AF616148

WATER AND IONS IN BIOLOGICAL SYSTEMS

WATER AND IONS IN BIOLOGICAL SYSTEMS

Edited by

Alberte Pullman
Institut de Biologie Physico-Chimique
Fondation Edmond de Rothschild
Paris, France

V. Vasilescu
Faculty of Medicine
Bucharest, Romania

and

L. Packer
University of California
Berkeley, California

PLENUM PRESS • NEW YORK AND LONDON

Library of Congress Cataloging in Publication Data

International Conference on Water and Ions in Biological Systems (2nd: 1982: Bucharest, Romania)
Water and ions in biological systems.

Proceedings of the Second International Conference on Water and Ions in Biological Systems, held Sept. 6–11, 1982, in Bucharest, Romania.
Bibliography: p.
Includes index.
1. Biological systems—Congresses. 2. Water—Physiological effect—Congresses. 3. Water in the body—Congresses. 4. Biological transport—Congresses. 5. Ions—Physiological effect—Congresses. I. Pullman, A. II. Vasilescu, Vasile. III. Packer, Lester. IV. Title.
QH501.I55 1982 574.19′212 85-3438
ISBN 0-306-41921-1

Proceedings of the Second International Conference on Water and Ions in Biological Systems, held September 6–11, 1982, in Bucharest, Romania, under the auspices of UNESCO, the Union of Societies of Medical Sciences in the Socialist Republic of Romania, the Romanian Biophysical Society, and the Academy of Medical Sciences, in cooperation with IUPAB

A Division of Plenum Publishing Corporation
233 Spring Street, New York, N.Y. 10013

Printed in the United States of America

PREFACE

As the First International Conference on Water and Ions in Biological Systems (Bucharest, June 25-27, 1980) was appreciated as a success, a second one was organized in the fall of the year 1982 under the sponsorship of the United Nations Educational, Scientific and Cultural Organization (UNESCO), the Romanian Academy of Medical Sciences, the Romanian Biophysical Society (Union of Societies for Medical Sciences in the Socialist Republic of Romania) and in co-operation with the International Union for Pure and Applied Biophysics (IUPAB).

The responsibility for the scientific program and organization of the Second Conference on Water fell on an International Scientific Committee which included Prof. J. Tigyi (Pécs), President of the UNESCO Expert Committee on Biophysics, Prof. K. Wüthrich, Secretary General of IUPAB and Prof. H. Eisenberg, (member of the IUPAB Council) under the guidance of an Executive Board whose members were Prof. J. Jaz (representative of UNESCO), Prof. B. Pullman (Vice-President of IUPAB) and Prof. V. Vasilescu (President of the Romanian Biophysical Society).

The Meeting was attended by more than 250 specialists including 150 Romanian participants and others from Bulgaria, Czechoslovakia, England, the Federal Republic of Germany, the German Democratic Republic, Greece, Hungary, India, Israel, Italy, Japan, the Netherlands, Nigeria, Poland, Sweden, Switzerland, USSR, USA, Venezuela, Yugoslavia.

The proceedings of the Conference took place in the Medical Faculty of Bucharest. The theoretical and practical importance of the Meeting was pointed out by the speakers, among whom were Prof. I. Ursu (Prime-Vice-President of the National Council for Science and Technology in Romania), Prof. Leonida Gherasim (Rector of the Institute of Medicine and Pharmacy in Bucharest), Prof. J. Jaz (Head of the Scientific Cooperation Bureau for the European and North American Region (UNESCO), Prof. H. Eisenberg (representative of IUPAB), Prof. C. F. Hazlewood (Baylor College of Medicine, Houston), Prof. E. Andronikashvili (Director of the Institute of Physics in Tbilisi),

Prof. M. B. Palma-Vittorelli (Institute of Physics, Palermo), Prof. A. Kotyk (Institute of Microbiology, Praga), Prof. V. Vasilescu (President of the Romanian Biophysical Society), Prof. A. T. Balaban (Polytechnic Institute, Bucharest).

The scientific program contained 4 plenary lectures, 15 symposia and 12 poster sessions.

From the lectures on biophysics, bioelectrochemistry and other scientific disciplines those who attended the Conference were able to understand better such fundamental problems as:

- role of water in the cell structure and function; physical state of water in the living cell; molecular mechanisms affected by the intramolecular water, as reflected in the cell function; the relation between cell function and the phenomena at aqueous biological interfaces; hydrophobic interactions and their role in the organization of the supramolecular structure; role of water in the evolution of biological systems;
- proton translocation and bioenergetic phenomena; specific proton-membrane interactions; proton translocation across biomembranes; mechanisms of proton translocation in the respiratory chain;
- role of ions in the cell structure and function; the state of K^+ in the living cell; calcium and regulation of cellular processes; metal ions and intramolecular interactions; the state of ions near biological interfaces; molecular mechanisms involved in the behavior of the ionic channels of excitable membranes.

Outstanding examples of theoretical approaches, of technical methods of high refinement, as well as of sophisticated experiments can be found in over 200 contributed papers to the conference.

The discussions maintained a high scientific level which was due, no doubt, to the presence of several senior specialists. This was a further reason why participants were able to appreciate the usefulness of the meeting and were anxious to make new contributions to such a tempting field of research.

The organizers wish to express their thanks for the financial support received from the Union of Societies for Medical Sciences in Romania and from the United Nations Educational, Scientific and Cultural Organization (UNESCO).

We are grateful also to the Romanian authorities and particularly to the National Council for Science and Technology, for assistance given, as is reflected by its continuing promotion of scientific research and international scientific cooperation.

Our thanks go equally to all those who contributed in some way or other to the success of this Meeting, which prompts us to continue the cycle of Conferences on Water and Ions in Biological Systems, encouraging the hope that a Third Conference will enjoy the same favorable reception.

V. Vasilescu

CONTENTS

ORGANIZATION AND FUNCTION OF WATER IN NATIVE CONFORMATION OF MACROMOLECULES

HYDROPHOBIC INTERACTIONS; THEIR ROLE IN THE ORGANIZATION OF SUPRAMOLECULAR SYSTEMS

WATER AND IONS IN HETEROGENEOUS SYSTEMS

WATER AND SOLUTE TRANSPORT

ROLE OF WATER AND IONS IN MEMBRANE STRUCTURE AND FUNCTION. WATER AND IONS IN EVOLUTION OF BIOLOGICAL SYSTEMS

PROTON TRANSLOCATION AND BIOENERGETIC PHENOMENA

CALCIUM AND THE REGULATION OF CELLULAR PROCESSES. WATER AND IONS IN MUSCLE

WATER, IONS AND DRUG ACTION, INCLUDING ANESTHETICS

IONIC CHANNELS

INTRODUCTORY REMARKS

B. Pullman

Council of IUPAB

When the European Expert Committee for Biophysics was created six years ago, at a constituent meeting held in Budapest, on June 2-4, 1976, hosted by Professor Tigyi, I have defined its scientific mission at the opening session by specifying what I considered to be our two fundamental duties: the first one, to survey and evaluate the most important and promising directions of research in the field of modern biophysics and, the second one, to select the subjects for possible scientific collaboration between eastern and western laboratories and to implement such a collaboration. I also added, and here I am quoting myself:

"Under UNESCO's sponsorship, the european meetings have the extremely important mission of enhancing scientific collaboration between its two principal groups of developed countries which differ in their political, economical and social structure. The enhancement of collaboration, human relations and friendship between these two groups is a primordial goal of these meetings, and should contribute to the atmosphere of "detente and entente" between east and west which is so essential for the peaceful future and development of mankind. I feel that we must consider ourselves as particularly privileged to be able to contribute here both to the development of science and to the improvement of relations between scientists and nations".

It was in Budapest, at this meeting, that we have established the basic structure and outlined the broad scientific goals of our Committee by creating two commissions within the frame of which we are performing till this day our collaborative research: the commission of Cell Biophysics under the chairmanship of Professor Tigyi and the Commission on Molecular Biophysics of which I had the honor to be designated as chairman. A further essential step forward in our collaboration occurred at a meeting of, I would say, the most active scientists of our group, in Paris, on May 25th 1977, held at the occasion of the International Symposium on Frontiers in Physico-Chemical Biology, celebrating the 50th anniversary of my institute, at which we have made the important choice of three main themes for the collaboration between eastern and western laboratories in the immediate and close future. These themes were:

1) Intra- and intermolecular interactions involving nucleic acids, proteins and their constituents;
2) Structure and function of biomembranes; and
3) The role of water and ions in molecular and cell biology.

Three impressive subjects, corresponding to three fundamental lines of research in modern biophysics, actively developed in many laboratories all around the world.

It seems to me that enough time has now elapsed since these first constituent meetings for us to be able to emit a judgement on the activity and achievement of our Committee. I, personally, believe that we have good reasons to be satisfied. It seems to me possible to ascertain, and with great satisfaction indeed, that the researches developed in our different laboratories and the collaborations established between them on these three topics have been fruitful, far beyond the reasonable expectation that we could have six years ago. They have permitted a very important development of deep and profitable contacts between the members of this Committee and their laboratories, an impressive exchange of collaborators, a large number of mutual visits and an important scientific production, consisting sometimes of joint publications from different laboratories. But, dominating this scientific success, is the materially invisible but not less real achievement of having created a most friendly and strongly linked international group of scientists. Although engaged in many international activities, I know of no other group collaborating in such a devoted and open-minded spirit with such an obvious pleasure and enthousiasm. We all know that since 1976 the international situation had somewhat deteriorated. It is our pride that this unfortunate evolution of world's affairs has had no damaging effect whatsoever on our structure, activity and on our intimate links. On the contrary, since last year we have added a new dimension to our existence, and an important one, by deciding to extend our activities to developing countries with whom we would like to share now what we have become a solid, well organized, well functioning body, our knowledge and our experience in a friendly, disinteresed and, we hope, helpful collaboration.

Among the different achievements of this Committee none illustrates as well the success obtained as the conference on Water and Cations in Biological Systems of which we are attending today the second one. The organization of activities on this Theme of our collaboration has been entrusted, since the Paris 1977 meeting, to Professor Vasilescu and the Committee certainly could not have made a better choice. The holding of the first meeting, 2 years ago, in this very surroundings, turned out to be a memorable event of International significance which crystallized the need of a more continuous collaboration on this most fundamental topic of modern molecular biophysics. At the same time, the excellency of the organization and the cordiality of the reception offered in Bucharest at that time naturally designated this beautiful City to become the center and the seat of this activity. This is an honour but it is also a heavy burden. I would like to express to Professor Vasilescu our deep thanks and appreciation for his indefatigable effort to organize and carry out these meetings in such a remarkable way and I would like to extend these thanks and sentiments of appreciation to the Roumanian Scientific and State Authorities, including the highest ones, for the substantial help and the open hospitality presented to our gatherings.

Of course, behind the individual efforts of persons or groups stands the fundamental support of UNESCO, which makes these reunions possible. Our deep thanks are thus due to Professor Jaz, the head of the European Scientific Bureau of UNESCO for this continuous activity on behalf of the Committee on Biophysics, his perseverent efforts of obtaining the maximum support for it and his deep comprehension of its needs and goals. Without his imaginative, brilliant and ceaseless activity nothing of all this could have been achieved.

The privilege which was offered me to address my greetings to you today is due, I believe, to my quality of former and now honorary chairman of this Committee, whose destiny is at present in the very able hands of Professor Tigyi. I would like, however, to add here today also the greetings on behalf of the Council of IUPAB of which I have the honor of being Vice-President.

This meeting is the second step on what we expect to be a long road of collaboration on a very important theme of modern biophysics for the benefit of scientific progress and human interrelations. We have a good start which implies an obligation of doing better in the future.

OPENING ADDRESS

J. Tigyi

Chairman of Biophysics Expert Group of UNESCO
European-North American Region

It is a great honour and privilege for me to greet all the participants and guests of the Second International Conference on Water and Ions in Biological Systems.

First of all I should like to express our great acknowledgement to the science section of UNESCO and its head, Assistant Director-General Professor A.R. Kaddoura and to European-North American Region of UNESCO and to its head Prof. J. Jaz for their continuous and effective support towards the biophysical research. Without their permanent help we would not be able to organize a similar scientific meeting at such a high level.

I should like to emphasize the very generous support of the Rumanian Academy of Sciences and the National Council of Science and Technology of the Rumanian Socialist Republic, which gave the condition sine qua non of this symposium. The special attention and contribution of President, Academician, Professor Elena Ceausescu and Academician J. Ursu was very precious. Thanks to the Institute of Medicine and Pharmacy and to Professor Rector L. Gherasim for the kind hospitality. Our appreciation should be expressed to the Union of Societies of Medical Sciences and to Prof. M. Voiculescu for their permanent help.

Professor V. Vasilescu and all the members of the Biophysics Department made a tremendous effort in organizing this symposium.

Their special merit was to organize this symposium second time also in Bucharest. The success of the first meeting was highly estimated internationally. The presence of the best experts of this field from 26 countries and famous experience and hospitality of the Rumanian colleagues is a complete guarantes of the success of this second symposium.

It is regrettable that the spiritus rector and founder of the UNESCO biophysics collaboration Prof. B. Pullman cannot be with us, his contribution to bringing into existence of this meeting was decisive.

I should like to wish successful discussions and pleasant friendly time together to all participants.

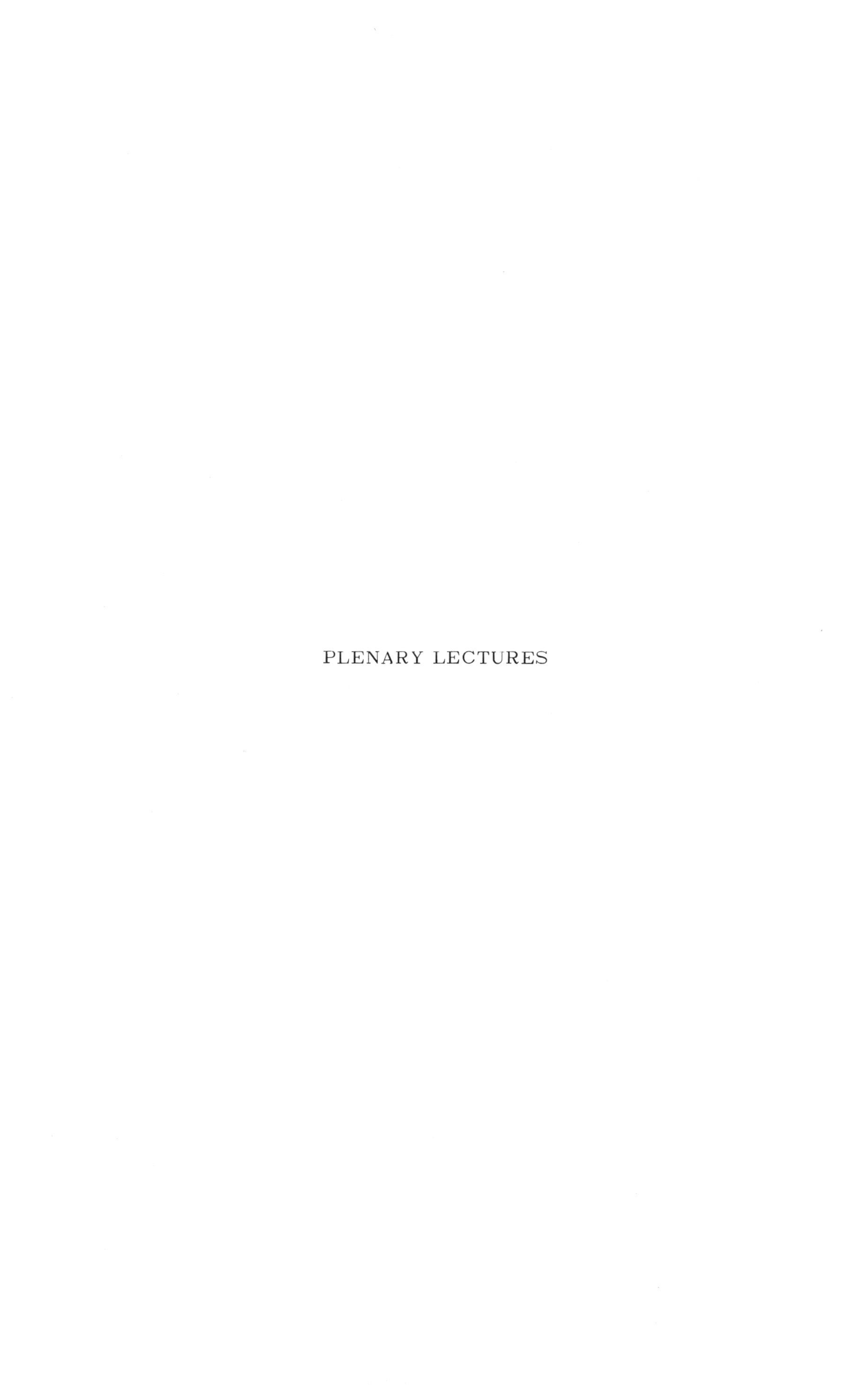

PLENARY LECTURES

WATER IN BIOLOGICAL SYSTEMS AS STUDIED BY NMR AND NEUTRON SCATTERING TECHNIQUES

Carlton F. Hazlewood

Department of Physiology
Baylor College of Medicine
Texas Medical Center, Houston Texas 77030

INTRODUCTION

Although water constitutes 60-90% of the weight of most biological tissues, its physical state within the cell has not yet been clearly characterized. Therefore, all attempts to understand the unequal distribution of ions between the inside and the outside of cells, the dynamic aspects of solute and water movement into and out of cells, and the bioelectric phenomena of cells, require fundamental assumptions concerning the physical state of cellular water. Ultimately, an understanding of cellular function will require a detailed characterization of the physical state of water.

The conventional assumption made by biologists is that the aqueous cytoplasm of a living cell is a dilute solution surrounded by a semipermeable membrane (1-4). Bioelectric phenomena of cells are also described classically by equations evolving from dilute solution theory. Osmotic theory, which is also entwined in the theory of dilute solutions, has earlier roots in the 18th century (1). Alternate views, however, proposing that cytoplasmic water and ions are not free, but rather are bound or adsorbed to macromolecular constituents have been proposed since the mid 19th century (5). These divergent views have not yet been reconciled, and reviews addressed to this controversy may be found elsewhere (6-8).

These divergent views of cellular function arise basically from different assumptions concerning the physical state of water and ions, and a resolution of these fundamental differences is needed. Other papers within this volume address the problem of the physical state of cellular ions and other solutes in biological systems, whereas this report focuses on new findings on the physical properties of water evolving from nuclear magnetic resonance (NMR) spectroscopy and from quasielastic neutron scattering studies of water protons. We especially emphasize the importance of the diffusive motion of water in biological systems.

NUCLEAR MAGNETIC RESONANCE

Models for NMR parameters in biological systems

The relaxation times of water protons in biological systems are substantially shorter than those for bulk water or water in dilute electrolyte solutions (9-12). Two general types of interpretations have been proposed to explain the reduction of the relaxation times: (1) "bound-free" models (13-16), and (2) the "ordered" water concepts (5,17-20). These two schools of thought have survived for more than 10 years without resolution.

Within the context of the "bound-free" models, equations have been developed which allow some explicit testing of the models. In this general set of models the relaxation times are space averages over a wide range of proton environments. Therefore, the presence of a small, rapidly relaxing fraction of protons in rapid exchange (even with another much larger "free" fraction) will dominate the average relaxation time observed. The diffusion coefficient on the other hand, will be dominated by the larger ("free") fraction of water. Furthermore, any reduction in D should be due to compartmentilization and/or obstruction effects within the cell.

In the case of the "ordered" water models, the polarized multilayer concept is the most specific but, at present, no formalized equations have been developed to describe the relaxation times. The polarized multilayer hypothesis, however, evolves from the fundamental assumptions that both the translational and rotational motions of water molecules are reduced. Therefore, in this view, the D of a substantial fraction, perhaps all, of the water should be reduced, and the reduction should be greater than that due to compartmentalization or obstruction effects. This distinction concerning the diffusion coefficient between the two general interpretations provides an opportunity to experimentally evaluate the models.

Diffusion coefficient

The diffusion coefficient (D) for water in biological systems is moderately reduced when compared to bulk water, whereas the relaxation times are reduced by an order of magnitude or more. The reduction in D (in many biological systems) is by a factor of the order of 2-6. Explanations proposed to account for this reduction generally fall into one of the following categories: (1) membrane compartmentalization of cellular water; (2) macromolecular structures within cellular compartments serving as local obstructions to diffusion; and (3) the induction of intrinsic changes in the water-water interactions in a substantial fraction of the intracellular water due to water macromolecule interactions. Clearly distinguishing between these alternatives has been difficult. Much of the difficulty arises from the heterogeneous character of the cell as well as to the fact that a single NMR measurement is carried out over a period of time, t, of the order of a few milliseconds or more. During this measuring time, the water molecules are diffusing within the cell, and their NMR parameters are determined by averages over distances $\sqrt{2\,Dt} \simeq 1$ micron, where D is the diffusion coefficient of bulk water and t is time over which the measurements are made. Interpretation of the results, therefore, requires detailed modeling of the heterogeneous environment of the cell.

If the large protein structures, including membranes, within cells do form impermeable compartments, D, measured by NMR techniques, would not only be reduced, but also would be a strong function of the measuring time (21). This possibility has been investigated and the results of the study indicated that it is unlikely that compartmentalization is the cause for the reduced value of D in biological systems (22). More recently measurements of D by way of pulsed field gradient techniques in which the measuring times can be varied have been used to evaluate this problem and it was found that cellular macromolecules serve as obstructions and can account for some, but not all, of the reduction of D in living cells (23).

The macroscopic scale of the measurement of D by NMR methods allows for the development of models based on the hydrodynamic interaction of water molecules with macromolecules. One of the earliest applications of the hydrodynamic method in biology was that of Wang (Wang, 1954). In this model the reduction of D for water in a dilute protein solution is due to interaction with the proteins (or macromolecules in general), which are impermeable for diffusion. This interaction is termed the obstruction effect and may be seperated into two contributions: (1) the cross-section for diffusion is reduced due to the volume occupied by the obstructions; and, (2) the path length for diffusion is lengthened.

Al alternative calculation termed the cellular method has been developed for two geometries likely to be found in various cells: (1) an ordered array of cylinders as would be found in striated muscle; and, (2) an order array of spheres as might be encountered within the interfilament spaces of striated muscles and generally in all cells. The development of the model, and its use in experimental analysis may be found elsewhere (24-27). In summary, the development of the cellular method yields approximate solutions which are relatively simple and which can be experimentally tested.

The following equations present the approximate solutions for an hexagonal array of cylinders (1) and a cubic array of spheres (2)

$$D' = \frac{D}{1 + 0.8\,\emptyset} \qquad (1)$$

$$D' = \frac{D}{1 + 0.63\,\emptyset} \qquad (2)$$

where D is the diffusion coefficient of bulk water, D' is the diffusion coefficient of water in the biological system, and $\emptyset$ is the volume fraction occupies by the obstructions (i.e., cylinders or spheres).

The solutions to this hydrodynamic model provide a means to test for obstructions in cellular systems. Equations (1) and (2) have been applied to skeletal muscle (25) to evaluate D as function of fiber orientation relative to the magnetic field (i.e., radial and longitudinal diffusion), to cysts of the brine shrimp to evaluate D as a function of hydration (26), and to barnacle muscle and macromolecular solutions to evaluate the overall effects of obstructions as a cause of the reduction of D (27). The conclusions reached in these studies, and in those of Tanner (23), are that macromolecular obstructions cannot fully account for the reduction of the diffusion coefficient in the biological systems studied. Therefore, the intrinsic translational diffusion coefficient of most of the cellular water is reduced.

QUASIELASTIC NEUTRON SCATTERING (QNS)

The QNS method provides information on the properties of scattering atoms which depend directly on the microscopic environment of the atoms. The correlation function for the atomic motion is evaluated over a space domain of a few angstroms and during a time interval of approximately 10^{-12} second. In the QNS method, the positional correlation of the atom determines the scattering whereas, in the case of NMR measurements, the relevent correlation function is that of the orientation of the protonic moments.

The method and results of our QNS studies are given in more detail in the paper of Rorschach, et. al. which is published in this volume and elsewhere (28). This study was undertaken in an attempt to determine the diffusive motion of water protons in a space domain very short compared to intermacromolecular distances within biological systems (i.e. its measurement of D made before the water molecules have moved more than a few angstroms). The study has thus far been restricted to the cysts of the brine shrimp (*Artemia*) because of the long time required to acquire the data needed for analysis (i.e. approximately 5 days). The *Artemia* cysts are well suited for this study, since the viability can be easily determined (29). The viability was 90 ± 2% in our studies and was unchanged by neutron irradiation.

A summary of the results of this study is given in Table 1.

Diffusion Parameters Determined from QNS Data

System	D/D_w	τ/τ_w	D_r (10^{-9} sec^{-1})
Pure water	1	1	*
Fully hydrated *Artemia* cysts (1.2 gms H_2O/gm dry cysts)	0.29	3.9	6.0

*8.1 x 10^{10} sec^{-1} obtained from dielectric relaxation data
D_w = 2.4 x 10^{-5} cm^2/sec--the diffusion coefficient of bulk water
τ_w = 1.2 x 10^{-12} sec---the correlation time of bulk water

In Table 1 the data from the QNS study on *Artemia* are compared to the data obtained on pure water. These data have been compared also to data obtained by other technologies (see Rorschach et al, this volume). The conclusion from this study is that water in the *Artemia* cysts has strongly reduced translational and rotational diffusion coefficients that are not due to obstructions, compartments or exchange with minor phases but are instead an intrinsic feature of the cellular water.

SUMMARY

The general findings of these studies indicate that the physical properties of water within cells are not equilivalent to those of bulk water or dilute aqueous solutions. Furthermore, the findings indicate that physical models that are developed to describe the relaxation processes of water protons in biological systems must be consistent with the observed decrease in both rotational and translational motion.

ACKNOWLEDGMENTS

United States Office of Naval Research Contracts N000-14-79-C-0492 and N00014-76-C-0100 and R.A. Welch Grant Q-390. Grants from the National Science Foundation, the Romanian National Council for Science and Technology and United States Office of Naval Research Grant N00014-82-G0099 are also acknowledged. The author is grateful for the secretarial assistance of Ms. Gerrie Hazlewood.

REFERENCES

1. H. Smith, A knowledge of the laws of solutions, Circulation 21:808 (1960).
2. D.A.T. Dick, "Cell Water," Butterworth, Inc., Washington D.C., (1966).
3. D.A.T. Dick, Osmotic properties of living cells, Int. Rev. Cyto. 8:387 (1959).
4. E.G. Olmstead, "Mammalian Cell Water," Lea and Febiger, Philadelphia (1966).
5. G.N. Ling, Hydration of macromolecules, in: "Water and Aqueous Solutions: Structure, Thermodynamics, and Transport Processes," R.A. Horne, ed., Wiley Interscience, New York (1972).
6. O. Hechter, Intracellular water structure and mechanisms of cellular transport, Ann. New York Acad. Sci., 125:625 (1965).
7. C.F. Hazlewood, Physicochemical state of ions and water in living tissues and model systems, Ann. New York Acad. Sci., 204:1 (1973).
8. C.F. Hazlewood, A view of the significance and understanding of the physical properties of cell-associated water, in: "Cell Associated Water", W. Drost-Hansen and J.S. Clegg, eds., Academic Press, New York, (1979).
9. R. Cooke and I.D. Kuntz, The properties of water in biological systems, Ann. Rev. Biophys. Bioeng.,3:126 (1974).
10. R. Mathur-DeVré, The nmr studies of water in biological systems, Prog. Biophys. Mol. Biol.,35:103 (1979).
11. S. Ratković, NMR studies of water in biological systems at different levels of their organization, Sci. Yugo., 7:19 (1981).
12. P.W. Kuchel, Nuclear magnetic resonance of biological samples, in: "Critical Reviews in Analytical Chemistry", B. Campbell and L. Meites, eds., CRC Press, Boca Raton, Florida, (October) (1981).
13. J.R. Zimmerman, and W.E. Brittin, Nuclear magnetic resonance studies in multiple phase systems: lifetime of a water molecule in an adsorbing phase on silica gel, J. Phys. Chem., 61:1328 (1957).
14. S.H. Koenig, R.G. Bryant, K. Hallenga, and G.S. Jacob, Magnetic cross-relaxation among protons in protein solutions, Biochem., 17:4348 (1978).
15. B.D. Sykes, W.E. Hull, and G.H. Snyder, Experimental evidence for the role of cross-relaxation in proton nuclear magnetic resonance spin lattice relaxation time measurements in proteins, Biophys. J., 21:137 (1978).
16. H.T. Edzes, and E.T. Samulski, Cross-relaxation and spin diffusion in the proton nmr of hydrated collagen, Nature, 265:521 (1977).

17. G.N. Ling, The physical state of water in living cell and model systems, Ann. New York Acad. Sci., 125:401 (1965).
18. F.W. Cope, Nuclear magnetic resonance evidence using D_2O for structured water in muscle and brain, Biophys. J., 9:303 (1969).
19. C.F. Hazlewood, B.L. Nichols and N.F. Chamberlain, Evidence for the existence of a minimum of two phases of ordered water in skeletal muscle, Nature, 222:747 (1969).
20. G.N. Ling, The polarized multilayer theory of cell water and other facets of the association-induction hypothesis concerning the distribution of ions and other solutes in living cells, in: "The Aqueous Cytoplasm," A.D. Keith, ed., Marcel Dekker, Inc., New York, (1979).
21. B. Robertson, Spin-echo decay of spins diffusing in a bounded region, Phys. Rev., 151:273 (1966).
22. D.C. Chang, H.E. Rorschach, B.L. Nichols and C.F. Hazlewood, Implications of diffusion coefficient measurements for the structure of cellular water, Ann. New York, Acad. Sci., 204: 434 (1973).
23. J. Tanner, Paper accepted for publication in Biochim. Biophys. Acta. (1983).
24. H.E. Rorschach, D.C. Chang, C.F. Hazlewood and B.L. Nichols, The diffusion of water in striated muscle, Ann. New York Acad. Sci., 204:444 (1973).
25. G.G. Cleveland, D.C. Chang, C.F. Hazlewood and H.E. Rorschach, Nuclear magnetic resonance measurements of skeletal muscle: anisotrophy of the diffusion coefficient of the intracellular water, Biophys. J., 16:1043 (1976).
26. P.K. Seitz, D.C. Chang, C.F. Hazlewood, H.E. Rorschach and J.S. Clegg, The self-diffusion of water, in: "Artemia Cysts", Arch. Biochem. Biophys., 210:517 (1981).
27. M.E. Clark, E.E. Burnell, N.R. Chapman and J.A.M. Hinke, Water in barnacle muscle IV. Factors contributing to reduced self-diffusion, Biophys. J., 39:289 (1982).
28. E.C. Trantham, Diffusive motion of water in biological and model systems, Dissertaion, Rice University, Houston, Texas (1981).
29. J.S. Clegg and Z.P. Conte, A review of the cellular and developmental biology of Artemia, in: "The Brine Shrimp Artemia", Vol. 2, eds., G. Persoone, P. Sorgeloos, O. Rolls and F. Jaspers, Universa Press, Wettern, Belgium (1980).

THEORETICAL MODELS FOR CALCIUM CHANNELS IN NERVE CELLS

Platon Kostyuk and Sergei Mironov

A. A. Bogomoletz Institute of Physiology
Ukrainian Academy of Sciences, Kiev, USSR

One of the important problems in modern biophysics is determination of the structure and function of different ionic channels in various excitable tissues. During the past three decades it became well established that there are three main types of ionic channels which differ in their kinetic, selective and pharmacological properties. These are sodium, potassium and calcium channels. Among them calcium channels appear to play the most significant role in the functioning of nerve cells (see, e.g. Llinas, 1979; Kostyuk, 1981). The reason for this lies in the fact that these channels are responsible not only for the electrical behaviour of nerve cells, but also provide a pathway for calcium ions into the cell. Entering the nerve cell Ca ions can trigger many different biochemical reactions related to the functioning, differentiation and proliferation of nerve cells (Llinas, 1979; Kostyuk, 1981).

Thus, it would be of great interest to develop theoretical quantitative models for calcium channels.

To describe quantitatively calcium channels it is necessary to investigate the potential of membrane surfaces. The reason for this can be illustrated by Fig. 1. It can be seen that even at constant voltage applied the value of membrane potential depends on the surface potential value. Moreover, the surface potential also affects the value of near-membrane concentration of permeant ions. According to the Boltzmann distribution, the more negative is the value of surface potential, the larger is the near-membrane concentration of permeant ions, hence, the larger is the conductance of the corresponding ionic channel (Kostyuk et al., 1982).

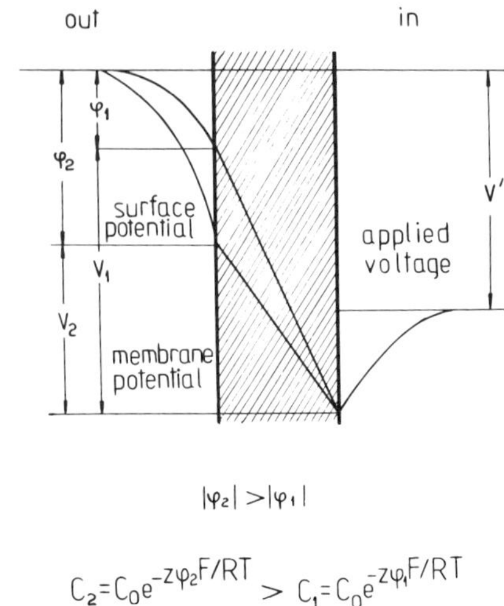

$$|\varphi_2| > |\varphi_1|$$

$$C_2 = C_0 e^{-z\varphi_2 F/RT} > C_1 = C_0 e^{-z\varphi_1 F/RT}$$

Fig. 1. Schematic description of the membrane potential profile.

The neglect of this effect may lead to serious errors in the interpretation of the experimental observations. To illustrate this point, let us consider the effect of hydrogen ions on the sodium conductance. Upon lowering pH of the extracellular solution we observed a decrease in sodium conductance which is usually attributed to the process of protonation of some acid group in the sodium channel (Hille, 1979). However, if we determine the value of surface potential change from the shift of the conductance curve along the potential axis it appears that the decrease in the limiting conductance g_{Na} is connected with the change of the surface potential only (Fig. 2, Mironov, 1982). In fact, for the example shown in Fig. 2 the values of g_{Na} are related as 1:0.80:0.55 (pH = 9.6 and 5, respectively). Taking into account the change of the surface potential we obtain the following sequence of g_{Na} values 1:0.88:0.63. Thus, we see that there is no need to associate this decrease in sodium conductance with the presence of some acid groups in the sodium channel which has been earlier postulated (Hille, 1979).

During our investigation of the surface potential of nerve cells it appeared that divalent ions interact quite strongly with those acid functional groups which determine the value of the surface potential. This effect has been described on the basis of double electric layer theory that takes into account the effects of cation adsorption (Kostyuk et al., 1982b). The values of binding constants obtained for alkaline-earth cations (Kostyuk et al., 1982b) are close to the analogous

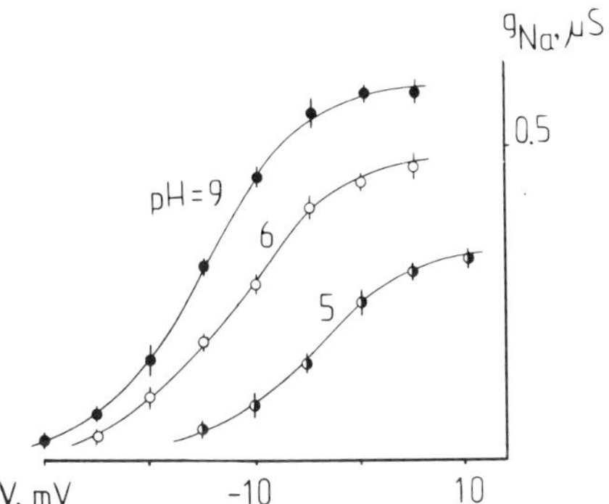

Fig. 2. Effect of pH of the extracellular solution on the sodium conductance of molluscan neurons.

characteristics of phosphatidylserine (Ohki and Kurland, 1981) which is the main charge constituent of natural membranes. Thus we can suppose that just phosphatidylserine is responsible for the surface potential of the neuronal membranes. These results we have used for description of ionic transport processes in the calcium channel.

We described ionic transport processes in the calcium channel on the basis of a three barrier model of its energy profile (Kostyuk et al., 1982a). Fig. 3 shows experimental current-voltage characteristics of the calcium channel in mammalian neurons and their approximation on the basis of our model. Fig. 3a clearly demonstrates the existence of saturation phenomena for calcium current, i.e. the magnitude of the calcium current approaches the limiting value when the concentration of calcium ions in the extracellular solution is increased. From the data shown in Fig. 3b we see that Ba ions are more permeable than Ca ions, however, the transition metal ions are less permeant and they block calcium conductance.

This blocking effect as well as the saturation phenomena can be explained from the energy profiles for different ions in the calcium channel (Fig. 4). According to our calculations the energy pr ofiles for different ions are practically identical except the first potential well (see Fig. 4). When the concentration of permeant ions in the extracellular solution is increased, this potential well is almost completely

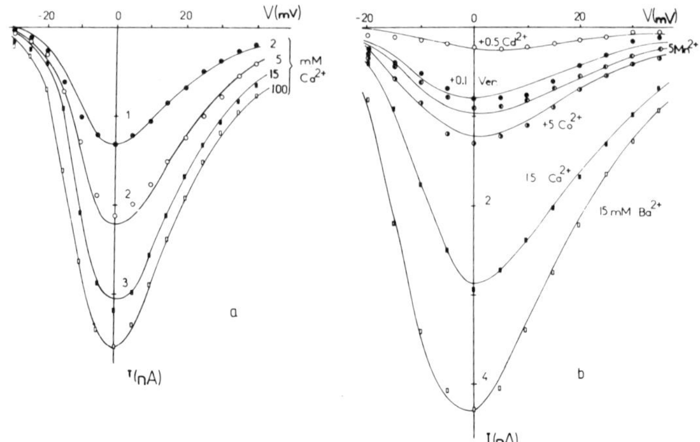

Fig. 3. Current–voltage characteristics of the calcium channel in mammalian neurons. Experimental data (points) were approximated by the curves calculated on the basis of a three barrier model of the energy profile for the calcium channel. Concentrations of divalent cations in the extracellular solution are indicated in the Figure.

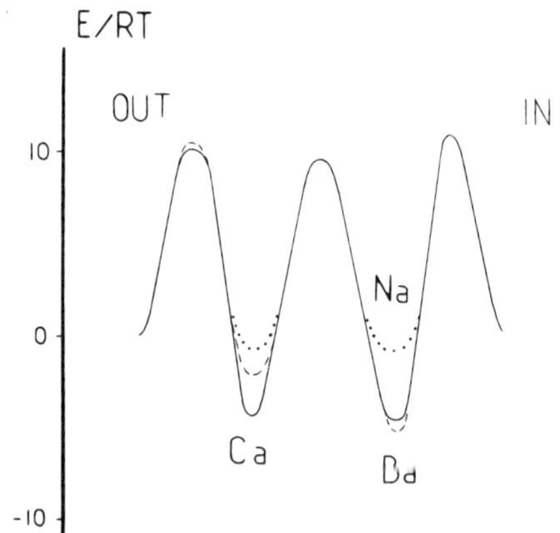

Fig. 4. Calculated energy profiles for Ca, Ba and Na ions in the calcium channel of molluscan neurons.

filled, thus the magnitude of the current tends to the maximal value. The deeper is the potential well, the lower is the value of the corresponding maximal current. In fact, calculated maximal single channel currents for molluscan neurons carried by Ca, Ba and Na ions agree with the experimental data (Kostyuk et al., 1982a). According to our calculations, transition metal ions possess the deepest potential well, thus they occupy the calcium channel and block the corresponding current if present even in small amounts in the extracellular solution.

From the physico-chemical point of view this potential well corresponds to some active centre located inside the calcium channel and its depth equals to free energy of a complex formed by this centre and cation. Knowing the energy profiles for different ions we calculated the values of dissociation constants of complexes formed by them (Table 1). These data demonstrate that the selectivity properties of the calcium channels determined by pK values are close for molluscan and mammalian neurons. These pK values also strongly resemble the data obtained for complexes of these ions with the carboxylic group in an aqueous solution (Table 1). Therefore we can suppose that this active centre of the calcium channel probably contains only one carboxylic group.

From these data it is also clear that the value of pK calculated for the calcium channel provides a suitable representation of its permeability. For example, transition metal cations form strong complexes with the carboxylic group (the values of pK are greater than 3) so they block the calcium conductance. On the contrary, for the alkaline earth metals the value of pK is lower than 2 and they are permeant cations. Thus, the presence of one carboxylic group is enough to provide the selectivity of the calcium channels for different divalent cations.

Recently it has been shown (Kostyuk & Krishtal, 1977) that calcium channels can become permeable to Na ions if EDTA is introduced into the extracellular solution. Together with Dr. Ya.M.Shuba we have performed a detailed investigation of this effect for molluscan neurons and have shown that it is connected with an abrupt decrease in the extracellular concentration of Ca ions down to a micromolar range leading probably to removal of these ions from a very specific Ca-binding site of the channel. This effect is not dependent on the membrane potential that indicates that this Ca-binding centre is located on the outer surface of the membrane.

Table 1. pK* of dissociation of complexes formed by different ions with calcium channels and with carboxylic group in an aqueous solution

Ligand**	Ba^{2+}	Sr^{2+}	Ca^{2+}	Co^{2+}	Mn^{2+}	Cd^{2+}
I	1.0	1.7	2.0	3.2	3.5	4.3
II	0.7	-	1.3	2.5	2.6	4.3
III	0.8	0.9	1.4	3.2	3.2	4.8

* pK = -log K, where K is the corresponding dissociation constant; ** Calcium channels in molluscan /I/ (Kostyuk et al., 1982a) and mammalian neurons /II/ (Kostyuk and Mironov, 1982); III - carboxylic group.

Other divalent cations produce similar effect, but their acting concentration is much greater than that for Ca ions. This effect can be approximated by Langmuir's isotherm with the following pK values (pK_{Ca}: pK_{Sr}: pK_{Ba}: pK_{Mg} = 6.6 : 5.5 : 4.8 : 4.2). This sequence fits well the corresponding pK values for EDTA whose binding capacity is mainly determined by its four carboxylic groups. The value of pK obtained for Ca ions is also close to pK values for different Ca-binding proteins such as parvalbumin, troponin C, calmodulin, etc. (Kretzinger and Nelson, 1976). It is thought that their Ca-binding capacity is mainly determined by the presence in their active centre of several carboxylic groups belonging to aspartic and glutamic amino acids. Thus, we can suppose that the high affinity of this Ca-binding site of the calcium channel is also connected with the presence of several carboxylic groups.

Thus, at present we can draw the following schematic description of the calcium channel (Fig. 5). It is a pore embedded in the lipid matrix possessing different Ca-binding sites. The outer one probably contains several carboxylic groups and control the selectivity of the calcium channel according to the charge of the permeant cations. When the concentration of calcium ions in the extracellular solution exceeds 1 μm, this centre becomes occupied and the calcium channel can pass only divalent cations, but if the concentration of Ca ions is lower than 1 μm this complex is

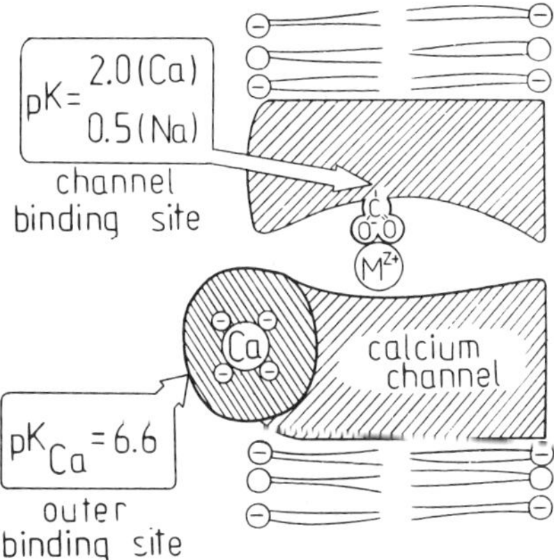

Fig. 5. Schematic description of the calcium channel.

destroyed and monovalent cations can pass through the calcium channel.

Permeant and blocking cations entering the channel form complexes with the channel binding site which is placed inside the channel and probably contains only one carboxylic group. This site has a much weaker binding capacity for Ca ions than the outer one.

Intracellular Ca ions in micromolar concentrations can also block Ca-conductance (Kostyuk and Krishtal, 1977). The mechanism of their action is not yet clear, however, one can suppose that Ca ions may interact directly with some Ca-inactivating binding site or indirectly through the system of cyclic nucleotides. Recent data obtained in our laboratory (Doroshenko et al., 1982) are in favour of the latter mechanism of the calcium channel inactivation. This effect may illustrate the coupling between the electrical behaviour of nerve cells and the intracellular biochemical processes.

Slow wave membrane potential oscillations observed for molluscan neurons (Meech, 1979) are an another example of such electro-biochemical coupling. The mechanism of this process can be described as follows (Fig. 6): the voltage-dependent calcium current leads to the membrane depolarization. In this phase of oscillations Ca ions enter the cell and activate Ca-dependent potassium conductance which then repolarizes the membrane. After this intracellular Ca ions are absorbed by some Ca-buffer systems. The increasing calcium current starts to exceed the potassium current and the next cycle of oscillations occurs. These curves (Fig. 7) were

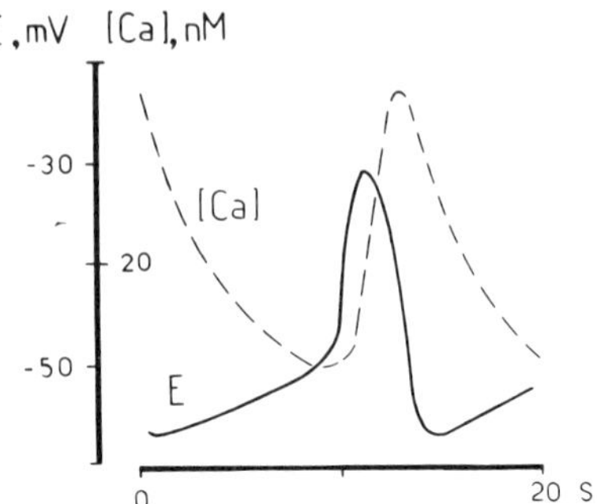

Fig. 6. Changes in the membrane potential and intracellular Ca concentration during slow wave oscillations. These curves were calculated on the basis of corresponding mathematical model developed to describe the slow-wave membrane oscillations observed in molluscan neurons (Mironov, 1983).

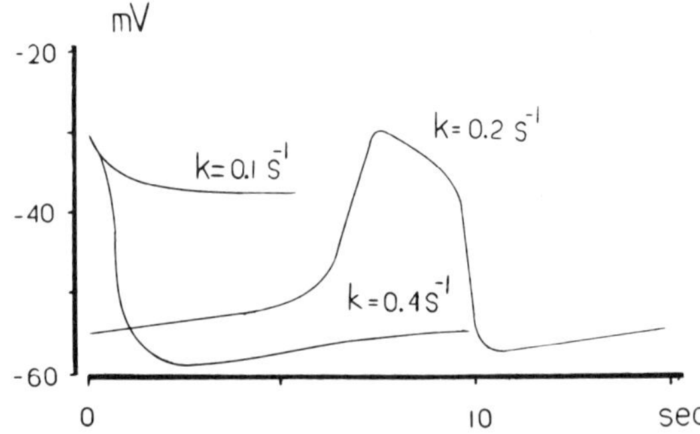

Fig. 7. Time-dependence of the membrane potential for different values of parameter k which enters the model of slow-wave membrane potential oscillations (Mironov, 1983).

calculated on the basis of corresponding mathematical model. They agree well with the observed changes of the membrane potential and intracellular Ca concentration during the oscillations measured experimentally on molluscan neurons by Gorman and Thomas (1978).

It should be noted that this oscillatory behaviour has a crucial dependence on the values of model parameters. Fig. 7 demonstrates the time-dependence of membrane potential for different values of parameter k which corresponds to the rate of Ca-uptake by intracellular buffer systems. When this process is fast the potassium current does not reveal a full-scale development and the membrane is depolarized. The value of membrane potential for this state is close to a threshold of the action potential generation. Therefore, in this state the neuron is expected to demonstrate a repetitive firing behaviour. On the contrary, when k is small, the process of intracellular Ca absorption slows down, the potassium current dominates and the system is at rest. For the intermediate values of k the processes of Ca entry and its uptake are well balanced, so the slow wave oscillations of the membrane potential with the period of about 15 sec are observed.

Thus, the model predicts that the neuron can be present in one of the three states: the repetitive firing state, the slow oscillation state or at rest. All these states are actually observed in experiments. Moreover, by changing some external conditions we can transfer neurons from one state to another one. For example, raising the external concentration of Ca or K ions we transfer the neurons from the resting state to the repetitive firing state through the state of stable oscillations. Similar effects can be induced by intracellular injection of cAMP which is known to change both the calcium conductance and the intracellular Ca level.

In summary, it should be noted that application of theoretical methods to quantitative description of the experimental data allows us not only to obtain the necessary characteristics of the system under study and to predict the behaviour of this system during the changes in external conditions but to provide the molecular mechanisms of underlying biophysical processes and to establish their possible functional role.

REFERENCES

Doroshenko, P. A., Kostyuk, P. G., and Martynyuk, A.E., 1982, Intracellular metabolism of adenosine 3',5'-

cyclic monophosphate and calcium inward current in perfused neurones of Helix pomatia, Neuroscience, 7 : 2125.

Gorman, A. L. F., and Thomas, M. V., 1978, Changes in the intracellular concentration of free calcium ions in a pacemaker neuron measured with the metallochromic indicator dye arsenazo III, J. Physiol., 275:357.

Hille, B., 1979, Ionic selectivity of Na and K channels of nerve membranes, in: "Membranes. A Series of Advances", G. Eisenman, ed., Marcel Dekker, New York.

Kostyuk, P.G. , 1981, Calcium channels in the neuronal membrane, Biochim. Biophys. Acta, 650:128.

Kostyuk, P. G., and Krishtal, O. A., 1977, Effects of calcium and calcium-chelating agents on the inward and outward currents in the membrane of mollusc neurones, J. Physiol., 270:569.

Kostyuk, P. G., and Mironov, S. L., 1982, Theoretical description of calcium channels in the neuronal membrane, Gen. Physiol. Biophys., 1:289.

Kostyuk, P. G., Mironov, S. L., and Doroshenko, P. A., 1982a, Energy profile of the calcium channel in the membrane of mollusc neurons, J. Memb. Biol., 70:181.

Kostyuk, P. G., Mironov, S. L., Doroshenko, P. A., and Ponomarev, V. N., 1982b, Surface charges on the outer side of mollusc neuron membrane, J. Memb. Biol., 70:171.

Kretzinger , R. H., and Nelson, D., 1976, Calcium in biological systems, Coord. Chem. Revs., 18:29.

Llinas, R., 1979, The role of calcium in neuronal function, in: "Neurosciences: Fourth Study Program", F. O. Schmitt, and F. G. Worden, eds., MIT Press, New York.

Meech, K. W., 1979, Membrane potential oscillations in molluscan burster neurons, J. Exp. Biol., 81: 93.

Mironov, S. L., 1982, The comparison of the selective filters of sodium and calcium channels in excitable membrane, Dokl. Akad. Nauk SSSR (Moscow), 268:731.

Mironov, S. L., 1983, Modeling of oscillatory electric activity of molluscan neurons, Neurophysiology (Kiev), 15:3.

Ohki, S., and Kurland, R., 1981, Surface potential of phosphatydylserine monolayers, Biochim. Biophys. Acta, 645:170.

DEVELOPMENT OF WATER RESEARCH IN BIOLOGY

J. Tigyi

Biophysical Institute of Medical University
Pécs, Hungary

At the beginning of the twentieth century some ingenious scientists[1,2] started to study the role of water in biological tissues but unfortunately this direction was not continued. In my opinion the main reason for the discontinuity was the lack of sophisticated physical and chemical methods which could have given exact results.

However, Ernst[3] developed a very simple and accurate method for studying the behavior of water in muscle and recognized in 1926 the essential role of water in muscle contraction. But his studies remained unaccepted because his results were incompatible with those of the leading scientists. A.V. Hill[4], around the thirties, announced that water in muscle exists in a free diffusible state. In 1950, E. Ernst[5] suggested the development of more exact methods for measuring the state of water in muscle. We developed three independent methods using the determination of a very characteristic thermodynamic parameter - the vapor pressure of water in the tissue as a function of water content.

Meanwhile, development of physics offered many very exact methods so the time became right to enter a new field of water research. The Symposium of NYASCI in 1973[6] can be considered as the real renaissance of modern water research. A large number of scientists and laboratories are active in this field and interest seems to be increasing.

I should like to make some general comments before I turn to details.

1. During the last three decades it has been recognized that the water and ion problems have to be studied in close correlation because in biological tissues these two systems always appear together and they have a lot of interrelations.

2. I should like to deal mainly with the muscle-water problem but the phenomena and relations of the muscle can be transferred - mutatis mutandis - to other tissues.

3. The problem of water in biological systems was originally a typical basic (pure) scientific problem but today - thanks to the exact results gained over the last decade - has become one of the most popular applied and practical fields of the sciences. Three fields of applied sciences have a constantly increasing interest towards water research: (a) medicine, (b) agriculture, (c) food industry.

Aging (e.g. cataract, Rácz[7]) and NMR imaging are among the most important problems of medicine. Frost and draft resistance in agriculture and quality preservation during the storage of food are among the most prominent questions in the food industry.

In this paper I should like to give an overview of the methodology of the water research using examples of work from our own laboratory. Table 1 gives an overview of the most frequently used methods.

Table 1. The most significant methods of studying muscle water and ions.

1. Exchange experiments
   a) extraction
   b) perfusion
2. Thermodynamic parameters
   a) vapor pressure
   b) diffusion coeff.
   c) heat of evaporation
3. Electronic
   a) potentials generated (RP. AP.)
   b) conductivity DC. AC. HF.
4. Magnetic EPR. NMR
5. ELMI
   a) staining
   b) radioautography
6. X-ray microanalysis
7. LAMMA (laser microprobe mass anal.)
8. Photoacoustic spectrometry

EXCHANGE EXPERIMENTS

Observing the time course of water exchange was one of the oldest methods used in the water research. Fig. 1 shows an experiment in which the sartorius muscle of the frog (Rana esculenta) was put in a 2.5 times hyperosmotic Ringer's solution[9]. The analysis of the time course of water loss can be explained by supposing the presence of a minimum of two components of the muscle water. Using labelled water, this kind of experiment can be made more exact and more informative. However, this experiment proved that the muscle is not an osmotic sack, because in the equilibrium state the muscle contains about two times more water than it should contain in an osmotic system.

Using different perfusion methods, we studied the exchange curves of different tissue compartments. Analyzing the exchange curves which are usually negative exponentials, or are composed of a few exponentials, we can estimate exactly the exchange rates of a certain material[10].

Measuring the specific density of the water is an interesting method for determination of the state of water in muscle[11]. Table 2 shows the change of specific density of water as a function of water content of the muscle. In this experiment measurement of specific density was performed by the hydrostatic method.

DETERMINATION OF SOME THERMODYNAMIC PARAMETERS

a) The most adequate thermodynamic parameter for determination of the state of water in biological tissues is the vapor pressure. The shape of the curve in Fig. 2 shows clearly the type of water binding. That is when the system studied is a simple solute, then the function mentioned above is a linear one; in the case of a swollen system the

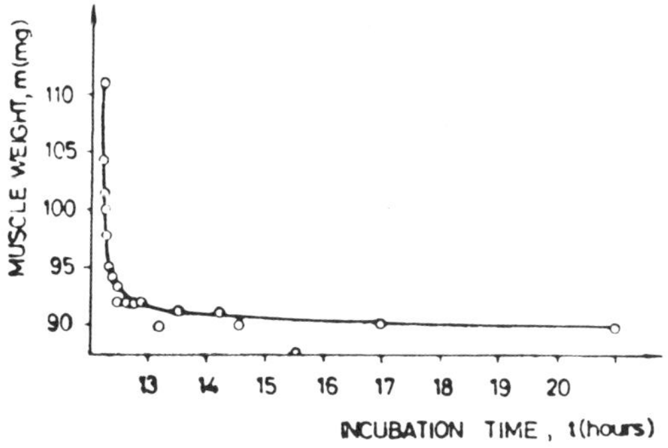

Fig. 1. The muscle weight (m) plotted against the time of soaking in a 2.5 hyperosmotic Ringer's solution.

Table 2.

N	i	d_i (g/cm^3)
1	4.01	1.012
2	2.99	1.014
3	1.69	1.025
4	1.40	1.032
5	0.58	1.076
6	0.30	1.141
7	0.19	1.192
8	0.14	1.234
9	0.12	1.26
10	0.10	1.28

curve has an S-shape. We have constructed differential manometers which offer a relatively exact determination of vapor pressure of water of muscle and other tissues. Fig. 3 shows the differential manometer of the author[5]. Fig. 4 shows a modified version made by Pócsik[13]. With this differential manometer, we could achieve an accuracy of 1% using 100 mg of tissue. Apparatus for direct determination of weight (water content) and vapor pressure is shown in Fig. 5 [9].

b) Determination of diffusion coefficient is a very useful method. However, there are a few cases when a simple indicator method is applicable[12].

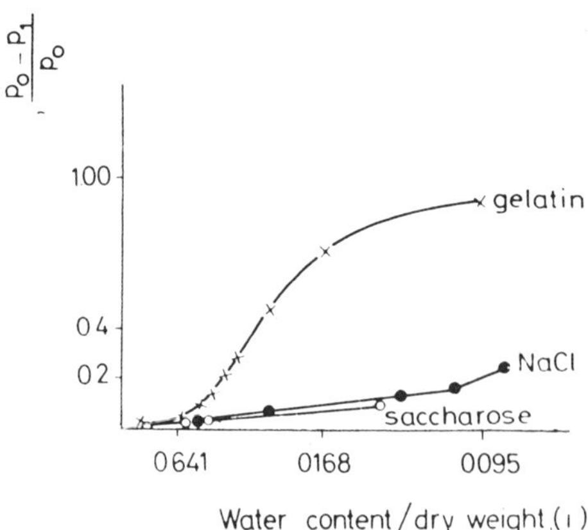

Fig. 2. Dependence of the relative vapor pressure of water on the water content in solutes (NaCl, sacharose) and a swollen (gelatin) system.

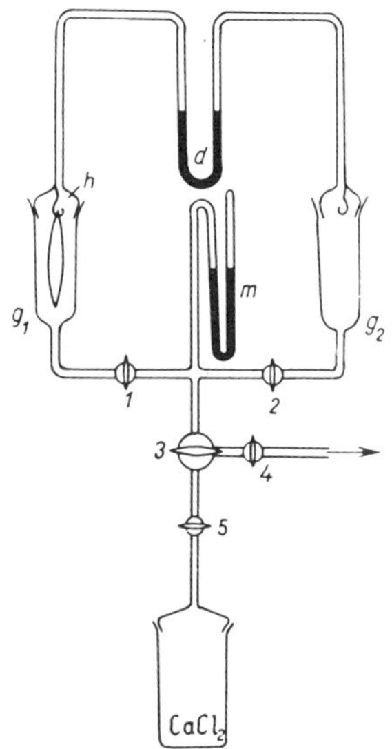

Fig. 3. Differential manometer for measuring the vapor pressure of water in biological tissues.

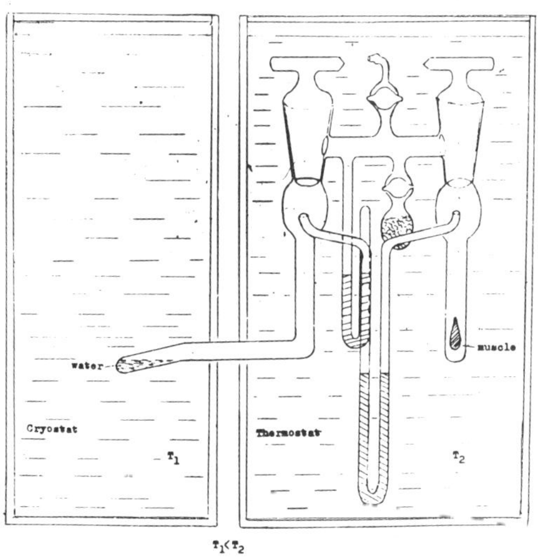

Fig. 4. A differential manometer modified by Pócsik[13].

c) Determination of evaporation heat is a very exact method because this parameter is in a very close relationship with the molecular state of water. Fig. 6 shows the results of Pócsik[14].

The evaporation heat of water increases successively with the decreasing water content. This method also gives an opportunity to calculate the binding energy of water.

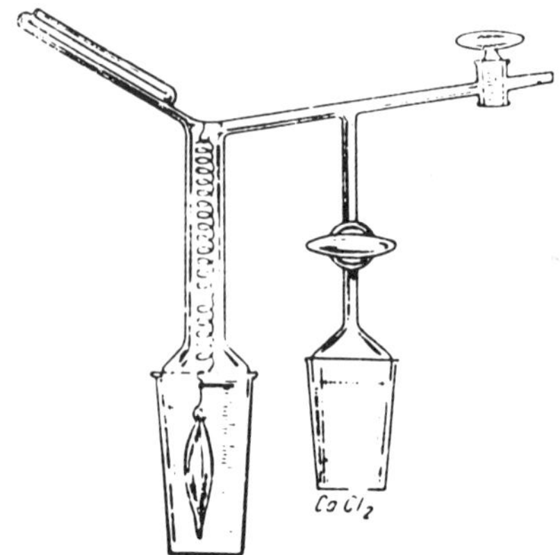

Fig. 5. Apparatus for simultaneous measurement of vapor pressure and the weight of the tissue.

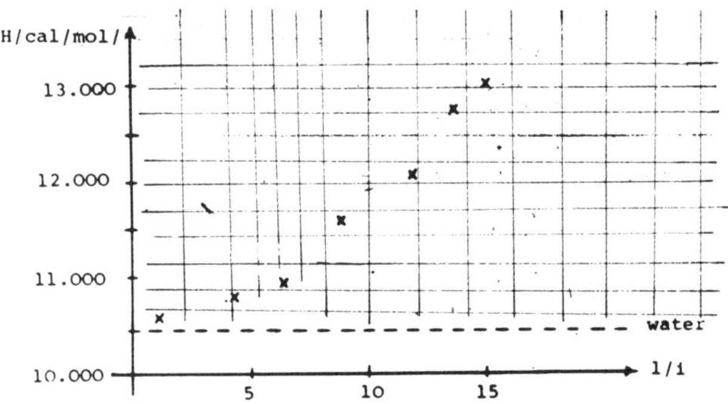

Fig. 6. The evaporation heat (H) of muscle water plotted against the reciprocal value of the relative water content ($\frac{1}{i}$).

ELECTRONIC METHODS

Electronic methods are suitable first of all to draw conclusions about the ionic concentration but they can also be used for obtaining indirect information about the state of water on the basis of the Nernst and Goldman equations.

a) The resting potential (RP) and action potential (AP) are widely used as indicators of muscle function. According to the membrane theory[15] the concentrations and the permeability constants determine the value of RP and the change of permeability coefficient is the reason for AP.

It is very interesting that the water content of the muscle influences AP and RP vary significantly. If we treat the muscle with

2.5 times hypertonic solution[16] or with D_2O Ringer solution the amplitude of an AP decreases to its half value without significant shift of ionic content (Fig. 7). The muscle looses only about 25% of its total water content, nevertheless the excitation-contraction coupling is totally blocked. So these experiments proved also the fundamental importance of studying the binding properties of water in muscle.

b) The measurements of the electrical conductivity can also give a lot of information about the state of ions and water in biological tissues. The DC and low frequency AC conductivity measurements are very much influenced by the structure of a tissue but using an ultra-high frequency current the hindrances of the structure can be avoided. The displacement of a K^+ ion e.g. at a frequency a 10^{10} Hz is less than 10^{-8} cm. The dipole rotation of water also can be measured by the microwave method. Masszi has elaborated such a method in our laboratory[17].

c) The use of ion specific electrodes offers a new way of studying the behavior of ions in biological tissues. A decade ago the smallest electrodes had a tip of about 1 μm which allowed a fundamental localization only in large cells. However, nowadays the tips of microelectrodes have diameter of 0.1 μm, so the future of the ion-specific electrodes seems to be very fruitful.

MAGNETIC RESONANCE METHODS

The use of magnetic field offers a very useful aid in water research. They are called radiospectrometric methods because of the range of the electromagnetic spectrum used.

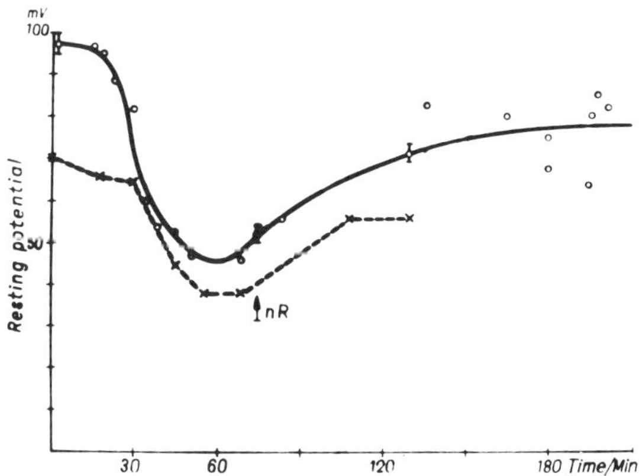

Fig. 7. The change of the resting potential of muscle kept from the zero time in a 2.5 hypertonic solution. Solid line: Rana pipiens dotted line: Bufo asiaticus. Arrow indicates the change of solution for normal Ringer's.

a) The nuclear magnetic resonance method (NMR) utilizes the very tiny change of magnetic dipole momentum and the change in the relaxation time(s) of the proton under different conditions. Hazlewood[6] was among the first who developed this procedure and his well-known and widely accepted results showed that in the muscle the water molecules are three different states. Rácz[2] in our laboratory succeeded in determining differences between the amount of bound water of cataractic and that of normal eye lenses. The medical significance of these experiments is evident. The differences in water binding in living tissues gave the opportunity of elaboration of NMR imaging which certainly helps medical diagnosis in a revolutionary manner. The development of NMR imaging is a good example of an unexpectedly prompt application of a fundamental procedure in practical medicine[8].

b) Another radiospectrometric method is the electron paramagnetic resonance spectrometry (EPR) which utilizes the microwave absorption properties of free radicals in a magnetic field. This procedure became the favorite method in membrane studies by the aid of stable free radicals (spin labels), e.g. using maleimide spin labels we can show details of the gating mechanism of ionic channels.

THE ELECTRON MICROSCOPE

An effective research tool in distribution of inorganic matter in biological tissues is the ELMI, and there are some promising procedures with high voltage electron microscopy which offer the possibility of imaging of tissues without dehydration. One of the most significant in these studies was offered by the ELMI radioautography. Fig. 8 shows the localization of K^{42} in the cross-striated muscle[18]. An essential requirement of this method was to prevent the displacement of water soluble ions during preparation. Fixation with ozmium tetroxide vapor seemed to solve this problem.

THE X-RAY MICROANALYZER

The combination of ELMI and X-ray spectrometer offers a revolutionary new possibility for determination of inorganic matter quantitatively in tissue sections. The sensitivity of the method is 10^{-14}g and the surface covered by the electron beam is 0.1 square μm.

Fig. 9 shows the difference in the concentration of potassium in the I and A band of cross-striation of a muscle fibril.

The difference between potassium concentration is different bands of the fibril is very strong proof in favor of the concept of bound potassium in the muscle.

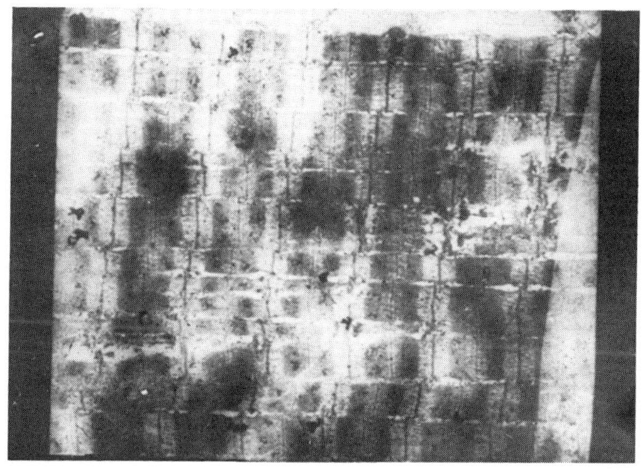

Fig. 8. Radioautographic localization of potassium (K^{42}) in cross-striated muscle.

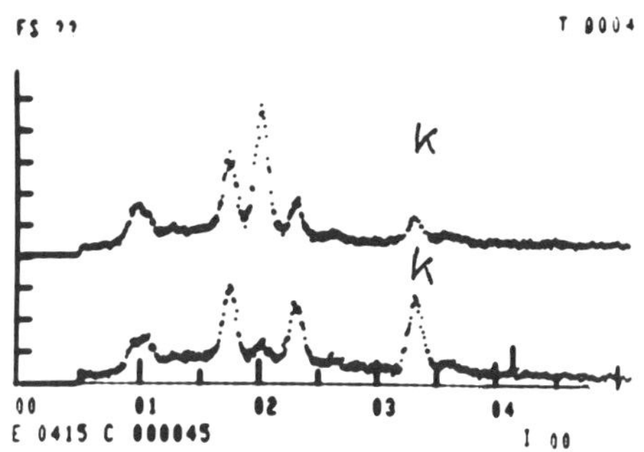

Fig. 9. X-ray microanalytic curve of a single muscle fibril.

THE LASER MICROSCOPE MASS ANALYZER

The laser microscope mass analyzer is a mircospectrophotometer in an improved form in which the advantages of homogeneous high intensity laser beams are utilized. Initially, it helps the high accuracy determination and localization of ions with a resolving power of 1 μm.

OPTOACOUSTIC SPECTROMETRY (OAS)

Optoacoustic spectrometry (OAS) is a new method in the study of structure of different substances. Rosencwaig[19] has developed the method which is based on the observation of very weak sound effects in a given sample caused by external light pulses. A small amount of biological tissue, e.g. 10 mg, is put in a photoacoustic chamber; an

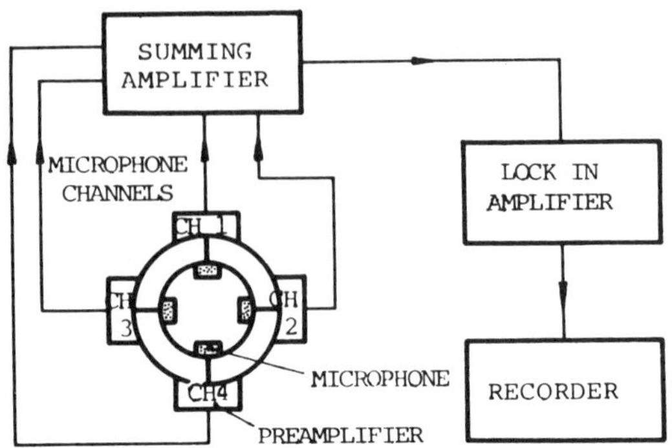

Fig. 10. Scheme of the optoacoustic chamber.

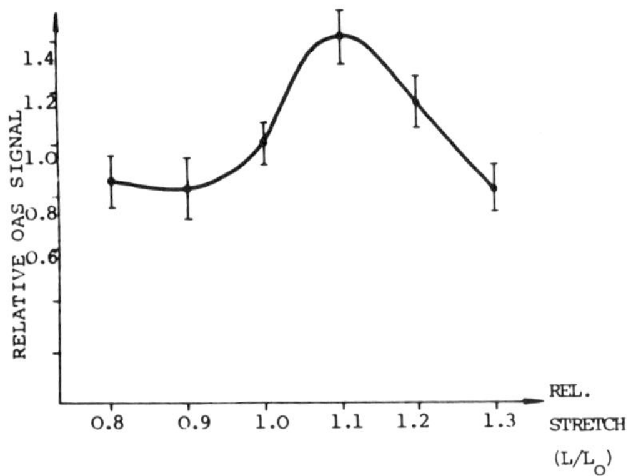

Fig. 11. The optoacoustic signal plotted against the length of the muscle (m. sartorius of the frog, Rana esculenta).

intense laser or other monochromatic light pulse of 30-100 Hz causes thermodilation and thermodiffusion effects periodically, which can be detected as a sound. The intensity of the optoacoustic signal is in a close correlation with the structure of the material. Hence one can detect the structure (even water binding) of the specimen. Fig. 10 shows the optoacoustic chamber made at our Institute.

Fig. 11 shows the relationship between the length and optoacoustic signal of the frog muscle sartorius[20]. It is very interesting that this function is very similar to the curve which was found by Huxley between the intensity of a certain synchroton X-ray diffraction line and the muscle length[21].

The complete interpretation and methodology of OAS is a task for the near future but it seems to be a hopeful method.

I am aware that I was unable to give a complete picture of all the methods of water and ion-research in biology. My subjective picture was dominated mostly by the practice used in our laboratory.

The methodology of research in this field has developed during the last decade very rapidly, offering new and more detailed information about basic relations of living matter.

REFERENCES

1. J. Loeb, Physiologische Untersuchungen über Ionenwirkungen, Arch.Ges.Physiol., 69:1 (1898).
2. E. Overton, Beiträge zur allgemeinen Muskel- und Nervenphysiologie, Arch.Ges.Physiol., 105:176 (1904).
3. E. Ernst and K. Czimber, Die Bindungsart des Wassers im Muskel, Arch.Ges.Physiol., 228:68 (1931).
4. A. V. Hill, The state of water in muscle and blood and the osmotic behavior of muscle, Proc.Roy.Soc.London B., 106:477 (1930).
5. E. Ernst, J. Tigyi, and A. Zahorcsek, Bindungszustand des Wassers und der Elektrolyte im Muskel, Acta Physiol.Acad.Sci. Hung., 1:5 (1950).
6. C. F. Hazlewood, ed., Annals of the New York Academy of Sciences 204 (1973).
7. K. Tompa and P. Rácz, Proton relaxation in normal and cataractous crystalline lenses, Proceedings of the XIX Congr. Ampere, Heidelberg, (1976).
8. H. R. Pykett, I. L. Mansfield, and P. Morris, NMR Imaging, in: "Lecture Notes in Physics," M. Schenkler, M. Fink, eds., New York, Vol. 112, 453 (1980).
9. J. Tigyi, The Role of Water in Muscular Activity, Periodicum Biologorum, 80:74 (1978).
10. J. Tigyi, "A Kisérleti Orvostudomány Vizsgáló Módszerei," 7:219, Budapest, (1965).
11. S. Pócsik, Bound water in muscle, Acta Biochim.Biophys.Acad.Sci. Hung., 2:149 (1967).
12. C. F. Hazlewood, Diffusion of Water in Striated Muscle, Ann.NY. Ac.Sci., 204:444 (1973).
13. S. Pócsik and L. Nagy, Relationship between stimulusthreshold and Water Content in Muscle, Studia Biophysica, 91:75 (1982).
14. S. Pócsik and J. Tigyi, Water Binding in Biological Systems Measured by Thermodynamic Methods Studia Biophysica, 84:35 (1981).
15. A. F. Huxley, CIBA Foundation Symposium, 31:171 (1975).
16. J. Tigyi, Relationship between the Resting Potential, Ion Content and Extracellular Space in Striated Muscle Treated with Hypertonic Solution, Acta Biochim.Biophys.Acad.Sci.Hung., 2:169 (1967).

17. G. Masszi, Dielectric Relaxation and Water Structure in Gelatine Solutions, Acta Biochim.Biophys.Acad.Sci.Hung., 7:349 (1972).
18. J. Tigyi, N. Kállay, A. Tigyi-Sebes, and K. Trombitás, Distribution and Function of Water and Ions in Resting and Contracted Muscle, *in*: "Internat. Cell Biology," 1980-81, H.G. Schweiger, Springer (1981).
19. A. Rosencwaig, "Photoacoustics and Photoacoustic Spectroscopy," Wiley, New York (1980).
20. G. Tigyi, J. Hegedus, and Ildiko Gargitai, Opoacoustic Spectrometry as a New Tool in the Structural Studies of the Striated Muscle, Studia Biophysica, (1983) in press.
21. H. E. Huxley, High Speed Time Resolved X-Ray Diffraction Studies on Muscle Contraction using Symchrotom Radiation, VII, Internat.Biophys.Congr.Abstracts, p. 17 Mexico City (1981).

WATER AND IONS IN NEUROBIOPHYSICAL PROCESSES

Vasile Vasilescu, Eva Katona and Cornelia Zaciu

Department of Biophysics
Faculty of Medicine
Bucharest - Romania

INTRODUCTION

Deeper understanding of phenomena involved in the functioning of the nervous system required the knowledge of the underlying physico-chemical mechanisms. This entailed the development of a new out-standing direction in neurobiology: neurobiophysics[1]. In spite of the spectacular findings in this field, recorded during the latest decades, many questions remained still unanswered. Thus the mechanisms of generation and propagation of excitation in various nervous structures or the mechanisms of excitation transmission between cells and of potential generation as response to external stimuli are rather poorly elucidated at the molecular level. The study of ions and of water role in the functioning of nervous system seems to be promising for a better understanding of all these mechanisms.

Knowledge of the importance of water and protons in the structure and function of various excitable systems has been a major concern in our Laboratory for many years[2-25]. The present work is a synthesis of some relevant data with emphasis on those recently obtained concerning the involvement of water and protons in the basic physico-chemical mechanisms which underpin the functioning of frog peripheral nerve and retina.

In order to preserve and control the function of various nervous structures non-destructive methods were mainly used. The dynamics of water molecules in the nerve and its dependence on the nerve state were investigated by nuclear magnetic resonance (NMR) measurements and by deuteration kinetics studies. There were followed up the effects of deuteration and of some neurotropic substances with reversible action, thus revealing the role of water in nerve anesthesia and in the cellular bioenergetics of nerve and retina. The action of partial and quasi-total deuteration on excitability characteristics of rythmically stimulated isolated nerve fibres was also investigated, disclosing the role of intracellular protons in the functioning of nerve fibre.

MOLECULAR DYNAMICS OF WATER IN THE PERIPHERAL NERVE

Water in the peripheral nerve, representing up to 76% of the total fresh weight, is constituted of several different compartments having different motional properties [10,12-13,21,24,26]. Analysis of our data concerning deuteration kinetics of frog sciatic nerve [10,12] revealed the existence of at least 3 distinct tissue water compartments, of different diffusion accessibility; their sizes and properties are dependent on the nerve state as illustrated in Fig. 1a. Our NMR studies concerning the relaxation properties of water protons in nerve [13,21,24] allowed to disclose at least 3 distinct relaxation components reflecting the existence of at least 3 subpopulations of water molecules having different mobility properties. The sizes and properties of these water compartments also appeared to be dependent on the nerve state, as illustrated in Fig. 1b.

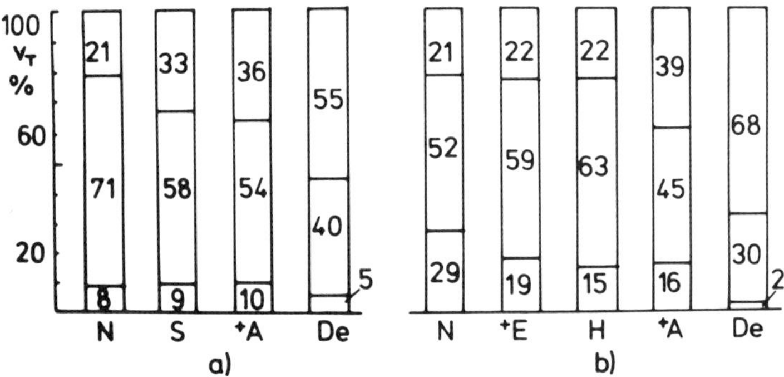

Fig. 1. Water compartments in a frog sciatic nerve. Figures from top to bottom represent the volumes of the fast exchanging, slowly exchanging and non-exchangeable water (a) and the volumes of slowly relaxing, of intermediate relaxation rate and fast relaxing water (b) respectively, expressed as % of the total water content N - normal, resting nerve, S - continuously stimulated, +A - acethylcholine treated, + E - anesthetized by ethyl ether, +H - halothane treated, De - "dead" nerves (kept 48 h at room temperature)

In order to follow up influence of the tissue deuteration degree on the mobility properties of water and on the state of nerve solid components, as well as influence of the state of nerve solid components on the tissue water properties, we investigated the relaxation behaviour of water protons in normal, in partially or quasi-totally deuterated and in "re-protonated" frog sciatic nerves.

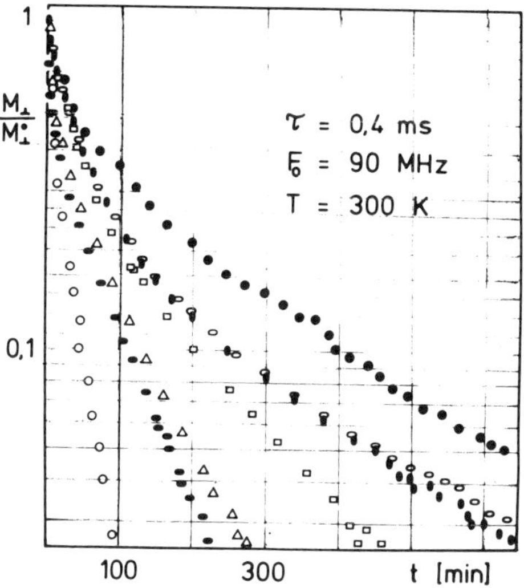

Fig. 2. Dependence of the spin-spin relaxation behaviour of water protons on the deuteration degree of frog sciatic nerves. The measurements were made at room temperature, by using Carr-Purcell-Meiboom-Gill technique with sequence pulse spacing τ = 0.4 ms. Different symbols correspond to different light water contents: ▮ - fresh nerve, total water volume = V_T; □ - deuterated nerve, light water content = 0.30 V_T; Δ - deuterated nerve, light water content = 0.12 V_T; ⬬ - quasi-totally deuterated nerve, light water content = 0.08 V_T. ⬭ - the same nerve, after rehydration, light water content: 1.01 V_T; o - deuterated and dehydrated nerve light water content = 0.08 V_T; • - deuterated, dehydrated and rehydrated nerve, light water content = 0.96 V_T.

A BKR-SXP Bruker resonance spectrometer was used for pulsed proton NMR measurements at 90 MHz. The spin-lattice relaxation times were determined by the inversion-recovery method, while the spin-spin relaxation behaviour by the Carr-Purcell-Meiboom-Gill method. With the view of obtaining different degrees of deuteration, the nerves were immersed for various time intervals in Ringer solutions prepared with 99.8% heavy water. Extent of the nerve deuteration was determined from extrapolation at the moment zero of the free-induction decay obtained after a 90° pulse. For "re-protonation" the deuterated nerves were immersed in normal Ringer solution.

Table 1. Variation of the spin-spin and spin-lattice relaxation behaviour of water protons in the frog peripheral nerve as a function of the deuteration degree (T = 300 K)

v ($\%v_T$)	T_1 (ms)	T_{2i} (ms) 1	2	3	$P_i^{(a)}$ ($\%v_T$) 1	2	3
100	1047	310	65	15	21	51	28
60	879	270	70	12	12	24	24
30	750	214	74	10	6	8	12
12	608	140	58	8	2,6	3	6,4
8	704	113	51	7	1,7	2,2	4,1
	v' ($\%v_T$)						
⊖ 8	101	320	66	15	21,5	50,5	28
○ 8			43	7		3,4	4,6
● 8	96	380	80	20	29	50	21

v – content of ordinary water in the deuterated nerve

v_T – total water content in the freshly dissected nerve

v' – ordinary water content following "re-protonation" or rehydration of the deuterated or deuterated and dehydrated nerves respectively

T_{2i}, p_i – spin-spin relaxation times and apparent fractional populations of the components obtained by graphical analysis of relaxation decays

T_1 – spin-lattice relaxation time

⊖ – quasi-totally deuterated nerve after "re-protonation";

○ – dehydrated deuterated nerve; ●–dehydrated deuterated nerve after rehydration

The relaxation behaviour of protons in the water of deuterated and "re-protonated" nerves is shown as varying with the deuteration degree of the nerve in Fig. 2 and Table 1.

The decrease in the ordinary water content of the nerve is observed to be accompanied by a gradual decrease of the spin-lattice relaxation time and by an apparent redistribution of the remaining ordinary water in the compartments initially observed. Thus, even in the quasi-totally deuterated nerve, the ordinary water for which heavy water cannot be exchanged through diffusion is divided into 3 components that is, besides the water associated to tissue macromolecules small fractions of slowly relaxing components are present. The changes induced by deuteration in the state of solid components of the nerve are reversible. Following "re-protonation", not only the total water content, but also the sizes and properties of water compartments appear virtually unmodified. Dehydration of the deuterated nerve results in disappearance of the slowly relaxing water fraction and occurence of irreversible changes in the state of solid components. After rehydration the dehydrated deuterated nerves have actually the same total water content as control nerves, but the properties of different fractions are modified.

WATER IN NERVE ANESTHESIA

Analysis of pulsed NMR data from anesthetized and control myelinated nerves shows that water is involved also in the mechanisms of anesthesia. A comparison of spin-spin relaxation decays of water protons in the peripheral nerve anesthetized by ethyl ether or by halothane with those obtained for the same nerve before anesthesia - as illustrated in Fig. 3 - reveals a slowing of the relaxation process as a whole, in the anesthetized nerves. This fact may hardly agree to the theory of water ordering during the anesthetic action. Graphical analysis of relaxation decays shows that the apparent relaxation times of the slowly relaxing component, and of that having intermediate relaxation rate, are indeed somewhat increased, the relaxation time of the fast relaxing component being practically unchanged. More pronounced changes however occur in the fractional populations of components, the component of intermediate relaxation rate enhancing while the fast relaxing component shows a decrease. The amount of "bound water" found to be 0.9 g H_2O/g dry mass in the normal peripheral nerve is 0.6 g H_2O/g dry mass in the ethyl ether anesthetized nerve and 0.5 g H_2O/g dry mass in a nerve anesthetized by halothane. These facts constitute a proof supporting Eyring's idea[27] that general anesthetics act by inducing protein conformation changes similar to those leading to reversible thermal inactivation of enzymes, which changes are usually accompanied by release of some amount of "bound water".

This mechanism of action of general anesthetics is also supported by our results concerning the kinetics of anesthesia settling in normal and deuterated peripheral nerves[17,28].

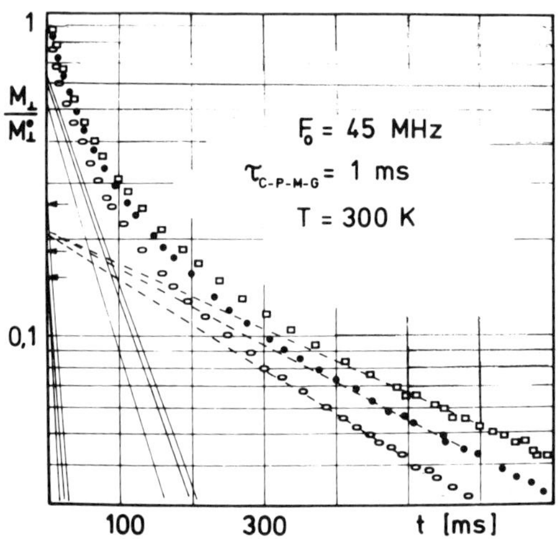

Fig. 3. Spin-spin relaxation decay of water protons in the frog sciatic nerve anesthetized by ethyl ether (●) or by halothane (□) as compared to that of the same nerve before treatment (○) and the relaxation components as revealed by graphical analysis.
Arrows indicate fractional populations corresponding to "bound water"

Using general anesthetics a significant increase of the time lag necessary to the settling of anesthesia was found in deuterated nerves as compared to controls (Fig. 4). This increase appears to be temperature-dependent (Fig. 5). As in quasi-totally deuterated nerves this time lag increase amounte to about 50% at room temperature, a diminution in the ion mobility cannot explain alone the observed phenomenon; most probably the enhanced stability of the native, more compact conformation of membrane proteins in deuterated nerves must equally be taken into account.

The true site and the mechanism of action of general anesthetics are not yet known. Either the membrane lipid phase or the protein hydrophobic region is their site of action, a change in the conformation

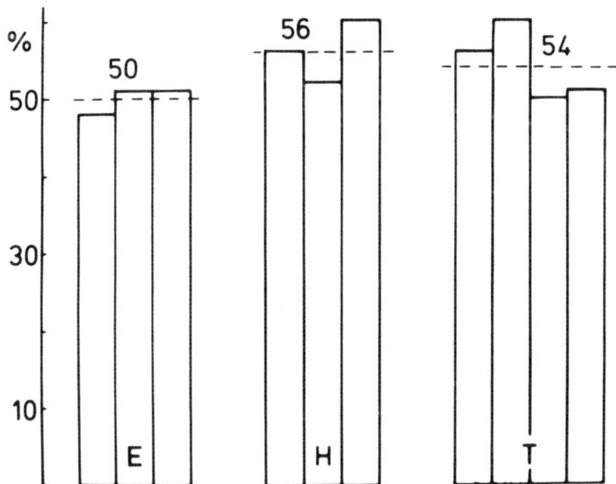

Fig. 4. Increase of time lag till settling of anesthesia in deuterated frog nerves as compared to control pairs, at T = 292 K. General anesthetics used: 2.5% ethyl ether (E); 2.5% halothane (H); and 2.5% thiopenthal (T)
Solvent vehicle: H_2O-Ringer and D_2O-Ringer respectively

of protein molecules seems to be involved. This change may occur under the direct action of anesthetics or as a result of the modification of the lipid phase. The fact that the presence of D_2O, stabilizing the more compact native conformation of proteins, delays settling of anesthesia also supports the theories according to which anesthetics act by making some essential proteins pass to a more unfolded, inactive conformation[27].

WATER AND PROTONS IN CELLULAR BIOENERGETICS

Dynamics of energy pools of excitable cells as well as its variation under deuteration and on continuous activity were followed up.

We used symmetric nerves and retina taken from the same frog. ATP concentrations in tissues were measured by Strehler's

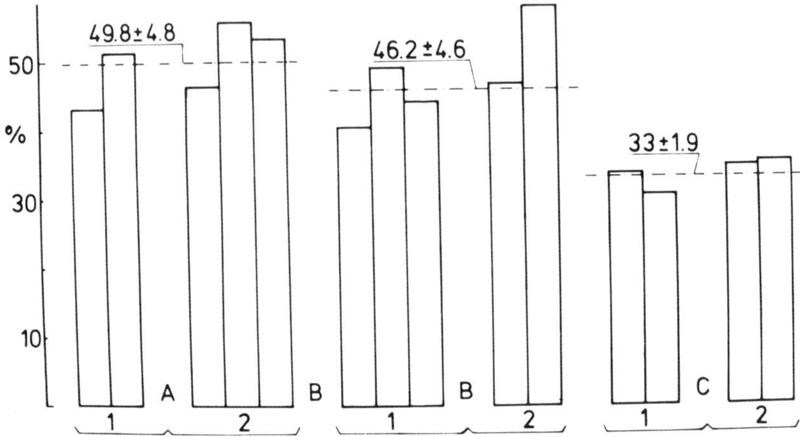

Fig. 5. Increase of time lag till settling of anesthesia in deuterated frog nerves as compared to control pairs at three different temperatures
T = 292 K (A), 295 K (B) and 300 K (C)
General anesthetic used: 2.5% ethyl ether
Solvent vehicle: H_2O-Ringer (1) and D_2O-Ringer respectively

bioluminescence method[29] modified by us[30], with the view of higher sensitivity.

The cellular ATP pool was revealed to be function of the tissue state. ATP consumption of deuterated systems, even in resting state, appears substantially enhanced as compared to that of the control ones. The difference is still more pronounced in the functioning systems, as illustrated in Fig. 6.

Simple immersion of the nerve in H_2O-Ringer solution results in a decrease of the cellular ATP pool by ∿ 4% in 90 min and its immersion in D_2O-Ringer, to a diminution reaching ∿ 22% in 90 min. Rhythmic stimulation induces a decrease of ∿ 16% in the ATP pool of control nerves and of ∿36% in those deuterated. ATP consumption of the deuterated resting nerve is increased by ∿ 21% in 90 min as against the control and, in case of continuous functioning, an additional

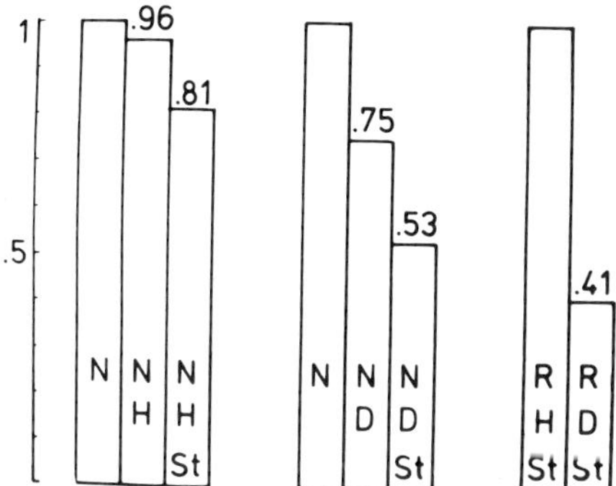

Fig. 6. Effects of rythmic stimulation and deuteration on cellular ATP content in peripheral nerve and retina
N - fresh nerve; N-H and N-D - resting nerves kept 90' in H_2O-Ringer and D_2O-Ringer respectively; N-H-St - nerve pairs after 90' of 50 Hz rythmic stimulation in H_2O-Ringer and D_2O-Ringer respectively; R-H-St and R-D-St - retina pairs kept 30' in dark and then 10' illuminated (white light, 40 lx) in H_2O-Ringer and D_2O-Ringer respectively
The figures represent the population mean values obtained for 4-6 groups of experiments on 6-8 pairs of nerves and retina respectively. Standard errors never exceed 12% of the mean values

increase of ∼ 14% in 90 min of ATP consumption is found in the deuterated nerve versus the control.

In retina the decrease of cellular ATP contents, on deuteration, is even more marked, namely up to ∼ 60%, as appears from a comparative experiment performed on retinae immersed for 30 min in D_2O-Ringer and H_2O-Ringer respectively, in the dark and then illuminated for 10 min.

Longer periods of immersion in D_2O-Ringer leading to a greater deuteration degree cause even more striking differences between the cellular ATP pools of deuterated and control systems.

The observed modifications are possible to be accounted for by the differences between the physical properties of protons and deuterons. Changes in the ion mobilities and also in the strength of intermolecular forces, entailing differences in the hydration degree of various species, can affect the kinetics of ATP synthesis and the ATP $\rightleftharpoons$ ADP equilibrium as well.

WATER AND PROTONS AS INVOLVED IN EXCITABILITY OF ISOLATED NERVE FIBRE

We investigated deuteration effects on the excitability of the Ranvier node membrane in the stationary state, by using voltage-clamp technique.

Another series of studies dealt with the effects of deuteration and of rehydration on the dynamical behaviour of the Ranvier node membrane. By dynamical behaviour we mean the way in which, the membrane functions during rhythmic stimulation and, as a result, the action potential appears and may be recorded.

The action potential (AP), along with the stimulus amplitude (P_{μ}) and the stimulus duration (D) were established by us as excitability parameters. As the stimulus energy is proportional to $P_{\mu}^2 D$, in order to meet the excitability conditions one obviously will change either the amplitude (P_{μ}) or the duration (D) of the stimulus.

Experiments were performed under external and quasi-total deuteration. For external deuteration the compartment of the node to be investigated was continuously flooded by D_2O-Ringer[18]. In order to realize a quasi-total deuteration of the fibre, we introduced D_2O-Ringer since the last stage of dissection and immersed afterwords the fibre 3-4 times in D_2O-Ringer; all the compartments of the chamber where the fibre was to be placed were also filled with D_2O-Ringer.

Our results can be summarized as follows:

Voltage-clamp experiments on the Ranvier node membrane at 12-15°C revealed that, 30 min after its immersion in D_2O, the ionic currents (both Na^+ and K^+) decrease by a factor of 30-40%, while the time-to-peak inward current increases by 10-20%[20].

When measuring ionic currents a second order phenomenon must be taken into account, namely the nodal gap resistance which modifies under deuteration. Our computer simulation shows that this error might be minimized by fitting the maxim-maximorum value of the ionic currents. In this case computer simulation[23] indicates minimal errors facing to the currents of the negative slope curve of the current-voltage characteristics.

As concerns the effects of external deuteration on the dynamical behaviour of the Ranvier node membrane, the following may be mentioned:

- The possible maximum frequency of stimulation at constant threshold decreases to half; the phenomenon was found to have rapid kinetics (2-3 min).

- The energetic condition of the stimulus still inducing excitation, by a variation either of its amplitude (P_μ) or of its duration (D), is expressed in the ratio $R_S = (P_\mu^2 D)_{D_2O} / (P_\mu^2 D)_{H_2O}$. This ratio found initially to be of only 1.62, increases up to 2.5 after 30 min of stimulation (Fig. 7).

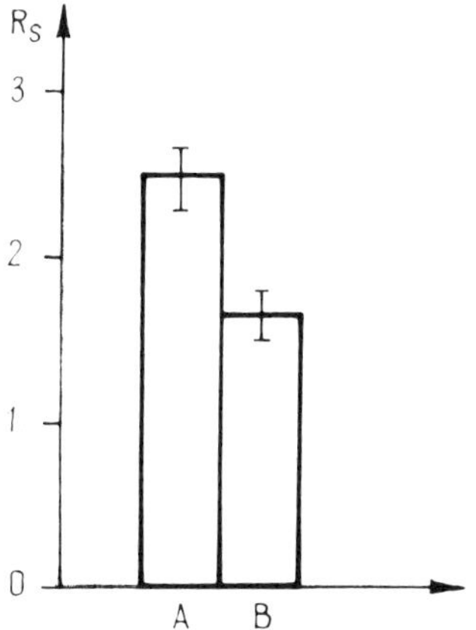

Fig. 7. Ratios (R_S) between the energy necessary to stimulus for keeping excitable the membrane immersed in D_2O-Ringer and that found in the case of the membrane immersed in H_2O-Ringer
A - ratios calculated after 30 min of stimulation
B - ratios calculated at the begining of experiment (considered as the moment $t = t_o$)

The above clearly show an interference of deuteration with the mechanisms involved in the excitability process during long-term stimulation.

From our studies related to the effect of quasi-total deuteration on the dynamical behaviour of the myelinated nerve fibre it appears that:

- The possible maximum frequency of rhythmic stimulation at constant threshold diminishes to a quarter of that found on membrane immersion in H_2O-Ringer, which is about 200 Hz (Fig. 8).

- Following 10-15 min of rhythmic stimulation with 150-200 Hz, the quasi-totally deuterated membrane loses its excitability, while the nondeuterated membrane keeps it for a long time (more than one hour). Coming back to a low frequency of stimulation (10 Hz), the nerve fibre submitted to the same quasi-total deuteration is observed to recover its excitability.

Rehydration effects, after excitability loss induced by quasi-total deuteration and high-frequency stimulation, are as follows:

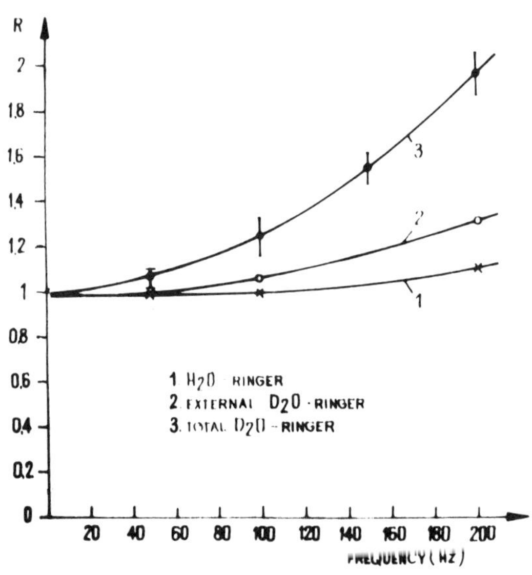

Fig. 8. Variation of the excitation threshold as a function of the stimulation frequency, under different conditions of fibre immersion. Curve (1) and (2) are obtained for the same membrane. Curve (3) is a plot of averages of 10 measurements; each R represents the ratio of the threshold at a given frequency f to that at the frequency f = 10 Hz.

- On external rehydration, when only the node under investigation is immersed in H_2O-Ringer, the other compartments, excepting that of the node, being maintained in D_2O-Ringer, excitability continues to be inhibited during stimulation with high frequencies.

- During a total rehydration, when the whole fibre is kept in H_2O-Ringer, excitability strongly reappears at high frequencies of stimulation; the phenomenon has a rapid kinetics.

Finally we concluded that reappearance of excitability, at low frequencies of stimulation and on the same quasi-total deuteration of nerve fibre, is indicative of the existence of one or more mechanisms strongly coupled with the excitability phenomena at high frequencies. Excitability abolishment induced by quasi-total deuteration and rhythmic stimulation becomes reversible following internal rehydration; therefore the mechanism(s) mentioned above is (are) drastically affected by proton-deuteron exchange in the intracellular medium.

Taking into account also our data revealing a substantial decrease of ATP pool in the deuterated nerve during stimulation, we can conclude that the results obtained are consistent with the idea of excitation-energy coupling which is more evident at high frequencies of stimulation and strongly affected by the proton-deuteron exchange in the intracellular medium.

CONCLUSIONS

- Water distribution in the peripheral nerve is dependent on the tissue state. Sizes and properties of water compartments in resting state systems are different from those found in the functioning ones. Reversible or irreversible abolishment of nerve function determines changes in tissue water distribution. Anesthesia settling leads to a decrease in the amount of the so-called "bound water"; continuous stimulation or acethyl-choline treatment results in significant enlargement of the extracellular fast-relaxing fast-exchanging nerve-water component. After death changes in nerve seem to bring about a gradual loss of compartmentation.

- Substitution of heavy water for tissue water modifies the kinetics of various processes: it alters most of the electrical parameters related to the kinetics of excitability processes and prolongs the time lag necessary to anaesthesia settling in the nerve.

- The forces responsible for the total water content of peripheral nerve are not significantly modified by deuteration; deuteration-induced changes in the state of nerve solid components are almost totally reversible; the water for which heavy water is not exchanged through diffusion in deuterated nerves is heterogeneous and seems to have mobility properties different from those of the water in the control nerve and from those of the water remained in dehydrated nerves.

- Heavy water for light water substitution interferes with cell energy mechanisms affecting ATP ⇌ ADP equilibrium and modifying the dynamics of cellular ATP pools.

- Deuteration acts also on bioelectrogenesis processes, particularly on excitation-energy coupling and on the early stages in the generation of action potential.

- Under voltage clamp conditions, replacement of light water by heavy water entails a decrease of the ionic currents measured in the Ranvier node membrane.

- The dynamical behaviour of the Ranvier node membrane is found to be affected even by only an external deuteration, but much more by a quasi-total deuteration: a higher energy is necessary for maintaining the excitability conditions in the deuterated membrane as compared to the control; on continuous stimulation quasi-total deuteration, leads to abolishment of excitability of the Ranvier node membrane. Total rehydration of the nerve fibre results in a rapid resumption of its function.

REFERENCES

1. V. Vasilescu and D. - G. Mărgineanu, "Introduction to Neurobiophysics", Abacus Press, Tunbridge Wells, Kent (1981)
2. V. Vasilescu, D. - G. Mărgineanu, D_2O Effect on Active Transport through Frog Skin, Naturwiss, 57:126 (1970)
3. V. Vasilescu, D. - G. Mărgineanu, Investigations on the Effects on Deuteration on Some Functions of Biological Membranes, Rev. Roum. Physiol. 8:217 (1971)
4. V. Vasilescu, D. -G. Mărgineanu, The Role of Water in Biological Membrane Phenomena as Revealed by Deuterium Isotope Effects, Rev. Roum. Physiol., 11:167 (1974)
5. V. Vasilescu, D. -G. Mărgineanu, Eva Katona, Kinetics of H_2O-D_2O Exchange in Muscle, Naturwiss, 62:187 (1975)
6. V. Vasilescu., Eva Katona, D. -G. Mărgineanu, Kinetics of D_2O Intake in the Striated Muscle, Studia Biophys. 52:223 (1975)
7. V. Vasilescu, Cornelia Zaciu, E. Truţia, ^2H_2O Effects on ATP Pool in Frog Stimulated Nerve, Rev. Roum. Morphol. Embriol. Physiol., 13:35 (1976)

8. Eugenia Chirieri , Constanţa Ganea, Ioana Aricescu, V. Vasilescu, Investigations concerning the Changes Induced by Deuterium for Hydrogen Substitution in Bioelectric Activity of the Frog Retina, Rev. Roum. Morphol. Embriol. Physiol., 14:119 (1977)
9. Eugenia Chirieri, Ioana Aricescu, Constanţa Ganea, V. Vasilescu, The Effects of Deuteration on the Frog Retina Bioelectrogenesis, Naturwiss, 64:149 (1977)
10. V. Vasilescu, D. - G. Mărgineanu, Eva Katona, Heavy Water intake in Tissues. II. H_2O-D_2O Exchange in the Myelinated Nerve of the Frog, Experientia, 33:192 (1977)
11. V. Vasilescu, A. Ciureş, Influence of Heavy Water on the Cardiac Automatism, Studia Biophys., 71:81 (1978)
12. V. Vasilescu, Eva Katona, D. - G. Mărgineanu, Recherches sur les Etats et le Role de l'eau dans les tissus excitables, Rev. Roum. Méd. Neurol. Psychiat., 16:168 (1978)
13. V. Vasilescu, Eva Katona, V. Simplăceanu, D. Demco, Water Compartments in the Myelinated Nerve. III. Pulsed NMR Results, Experientia, 34:1443 (1978)
14. Eva Katona, D. -G. Mărgineanu, V. Vasilescu, Water Compartments in Living, Glycerinated and Fixed Skeletal Muscles of the Frog, Cell Tiss. Res., 203:331-338 (1979)
15. Constanţa Ganea, V. Vasilescu, Heavy Water Effects on Certain Energetic Processes in Retina, Rev. Roum. Morphol. Embryol. Physiol. PHYSIOL., 16:59-62 (1979)
16. V. Vasilescu, Eva Katona, Heavy Water Effects on Excitable Membranes, in Frontiers of Bioorganic Chemistry and Molecular Biology (ed. S. N. Ananchenko), Pergamon Press, Oxford, New York, 445-464 (1980)
17. V. Vasilescu, Eva Katona, New Data concerning the Role of Water in Biosystems, Rev. Roum. Physiol., 17:3-17 (1980)
18. V. Vasilescu, Cornelia Zaciu, Deuteration Effect on Dynamical Behaviour of the Ranvier Node Membrane, Studia biophys., 78:107-118 (1980)
19. V. Vasilescu, Water in biosystems, Studia biophys., 84:1, (1981)
20. Cornelia Zaciu, V. Vasilescu, Excitability properties of the Ranvier node membrane as investigated by deuteration, Studia biophys., 84:71-72 (1981)
21. Eva Katona, V. Vasilescu, Pulsed NMR data on water state in nerve, Studia biophys., 84:79-80 (1981)
22. Constanţa Ganea, V. Vasilescu, Heavy Water Effects on the kinetic and thermodynamic parameters of bioluminescence, Studia biophys., 84:59-60 (1981)

23. Cornelia Zaciu, Mioara Tripşa, V. Vasilescu, Effect of the Nodal gap resistance on ionic current measurements in the Ranvier node membrane as revealed by computer simulation Biophys. J., 36:797-802 (1981)
24. Eva Katona, Constanţa Ganea, A. Popescu, V. Vasilescu, Dynamics of Water in hthe Peripheral Nerve in "Biophysics of Water" F. Franks, ed. John Wiley and Sons Limited, Chichester, New York (1982)
25. A. Popescu, Eva Katona, Constanţa Ganea, V. Vasilescu Kinetics of Vacuum Dehydration in the Study of Tissue, Water in "Biophysics of Water", F. Franks, ed., John Wiley and Sons Limited, Chichester, New York (1982)
26. D. C. Chang and C. F. Hazlewood, Nuclear Magnetic Resonance Study of Squid Giant Axon, Biochim. Biophys. Acta 630:131-136 (1980)
27. H. Eyring, J. W. Woodbury and J. S. D'Arrigo, A Molecular Mechanism of General Anesthesia, Anesthesiology 38:415-424 (1973)
28. V. Vasilescu, Eva Katona, A. Popescu, Cornelia Zaciu. Constanţa Ganea, Some Problems concerning the Role of Water and Protons in the Function of Biological Membranes in "Membrane Processes: Molecular Biological Aspects and Medical Applications", G. Benga, H. Baum and F. Kummerow, eds., Springer Verlag, New York, in press
29. B. L. Strehler, Adenosine-5'-triphosphate (Determination with Luciferase) in "Methods of Enzymatic Analysis, Haus-Ulrich Bergmeyer, ed., Academic Press, London, pp. 558-573 (1963)
30. V. Vasilescu, E. Truţia and Cornelia Zaciu, Contributions to the perfecting ofthe enzymatic method used for the determination of ATP in biological systems, Rev. Roum. Morphol. Embryol. Physiol., Physiologie, 12:87-90 (1975)

SYMPOSIA

INTERACTIONS IN TERNARY SYSTEMS - WATER, IONS, BIOPOLYMERS

DYNAMIC INTERACTION OF WATER MOLECULES, INORGANIC IONS AND CHROMATIN STRUCTURES IN ISOLATED THYMUS NUCLEI

Miklos Kellermayer* and Carlton F. Hazlewood

Department of Physiology, Baylor College of Medicine
Houston, Texas 77030
*Present address: Department of Clinical Chemistry
University Medical School, Pécs, Hungary 7624

In the last decade, our knowledge has tremendously improved as far as the molecular anatomy of the genes concerned, but the mechanism of the structural and functional organization of the genetic material within the nucleus of the living cell is still obscure. What we know is that the water molecules, inorganic ions and certain mobile proteins are involved in the dynamic supramolecular organization of the genetic material (Agutter and Richardson, 1980; Belmont and Nicolini, 1981; Bonner, 1978; Cameron et al., 1977; Kellermayer et al., 1974; Kellermayer, 1981). The relationship between hydration of chromatin, volume changes of nuclei and the gene activity can be well documented with one of our earlier observations on nuclei of chicken erythrocytes hybridised with SV 40 virus induced tumor (H-50) cells (Kellermayer et al., 1978). Nuclei of the chicken erythrocytes are inactive as far as their DNA and RNA synthesis concerned, but after insertion into actively growing cells, they undergo a rapid reactivation (Harris, 1967). Twenty-four hours after fusion, parallel with the activation of the gene functions, the total dry mass of the inserted nuclei increases, but the mass concentration decreases (Table 1.). The increased dry mass of the active nuclei is mainly due to the intranuclear accumulation of mobile proteins, but there is no reasonable explanation why the water content of the chromatin structures are higher in the larger, active than in the smaller, inactive nuclei.

Using whole intact cells, we are unable to answer the question of chromatin hydration, we can not get any better insight into the mechanism of the volume regulation of the cell nucleus and we can not test the involvement of water molecules, inorganic ions and mobile proteins in these complex processes. These are the major reasons why we selected isolated nuclei for our present study. At the beginning, our experimental requirements were: 1./ the removal of all of the lipoid membranes; 2./ no DNA loss; 3./ maintained structural integrity of chromatin; 4./ sufficient quantity of isolated nuclei for water and inorganic ion measurements.

Table 1. Interference microscopic measurements on nuclei of chicken erythrocytes fused with SV 40 virus induced tumor cells

	Erythrocyte nuclei in H-50 cell hybrid	
	4 h	24 h
n	18	17
Area (μ^2)	17.3 ± 4.1	36.2 ± 4.7
Dry mass ($g \times 10^{-12}$)	18.9 ± 4.3	27.1 ± 3.6
Concentration ($g \times 10^{-12}/\mu^2$)	1.10 ± 0.08	0.75 ± 0.10

Calf thymus nucleiisolated in a buffered sucrose solution containing Ca and Mg divalent cations (TSCM) and Triton X100 nonionic detergent in 0.2% concentration fulfilled these requirements. The detergent solubilised all of the lipoid membranes and the nuclei remained stabile in the TSCM for more than 50 hours (Fig.1.A., and Fig.2.). One part of the isolated nuclei were transferred and equilibrated with a 75-75 mMol/l Na and K buffer containing Ca and Mg divalent ions (TNKCM). The other part was treated with a 1 to 5 mixture of TNKCM and TSCM solutions. This solution was named as 1/5 TNKCM. The nuclei are smaller in the medium containing 15-15 mMol/l Na and K than in TSCM with no Na and K (Fig.1.B., and Fig.3.). On the other hand the nuclei are completely disrupted in a medium containing 75 mMol/l Na and K, that is at 150 mMol/l monovalent cation. I want to remind you that the 150 mMol/l monovalent cation concentration is equal or even a slightly lower than that is supposed to be in the nuclear sap of the living cell nucleus (Fig. 1.C.,and Fig.4.) .This disruption of the chromatin structures is one of the major arguments that the chromatin can not be exposed to a 150 mMol/l or higher free electrolyte concentration inside the living cell nucleus.Consequently,this means that some or most of the intranuclear cations must be adsorbed or sequestered within the nucleus of the living cell (Cameron et al.,1977;Kellermayer and Hazlewood,1979;Kellermayer,1981; Negendank,1982).

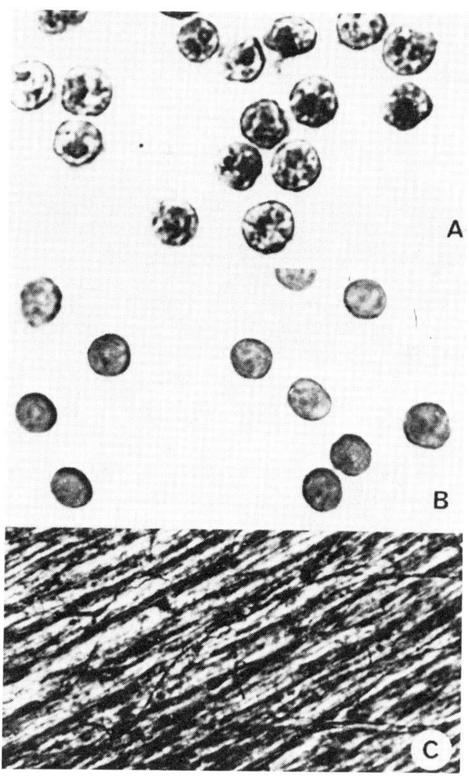

Fig.1. Light micrographs of isolated calf thymus nuclei in A.TSCM;B.1/5 TNKCM;C.TNKCM solutions and stained with crystal violet.The divalent Ca,Mg concentration is 3-3 mMol/l in each solution,while the TSCM is Na and K free,the 1/5 TNKCM contains 15-15 and the TNKCM 75-75 mMol/l Na and K.The average size of nuclei treated with TSCM is $\bar{x}$=23.8 ±3.8 (n=42), and with the 1/5 TNKCM solution the same parameter is $\bar{x}$=12.8 ± 1.7 (n=32) . A and B x1.600; C x 140.

The nuclei equilibrated to the three different solutions were pelleted down at different forces and their water content measured (Table 2.) .

Table 2. Water content of isolated thymus nuclei at different g forces %-water in pellet

	g forces						
	500	1000	5000	20 000	40 000	100 000	150 000
TSCM	81.6 ±2.3 n:7	77.9 ±1.1 n:15	72.3 ±1.0 n:5	70.6 ±1.2 n:7	67.8 ±0.6 n:5	65.3 ±0.5 n:12	65.7 ±0.5 n:7
1/5TNKCM	76.1 ±1.5 n:7	73.6 ±0.6 n:7	72.3 ±1.1 n:6	71.5 ±1.1 n:5	69.0 ±1.2 n:6	67.6 ±0.8 n:8	67.4 ±0.9 n:6
TNKCM	81.4 ±1.3 n:5	80.4 ±1.1 n:5	78.7 ±0.9 n:5	78.5 ±0.7 n:5	77.3 ±0.7 n:5	76.2 ±1.2 n:8	76.3 ±1.0 n:5

The divalent Ca^{++} and Mg^{++} content is 3-3 mMol/l in each solution but the monovalent Na^{+} and K^{+} content is 0 in TSCM, 15-15 mMol/l in 1/5 TNKCM and 75-75 mMol/l in TNKCM solution(for details, see Materials and Methods). The water content ratios of nuclei equilibrated with TSCM and with 1/5 TNKCM solutions are: 1.07 and 1.06 at 500 and 1,000, but 0.96 at 100,000 and 150,000 g forces.

At 100.000 g, the water content of the pelleted nuclei reached a plateau value. We call this value, the water holding capacity of the nuclei. The water holding capacity was the lowest with the TSCM solution (no monovalent ion), higher with the 1/5 TNKCM and highest with the TNKCM solution. It is interesting to note that at low speed, tha TSCM treated nuclei contained more water than the nuclei treated with the 1/5 TNKCM solution (Table 2.).

Studying the motional characteristics of water molecules, the NMR water proton signals, the T_1 and T_2 values were shorter in each case than those for ordinary water but proportional to the water content of the nuclear pellets (Table 3.).

The electron microscopic procedure we used here supplied us with good information about the structure of chromatin and the removal of the lipoid membranes, but it was incapable to show the nuclear matrix elements especially in the interchromatin spaces. The nuclei isolated and equilibrated to TSCM are not surrounded with membranes (Fig.2.). The chromatin bodies are highly condensed and the interchromatin spaces emptied. The nuclei incubated in the 1/5 TNKCM are smaller but the chromatin substance is less condensed (Fig.3.) . On the other hand the interchromatin spaces narrowed. Isolated nuclei washed and equilibrated with TNKCM solution aggregate forming a gluey clump (Fig.4.) . The individual nuclei are not discernible and the

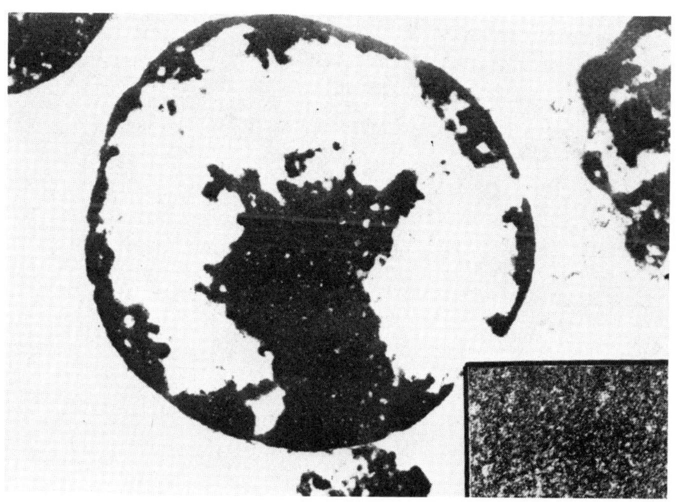

Fig. 2. Electron microgrph of isolated calf thymus nuclei equilibrated to TSCM solution x 30.000; x 142.000 for the inset.

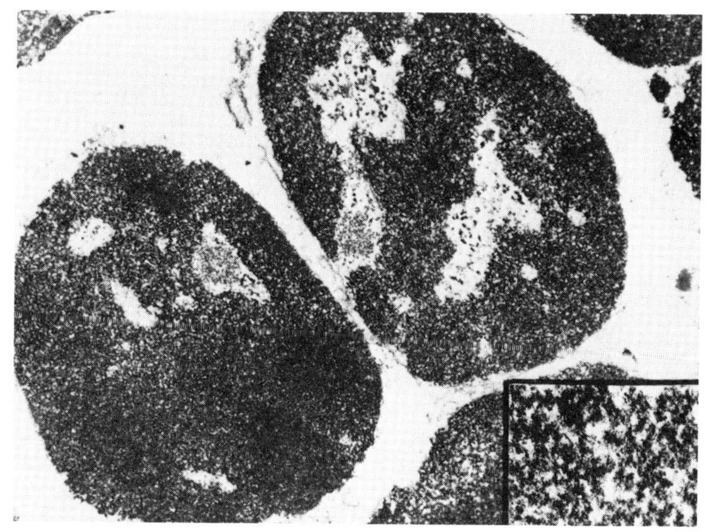

Fig. 3. Electron micrograph of isolated calf thymus nuclei equilibrated to 1/5 TNKCM solution x 30.000; x 142.000 for the inset.

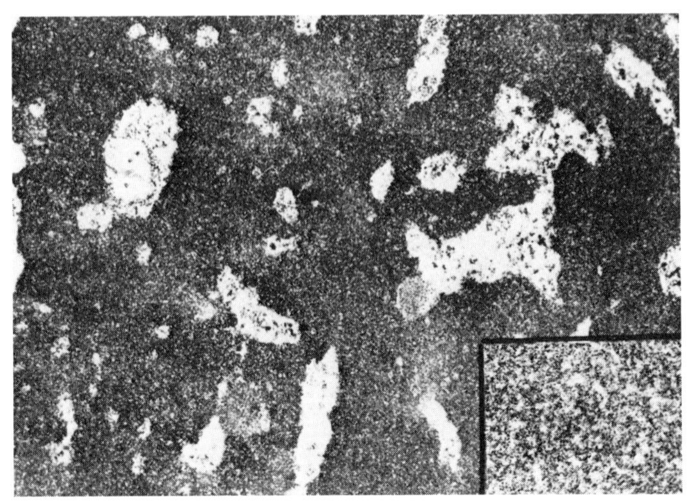

Fig. 4. Electron micrograph of isolated calf thymus nuclei equilibrated to TNKCM solution. No separated nuclei are visible and the chromatin bodies are confluent. x 30.000; x 142.000 in inset

chromatin appears to be loosely arranged.

We also measured the Na and K content of the nuclear pellets but instead to give the absolute values, the distribution coefficient, the ρ-values were calculated (Table 3.) . The ρ-value was introduced by Ling and it designates the ratio of a given solute concentration inside and outside (Ling et al., 1980). If this ratio is bigger than 1 accumulation and if smaller than 1 exclusion takes place. In case of intact living cells, the K is accumulated and Na is excluded. At the beginning of these experiments, we were anxious to know whether our isolated nuclei having all of the fix polymers, DNA with its tight associates i.e. histones and certain non histone proteins and also the nuclear matrix, would or would not accumulate and/or exclude monovalent K and Na ions. As it can be seen in Table 3, they accumulate both cations but much less extent as the living cells do. At this point, we must agree that something has been lost during the isolation. Something which is responsible for those distribution values characteristic for living cells becouse our nuclei at equilibrium can not keep the same gradients as the living cells do. In other words, the isolated nuclei can not simulate perfectly the nuclei of living cells. It seemed to be logical to test the two major cellular components lost during the isolation. These are the membranes and the proteins solubilised by the detergent. Unfortonately, the proteins mobilised from the cells showed no tendency at all to accumulate or ex-

Table 3. K and Na content, T_1 and T_2 values of isolated thymus nuclei

	TSCM		1/5		TNKCM	
	1.000 g	100.000 g	1.000 g	100.000 g	1.000 g	100.000 g
H_2O %	77.4 ±0.92 n=12	64.5 ±0.89 n=12	72.6 ±0.94 n=12	68.1 ±1.03 n=12	80.4 ±1.13 n=12	76.2 ±1.7 n=12
ρ K	----	----	1.64 ±0.04 n=3	1.86 ±0.06 n=3	1.42 ±0.01 n=3	1.49 ±0.02 n=3
ρ Na	----	----	2.00 ±0.02 n=3	2.22 ±0.04 n=3	1.48 ±0.01 n=3	1.52 ±0.03 n=3
T_1	643.8 ±25.8 n=12	373.6 ±10.2 n=8	594.2 ±8.0 n=4	460.9 ±7.4 n=3	789.8 ±1.9 n=5	646.7 ±1.8 n=3
T_2	65.1 ±7.2 n=8	30.3 ±1.9 n=4	55.5 ±2.8 n=4	42.0 ±1.9 n=4	95.6 ±3.5 n=4	72.9 ±2.0 n=4

TSCM = 0.01 M Tris-HCl buffer (pH 7.2), 0.25 M sucrose, 3 mM $CaCl_2$ and 3 mM $MgCl_2$

1/5 = 0.01 M Tris-HCl buffer (pH 7.2), 0.25 M sucrose, 15 mM NaCl, 15 mM KCl, 3 mM $CaCl_2$, 3 mM $MgCl_2$

TNKCM = 0.01 M Tris-HCl buffer (pH 7.2), 75 mM NaCl, 75 mM KCl, 3 mM $CaCl_2$, 3 mM $MgCl_2$

ρ K = concentration of K in water of pelleted nuclei divided by K concentration of supernatant

ρ Na = concentration of Na in water of pelleted nuclei divided by Na concentration of supernatant

T_1 = NMR Spin - lattice relaxation time of water hydrogen proton determined on nuclear pellet

T_2 = NMR longitudinal relaxation time of water hydrogen proton determined on nuclear pellet.

clude K and Na ions, but the 150 mMol/l or higher intracellular free ion concentration suggested by the membrane theory has been strongly questioned by our present results as well.

At this point, the application of a new detergent turned out to be a promising experimental approach. Brij 58 (Sigma) was found to solubilise the lipoid membranes of the cell with the same speed as the Triton X100 detergent does but the mobilization of the proteins seemed to be much slower (Schliwa et al., 1981). For Brij and Triton experiments the cal thymus cells were first separated in serum. When these cells were exposed to 0.2% Brij 58 for 5 minutes both the cellular and nuclear membranes disappeared (submitted to Nature). As it was shown before on the electron microscopic pictures, the same happened with the membranes after the 0.2% Triton X100 treatment. After the detergent treatment, 10 more minutes were needed for centrifugation and separation of the samples for the Na and K measurements. The exposition times (detergent treatment times) were 5, 10 and 20 minutes, which means that our time points are 15, 20 and 30 minutes (Fig. 5., and Table 4.). At the first time point (15 minute), the K ρ-value dropped to the equilibrium

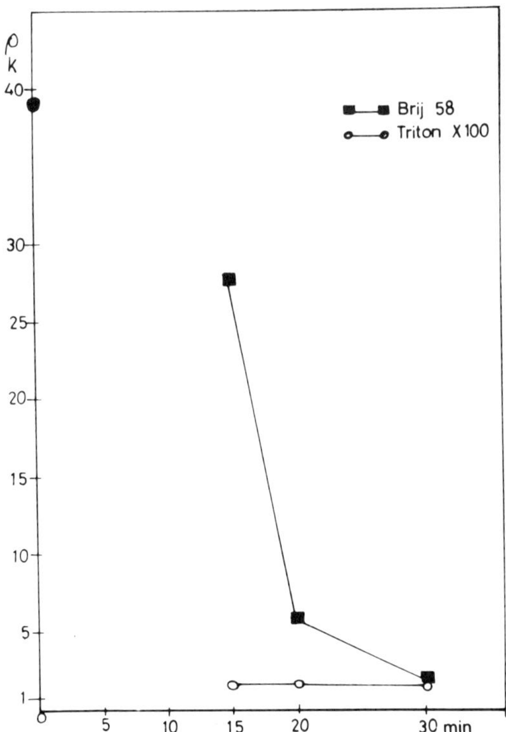

Fig. 5. Kinetics of K release fro detergent "opened" thymus lymphocytes.

value (close to 2) in case of Triton treatment but it remained high (close to 30) in case of Brij 58 treatment. You have to remember that by this time all of the lipoid membranes are solubilized by both detergents. Consequently, the high potassium gradient in the detergent "opened" cells must have been maintained by different intracellular/intranuclear components than the lipoid membranes. The mobilisation of K from the detergent "opened" cells is rapid in case of Triton X100, but slow in case of Brij 58 and it shows a sigmoid-like kinetics (Fig. 5.). The delayed release of K and this sigmoid behaviour suggest a cooperative process involved in the intracellular sequestration of K. We think that the "loosely associated" or "loosely bound proteins (known as mobile proteins) are responsible for the intracellular or intranuclear accumulation of K. It seems to be logical to assume that mobile proteins which can be as much as 70-75% of the total cellular protein, undergo a phase-transition when they associate to fix structural elements like

Table 4. The mobilisation of K from detergent "opened" thymus lymphocytes

Water content and K_ρ-values of detergent treated thymus lymphocytes		n	H_2O %	K_ρ
Control (Intact thymocytes)		8	76.6 ±0.4	39.5 ±0.9
15'	Brij 58	18	76.8 ±0.5	28.2 ±1.4
	Triton X 100	18	73.5 ±0.2	1.95 ±0.04
20'	Brij 58	8	74.5 ±0.13	6.65 ±0.2
	Triton X 100	8	68.1 ±0.2	2.18 ±0.06
30'	Brij 58	8	69.0 ±0.2	2.02 ±0.1
	Triton X 100	8	67.8 ±0.2	1.96 ±0.03

chromatin, nuclear matrix or cytoskeleton. This phase transition makes the mobile proteins capable to adsorbe or seqestrate inorganic ions and the numbers of the polarised layers of water molecules (Hazlewood, 1979; Ling, 1979) are also supposed to change sensitively with this globular-fibrillar phase transition.

Summary

Isolated thymus nuclei were exposed to buffer solutions containing Ca and Mg ions in constant and monovalent Na, K ions in increasing concentrations. The morphology of nuclei was studied by light and electron microscopy. The nuclei exposed or equilibrated to different solutions were centrifuged with increasing g forces and their K, Na, water content, T_1, T_2 relaxation times of water protons measured. The hydration of chromatin named the water holding capacity was higher at higher monovalent cation concentrations. The T_1 and T_2 values were shorter than for ordinary water but proportional to the water content of the nuclei. The isolated nuclei accumulated both K and Na but the accumulation index (ρ-value) didn't exceed 2 at equilibrium. In the last series of experiments, the K content of the nuclei were measured short (15,20,30 minutes) after the detergent (Triton X100 and Brij 58) attack.Although,both detergents removed all of the lipoid membranes in 5 minutes,a 30 time gradient was found for K between the inner and external concentrations.The easily mobilisable proteins are supposed to be responsible for the observed gradient and in a broader sense,these proteins are supposed to be the major regulators of the intracellular ionic milieu.

References

Agutter,P.S.,and Richardson,J.C.W.,1980,Nuclear nonchromatin proteinaceus structures:Their role in organization of interphase nucleus,J.Cell Sci.,44:395.

Belmont,A.,and Nicolini,C.,1981,Polyelectrolyte theory and chromatin-DNA quaternary structure:Role of ionic strength and H_1 histone,J.Theor.Biol.,90:169

Bonner,W.M.,1978,Protein migration and accumulation in nuclei, in:The Cell Nucleus,Vol.VI.,H.Busch,ed.,Acad.Press,New York,p.97.

Cameron,I.L.,Sparks,R.L.,Horn,K.L.,and Smith,N.R.,1977,Concentration of elements in mitotic chromatin as measured by X-ray microanalysis,J.Cell Biol.,73:193.

Harris,H.,1967,The reactivation of the red cell nucleus, J.Cell Sci.,2:23.

Hazlewood,C.F.,1979,A view of the significance and understanding of the physical properties of cell associated water, in:Cell-Associated Water,W.Drost-Hansen,and J.Clegg,eds., Acad Press,New York,p.165.

Kellermayer,M.,Olson,M.O.J.,Smetana,K.,Daskal,Y.,and Busch,H., 1974,Effect of various ionic media on extraction of soluble nuclear proteins and on nuclear ultrastructure, Exp.Cell Res.,85:191.

Kellermayer,M.,Jobst,K.,and Szucs,G.,1978,Inhibition of internuclear transport of SV 40 induced T antigen in heterokaryons,Cell Biol.Int.Rep.,2:19

Kellermayer,M.,and Hazlewood,C.F.,1979,Dynamic inorganic ion-protein interactions in structural organization of DNA of living cell nuclei,Cancer Biochem.Biophys.,3:181.

Kellermayer,M.,1981,Soluble and "loosely bound" nuclear proteins in regulation of the ionic environment in living cell,in:International Cell Biology 1980-1981,H.G.Schweiger, ed.,Springer-Verlag,Heidelberg,p.915.

Ling,G.N.,1979,The polarized multilayer theory of cell water according to the asstiation-induction hypothesis,in:Cell-Associated Water,W.Drost-Hansen,and J.Clegg,eds.,Acad Press,New York,p.265.

Ling,G.N.,Ochsenfeld,M.M.,Wqlton,C., and Bersinger,T.J.,1980, Mechanism of solute exclusion from cells:The role of protein-water interaction,Physiol.Chem.Physics,12:3.

Negendank,W.,1982,Studies of ions and water in human lymphocytes,Biochim.Biophys.Acta,694:123.

Schliwa,M.,van Blerkom,J.,and Porter,K.R.,1981,Stabilisation of the cytoplasmic ground substance in detergent opened cells and a structural and biochemical analysis of its composition,Proc.Natl.Acad.Sci.USA.,78:4329.

Acknowledgement

This work was supported by contracts No0014-76-C-1100 from the United States Office of Naval Research, No: 0-390 from Robert A. A.Welch Foundation and No: 05/3-16/255 from the National Health Ministry of Hungary.

METAL ION EFFECT ON MOLECULE SIZES AND INTRAMOLECULAR INTERACTION IN DNA

S. V. Kornilova, Yu. P. Blagoy, and A. G. Shkorbatov

Physico-Technical Institute of Low Temperatures
Ukrainian Academy of Sciences
Kharkov 310164, USSR

INTRODUCTION

The study of metal ion effects on the structure and conformational stability of DNA is an important problem of the nuclear acid physics connected with biological functions of DNA-macromolecules. As shown by a number of viscometric studies[1-3], the addition of bivalent metal ions at very small concentrations cause a considerable reduction of intrinsic viscosity of DNA $[\eta]$. In this paper we study the effect of Mn^{2+} and Cu^{2+} ions on DNA of high ($\sim 10^7$) and low ($\sim 10^5$) molecular weight in solutions of different ionic strength. Concentration dependences are calculated for the excluded volume parameter (ε), segment interaction parameter (Z) and expanding coefficient (α) for DNA macromolecules. These dependences permit one to judge upon the contributions of the electrostatic long- and short-range effects characterized by the values of ε and statistical segment A during the DNA interaction with metal ions. The effect of different degrees of ion binding (ι) upon DNA intrinsic viscosity is analysed in terms of the theory of equilibrium ion-DNA binding. This approach may prove efficient in estimating the ion-DNA constants from experimental results on viscosity.

EXPERIMENTAL TECHNIQUE AND MATERIALS

The intrinsic viscosity of DNA was measured using a Zimm-Crothers low shear rotating cylinder viscometer[4]. The studies were carried out on calf thymus DNA. Its protein content as estimated by the method of Lowry[5] was <0.5%; the hypochromicity was 40%; molecular extinction coefficient 6450; molecular weight 1.9 10^7, 1.2 10^7 and 2.1 10^5 was determined from intrinsic viscosity[6]. Fragmentation of high (1.9 10^7) molecular weight DNA to the low molecular weight (2.1 10^5) was effected by the method[7]. The temperature of solutions under study was kept constant, equal to 23.7°C at 0.2 to 10^{-3}M NaCl and 16°C at 2 10^{-4}M NaCl. Taking into account strong gradient dependence of η at 2 10^{-4}M NaCl, $[\eta]$ was determined at $g<0.07\ s^{-1}$ and DNA concentration less than 20 μg/ml. The intrinsic viscosity was measured at Na^+ concentration increasing from 2 10^{-4}M to 0.2 M. The $[\eta]$ was also considered as a function of concentration of bivalent ions Mn^{2+} and Cu^{2+} with NaCl contents of 2 10^{-2}M, 10^{-3}M and 2 10^{-4}M. The detailed description of the DNA solution preparation is given in Reference 8.

RESULTS

The dependences of $[\eta]$ on bivalent ion concentrations for high-molecular-weight DNA at different NaCl contents in the solution are shown in Figure 1. A sharp increase in viscosity at small ion concentrations (10^{-6} to 10^{-5}M) and low ionic strengths (10^{-3}M Na^+ and 2 10^{-4}M Na^+) is of particular interest. At high ion concentrations the viscosity values are close for all the curves. Figure 2 compares the dependences of DNA viscosity with two essentially different molecular weights 1.9 10^7 and 2.1 10^5.

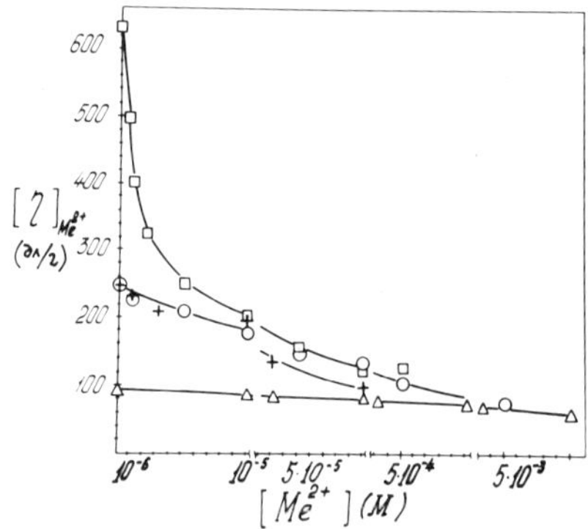

Fig. 1. DNA $[\eta]$ as a function of bivalent ion concentration for high-molecular-weight DNA at different NaCl contents. △ - Mn^{2+}, [NaCl] = 2 10^{-2}M, M = 1.2 10^7; ⊙ - Mn^{2+}, + - Cu^{2+}, [NaCl] = 10^{-3}M, M = 1.9 10^7; ⊡ - Mn^{2+}, [NaCl] = 2 10^{-4}M, M = 1.9 10^7.

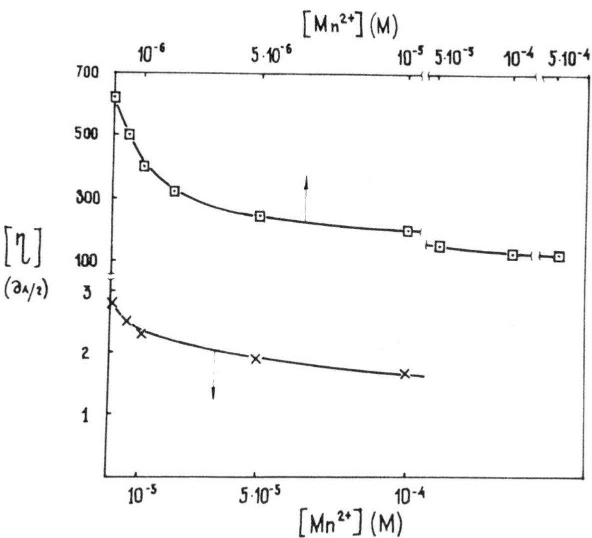

Fig. 2. Dependences of DNA [η] with two different molecular weights on Mn^{2+} ion concentration at [NaCl] = 2 10^{-4}M. ⊡ - M = 1.9 10^7; X - M = 2.1 10^5.

Theoretical consideration of the polyelectrolytic expansion of DNA gives a linear dependence [η] vs $1/\sqrt{J}$ (J - is the ionic strength). But this dependence agrees with experiment only for monovalent ions and in the region $J>10^{-3}$M [9]. Bivalent ions cause a stronger concentration dependence of [η] (Figure 3), which cannot be accounted for only by an enchancement of the Debye-Hückel screening of DNA negative charges at an increasing ionic strength. The data for bivalent ions might be interpreted on the basis of earlier results[10] taking into account that in the process of binding with a DNA molecule a bivalent ion neutralizes its two charges. Under this condition the mean negative charge of the monomeric DNA link q is linearly dependent on the degree of ion binding τ:

$$q = q_o(1-2\tau),\quad 0\leqslant\tau\leqslant0.5 \tag{1}$$

According to Fixman's theory[10] viscosity has approximately linear dependence on (1-2τ):

$$[\eta] = [\eta]_o + \frac{\text{const}}{\sqrt{J}}\, q_o(1-2\tau) \tag{2}$$

where $[\eta]_o$ is the intrinsic viscosity of nonperturbed molecules (at θ-conditions) and J is calculated as

$$J = [Na^+]z^2_{Na^+} + [Me^{2+}]z^2_{Me^{2+}} \tag{3}$$

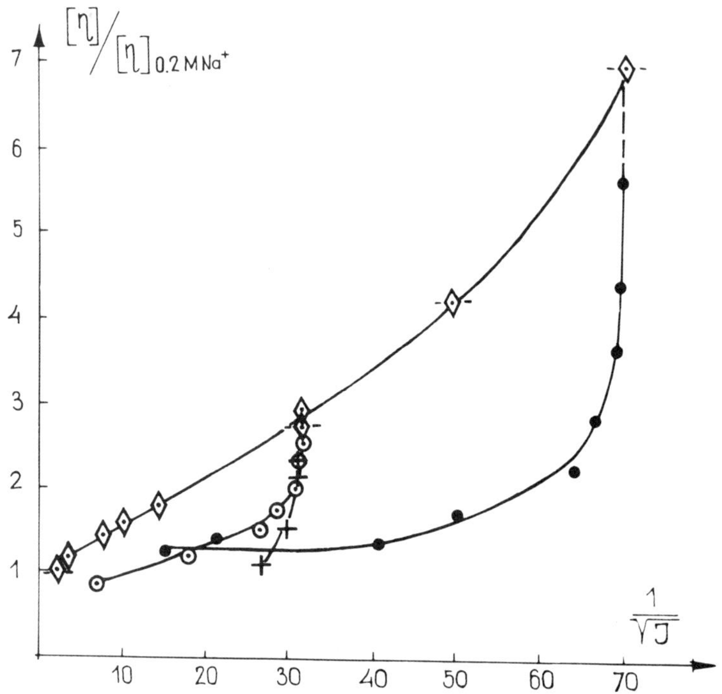

Fig. 3. Dependences of DNA relative intrinsic viscosity on $1/\sqrt{J}$, ● – Mn^{2+}, [NaCl] = 2 10^{-4}, M = 1.9 10^7; ⊙ – Mn^{2+}, + – Cu^{2+}, [NaCl] = 10^{-3}M, M = 1.9 10^7, ◊ – Na^+, M = 1.2 10^7; -◊- – Na^+, M = 1.9 10^7.

The general effect of the monovalent (Na^+) and bivalent (Me^{2+}) ions binding is described by the parameter τ_t:

$$\tau_t = \tau_{Me^{2+}} + 0.5\tau_{Na^+}, \quad 0 \leqslant \tau_{Na^+} \leqslant 1 \tag{4}$$

The degree of ions binding τ_t was calculated using the system of Scatchard equations for the case of competition between Na^+ and Me^{2+} ions:

$$\begin{cases} \dfrac{\tau_{Me^{2+}}}{A_{fMe^{2+}}} = K^\circ_{Me^{2+}} e^{-W_{Me^{2+}}\tau_t} (0.5-\tau_t) \\ \dfrac{\tau_{Na^+}}{A_{fNa^+}} = K^\circ_{Na^+} e^{-2W_{Na^+}\tau_t} (1-2\tau_t) \end{cases} \tag{5}$$

where A_f is the free ion concentration, K° – the apparent constant of ion binding, W – parameter of electrostatic ion interaction. For Mn^{2+} the K° and W values were taken from literature[11], for Na^+ – from [Reference 12]. Taking into account that the appropriate constants for Cu^{2+} must be close to those for Mn^{2+}[13], the same values were used for both types of ions.

The dependence of relative intrinsic viscosity versus ($n-\tau_t$) (or τ_t) is almost a universal plot for $(n-\tau_t)<0.2$, where n- is the total number of binding sites per nucleotide (Figure 4). As seen from Figure 4, at 2 $10^{-4}Na^+$ the Na^+ ions compensate for only about 20% of negative charges. As a result, in the process of bivalent ion binding, the viscosity experiences great changes. At $10^{-3}Na^+$, Na^+ ions occupy about 40% binding sites and at 2 $10^{-2}Na^+$ monovalent counterions neutralize 80% of charges. The appropriate change of $[\eta]$ in the bivalent ion case is rather small (triangles).

A great increase of $[\eta]$ observed in high-molecular weight DNA (19 10^6) at 2 10^{-4}M NaCl can be attributed mainly to a strong variation of statistical segment observed in this region of NaCl concentrations[14,15] which is not accounted for by Fixman's theory. The existing theories fail to explain properly a rise in the DNA viscosity at low molecular weight (2.1 10^{-5}) for which the excluded volume effect is small (Figure 2). It should be noted, that the increase in the statistical segment length or the DNA stiffness alone cannot be responsible for the experimentally observed rise of viscosity in short DNA chains. According to the Yamakawa and Fujii theory[16] the DNA $[\eta]$ cannot increase by more than a factor of 1.5.

Considering the contribution of long-range electrostatic interaction in chain expansion we calculate the excluded volume parameter ε and expanding coefficient α as a function of τ_t. The general relations between the meansquare end-to-end distance H^2, ε and α were employed.

$$H^2 = A^{1-\varepsilon} L^{1+\varepsilon} \tag{6}$$

$$H^2 = \alpha^2 H_\theta^2 \tag{7}$$

$$\alpha^3 = [\eta]\phi_\theta/([\eta]_\theta\phi) \tag{8}$$

where A is statistical segment of the chain, L is contour length of the chain, index θ marks the values at θ-point. ϕ is the Flory parameter, dependent on ε according to Ptitsyn's theory[17].

$$\phi = \phi_\theta(1-2.63\varepsilon + 2.86\varepsilon^2) \tag{9}$$

The dependence of the Flory parameter on A and L was taken into account using Yamakawa and Fujii work[16]. As a result ε values were found solving the equation:

$$\left[\frac{L^{1+\varepsilon}A^{1-\varepsilon}}{LA_\theta}\right]^{3/2} = \frac{\phi_\theta(L,A)[\eta]}{\phi(L,A)(1-2.63\varepsilon+2.86\varepsilon^2)[\eta]_\theta} \tag{10}$$

We have calculated ε for sodium ions using Hagerman's data[15] on the statistic segment length A. Figure 5 shows the variation of ε at

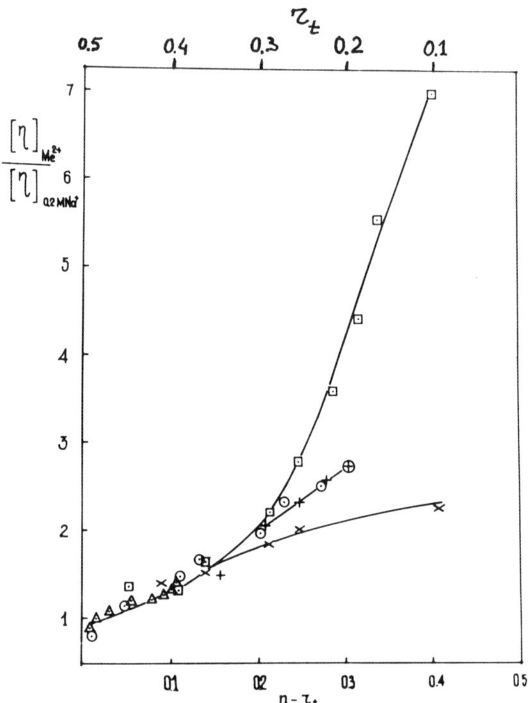

Fig. 4. Dependence of DNA relative intrinsic viscosity on ($n - \tau_t$) at different NaCl contents, ▵ - Mn^{2+}, [NaCl] = 2 10^{-2}M, M = 1.2 10^7; ⊙ - Mn^{2+}, + - Cu^{2+} [NaCl] = 10^{-3}M, M = 1.9 10^7; X - Mn^{2+}, [NaCl] = 2 10^{-4}M, M = 2.1 10^5; ⊡ - Mn^{2+}, [NaCl] = 2 10^{-4}M, M = 1.9 10^7.

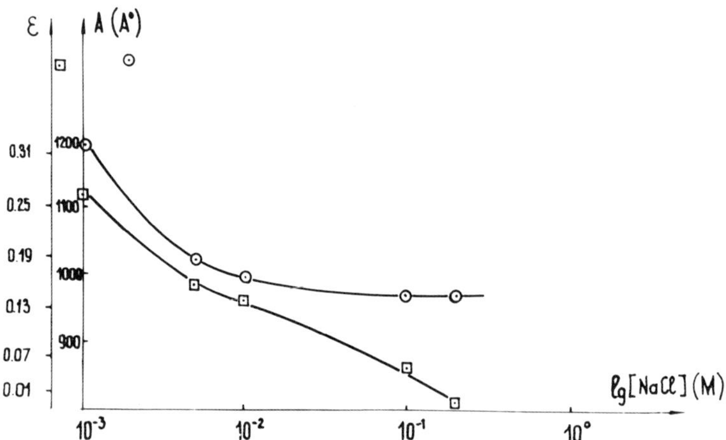

Fig. 5. Variation of ε and statistical DNA segment A with NaCl concentration. ⊙ - A by Hagerman[15] ⊡ - ε.

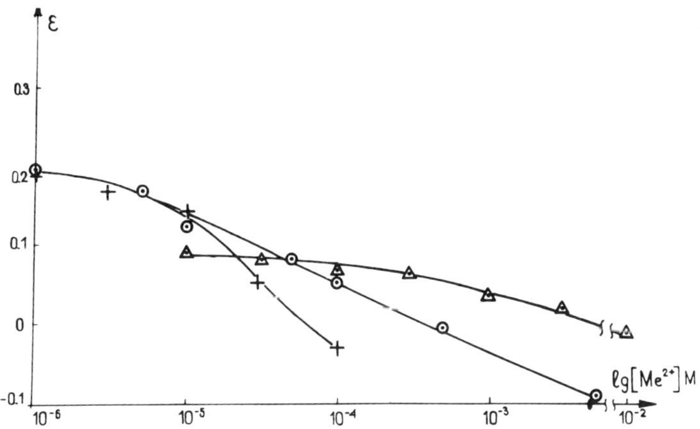

Fig. 6. ε as a function of bivalent ion concentration [Me^{2+}], + - Cu^{2+}, ⊙ - Mn^{2+}, [NaCl] = 10^{-3}M, M = 1.9 10^7; △ - Mn^{2+}, [NaCl] = 2 10^{-2}M, M = 1.2 10^7.

different concentrations NaCl and A vs [NaCl] by Hagerman. In the region 10^{-3}M<[NaCl]<10^{-2}M the contributions of both effects of electrostatic long (ε) and short (A) range interactions upon [η] are comparable. But if [NaCl]>10^{-2}M the statistical segment A becomes constant, thus the viscosity variation depends on excluded volume effect only. The dependences [η] vs [Me^{2+}] were studied at sufficiently high concentrations [NaCl]=10^{-3}M, 2 10^{-2}M, when A = const[14,15]. We have used the last assumption in the presence of bivalent ions and calculated dependence ε vs [Me^{2+}] by Equation (10) (Figure 6).

Another description of long-range interaction may be done by the parameter $\mathcal{Z}$, which is proportional to the mean effective potential of segment interation:

$$\mathcal{Z} = \left(\frac{3\pi}{2}\right)^{3/2} \sqrt{N}\, A^{-3}\, \nu_0 \tag{11}$$

where N - the number of chain segments, ν_0 - effective excluded volume of segment. We have calculated $\mathcal{Z}$ using the relation with the expanding coefficient α [18]:

$$\alpha^5 = 1 + \frac{20}{3}\mathcal{Z} + 4\pi\mathcal{Z}^2 \tag{12}$$

Figures 7 and 8 show the dependences ε and $\mathcal{Z}$ vs τ_t, for the cases of Cu^{2+} and Mn^{2+} binding. As seen in Figure 7, the parameter ε becomes negative at τ_t>0.3. Possible values of ε belong to the interval $-1/3<\varepsilon<0.2$ [19] and negative ε correspond to the compression of DNA chain (relative to θ-point). The dependence Z vs τ_t presents a similar picture. As a whole, the dependences of ε on full concentration of bivalent ions are of individual character (Figure 6), whereas the τ_t dependence are universal (Figure 7).

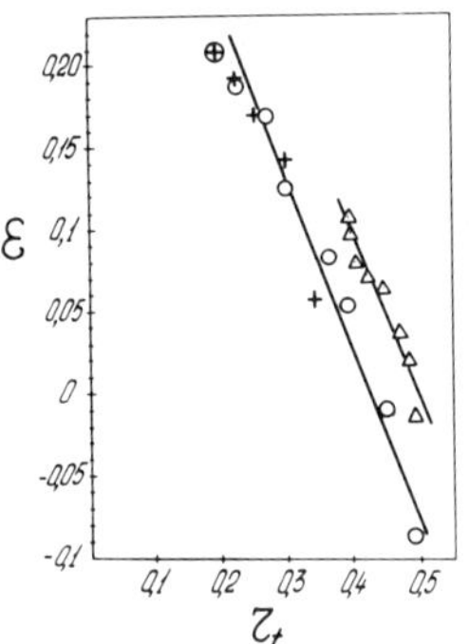

Fig. 7. ε vs degree of ion binding τ_t. + - Cu^{2+}, ⊙ - Mn^{2+}, [NaCl] = 10^{-3}M, M = 1.9 10^7; ▵ - Mn^{2+}, [NaCl] = 2 10^{-2}M, M = 1.2 10^7.

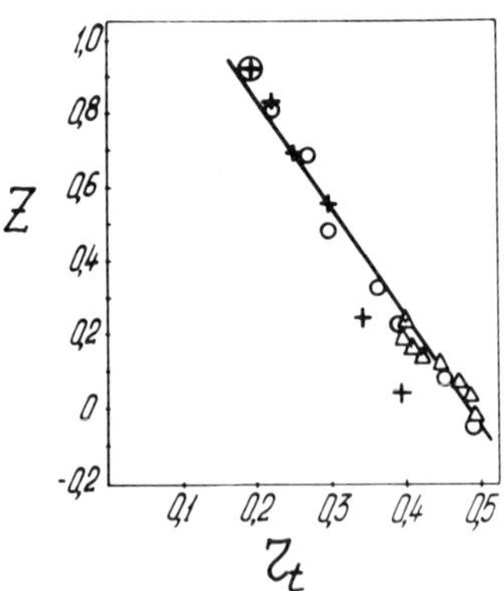

Fig. 8. Z vs degree of ion binding τ_t. + - Cu^{2+}, ⊙ - Mn^{2+}, [NaCl] = 10^{-3}M, M = 1.9 10^7, ▵ - Mn^{2+}, [NaCl] = 2 10^{-2}M, M = 1.2 10^7.

REFERENCES

1. K. Baxter-Gabbard and D. Freser, The Effects of Cations and Diamines on the Viscosity of T2 DNA, Biopolymers, 13:207 (1974).
2. M. Daune, Interactions of Metal Ions with Nucleic Acids, Metal.Ion.Biol.Syst., 3:1 (1974).
3. C. K. Pillai and V. S. Nandi, Binding of Gold (III) with DNA, Biopolymers, 12:1431 (1974).
4. B. N. Zimm and D. M. Crothers, Simplified Rotating Cylinder Viscometer for DNA, Proc.Nat.Acad.Sci.USA, 48:905 (1962).
5. O. H. Lowry, N. Rosebrough, A. Farr, and R. Randall, Protein Measurement with the Folin Phenol Reagent, J.Biol.Chem., 193:265 (1951).
6. J. Eigner and P. Doty, The Native, Denatured and Renatured States of Deoxyribonucleic Acid, J.Mol.Biol., 12:549 (1965).
7. A. V. Shugalii and A. B. Fonarev, On the Application of DNA Sonication in Study of Reassociation Kinetics, Biochimia, 40:598 (1975).
8. Yu.P. Blagoy, S. V. Kornilova, and V. I. Sokhan, Study of Changing Characteristic Viscosity of DNA Interacting with Cu^{2+} and Mn^{2+} Ions, Molekularnaya Biologia, 16:210 (1982).
9. S. V. Slonitskii, E. V. Frisman, A. K. Valeev, A. E. Eliashevich, and A. M. Valeev, Intrinsic Viscosity Evaluation of Synthetic and Biological Polyelectrolytes of Different Stiffness, Molekularnaya Biologia, 14:484 (1980).
10. M. Fixman, Polyelectrolytes: A Fuzzy Sphere Model, J.Chem.Phys., 41:3772 (1964).
11. J. Reuben and E. Gabby, Binding of Manganese (II) to DNA and the Competitive Effects of Metal Ions and Organic Cations. An Electron Paramagnetic Resonance Study, Biochemistry, 14:1230 (1975).
12. R. Clement, J. Sturn, and M. Daune, Interaction of Metallic Cations with DNA. VI. Specific Binding of Mn^{2+} and Mg^{2+}, Biopolymers, 12:405 (1973).
13. Yu.P. Blagoy, V. Sorokin, V. Valeev, and G. Gladchenko, Bivalent Copper Ion Effect on Thermal DNA Denaturation, Molekularnaya Biologia, 12:795 (1978).
14. S. V. Slonitskii and E. V. Frisman, Ionic Strength Effect on the Thermodynamical Stiffness of DNA Molecule in Aqueous and Aqueous-Organic Solvents, Molekularnaya Biologia, 14:496 (1980).
15. P. J. Hagerman, Investigation of the Flexibility of DNA using Transient Electric Birefrigence, Biopolymers, 20:1503 (1981).
16. H. Yamakawa and M. Fujii, Intrinsic Viscosity of Worm-like Chains. Determination of the Shift Factor, Macromolecules, 7:128 (1974).
17. O. B. Ptitsyn and Yu. E. Eizner, Hydrodynamic of Polymer Solutions. II Hydrodynamic Properties of Macromolecules in Good Solvents, Zhurn.Techn.Fiz., 29:1117 (1959).

18. I. M. Lifshits, A. Yu. Grosberg, and A. R. Khokhlov, Volume Interactions in the Statistical Physics of a Polymer Macromolecule, Uspekhi Fiz.Nauk, 127:353 (1979).
19. B. A. Fedorov, Theory of Small-Angle X-Ray Scattering of Compact Macromolecules in Solution, Branched Gauss Chains, Vysokomolekularnye Soedinenia, 12-A:810 (1970).

EXPERIMENTAL CONFIRMATION OF THE POLARIZED MULTILAYER THEORY OF CELL WATER INCLUDING DATA THAT LEAD TO AN IMPROVED DEFINITION OF COLLOIDS

Gilbert N. Ling

Department of Molecular Biology
Pennsylvania Hospital
Philadelphia, Pa

SUMMARY

The polarized multilayer theory of cell water, as part of the association-induction hypothesis, and its rapidly gathering supportive experimental evidence were reviewed. It was shown that the new insight offered by this theory reconciles many hitherto unexplained phenomena and that new experimental data also suggest an improved definition of the concept of colloids.

I. GELATIN: THE NAMESAKE OF COLLOIDAL CHEMISTRY

Thomas Graham, who introduced in 1861 the concept as well as the name of colloids, explained in these words, "The plastic elements of the animal body are found in this class. As gelatine appears to be its type, it is proposed to designate substances of the class as colloids..." (κολλοσ, the Greek word for glue, is largely gelatine) (Graham, 1861, p. 182). The inventor of dialysis, Graham also stated in the same article (p. 185), "The water of the gelatinous starch is not directly available as a medium for the diffusion of either the sugar or gum, being in a state of true combination, feeble although the union with starch may be..." Beside gelatine, starch and other colloidal materials, Graham also investigated copper ferrocyanide, which is a highly gelatinous reddish-brown precipitate formed when copper sulfate is mixed with K ferrocyanide. It is the study of a membrane of this gelatinous precipitate by M. Traube (1867) and by W. Pfeffer (1877) that led to the founding of the membrane theory. Unfortunately, Traube did not follow the clues Graham revealed but chose to champion the so-called "atomic sieve theory" to explain the impermeability of the copper ferrocyanide gel membrane to sugars and other solutes. Later x-ray and electron diffraction studies have shown that the interstices of the copper ferrocyanide lattice (100-200 Å) is much larger than that of sucrose (diameter, 9 Å), thus disproving the atomic sieve idea (Fordham and Tyson, 1937; Glasstone, 1946). The semipermeable property of this famous membrane model has remained unexplained until quite recently (see below).

That interaction of biological materials causes changes of property of water was recognized even before Graham's paper of 1861. For example, in 1849, Carl Ludwig, known as the father of modern physiology, demonstrated that dried pig bladder when immersed in a salt solution takes up a much more dilute salt solution than the bathing solution in which the bladder was immersed. This " water of imbibition" or " Schwellungswaser" clearly has different solvent properties from normal water.

In 1934 Holleman, Bungenberg de Jong and Modderman carried out similar study as that described above by Carl Ludwig. However, instead of dried pig's bladder, gelatine was used. Like Ludwig, they also found that water in gelatine gel accommodated, at equilibrium, less Na_2SO_4 than that present in the incubation solution. These authors interpreted their data in terms of " negative adsorption." Similarly when Mc Mahon et al (1940) found lower sucrose concentration in the water taken up by copper ferrocyanide gel than in the surrounding solution they too interpreted the data in terms of " negative adsorption," a term, which I think was not well chosen. Conceptually it is obtuse (see below).

Beside the unusual solvency property, gelatine water system exhibits another distinctive feature: resistance to freezing. In a gel containing 35% water, and 65% gelatine, the water will not freeze even at the temperature of liquid nitrogen (-195.8° C) (Moran, 1926).

In summary, since its very inception, gelatine was seen as having the property representative of colloids including the living matter. These colloidal systems exhibit unusual solvent properties (and have been historically explained in terms of " negative adsorption") and are resistant to freezing.

II. EARLY CONCEPTS OF WATER IN LIVING CELLS

A. Ideas of Pfeffer and Overton

According to the conventional membrane-pump theory of the living cell, all or nearly all water in living cells is free. It is therefore interesting to note that those who first clearly suggested that a substantial portion of the water in living cells is not free but exists as "Schwellungswaser" were in fact the founders of the membrane theory, W. Pfeffer and one of its great supporters, E. Overton (1902). Overton found that muscle cells when immersed in a solution half of the osmotic strength of a Ringer solution, do not swell to twice their natural size, as would be the case, according to the membrane theory and the van't Hoff equation of osmosis. Overton suggested that a sizable fraction of the water in frog muscle cells is " Schwellungswaser" which was considered to be osmotically inactive. In later quantitative assessment of the amount of this " osmotically inactive" water, the water content of cells is plotted against the reciprocal of the concentration (C) of solutes in the medium. Extrapolating to $C^{-1} = 0$ yields the amount of this fraction of osmotically inactive water. Ling and Negendank (1970) showed that this concept is strictly speaking, incorrect, because it, in fact, says that part of the cell water does not leave the cell even at zero water vapor pressure in the environment. Actual measurement showed that, under this condition, no significant amount of water remains in frog muscle. These findings showed that the kind of cell water behavior contrary to the assumption of the membrane theory as observed by Overton, cannot be reconciled by assuming a portion of it to be osmotically inactive. A better explanation has to be found.

B. Ideas of Gortner

Other proponents of unusual water in living cells included Fisher and Suer (1939) and Gortner (1930). Gortner's work represents a number of physico-chemical studies by him and others of water associated with biological materials. His criteria of " bound water" were based on its nonsolvency for solutes and nonfreezing at -20° C. From 1940 on, the bound water idea eventually almost disappeared from the literature in consequence of two types of contradictory experimental findings: (i) urea was found to distribute equally between muscle cell water and the surrounding medium (Hill, 1930); ethylene glycol was found to distribute equally between erythrocyte water and the surrounding medium (Mac Leod and Ponder, 1936). These studies led to the conclusion that no bound water exists in living cells. (ii) Blanchard (1940) in his review on " Bound Water" refuted the non-freezing water by pointing out that pure water can be supercooled to temperature lower than -20° C. This and other reasons led Blanchard to reach the conclusion: No bound water exists in living cells.

C. Troshin's Coacervate Theory of Cell Water

A. S. Troshin (1966) from the Soviet Union, was impressed by the findings of Holleman, Bungenberg de Jong and Modderman (1934) mentioned above, confirmed their experiments and suggested that the low levels of sugar, free amino acids, and ions found in living cells have a similar origin: both gelatin and living cell protoplasm represent colloidal " coacervates" systems in which water has different solubilities for solutes (Bungenberg de Jong, 1949). Regretably this highly talented and productive scientist ceased publishing on this subject in the sixties and he did not further elaborate on how and why coacervates exclude solutes.

III. THE POLARIZED MULTILAYER THEORY OF CELL WATER AS PART OF THE ASSOCIATION-INDUCTION HYPOTHESIS

In 1965 Ling expanded his theory of the living cell, the association-induction (AI) hypothesis to include a more detailed theory of the physical state of the bulk of cell water (Ling, 1965). This theory attempts not only to answer the how and why the bulk of water in a resting cell exists in a different physical state, but also why and how solutes are excluded from water in this state. Before discussing this theory, I shall first briefly sketch its historical background.

A. Historical Background of the Polarized Multilayer Theory of Cell Water

1. Contribution from Physicists. Gases condense on solid surface often in a characteristic manner: a more or less flat plateau is reached at low gas concentrations in the external phase, followed by a steep rise at a higher gas concentration range. Thus this type of adsorption isotherm is ~ -shaped. Physicists De Boer and Zwikker (1929) developed a theory for this phenomenon in terms of multilayer adsorption. This theory was severely criticized by Brunauer, Emmett, and Teller (1938) who pointed out even in the case of very large and hence polarizable atoms like argon or iodine, the solid surface cannot polarize more than one layer of the gaseous molecules. Instead they proposed their own theory (later known as the BET theory) in which additional layers of gases taken up beyond the first adsorbed layer are condensed in the same way as in liquid liquids. However, Brunauer, Emmett, and Teller also specifically pointed out that their criticiam was limited to gaseous molecules that have no permanent dipole moments. For gaseous molecules with permanent dipole moments, as in the case of water molecules, polarized multilayer like that suggested by De Boer and Zwikker are quite feasible. Indeed, this case had already been treated by Bradley (1936) who presented a polarized multilayer adsorption isotherm (Eq. 1) in a form quite similar to the De Boer-Zwikker isotherm,

$$\log \left(\frac{p_o}{p}\right) = K_1 K_3^a + K_4, \quad (1)$$

where p is the vapor pressure of the gas under the experimental condition and p_o is p at full saturation. Thus p/p_o is the relative vapor pressure. a is the amount of gas taken up. K_1, K_3, and K_4 are constants under a defined condition. Equation 1 can be written in a double log form:

$$\log\left(\log\left(\frac{p_o}{p}\right) - K_4\right) = a \log K_3 + K_1 \quad . \tag{2}$$

2. Contribution from Industrial Physico-chemists. In the textile industry, sorption of water is an important subject of concern. Large uptake by cotton and wool fiber had been for a long time explained as due to capillary condensation; i.e., the bulk of water is held as normal liquid water in pores and narrow channels in the fibers. Benson and Ellis (1948, 1950), in their study of N_2, O_2 and CH_4 sorption on dry proteins could not find evidence for such pores and channels. In fact they found that the total amount of sorbed gases depends only on the state of subdivision of the proteins and the total surfaces exposed. In sharp contrast, extensive alteration of the physical state of wool, egg albumin, and silk did not alter the amount of water sorbed (Mellon, Korn, and Hoover, 1949). In these cases water uptake depends only on the specific sites available in a protein; the state of subdivisions makes little difference (Benson, Ellis, and Zwanzig, 1950).

From this group of scientists also came the important demonstration that beside polar side chains, the polypeptide chains of proteins offer important sites of hydration, as indicated by the sorption of large amount of water by amorphous poly-glycine ester (Mellon, Korn, and Hoover, 1948) and by polyvinylpyrrolidone $\left[\overset{\text{N}=\text{O}}{\text{-CH-CH}_2\text{-}}\right]_n$ which contain no polar side chains (Dole and Faller, 1950). Furthermore, this water uptake by the fibrous proteins and polymers follows Bradley's polarized multilayer adsorption isotherm described by Equation 2.

B. Theoretical Reasons for the Introduction of the Polarized Multilayer Theory of Cell Water

The logical sequence for the polarized multilayer theory of cell water includes the following:

1. In the water of living cells various solutes (e.g., Na^+, sugars, free amino acids) are found in much lower concentration than in the surrounding medium. There are only three basic types of mechanism for the maintenance of such a difference in concentration between the cell and its aqueous environment. Two of these, absolute membrane impermeability and pumping, have both been ruled out (Ling, 1962, 1983), leaving a difference in the

physico-chemical nature or solubility of the cell internal environment (i.e., water) as the only class of mechanism still tenable.

2. Virtually all cell water must be different, because the concentration of Na^+ and other solutes in the water of many cells may be only 5% of that in the external environment. This sets a limit of normal liquid water in the cells at 5%. Most likely this limit is even lower because it is unlikely that the affected water can exclude Na^+ completely as assumed in this 5% calculation.

3. Cell water comprises from 60 to 85% of the weight of most cells. To convert 95 to 100% of this cell water to something with distinctly different properties, there must be another substance or substances that are at once ubiquitous enough and abundant enough in all cells. Since some cells (e.g., human erythrocytes) contain no DNA or RNA in significant amount, this substance can only be proteins. (This does not rule out a similar role for the nucleic acids and even polysaccharides in some cells to serve a similar role assigned primarily to proteins.)

4. If all the cell proteins are stretched out and uniformly distributed throughout the cell, the space between nearest neighboring chains would be about 20 Å (Ling, 1962), which is roughly equal to $\frac{20}{2.8}$ = 7 water molecules diameters wide. Since not all cells proteins participate, the number of water molecules affected by the involved protein must be greater (but not vastly greater). Thus multilayers of water must be somewhat affected by the cell proteins.

5. One concept introduced of such an influence of proteins on water was the " iceberg" concept (Jacobson, 1955; Klotz, 1958). In this theory protien surfaces offer regular sites that have the geometry of tridymite-like Ice-I structure. This then in some unspecified way induces the formation of more ice layers (at room or body temperature). This theory has been ruled out by the fact that cell water can be supercooled and maintained in that condition but not when the cell water is touched with a seeding ice crystal through a cut end of the cell. Rapid ice formation in cells then follows (Ling and Miller, 1970). This finding shows that no ice could be present in normal cells to begin with (see below).

6. Theoretical considerations discussed in Sect. II, 1. and 2. offer sound basis for the consideration that the bulk of cell water is polarized by cell proteins in the form of multilayers.

C. The Polarized Multilayer Theory of Cell Water and the Theory for Solute Exclusion from Water in This State

The polarized multilayer theory of cell water first proposed in 1965 as part of the AI Hypothesis (Ling, 1965) was amplified in 1973 and 1975 (Ling, 1973, Ling and Sobel, 1975), includes the following postulations.

1. In the intracellular space of all living cells not occupied by other matter, a network of regularly and finely dispersed proteins, referred to as " matrix proteins" exists.

2. The matrix proteins exist in an extended conformation with their backbone NHCO groups directly exposed to the bulk-phase water. The NH and CO group, being positively charged (P) and negatively charged (N) respectively constitute a NP-NP-NP system.

3. Each of these CO groups and its neighboring NH group orient rows of water molecules with opposite orientations. As a result there is not only radical polarization of the water molecules along each row but also lateral interaction between water molecules in immediately neighboring rows. The entire assembly thus assumes 3-dimensional cooperative characteristics.

4. As a result of the multilayer polarization, the water molecules have reduced rotational as well as translational motional freedom.

5. The electrical polarization of the assembly of water molecules and the motional restriction thus created provides two basic mechanisms for the exclusion of large solute molecules from this polarized water: enthalpic and entropic. In both mechanisms, small solutes and solutes that can fit into the water lattice may not be excluded at all or may even be preferentially taken up. The degree of exclusion as measured by the equilibrium distribution coefficient (q-value) increases with increasing molecular sizes and complexity (the size rule).

6. The degree of motional restriction considered is very modest (Ling, 1979). Thus to produce a q-value of 0.1 for a solute (which is quite low for q-values), the required motional restriction, represents no more than a factor of 10 in the " partition function" ratios. This type of water structure is dynamic, the long-range order can be revealed only after many repeated photographic exposures are taken.

IV. EXPERIMENTAL TESTING OF SEVERAL OF THE PREDICTIONS OF THE THEORY

A. Solvency

A corollary of polarized multilayer theory of cell water is that when a protein exists in an α-helical or other intramacromolecularly H-bonded structure, the bulk phase water will not be influenced. As a result, even large solutes like hydrated Na^+, sugar, etc., will either not be excluded at all or minimally excluded. This theoretical corollary was confirmed by Ling, Ochsenfeld, Walton, and Bersinger (1980).

However, when either due to specific non-helical molecular structure, or exposure to denaturants like urea or guanidine HCl, the backbone NHCO groups of a protein are exposed to the bulk phase solvent, solute exclusion then is observed.

From these studies, the conclusion was also reached that the reason water associated with gelatin excluded Na^+ and other solutes, normally excluded from living cells, is due to gelatin's existence in an extended conformation, as a result of the possession of the repeated triads of non-helical forming amino acid residues: glycine, proline, and hydroxyproline. In full agreement, a number of synthetic polymers including polyvinylmethyl ether (PVME) $(-CH-C(H)(O-CH_3)-)_n$; polyvinylpyrrolidone (PVP); poly(ethylene oxide) (PEO) $(-CH_2-O-CH_2-)_n$; and methylcellulose (Ling, Walton, and Bersinger, 1980) which do not contain an H-donating group like NH in the polypeptide chain also alter the solvency of the bulk-phase water. These findings show that all that is needed to produce the solvency effect on the bulk-phase water is the presence of unshielded oxygen atoms with its lone pair electrons along the chain at distance roughly equal to that of two water diameters.

In agreement with this theory it was found that water associated with PVME, does not exclude methanol but excludes larger hydroxylic compounds in rough proportion to the sizes and complexities in full agreement with the size rule long ago reported for living cells (Ling, Miller, and Ochsenfeld, 1973).

B. Degree of Motional Restriction of Polarized Water in Model Systems

The NMR relaxation time, T_1 and T_2 of the proton of water associated with PVP, PVME, and PEO were studied (Ling and Murphy, 1982). The T_1/T_2 ratios are very close to that of pure water (i.e., near unity). This equality of T_1 and T_2 and the actual values of T_1 and T_2 at the magnetic field used, permits a rough estimate of the rotational correlation time (τ_c) which turns out to be no more than 1 order of magnitude longer than normal liquid water and thus vastly shorter than that of ice.

C. Swelling and Shrinkage of Polymer-Water Systems and of Living Cells Without Intact Membrane

Dialysis sacs filled with a 30% solution of the neutral polymer (PEO) maintain a swollen, unchanged, or shrunken volume depending on the concentrations of Na-citrate in the external solution even though the dialysis sac is fully permeable to Na-citrate (Ling, 1980). This confirms the theoretical expectation that it is primarily the multilayer polarization and hence lowered activity of the polymer-dominated water in conjunction with the low equilibrium distribution coefficient or q-value of Na-citrate in the polymer-oriented water that determine the volume of the sacs. A similar mechanism was suggested in the AI Hypothesis for the maintenance of the size of living cells, which have been shown to swell or shrink regardless of the presence or absence of an intact cell membrane (Ling and Walton, 1976).

D. Freezing Properties

Ling and Zhang in work yet to be published showed that the non-freezing water in gelatin-gel is also found in water associated with PEO and other polymers. They came to the conclusion non-freezing is yet another trait of the water polarized in multilayers. These data also clearly showed that the polymer-oriented water is definitely not ice as postulated in the iceberg theories.

E. Adherence to the Bradley Adsorption Isotherm

Ling and Negendank (1970) showed that 95% of the water in isolated frog muscle cells follow Bradley's adsorption isotherm (Equations 1 and 2). In the last six years, three different laboratories across the world using four modern methods, showed that the bulk of the major intracellular cation K^+ is adsorbed (Edelmann, 1977, 1980-81, 1981; Ling, 1966; Trombitas and Tigyi-Sebes, 1979). Yet the cells are in osmotic equilibrium with a Ringer solution which has an osmotic strength equal to that of a 0.118 M NaCl solution in which both Na^+ and Cl^- are free. This osmotic equilibrium with a Ringer solution is maintained whether the muscle is in direct contact with the Ringer solution or separated from the Ringer solution by an air space as was the case in the experiments described by Ling and Negendank (1970). In terms of the AI Hypothesis, these observations indicate that the osmotic balance is maintained primarily by the matrix proteins. Although not yet clearly identified, it is believed that actin, tubulin, and other cytoskeletal proteins may play major roles. However, it is also believed that they exist in a finer state of dispersion than seen in EM pictures. In other words we expect that the matrix proteins have properties illustrated, though less intensely, by gelatin, while other proteins like certain enzymes which normally exist in a highly helical form contribute less to the cell's activity.

If this interpretation is correct, one would expect that at a partial vapor pressure of 0.996 (i.e., the vapor pressure of a Ringer solution), proteins which, (existing in an extended conformation) cause solvency change of the bulk phase water will take up large amount of water while native globular proteins (existing in highly α-helical conformations) will take up much less water. In fact Ling has shown that this is precisely the case. Among those that take up a huge amount of water is once more gelatin (Ling, 1973, 1983).

F. Quasi-elastic Neutron Scattering

Trantham, Rorschach, Clegg, Hazlewood, and Nicklow (1981), (see also Rorschach et al, this volume) presented their work in this symposium on the quasi-elastic neutron scattering of water in brine shrimp (Artemia) cysts. From these studies they reached the conclusion that the majority of cell water has strongly reduced translational and rotational motional freedom. Their findings are in agreement with the predictions of the polarized multilayer theory of cell water (Ling, 1965, 1967; Ling, Ochsenfeld, and Karreman, 1967). Just as exciting was the finding of Rorschach and coworkers that similar neutron-scattering studies revealed water associated with 36% PEO exhibits properties of translational and rotational motional restrictions similar to the water in the Artemia cysts (Rorschach, personal communication).

V. A BACKWARD GLANCE AT HISTORY: ATTEMPTS TO SOLVE SOME SO-FAR UNRESOLVED QUESTIONS

A. The Virtual Impermeability to Sucrose of Copper-ferrocyanide Gel Membranes and Cellulose Acetate Membranes Which Have Pores Many Times Larger Than the Diameter of Sucrose Molecules

Cellulose, when existing in crystalline form (microcrystallin cellulose) reacts weakly or not at all with water. Only when bulky groups are introduced between the glucose chains, as in methyl cellulose, do they polarize water in multilayers, endowing this water with reduced solvency for sugars, hydrated ions, etc. (Ling, Walton, and Bersinger, 1980). Apparently (activated) cellulose acetate behaves in a similar manner. Thus Ling (1973) showed that cellulose acetate sheet shows similar semipermeability properties as the living cell "membrane" (inverted frog skin) for 11 hydroxylic compounds at 3 different temperatures. Since the pores of these activated cellulose acetate membranes have pores with an average diameter of 45 Å (Ling, 1973) which is 5 times wider than the diameter of sucrose, to which the membrane is virtually impermeable, clearly the situation closely resembles the case of copper-ferrocyanide gel membrane. In both, a low q-value of sucrose for the water polarized by either the cellulose acetate, or by the copper-ferrocyanide, in addition to greatly reduced diffusion coefficient (D) for sucrose in this polarized water accounts for its low permeability (P = qD). Of course, q-value as well as D-value decreases as the size and complexity of the hydraulic compound decreases. Therefore the permeability of these hydroxylic compounds are progressively higher for the compounds with smaller size. Water, having a q-value of 1.0 and apparently a correspondingly high D, is the most permeable through both models and the living cell membrane.

B. An Amplification of the Term, Negative Adsorption

With multilayer adsorption of water established, the term that a solute is " negatively adsorbed," can be better understood. It signifies that like Na^+, glucose and free amino acids, a negatively adsorbed solute is that which is excluded from the polarized multilayers. However, the term remains a poor one, first, in the sense the solute excluded does not have direct contact with the polarizing surface and thus is not at all AD-sorbed. Second, negative adsorption suggests that the surface sites actually repel it, which they don't.

C. Overton's Schwellungswaser in Living Cells

Overton was forced to postulate that a portion of the cell water to be different and in later investigator's lingo, osmotically inactive. I already pointed out that this is not a sound idea and is refuted by actual measurements. However, the basic trouble originates from the assumption that the bulk of cell water is normal liquid water and in which free solutes like K^+ determine its osmotic strength according to van't Hoff's law. Since these assumptions have proven invalid and since the volume change is primarily a matter of multilayer adsorption of water (Eq. 1, see also Ling and Peterson, 1977) clearly no postulate of part of the water being osmotically active can explain the full behavior pattern.

D. Urea and Ethylene Glycol Are Not the Proper Probe for Water Existing in the State of Polarized Multilayers

The " bound water," as used by earlier investigators, was given attributes that were incorrect. Among those was the attribute of categorical non-solvency - i.e., non-solvency for any solutes. It was by siezing this incorrect assumption, that the bound water concept of Gortner and others was " disproven" by Hill, and Mac Leod and Ponder mentioned earlier. According to the polarized multilayer theory, small molecules and molecules that can fit into the water lattice are not excluded. These non-excluded solutes experimentally demonstrated include urea and ethylene glycol (Ling, Walton, and Bersinger, 1980), which were respectively that used by Hill (1930) and by Mac Leod and Ponder providing the major evidence against the bound water ideas of the 1930's.

VI. HOW CAN WE DELIBERATELY FORGET THE MAJOR ADVANCEMENT MADE IN UNDERSTANDING WATER STRUCTURE ON THE BASIS OF ITS TETRAHEDIAL H-BONDED STRUCTURE?

Bernal and Fowler (1933) proposed the first modern structural theory of water. That liquid water retains short-range order is now well established. The tetrahedial structure of water molecule underlies a natural tendency to form 4 pairs of H-bonds with neighboring water molecules like in Ice I. Although no concensus of liquid water structure has been reached yet, this basic element of tetrahedial structure is universally accepted (Eisenberg and Kauzmann, 1969). The question may be raised, " Why should we now propose a modified liquid water structure, which seems to ignore these important detailed knowledge?"

At the time when French impressionists started to make their unorthodox artistic renditions of nature, they might well have been asked a similar question, " Why do you ignore the knowledge we already have gained in detail painting, in preference for the crude pictures you seem to prefer?"

My answer to the criticism of the choice of the dipole-dipole model over the more detailed H-bonded model may be similar to one that could have been offered by the French impressionists: " In order to see the whole forest, details must be deliberately ignored."

Equally important, the dipole-dipole interaction is preferred over the tetrahedially-H-bonded water molecule interaction because at this moment we can predict certain properties of water on the basis of the dipole model which we cannot when based on the complicated tetrahedial structure. Thus dipole-dipole interaction may predict in one way of orientation, an attraction between two water molecules, and in another orientation, there can be repulsion. The present-day model of the H-bonded structure predicts attraction all right but would be strained to describe less favorable interaction than no interaction. In short, the dipole-dipole interaction model is, at this time at least, correct, highly useful, and informative in the same sense, an impressionist picture of a landscape is more representative than a picture in which every leaf of a tree has been painted.

VII. A REVISED DEFINITION OF COLLOIDS

Graham defined colloids as materials having the properties of gelatine. In years following, the development of colloidal chemistry emphasized the large molecular size of colloidal materials. Yet we have now amply shown that gelatin does indeed stand apart from native proteins with similar molecular weights and that gelatin owes this unusual property not to large molecular weights (witness copper ferrocyanide is a colloid, like gelatin but having a very small molecular weight) but to its ability to orient multilayers of water on its extended polypeptide chains. I would therefore like to suggest the colloids as envisaged by its founder Thomas Graham are distinguished by their ability to achieve multilayer interaction with the bulk phase solvent, water.

ACKNOWLEDGMENT

The foregoing work was supported by NIH Grants 2-R01-CA16301-3 and 2-R01-GM11422-13, and by Office of Naval Research Contract N00014-79-C-0126.

REFFERENCES

Benson, S. W., and Ellis, D. A., 1948, Surface areas of proteins. I. Surface areas, and heats of absorption, J. Amer. Chem. Soc. 70:3563.

Benson, S. W., and Ellis, D. A., 1950, Surface areas of proteins. II. Adsorption of non-polar gases, J. Amer. Chem. Soc. 72:2095.

Benson, S. W., Ellis, D. A., and Zwanzig, R. W., 1950, Surface areas of proteins. III. Adsorption of water, J. Amer. Chem. Soc. 72:2102.

Bernal, J. D., and Fowler, R. H., 1933, A theory of water and ionic solution, with particular reference to hydrogen and hydroxyl ions, J. Chem. Phys. 1:515.

Blanchard, K. C., 1940, Water, free and bound, Cold Springs Harb. Symp. Quant. Biol. 8:1.

Bradley, S., 1936, Polymolecular adsorbed films. Part II. The general theory of the condensation of vapors on finely divided solids, J. Chem. Soc. 1467:1799.

Brunauer, S., Emmett, P. H., and Teller, E., 1938, Adsorption of gases in multimolecular layers, J. Amer. Chem. Soc. 60:309.

Bungenberg de Jong, H. G., 1949, Crystallization-coacervation-flocculation *in*: "Colloid Science," Vol. 2, H. R. Kruyt, ed., Elsevier Publ. Co., Inc., New York, p. 232.
De Boer, J. H., and Zwikker, C., 1929, Adsorption als Folge von Polarisation; Die Adsorptionsisotherme, *Z. Physik. Chem.*, B3:407.
Dole, M., and Faller, I. L., 1950, Water sorption by synthetic high polymers, *J. Amer. Chem. Soc.* 72:414.
Edelmann, L., 1977, Potassium adsorption sites in frog muscle visualized by cesium and thallium under transmission electron microscope, *Physiol. Chem. Phys.* 9:313.
Edelmann, L., 1981, Electron microscopic demonstration of potassium binding sites in muscle, *in*: "Intern. Cell Biol. 1980-1981" H. G. Schweiger, ed., Springer-Verlag, Berlin, p. 941.
Edelmann, L., 1981, Selective accumulation of Li^+, Na^+, K^+, Rb^+, and Cs^+ at protein sites of freeze-dried embedded muscle detected by LAMMA, *Fresnius A. Anal. Chem.*, 308:218.
Eisenberg, D., and Kauzmann, W., 1969, "The Structure and Properties of Water," Oxford Press, Oxford.
Fischer, M. H., and Suer, W. J., 1939, Base-protein acid compounds prepared from fibrin, *Arch. Pathol.*, 27:811.
Fordham, S., and Tyson, J. T., 1937, Structure of semipermeable membrane of inorganic salts, *J. Chem. Soc.*, 31:483.
Glasstone, S., 1946, "Textbook of Physical Chemistry," 2nd. ed., Van Nostrand, New York.
Graham, T., 1861, Liquid diffusion applied to analysis, *Phil. Trans. Roy. Soc.*, 151:183.
Gortner, R. A., 1930, The state of water in colloidal and living systems, *Trans. Farad. Soc.*, 26:678.
Hill, A. V., 1930, State of water in muscle and blood and the osmotic behavior of muscle, *Proc. Roy. Soc. B.*, 106:477.
Holleman, L. W., Bungenberg de Jong, H. G., and Modderman, R. S. T., 1934, Zur Kenntnis-der lyophilen Kolloide. XXI Mitteilung über Koazervation. I: Einfache Koazervaten von Gelatinesolen, *Koll. Beihefte*, 39:334.
Jacobson, B., 1955, On the interpretation of dielectric constants of aqueous macromolecular solutions. Hydration of aqueous macromolecular solutions, *J. Amer. Chem. Soc.*, 77:2919.
Klotz, I., 1958, Protein hydration and behavior, *Science*, 128:815.
Ling, G. N., 1962, "A Physical Theory of the Living State: The Association-Induction Hypothesis," Blaisdell, Waltham.
Ling, G. N., 1965, The physical state of water in living cell and model systems, *Ann. N.Y. Acad. Sci.*, 125:401.
Ling, G. N., 1967, Effects of temperature on the state of water in the living cell, *in*: "Thermobiology," A. Rose, ed., Academic Press, New York, p. 5.
Ling, G. N., 1973, What component of the living cell is responsible for its semipermeable properties? Polarized water or lipids? *Biophys. J.*, 13:807.

Ling, G. N., 1977, K^+ localization in muscle cells by autoradiography, and identification of K^+ adsorbing sites in living muscle cells with uranium binding sites in electron micrographs of fixed call preparations, Physiol. Chem. Phys. 9:319.
Ling, G. N., 1979, The polarized multilayer theory of cell water and other facets of the association-induction hypothesis concerning the distribution of ions and other solutes in living cells, in: " The Aqueous Cytoplasm," Alec D. Keith, ed., Marcel Dekker, Inc., New York, p. 23.
Ling, G. N., 1980, The role of multilayer polarization of cell water in the cell swelling and shrinkage of living cells, Physiol. Chem. Phys. 12:383.
Ling, G. N., 1983, " In Search of the Physical Basis of Life," Plenum Publishing Corp., New York.
Ling, G. N., and Miller, C., 1977, Structural changes of intracellular water in caffeine-contracted muscle cells, Physiol. Chem. Phys. 2:495.
Ling, G. N., Miller, C., and Ochsenfeld, M. M., 1973, The physical state of solutes and water in living cells according to the association-induction hypothesis, Ann. N.Y. Acad. Sci. 204:6.
Ling, G. N., and Murphy, R. C., 1982, NMR relaxation of water protons under the influence of proteins and other linear polymers, Physiol. Chem. Phys. 14:xxx (in press)
Ling, G. N., and Negendank, W., 1970, The physical state of water in frog muscles, Physiol. Chem. Phys. 2:15.
Ling, G. N., Ochsenfeld, M. M., and Karreman, G., 1967, Is the cell membrane a universal rate-limiting barrier to the movement of water between the living cell and its surrounding medium? J. Gen. Physiol. 50:1807.
Ling, G. N., Ochsenfeld, M. M., Walton, C., and Bersinger, T. J., 1980, Mechanism of solute exclusion from cells: The role of protein-water interaction, Physiol. Chem. Phys. 12:3.
Ling, G. N., and Peterson, K., 1977, A theory of cell swelling in high concentrations of KCl and other chloride salts, Bull. of Math. Biol. 39:721.
Ling, G. N., and Sobel, A. M., 1975, The mechanism for the exclusion of sugars from the water in a model of the living cell: The ion-exchange resin: Pore size or water structure? Physiol. Chem. Phys. 7:415.
Ling, G. N., Walton, C., and Bersinger, T. J., 1980, Reduced solubility of polymer-oriented water for sodium salts, amino acids, and other solutes normally maintained at low levels in living cells, Physiol. Chem. Phys. 12:111.
Ludwig, C., 1849, Ueber die endosmotischen Aequivalente und die endosmotische Theorie, Z. rationelle Medizin von Henle 8:1.
Mac Leod, J., and Ponder, E., 1936, Solvent water in the mammalian erythrocytes, J. Physiol. 86:147.
Mc Mahon, B. C., Hartung, E. J., and Walban, W. J., 1940, Studies in membrane permeability. II. The adsorption of sucrose and two salts on cupric ferrocyanide, Trans. Farad. Soc. 36:515.

Mellon, E. F., Korn, A. H., and Hoover, S. R., 1949, Water absorption of proteins. IV. Effect of physical structure, J. Amer. Chem. Soc. 71:2761.

Moran, T., 1926, The freezing of gelatin gel, Proc. Roy. Soc. A 112: 30.

Overton, E., 1902, Beiträge zur allgemeinen Muskel und Nervenphysiologie, Pflügers Arch. ges Physiol. 92:115.

Pfeffer, W., 1877, " Osmitische Untersuchungen, Studien zur Sell Mechanik," (1st ed.), W. Engelmann, Leipzig.

Rorschach, H. E., Trantham, E. C., Heidorn, D. B., Hazlewood, C. F., Clegg, J. S., Nicklow, R. M., and Wakabayashi, N., this volume. The diffusive motion of protons in pure water, agarose gel and Artemia cysts as measured by quasi-elastic neutron scattering.

Trantham, E. C., Rorschach, H. E., Clegg, J. S., Hazlewood, C. F., and Nicklow, R. M., 1981, QNS measurements on water in biological and model systems, Amer. Inst. Phys. Conf. Proc. 89:264.

Traube, M., 1867, Experimente zur Theorie des Zellenbildung und Endosmosis, Arch. (Anat. u) Physiol. 87:165.

Trombitas, C., and Tigyi-Sebes, A., 1979, X-ray microanalysis studies on native myofibrils and mitochondria isolated by microdissection from honey-bee flight muscle, Acta Physiol. Acad. Sci. Hung. 14:271.

Troshin, A. S., 1966, " Problems of Cell Permeability," (English transl. by M. G. Hall) (W. F. Widdas, ed.) Pergamon Press, London.

THE INFLUENCE OF IONS ON WATER STRUCTURE AND ON AQUEOUS SYSTEMS

Werner A. P. Luck

Fachbereich Physikalische Chemie
Philipps Universität Marburg
3550 Marburg, West Germany

INTRODUCTION

Life would not be possible on our planet without the anomalous properties of water[1,2]. Water should melt at -90°C and boil at about -80°C if water did not have the strong H-bond interactions. In biochemical systems water structure is the second interesting problem of water research. We will give a review on the present status of knowledge in the sequence:

I. STRUCTURE OF PURE LIQUID WATER

The content of closed H-bonds is one of the dominating factors of the properties of liquid water. It can be determined by infrared spectroscopy, especially by precise overtone spectroscopy [2,3,4]. The comparison of the concentration dependency of the solution spectra of alcohols in CCl_4 with the temperature dependence of spectra of pure liquids (Figure 1)[2] demonstrates that a T-increase (Figure 2) changes the intensity of the sharp OH stretching overtone band at 7100 cm^{-1} similar to a concentration decrease in solutions (Figure 1). As can be identified from solutions' spectra this band belongs to OH group not interacting with H-bonds[3]. Only the half-width of this band is larger in liquids and its intensity at the maximum is smaller but the band area $\int \varepsilon d\nu$ is equal[4]. This means that we could determine the measurement of the heights, or better, the band areas at 7100 cm^{-1} the content of non H-bonded (so called "free") OH groups (Figure 3). The comparison between alcohols and water spectra especially with HOD spectra - [5] - allows application of this useful method on water spectra too.

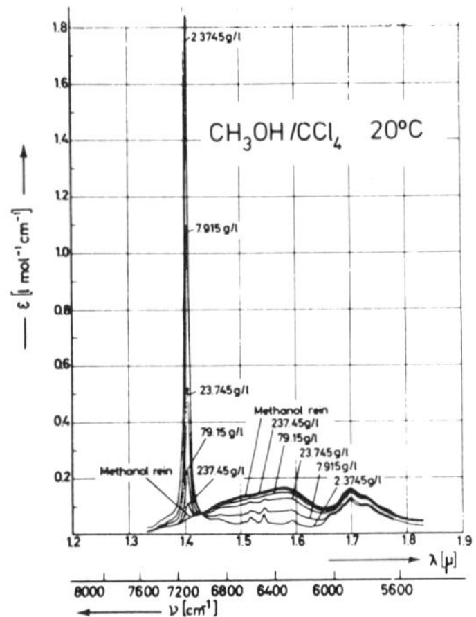

Fig. 1. Concentration dependence of OH stretching overtone of CH_3OH in CCl_4 solutions at 20°C. The sharp band at 7100 cm^{-1} is a measure of non H-bonded OH.

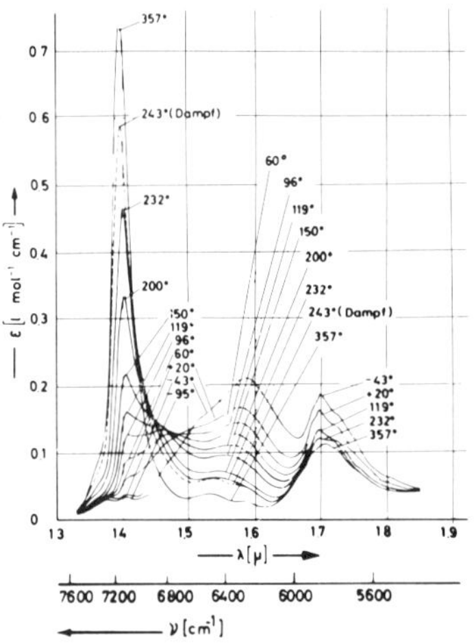

Fig. 2. The temperature dependence of OH stretching overtone of pure liquid CH_3OH in equilibrium with its vapor. The 7100 cm^{-1} band is a measure of non H-bonded OH.

We have to stress that this simple method still needs some study, especially with the seldom-used precise quantitative spectroscopy. Some water spectra are published without enough research and it seems, therefore, that the literature on water spectra is inconsistent.

The difference between the total concentration of OH groups to figure 3 gives H-bonded OH groups. There may exist a more or less distribution of H-bond angles[2,6,7] or distances.

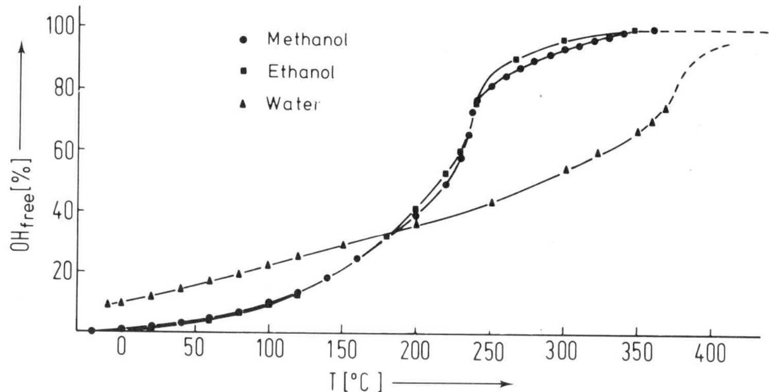

Fig. 3. The content of non H-bonded OH (OH_{free}) of liquid H_2O and alcohols in equilibrium with its vapors as function of temperature.

The caloric properties of liquid water can be described approximately[2,8] with reasonable exactness by the simple equilibrium:

$$OH_f + \Theta_f \rightleftharpoons OH_{bond} \quad (1)$$

OH_f: non H-bonded OH concentration
Θ_f : concentration of non H-bonded lone pair electrons
OH_b: H-bonded OH concentration

Equation (1) gives the fundamental understanding of aqueous systems. The success of the first approximation of (1) could mean that distributions of H-bond angles and distances show only small differences in interaction energies, because by breaking the H-bonds the oxygen distance could decrease and therefore the van de Waals interaction between water molecules would increase. The size of normal van der Waals interaction may be of about 30% of the total interaction energy of ice or liquid water near melting point[2,9]. The differences between water and alcohol in Figure 3 are caused by the 50% excess of single pair electron of alcohol and the different volume concentrations of OH.

II. Structure of Electrolyte Solutions

II. 1. Structure Temperature

Spectroscopic attempts to describe aqueous solutions are more difficult than pure water and only possible by simplification.

The first experiments to compare water spectra of electrolyte solutions with T-dependence of pure water showed the following results[10]:

1. Ions disturb the water little. We needed concentrations higher than about 0.1 mol.liter to find disturbances of water spectra.

2. The disturbances of water spectra by ions are very similar to a T-change of pure water. Bernal and Fowler[11] have given the nomenclature: structure temperature T_{str} as temperature T which pure water should have with similar H-bonds. Our spectroscopic method has given at first a method to determine T_{str} quantitatively as a T of pure water with similar absorbances in the frequency region of "free" OH. T_{str} determinations in the region of H-bonded OH would give little different T_{str}-values[12,13] because of the disturbances of OH-bonds depend on the charge and size of the ions. To describe ternary systems: water/ions/ solute our definition of T_{str} with the absorbance in the region of free OH is preferred because the solubility of third molecules depend strongly on the content of free OH (see equilibrium)[1].

3. The ion series obtained by the spectroscopically determined change of the water spectra[10] is the same (Figure 4) as the Hofmeister or lyotropic ion series of colloid chemistry. Therefore our spectroscopic method[12] was able to find the cause of the Hofmeister ion series: as a series of disturbance of water structure.

4. The position of every salt in the ion series depends mainly on the anion and less on the cations in the first instance.

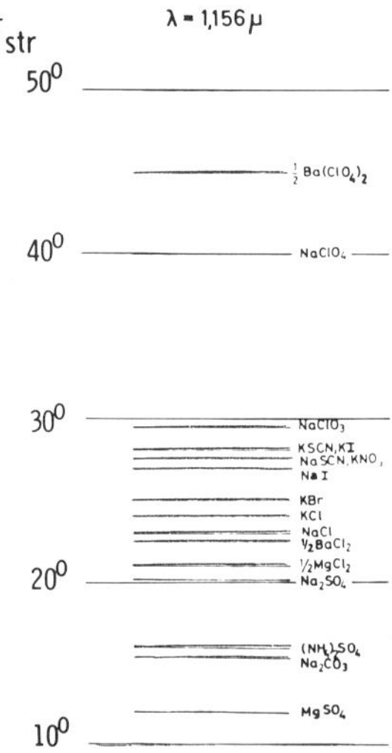

Fig. 4. The spectroscopic determined structure temperature T_{str} as rough heuristic measure of the changed water structure by ions. (Solution T = 20°C) C_{Anion} =1 mole/liter). The ions series obtained by T_{str} is similar to the Hofmeister ion series and the recognized Hofmeister series as a series of disturbed water structure.

This factor cold be a property of our spectra: the OH stretching vibration can be influenced easier by anion-proton interactions than cation-lone pair electron interactions. But other properties too demonstrate a special role of anions in electrolyte solutions. Figure 5 shows the partial molar volume V_1 of water minus its molar volume V_{01} in pure state as the function of salt concentration (full lines or dotted line in Figure 5). If we plot the $MgCl_2$ solution data as function of anion concentration the V_1 values are the same as in NaCl-solutions. There seems to be no difference between the influence of Na^+ or Mg^{++} on the partial molar volume of water.

5. The ion series in Figure 4 is split in two parts: one has a structure T higher than the solution T, this group is called structure breaker because the H-bond structure is weaker than in pure water; the second group has a $T_{str} < T_{solution}$, this group is called structure maker ions. These two groups can be also observed with NMR measurements[13,14]. $T_{str} > T_{solution}$ means: water has more "free OH", it is more hydrophilic, and can give "salt-in" effects for organic solutes. $T_{str} < T_{solution}$ means: water has less "free OH", it has "salt-out" effects or organic solutes.

II. 2. EXAMPLE OF HOFMEISTER SERIES

"Salt-in" and "salt-out" effects can be easily studied by solutions of polyethylenoxides with hydrophobic end groups, for instance of the type:

$$\text{iso-}C_8H_{17} - \langle\bigcirc\rangle - (O\text{-}CH_2\text{-}CH_2\text{-})_n - OH \text{ (abbreviation: POIP-n)}$$

At high temperatures - dependant from the hydrophilic/hydrophobic balance given by the number n of ethylenoxide groups - the H-bonds of the ether oxygen atoms are not strong enough to carry the hydrophobic end group and two phases are formed: a PIOP reach organic phase and an aqueous phase. The turbidity point T_k (one phase is observed at $T < T_k$) is sensitive on salt additives, indicating a change of H-bond structure of water (Figure 6).

Figure 6 demonstrates -like the Hofmeister ion series - the bigger effect of anions as cations and a split in two groups of ions with $T_k < T_{ko}$ (pure water) and $T_k > T_{ko}$ (pure water).

The turbidity points T_k are correlated to the spectroscopic determined T_{str}[10].

The boundary between "salt-in" effects and "salt-out" effects depends a little on the solution temperature and on the solutes.

The solute effect on this border is demonstrated in Figure 7. The partition coefficients K_{salt} of different solutes between cyclo-

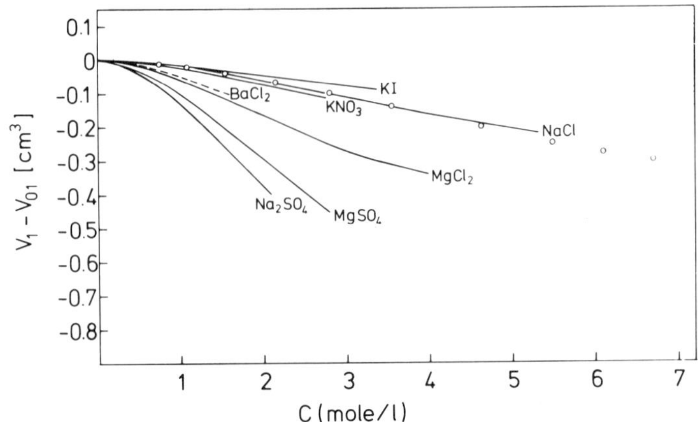

Fig. 5. Partial molar volume V_1 of water in salt solutions at 20°C as function of salt concentration (full and dotted lines). O: $MgCl_2$ datas as function of anion concentration.

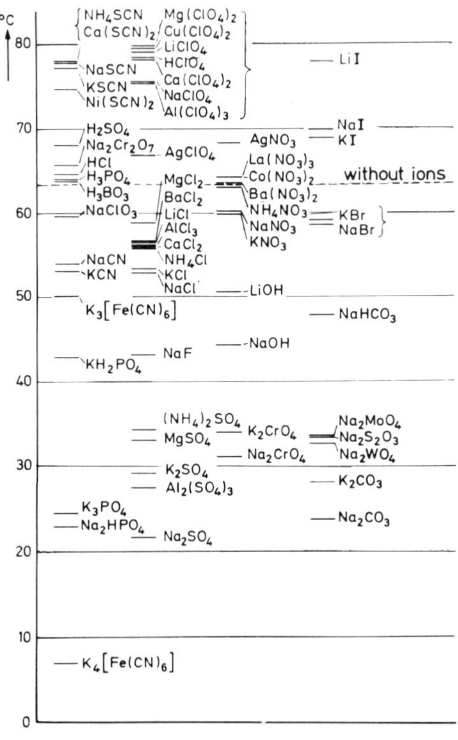

Fig. 6. Turbidity point T_k of PIOP-9 in aqueous solutions with 0.5m salt. The anion effect is bigger than the cation one.

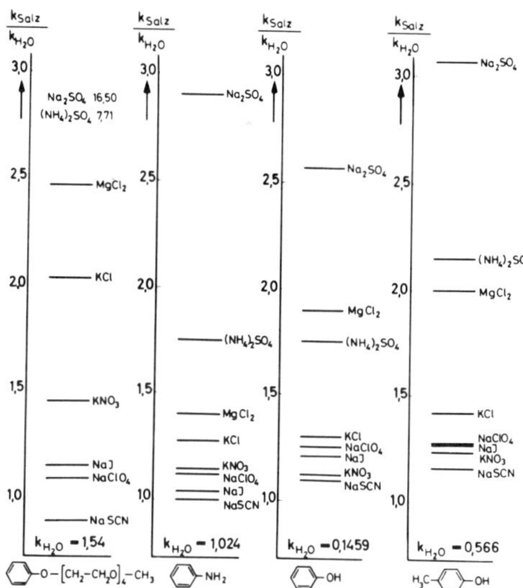

Fig. 7. Partition coefficients k_{salt} of different solutes between cyclohexane and salt solution (1 m, 20°C) over the coefficient with pure water k_{H2O}. Border between salt-in/ salt-out effects depend on rivalry H-bonds water-water, water-solute, water-ions.

hexane and electrolyte solution (salt concentrations 1 mol/liter) are divided by the partition coefficient k_{H2O} between cyclohexane/water. This change of the boundary $k_{salt}/k_{H2O} \gtrless 1$ may depend on the strength of the H-bonds between the solute and water.

II. 3. UNDERSTANDING THE STRUCTURE MAKER EFFECT

Debye[16,17] has described the structure maker effect of ions by the strong Coulomb fields around ions. Debye assumed that the water dipoles are orientated around every ion until saturation. Therefore organic solutes could not enter such areas of total orientated water dipoles. Debye has estimated that this effect orientates water around every single charged ion until a distance of about 11.10^{-8}cm. If water had an ice-like structure in this area this would mean every ion orientates water in a sphere of about 190 H_2O molecules. If organic solutes need free OH groups of water or distrubed H-bond structures the salt-out effect of ions could be easily understood by this orientation of water around ions.

II. 4. UNDERSTANDING OF THE STRUCTURE BREAKING IONS

Debye's theory cannot be used to understand the salt-in effect. The interaction of ions-water could be studied spectroscopically by studies of the water Raman spectra of crystalline hydrates. Figure 8 demonstrates some examples with the result that we found the maxima of water stretching bands on solid hydrates mainly between the vapor spectra and ice spectra[18,19]. Studies of H-bond spectroscopy gave the so-called Badger-Bauer rule: the frequency shift $\Delta\nu$ of OH stretching bands by H-bonds is proportional to the H-bond interaction energy Δ H:

$$\Delta\nu \sim \Delta H_H \qquad (2)$$

If we apply this rule to anion-water interactions, too, we could ask: is it possible that some anions interact with one OH of water not stronger than two water molecules?

This question seems to be foolish if we remember that hydration energies of ions are given in the literature in the size order of 100 kcal/mol or more.

But the contradiction between both studies can be reduced if we remember that hydration energies are given per mol and that the limited solubility of salts in water indicate that ions need a group of water molecules to be solved or to produce the hydration energy. If we assume that the solubility gives at a certain temperature a minimum n of water molecules which produces the hydration energies H^+ of cations and H^- of anions we see in Table 1 this water number n per mol salt at saturated solutions. If we remember now that every water molecule has four centers of Coulombic interactions: the two protons and the two lone pair electrons we could divide the total

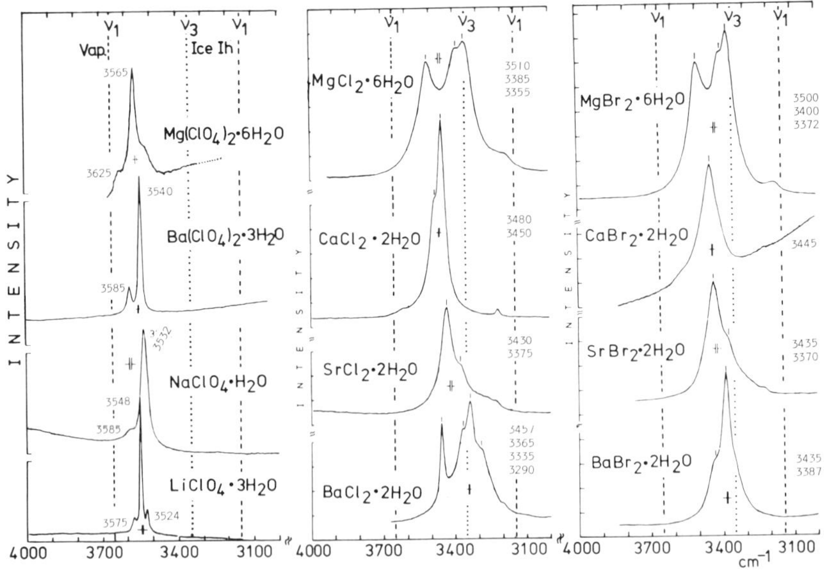

Fig. 8. Raman spectra of hydrate water in crystalline hydrates. Band maxima are mainly between vapor and ice spectra.

hydration energy by 4 n to estimate the mean energy per active water group. Table 1 demonstrates that a lot of ions give a size of order of 5 kcal/mol or lower for this quotient. This is the size of the H-bond energy per OH group. Higher values in the last column of Table 1 could mean that ion pairs are formed or that the cation-water interaction - which is insensitive to the water spectra - could be larger than the anion-water one.

A third method gives us a similar result: Kleeberg in our laboratory[23] has measured water spectra in CH_2Cl_2 and in ternary mixtures: water/CH_2Cl_2/crytands (Figure 9). In these fundamental regions of Figure 9 we can observe that the cryptands ν_3 vibration of water and the symetric ν_1. ν_1 is shifted if one OH is H-bonded [20,21,22]. All spectroscopic results demonstrate that the anion - water interaction changes the water spectra similar to organic bases.

Therefore the bands around 3400 cm^{-1} can be correlated to OH groups interacting with different anions. This method is very useful because the cryptands protect the cations and we could assume that we are observing mainly the water-anion interaction.

Table 1. Mol water n per mol salt in saturated salt solutions at 25°C. Sum of hydration energies of cation ΔH_{h+} and of anions ΔH_{h-} divided by the active water groups 4n at saturation gives values not far from the size of order of water.... water interaction.

solute	$n = \frac{\text{mole } H_2O}{\text{mole solute}}$ 25° C	$\frac{(\Delta H_{h+} + \Delta H_{h-})}{4n}$ [kcal]
CsF	2.30	- 20.5
LiCl	2.79	- 19.2
LiBr	2.83	- 18.6
KF	3.21	- 15.5
RbF	4.43	- 10.9
NaI	4.50	- 9.4
CsCl	4 77	- 8.2
LiI	4.94	- 10
NaBr	6.04	- 7.6
KI	6.22	- 6
RbI	7.07	- 5.1
RbCl	7.12	- 5.7
RbBr	7.82	- 5
NaCl	8.38	- 5.5
CsBr	9.66	- 3.9
KBr	9.88	- 4.1
CsI	16.59	- 2.1
NaF	55.5	- 2
LiF	959.7	- 0.06

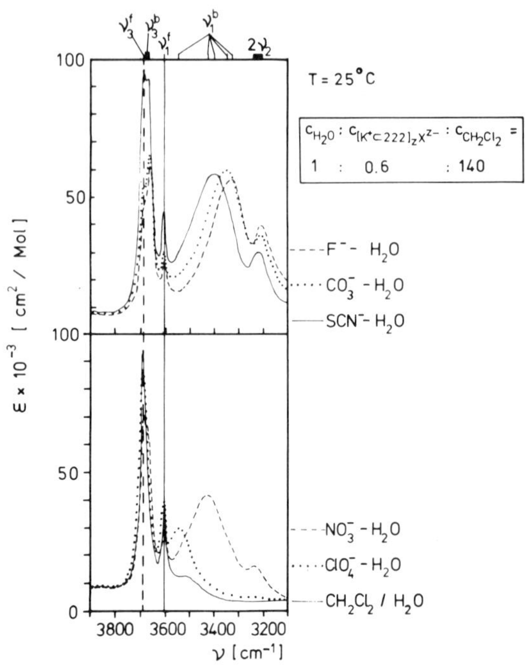

Fig. 9. Fundamental vibrations of water in solutions: CH_2Cl_2/cryptand/ions.
ν_1^f : band of water non interacting with anions
ν_1^b : band of water interacting with anions.

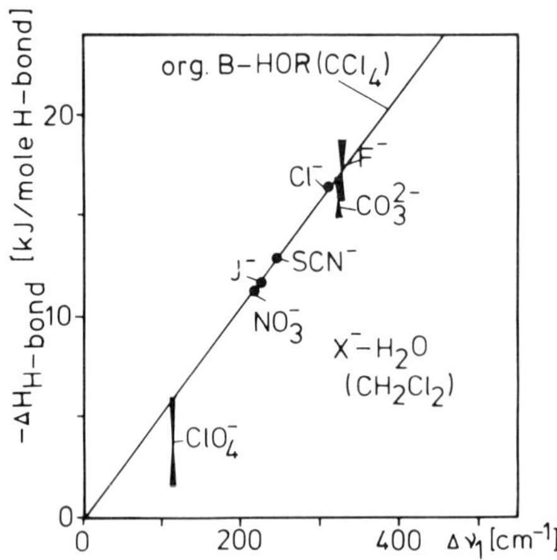

Fig. 10. Interaction energy ΔH_{H-bond} of water-anions as function of frequency shift $\Delta\nu_1$ determined by the T-dependence of ν_1^f and ν_1^b of figure 9. Validity of the Badger-Bauer rule for water-anion interactions[23].

The shifts are observed in the series (Figure 10)

$ClO_4^- < NO_3^- < I^- < SCN^- < Cl^- < CO_3^{--} < F^-$

a similar series to the Hofmeister ions series or the T_{str} - series. By studying the T-dependence of such spectra Kleeberg has determined the interaction energies ΔH_H of these water-anion complexes and obtained results of Figure 10. This method, too, gives ΔH_H values in the size of the water-water interaction of about 15.5 kJ/mol (3.7 kcal/mol) [24].

All three methods: 1. structure T measurements with aqueous solutions, 2. the spectra of crystalline hydrates, or, 3. of ternary systems water/CH_2Cl_2/cryptands, give similar results:

The interaction between a number of anions and water seems to be in the similar size of order to the water-water interaction per one OH group.

If this result is true and not an artefact of our indirect conclusions we could assume that the H-bonds of water or alcohols are as strong as the ion-dipole interactions by the strong Coulomb forces of protons or lone pair electrons, because both are near the surface of water molecules and Coulomb forces depend on charge and distance.

A difference between ion interaction is that we could have bigger coordination numbers[25] but H-bonds with organic basis have only a coordination number of one.

Figure 11. demonstrates the positions of the water bands ν_1 and ν_3 in presence of organic bases, Figure 12 in presence of different anions in the cryptand solutions of CH_2Cl_2. Both figures demonstrate:

The anions-water interaction are similar to H-bond water-organic bases.

Acids and bases give stronger influences on the water spectra than neutral salts[18,26], indicating the important role of pH.

III. BINARY INTERACTION WATER-POLYMERS

The interaction between water and hydrophilic organic molecules depend on the base strength of the hydrophilic group and the acidity of its protons. The action between water and organic molecules can be discussed as two types: H-bond and van der Waals interaction. Small hydrophobic molecules like CH_4, H_2 etc. can be solved in water by dispersion interaction[31]. The solubility of rare gases in H_2O increases with the molecular weight M because the dispersion inter-

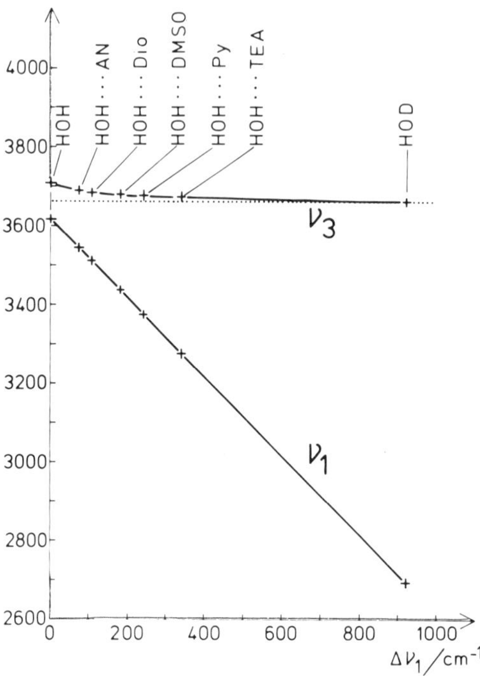

Fig. 11. AN: Acetonitrile Dio: Dioxane DMSO: Dimethylsulfoxide
Py: Pyridine TEA: Triethylamine
Frequencies of the asymetric water vibration ν_3 and the ν_1 of water base complexes (1:1) as function of the shift $\Delta\nu_1$ (complex vibration - vapor vibration). H-bonds with 1 OH shifts ν_1 proportional to ΔH and decoupled ν_3 to the vibrations of the non-bonded OH vibration $(\nu_1 + \nu_3)$free/2 [21].

action increases with M. But the solubility in water as function of M has a maximum and depends not only on the strength of dispersion interaction but on the size of the solute[23,32]. We could look for the cause in the H-bond disturbance if we have to form holes in water for big solute molecules.

The solubility of hydrophilic organic molecules depends on their H-bond acceptor and donor strength. Both mechanisms, interaction water-hydrophobic groups and H-bonds water-hydrophilic groups can be differentiated by spectroscopy. The OH or NH stretching modes are more sensitive on H-bonds than van der Waals interactions[9,23]. FIgure 13 gives as example the heat of vaporization ΔU_{vap} of some compounds from C_2H_5 as function of the alcohol OH frequency. The hydrophobic interaction induces only small shifts, but H-bonds induce stronger ones. A similar relation is obtained for aqueous solutions [23]. The heat of solutions ΔH_{AB} in water is the difference

$$\Delta H_{AB} = \Delta H^*_{AB} - \Delta H_{AA} - \Delta H_{BB}$$

ΔH^*_{AB} : Interaction energy water-solute
ΔH_{AA} : Interaction energy water-water
ΔH_{BB} : Interaction energy solute-solute

If ΔH_{BB} of hydrophilic compounds like phenol is in a similar size of order to ΔH^*_{AB}, two phases can be formed: a water rich phase with some organic molecules and an organic rich phase with some water in it. Such two-phase formations with compounds of medium or high molecular weight are called coacervate formations and the organic phase is called the coacervate.

In ternary mixtures properties could depend on the difference of the H-bonds of the two solutes to water.

III. 1. HYDRATION OF POLYMERS

Polyethenoxide derivatives (PEO) are interesting in the study of the hydration of hydrophilic polymers. The high surface tension of water forms droplets of liquid water. But small amounts of high molecular polyethylenoxide additives allow the drawing of hydrated

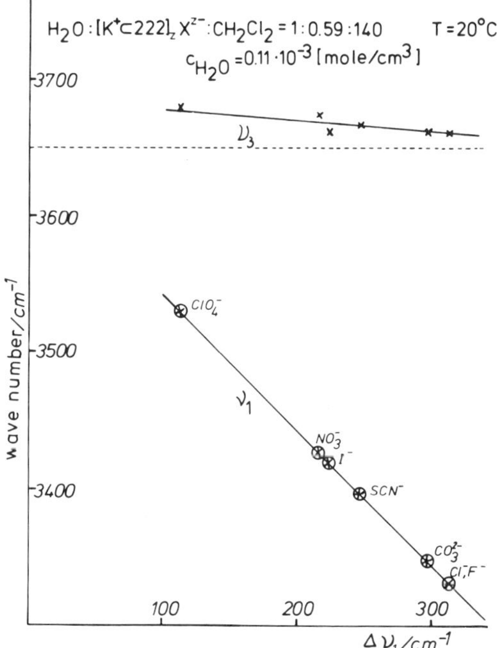

Fig. 12. Frequencies ν_3 and ν_1 of water in complexes with different anions in systems: water/CH_2Cl_2/cryptand/ions. Figure 11 and 12 demonstrate the similarity of water-anion interactions to water-base H-bonds.

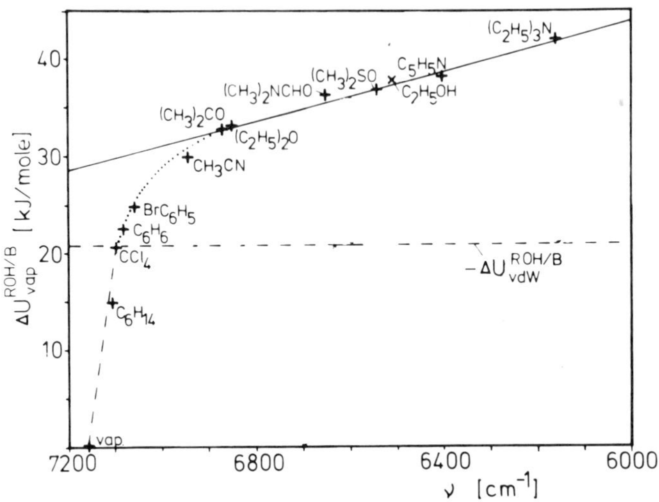

Fig. 13. The heat of vaporization ΔU_{vap} at 25°C from ethanolic solutions as function of the ethanol OH frequency (25°C).

PEO like gossamer from the water, demonstrating that hydration of PEO has higher interaction as the water-water inter-action.

PEO products with hydrophobic groups like PIOP-n (see chapter II.2) give two phase formation at higher temperatures at a sharp turbidity point T_K (see Figure 6). The organic phase of PIOP has a water content of about 20 H_2O per ether oxygen at T_K. This water content decreases with increasing T until 2 H_2O per oxygen, showing a special stable dihydrate[26]. The turbidity point depend hardly on the PIOP concentration, in the concentration region C_0 between 1 g/l until 100 g/l (0-0.2 mol/l). T_K changes only about 3° with a minimum at about 0.08 mol PIOP/l. Figure 14 demonstrates that the size of T_K is an indicator of the strength of the H-bonds between water and the ether oxygen atoms. This interaction is able to carry certain amounts of hydrophobic groups. The hydrophobic/hydrophilic balance, the ratio between n and the length of the hydrophobic chain, determines the turbidity temperature:

Table 2. Turbidity point of PIOP - n in °C. for 0<C_0L0.2 mol.l

n	7	9	10	17
T	49°-52°	63°-64°	75°-77°	91.5°-95°

The heuristic understanding of the ion effects by the structure temperature T_{str} can be demonstrated at the transition temperatures of ribonuclease in electrolyte solutions (Figure 15). The points

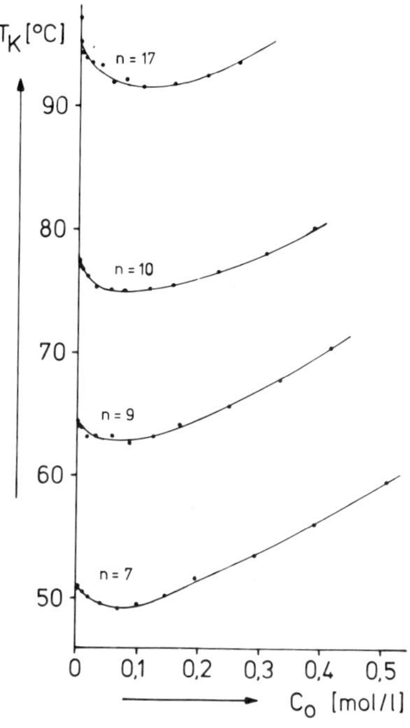

Fig. 14. The turbidity temperature T_K of iso-C_8H_{17}-⟨◯⟩-$(O-CH_2-CH_2-)_n$-OH in water as function of its concentration and the content of ethylenoxide n.

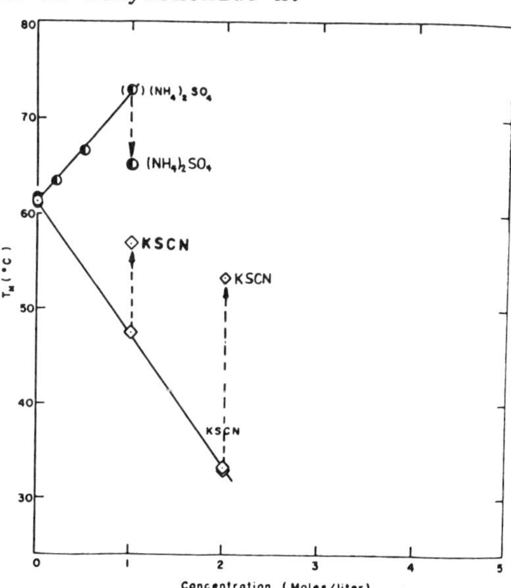

Fig. 15. Full line transition temperature T_m of 0.5% ribonuclease in water (measurements of Hippel and Wong 1965). Arrows indicate the spectroscopically determined structure temperatures T_{str}, indicating that T_m in salt solutions mean similar H-bond structures of the solutions at T_m.

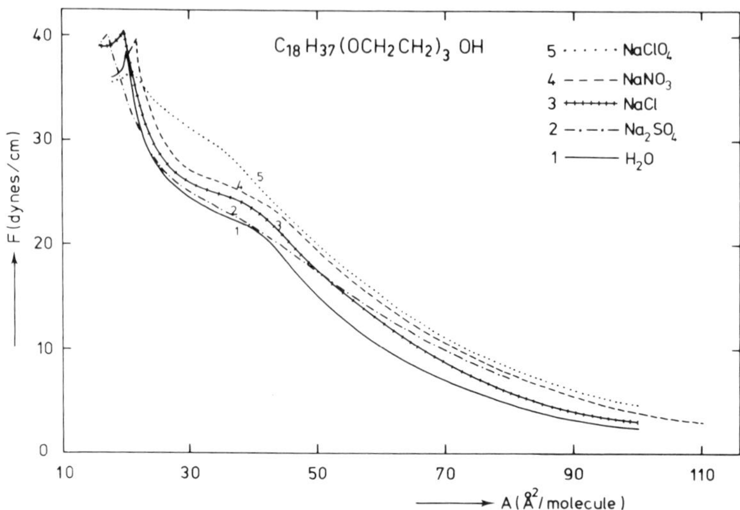

Fig. 16. Force (F) area (A) diagrams of monolyers of $C_{18}H_{37}$-$(OCH_2CH_2)_3$-OH (TEO) at 24°C on water and 1 m electrolyte solutions demonstrate influence of the changed water structure on F-A [27].

connected with arrows give the T_{str} determined at the transition temperatures of ribonuclease with 1 mol/1, 2mol/1 $(NO_4)_2SH_4$ or KSCN. The T_{str} values are not far from the transition T in pure water, which means in ion solutions the transition T is found at a T of common H-bonded state of water.

A second example of the role of T_{str} on the hydration of polymers is found in measurements of the force-area diagrams of monolayers of non-water soluble ethylenoxide products like $C_{18}H_{37}$-$(OCH_2CH_2)_3$-OH (TEO) Figure 16. The structure-maker $Na(SO_4)_2$ decreases the diagram as a T decrease[27]. Structure breaking ions change the diagram as a T increase.

If we plot the force F needed to get an area of 30 A^2 per molecules of TEO against the spectroscopic T_{str} values we get a proportionality (Figure 17). The same is valid if we plot the whole area $\int F\,dA$ against T_{str}. Figures 16 and 17 demonstrate that similar ion effects happen at surfaces. Force-area diagrams depend on the interaction of the monolayers to the aqueous subphase.

III. 2. MICELLE FORMATION

The U.V. spectra of the aromatic ring in the PIOP-n products are sensitive to the environment[28]. A small band at 283 mμ in presence of hydrophobic groups appears contrary to water. With increasing concentrations of PIOP-n in water the absorbance of this 283 mμ band increases indicating an equilibrium[28].

$$m \text{ monomers} \rightleftharpoons \text{micelle} \tag{3}$$

with m about 6. This spectroscopic method determines equilibrium constant:

$$K = \frac{(\text{monomer})^m}{(\text{micelle})} \tag{4}$$

It demonstrates the understanding of the so-called critical concentration C_k of micelle formation. As far as C_k PIOP-n solves in water mainly as monomers, above C_k added PIOP-n amounts increase mainly the micelle concentrations and the monomer concentration keeps more or less constant[28].

These micelles are able to solve organic molecules (solubilisation). For instance dyestuff molecules solve inside the micelles giving an equilibrium

$$\text{free dye} + \text{micelle} \rightleftharpoons \text{complex (dye-micelle nucleus)} \tag{5}$$

This equilibrium is applied to dyeing processes of textile chemistry[28,29]. The PIOP products act like a buffer. Only the free dye is in equilibrium with fibres. By the complex formation equation (5) the free dye concentration changes less during a dyeing process. The hydrophobic polyethylenoxides act therefore as levelling agents[29,30].

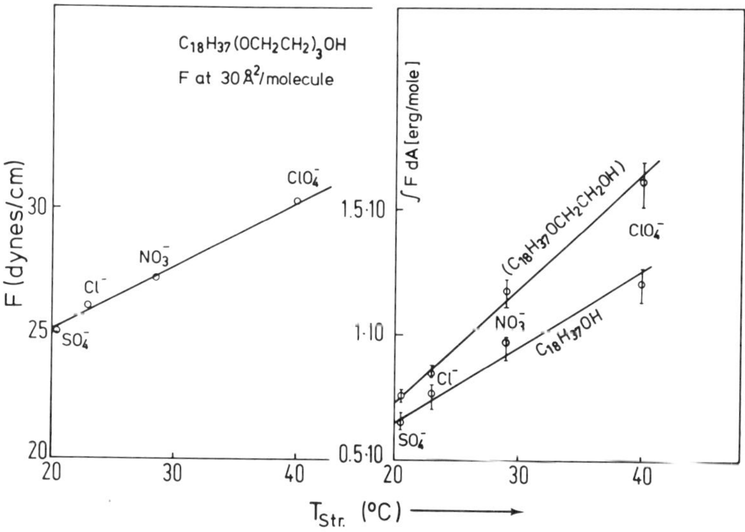

Fig. 17. Left: Force (F) to get an area of 30 A^2/molecule in Figure 16 in presence of different anions as function of the subphase structure temperature T_{str}[27]. Right: $\int$FdA of Figure 16 and of $C_{18}H_{37}OH$ as function of T_{str}.

III. 3. WATER IN BIOPOLYMERS

That water exists in different types in biologic systems is a discussion point[31-33], and the terms "bonded" and "free" are the terms used to describe water in cells. The spectroscopic method could establish the question. Figure 18 gives at the bottom the desorption isotherms of water in tendon or gelatine as function of the relative humidity[34,35]. At about 50% relative humidity is found a plateau which can be understood as monohydrate water/polymer. Figure 18 shows in the middle the band maxima of the combination band ($\nu_2 + \nu_3$) of water in the same polymers. In the region of the hydrate formation we found[34,35,36] the level of the position of the water band shifted to a stronger H-bonded water form. The secondary hydrate above 50% relative humidity indicates a transition from this hydrate water to a liquid-like structure.

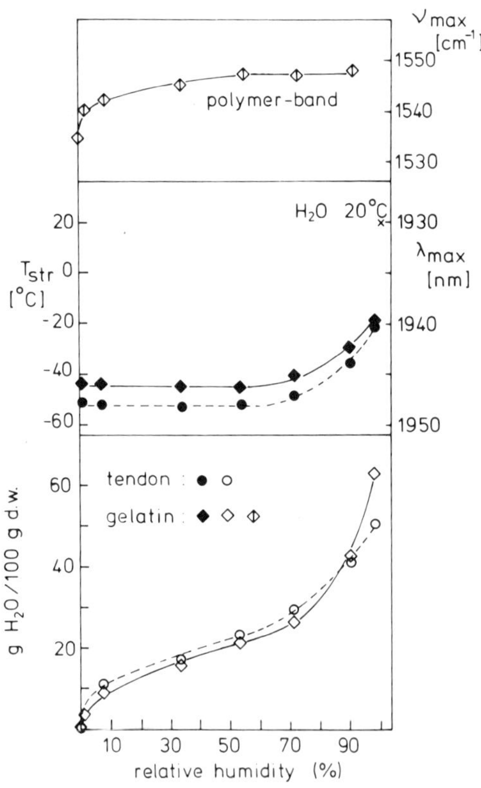

Fig. 18. Bottom: Desorption isotherms of water as function of relative humidity. Top (right scale): band maxima of combination water band. Top (left scale) : structure temperature of water in tendon or gelatine.

The stronger bonded water of the hydrate may be caused by the stronger H-bond acceptor groups of the peptides. But the difference from liquid water is small. Therefore we prefer the nomenclature "hydrate water" and "liquid-like water", remembering that liquid-like water is very nearly bonded and not free.

The small difference in the H-bond structure of the hydrate water seems at first contrary to the very different properties of biological systems at different humidities. For instance seed can be stored for a long time without growing at low humidities of about 65% but starts to grow at 100% humidity; food can be stored by small reduction of the water activity by drying or by salt and sugar additions, bacteria does not grow at small reductions of relative humidities. These big effects can be understood easier if we look at the left scale of the water band maxima in Figure 18. The structure temperatures of pure water are given with similar band maxima. The hydrate water of both biopolymers corresponds to H-bond strength which supercooled water would have at temperatures below - 40°C, at temperatures where the kinetics of bioreactions would be reduced.

With similar spectroscopical measurements of hydrate water in membranes for sea water desalination this membrane mechanism could be described[37.38].

IV. TERNARY SYSTEMS

IV. 1. WATER / ORGANIC / ORGANIC MOLECULES

In ternary systems; water/PIOP-n/organic molecules, the turbidity point is measured as the sum of interaction changes dependant on the hydrophilic/hydrophobic properties of the organic molecules (Figure 19). Adding 1 mol.liter of C_2H_5OH increases T_k on 11.5° similar to the increase of PIOP-9 on one ethylenoxide group to PIOP-10 (Figure 14). The comparison between the action of different alcohols demonstrates that the increase of hydrophobic groups increases the lift of T_k, indicating that the rise of hydrophilicity of the micelle nucleus occurs by adsorbing the alcohol as it takes part in the T_k-change. The T_k-increase by adding hydrophilic molecules is a good model for demonstrating the effect of protecting the flocculation of colloids to increase their stability. Adding the water bad soluble n-butanol decreases T_k (1 m at 18°C). Moderate water soluble molecules like cyclohexanole or aniline reduce T_k strongly.

Figure 19 demonstrates solubilisation or antisolubilisation effects of small molecules on hydrophilic polymers of medium molecular weight.

We could assume that the increase of T_k by high concentrations of PIOP-n (Figure 14) may be induced by a solubilisation of micelle

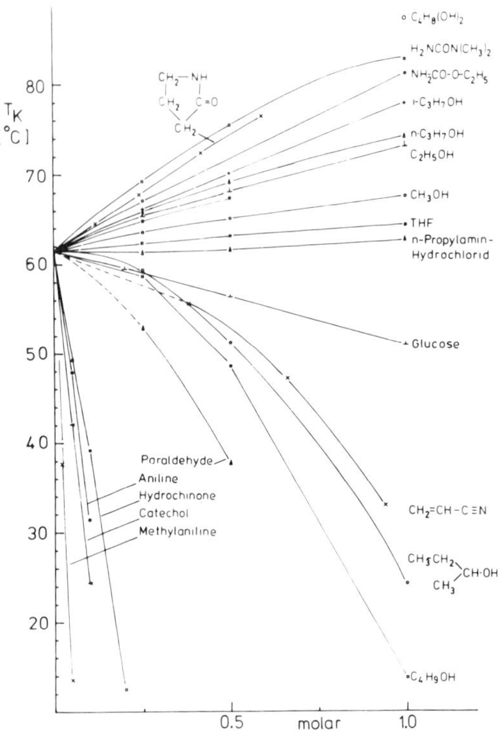

Fig. 19. Change of the turbidity of PIOP-9 by organic additives. A comparison with Fig. 6 demonstrates the similarity between anion and H-bonds effects of hydrophilic groups.

nucleus by the ethylenoxide groups of a neighbor micelle. In addition the change of the water structure by hydration of ether oxygens should be discussed.

IV. 2. WATER / ORGANIC MOLECULES / IONS

A comparison between Figure 19 and Figure 6 demonstrates again a similarity between the effects or organic hydrophilic groups and anion effects on H_2O. (As in figures 11 and 12). The ion effects on interaction water/hydrophilic molecules can be studied too by measurement of the ion influence on the micelle formation of PIOP-n compounds[26,28,39,40]. The micelle equilibrium constant K_{mol} of equation (4) is very sensitive to ion concentration and on the type of ions corresponding to the Hofmeister series or to T_{str}.[26,28,33]. Figure 20 shows as an example K in molar units in aqueous solutions of 0,5 mol salt per liter.

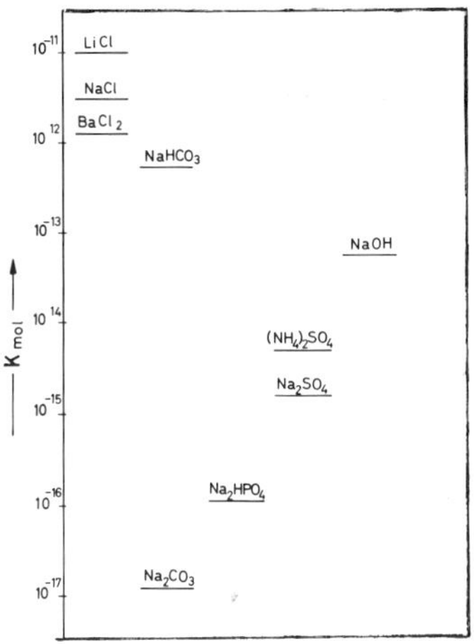

Fig. 20. Change of the equilibrium constant k mol of micelle formation of PIOP-40 with salts (0.5 mole/liter, 20°C).

The composition of the coacervate of polyethylenoxides above the turbidity point can be influenced intensively by added ions. Figure 21 demonstrates the water content of the organic phase of PIOP-9 at 80° in presence of different ions (given in mole H_2O per mole ethylenoxide groups EO) with salt concentration 0.5 mole/liter below T_k). Salts which disturb the H-bond structure of water such as NaSCN, $NaClO_4$ or NaI increase the hydration of PIOP whilst salts which reduce the content of "free" OH-like sulfates decrease it. A similar influence of salts corresponding to its Hofmeister series or its T_{str} can be observed by the PIOP-9 content x of the aqueous phase (figure 22). Both figures 22 and 23 indicate a changed hydration of PIOP-9 in presence of ions caused by the change of water structure by the ions and the similar properties of the aqueous and organic phases.

The ion effects on the water structure observed by OH spectra is relatively small but solubilities of organic molecules relatively large. In a heuristic method we could estimate the ion effects on solubility with the simplified assumption that the hydrate sphere around every ion is excluded for other solubilities. At partition coefficient k determinations of organic solutes like p-cresol between cyclohexane/water we could assume too that k will not be changed by ion outside its hydrate sphere. With the observed change of k by ions we could estimate the hydration numbers H.N. of different salts-[26,33]. Figure 23 shows such an example with function of Na_2SO_4 concentrations and temperature as parameter. These apparent hydration numbers should only demonstrate the large ion effects. They start corresponding to 100 H_2O molecules per ion pairs at low con-

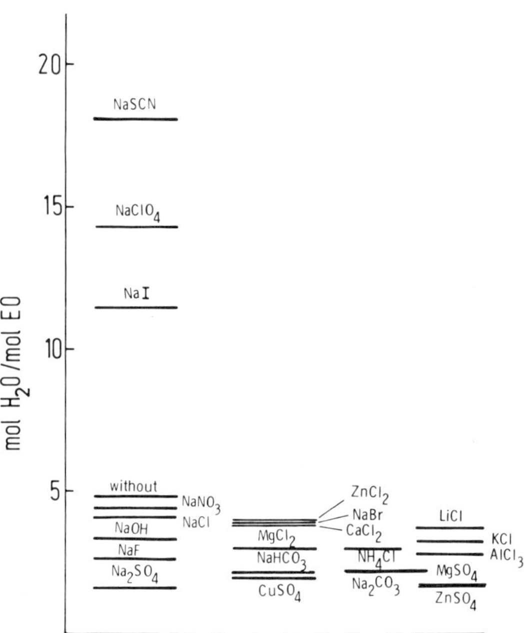

Fig. 21. Water content per ethylenoxide unit (EO) of PIOP-9 at 80°C, without and in presence of 0.5 mole/l salts.

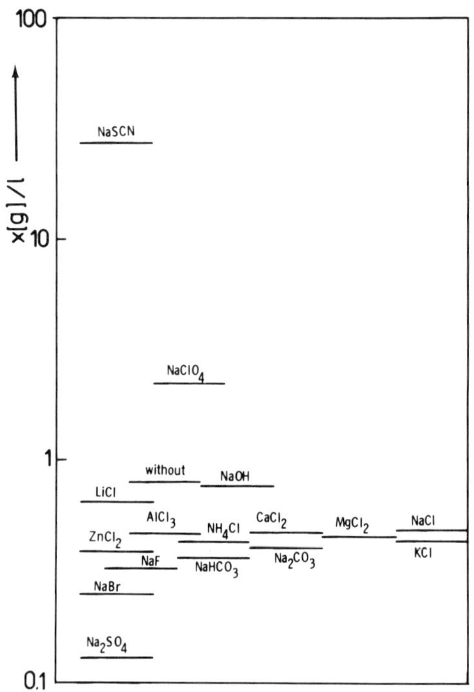

Fig. 22. PIOP-9 content x of the aqueous phase of PIOP-9 at 80°C, without and in presence of 0.5 mole/l salts.

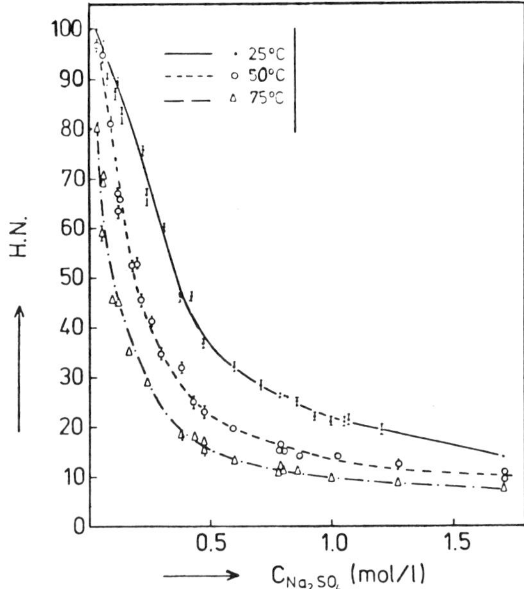

Fig. 23. Apparent hydration numbers H.N. of ions determined by the partition coefficients of p-cresol between cyclohexane and electrolyte solutions at different ion concentrations and temperatures.

centration. This size agrees with Debye's estimation of the sphere strongly oriented by the ion Coulomb field[16,17]. This sphere is so large that its decrease with ion concentration by disturbance of a secondary hydration sphere (Figure 23), seems to be a probable.

The discrepancy between the big effects of Figure 23 and the small disturbance of the OH spectra becomes smaller if we estimate the extension of H-bonds in liquid water by the determined "free" OH groups[2,4,5,8]. At 25°C we could estimate that the measured content of 12% free OH means that the network of H-bonds extends about 200 water molecules. The recalculation starting with the changed content of "free" OH in lm/l Na_2SO_4 solutions gives an enlargement of the H-bond network by 20 water molecules. This value agrees with the result of Figure 23 fairly well.

Similar hydration numbers like the rough estimation of figure 23 we obtained by the change of micelle equilibrium constants of PIOP compounds[26] or by the changes solubility of simple gases such as O_2,H_2[26]. This agreement demonstrates that this simplified estimation is a useful method to predict salt effects on solubility of solutes in electrolyte solutions. The usefulness of a hypothesis in science can be proved by prediction of unknown properties at the base of such a hypothesis. This we demonstrated by the agreement between the temperature effect on the hydration numbers by Na_2SO_4 in Figure 23 and the spectroscopic change of "free" OH content. Spectroscopy of NaCi solutions has found, that hydration numbers of NaCl do not change much in the range 20°C < T < 120°C. If the simplified hydration numbers are used we could predict that the salt effect on partition coefficients by NaCl does not change much with T either. Figure 24 demonstrates that our measurements - calculated after this prediction - agree with it[26,32]. The big ion effects on solubilities may correlate with the extended H-bond system in liquid water. Solutes interact at first with the small content of defects of free OH groups. Ions and other solutes rival on the free OH groups of water.

The description of ion effects in ternary systems: water/organic solutes/ions based on our I.R. spectroscopy results seems to be generally useful for any system in which the Hofmeister ion series is observed. For instance Figure 25 shows the similar ion series observed at the melting temperature of agarose or of gelatine[23]. The melting temperatures T_m of gelatine in the presence of 1 mol/liter salts is correlated with the spectroscopically determined structure temperature T_{str} as a rough measure of the water structure.

Our rough approximate description of ternary systems attempts to combine the hydration phenomena of ions and organic solutes. Frank and Wen[41] or Lilley[42] have described the ion pair formation in aqueous solutions by an alliance of the primary hydration spheres of anions and cations, forming a common secondary hydrate sphere.

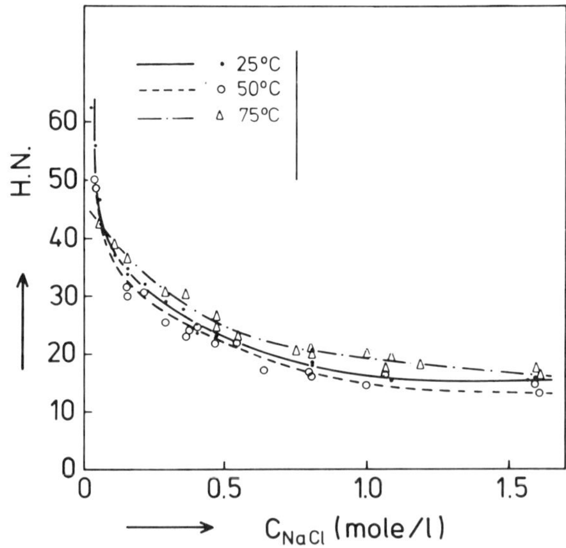

Fig. 24. The apparent hydration numbers of NaCl determined with same method in figure 24 are nearly T-independent as spectroscopic results had predicted.

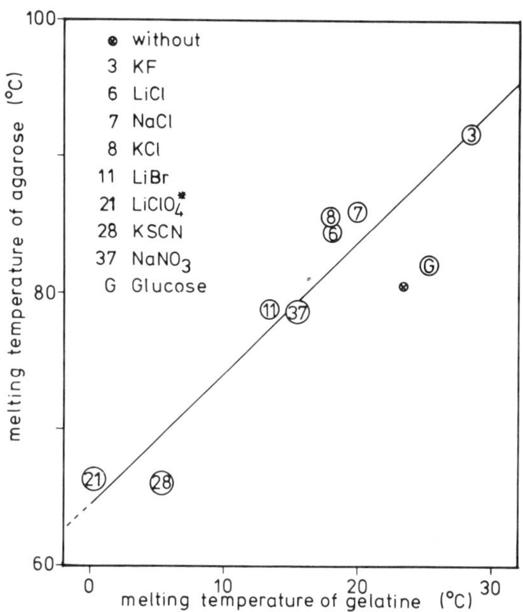

Fig. 25. Parallel effects of ions on the melting temperatures of agarose and gelatine [23].

(Figure 26(a)). The usefulness of the differentiation primary and secondary hydration we could extend on organic hydrations by the observation that the secondary hydration of PIOP above the turbidity temperature of two phase formation measured by the water content of the organic phase can be reduced by increasing the separation temperature of the two phases or by the ion concentration of the aqueous phase [26]. But with both methods we could not reduce the water content below two H_2O per ether oxygen atom[26]. At this dihydrate water content the viscosity has a maximum and x-ray scattering indicates a special structure[26].

We could apply the hypothesis: during coacervate formations the secondary hydrate spheres of organic molecules combine.[43]. (Figure 26). The idea to differ the primary hydration and the diffuse secondary hydration has been already described by Wolfgang Ostwald[44].

For example - the hypothesis that the ternary systems can be described approximatively by the hydration of both components and predict some ion effects by the structure change series - figure 27 - of gives a general view of the complicated question of the ternary systems: water/ions/solvents could form two phases or not at room temperature. This possibly depends on the H-bond strength of the solvent and the position on the ion series[23]. The solvent series of its concentration of hydrophilic groups C_A is seen (row 2 in figure 27).

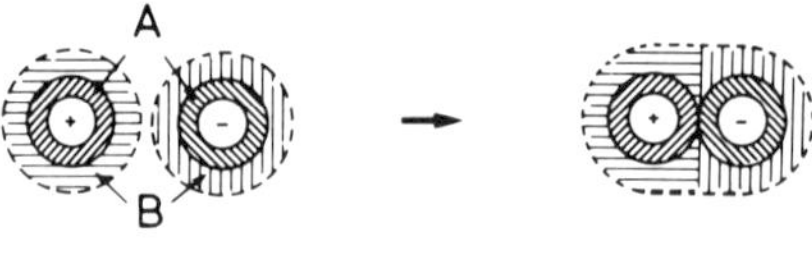

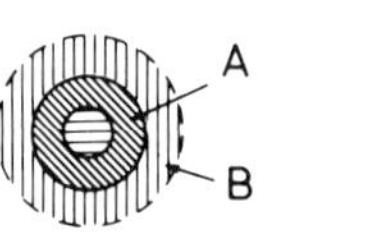

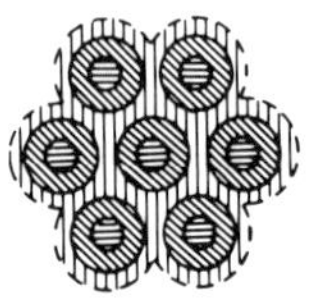

Fig. 26. a) model of ion pair formation and its hydration; b) model of coacervate formation by the network of its secondary hydrates

\+ possibility to form coacervates

(+) during coacervate formation 3 phases

C_A : concentration in mol/l of hydrophilic groups in the pure solvent

	C_A[mol/l]	K_2CO_3	Na_2CO_3	KF	K_2CrO_4 Li_2SO_4	Glucose	$NaNO_3$	$CuSO_4$ Na_2SO_4	KSCN	KCl	KNO_3	KI
Formamid	50.33	-	-	-	-	-	-	-	-	-	-	-
Glycerin	41.08	-	-	-	-	-	-	-	-	-	-	-
Methanol	24.70	+	-	-	-	-	-	-	-	-	-	-
Ethanol	17.13	+	(+)	+	-	-	-	-	-	-	-	-
Dimethylsulfoxid	14.10	+	(+)	+	-	-	-	-	-	-	-	-
Aceton	13.60	+	+	+	+	(+)	-	-	-	-	-	-
iso-Propanol	13.07	+	+	+	+	(+)	+	(+)	+	-	-	-
Tetrahydrofuran	12.33	+	+	+	+	+	+	(+)	-	+	(+)	-
n-Propanol	13.37	+	+	+	+	+	+	(+)	+	+	-	-
tert.-Butanol	10.64	+	+	+	+	(+)	+	(+)	+	+	(+)	+

Fig. 27. +: possibility of formation of two phases at room temperatures in ternary systems: water/ions/solvent [23] -: two phases not observed. (+); during two phase formation at first 3 phases. C_A: concentration of hydrophilic groups in Mole/liter pure solvent.

IV. 3. EXTENSION OF TERNARY OBSERVATIONS ON SOLID POLYMER PHASES AS ONE COMPONENT.

The paralellism between the properties of two coacervate phases induces the hypothesis that the ion effects observed with these methods could be expected, too, on ternary systems. That is electrolyte solutions in contact with solid polymers.

We can demonstrate that the Na ClO_4 which increases the non H-bonded OH groups in liquid water increases too the water uptake of 6-polyamide[32]. The water content of 6-polyamide decreases in contact with Na_2SO_4 solutions in which the content of "free" OH is reduced. Figure 28 shows another experiment of this type[23]. The water uptake of bovine nasal cartilage at 4°C can be reduced in contact with 0.8% NaCl solution instead of pure water (Figure 28).

We would expect that similar results would happen in living cells too. In some cases it is to be expected that some special ion adsorption by polymers could happen in addition to the normal Hofmeister effects between change of water structure by ions and hydration effects. In such cases deviations from the Hofmeister ion series could be observed.

IV. 4. ADDITIVITY AND RIVALRY OF DIFFERENT ION HYDRATES

The ion effects on the water structure change seem to be additive at medium concentrations. One example shows that the changes of the turbidity points of PIOP-9 in 1m mixtures Na_2SO_4 solutions can be calculated by the sum of SO_4 decrease and ClO_4 increase effects [32].

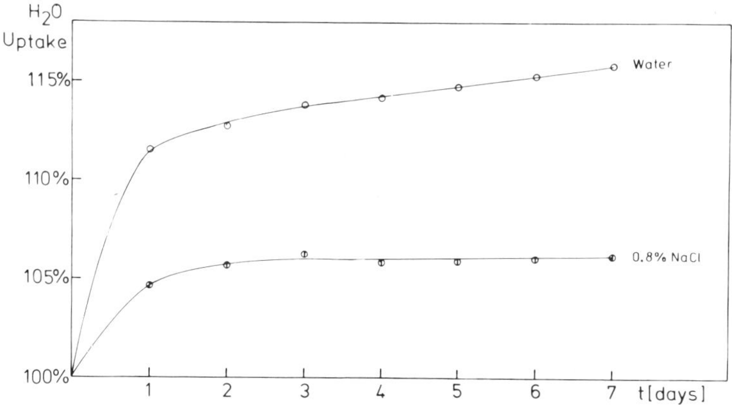

Fig. 28. Uptake of water by bovine nasal cartilage is influenced by ions[23]. (100%: fresh weight 4°C).

Figure 29 gives another example of the nearly additivity of ion effects and effects of added solvents on the turbidity point of PIOP-9. Kleeberg shows in his thesis, too, that such an additivity can be observed by the change of water spectra in ternary or quaternary systems too[23]. At higher concentrations a rivalry between different hydrations can happen. The concentration dependence of the hydration numbers in figures 23 and 24 may indicate a coupling of the different hydration spheres. Schiöberg in our laboratory showed spectroscopically a concentration dependence of the hydration numbers of ClO_4^- anions[45].

Similarly we could assume that the decrease of the turbidity point of PIOP-9 by KI at concentrations higher than 2 mol KI/liter would be caused by the rivalry between the hydrations of ions and PIOP-9. (Figure 31). This hypothesis could be established by the fact that in this region T_k depends on the PIOP-9 concentration, which is contrary to low ion concentrations and contrary to the observation that without the presence of ions an increase of PIOP-9 concentration increases T_k (Figure 14). Corresponding to our assumption that, in solutions higher than 2 mol KI.liter or 1 mole ClO_4/liter, the hydration of PIOP-9 in the organic phase decreases. (Figure 31).

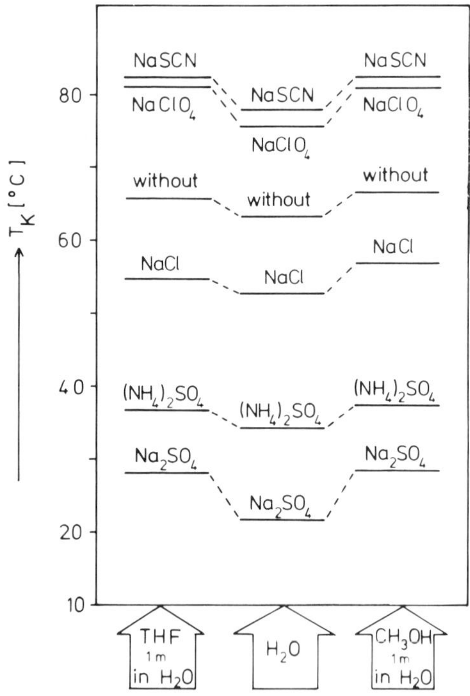

Fig. 29. Turbidity point changes of PIOP-9 (10g/l) by 0.5 m salt or 1 m added tetrahydrofuran(THF) or CH_3OH are nearly additive, indicating the model of nearly additivity in ternary systems.

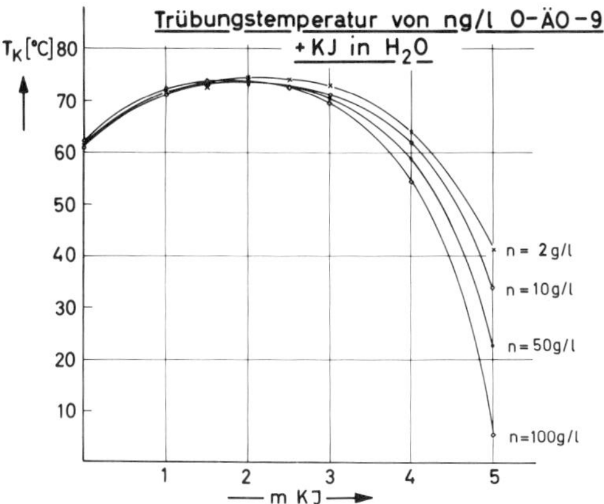

Fig. 30. Turbidity point of PIOP-9 as function of molar KI concentration m and amount of n of PIOP-9 in g/l. Rivalry between ion- and solute-hydration.

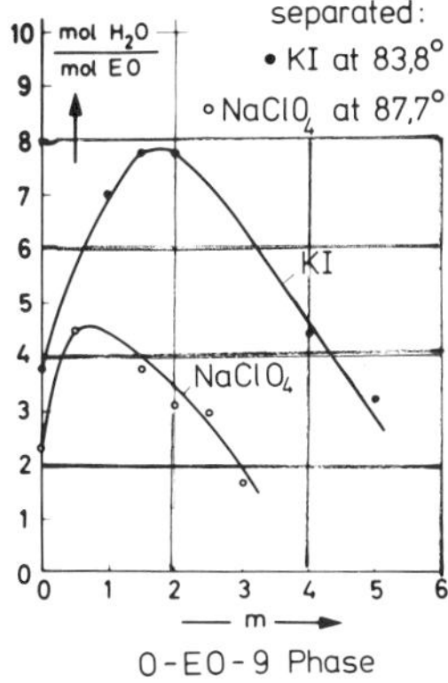

Fig. 31. Water content per ethylenoxide unit EO as function of molar concentration of KI (83.8°C) and $NaClO_4$ (87.7°C).

The salt solubility in water/solvent mixtures also indicates hydration rivalries. Figure 32 shows the reciprocal salt solubilities given in mol water per mol salt as function of the water content in water/acetone or water/ethanol mixtures. In water rich mixtures the salt solubilities correspond to their values in pure water. At low water contents the solubility is lower than at higher ones. The decrease of salt solubility becomes stronger in a region where the amount of water molecules per solvent is lower than the hydration sphere of saturated salt solutions in pure water. This may indicate that ions need a group of liquid-like water molecules to be solved. A similar conclusion was obtained by the investigations of desalination membranes[37,38]. The are efficient if their holes are smaller than the diameter of the group of water molecules which NaCl need to be solved and in this case the membrane hydration is preferred by steric hindrance. The curvature of the solubility diagram in Figure 32 starts a little earlier than these values by the hydration of the solvents. The decrease in salt solubilities at low water concentrations could be described as rivalry between the salt hydration and the solvent hydration. With KCl or NaCl the salts seem to loose this struggle because their solubility decreases but not the miscibility of solvents in water. This indicates too that the ion-water interactions are not stronger compared with H-bonds of the solvents. A conclusion which we obtained by spectroscopic results of hydrates. But in presence of K_2CrO_4 the ion interaction may be stronger than the interaction water-solvent. In the region where the amount of needed water per ion is bigger than the number of water molecules between 2 solvent molecules, two phases are formed indicating a higher affinity of water to ions than to solvents (Figure 32 compare Figure 27).

Conclusions

Properties of water can be described by its content of H-bonded or non H-bonded OH groups. Hydrophilic molecules or ions are hydrated in water depending on its H-bond acceptor strength. The interaction between anions and water could be described as medium

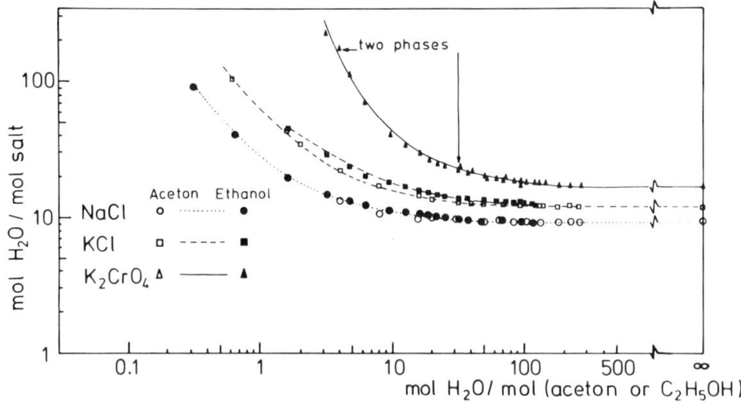

Fig. 32. Reciprocal solubility of different salts given in mole H_2O per mole salt as function of the water content in mixtures with acetone or ethanol (T=23°C).

strong H-bonds too. In ternary mixtures the hydration of ions and other solutes are additive with hydration structures of the similar binary systems. A rivalry between the hydration of ions and other solutes begins at higher concentrations only. The strength of interaction energies, water-ion or water-solute, determines the properties of ternary or quaternary systems. The anion-water interactions per one OH group of water may be in a similar size of order water.... water interactions. This experience could help to understand aqueous systems. Properties of two phases in equilibrium with an aqueous one are correlated. Therefore with the knowledge of one phase one could predict properties of the other. This experience may help to understand properties of biochemical systems studying its aqueous phase.

I thank my coworkers W. Ditter, R. Ghosh, H. Kleeberg, Ch. Buanom-Om and D. Schiöberg for their cooperation during our research.

REFERENCES

1. W. A. P. Luck, Phys.Bl., 22:347-357 (1966).
2. W. A. P. Luck, Angew.Chem., 92:29-42 (1980).
3. W. A. P. Luck, Infrared studies of hydrogen bonding in pure liquids and solutions, in: "The Hydrogen Bond in Water: A Comprehensive Treatise," Plenum Publ. Corp. New York, Chpt. 4 Vol.II, 225-320 (1973).
4. W. A. P. Luck and W. Ditter, Ber.Bunsenges.Phys.Chem., 72:365 (1968).
5. W. A. P. Luck and W. Ditter, Z.Naturforschung 24b:482 (1969).
6. W. A. P. Luck, Naturwissenschaften 52:25-31, 49-52 (1965).
7. W. A. P. Luck, The angle dependence of hydrogen-bond-interactions, in: "The Hydrogen Bond," Schuster-Zundel-Sandorfy, Verlag, North Holland Publ. Vol. II, chapt. 11 p. 527-562 (1976).
8. W. A. P. Luck, Discuss. Faraday Soc. 43:115-127 (1967).
9. H. Kleeberg, O. Kocak, and W. A. P. Luck, J.Solution Chem., 11:611-624 (1982).
10. W. A. P. Luck, Ber. Bunsenges. Phys. Chem., 69:69-76 (1965).
11. J. D. Bernal and R. H. Fowler, J.Chem.Phys., 1:516 (1933).
12. W. A. P. Luck and A. P. Zukovskij, Lilijanie ionver na IK-spektr wody w obertonni oblasti, in: "Molecular Physics and Biophysics of Water Systems, Vol. 2, p. 131-140, A. J. Sidorovo, ed., Leningrad University Press (1974).
13. J. Paquette and C. Joliceur, J.sol.chem., 6:403 (1977).
14. J. N. Shoorly and B. J. Alder, J.Chem.Phys., 23:805 (1955).
15. G. Hertz, in:"Structure of Water and Aqueous Solutions," W. A. P. Luck, ed., Verlag Chemie, Weinheim (1974).
16. P. Debye, Polare Moleküle, Leipzig, Hirzel (1929).
17. H. Sack, Phys.Ztschr., 28:199 (1927).
18. C. Buanam-Om, W. A. P. Luck and D. Schiöberg, Z.Phys.Chem., Neue Folge 117:19-35 (1979).

19. Thesis C. Buanam-Om, Marburg (1981).
20. D. Schiöberg and W. A. P. Luck, Spectroscopy Letters 10 (8), 613-618 (1977).
21. W. A. P. Luck and D. Schiöberg, Advan.Mol.Relaxation Processes, 14:277-296 (1979).
22. D. Schiöberg and W. A. P. Luck, J.Chem.Soc.Faraday Trans., I, 75:762-773 (1979).
23. Thesis H. Kleeberg, Marburg (1983).
24. W. A. P. Luck, Infrared overtone region, in: "Structure of Water and Aqueous Solutions," Kap. III, 3., Verlag Chemie/Physik Verlag Weinheim (1974).
25. D. Schiöberg, Ber.Bunsenges.Phys.Chem., 85:513 (1981).
26. W. A. P. Luck, Fortschr.chem.Forsch., 4:653 (1964).
27. W. A. P. Luck and S. S. Shah, Prog.Colloid & Polymer Sci., 65:53 (1978); Thesis S. S. Shah, Marburg (1978).
28. W. A. P. Luck, Angew.Chem., 72:57 (1969).
29. W. A. P. Luck, Melliand 41:315 (1960).
30. W. A. P. Luck, J.Soc.Dyers and Colourists., 74:221 (1958).
31. W. A. P. Luck, Schriftenreihe des Deutschen Wollforschungs-institutes Aachen 84:215 (1981) and in: "Flüssige Arzneiformenund Arzneimittelsicherheit, hrsg. von D. Essig, P. Schmidt und H. Stumpf, Wissenschaftl. Verlagsges. mbH, Stuttgart S. 17-47 (1981).
32. W. A. P. Luck, Topics in Current Chemistry, Vol. 64 Fortschritte der chem. Forschung S. 113-180 (1976).
33. C. F. Hazlewood, Ann.New York Academy Sci., Vo. 204 (1973).
34. H. Kleeberg and W. A. P. Luck, Naturwissenschaften 64:223 (1977).
35. W. A. P. Luck and H. Kleeberg, Structure of Water and Aqueous Systems in: "Photosynthetic Oxygen Evolution, H. Metzner ed., Academic Press, London-New York, San Francisco, S. 1-29 (1978).
36. H. Kleeberg and W. A. P. Luck, Studia biophysica 84.19 (1981).
37. W. A. P. Luck, D. Schiöberg and U. Siemann, J. C. S. Faraday II 76:136 (1980).
38. W. A. P. Luck, D. Schiöberg and U. Siemann, Ber.Bunsenges.Phys. Chem., 83:1085 (1979).
39. W. A. P. Luck, II. Internat.Kongr.f. Grenzflächenaktive Stoffe, Köln, Bd. I Sekt. A., S. 264 (1960).
40. P. Becher, J.coll.science 17:325 (1962); 18:196 (1963).
41. H. S. Frank and W. V. Wen, Disc.Faraday Soc., 24:133 (1957).
42. T. H. Lilley, in: "Water A Comprehensive Treatise," F. Franks, ed., Plenum Press, New York Vol. 3 p. 266 (1973).
43. W. A. P. Luck, Structure of aqueous systems and the influence of electrolytes, in: "Water IN Polymers," St. P. Rowland ACS, ed., Symposium Series Nr. 127, S. 43-71 (1980).
44. W. Ostwald and R. H. Hertel, Koll.Ztschr., 47:258, 357 (1929).

ORGANIZATION AND FUNCTION OF WATER IN NATIVE CONFORMATION OF MACROMOLECULES

DIVALENT METALS AND CANCER

E. L. Andronikashvili

Institute of Physics
Academy of Sciences of the Georgian SSR
Tbilisi, USSR

The carcinogenic action of some metals was known long ago. However, their specific role remained obscure. The enormous role of metals in cancer development as all levels of biological functioning has been revealed during the last ten years. The beginning of these investigations should be dated back to 1969[1,2] when our papers devoted to the additional incorporation of metal ions in nucleic acids of transformed cells as compared to the norm were published.

However, setting the aim to demonstrate that the malignant tumor development is accompanied by the change of metal concentration upon all organization levels without exception, we begin with the organ level - with mammary gland cancer. The data presented in Table 1 taken from the paper by Santaliquido et al. (1976) show that tumorous mammary gland needs many metals, both non carcinogenic and carcinogenic, as compared with non-tumorous one.

Why do we call certain metals carcinogenic? The answer may be obtained using the data taken from the Table given by C. P. Flessel et al.[3] (Table 2). Is there any correlation between the amount of metal in an organ, e.g., that of zinc in blood, and the course of the disease? Our experiments carried out with the blood of leukemic patients have shown that such correlation exists (Figure 1)[4]. Even more effective are the experiments conducted by Hrgovcic et al.[5] who have studied Hodgkin's disease and shown that copper concentration in blood serum at the height of the disease increases 5 times with respect to the norm. Despite good clinical data, one cannot consider remission until copper concentration becomes normal. The relapse can take place even in 2 or 3 months.

Table 1. Metal Redistribution Between Healthy and Cancerous Tissues. Mammary Gland Cancer.

μg/g dry weight	Cancer	Norm	Cancer to Norm Ratio
Zn	23.5	5.5	4.3
Fe	234	72	3.25
Ca	1173	270	4.3
Mg	386	77	5.00
Mn	0.59	0.29	2.0
Cr	0.38	0.20	1.9

Table 2. Carcinogenicity Index.

Cr 21	Fe 11	Hg 3
Ni 21	Mn 8	Mo 3
Cd 15	Pt 8	Se 3
Be 15	Co 7	Ti 3
As 15	Zn 7	Sb 3
Pb 15	Ag 3	Al 1
	Cu 3	Te 1

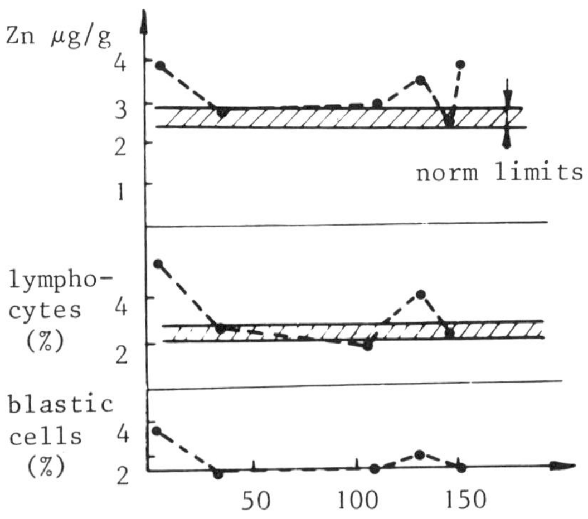

Fig. 1. Presence of correlation between zinc content in blood, quantity of lymphocytes and blastic cells in acute leukemia dynamics (shading corresponds to healthy donor blood).

Microelements do not only control the course of the disease. Our experiments[6] have shown that the chance of tumor appearance can be determined according to the ratio of iron and zinc concentrations in the human stomach mucous membrane changing in the average from 9.0 in the norm to 1.6 in case of cancer. In other words, it is possible to diagnose the pre-cancerous state. Similar experiments have been carried out on rats with the transplanted cancer of the sciatic nerve. In this case the clinical diagnosis has been two months behind the "microelemental" one.

Now let us pass from the whole organism and separate organ level to the cellular level. As an example, we present nickel and manganese concentrations in healthy and leukemic lymphocytes[7]. Table 3 shows that the amount of microelements Ni and Mn in cells in the case of the malignant transformation exceeds the norm.

As for the subcellular level, in the process of the malignant disease, metal redistribution between cell organellae also takes place[8]. The redistribution of metals (in the present case, that of zinc) between the organellae of the sarcomatous animal liver tissue cells observed as a result of X-ray irradiation in vivo[9] is especially interesting. As Table 4 shows, zinc leaves the most part of organellae and concentrates in mitochondriae. During the subsequent hours, a partial reparation occurs for the most part of organellae, and zinc concentration tends to its normal value, although very slowly.

What physical properties of tumor chromatin are changed? In case of cancer, metals play an important (and sometimes determining) role in such structures as chromatin. Figure 2 shows the heat absorption curve of chromatin isolated from the liver of C-3 HA line mice obtained by J. Monaselidze et al. (1982). This curve is compared with the heat absorption curve of chromatin from hepatoma 22A. While the curve of normal chromatin has two maxima, that of hepatoma chromatin has only one maximum. However, after the removal of metals out of hepatoma chromatin using EDTA, the second maximum appears on the curve again, while after incubating the normal chromatin in $MgCl_2$ solution, the above maximum disappears again. Hence, the excessive amount of metals in cancer chromatin influences pure physical properties, such as the melting temperature and the shape of the melting curve of this overmolecular structure.

Table 3. The Change of Metal Concentration in Cells (Lymphocytes).

$\mu g/10^8$ cells	Norm	Acute leukemia
Ni	0.146	0.29
Mn	0.043	0.11

Table 4. Subcellular Organellae of M-1 Tumor After X-Ray Irradiation to the Dose of 1 000 R (g/g dry weight x 10^{-5}).

Object under investigation: Zn in:	Non-Irradiated Tumor	Tumor after the Irradiation (time in hours)		
		1	4	24
nuclei	1.4	0.4	0.6	0.93
mitochondriae	3.4	18.5	5.2	5.7
microsomes	4.8	4.2	4.1	4.8
supernatant	3.2	2.9	2.6	1.6
lysosomes	4.8	1.1	2.1	2.8
lysosomal membranes	3.3	0.5	1.4	2.0

The change of metal concentration takes place in macromolecules, as well. The content of such biogenic metals as calcium, magnesium and zinc in histone proteins has been studied at our Institute. Table 5 shows that in all histone fractions the amount of the investigated metals in case of hepatoma 22A are higher than in the norm. Besides the fact that cancer histones contain a larger amount of metals, the physical properties of these proteins are changed, which is seen clearly from the comparison of circular dichroism spectra.

Calcium addition to the normal H_{2b} histone or the removal of calcium out of the same histone isolated from tumor significantly change the circular dichroism spectra. This experiment was carried out jointly by Sichinava (Tbilisi), Esipova (Moscow) and Ramm (Leningrad) (1982).

The presence of divalent metals incorporated in DNA and RNA molecules was discovered by Walker and Vallee[10] as early as in 1959. As it has been mentioned above, the beginning of our work dates back to 1969 when Andronikashvili, Belokobyl'ski et al.[1,2] discovered that in case of malignant transformation the excessive amount of metals penetrates nucleic acid molecules. These data are presented in Table 6.

One may ask: how do metal ions penetrate into DNA? One of the mechanisms was established in "in vitro" experiments by Sirover and Loeb in 1976[11]. Those metals which introduce errors into the synthesis fidelity "in vitro" during the formation of polynucleotide chains, behave "in vivo" as mutagens and carcinogens. Here is an extract from the paper by these authors (Table 7).

Three metals (Fe, Cu and Zn) among the investigated ones represent an exception: being non-mutagenic, they are slightly carcinogenic. Thus, certain metal ions introduce errors into the structure of polynucleotide chains. The most dangerous error is the non-

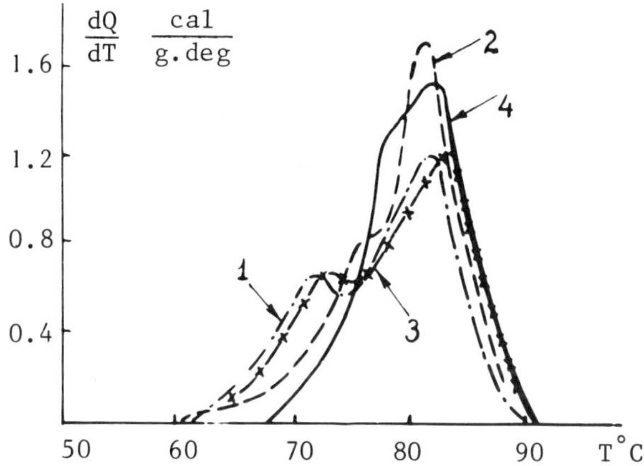

Fig. 2. Heat absorption curves of chromatin: 1 - chromatin from liver, 2 - chromatin from hepatoma 22A, 3 - chromatin from hepatoma 22A after its treatment with EDTA, 4 - chromatin from liver after the incubation in $MgCl_2$ solution.

Table 5. Chromatin Histone Proteins.

	Ca	Mg	Zn
norm	1 400	1 000	55
hepatoma	1 800	1 700	68
norm	480	240	42
hepatoma	840	320	52
norm	1 000	1 100	40
hepatoma	1 900	1 450	120
norm	1 200	1 100	77
hepatoma	2 300	2 000	83

Table 6. The Increase of Divalent Metal Concentration in Tumor DNAs.

DNA isolated from	Fe	Zn	Co	Sb
Normal tissue	16	0.5	<0.07	∿1
Sarcoma M-1	24	2.0	0.31	1.4
Carcinosarcoma W-256	32	2.1	0.34	2.5

Table 7. What Is the Difference Between Carcinogenic and Non-Carcinogenic Metals.

Compound	Template	Error frequency (metal concentration)	Fidelity decrease	Carcinogenic mutagen
$BeCl_2$	poly (A)	15 (10mM)	+	+
$CoCl_2$	poly (C)	8.37 (4mM)	+	+
CrO_3	poly d (A-T)	3.83 (16mM)	+	+
$NiCl_2$	poly (C)	1.92 (8mM)	+	+
$MnCl_2$	poly (C)	3.78 (10mM)	+	+
NaCl	poly d (A-T)	0.81 (120mM)	-	-
KCl	poly d (A-T)	1.11 (150mM)	-	-
$MgSO_4$	poly (C)	1.11 (10mM)	-	-
$CaCl_2$	poly (A)	1.28 (5mM)	-	-
$ZnCl_2$	poly d (A-T)	1.06 (0.4mM)	-	±

complementary G-T mutation which introduces metal ion into DNA due to the high chemical activity of guanine. According to the papers of Zavriev et al.[12] metal ion binding with guanine makes its electronic structure similar to that of adenine which leads to a number of very important consequences.

What physical properties of DNA change due to the increase of divalent metal ion concentration? Monaselidze with co-workers have shown using the method of differential microcalorimetry[13] that DNA from transformed cells melts at the temperatures 1-0.7° below the norm (Figure 3). We have proved that this effect is directly connected with the presence of metal ions in DNA.

A question arises: what happens if metal ion reacting with guanine and thus making its electronic structure similar to that of adenine, becomes incorporated in the promoter? Such promoter will fail, since it cannot be recognized by RNA- or DNA-polymerases. The same can be applied to gene-regulator coding a repressor-protein with a wrong amino acid sequence under similar conditions. Then the repressor will be unrecognizeable by the promoter. Hence, the initiation of RNA synthesis will be beyond the control realized by repressor-proteins and will proceed in a random continuous manner. Thus it is the case of the malignant transformation, with an uncontrollable and random proliferation of cells.

It turns out that the presence of metal ions in DNA directly determines a number of important phenomena at the cellular level. Thus, for example, it has been found in[14] that leukemic human bone marrow cells are denaturated at the temperature 10°C above that of

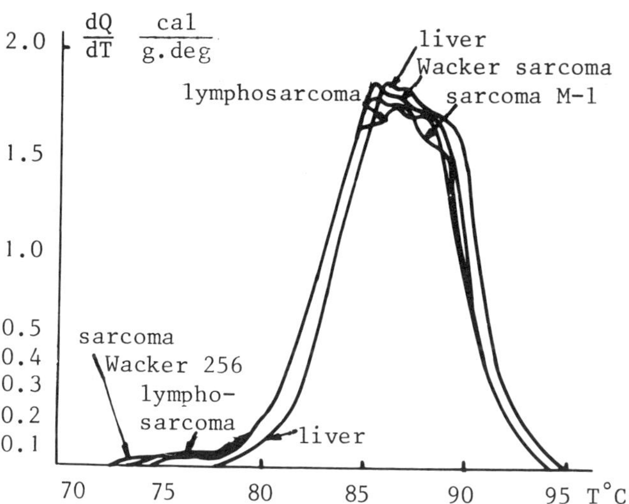

Fig. 3. Influence of metals upon DNA physical properties. Tumorous DNA melts at the temperatures 1 - 0.7° below the norm.

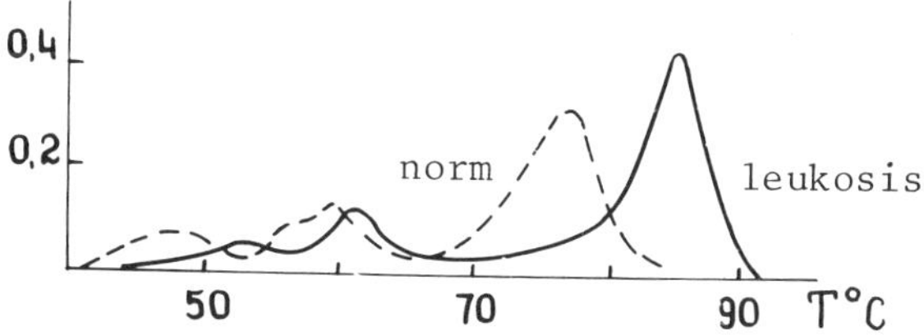

Fig. 4. Heat absorption curves of cells from human bone marrow.

bone marrow cells of a donor (see Figure 4). This new phenomenon discovered at our Institute represents the basis of malignant disease diagnosis.

REFERENCES

1. E. L.Andronikashvili, Preprint of the Institute of Physics, Acad.Sci.of Georgian SSR, (1969).
2. E. L. Andronikashvili, L. M. Mosulishvili, A. I. Belokobylski, V. P. Manjgaladze, N. E. Kharabadze, E. Yu. Efremova, Doklady Akademii Nauk SSSR, 195,4,979-982 (1970).
3. C. P. Flessel, A. Furst, S. B. Radding, Comparison of carcinogenic metals, in: "Metal Ions in Biological Systems," Marcel Dekker Inc., N. Y. and Basel, v.10, pp.24-54 (1980).
4. E. L. Andronikashvili and L. M. Mosulishvili, Human leukemia and trace elements, in: "Metal Ions in Biological Systems," Marcel Dekker Inc., N. Y. and Basel, v.10, pp.167-206 (1980).

5. M. Hrgovcic, C. F. Tessmer, F. B. Thomas, P. S. Ong, J. F. Gamble, and Shullenberger, Cancer, 32:1512 (1979).
6. E. L. Andronikashvili, L. M. Mosulishvili et al., J.Radioanal. Chem., 58:1 49-59 (1980).
7. L. V. Popova and S. V. Mazukha, Proc. of Voronezh Medical Institute, 94:58 (1975).
8. L. K. Tkeshelashvili, I. S. Mepisashvili, L. M. Mosulishvili, and N. E. Kuchava, in: "Radiation Investigations,"Tbilisi, "Metsniereba," 11:25-31 (1975); L. K. Tkeshelashvili, I. S. Mepisashvili, L. M. Mosulishvili, N. E. Kuchava, and J. N. Novikova, in: "Radiation Investigations," Tbilisi, "Metsniereba," 11:33-50 (1975).
9. E. L. Andronikashvili, L. M. Mosulishvili, L. K. Tkeshelashvili, A. N. Rcheulishvili, and A. E. Matevosyan, Bull.Experim. Biologii i Meditsiny, 82:11 (1976).
10. W. E. Walker and B. L. Vallee, J.Biol.Chemistry, 234:3257 (1959).
11. M. A. Sirover and L. A. Loeb, Science, 194:1434 (1976).
12. S. K. Zavriev, L. E. Minchenkova, M. Vorlickova, A. M. Kolchinsky, M. V. Volkenstein, and V. I. Ivanov. BBA, 564:212 (1979).
13. J. R. Monaselidze, Z. I. Chanchalashvili, G. N. Mgeladze, G. V. Majagaladze, and D. K. Tsitsishvili, Proc.VII All-Union Conf.on Calorimetry, Chernogolovka, 415 (1977).
14. Z. I. Chanchalashvili, G. V. Majagaladze, G. N. Mgeladze, and J. R. Monaselidze, Conformational Changes of Biopolymers in Solutions, Proc.III All-Union Conf., Tbilisi, "Metsniereba," p.86 (1975).

COMPUTER SIMULATIONS OF STRUCTURAL ORGANIZATION OF WATER AROUND BIOMOLECULE

Chen Run-sheng[1], Ni Xiang-shan[1] and Shi Xiu-fan[2]

[1]Institute of Biophysics, Academia Sinica
Beijing, China
[2]Kunming Institute of Zoology
Academia Sinica, Kunming, China

In order to elucidate a life system at molecular and submolecular levels, it is necessary to apply physical concepts and methods to biology. To perform computer simulations of motions of biomolecules, we must calculate the intra- and interaction energies in terms of empirical, semiempirical or ab initio methods. Using CNDO/2, we have studied the interaction between chiral molecules[1,2] and the rotation barrier of bases in nucleic acid[3,4]. Owing to the difficulties in applying CNDO and ab initio methods to macromolecules, we adopted the atom-atom potentials developed by E. Clementi to deal with polypeptide and oligonucleotide[5,6]. As neuropeptides are of interest, we calculated the interaction between leu-enkephalin and water molecule.

THE INTERACTION BETWEEN LEU-ENKEPHALIN AND WATER MOLECULE

Two pentapeptides separated by Huges et al. (1975) from extracts of pig brains exhibit morphine-like effects in both central and peripheral nervous systems. Their amino acid sequences are as follows:

H-Tyr-Gly-Gly-Phe-Met-OH
H-Tyr-Gly-Gly-Phe-Leu-OH

The former is termed methionine-enkephalin, the latter leucine-enkephalin. In order to understand how the pentapeptide enkephalin can mimic the action of non-peptidic substance, morphine, it is necessary to determine their conformations, especially the conformations in aqueous solution. In recent years many studies on enkephalin conformation have been carried out by means of NMR spectroscopy in many laboratories. We hoped to obtain some information about the interaction between enkephalin and water by quantum chemical calculation.

The molecules of enkephalin and water were held rigid during the calculation of interaction energy. The dihedral angles of leu-enkephalin were taken from the data of X-ray diffraction[7]. The structure parameters of water molecule[8] are: R_{OH}=0. 9572Å, ∠HOH=104.52. The procedure for calculating the interaction energy between two molecules lies in the ability to convert the calculation into a summation of these interaction energies between every atom-atom, respectively, constituting the two molecules. The formula is as follows:

$$I(E,W) = \sum_i \sum_{j \neq i} I_{ij}^{ab}(E,W) = \sum_i \sum_{j \neq i} (-A_{ij}^{ab}/r_{ij}^6 + B_{ij}^{ab}/r_{ij}^{12} + C_{ij}^{ab} q_i q_j / r_{ij})$$

where a,b denote the classes, which the atoms i and j belong to, respectively, A_{ij},B_{ij},C_{ij} fitting constants, r_{ij} the distance between i and j, q_i,q_j the net charges of atoms i and j, respectively. Taking the geometric center of leu-enkephalin as the spherical center, we constructed 15 spheres, the radii of the first twelve are from 10 to 21 a.u., the spherical interval is 1 a.u., and the radii of the last three are 23, 25 and 30 a.u., respectively. On each sphere we constructed 62 points. In addition to those points more points were added in proper positions. At each point the oxygen atom of a water molecule is fixed, then the hydrogens are rotated so obtain the minimum of interaction energy. The iso-energy surfaces are obtained by connecting these points marked by identical energy.

Three cross sections, perpendicular to one another, of the total energy surfaces are shown in Figures 1-3, which are the iso-energy contour maps on the X-Y, Y-Z and X-Z planes, respectively. The innermost contours are 0, the outermost (in Figure 1 and Figure 3) are -0.1 kcal/mol, in Figure 2 is -0.5 kcal/mol. From-1.0 kcal/mol. In Figure 1 the hydrophilic and hydrophobic regions of leu-enkephalin can be seen clearly. There is an intermediate, near strong (up to -13 kcal/mol), hydrophilic area around the -OH group of the tyrosine residue, and a strong hydrophobic region is formed around the leusine residue. Along the peptide chain there is a weak (to -7 kcal/mol) hydrophilic region. From the total interaction energy surface between leu-enkephalin and a water molecule, a general idea can be obtained of how the water molecules are positioned and orientated around the enkephalin in aqueous solution.

It is established that the N atom in the terminal amino group of enkephalin corresponds to the N in morphine. The electrostatic molecular potential contour maps around the N atoms of morphine and nalorphine in the lone pair directions are almost identical both in

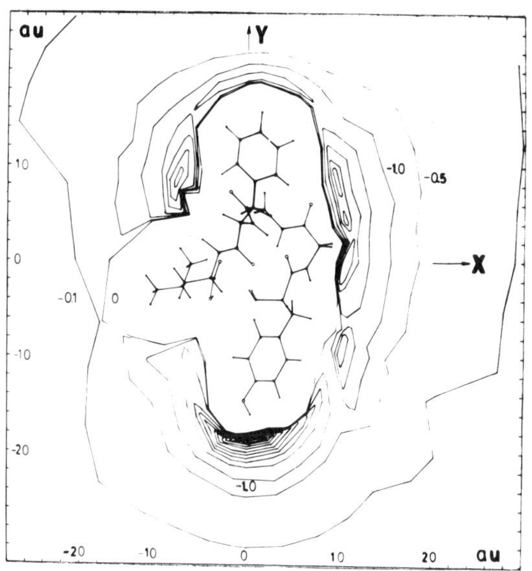

Fig. 1. Iso-energy contour map on X-Y plane (Z=0) for the interaction energy of leu-enkephalin with one water molecule.

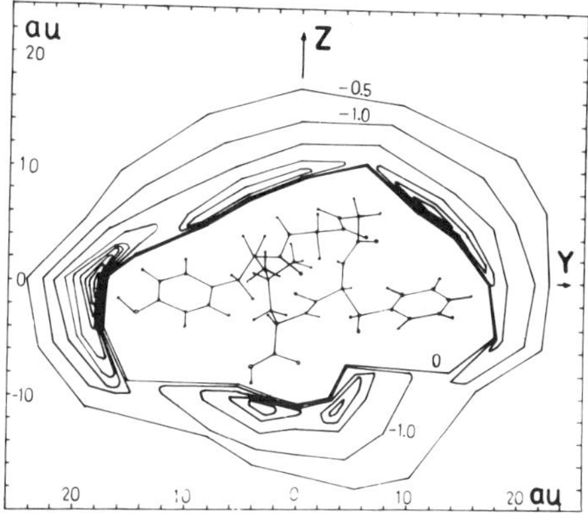

Fig. 2. On Y-Z plane (X=0).

shape and in numerical value[9]. N atom in these molecules are pharmacophores. In the crystal structure of leu-enkephalin, between the N atom in the terminal amino group and the carbonyl oxygen of Phe an intramolecular hydrogen bond is formed[7]. Hence we speculate that in solution this hydrogen bond may be broken, making the N atom exposed and in a state of exhibiting its function. There is also a probability that by the hydrophobic force the torsion angles of the

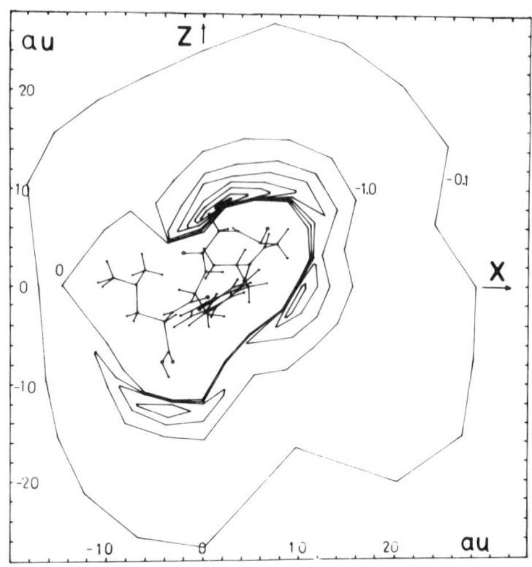

Fig. 3. On X-Z plane (Y=0).

leucine residue may be changed so it may combine with the hydrophobic region of the receptor in a proper conformation.

Having calculated various interaction energies between biomolecule and water molecules, it becomes feasible to make computer simulations of aqueous solutions of biomolecules by Monte Carlo approach, which is useful for discussing the equilibrium structure and thermodynamic properties of fluids. We have carried out MC simulations of some amino acid solutions.

MONTE CARLO SIMULATION OF AQUEOUS SOLUTION OF ALANINE

The simulation temperature is 298K, and the concentration of the solution is 8.11% corresponding roughly to the half of that in saturated solution. The density of the solution is given as 1.00 g/cm[3]. Periodic boundary condition is used to remove the effect of the surface. There is one neutral alanine molecule surrounded by 56 water molecules in the central cube with edges of 12.2166 Å in length. All molecules in the system are regarded as rigids. According to our previous calculation the interaction energy between two alanine molecules in distance with 12.2166 Å may be neglected even if they are in the most favorable bonding configuration. So we calculate the alanine-water and water-water two-body interactions only, taking no account of the alanine-alanine interaction.

The water-water interaction potential was obtained by O. Matsouka et al., from CI calculation and fitted to an approximately analytical solution[10]:

$$Y_{w-w} = q^2\left(\frac{1}{\gamma_{13}} + \frac{1}{\gamma_{14}} + \frac{1}{\gamma_{23}} + \frac{1}{\gamma_{24}}\right) + \frac{4q^2}{\gamma_{78}}$$

$$-2q^2\left(\frac{1}{\gamma_{18}} + \frac{1}{\gamma_{28}} + \frac{1}{\gamma_{37}} + \frac{1}{\gamma_{47}}\right) + a_1\exp(-b_1\gamma_{56})$$

$$+a_2[\exp(-b_2\gamma_{13}) + \exp(-b_2\gamma_{14}) + \exp(-b_2\gamma_{23}) + \exp(-b_2\gamma_{24})]$$

$$+a_3[\exp(-b_3\gamma_{16}) + \exp(-b_3\gamma_{26}) + \exp(-b_3\gamma_{35}) + \exp(-b_3\gamma_{45})]$$

$$-a_4[\exp(-b_4\gamma_{16}) + \exp(-b_4\gamma_{26}) + \exp(-b_4\gamma_{35}) + \exp(-b_4\gamma_{45})]$$

where a_1, a_2, a_3, a_4, b_1, b_2, b_3, b_4 are the fitting constants, the distance r is designated in Figure 4. The point M in Figure 4 is positioned on the symmetry axis of c_{2v} of the water molecule, apart from the oxygen atom 0.2677 Å, its net charge is -2q. The geometry of water molecule was taken from W. S. Benedict's result. The geometry of alanine molecule and the alanine-water interaction potential was based on E. Clementi[5]. Metropolis sampling scheme was adopted. [11]. The position and orientation of alanine in the cube are as follows: C^α is put on the center of the cube (the origin of coordinates), the C^α-H^α is set towards the direction of the vector (-1,-1,-1) and the coordinates X and Y of C' are set to be equal and positive so that the bonds C^α-C', C^α -N and C^α-C^β are nearly directed to the directions of lines OB, OE and OG, respectively (see Figure 5). The position of a water molecule is defined by six variables (X,Y,Z, θ,ϕ,ψ). Because H^α is surrounded by three groups which are much bigger than H^α in size, one could not expect to get significant information about it. Therefore the cube is equally divided into three pentahedral regions by three planes FDA, FDC and FDH, so that the three groups (carboxyl, amino and methyl), in which we are interested, lie respectively in one of the regions. 101 concentric spheres (r_{min} =1.532 Å, r_{max} =7.378 Å) were drawn with their centers at C', N and C^β, respectively, to form 100 spherical shells for the related statistical treatment.

All together we have calculated 10240 cycles (one cycle means every water molecule moves once successively). The first 5120 cycles were discarded. The following results were averaged over the last 5120 cycles. Subaverages were taken over macrosteps consisting of 256 cycles. The rule of thumb was used[12] to choose suitable step length so that about half of the moves in Markov chain were accepted. For standardization and convenience of comparison, in following, the interaction energy between two molecules implies half of the interaction energy, while the bonding energy between two molecules represents the whole interaction energy.

Energy Results (in kcal/mol)

The average water-water interaction energy: -8.9±0.1 The average alanine-water interaction energy: -13.4±1.1. The variation of water and alanine-water interaction energies with the distances R between the oxygen atom in a water molecule and the central atom in

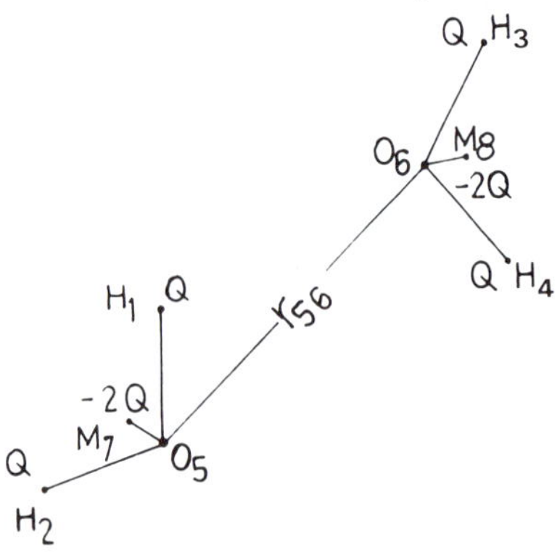

Fig. 4. Definition of the used distances in water-water potential.

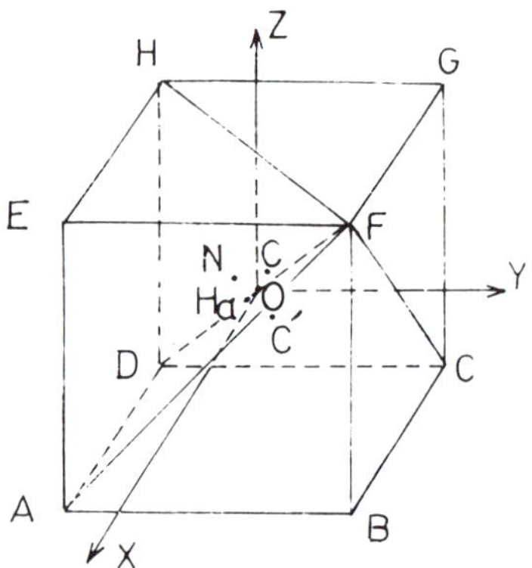

Fig. 5. Location of alanine molecule in cube and division of three regions.

each region are shown in Figures 6 and 7, respectively. In Figure 7 it can be seen clearly that in carboxyl region, the energy valley is as deep as -5.1, in amino region and the depth of its valley is about a quarter of the former and in methyl region the depth of its valley is as shallow as one percent of the first. This shows that the hydrophilicity of carboxyl is the strongest in the three groups, while that of methyl is the weakest.

Structural Results

The orientational correlation functions of the dipole moments of water molecules, which are defined by the projection of a unit vector in the direction of $\overline{OM}$ (see Figure 4) on the direction from the central atom to the oxygen atom of a water molecule as a function of the distance R, in three regions are given in Figure. The distri-

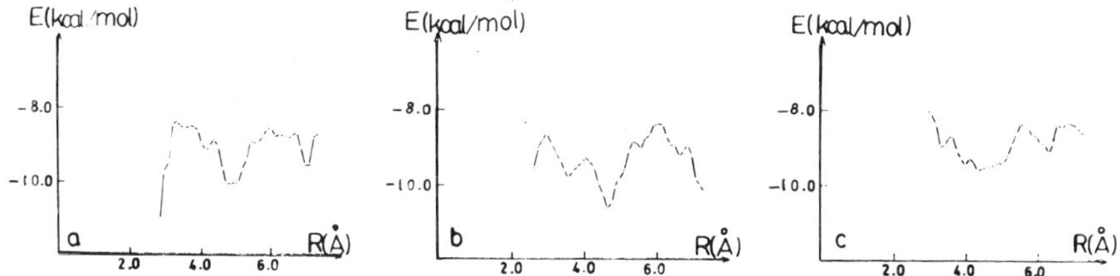

Fig. 6. Interaction energies between a water molecule whose oxygen atom located in a region and all the other water molecules as functions of R. (a) for carboxyl, (b) for amino group, (c) for methyl group.

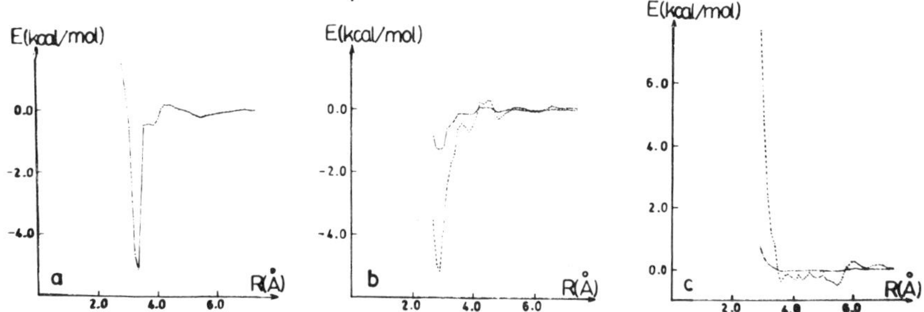

Fig. 7. Interaction energies between a water molecule whose oxygen atom locates in a region and the alanine molecule as functions of R. (a) for carboxyl, (b) for amino group (the dotted curve is the fourfold enlarged one of the solid), (c) for methyl group (the dotted curve is the tenfold enlarged one of the solid).

bution functions for oxygen and hydrogen atoms of water molecules, which are defined by the ratio of the particle number in a spherical shell to the average water molecule number in the corresponding shell as a function of the radius R of the spherical shell[14], in three regions are given in Figure 9.

According to the first minima of the radial distribution function for oxygen atoms in three regions in Figure 9, the number of water molecules in the first solvation shell of each region could be determined. They are 10.0 for the C'-region, 8.6 for the N-region and 9.4 for the C^{β}-region. Summing up the numbers of these three regions, we found that the number of water molecules in the first solvation shell surrounding the whole alanine molecule is 28, in which there are 3-5 molecules, their bond energy with alanine molecule is more than 2 kcal/mol, the rest formed a hydrogen-bond clathrate depending on water-water interaction. There is only one water molecule whose bond energy is above 11 kcal/mol. We thought this one may be the bond-water, therefore estimated the amount of bond-water of alanine is about 20% (w/w).

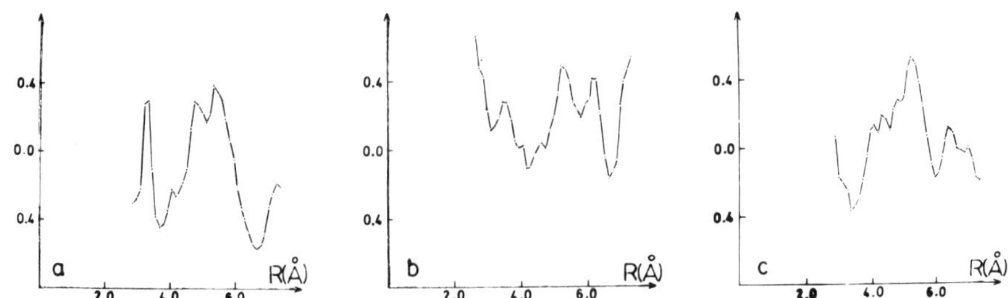

Fig. 8. Orientational correlation functions in (a) the C', (b) the N and (c) the C^{β} regions of alanine.

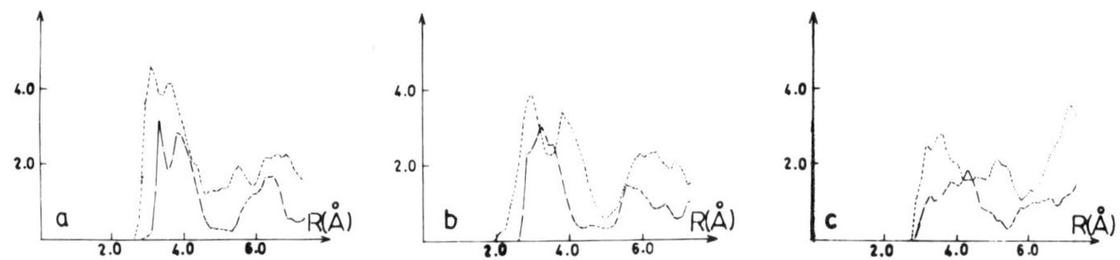

Fig. 9. Radial distribution functions in (a) the C', (b) the N and (c) the C^{β} regions of alanine. The solid lines are for the oxygen atoms and the dotted lines for the hydrogen atoms in the water molecules in each region.

A model of the alanine molecule and 29 water molecules in the first solvation shell is given in Figure 10, where only the oxygen atoms of water molecules was drawn and the connecting sticks means that the bond energy between the two molecules is over 1.5 kcal/mol. In this model the water molecules No. 6,7,10,11,16,22,24,26,27,36 and 40 belong to the C'-region, No. 18,19,30,31,32,34,43,47 and 51 belong to the N-region and No. 21,33,35,37,38,39,45,49 and 50 belong to the C^{β}-region. Among these 29 water molecules there are only five water molecules between one of which and the alanine molecule the bond energy is more than 1.5 kcal/mol. They are water molecules No. 11, 19,22,24 and 47. The water molecule No. 11 combined fast, its bond energy reached 16.4 kcal/mol.

The interaction energy between the alanine molecule and these molecules of the first solvation shell in corresponding region are: in C'-region: -9.70 kcal/mol, in N-region: -3.05 kcal/mol, in C^{β}-region: 0.03 kcal/mol, the sum of them is -12.8 kcal/mol, which accounts for 95% of the interaction energy between the alanine molecule and the total water molecules, while the number of water molecules in the first solvation shell is only 50% of the total water molecules. So principally there is only one solvation shell surrounding a neutral alanine molecule.

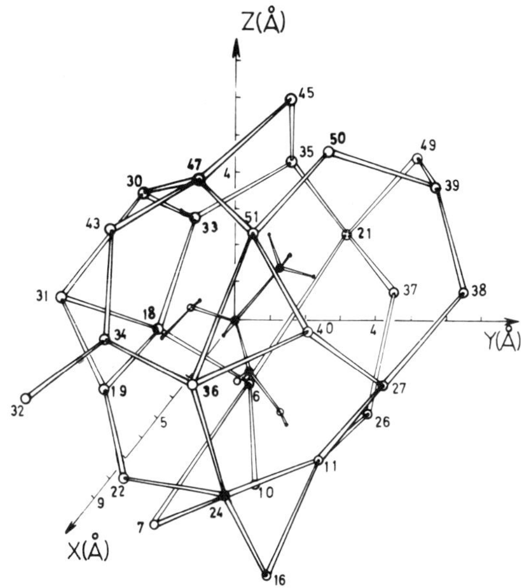

Fig. 10. Perspective drawing of the distribution of the alanine molecule and the water molecules in the first solvation shell.

Because the three groups are similar in size and the division of regions is proper, the structure results obtained is clear. The results for the two hydrophilic groups are comparable with those in refs. 13,15, and for the hydrophobic group it agrees with that for $-CH_3$ group in ref. 16. The proposal of G. Bolis et al., in that paper was verified.

In order to study the macroscopic dynamic character of biomolecular system, we can apply dissipative structure theories to analysis of the solution structure in life system with the help of computer. We have proposed a model of protein synthesis process and studied the relaxation process of forming the structure and the effect on the asymmetry stationary state structure of protein.

DYNAMIC MODEL OF PROTEIN SYNTHESIS[17,18,19]

Making use of the dissipative theory[20.21,22] to the protein synthesis process on mRNA template, we proposed the reaction diffusion equations of non-spherical molecules as follows:

$$
\begin{cases}
\frac{\partial}{\partial t}(M) = -KK_1M \cdot a \cdot z & + KMa^{s-1} \cdot a. \\
\frac{\partial}{\partial t}(Ma) = KK_1M \cdot a \cdot z - KMa \cdot a. \\
\frac{\partial}{\partial t}(Ma^2) = \quad + KMa \cdot a - KMa^2 \cdot a. \\
\vdots \quad\quad \vdots \\
\frac{\partial}{\partial t}(Ma^{s-1}) = & KMa^{s-2} \cdot a - KMa^{s-1} \cdot a. \\
\frac{\partial a}{\partial t} = -K\sum_{i=1}^{s-1}(Ma^i \cdot a) - KK_1M \cdot a \cdot z + D_\perp^A \frac{\partial^2 a}{\partial x^2} + D_\perp^A \frac{\partial^2 a}{\partial y^2} + D_\parallel^A \frac{\partial^2 a}{\partial z^2}, \\
\frac{\partial P}{\partial t} = KMa^{s-1} \cdot a & + D_\perp^p \frac{\partial^2 P}{\partial x^2} + D_\perp^p \frac{\partial^2 P}{\partial y^2} + D_\parallel^p \frac{\partial^2 P}{\partial z^2},
\end{cases}
\tag{1}
$$

where $a, P, M, Ma, \ldots, Ma^{s-1}$ denote the concentrations of amino acid, protein, mRNA template and mRNA template combined by amino acids, respectively, K is the reaction rate constant, K_1 the positional factor, s the number of amino acid residues in protein, the superscripts A and p of D denote the diffusion coefficients of amino acid and protein, respectively.

Because of the conservation of the total number of the template, the stationary state equation is yielded from Equation (1):

$$
\begin{cases}
0 = -KR_m \cdot a_f + \frac{KR_m}{S} \cdot a_f - \frac{KK_1R_m}{S} \cdot a_f \cdot z + D_\perp^a \frac{\partial^2 a_f}{\partial x^2} + D_\perp^a \frac{\partial^2 a_f}{\partial y^2} + D_\parallel^a \frac{\partial^2 a_f}{\partial z^2} \\
0 = \frac{K}{S} R_m \cdot a_f + D_\perp^p \frac{\partial^2 P_f}{\partial x^2} + D_\perp^p \frac{\partial^2 P_f}{\partial y^2} + D_\parallel^p \frac{\partial^2 P_f}{\partial z^2},
\end{cases}
\tag{2}
$$

where a_f, P_f denote respectively the concentration of amino acid and protein at the point $\underline{R}$ in stationary state, R_m the total number of the templates.

When considerating the shape of the cell, we selected a cylinder with 1 mm in diameter and 1 mm in height as the reaction region. Let $-\frac{KK_1R_m}{S} = \beta'$, $-\frac{KR_m}{S}(S-1) = \gamma'$, $\frac{KR_m}{S} = \delta'$ and considering the symmetry of this model, the cylindrical coordinate is used and then Equation (2) is expressed as follows:

$$
\begin{cases}
\left[D_\perp^a \frac{\partial^2}{\partial r^2} + D_\perp^a \frac{1}{\gamma}\frac{\partial}{\partial r} + D_\parallel^a \frac{\partial^2}{\partial z^2} + \beta' z + \gamma'\right] a_f = 0, \\
\left[D_\perp^p \frac{\partial^2}{\partial r^2} + D_\perp^p \frac{1}{\gamma}\frac{\partial}{\partial r} + D_\parallel^p \frac{\partial^2}{\partial z^2}\right] P_f + \delta' a_f = 0.
\end{cases}
\tag{3}
$$

Taking the first boundary condition, using the finite difference method, the numerical computations of the partial differential equations in Equation (3) are performed and the resulting large linear algebraic equations are solved by means of the over-relaxation method. Suitable over-relaxation factor is obtained by optimum seeking method. The constants used in the calculations were taken from Sober and Harte (1970).

The results showed that in translation phase the protein synthesized on mRNA template are not distributed homogenously in cell, but its concentration formed a certain spacial structure. In Figure 11 a steady-state concentration distribution of protein is shown.

We also found that the molecular configurations can influence the asymmetry of the stationary state structure of protein, but there was no linear relationship between the effect and the deviation degree from sphere of the molecular shape. In contrast, as the molecular shape varies near a sphere, the effect produced by its configuration change on the stationary state concentration distribution of protein becomes more obvious than that of far from a sphere. In addition, the molecular configuration also influenced the total amount of synthesized protein.

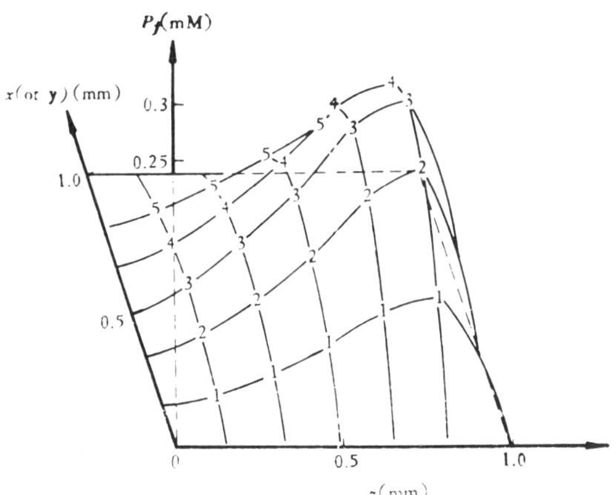

Fig. 11. Stationary state of protein (x-z sectuibm y = 1mm or y-z section, x = 1mm).

Analyzing the relaxation process of Equation (1) by means of alternative direction implicit method (A.D.I.M.), we found that the relaxation time of protein synthesis was 300 sec. and that the two patterns of the concentration distribution of protein were fundamentally consistent when during the relaxation process and when in a stationary state.

In summary, all these results mentioned above show that computer simulation is useful in studying the solution structure of a life system.

Acknowledgements

We wish to express our gratitude to Prof. E. Clementi who has provided us with a great deal of valuable material which was useful for analyzing the results of this paper. We also want to thank Prof. Xu Jin-hua and Ding Da-fu for their guidance and support.

REFERENCES

1. Chen Run-sheng and Ni Xiang-shan, Interaction of chiral molecules 1. L-L and L-D alanine two-body system, J.Mol.Sci., (in chinese) 4:191 (1982).
2. Chen Run-sheng, Ni Xiang-shan, and Peng Zao-yuan, Interaction of chiral molecules 2. L-L and L-D alanine system: net charge, population and dipole moment, Biochem.Biophy.Prog., (in Chinese) 5:35 (1982).
3. Chen Run-sheng, Rotational barrier of guanosine nucleoside about Z-DNA, J.Mol.Sci., (in Chinese) 3:81 (1982).
4. Chen Run-shen, DNA unwinding and control of information expressions, J.Mol.Sci., (in Chinese) 1:81 (1982).
5. E. Clementi, F. Cavallon, and R. Scordamaglia, Analytical potentials from ab initio computation for the interaction between biomolecules, 1. Water with amino acids, J.Am.Chem. Soc., 99:5531 (1977).
6. M. Ragazzi, D. R. Ferro, and E. Clementi, Analytical potentials from ab initio computation for the interaction between biomolecules 5. Formyl-triglycyl amide and water, J.Chem.Phys., 70:1040 (1979).
7. G. D. Smith and J. F. Griffin, Conformation of leu^5-enkephalin from X-ray diffraction, Science., 199:1214 (1978).
8. W. S. Benedict, N. Gailar, and E. K. Plyler, Rotation-vibration spectra of deuterated water vapor, J. Chem.Phys., 24:1139 (1956).
9. J. J.Kaufman, Quantum chemical and physiocochemical influences on structure-activity relations and drug design, Int.J. Quantum Chem., 16:221 (1979).
10. O. Matsuoka, E. Clementi, and M. Yoshimine, CI study of the water dimer potential surface, J.Chem.Phys., 64:1351 (1976).
11. N. Metropolis, S. W. Rosenbluth, M. N. Rosenblutf, A. H. Teller, and E. Teller, Equation of state calculations by fast computing machines, J.Chem.Phys., 21:1087 (1953).
12. J. P. Valleau and S. G. WHittington, "Statistical mechanics part A: equilibrium techniques," B. J. Berne ed., Plenum Press, N.Y., (1977).
13. S. Romano and E. Clementi, Monte Carlo simulation of the interaction between water and biomolecules: glycine and the corresponding zwitterion, Gazz.Chim.Ital., 198:319 (1978).
14. D. Eisenbegr and W. Kauzmann, "The structure and properties of water," Oxford U.P., N.Y., (1969).

15. S. Romano and E. Clementi, Monte Carlo simulation of water solvent with biomolecules: serine and the corresponding zwitterion, Int.J.Quantum Chem., 17:1007 (1980).
16. G. Bolis and E. Clementi, Methane in aqueous solution at 300 K, Chem.Phys.Lett., 82:147 (1981).
17. Chen Run-sheng, Dynamic model of translation phase, Scientia Sinica (series B) 10:1044 (1982).
18. Chen Run-sheng and Ni Xian-shan, Molecular configuration and macroscopic dynamic structure of protein, KEXUE TONGBAO 28:96 (1983).
19. Ni Xian-shan and Chen Run-sheng, Studies on evolution of dynamic structure of protein synthesis, KEXUE TONGBAO 28:244 (1983).
20. G. Nicolis and I. Prigogine, "Self-organization in nonequilibrium system," John Wiley & Sons, N.Y. (1977).
21. P. Glansdorff and I. Prigogine, "Thermodynamics of structure, stability and fluctuations," Wiley-Interscience, N.Y. (1971).
22. Xu Jin-hua and Ding Da-fu, Replication of biological macromolecules and formation of dissipative structures, Scientia Sinica 22:1206 (1979).
23. "CRC Handbook of Biochemistry, selected data for Molecular Biology,"H. A. Sober and R. A. Harte, ed., CRC Press (1970).

HYDRATE-H_2O AND LIQUID LIKE H_2O

H_2O INTERACTIONS WHICH ARE NECESSARY FOR BIOLOGICAL ACTIVITY

Hubertus Kleeberg and Werner A.P. Luck

Department of Physical Chemistry

University of Marburg, FRG

SUMMARY

H-bond interactions of H_2O can be investigated with high accuracy in the region of the H_2O combination vibration (near infrared (NIR): 1900-2000 nm or 5300-5000 cm^{-1}).

Below 50 % relative humidity H_2O forms H-bonds with the polar groups of biopolymers (Hydrate-H_2O). This H_2O is necessary for the stability of biomolecules but not sufficient for them to develop activity.

Above 50 % relative humidity liquid like H_2O is present in addition to Hydrate-H_2O.

The IR-spectrosopically determined amount of H_2O which is changed in its intermolecular interactions in comparison to pure water is in agreement with the results of other methods. The consequences of biopolymer-H_2O interactions are discussed with respect to enzymatic activity and partition coefficients in biological systems.

INTRODUCTION

It is generally agreed that investigations on the hydration of biopolymers are not only important for the production of foodstuff, cloth, leather, and so on (1) but also for the understanding of the polymerstructure or the energy balance during interactions of biologically active molecules as enzyme-substrate for example.

There is, however, a large number of different "termini technici" (for example: bound, non-freezable, vicinal, interstitial, adsorbed water) for H_2O which is changed in its properties in comparison to pure water by solutes (2-4). The nomenclature is by no means uniforms, since it largely depends on the methods and concepts used. This situation is very confusing at least for those who are not directly involved in H_2O-research, but who are interested in the results.

Consequently it is useful to find methods which permit to investigate the different interactions of H_2O molecules. This is possible to some extent by means of near infrared (NIR) spectroscopy (3,5-7) especially in systems containing polymers (8-10). As will be shown below, H_2O interactions in aqueous systems can be divided into two main groups:

HYDRATE-H_2O: this H_2O forms H-bonds with the polar groups of the solute;

LIQUID LIKE H_2O: this H_2O forms H-bonds with other H_2O molecules. The H-bond equilibrium of this H_2O:

$$OH_f + \theta_f \rightleftharpoons \text{H-bond}$$

(OH_f = not H-bonded OH ("free OH"); θ_f = lone electron pair -not interacting with an electron acceptor- ("free" lone pairs) needs not necessarily be the same as in pure water at the same temperature. This may be due to changes in the volume concentration of OH_f and/or θ_f for example.

The separation of H_2O interactions with respect to the H-bond acceptor molecule seems useful, since H-bonding is very important for an understanding of the properties of aqueous systems, although van der Waals forces should not be underestimated (7,11,12). The given nomenclature does not imply any statemant on the strength of interactions (like bound or free water suggests).

With respect to IR investigations it should be stressed that the time scale of the method is shorter than the lifetime of H-bonds (about 10^{-12} sec. (2)). Thus the amount of free OH and different H-bonds in the sample space can be separated experimentally.

With methods which are working on a longer time scale, information about the average H_2O properties are always gained i.e. H_2O involved in different H-bond interactions (see ref. 2,5-7). It will be shown that the results of methods which average over times longer or shorter than the life-time of an H-bond do not disagree.

PROTEIN HYDRATION

Concentration dependence

Although the triple helical structure of collagen (13,14) is well established, the results on the hydration of the protein are contradictory: x-ray structure analysis (14) indicated that about 18 g H_2O/100 g dry mass (d.m.) are an integral part of the polymer structure; calorimetric measurements (15-17) indicate that about 25-30 g H_2O/100 g d.m. have clearly different properties than pure water and NMR investigations show that about 50 g H_2O/100 g d.m. are not freezable (18-20).

On successive reduction of the H_2O-content of collagen or gelatin H_2O-spectra in the region of the H_2O combination band (NIR: 1890 - 2000 nm or 5300-5000 cm^{-1}) indicate the following:

- the position of the band maximum (ν_{max}) is shifted to shorter wave numbers until H_2O contents of 23 g H_2O/100g d.m. (corresponding to samples equilibrated at 50 % relative humidity (r.h.)). Below this H_2O content ν_{max} remains constant (see figure 1).
- the differences between the spectra at adjacent H_2O contents may be separated into three groups: a: Hydrate-H_2O: below 50 % r.h. the H_2O spectra are closely identical to those of H_2O in N-ethyl-acetamid. This indicates that H_2O is H-bonding predominantly to the peptide groups of the protein. About 1.2 H_2O/amino acid residue is present at this r.h.. This indicates that all lone electron pairs of the protein act as H-bond acceptors of H_2O. b: Liquid like H_2O with changed H-bond equilibrium: between 23 and 280 g H_2O /100 g d.m. difference bands are very similar to the spectrum of pure water at the same temperature. Slightly less free OH of H_2O, however, seems to be present in the collagenous system than in pure water. c: Liquid like H_2O: above 280 g H_2O/100 g d.m. the difference spectra are indistinguishable from that of pure water at the same temperature.

In figure 1 different properties of collagenous systems from literature data are plotted as a function of H_2O content. In all cases the change in the properties is most pronounced at about 23 g H_2O/100 g d.m. (as indicated by a vertical line in the figure). This H_2O content corresponds to approximately 50 % r.h.

The IR spectrum of the polymer gives information about the concentration dependence of changes of the protein (13,21,22). For example the position of the amide II band (see figure 1 and ref. 10,13,21,22) is constant above 50 % r.h. and changes below

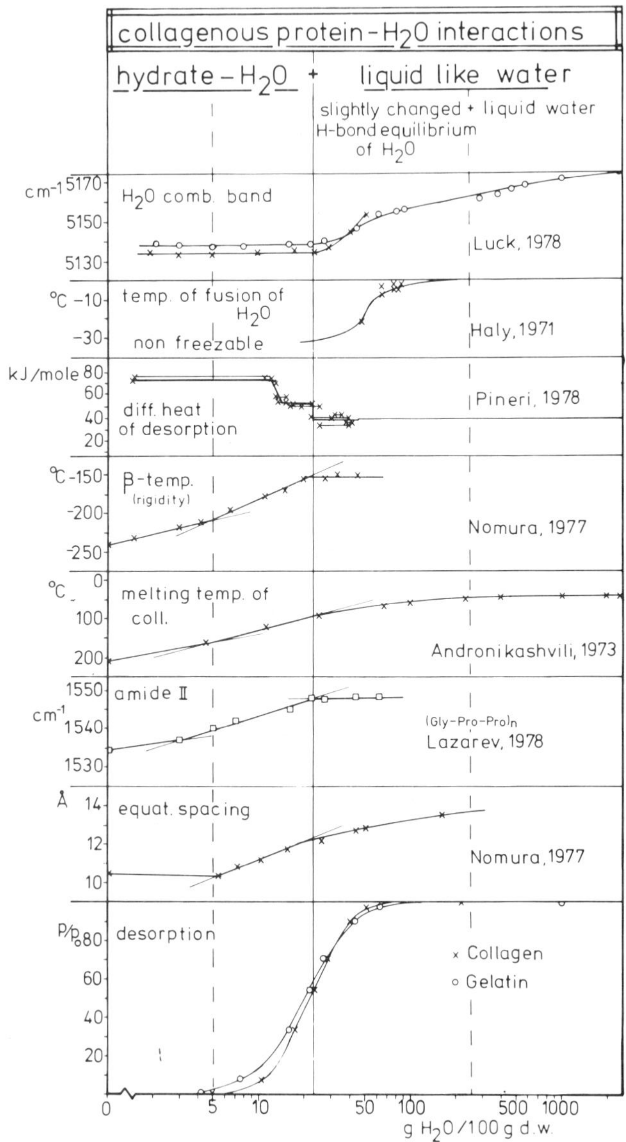

Figure 1: Dependence of different properties of collagenous systems on the H_2O content. The change in the properties is most pronounced at the H_2O content which corresponds to hydrate-H_2O (23 g H_2O/100 g dry weight, indicated by a vertical line). Values were taken from ref.10,16,17,22,27-29 (see also: Todireanu S., Angel G. and Nahorniak V., this volume)

this value. This indicates that the state of the polymer depends predominantly on the presence of Hydrate-H_2O.

Temperature dependence

In collagenous protein containing 25 - 100 g H_2O/100 g d.m. the melting temperature of ice is reduced (16,17). With a sample containing 80 g H_2O/100 g d.m. this phase transition is spectroscopically observed between -10 and - 20 °C. The spectrum at -50°C (figure 2) can not be reproduced quantitatively by an addition of the spectra at 50% r.h. and ice (both at -50°C). Taking into account a third band (indicated by ΔA in figure 2) a quantitative agreement between the band addition and the experimental spectrum is possible.

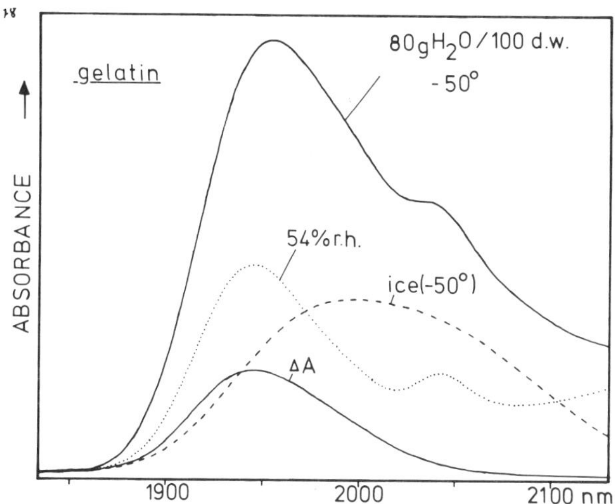

Figure 2: The near infrared spectrum of 80g H_2O in 100 g gelatin at -50°C can be decomposed into three components: the spectra of ice, gelatin at 50 % r.h. and a third component (indicated as ΔA).

From the relative area of the ΔA-band the corresponding amount of H_2O may be estimated to 15- 20 g H_2O/100 g d.m. The position of this band and stoichiometric considerations (see above) indicate that the band may be due to H-bonds between H_2O molecules which resemble more to "supercooled" water than to ice. Possibly this H_2O forms H-bonds with the lone pairs of Hydrate-H_2O (10).

It seems reasonable to assume that this H_2O (ΔA in figure 2) has a contribution to the heat of melting like ice (compare ref. 16,17), but a contribution to NMR measurements like Hydrate-H_2O.

This interpretation would explain the differences in the determination of non-freezable H_2O by the methods mentioned.

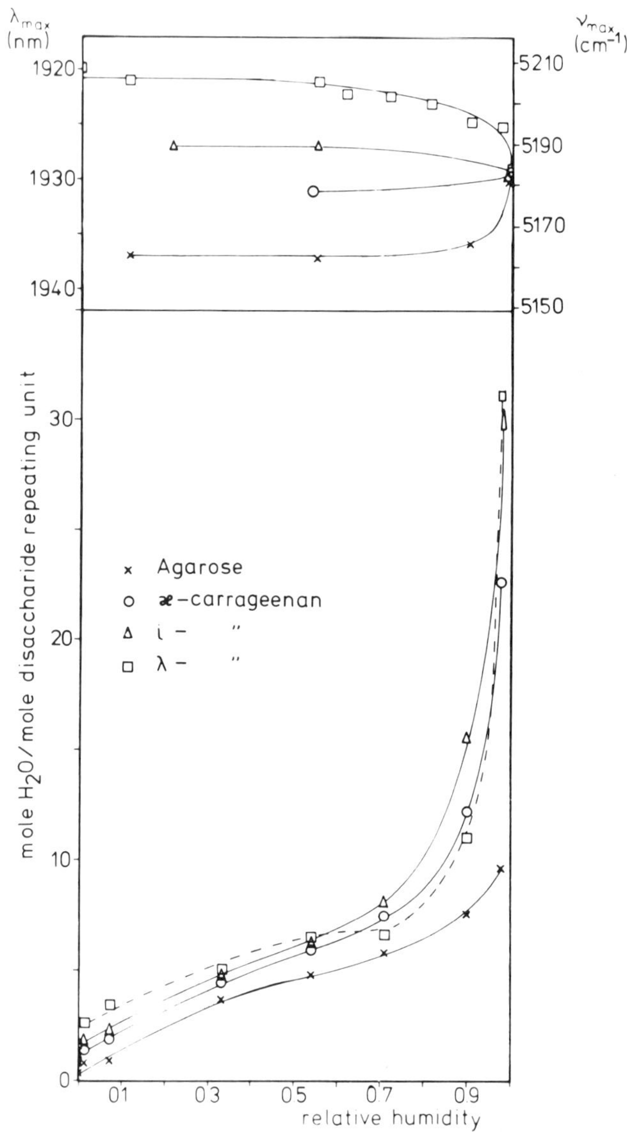

Figure 3: The position of the H_2O combination band and H_2O content of agarose, χ-, i- and λ-carrageenan at different relative humidities are compared.

POLYSACCHARIDE HYDRATION

In figure 3 the IR spectroscopic results and desorption isotherms for agarose, χ-, i- and λ-carrageenan are compared. In each case ν_{max} of the H_2O band changes above and remains constant below 50 % r.h. (8-10). At 50 % r.h. the H_2O content of the samples increases and ν_{max} of the H_2O band is shifted to larger wave numbers (weaker average H-bond strength of Hydrate-H_2O (7,10-12)) with increasing degree of sulfatation i.e. in the series:

agarose <χ-carrageenan < i-carrageenan < λ-carrageenan

Analogous results are found for different Glycosaminoglycans (8,10).

Thus we can conclude:

- at 50 % r.h. H_2O forms H-bonds to the polar groups of polysaccarides; if this Hydrate-H_2O is desorbed the state of the polymer changes;
- the sulfateester group is a comparatively weak H-bond acceptor in comparison to OH (like in agarose for example);
- on substitution of an OH-group by an SO_4^--group the amount of Hydrate-H_2O increases, since the number of lone electron pairs is enlarged;
- above 50 % r.h. liquid like H_2O is present in the sample in addition to Hydrate-H_2O.

Analysis of spectra of 5-23 % solutions of polysaccharides show that Hydrate-H_2O is present with equal amount and H-bonding like at 50 % r.h. (8,10).

INFLUENCE OF HYDRATE-H_2O ON BIOLOGICAL ACTIVITY

In figure 4 the biological activity in different systems is compared with the typical sorption isotherm.

The enzymatic activity of ß-amylase, phospholipase or the growth rate of microorganisms (Xeromyces bisporus) (23) increases above 50 % r.h.. The inactivity below 50 % r.h. obviously is due to the lack of liquid like H_2O.

If we take into account that the desorption of Hydrate-H_2O may irreversibly inactivate biological substances we may conclude that:

> Hydrate-H_2O is necessary for the stability of bio-molecules but not sufficient for their action. At least small amounts of liquid like H_2O have to be present for the activity of biological systems.

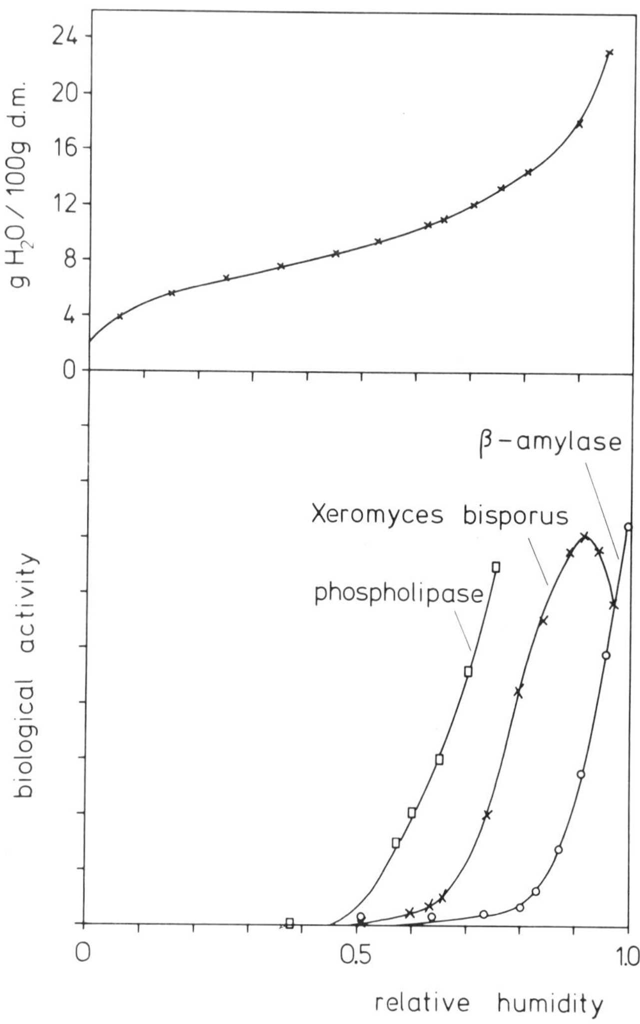

Figure 4: The upper part of the figure shows a typical sorption isotherm (of lecithin). The lower part shows that the biological activity depends on the H_2O content. At least small amounts of liquid like H_2O have to be present for biological activity; this is the case above 50 % relative humidity. As examples the enzymatic activity of phospholipase (from barley malt flour) on lecithin, ß-amylase on starch and the growth of Xeromyces bisporus are given (taken from ref. 23).

HYDRATION IN TERNARY SYSTEMS

IR investigations in low molecular weight ternary systems indicate that the solubility of organic molecules in aqueous solutions of solids (salts, sugars etc.) or the solubility of solids in water-organic solvent mixtures is a competitive process.

The solubility in these ternary systems is usually reduced on behalf of solute-H_2O interactions. The separation of an organic and/or solid phase from aqueous solutions on addition of solute seems to depend largely on the strength and amount of H_2O-solute interactions. Hydrate-H_2O does not seem to have the property to solvate another solute in the aqueous phase (10).

The situation in ternary systems containing polymers seems to be very similar. The composition of coacervate (24,25) which is in equilibrium with an aqueous solution depends on the solute and its concentration. In the polymer rich coacervate the content of solute is usually smaller than in the equilibrium liquid (24). About 10 % of the H_2O of chondroitinsulfate-gelatin-coacervates (approx. 2000 g H_2O/100 g d.m.; the H_2O content depends on the solute added) does not behave as a solvent. This amount (approx. 200 g H_2O/100 g gelatin corresponds to the amount of Hydrate-H_2O and Liquid like H_2O with a changed H-bond equilibrium of gelatin (see above).

A reduction of the partition coefficient is frequently found in biological tissues as well (26 and references cited therein). In both cases mentioned the reduced solubility of solutes seems to have the same reasons as in low molecular weight solutions: Hydrate-H_2O and liquid like H_2O with a changed H-bond equilibrium do not act as solvent for other solutes because of different intermolecular interactions.

REFERENCES

1. Rockland L.B. and Stewart G.F. (eds.), Water Activity: Influences on Food Quality, Academic Press, New York (1981)
2. Eisenberg D. and Kauzmann W., The structure and Properties of Water, Oxford Univ. Press, London (1969)
3. Luck W.A.P. and Kleeberg H., p. 421 in ref. 1.
4. Drost-Hansen W. and Clegg J. (eds.), Cell-Associated Water, Academic Press, New York (1979)
5. Luck W.A.P. and Schiöberg D., Advan. Mol. Relaxation Processes 14, 277 (1979)
6. Schiöberg D. and Luck W.A.P., J.C.S. Faraday Trans. I, 75, 762 (1979).

7. Kleeberg H., in: Intermolecular Forces, Vol.14, p. 465, (Pullman B., ed.) D. Reidel Publ. Co., Dordrecht (1981).
8. Kleeberg H. and Luck W.A.P., in: Glycoconjugates (Schauer R. et al., eds.), p. 98, G. Thieme Publishers, Stuttgart (1979).
9. Luck W.A.P. and Kleeberg H., in: Photosynthetic Oxygen Evolution (Metzner H., ed.), Academic Press, London (1978).
10. Kleeberg H. and Luck W.A.P., in press; Kleeberg H., Thesis, University of Marburg, in preparation.
11. Luck W.A.P., in: see ref. 7, p. 199.
12. Kleeberg H., Kocak Ö. and Luck W.A.P., J. Solut. Chem. 11 (1982).
13. Fraser R.D.B. and MacRae T.P., Conformation in Fibrous Proteins Academic Press, New York (1973).
14. Ramachandran G.N. and Chandrasekharan R., Biopol., 6, 1649 (1968)
15. Berlin E., Kilman P.G. and Pallansch M.J., J. Coll. Interface Science, 34, 488 (1970).
16. Mrevlishvili G.M., in: L'Eau et les Systèmes Biologiques (C.N.R.S., ed.), no 246, Paris (1976).
17. Haly A.R. and Snaith J.W., Biopol., 10, 1681 (1971).
18. Kuntz I.D. Jr., Brassfield T.S., Law G.D. and Purcell G.V., Science, 163, 1329 (1969).
19. Dehl R.E., Science, 170, 738 (1970).
20. Hoeve C.A.J. and Tata A.S., J. Phys. Chem., 82 1660 (1978).
21. Susi H., Ard J.S. and Carroll R.J., Biopol., 10, 1597 (1971).
22. Lazarev Yu.A. et al., Biopol. 17, 1197 (1979).
23. Tome D. et Bizot H., Les Aliments à Humidité Intermèdiaire, Serie Synthèses Bibliographique No. 16, APRIA, Paris (1978)
24. Troshin A.S., Das Problem der Zellpermeabilität, VEB G. Fischer Verlag, Jena (1958)
25. Kruyt H.R. (ed.), Colloid Science II, Elsevier Publ. Co., New York (1949).
26. Garlid K.D., in: see ref.4, p.293; and ref. 16, p.317.
27. Pineri M.H., Escoubes M. and Roche G., Biopol., 17,2799 (1978).
28. Nomura S., Hiltner A., Blando J.B. and Baer E., Biopol., 16, 231 (1977).
29. Andronikashvili E.L., Preprint 04-BP, Inst. of Physics, Acad. of Sci. of the georgian SSR, Tbilissi (1972)

HYDRATION OF DNA AND POSSIBLE ROLE OF WATER IN THE TRANSITION BETWEEN RIGHT- AND LEFT-HANDED DOUBLE HELIX

G. M. Mrevlishvili, G. Sh. Japaridze, V. M. Sokhadze,
Z. I. Chanchalashvili and D. A. Tatishvili

Institute of Physics
Academy of Sciences of the Georgian USSR
Tbilisi, 380077, USSR

INTRODUCTION

As we already know from the classical X-ray studies, the conformation of DNA double helix depends to a great extent on water and different cation content[1,2]. Unfortunately, until now we have been forced to consider structural parameters of DNA molecules only in terms of spatial molecular models[3]. Only recently a noticeable progress has been made in this field[4-10]. The obtained data show[5] that structural and dynamic properties of polynucleotide chains with definite nucleotide sequence lead to the forms of double helix radically differing from B-DNA conformation. We are considering so-called Z-DNA, or left handed DNA with the structure determined on the basis of X-ray analysis of monocrystals of oligonucleotides with alternating pur-pyr sequences[4-7].

Important results were obtained by Dickerson and Drew[8,9] who were the first to establish structural parameters of dodecamer d(C G C G A A T T C G C G) in B-form and the coordinates of water molecules forming hydrate layers in the wide and narrow grooves of the molecule. Putting aside the question of the possible biological role of the Z-form DNA (Discussed in[4,11,12]) the elucidation of the nature of the transition from the right-handed DNA helix into the left-handed conformation is extremely important. The most pronounced transition has been observed in aqueous solutions of oligonucleotides d(CpGpCpGp;...)[13,14] These transitions occur in aqueous solutions of alternating oligo- and polynucleotides at high salt concentration, in particular, with the increase of NaCl (2.5M) and $MgCl_2$ (0.7 M) molarity. Thus, Z-form of DNA has been discovered, and the conditions forcing the double helix to transform from the right-handed form into the left-handed one have been established[14-17]. What is the main cause responsible for such a transition? We have tried to answer this question using the method of low-temperature differential microcalorimetry on such objects as synthetic alternating polynucleotides poly-[d(A-T)] and poly-[d(G-C)] and natural DNA[19,20]. The hydration of natural DNAs has been shown to depend on GC-content [18-20] and to fit the equation $n_{DNA} = \{28 - 0.12 \cdot (\%GC)\}$ (Figure 1). As well as this it has been shown that the quantity of water molecules in the inner layer ("structural" part of the hydration shell) also depends on G-C content ($n_{str} = \{12 - 0.06 \cdot (\%GC)\}$). The experimental data presented in Figure 1 are extremely important as they imply that B-Z transition does not take place in aqueous solutions of AT alternating polynucleotides, while such transition occurs for GC alternating polynucleotides[13,14]. (It has been recently shown that (dG-dC) polymer methylation leads to Z-conformation under the physiological environmental conditions)[11]. The questions arise: 1. What is the role of AT- and GC-base pair hydration heterogeneity in the conformational transitions within the double-stranded state? 2. What is the effect of salt (the range of B-Z transition) on thermodynamic parameters of water?

EXPERIMENTAL

Materials

Na-salts of synthetic alternating polynucleotides poly-[d(A-T)] and poly-[d(G-C)] produced by "Boehringer Mannheim GmbH" were used in

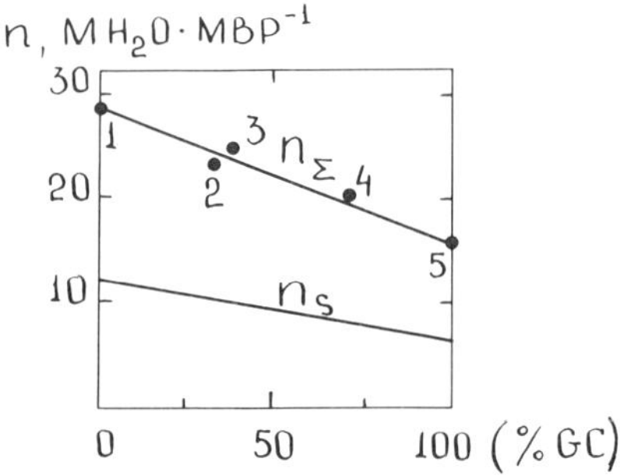

Fig. 1. The dependence of DNA hydration parameter (n) on G-C content ($n = M\ H_2O/MBP$): 1 - Na-poly-[d(A-T)]; 2 - Na-DNA phage T2; 3 - Na-DNA calf thymus; 4 - Na-DNA M. Lysodeikticus; 5 - Na-poly-[d(G-C)].

the experiments. All solutions were prepared with distilled deionized water. The concentration of $MgCl_2$ salt varied over the range of 0.2-1.0 M by adding $MgCl_2$ solutions of high concentration into aqueous solutions of polynucleotides in order to achieve the necessary molarity, while both polynucleotide concentration and pH remained constant. The initial and final concentrations of polynucleotides were determined spectrophotometrically and by weighting directly in microcalorimetric cells after the experiment. The cells with the volume of 0.1 ml were used (see "Method"). The initial weights of polynucleotides in cells were 1.63 mg for poly-[d(A-T)] and 2.5 mg for poly-[d(G-C)].

Method

Thermodynamic parameters of solvent phase transitions, eutectic ($MgCl_2$-H_2O) mixture melting, and polynucleotide thermal denaturation process have been measured on a low-temperature scanning microcalorimeter designed at our laboratory[21,22]. The principal technical characteristics of the microcalorimeter are as follows: working temperature range 80-400 K, heating rate 2°/hour-15°/hour, working volume of the cells 0.05-0.5 ml, sensitivity $\sim 10^{-7}$W, accuracy of temperature registration ± 0.1 K. Figure 2 shows an example of the record of heat absorption process observed in low-temperature region. The areas under heat absorption peaks correspond to the melting heats of eutectic H_2O-$MgCl_2$ (ΔH_u) and ice (ΔH_{H_2O}) for a given concen-

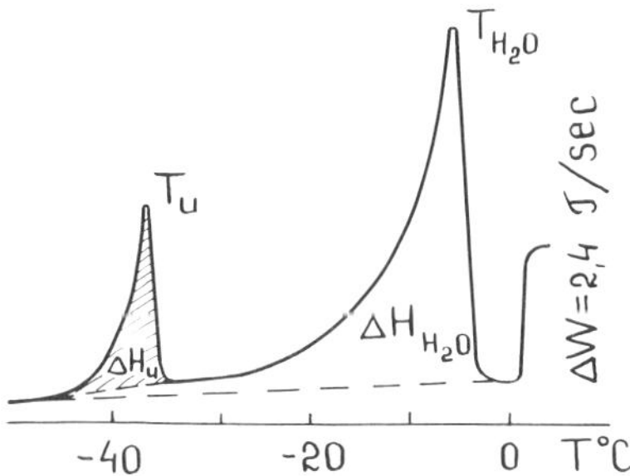

Fig. 2. Records of heat absorption processes at the heating of poly-[d(A-T)] - H_2O-$MgCl_2$ freezed water solutions. (See explanations in the text). 1.63 mg of the polymer has been in the calorimetric cell; the total weight of the specimen is 67.22 mg; $MgCl_2$ molarity is 0.66 M; heating rate V = 14°/hour; ΔW is the value of the calibration mark.

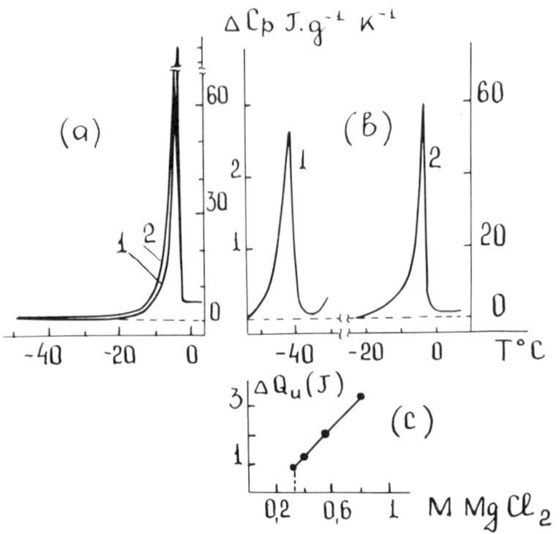

Fig. 3. Temperature dependence of the excess heat capacity in the ternary system of poly-[d(G-C)] -H_2O-$MgCl_2$ over the low-temperature range at various $MgCl_2$ concentrations; (a) (1) - 0.17 M $MgCl_2$, (2) - 0.24 M $MgCl_2$; (b) (1) - melting peak of $MgCl_2$-H_2O eutectic in the ternary system with $MgCl_2$ concentration 0.31 M, (2) - ice melting peak; (c) - the dependence of the eutectic melting heat on $MgCl_2$ concentration in the ternary system.

tration of polynucleotide and $MgCl_2$. Varying salt concentration in desirable range of molarities, we determine important thermodynamic parameters T_{H_2O}, T_u - the temperatures, ΔH_{H_2O}, ΔH_u - the enthalpies, and the entropies $\Delta S_{H_2O} = \Delta H_{H_2O}/T_{H_2O}$, $\Delta S_u = \Delta H_u/T_u$ of ice and eutectic melting, respectively. Changes of absolute values of these thermodynamic parameters compared to the values observed in pure water and during pure eutectic mixture melting, testify the changes of both bulk water properties and the mechanisms of ion interaction with macromolecules[21,22].

RESULTS AND DISCUSSION

The temperature dependences of heat capacity of poly-[d(G-C)]-H_2O-$MgCl_2$ system at different $MgCl_2$ concentrations are given in Figure 3. As one can see, the increase of salt concentration up to 0.32 $MgCl_2$ does not lead to the emergence of eutectic H_2O-$MgCl_2$ melting peak (the temperature of pure H_2O-$MgCl_2$ eutectic melting equals -37.6°C). That means that all Mg^{++} ions are bound to the polynucleotide chains[21]. The sharp heat absorption peak is ob-

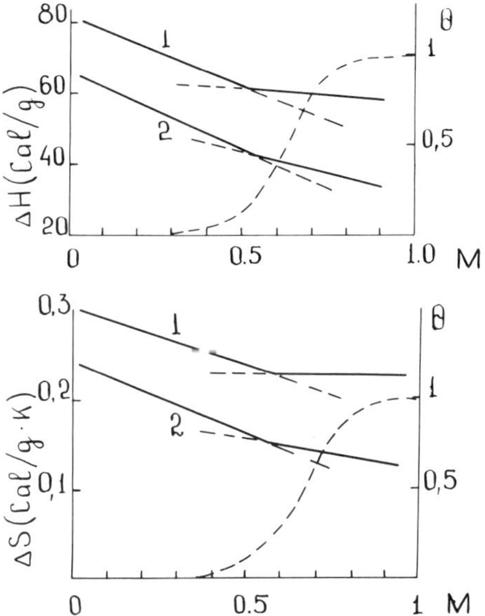

Fig. 4. The dependence of the enthalpy (ΔH) and the entropy (ΔS) of water melting on $MgCl_2$ concentration (in moles) in poly-[d(A-T)] - H_2O-$MgCl_2$ (1) and poly-[d(G-C)] - H_2O-$MgCl_2$ (2) systems. Dotted line (Θ) corresponds to Pohl-Iovin transition.[13] (Here and later 1 cal = 4.184 J).

served at $MgCl_2$ concentrations of 0.32-0.4 M over the temperature range from -50 to -37°C, which points to the appearance of free Mg^{++} and Cl^- ions forming eutectic mixture with ice crystals at low temperatures. Such dependences of the eutectic melting enthalpy on $MgCl_2$ concentration are given in Figure 3. The increase of ΔQ_u testifies that the increase of salt concentration does not lead to the additional binding of ions. However, enthalpy and entropy of ice melting (i.e., properties of bulk water) change abruptly with the increase of $MgCl_2$ concentration. Enthalpy and entropy of ice melting decrease (Figure 4) abruptly until $MgCl_2$ concentration equal to 0.5 M. As it has been mentioned, only a part of Mg^{++} ions equal to

0.75-1.0 M of Mg^{++} per phosphate, is bound to polynucleotide chains (figure 3c). On the other hand, according to Pohl-Jovin data[13] (see also [14]), just at these molarities the double helix transition from the right-handed conformation into the left-handed one begins. The sharp decrease of thermodynamic parameters of water melting points to a significant rearrangement of the total solvent structure.

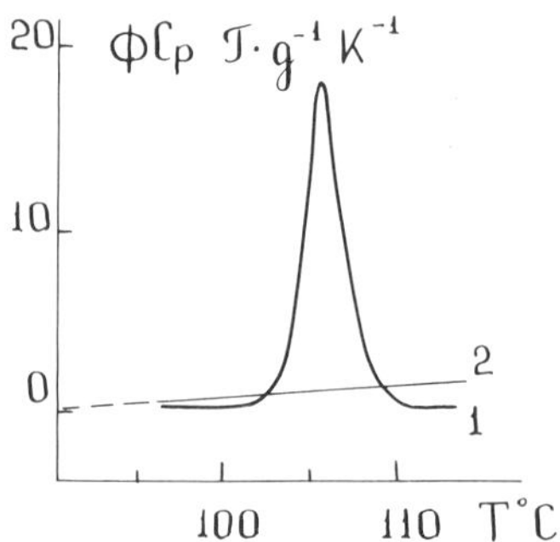

Fig. 5. The temperature dependence of the excess heat capacity in poly-[d(G-C)]-H_2O-$MgCl_2$ system. $MgCl_2$ molarity is 0.85 M. 1 - the curve of poly-[d(G-C)] heat denaturation, 2 - the repeated measurement of $\Delta C_p=f(T)$ after the slow cooling of the solution (polymer concentration is 2.5 percent by weight).

Thus, B-Z transition is characterized by the abrupt change of thermodynamic parameters of the solvent reflecting the state of bulk water. Hence, the role of ions in the transition consists not only in the necessity of the direct binding of polynucleotide chains to the screening phosphate charges, but in the change of chemical potential of water and the perturbation of the whole solvent structure, too. The curve of the heat denaturation of poly-[d(G-C)] obtained under the conditions where B-Z transition should be entirely completed, is given in Figure 5. The melting thermodynamic parameters are as follows: $T_d=106°C$, $\Delta T_d^{1/2}=2°$, $\Delta H_{d,Z}=34.1$ kJ/MBP. As it is seen, the enthalpy of poly-[d(G-C)] denaturation under the Z-form existence conditions differs slightly from the enthalpy of B-DNA unwinding[24]. The complete irreversibility of the heat denaturation process should be noted (Figure 5, curve 2).

REFERENCES

1. R. E. Franklin and R. G. Gosling, Acta Crystallogr. 6:673 (1953).
2. M. H. F. Wilkins, W. E. Seeds, A. R. Stokes, and H. R. Wilson, Nature, 172:759 (1953).
3. S. Arnott, 1st Cleveland symposium on macromolecules, in: "Secondary Structures of Polynucleotides," pp.87-104, Elsevier, Amsterdam (1977).
4. A. H. J. Wang, G. J. Quigley, F. J. Kolpak, J. L. Crawford, J. H. van Boom, G. van der Marel, and A. Rich, Nature, 282:680 (1979).
5. A. H. J. Wang, G. J. Quigley, F. J. Kolpak, G. van der Marel, J. H. van Boom, and A. Rich, Science, 211:171 (1981).
6. H. Drew, T. Takano, S. Tanaka, K. Itakura, and R. E. Dickerson, Nature, 286:567 (1980).
7. J. L. Crawford, F. J. Kolpak, A. H.-J. Wang, G. J. Quigley, J. H. van Boom, G. van der Marel and A. Rich, Proc. Natl.Acad. Sci.USA, 77:4016 (1980).
8. R. E. Dickerson and H. R. Drew, J.Mol.Biol. 149:761 (1981).
9. H. R. Drew and R. E. Dickerson, J.Mol.Biol., 151:535 (1981).
10. S. Arnott, R. Chandrasekharan, D. L. Birdsall, A. G. W. Leslie, and R. L. Ratliff, Nature, 283:743 (1980).
11. J. Nickol, M. Behe, and G. Felsenfeld, Proc.Natl.Acad.Sci.USA, 79:1771 (1982).
12. J. Klysik, S. M. Stirdivant, J. E. Larson, P. A. Hart, and R. D. Wells, Nature, 290:672 (1982).
13. F. M. Pohl and T. M. Iovin, J.Mol.Biol. 67:375 (1972).
14. D. J. Pathel, L. T. Canuel, and F. M. Pohl, Proc.Natl.Acad.Sci. USA, 76:2508 (1979).
15. T. J. Thamann, R. C. Lord, A. H.- J. Wang, and A. Rich, Nucleic Acids Res. 9:5443 (1981).
16. D. J. Pathel, S. A. Kozlowski, A. Nordheim, and A. Rich, Proc. Natl.Acad.Sci., 79:1413 (1982).
17. E. Minyat and V. Ivanov, in: "Symposium of Biophysics of Nucl.Acids and Nucleoproteins," Tallin, USSR (1981).
18. E. L. Andronikashvili, G. M. Mrevlishvili, G. Sh. Japaridze, V. M. Sokhadze, D. A. Tatishvili, and L. Orvelashvili, in: International Symp. on Biocalorimetry (Abstracts) p. 66, Tbilisi, USSR (1981).
19. G. M. Mrevlishvili, G. Sh. Japaridze, V. M. Sokhadze, D. A. Tatishvili, and L. V. Orvelashvili, Molec.Biol.(USSR) 15:336 (1981).
20. G. M. Mrevlishvili, Dokl.Akad.Nauk SSSR (USSR) 260:761 (1981).
21. G. M. Mrevlishvili, Biofizika (USSR) 1:180 (1977).
22. M. Mrevlishvili, Sov.Phys.Usp. (American Inst. of Phys.Publ.) 22 (6):433 (1979).
23. R. A. Robinson and R. H. Stokes, in: "Electrolitic Solutions," Butterworth, London (1959).
24. H. Klump and J. Ackermann, Biopolymers, 10:513 (1971).

ION PAIRS OF SIMPLE ELECTROLYTES AND OF POLYIONIC BIOPOLYMERS IN ELECTRIC FIELDS

Eberhard Neumann*, Kinko Tsuji and Dieter Schallreuter

Biophysical Chemistry Unit
Max-Planck-Institut für Biochemie
D-8033 Martinsried/München, F.R. Germany

INTRODUCTION

'I believe that the ultimate goal of biological study is to "translate" the phenomena of life into meaningful physical concepts'.[1] Aharon Katzir-Katchalsky (1914-1972) to the memory of whom this account is dedicated.

The dynamics of biological macromolecules and membranes appears to involve a special electrostatic interaction element: ion-pairing. Associations of cations and anions are well known from simple electrolytes, the Bjerrum pairs,or in the counterion association of linear polyelectrolytes such as the various nucleic acids.

Cation-anion pair configurations may be of particular functional relevance for membrane protiens, especially for the ion flow gating protiens in excitable membranes. Fixed ionic groups of membrane components are probably the preferred interaction partners of the electric field of biological membranes.

In this context it is recalled that all biomembranes under in vivo conditions appear to have an electric potential difference (membrane potential) and to experience changes of this electrical property. The electric potential difference is equivalent to an electric field across the membrane components.

In nerve excitation, the physiological action potential represents a transient change of the electric membrane field in

* Address of correspondence

magnitude and direction from approximately 70 $kVcm^{-1}$ to 40 $kVcm^{-1}$ in the opposite direction.[2] A further example of functionally important changes in the electric membrane field is encountered in the hyperpolarizing increase of the absolute value of the membrane potential during light-induced proton pumping of bacteriorhodopsin in halobacteria; this increase in the membrane field is concomitant with a decrease in the proton transport.[3] Indeed, there is a variety of epiphenomena where structure and function of membrane components appear to be strongly coupled to the electric field of biological membranes.

Recent experimental progress of chemical relaxation kinetics in high electric fields[4] has brought new information on ion pairing processes in simple inorganic salts and in linear polyelectrolytes[4] as well as on membrane transport proteins.[5-7] In the following these advances will be briefly summarized.

EXPERIMENTALS

A recently developed electric field-jump relaxation spectrometer permits the simultaneous measurements of field-induced changes in electric conductance and in optical properties of solutions in the ns-ms time range with high resolution.[4] Rectangular field pulses up to 100 kV cm^{-1} and of variable pulse length (0.5μs-ms) can be applied by cable discharge in a single pulse mode to solutions and suspensions.

ION PAIRS IN $MgSO_4$ SOLUTIONS

The electric conductance relaxation spectrum, as response to a rectangular field pulse, of an aqueous $MgSO_4$ solution exhibits a rapid (0.1μs) part and a slow phase in the 1μs time range. The slow conductivity increase represents a discrete normal mode, depending on the salt concentration and on the electric field strength. The electric data are consistent with familiar concepts of the ionic atmosphere relaxation and of the Onsager field dissociation effect. The analysis of the completely analyzable slow relaxation mode yields an electric dipole moment (~50 debye) which is consistent with the presence the outer-sphere complex $Mg^{2+} \cdot H_2O \cdot SO_4^{2-}$; an ion-pair involving at least one H_2O dipole shell between the ionic charge carriers of the ion-pair dipole. Since the treatment of polarizable dipoles in polar solvents is intricate and notoriously difficult, the experimental determination of dipole moments of ion-pairs in aqueous solution represents an important fundamental information.

COUNTERION FLOW COUPLING IN POLYELECTROLYTES

The simultaneous measurements of electric conductance and UV electric dichroism of the linear polyelectrolyte poly(ribo-adenylate K+) have shown that (1) the K+-conterions of the inner ionic atmosphere of poly(rA,K+) dissociate from the polyion with the same rate as the rod-like ends and the middle part of the polyelectrolyte rotate into the field direction,
(2) counterion dissociation induces increased destacking of the adenine bases that is also rate-limited by the orientation process,
(3) rotation rate and amplitudes of orientation, of counterion dissociation, and of base destacking strongly decrease with increasing concentration of free counterions (KCl),
(4) 'end effects' are important and counterion-polyion polarization causes a large induced dipole moment which is the origin of the orientation processes.

The electro-optic data of poly(rA,K^+) can be consistently described in terms of a flow concept developed by Katchalsky and Neumann,[8,9] involving saturation in the counterion polarization (electro-diffusion limit) and concentration effects in terms of a simple mass action formalism.[4].

Polarization flow model

According to the coupled flow concept the external electric field, E, causes a counterion flow $J_p=v[C^+]_p$ along the polyion; $v=u\cdot E$ is the velocity, u is the mobility and $[C^+]_p$ the local concentration of counterions in the inner (condensed) counterion layer. The counterion polarization and the large induced dipole moment finally result from flow imbalances at the two ends of the polyelectrolyte rod.

The 'polarization flow' J_p causes a local depletion of counterions at the one rod end (negative dipole charge) and a local accumulation at the other end (positive dipole charge). The local depletion range gives rise to an influx $J_-=v[C^+]$ of free counterions ($[C^+] \leq 1$ mM) from the solution. Because $[C^+] < [C^+]_p$, the negative dipole charge is caused by the flow difference $\Delta J_-=v\cdot([C^+]_p-[C^+])$. On the other hand the positive dipole charge arises from $\Delta J_+=v([C^+]_p-[C^+])$ because the outflow $J_+=v[C^+]$ is not balanced by the local influx J_p into the accumulation range.

When $[C^+]$ is increased (by KCl addition) all processes which are coupled to counterion polarization: orientations, counterion dissociation and base-destacking are reduced in electric fields. In the limit of $[C^+] = [C^+]_p$ neither polarization, nor any electro-optical and chemical-conformational reaction flows appear.

The anisotropy of the counterion movement along polyionic surfaces suggest that counterion exchange as well as influx and efflux of counterionic substrates or hormones occur preferably at the border lines of the ionic atmospheres which cover polyionic regions on macromolecular enzyme and receptor proteins and membranes. Once part of the ion cloud, such substrates and activator substances may reach the active sites via surface diffusion.[10,11]

In this model, the border regions of the counterion atmosphere serve as a preferable cross section for trapping counterionic substrates, probably in diffuse ion-pair configurations. In brief, the presence of surface charges and the dynamic element of ion-pair configurations coupled to surface diffusion will appreciably accelerate the effective rates of counterionic ligand binding.

MEMBRANE PROTEINS IN ELECTRIC FIELDS

Bacteriorhodopsin of purple membrane fragments

An instructive example for the action of electric fields on membrane proteins and ion transport processes is the bacteriorhodopsin of the purple membranes of halobacteria. The light-driven proton pump activity of this protein appears to be feedback regulated by a hyperpolarizing increase in the electric membrane field.[6]

Recent electro-optic data of aqueous suspensions of purple membranes demonstrate that high electric field pulses ($1-30 \times 10^5$ Vm^{-1}, 1-100 µs) cause structural transitions in two kinetically different phases.[7] The rapid mode (~1µs) represents a concerted change in the orientations of both the retinal and some tyrosine and/or tryptophan side chains concomitant with alterations of the local protein environment of these chromophores. The slower mode (~100 µs) involves changes in the microenvironment of aromatic amino acids, at fixed chromophor orientations and in the pK values of a least two types of proton binding sites leading to a sequential uptake and release of protons.[6]

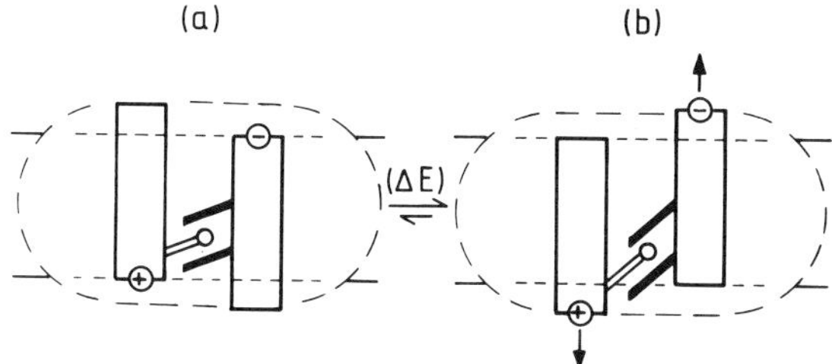

Fig. 1 Principle of the saturatable induced dipole mechanism causing positional changes of side chains in helical membrane proteins. In bacteriorhodopsin helical parts with different net charge may move transversal to the membrane plane in opposite directions when the electric membrane field is increased, (a) ➞ (b). The geometrically limited increase in the distance of the charge centers is equivalent to a saturable induced dipole moment.[7] The transversal displacement of at least one of the two helical parts can thereby cause a concerted rotational shift of the retinal (═O) and of aromatic amino acid side chains which may sandwich (T.H. Haines) the retinal chromophore.

Electric gating of ion transport

The dependence of the electro-optic relaxation amplitudes of purple membrane suspensions on the field strength suggests a saturable induced-dipole moment mechanism.[5,7] The electric field is apparently able to increase in a strongly cooperative manner, the mean distance between positively and negatively charged amino acid side groups (ion-pairs) such that the mean dipole moment increases; see Fig. 1. The kinetics of the electro-optic changes and the steep field dependence indicate a rather large reaction dipole moment, $\sim 1.1 \times 10^{-25}$ Cm ($=3.3 \cdot 10^{4}$ debye) per cooperative unit at a field strength of 1.3×10^{5} Vm^{-1} (which is the transition midpoint).[7]

These results are not only suggestive for a possible control function of the electric field of the bacterial membrane during the photocycle of bacteriorhodopsin, but also for a possibly general, induced-dipole mechanism for electric field dependent structural changes in membrane transport proteins such as the gating proteins in the excitable membranes of nerve and muscle cells. Analoguous to the scheme in Fig. 1, we may suggest an induced-dipole gating mechanism for the axonal Na^+ channel in a highly schematic form, also incorporating ion exchange Ca^{2+}/Na^+.[12]

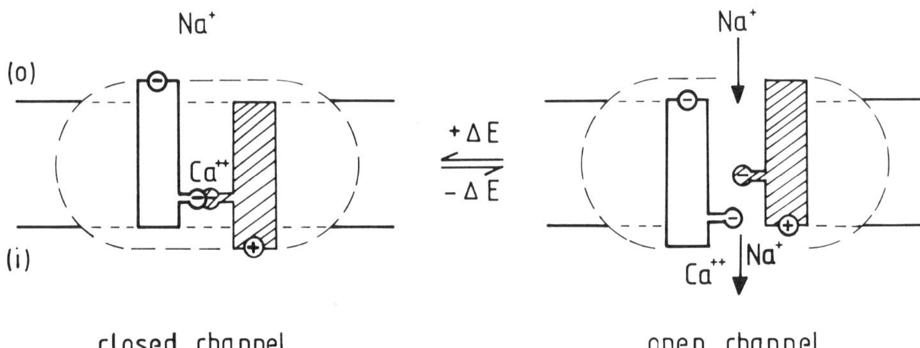

Fig. 2 Model for a gating element of the axonal Na^+ channel. At high electric field (resting membrane potential) between the outside (o) and the inside (i) of the excitable membrane, the gating protein is in the closed conformation. Ca^{2+} ions are bound to anionic side chains of amino acids of helical parts kept by the membrane field in a high dipole moment configuration (large distance between the cationic and anionic groups). Depolarization reduces the electric membrane field which, in turn, reduces the electric polarization by transversal movements of the helical parts, decreasing the distance between the fixed charges. The Ca^{2+} ion of the former bridge is exchanged by Na^+ ions which flow through the open, field-relaxed conformation into the cell interior.

Long-lived electric imprint into structure

A further kinetic aspect of the bacteriorhodopsin data may also be of general fundamental importance. Compared to the rapid induction of the electric field-mediated structural changes in the purple membranes (μs range), annealing of the changes after the electric impulse is very slow. The field-induced conformational transitions in bacteriorhodopsin are thus long-lived compared to the pulse duration; these transitions therefore exhibit memory properties.

Longevity of field-induced structural changes may also offer a basis for the interpretation of the ATP formation by thylakoid membranes[13,14] and by submitochondrial particles[15] after the exposure to the electric fields. When the lifetime of the electric field-induced ATPase activation (to synthetize ATP) is greater than the pulse duration, 'after-field effects' may occur in a way similar to that of the after-field pH changes[6] observed in bacteriorhodopsin of purple membranes. The electric field-induced energization[16] of ATP synthetase would thus be caused by an electric field-induced, long-lived structural transition to the active enzyme conformation.

It is recalled that electric field-induced, long-lived structural changes had been suggested as a mechanism for the initial, physical step of biological memory recording.[8]

THERMODYNAMICS OF CHEMICAL TRANSFORMATIONS IN ELECTRIC FIELDS

Structural transitions of the type $B_o \rightleftharpoons B_1$, induced by an electric field E at constant pressure P and Kelvin temperature T may be described by

$$(\partial \ln K/\partial E)_{P,T} = \Delta M/RT \qquad (1)$$

where $K = [B_1]/[B_o]$ is the apparent equilibrium constant (concentration ratio of the conformers) and R is the gas constant.[17] The apparent molar reaction moment ΔM is the difference between the field-parallel components of the molar (permanent and induced) dipole moments of the conformers: $\Delta M = M_1 - M_o$. Thus $\Delta M \neq 0$, if $M_1 \neq M_o$.

In terms of the mole fraction $\theta = [B_1] / [B_1] + [B_o]$ and with $K=\theta/(1-\theta)$ eq.(1) leads to:

$$(\partial \theta/\partial E)_{P,T} = \theta(1-\theta)\ \Delta M/RT \qquad (2)$$

With $\bar{\varepsilon}_o$ and $\bar{\varepsilon}_1$ being the (average) extinction coefficients of the conformers B_o and B_1, respectively,[5,7] the chemical transition may be measured as an absorbance change. Per cm light path, $\Delta A^{(ch)} = (\bar{\varepsilon}_1 - \bar{\varepsilon}_o) \cdot c\, \Delta\theta$, where $c = [B_1] + [B_o]$ is the total concentration of B. It is now readily shown that at a given field strength,

$$\Delta M = RT\{(1-\theta^\circ)/[\theta(1-\theta)]\}[\partial(\Delta A/\Delta A_s/\partial E]_{P,T}\ , \tag{3}$$

where θ° refers to E=O and ΔA_s is the saturation value at θ=1.

Dielectrochemical affinity

Chemical electric field effects of dipole equilibria can be rigorously treated in terms of the dielectrochemical potential of the dipolar charge configuration B_j with the dipole moment

$$\underset{\sim}{m}_j = |z_j|\ e_o \underset{\sim}{r}_j \tag{4}$$

where e_o is the (positive) elementary charge, z_j is the charge number of one of the charges of the dipole and $\underset{\sim}{r}_j$ is the charge distance vector. In a macromolecule, oppositely charged groups may be grouped together into ion-pairs m_j. The total dipole moment is then the vector sum over all the individual contributions of the various ion pairs.

For plate condensor geometry the measurable polarization M is the parallel component of the total macroscopic dipole component $\underset{\sim}{M} = \underset{\sim}{P} \cdot V$ where $\underset{\sim}{P}$ is the familiar polarization per unit volume and V is the volume. It is obvious that

$$M = (\underset{\sim}{M})_{\parallel} = N_A \sum_j n_j \langle m_j \cos\vartheta_j \rangle$$

$$= N_A \sum_j n_j m_j = \sum_j n_j M_j \tag{5}$$

where N_A is the Avogadro-Loschmidt constant, n_j is the mole quantity of the dipole type j, M_j is the average partial molar dipole moment given by $M_j = (\partial M/\partial n_j)_{n=n_j} = N_A \langle m_j \cos\vartheta_j \rangle$ and ϑ_j is the angle between the dipole vector $\underset{\sim}{m}_j$ and the electric field.[18]

The molar reaction moment in Eq.(2) for the chemical change $\Sigma\nu_j B_j$=O at constant T and P and at a given field strength E is defined by

$$\Delta M = \left(\frac{\partial M}{\partial \xi}\right)_E = \sum_j \nu_j M_j \quad , \tag{6}$$

where ν_j is the stoichiometric coefficient (with sign) of species B_j and $d\xi = dn_j/\nu_j$ is the differential advancement of the reaction.

The derivation of eq.(1) requires the introduction of Guggenheim's characteristic Gibbs function[19] defined as the difference

$$\tilde{G} = G - EM \tag{7}$$

of the ordinary Gibbs free energy at E and the electric work $\int dW_{el} = \int MdE = ME$.

With U and S denoting the inner energy anf the total entropy in the field eq.(7) is rewritten as [18]

$$\tilde{G} + U - TS + PV - EM \tag{8}$$

The general Gibbs equation for chemical changes in electric fields is:

$$dU = TdS - PdV + \sum_j \tilde{\mu}_j dn_j + EdM \tag{9}$$

where $\tilde{\mu}_j$ denotes the chemical potential of species B_j in the field. Eqs. (8) and (9) then yield

$$d\tilde{G} = dG - d(EM) = -SdT + VdP + \Sigma\tilde{\mu}_j dn_j - MdE, \tag{10}$$

that at constant P,T reduces to

$$d\tilde{G} = \Sigma\tilde{\mu}_j \, dn_j - MdE \tag{11}$$

Consistent with ordinary thermodynamics we obtain from eq.(11)

$$\tilde{\mu}_j = (\partial\tilde{G}/\partial n_j)_{P,T,E,n \neq n_j} \tag{12}$$

where all n except for n_j are constant. By cross differentiation eq.(11) yields

$$(\partial\tilde{\mu}_j/\partial E)_{P,T} = -(\partial M/\partial n_j)_{P,T,E,n\neq n_j} = -M_j \quad (13)$$

where eq.(5) was applied to obtain the partial molar dipole moment M_j.

According to Kirkwood and Oppenheim, integration of eq.(13) between E and E=O yields

$$\tilde{\mu}_j = \mu_j - \int M_j \, dE \quad (14)$$

where $\tilde{\mu}_j(0) = \mu_j$ is the ordinary chemical potential at E=O.

Analoguous to Guggenheim's electrochemical potential we may call $\tilde{\mu}_j$ the dielectrochemical potential[11,18] of the dipolar species. Similar to a previous treatment[18] we introduce a standard dielectrochemical potential μ_j^θ and write

$$\tilde{\mu}_j = \tilde{\mu}_j^\theta + RT \ln \tilde{a}_j \quad (15)$$

where $\tilde{a}_j = c_j\tilde{y}_j$ is the activity and $\tilde{y}_j$ the activity coefficient in the field; evidently $\tilde{\mu}_j^\theta = \mu_j^\theta - \int M_j^\theta \, dE$.[18]

For dipolar interaction partners according to $\Sigma\nu_j B_j = 0$ in electric fields, a dielectrochemical affinity[18] is defined by

$$\tilde{A} = -\Sigma\nu_j\tilde{\mu}_j = -\Sigma\nu_j\tilde{\mu}_j^\theta - RT\,\Sigma\nu_j \ln \tilde{a}_j \quad (16)$$

The characteristic Gibbs function for isothermal-isobaric chemical changes is then derived from eqs.(11) and (16),

$$d\tilde{G} = -\tilde{A}d\xi - MdE \leq 0 \quad (17)$$

and, consistently,

$$\tilde{A} = -(\partial\tilde{G}/\partial\xi)_{P,T,E} \geq 0 \quad (18)$$

yielding familiar expressions for reversible (equilibrium) processes ($d\tilde{G}=0$, $\tilde{A}=0$) and for irreversible (nonequilibrium) processes ($d\tilde{G}<0$, $\tilde{A}>0$). Note that the ordinary Gibbs free energy increases in the presence of E.

Using eq.(16) and the definition of the thermodynamic equilibrium constant $K^{\ominus}=\Pi\bar{a}_j^{\nu_j} = K\cdot Y$ with $K=\Pi\bar{c}_j^{\nu_j}$ (concentration ratio) and $\bar{Y}=\Pi\bar{y}_j^{\nu_j}$ together with the nonequilibrium activity ratio $Q^{\ominus}=\Pi a_j^{\nu_j}$ we obtain

$$\tilde{A} = RT\ (\ln K^{\ominus} - \ln Q^{\ominus}), \tag{19}$$

where $RT \ln K^{\ominus} = -\Sigma\nu_j\tilde{\mu}_j^{\theta}$; the bar denotes equilibrium. Following now the treatment of SCHWARZ[21] it can be shown that, using the equilibrium criteria $d\tilde{G}=0$ and $\tilde{A}=0$, eqs.(17) and (19) yield, respectively,

$$(\partial\tilde{A}/\partial E)_{\xi} = (\partial M/\partial\xi)_{E} = \Delta M \tag{20}$$

and, provided that the dependence of $\tilde{Y}$ on ξ is negligible, eq.(1) obtains.

Generally, the van't Hoff relationship for electric field effects on chemical processes is

$$(\partial \ln K^{\ominus}/\partial E)_{P,T} = \Delta M^{\ominus}/RT , \tag{21}$$

where

$$\Delta M^{\ominus} = \Delta M - RT(\partial \ln\tilde{Y}/\partial E)_{\xi,P,T} , \tag{22}$$

derived from[18] $\int(M_j-M_j^{\ominus})dE = -RT \ln \tilde{y}_j$ with $\tilde{y}_j=\tilde{a}_j/a_j$; a_j being associated with μ_j(at E=0) according to $\mu_j = \mu_j^{\ominus} + RT \ln a_j$.

For computer analysis of actual chemical transition curves $\Theta(E)$, see eq.(2), it is useful to apply the integrated form of eq.(21):

$$K^{\ominus}(E) = K^{\ominus}(o) \cdot \exp\left[\int\Delta M^{\ominus} dE/RT\right] \tag{23}$$

CONCLUSION

The thermodynamic formalism outlined here has been used to analyze the electric field effects observed in the proton transport protein bacteriorhodopsin of purple membrane fragments. It has been found that this protein is particularly sensitive to the electric field of the membrane. The apparent coupling of the light-induced proton-pumping cycle to the membrane field[3,6] represents an interesting bioelectric phenomenon which may be partially investigated in vitro. In summary, the study of bioelectric structures and processes on the molecular level of chemistry and physics may thus be classed as one of the efforts 'to "translate" (bioelectric) life phenomena into meaningful physical concepts'.

ACKNOWLEDGEMENTS

We thank Prof. Michel Mandel, Leiden, for instructive comments on chemical thermodynamics in electric fields. The financial assistance of the Stiftung Volkswagenwerk, grant I/34706, and of the Deutsche Forschungsgemeinschaft, grant Ne 227, is gratefully acknowledged.

REFERENCES

1. A. Katchalsky, Chemical dynamics of macromolecules and its cybernetic significance, in: "Biology and the physical sciences", S. Devons, ed., Columbia University Press, New York (1969), pp. 267-298.
2. E. Neumann, Molecular hysteresis and its cybernetic significance, Angew. Chem. internat. Edit. 12: 356-369 (1973).
3. H. Michel and D. Oesterhelt, Light-induced changes of the pH-gradient and the membrane potential in halobacterium halobium, FEBS Lett. 65: 175-178 (1976).
4. D. Schallreuter, Ionenassoziation in einfachen Elektrolyten und biologischen Polyelektrolyten, Thesis, Kostanz and Martinsried (1982).
5. K. Tsuji and E. Neumann, Structural changes in bacteriorhodopsin induced by electric impulses, Int. J. Biol. Macromolecules, 3:231-242 (1981).
6. K. Tsuji and E. Neumann, Electric field-induced pK-changes in membrane-bound bacteriorhodopsin, FEBS Lett. 128:265-268 (1981).
7. K. Tusji and E. Neumann, Conformational flexibility of membrane proteins in electric fields. I. UV absorbance and light scattering of bacteriorhodopsin in purple membranes, Biophys. Chem. 17 (1983).
8. E. Neumann and A. Katchalsky, Long-lived conformation changes induced by electric impulses in biopolymers, Proc. Natl. Acad. Sci. USA 69:993-997 (1972).
9. E. Neumann, Electric field effects in biopolymer structures and electrical chemical memory recording, in: "Ions in macromolecular and biological systems", D.H. Everett and B. Vincent, eds., Scientechnica, Bristol (1978), pp. 170-191.
10. T.L. Rosenberry and E. Neumann, The interaction of ligands with acetylcholinesterase. Use of temperature-jump relaxation kinetics in the binding of specific fluorescent ligands, Biochemistry 16:3786-3792 (1977).
11. E. Neumann, Dynamics of molecular recognition in enzyme catalysis, in: "Structural and functional aspects of enzyme catalysis", H. Eggerer and R. Huber, eds., 32. Mosbach Coll., Springer, Berlin (1981), pp. 45-58.

12. E. Neumann, D. Nachmansohn and A. Katchalsky, An attempt at an integral interpretation of nerve excitability, Proc. Natl. Acad. Sci. USA 70: 727-731 (1973).
13. H.T. Witt, E. Schlodder, and P. Graeber, Membrane-bound ATP synthesis generated by an external electric field, FEBS Lett. 69: 272-276 (1976).
14. Ch. Vinkler and R. Korenstein, Characterization of external electric field-driven ATP synthesis in chloroplasts, Proc. Natl. Acad. Sci. USA 79: 3183-3187 (1982).
15. J. Teissie, B.E. Knox, T.Y. Tsong, and J. Wehrle, Synthesis of ATP in respiration-inhibited submitochondrial particles induced by μs electric pulses, Proc. Natl. Acad. Sci. USA 78: 7473-7477 (1981).
16. E. Schlodder and H.T. Witt, Relation between the initial kinetics of ATP synthesis chloroplast ATPase studied by external field pulses, Biochim. Biophys. Acta 635: 571-584 (1981).
17. K. Bergmann, M. Eigen, and L.C.M. DeMaeyer, Dielettrische Absorption als Folge chemischer Reaktionen, Ber. Bunsenges. Phys. Chem. 67:819-826 (1963).
18. E. Neumann, Fundamentals of electric-chemical field effects in biological macromolecules, in: "Electric field effects in macromolecules and membranes", U. Zimmermann and R. Benz, eds., Springer, Berlin (1983).
19. E.A. Guggenheim, "Thermodynamics" 5th Ed., North Holland, Amsterdam (1967).
20. G. Schwarz, On dielectric relaxation due to chemical rate processes, J. Phys. Chem. 71: 4021-4030 (1967).
21. J.G. Kirkwood and I. Oppenheim, "Chemical Thermodynamics", McGrawhill, New York (1961), Chp. 14.

THE ROLE OF WATER OF HYDRATION IN ACYLATION OF BUTYRYLCHOLINESTERASE

Miloš R. Pavlič and Matjaž Zorko

Institute of Biochemistry
Medical Faculty
Ljubljana, Yugoslavia

INTRODUCTION

It was found recently in our laboratory (in preparation) that the acylation of butyrylcholinesterase (acylcholine acyl-hydrolase, EC 3.1.1.8) by methanesulfonylfluoride - a reaction analogous to the acylation of butyrylcholinesterase by its natural substrate - - is accelerated by tetramethylammonium and tetraethylammonium and that the accelerators improve the binding between the enzyme and the acylating agent. The same mechanism was found in our laboratory to be operative in the acceleration of acylation of acetylcholinesterase (acetylcholine hydrolase, EC 3.1.1.7)[1,2]; in this case, the acceleration was shown to result from hydration changes in the active site of acetylcholinesterase under the influence of accelerator[3-5].

It seemed interesting to elucidate a possible role of water in the acceleration of acylation of butyrylcholinesterase. In order to do this the influence of the monovalent cations of the Hofmeister series on the acylation of bytyrylcholinesterase by methanesulfonylfluoride was investigated.

METHODS AND MATERIALS

The effect of the cations on acylation was investigated by studying the methanesulfonylation of butyrylcholinesterase in the presence of each individual cation. The second-order rate constant for the methanesulfonylation in the presence of a cation, k_a, was determined in the following way[2]. Butyrylcholinesterase was incubated with methanesulfonylfluoride in a given concentration for 1-20 minutes and the remaining enzyme activity measured by the method of Ellman et al.[6]. Then the apparent first-order rate constant was calculated from the time dependence of methanesulfonylation; five first-order rate constants were determined from five concentrations of methanesulfonylfluoride. Subsequently the second-order rate constant was determined from the dependence of the first-order rate constants on the concentrations of methanesulfonylfluoride. Finally the second-order rate constant for the methanesulfonylation at saturating concentrations of the cation, k_A, was extrapolated.

Experiments were carried out at 25°C, pH 8.4. The enzyme used was purchased from Worthington, CHE 36N696F, 11 units/mg.

RESULTS AND DISCUSSION

The main results of our experiments, together with some other relevant data, are summarized in Fig. 1 and Table 1.

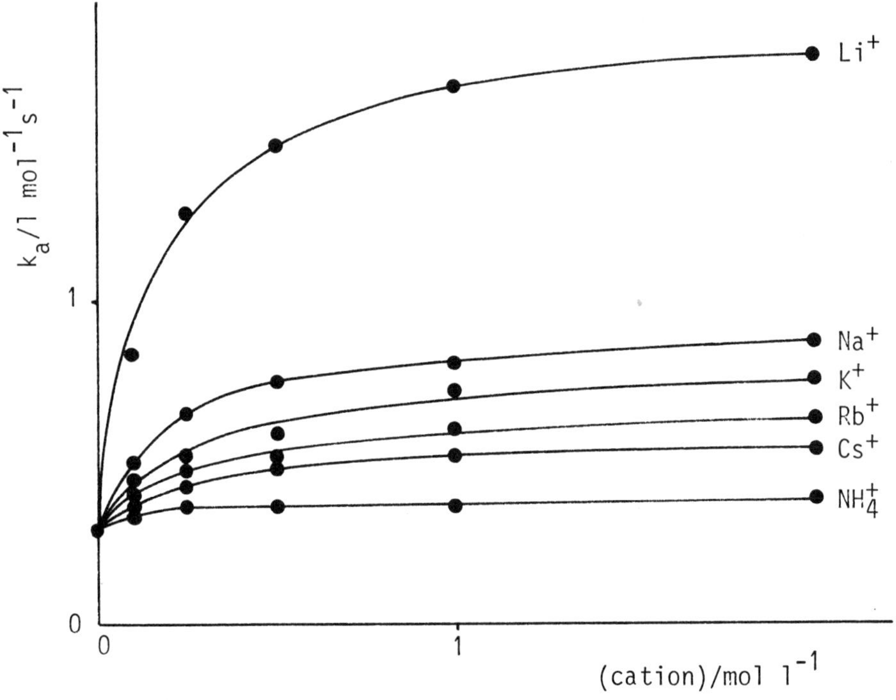

Fig. 1. Effect of the cations of the Hofmeister series on the acylation of butyrylcholinesterase by methanesulfonylfluoride. k_a is the second-order rate constant for the acylation in the presence of cation.

Table 1. Acceleration of the acylation of butyrylcholinesterase by the cations of the Hofmeister series and some other relevant data.

Cation	Accelera-tion k_a/k	r(cr) nm	r(s) nm	B	ΔS(ex) $JK^{-1}mol^{-1}$
NH_4^+	1.28	0.148	0.125	-0.007	-
Cs^+	1.90	0.169	0.119	-0.045	81.2
Rb^+	2.13	0.148	0.118	-0.030	74.9
K^+	2.57	0.133	0.125	-0.007	58.2
Na^+	3.17	0.095	0.183	0.086	42.7
Li^+	6.3	0.060	0.237	0.150	-14.7

k_a is the second-order rate constant for the methanesulfonylation of butyrylcholinesterase in the presence of cation; k, the second--order rate constant for the methanesulfonylation without cation; r(cr), crystal ionic radius (from Monk[7]); r(s), solvated ionic radius (from Monk[7]); B, viscosity coefficient (from Gordon[8]); ΔS(ex), extra entropy of hydration (from Gordon[8]).

It is evident from Fig. 1. that the monovalent cations of the Hofmeister series accelerate the acylation of butyrylcholinesterase by methanesulfonylfluoride and that they do it exactly according to their position in the series.

It can be seen from Table I that the acceleration is strikingly well correlated with the ionic radia ($r = 0.91$; $p < 0.001$ and $r = 0.90$; $p < 0.001$, respectively), the viscosity coefficients ($r = 0.90$; $p < 0.001$) and the extra entropy of hydration ($r = 0.98$; $p < 0.001$). Since these properties of the cations are undoubtly related to their hydration, it must be the same phenomenon, i.e. hydration, which is involved both in the above mentioned characteristics of the cations and in the process of acceleration. Consequently, the acceleration of the acylation of butyrylcholinesterase by cationic accelerators must involve hydration changes at the active site of butyrylcholinesterase. It seems safe to conclude (cf. ref. 5) that the accelerator loosens water of hydration at the nucleophilic serine - OH group in the esteratic site of the enzyme which results in a faster formation of the complex between the esteratic site and the acylating agent, thus improving the affinity between the two reactants.

REFERENCES

1. Pavlič, M.R. and Wilson, I.B. (1978) Biochim. Biophys. Acta 523, 101-108.
2. Pavlič, M.R. and Zorko, M. (1978) Biochim. Biophys. Acta 524, 340-348.
3. Pavlič, M.R., in "Enzyme Regulation and Mechamism of Action" (P. Mildner and B. Ries, eds.) FEBS Vol. 60, Pergamon Press, Oxford and New York, 1980, pp. 113-122.
4. Pavlič, M.R. in "Synaptic Constituents in Health and Disease" (M. Brzin, D. Sket and H. Bachelard, eds.) Mladinska knjiga and Pergamon Press, Ljubljana and Oxford, 1980, pp. 551-557.
5. Pavlič, M.R. (1980) Period. biol. 82, 449-453.
6. Ellman, G.L., Courtney, K.D., Andres, V.Jr. and Reatherstone, R.M. (1961) Biochem. Pharmacol. 7, 88-95.
7. Monk, C.B. "Electrolytic Dissociation" Acad. Press, London and New York, 1961, p. 271.
8. Gordon, J.E. "The Organic Chemistry of Electrolyte solutions" John Willey and Sons, New York, London, Sidney and Toronto, 1975, p. 186, p. 197.

CONFORMATIONAL CHANGES OF RHODOPSIN IN H_2O, D_2O AND T_2O

Al., Dinu, I. Aricescu, E. Chirieri-Kovács,
and V. Vasilescu
Department of Biophysics
Faculty of Medicine
Bucharest, Romania

The receptor layer of the retina contains the cells - rods and cones - where the light triggers the nerve impulse which carries the visual information all along the visual pathways. The outer segments of receptor cells are structures specialized to convert light signals into electrical ones. These segments consist of regular stac kings of thousands of membranes loaded with the visual pigment - rhodopsin[8]. (Fig. 1).

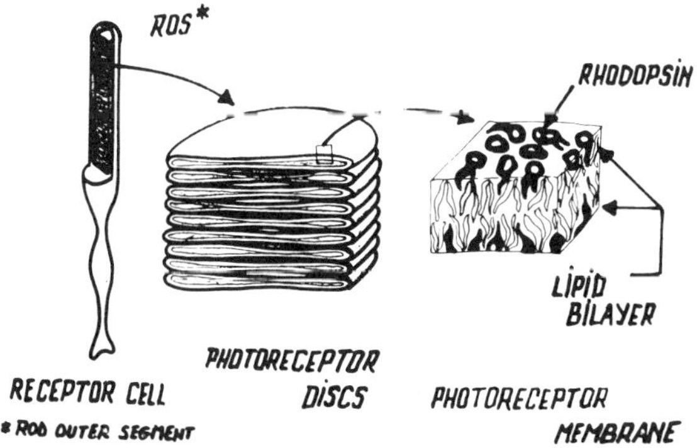

Fig. 1. Schematic representation of a rod, a photoreceptor disc a segment of the rhodopsin-containing membrane

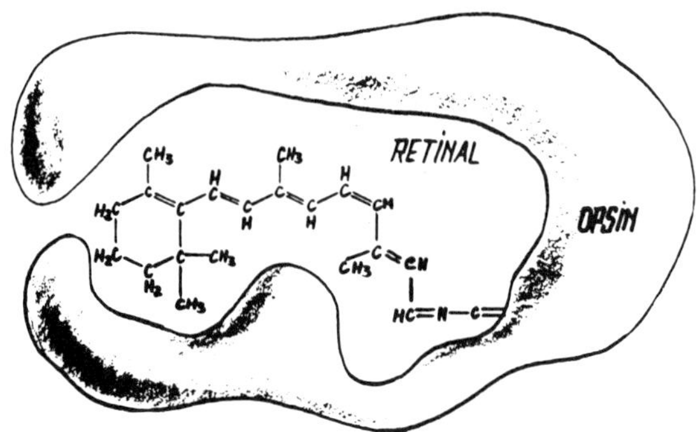

Fig. 2. Rhodopsin

Under illumination the chromophoric part of rhodopsin-the retinal (Fig. 2) is known to undergo isomerization which entails a conformational change of the whole rhodopsin molecule[10].

Finally, the chromophore breaks away from the opsin part of the molecule.

These changes trigger off hyperpolarization of the receptor cell, i.e. visual excitation[10].

The way in which the structural changes of rhodopsin induce hyperpolarization of the cell membrane is still unknown and constitutes at present the capital problem of visual reception.

We endeavoured to find out something about the primary mechanisms of visual excitation and tried to see which is the role of water and protons in these processes.

First we illuminated the isolated frog retina and recorded the ERG-transretinal variation of potential - as response to illumination - in H_2O and D_2O (Fig. 3). The latency of the ON response was found not to change at all in D_2O, while that of the OFF response rose remarkably[2].

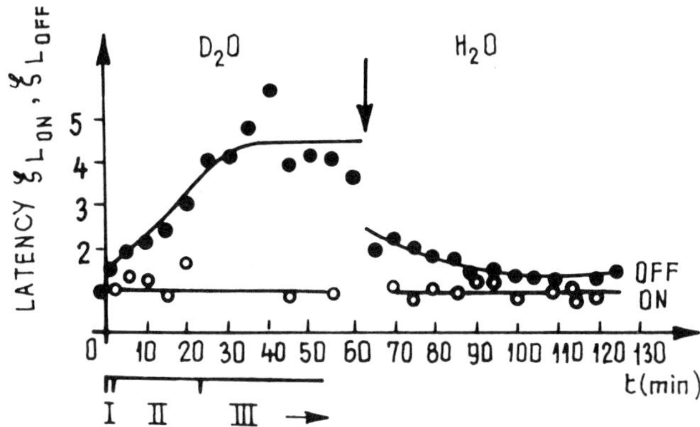

Fig. 3. The latency of the OFF response (o) increases in D_2O, while the latency of the ON response (o) does not change

This would mean that protons are strongly involved in the processes occurring during the latency time of the OFF response - a play in which rhodopsin too is an important character.

So we decided to extract rhodopsin from rod discs and to study its structural changes, under illumination, in H_2O, D_2O and T_2O solutions. To this effect, the specific conductivity of illuminated and unilluminated solutions of different concentrations was measured.

Rhodopsin was extracted from photoreceptor membranes by digitonin[16] which is a nonionic detergent ($C_{56}H_{92}O_{29}$). In such solutions each rhodopsin molecule is included in a complex micella formed by aggregation of 3 digitonin units[4]. Each digitonin unit is in fact itself a micella composed of 75 digitonin molecules.

Thus the rhodopsin solutions studied by us are suspensions of such complex micellae in water.

In the initial rhodopsin extract digitonin concentration was $8.04 \cdot 10^{-3}$ M.

That is perhaps the reason why our conductivity curves show "pockets" and why in the "pocket" region there is a much greater dispersion of the experimental points.

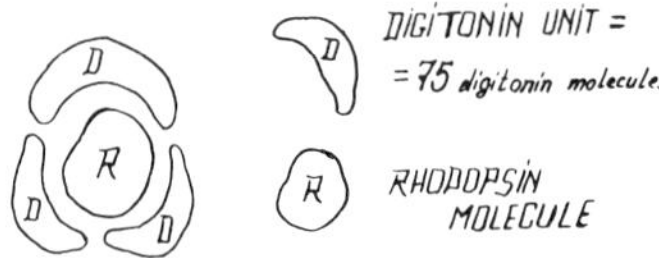

Fig. 4. Rhodopsin-digitonin complex

In order to obtain different solutions, the dilution was successively doubled with tri-distilled and deionized water. By plotting the variation of the specific conductivity as a function of concentration, we got the specific conductivity curves in H_2O, D_2O and T_2O (Figs. 5, 6, 7). The negative slopes of the conductivity curves, corresponding to the first dilutions, suggest that, following these dilutions, the micellar systems of complex solutions begin to disaggregate into digitonin micellae and rhodopsin molecules. We determined from the curves the critical micellization concentration which is $7 \cdot 10^{-4}$ M.

After the first dilution, our solutions contain a mixture of nonionic digitonin micellae with a very low conductivity (according to our measurements digitonin conductivity is very close to that of distilled water) and free rhodopsin molecules which are ionic. Consequently, the conductivity of the solution may be assumed to be due to rhodopsin molecules.

It appears from our results that the conductivity properties of illuminated rhodopsin are different from those found when rhodopsin is unilluminated. At high concentrations, before digitonin-rhodopsin complex disaggregates, the conductivity of the unilluminated solutions is greater than that of the illuminated ones. This is probably due to the stretched configuration of illuminated rhodopsin, which results in an increase of the micellar size.

At lower concentrations (when probably rhodopsin molecules have already been delivered from rhodopsin-digitonin complex) a "pocket" appears in the plot of unilluminated solutions, which is musch smaller or even absent in those which are illuminated.

As Ruth Hubbard observed in her early studies on rhodopsin solutions[4] when "digitonin leaves the micellae, the complexes become unstable and condense with one another to form aggregates of varying size, ... they never seem to reach a stable configuration".

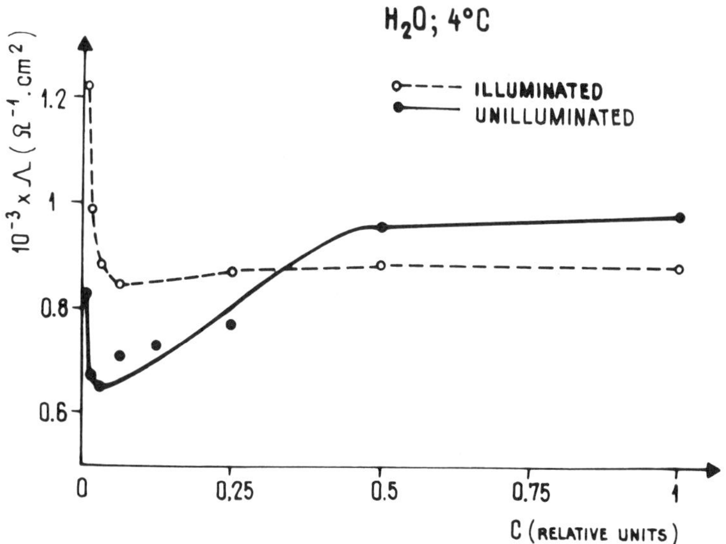

Fig. 5. Conductivity curves of rhodopsin solutions in H_2O. The equivalent conductivity $\Lambda = \frac{1000\ \gamma}{c}$ was plotted versus concentration, c. The initial concentration of the extract was conventionally taken as unit; thus the lower concentrations are fractions of that unit

Instead of classically rising with decreasing of concentrations, the conductivity of "dark"* rhodopsin drops because of the formation of aggregates and fluctuates on account of their instability.

Illuminated solutions have no pockets probably because the loosened configuration of opsin lowers its ability to form aggregates. And besides that, opsin molecules being more polar than unilluminated rhodopsin (as shown by Radding and Wald in 1955)[6] they are more time "light"** rhodopsin contains a double number of carriers (from each rhodopsin molecule a chromophore is bronken off) and thus it is more conductive than "dark"-rhodopsin.

When the concentration continues to decrease rhodopsin aggregates are dissociating and conductivity rises again; but every time at low concentrations, conductivity of "dark-rhodopsin" is lower than that of "light-rhodopsin"; this difference may also be due to the greater polarity of opsin and the double number of carriers.

In D_2O the tendency to form "pockets" is much smaller than in H_2O the two curves for "dark" and "light"-rhodopsin thus getting much closer (Fig. 6).

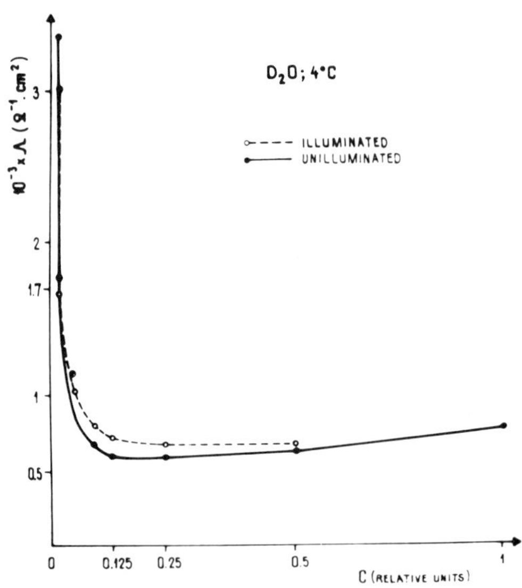

Fig. 6. Conductivity curves of rhodopsin solutions in D_2O

* unilluminated

** illuminated

Radding and Wald[6] assert that the trans-conformation of rhodopsin under illumination is similar to a denaturation process.

Many experiments showed that in D_2O proteins have an enhanced stability against denaturation. Thus in D_2O opsin may be more stable because some of its H-bonds are replaced by D-bonds which are stronger and, consequantly, it does not undergo the conformational changes usual under illumination. This would be an explanation for the closeness of the "light" and "dark" - rhodopsin curves. We however wonder why the conductivity curves in D_2O show no "pockets", as the tendency of rhodopsin to form aggregates should be stronger in D_2O.

Similar results were obtained also in the case of tritiated water (Fig. 7); althrough we expected a more pronounced isotopic effect in T_2O, this could not be revealed, because in T_2O solutions available, the concentration of tritium ions was relatively small namely 25%.

As concerns the practical value of our findings, we think that they could be useful for elucidating to a certain extent a problem much disputed today, namely: evidence available indicates that in vivo, in the ROS* membranes, rhodopsin is monomeric on a time average.

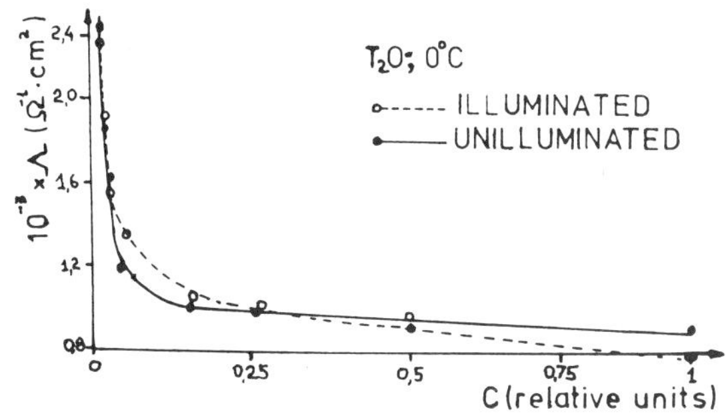

Fig. 7. Conductivity curves of rhodopsin solutions in T_2O

* Rod Outer Segment

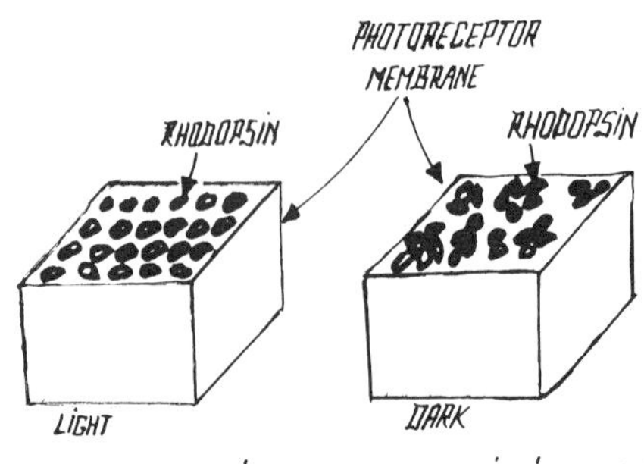

Fig. 8. Aggregation of rhodopsin in the dark

But in 1977[5], Montal et al. suggested that rhodopsin monomers interact upon light exposure to transiently form a multimeric channel. Since then, at least three laboratories have examined ROS membranes for obtaining evidence on light-induced association of rhodopsin; the results however, were not very conclusive.

In 1980, Shaw and coworkers[7] reported that the molecular weight of the cross- linked species of bovine rhodopsin are enhanced upon illumination of membranes; they showed this electrophoretically.

Our results suggest that rhodopsin is able to form multimers in the dark, while in the light this ability is much impaired, as shown in Fig. 8.

REFERENCES

1. Brett M. and Findley J. B. C. (1979): Biochem. J., 177, 215
2. Chirieri Eugenia, Aricescu Ioana, Ganea Constanţa, Vasilescu V. (1977): Naturwissenschaften, 64, 149
3. Downer N. and Cone R. A. (1978): Biophys. J., 21, 135a
4. Hubbard Ruth (1954): J. Gen. Physiol. 37, 381

5. Montal M., Darzon A. and Trissl H. W. (1977): Nature, 267, 221
6. Radding C. M. and Wald G. (1955-56): J. Gen. Physiol., 39, 909
7. Shaw A., Crain R., Marinetti G. V., O'Brien D. and Tyminski P. N. (1980): Biochim. Biophys. Acta, 603, 313
8. Sjostrand F. S. and Barajas L. J. (1968): J. Ultrastruct. Res., 25, 121
9. Wald G. (1938): J. Gen. Physiol., 21, 795
10. Wald G. (1968): Nature, 219, 800

HYDRATION AND THERMOTRANSITION OF COLLAGEN FIBRE

Zhang Jizhen, Zhang Zhenglian, Zhang Wending,
Fu Yazhen, Ye Guohue, and Nin Zeling

Institute of Biophysics
Academia Sinica
Beijing, China

ABSTRACT

The hydration and thermotransition behaviours of bovine achilles tendon collagen (BAT) in the water content range of 0.023 to 13 g/g have been studied by differential scanning calorimetry (DSC). The results are discussed in terms of seven steps involving different states of water and their effects on the fibrous structure and its thermotransition. For collagen without water, neither melting nor thermotransition can occur. For BAT with two water molecules per every three residues its crystallinity is complete. Every three residues contains 1.78 intramolecular H bonds. ΔH_H is equal to 1.43 Kcal/mole, and T_{tr} is not more than 376 K. The rest of the water has no effect on the melting of the molecule. We propose that the thermotransition of collagen swollen in water involves melting of the molecule and the gradual disengagement of molecular and fibrous structure; the enthalpy values in the three processes being 0.85, 0.52, and 0.36 Kcal/mole of residue. The transition temperature is not more than 332 K. We argue that water is an essential factor in the formation of intramolecular H bond and crystallinity of the molecule. The term " melting phase transition" usually used to describe collagen to gelatin transition or thermal shrinkage of collagen fibre, seems to be inaccurate and misleading. A proposal for the mechanism of the thermotransition of collagen fibre is made. The stability of collagen may depend on factors inherent at all levels of native conformational structure.

INTRODUCTION

Although the collagen to gelatin transition has been extensively studied, many problems, such as what forms stabilize the structure of collagen and what role water plays in maintaining its structure and stability, are still controversial (1-7). The collagen to gelatin transition generally is recognized as a melting phase transition (2-10), and the enthalpy related to this transition measured by various methods is considered to be its melting phase transition enthalpy. The values reported in the literature, however, are considerably diverse. An analysis of these data has led to different explanations of its stability.

Privalov and Tikopulo obtained an enthalpy value of 1530 cal/mole (of residue) for tropocollagen in solution (11, 12), which is too high to be accounted for by the single- or even double-bond model of collagen structure (13-17). These investigators suggested that a regular water structure near the collagen molecule or a fourth chain of water is responsible for stability of structure and a large enthalpy value (7, 18-21). Finch et al (22-24) acquired values of 510-1040 cal/mole for bovine achilles tendon collagen which is much less than those of tropocollagen. They proposed the hydrophobic interaction as a major factor in stabilizing collagen fibre (6, 21, 25). Mc Clain and Wiley (26) concluded that intermolecular crosslinks in fibre are much less liable for stabilizing the fibrous structure, based on their value of 1055 cal/mole for both soluble and insoluble collagen. A similar dissention also appears in theoretical considerations of the stability of collagen (5, 19, 27, 28).

The stability of collagen is closely associated with its imino acid residue content (1-3, 11, 12, 18); however, this kind of residue is known not to be able to form an intramolecular H bond, and thus the stability must be related to other factors than the hydrogen bond. On the other hand, it is well known that collagen fibre has a stable structure composed of molecules connected by intricate crosslinks. Its shrinkage temperature is higher by as much as 20-30° than that of the molecule in solution (3, 4, 29, 30). It performs all physiological functions within the body and does not disperse at all. All these points demonstrate that crosslinks may play a decisive role in stabilizing the fibrous structure, and may also show a great effect on the melting property of crystalline structure upon thermotransition.

The aim of this work is to study the hydration characteristics of collagen fibre and the effect of water on its stability and thermotransition over the whole hydration range using DSC method.

METHODS

Bovine achilles tendon collagen (CalBiochem, Lot 901627) was used in most studies. For comparison a threadline fibre (Sigma No. C-4387, Lot 89C-0113) and BAT prepared according to Einbiner's method (31) also were measured. A sample with a water content of 0.023 g(water)/g (dry BAT) was obtained by quickly sealing a piece of fibre of 1.6 $\pm$ 0.2 mg (without any crack) into a Perkin-Elmer (P/E) volatile aluminum pan after keeping it in a desiccator with P_2O_5 at 50° C over two months. Samples with water contents below 0.60 g/g were prepared by equilibrating the pieces in desiccators with various saturated salt solutions. The samples with water contents from 0.60 to 13 g/g were obtained by directly adding redistilled water. The water content and dry mass of samples were determined by puncturing the lid of capsule and keeping denatured samples at 80° C over 8 hours followed by drying them in desiccators with P_2O_5 at 120° C over 48 hours. Samples with ethylene glycol were obtained by directly adding glycol to the fibre piece, or equilibrating them with different water content as Flory did. Their water content and dry mass were calculated from an average water content of samples without considering the glycol presence. All the samples were kept at room temperature over 48 hours for further equilibrating after sealing. All weighings were made on a P/E Model AD-2Z autobalance. Ethylene glycol was an analytical grade reagent with water content less than one percent.

A P/E Model DSC-2 differential scanning calorimeter with a sub-ambient kit was used for all thermal measurements. The block of sample holder was kept at 220 K by means of Intercooler-II. It was operated at a sensitivity range of 0.5 or 1.0 mcal/sec full scale. The baseline was compensated by Auto Scanning Zero, the greatest shift of which was not more than $\pm$ 0.01 mcal/sec in range of 520 K. Both temperature and energy were calibrated at a rate of 5°/min, using P/E indium standard and redistilled water. All data were registered and calculated by Textronix-31 and the output of transition temperature was the point of maximum heat effect. Chart speed of recorder was 20 mm/min.

Normal measurements were performed at a rate of 5°/min from 275 K to a temperature which was 20° higher than the end of transition. In case of low temperature measurement the scanning started from 225 K. The cooling rate was 320°/min. While measuring ice fusion with a decreased melting temperature a cooling rate of 0.31°/min was used. A sealed empty pan served as the reference.

RESULTS

A series of thermograms of BAT recorded with increasing water content is shown in Fig. 1. At water contents higher than 0.59 g/g

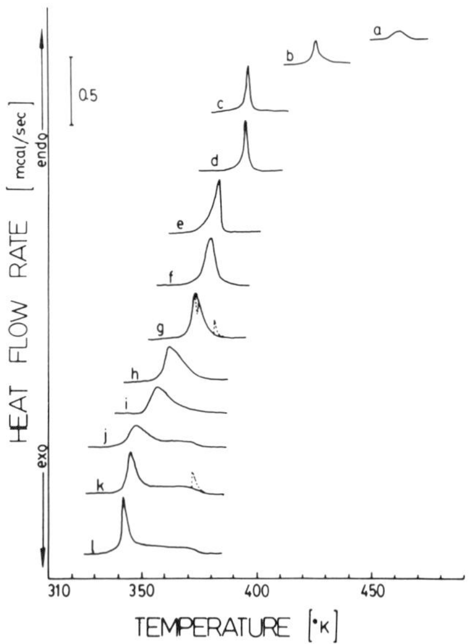

Fig. 1. DSC thermograms of hydrated BAT in 320-490 K. Sample weights are 1.62 $\pm$ 0.03 mg. Water contents (g/g) are: a, 0.026; b, 0.065; c, 0.13; d, 0.16; e, 0.25; f, 0.33; g, 0.43; h, 0.59; i, 0.72; j, 1.07; k, 1.18; 1, $\geq$ 2.0.

The thermograms begin gradually dividing into a sharp peak on the lower temperature side and a shoulder-tail peak in the higher. Above a water content of 2.0 g/g no change in the shape of thermograms was observed. All thermodynamic parameters obtained from the thermograms show a strong, but non-monotonic dependence on the degree of hydration, as shown in Fig. 2, A-D and Table 1. Above 2.0 g/g all parameters remain constant within experimental error. For the dry sample the lowest water content which could be obtained was 0.03 g/g. The half width $\Delta T_{\frac{1}{2}}$ of its thermogram is 6.3^{o} C, transition heat Q_{tr} is 0.8 cal/g (dry BAT) and the transition temperature T_{tr} is 493 $\pm$ 7 K. Above 510 K, irregular peaks were observed due to protein decomposition. Extrapolation of the curve to zero water content is difficult because of the rapidly rising curve at low hydration and the experimental errors. If such an extrapolation is attempted, the intercept is at 590 $\pm$ 20 K which is

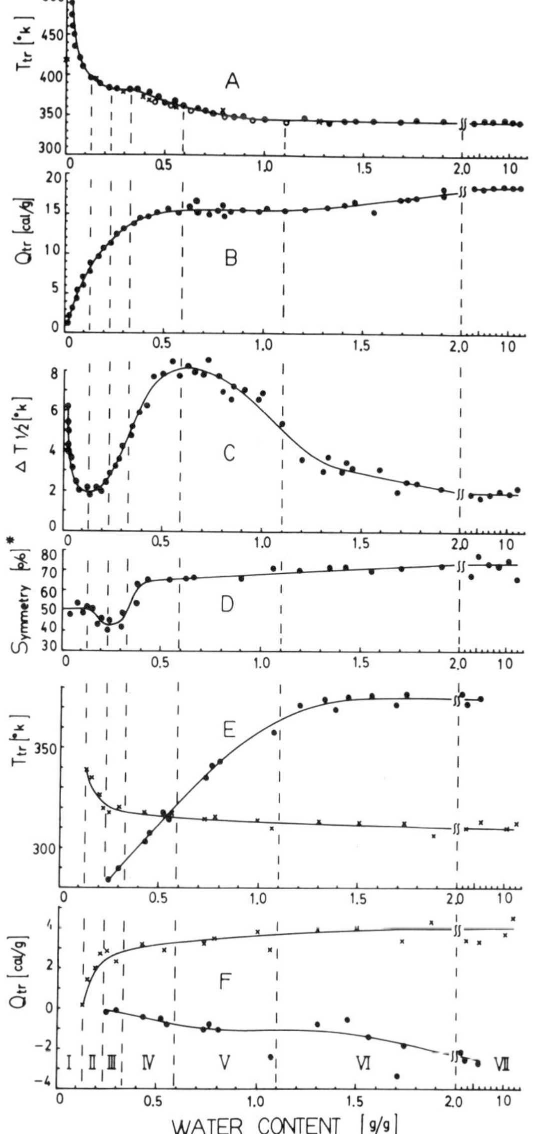

Figure 2. (Continued on following page)

Fig. 2. (Cont.) Dependence of thermodynamic parameters of denatured fibre of BAT on water content. T_{tr}, temperature of transition; Q_{tr}, heat value transition; $\Delta T_{\frac{1}{2}}$, half width of thermogram; *Symmetry: percentage of the area covered by half of a thermogram in the higher temperature side to its total area while a line drawing perpendicularly from the maximum heat absorption of the thermogram to the baseline divides it into two parts. In Curve A, symbols (x) and (o): T_{tr} caldulated from Flory's (9) and DiLisi (28) results respectively. In Curve E and F, symbol (●): T_{tr} and Q_{tr} of exothermal peak remeasured immediately after thermo-transition; Symbol (x): T_{tr} and Q_{tr} of endothermal peak measured after keeping denatured sample at room temperature for 10 days.

Table 1. Thermodynamic Properties of Hydrated BAT

	I_1	I_2	II	III	IV	V	VI	VII
water content (g/g)	0.07	0.15	0.23	0.33	0.59	1.10	2.0	13.0
Number of water molecules/triplet	1	2	3.5	5	9	17	30	130
T_{tr} (K)	424.1	393.5	383.2	380.7	359.4	345.3	342.4	342.4
Q_{tr} (cal/g)	5.21	9.29	11.6	13.4	14.9	15.1	18.8	19
$\Delta T_{\frac{1}{2}}$ ($^{\circ}$ K)	2.6	1.9	2.4	4.5	8.2	4.6	2.1	2.1
Symmetry	48	51	41	49	63	67	73	73

much higher than the 418 K and 473 K suggested by Flory (9) and Monaselidze et al (32, 33). The T_{tr} curve shows a plateau in the water content range of 0.23 to 0.33 g/g, where $T_{tr} = 382 \pm 1$ K. On the Q_{tr} curve a plateau is also seen in the water content range of 0.59 - 1.1 g/g, where $Q_{tr} = 15 \pm 1$ cal/g. $\Delta T_{\frac{1}{2}}$ reveals a complicated dependence on the water content. Figure 2-D shows the dependence on water content of the symmetry of the thermograms. At water contents of below 0.15 and of 0.33 g/g the thermograms are symmetrical, above 0.33 g/g the area on the higher temperature side becomes larger, and above 2.0 g/g the ratio of this part to the total area reaches to 73 percent.

An endothermal peak is observed between 310 - 350 K after keeping denatured samples at room temperature, as shown by all d curves of Fig. 3. Its area increases with time and becomes constant after

7-8 days. For water swollen samples this endothermal peak has T_{tr} = 311 K, and Q_{tr} = 4.2 cal/g when remeasuring denatured samples, immediately after transition. Above the water content of 0.23 g/g a melting peak of ice at 273 K and an exothermal peak in the range 280-380 K appear (Curves II-c and III-c) and disappear when samples are kept at room temperature over two hours. Above the water content of 2.0 g/g this exothermal peak shows T_{tr} = 375 K and Q_{tr} = 2-3 cal/g. For samples with the water content higher than 4.0 g/g the exothermal peak sometimes was not observed. Both endo- and exothermal effects are completely reversible. Their dependence on hydration degree is shown in Fig. 2-E and F. Above a water content of about 0.38 g/g, a melting peak of ice with a decreased melting temperature (34) appears in the range 230-273 K before and after denaturation as shown in Fig. 3, Curve III-a and d. Curve III-b indi-

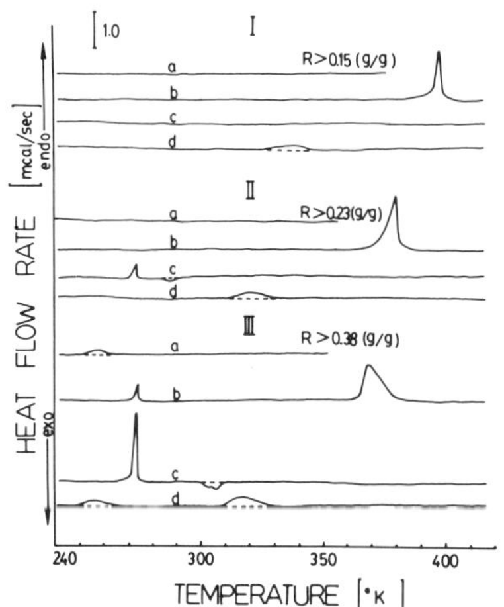

Fig. 3. Thermograms of BAT with various amounts of water measured before and after transition. Sample weights are 1.62 ± 0.03 mg. a: up to a temperature 5° C lower than the onset of the peak before transition; b: following Curve a; c: immediately after completing the transition; d: after keeping denatured sample at room temperature for ten days.

cates that this part of water can be released before transition by heating, but no heat effect can be observed above the ice melting point in the first run (Curve III-a). All melting enthalpy value of ice mentioned above increase with the degree of hydration (35).

A typical thermogram of BAT with the water content of 0.13-0.15 g/g is shown in Fig. 4-1, which is symmetrical, $\Delta T_{\frac{1}{2}} = 1.9^{o}$ C, $T_{tr} = 393.5$ K, $Q_{tr} = 9.29$ cal/g. Assuming that an amino acid residue is taken to have an average molecular weight of 91 (3), this water content corresponds to two water molecules per triplet, giving an enthalpy value of $\Delta H_{tr} = 850$ cal/mole (of residue).

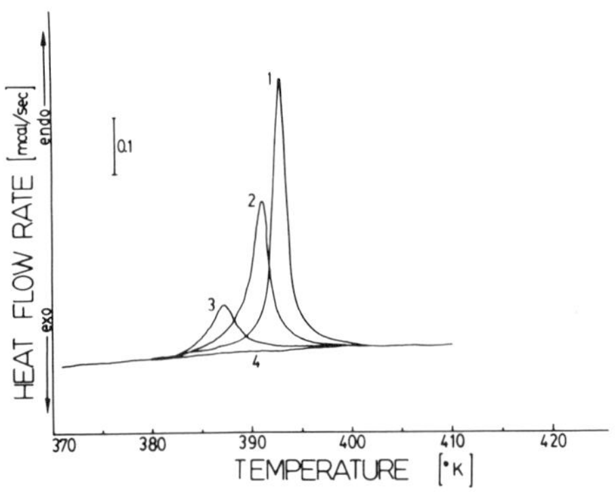

Fig. 4. Thermograms of BAT with water content of 0.148 ± 0.003 g/g measured in different scan procedures. Sample weights are 1.62 ± 0.03 mg. 1: Normal, 2-4: after eigher keeping sample at 373, 375, and 376 K for 15 hrs. or scanning up to 391, 392, and 393 K, respectively, and then cooling.

Fig. 5-1 shows a typical thermogram of a sample swollen in water. This is more complicated. It consists of a sharp peak at 342.4 K and a shoulder-tail on the higher temperature side, covering a 45° range. The sharp peak has $\Delta T_{\frac{1}{2}} = 2.1$. From Fig. 5 the following data were obtained by calculation: total transition heat $Q_{tr} = 19 \pm 2$ cal/g, $\Delta H_{tr} = 1730$ cal/mole; $Q_{abe} = 8.6$ cal/g;

Q_{abf} = 10.1 cal/g; Q_{abg} = 9.8 cal/g; ΔH_{tr} = 890 cal/mole. A preparation of BAT from our laboratory having an identical water content as the CalBiochem samples shows a thermogram the same as the CalBiochem sample in Fig. 5-1 but with $\Delta T_{\frac{1}{2}}$ = 2.2 ^{o}; T_{tr} = 341.8 K; Q_{tr} = 19.47 cal/g; Q_{abg} = 9.46 cal/g. The thermogram for the Sigma materials also has a sharp peak and a shoulder, T_{tr} = 341.6 K; Q_{tr} = 17.84 cal/g; $\Delta T_{\frac{1}{2}}$ = 3.2, the area of the sharp peak is slightly larger. Fig. 4 and 5 also show thermograms of BAT with two kinds of water contents recorded in different ways. A shoulder can only be observed (Fig. 5-3) after either keeping water-swollen BAT at 332 K over 15 hours or by scanning to 342 K, cooling and then rescanning. In this case, Q_{tr} = 9.50 cal/g. in the same way, for the sample with a water content of 0.13 - 0.15 g/g, but kept at 376 K or scanned to 393 K, the thermogram disappears completely, as shown in Fig. 4-4. The decrease of the peak area of the thermogram obtained by a second run demonstrates that some type of reaction happened while keeping the sample at a certain temperature. In regard to the sharp peaks of both samples, the sum of the areas of thermograms recorded in the two-step scan is only 30-40% of the normal. For both samples

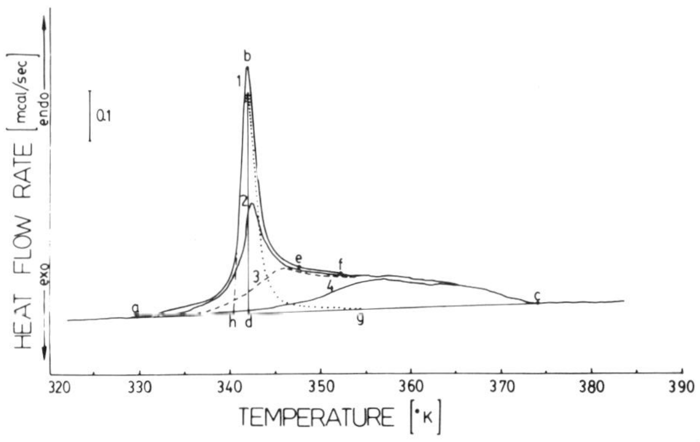

Fig. 5. Thermograms of BAT with water content of 2.3 ± 0.2 g/g measured in different scan procedures. Sample weights are 1.62 ± 0.03 mg. 1: Normal; 2-4: after either keeping sample at 330.5, 332, and 334 K or scanning up to 340.5, 342, and 350 K respectively, and then cooling. Temperature of Point f is T_{tr} + 10 K.

their T_{tr} increased by 0.8^o while the scan rate is doubled in the rate range of 0.31 - 20°/min, but for indium and ice, T_{tr} only increased by 0.06 - 0.1^o. For samples with water contents of 0.59 and 1.10 g/g, the sum of the areas obtained in the two-step scan can reach 80-90% of their normal. At water contents below 0.33 g/g the T_{tr} measured in the second scan decreases; on the contrary, in the case of water contents higher than 0.33 g/g, T_{tr} increases. This is clearly shown in Figs. 4 and 5.

Fig. 6 displays a series of water swollen BAT scans measured after transition or partial transition. Curves d and e indicate that after the transition related to the sharp peak comes just to the end, only a heat capacity change (possibly a glass transition) is observed at about 320 K. However, once the transition related to the shoulder takes place either completely or partially, a low temperature endothermal peak is recorded, as shown in Curve g. Conversely, samples with water contents of 0.59 and 1.10 g/g having undergone partial transition can show both exo- and endothermal peaks as well as a further transition peak of fibrous structure undenatured in the first scan. The latter is similar to Curve g and not a reversible transition.

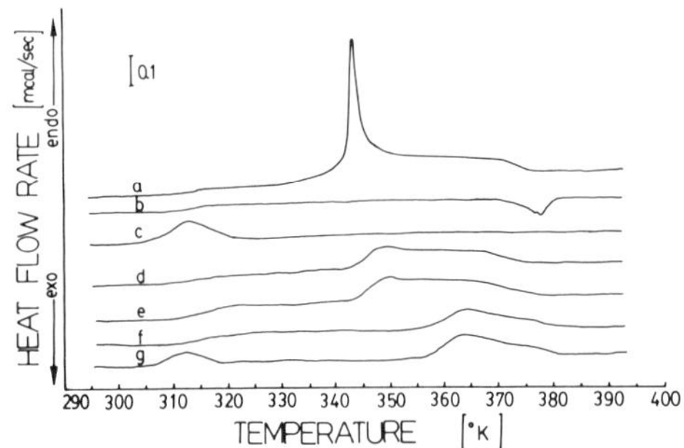

Fig. 6. Thermograms of BAT with water content of 2.3 ∓ 0.2 g/g measured after denaturation to different extent. Sample weights are 1.62 ± 0.03 g. a: normal; b: immediately after transition; e: after keeping the denatured sample at room temperature for ten days; d: immediately after scanning sample up to 342 K and then cooling; e: after either keeping sample at 332 K for 15 hrs. and then cooling or scanning up to 342 K and then keeping at toom temperature for ten days; f: immediately after scanning up to 350 K and then cooling; g: The sample was scanned up to 350 K then kept at room temperature for ten days.

In the hydration range of 0.15 to 13.0 g/g an irregular thermogram, a splitting peak or a small sharp peak are often observed overlying somewhere on thermograms, as shown in Fig. 1-g and k. Samples with a water content of about 0.59 g/g give the irregular peaks more often. A piece of fibre cracked by cutting or two pieces in the same pan display a splitting, or even two adjacent individual peaks; Q_{tr} however, is not affected by this behaviour.

DISCUSSION

Seven Steps of Hydration and Thermotransition

BAT calorimetric results allow resolution of its hydration and thermotransition into seven steps shown in Fig. 2 and Table 1. Hydration range I is a process in which water bridges the interchain H bonds and helix crystallinity is gradually formed and perfected. The denatured enthalpy of the samples reflects a degree of their perfection and the abrupt changes of other thermodynamic parameters are also connected with this process. It has been proved that two water molecules per triplet are indispensible for maintaining the crystalline structure of collagen (17). The results suggest that water is an essential factor for formation of intramolecular H bond and crystallinity in the molecule. X-ray results have shown that no collagen crystalline structure exists without water (17, 18). For thoroughly dried BAT we argue that since neither melting nor thermotransition appears to be observed, the melting temperature of pure collagen, T_{tr}, may lose its proper meaning. For BAT with two water molecules per triplet, ΔH_{tr} = 2540 cal/triplet and, as seen in Table 1, the ratio of their enthalpy values is 1:0.78, then H_{tr} = 1430 cal/mole H bond; this is in good agreement with the double-bond model of collagen both in H bond number and energy (5, 15-18, 27, 36-38).

It is interesting to notice that Hydration Range II corresponds to the slow-rising linear part of the water sorption isotherm. In this range water molecules may be gradually attached to peptide NH-bonds (39). Table 1 shows that the ratio of enthalpy values caused by the first, second, and the third with half of the fourth water molecule absorbed by a triplet is 2.6:2.0:0.8. Supposing that to some extent these enthalpy values are considered to be an expression of the strength of hydrogen bonding between water and peptide group, it may show that the first two water molecules are triple- and double-bonded as water bridges inside the helix (17), the last one and a half are bounded outside the helix via a single-H bond. Thus in Range II the hydration of collagen may not be affected by a change in pH. These water molecules do not act as a solvent for electrolytes and nonelectrolytes and therefore ions have no observable effect on denaturation of tropocollagen (1, 25). These waters are a key factor in gelatinization and reformation of the helix (Fig. 2 E and F) and may play a role as outside structural water for the helix.

In Range III the waters provide an aqueous medium for forming hydrophobic bonds among denatured peptide chains and is released during thermotransition but they dod not freeze even at 225 K (Fig. 3-II). Possibly they are binding at weak surface sites of the protein molecules and have no obvious effect on T_{tr} of molecule. A symmetric thermogram appears again (Fig. 2-A and D).

Further, in Range IV this water not only can be released before thermotransition and also can be frozen before and after transition (Fig. 3 III). This fact presumably shows that water clusters appear in fibrous structure and may be filled in hole zones between N- and C-terminals of molecules as well as among molecules and microfibrils. Owing to large gaps the appearance of clusters leads to another heterogeneity in fibrous structure which brings about a severe dispersion of T_{tr} of the molecules, and, thus, the thermograms become widest.

In Range V clusters turn into large area of water and the hydrated structure begins to become homogeneous once more, and so the two stages of thermotransition separate gradually as the degree of hydration increases. In this case, if the tail of the peak was neglected, a fall on the Q_{tr} curve might be observed, as reported by Leuscher et al (25). Moreover, in this range Q_{tr} remains more constant within the experimental error and is 15.1 ± 0.8 g cal/g, i.e., $H_{tr} = 1370 \pm 70$ cal/mole, which would be the expected enthalpy value of BAT molecules with approximately 216 imino acid residues per thousand residues in solution (40). At this stage the transition of molecules might be considered to be completed if the effect of the fibrous structure on it is ignored.

Range VI, in fact, is a process in which all factors in the fibre become more homogeneous. A slow rising of Q_{tr} appears to mean a beginning of gelatinization of the fibrous structure. The constancy of all thermodynamic parameters in Range VII indicates that a final homogeneity is reached.

Melting and Gelatinization

A series of intermediate stages were observed in chemical denaturation of collagen fibre (4). In solution, the denaturation enthalpy of tropocollagen includes two kinds of thermal effects (11, 41-44). Engel found that denaturation of soluble collagen consists of two steps and a decrease in its molecular weight occurs in the second (45). Also in solution with a certain concentration range of reagents, with increasing concentration of urea and SCN^-, etc., the sharp peak of swollen BAT becomes smaller and finally vanishes, but the shoulder still remains. With increasing concentration of linking agents, the area of the sharp peak increases; however, the total area of thermogram remains constant (46).

The melting of a single crystalline molecule of collagen may give a sharp peak. While the water environment of molecules and fibrous structure of protein are in a perfectly homogeneous state, T_{tr} of all the molecules are almost identical so that a sharp peak is also recorded as shown in Figs. 4-1 or 5-1. The rest of the water in excess of two water molecules absorbed by a triplet may have no effect on the melting of the molecule, but enhance the mobility of the peptide chains and weakens intermolecular interactions, thereby reducing T_{tr} of molecule. In addition, this water may provide an essential medium for gelatinization, the heat effect of which overlaps the higher temperature side of the melting peak and makes the peak become wider or even tailed. Insufficient water content may bring about a heterogeneity in fibrous structure, which leads to a dispersion of T_{tr} of molecules, and the thermogram will widen. The thermogram probably is an overlap of different transition peaks of molecules with different T_{tr}. In the hydration range from 0.023 to 13.0 g/g the fibrous structure goes through a process of heterogeneity-homogeneity-heterogeneity-homogeneity and also the degree of gelatinization increases greatly over this range, there being none at the initial point. This seems to be a more acceptable reason for a series of complicated variations in the shape of the thermograms in all thermal dynamic parameters and behaviours of denatured fibre with hydration degree shown in Figs. 1 and 2.

The sharp peak probably is attributed to a rupture of intramolecular H bonds or the so-called melting of the molecule, followed by a folding of the melted polypeptide chains. At a temperature below T_{tr}, reformation of a local helix from a coil can take place (1-6, 41); the endothermal peak in 310-350 K measured after keepint denatured samples with water contents higher than o.15 g/g at room temperature (Figs. 2 and 3) may be associated with a retransition of reformed helices. Hydrophobic bond rupture is known to be an exothermal process (47), and thereby, the exothermal peak in 280-380 K measured immediately after transition of BAT with water contents above 0.23 g/g (Figs. 2 and 3) is presumably related to a breakdown of hydrophobic bonds formed among polypeptide chains. The fact that a sample subject to a complete transition or sharp peak has given only a glass transition (Fig. 6d and e) may suggest that melting of the molecule probably is only a relaxation process of the helix, followed by a folding of polypeptide chains and that the melted helix is not untwisted. As this intermediate stable state polypeptide chains may not contact each other. In the shoulder step transition the molecular structure and the fibrous structure probably undergo a slow disintegration, such as a gradual disruption of intra- and intermolecular crosslinks and unwinding of the melted helix. Only when this second step occurs, can the helix-to-coil transition be completed, and the molecular weight decreases (45), which is a gelatinization process.

Cooperative Characteristic of Transition

A great number of reports have been contributed to the study on the mechanism of shrinkage of collagen (1, 48-49). It seems likely that melting of a molecule is initiated at a certain weak point with the lowest stability. Losing the hydrogen binding the helix structure probably relaxes, disperses, and folds back on itself locally and consequently creates a local disruptive force which is able to accelerate the melting process of the molecule. Besides, an additional tension must exert influence on the fibre stability too. This force may promote the melting of neighboring molecules by transmission of it via intermolecular crosslinks. On the other hand, neighboring stable molecules without weak points may, in turn, prevent further development of melting. Such intra- and intermolecular interactions could cause the melting of collagen to have a more striking cooperative nature than that of globulin. When the number of melted regions approaches a certain threshold, the fibrous structure becomes so weak that it is most likely that an avalanche-like disruption of molecules in the fibre would take place. Even at much lower temperature than T_{tr} the reaction will spontaneously spread to all molecules in the fibre. This reaction is dependent not on the melting phase transition temperature but on the number and stability of weak points in the molecules, their transition extent and the possibility of cooperative transference. It shows certain analogies to a chain reaction. For BAT with water contents of 0.13 - 0.15 or more than 2.0 g/g the reaction related to the sharp peak can be completed even at the temperature which is 10° or 17° C lower than their T_{tr} (Figs. 4 and 5). While measuring them up to 342 and 393 K respectively at a rate of 5°/min (0.083^{o}/sec) in the first scan, only approximately one third of the reaction related to their sharp peaks has been completed (it needs another five seconds to reach the midpoint of reaction), a 3° rise in temperature (i.e., 36 sec.) need to be taken to basically complete the reaction. When cooling the sample with 320°/min. five seconds is enough to reduce the temperature by 27° C. One must conclude that the part of the uncompleted reaction is carried out at a much lower temperature, a conclusion that cannot be explained in terms of the normal idea of a melting phase transition. A 2° rise in temperature leads to a doubling of the spread of shrinkage of fibre (1, 49); this, also, is quite different from the behaviours of an ordinary chemical reaction. However, with the above assumption, these results can easily be understood. The number and stability of weak points of molecules may be associated with the imino acid residue content and the distribution of intra- and intermolecular crosslinks. The fact that the shrinkage temperature of collagen fibre is 20-30° C higher than that of the molecule can be caused only by intermolecular crosslinks (3, 4, 29, 30). There is no doubt that all these factors bring about a great difference in the stability of various regions along the 3000 Å of the molecule, and show their own characters and need for energy upon denaturation. That is why the collagen to gelatin transition possesses its own complexity and peculiarity, basically differing from either ordinary phase transition reactions, or denaturation process of other proteins.

The transmission of cooperation probably depends on homogeneity of water and structure in collagen. The transition of a thin piece of fibre in chemical denaturation is faster (4). T_s of a compact sample is higher than that of a looser sample. Severe heterogeneity, or mechanical cracks in fibrous structure probably can inhibit the transmission and create two or even several cooperative regions with different stability, having their own melting process. Thus the splitting peaks or additive small ones, have been observed (Fig. 1-g and k). The former was reported by Finch and Ledward and explained in terms of an exothermal effect from a breakdown of hydrophobic bonds (22). The present results show that these two kinds of peaks are probably brought about by an identical cause. For BAT with water contents around 0.59 g/g, as mentioned above, severe heterogeneity in hydrated fibrous structure causes T_{tr} of molecules to be scattered over the widest temperature range. The $\Delta T_{\frac{1}{2}}$ and area of the Sigma material's sharp peak may just be caused by more cooperative regions of a threadlike value.

In the low hydration region the behaviour of the thermotransition of BAT is more similar to that of calfskin (25) than to that of rat (32, 33, 43, 44); this may be a species dependence.

T_{tr} calculated from results of BAT and rat tail tendon (RTT) (partial)-ethylene glycol by Flory and Garrett (9) and of hydrated BAT by DiLisi et al (28) are shown in Fig. 2-A. It is clearly seen that in spite of a different medium they coincide either on the same curve or on the point obtained from extrapolating the temperature (Point h in Fig. 5). However, they are lower than T_{tr} of the BAT-water (very little) glycol ternary system measured in this work. Here rises a question. Is there any possibility that some water content of the BAT-glycol system of Flory's experiment was neglected?

SUMMARY

The stability of collagen is related to factors operating at all levels of the native conformation structure. The collagen to gelatin transition probably includes melting of molecules, molecular gelatinization, and gelatinization of fibrous structure. The enthalpy of the transition may be represented as follows:

$$\Delta H_{tr} = \Delta H_{mm} + \Delta H_{mg} + \Delta H_{fg}$$

where mm represents the molecular melting, mg the molecular gelatinization, and fg the fibrous gelatinization. From our DSC measurements, we obtained a total enthalpy value ΔH_{tr} = 1.73 Kcal/mole, with ΔH_{mm}, ΔH_{mg}, and ΔH_{fg} being 0.85, 0.52, and 0.36 Kcal/mole, respectively. The rupture of the intramolecular H bond is considered as the melting of molecule and helix melting which is accompanied by a folding of polypeptide chains. The DSC results lead us to believe that the melting is the first stage not only of the collagen to gelatin transition but also a first stage of thermal shrinkage of collagen fibre. No exact or genuine thermodynamical phase transition point exists. Rather, the transition consists of a series of intermediate states. In addition, no obvious reversibility of the transition has been observed under the conditions studied. All these indicate that besides some characters of phase transition, the transition is a rate-dependent process too (1, 8, 9, 49). Moreover, each of the three stages is most likely to be a complicated process involving many factors, including breakdown and reformation of all types of bonds and interactions, and it is difficult to distinguish and to separate them using only thermodynamic methods. It seems to be oversimplified both theoretically and experimentally if the transition of the collagen to gelatin, or even the thermal shrinkage of collagen fibre is presumed only to be a melting phase transition. Recently a better linear dependence of the total enthalpy value of tropocollagen in solution on its imino acid residue content has been found (11, 12, 21). But the data for insoluble collagen are more confusing (21, 26). In addition, influence of ions and other reagents on the transition of collagen is attracting an increasing attention (1, 8, 10, 23, 24, 29); the results so far have shown more diversity (23, 24, 46). This may just be the result of a confusion of these several overlapping transition stages of collagen. The factors influencing thermostability of collagen may be interpreted more perfectly only if the mechanism of the collagen to gelatin transition at molecular level is clarified.

ACKNOWLEDGEMENT

We gratefully acknowledge the direction of Prof. Bei Shizhang; valuable discussion by Prof. Shen Shumin, Drs. J. L. Finnay, and R. S. Chen; linguistic assistance by J. L. Finney; preparation of BAT by Dr. S. Z. Linc; and manuscript preparation assitance by M. M. Ochsenfeld and J. F. Brogan

REFERENCES

(1) K. H. Gustavson, "The Chemistry and Reactivity of Collagen," Academic Press, New York (1956).
(2) W. Traub and K. A. Piez, Adv. Protein Chem. 25:243 (1971).
(3) W. F. Harrington and P. H. von Hippel, Adv. Protein Chem. 16:1 (1961).
(4) G. N. Ramachandran, in: "Treatise on Collagen," Vol 1, p. 103, Academic Press, New York (1967).
(5) P. H. Von Hippel, in: "Treatise on Collagen," Vol. 1, p. 253, G. N. Ramachandran, ed., Academic Press, New York (1967).
(6) P. Bornstein and W. Traub, in: "The Proteins," Vol. 4, p. 411, H. Neurath and R. L. Hill, eds., Academic Press, New York (1979).
(7) P. L. Privalov, in: "Biological Microcalorimetry," p. 413, A. E. Beezer, ed., Academic Press, London (1980).
(8) R. R. Garrett and P. I. Flory, Nature 177:176 (1956).
(9) P. J. Flory and R. R. Garrett, J. Am. Chem. Soc. 80:4836 (1958).
(10) A. Veis, in: "Treatise on Collagen," Vol. 1, p. 367, G. N. Ramachandran, ed., Academic Press, New York (1967).
(11) P. L. Privalov and E. I. Tiktopulo, Biopolymers 9:127 (1970).
(12) P. L. Privalov and E. I. Tiktopulo, Biofzika 13:955 (1968).
(13) A. Rich and F. H. C. Crick, Nature 176:915 (1955).
(14) A. Rich and F. H. C. Crick, J. Mol. Biol. 3:483 (1961).
(15) G. N. Ramachandran and G. Kartha, Nature 176:593 (1955).
(16) G. N. Ramachandran and V. Sasisekharan, Biochim. Biophys. Acta 109:314 (1965).
(17) G. N. Ramachandran and R. Chandrasekharan, Biopolymers 6:1649 (1968).
(18) P. L. Privalov, E. I. Tiktopulo and V. M. Tischenko, J. Mol. Biol. 127:203 (1979).
(19) A. Cooper, J. Mol. Biol. 55:123 (1971).
(20) E. L. Andronikashvili, G. M. Mrevlishvili, G. Sh. Japaridze, V. M. Sokhadze and K. A. Kvavadz, Biopolymers 15:1991 (1976).
(21) S. Menashi, A. Finch, P. I. Gardner and D. A. Ledward, Biochim. Biophys. Acta 444:623 (1976).
(22) A. Finch and D. A. Ledward, Biochim. Biophys. Acta 278:433 (1972).
(23) A. Finch and D. A. Ledward, Biochim. Biophys. Acta 295:296 (1973).
(24) A. Finch, P. J. Gardner, D. A. Ledward and S. Menashi, Biochim. Biophys. Acta 365:400 (1974)

(25) M. Luescher, M. Rüegg and P. Schindler, Biopolymers 13:2489 (1974).
(26) P. E. Mc Clain and E. R. Wiley, J. Biol. Chem. 247:692 (1972).
(27) W. F. Harrington, J. Mol. Biol. 9:613 (1964)
(28) C. DeLisi and M. H. Shamos, J. Polymer. Sci. Pt. A 10:673 (1972).
(29) P. Doty and T. Nishihara, in: "Recent Advances in Gelatin and Glue Research," p. 92, G. Stainsby, ed., Pergamon Press, London (1958).
(30) B. J. Rigby, Biochim. Biophys. Acta 133:272 (1967).
(31) J. Einbinder and M. Schubert, J. Biol. Chem. 188:335 (1951).
(32) D. R. Monaselidze and N. G. Bakradze, Dokl. Akad. Nauka USSR 189:899 (1969).
(33) E. L. Andronikashvili, D. R. Monaselide, N. D. Bakradze and R. S. Svanidze, in: "Konformatsionnie Izmeneniya Biopolymerov v Rastvorakr," p. 164, E. L. Andronikashvili, ed., Nauka Publishing House, Moscow (1973).
(34) A. R. Haly and J. W. Snaith, Biopolymers 10:1681 (1971).
(35) J. Z. Zhang, Z. L. Zhang and W. D. Zhang, in preparation.
(36) W. F. Harrington and P. H. von Hippel, Arch. Biochem. Biophys. 92:100 (1961).
(37) R. Y. Yee, S. W. Englander and P. H. von Hippel, J. Mol. Biol. 83:1 (1974).
(38) H. A. Scheraga, in: "The Proteins," Vol. 1, p. 477, H. Neurath, ed., Academic Press, New York (1963).
(39) H. Susi, J. S. Ard and J. R. Carroll, Biopolymers 10:1597 (1971).
(40) J. E. Eastoe, in: "Treatise on Collagen," Vol. 1, p. 1, G. N. Ramachandran, ed., Academic Press, New York (1967).
(41) P. L. Privalov and E. I. Tiktopulo, Biofizika 14:20 (1969).
(42) D. R. Monaselide and N. D. Bakradze, Dokl. Akad. Nauka USSR 183:1205 (1968)
(43) E. L. Andronikashvili, Biofizika 17:1068 (1972).
(44) E. L. Andronikashvili, D. R. Monaselidze, N. G. Bakradze, Z. L. Chanchalashvili, G. M. Mgeladze, E. L. Mikadze and G. M. Mrevlishvili, in: "Konformatsionnie Izmeneniya Biopolymerov v Rastvorakh," p. 171, E. L. Andronikashvili, ed., Nauka Publishing House, Moscow (1973).
(45) J. Engel, Arch. Biochem. Biophys. 97:150 (1962).
(46) J. Z. Zhang and W. D. Zhang, in preparation.
(47) H. Boedtker and P. Doty, J. Am. Chem. Soc. 78:4267 (1957).
(48) K. G. A. Pankhurst, Nature 159:538 (1947).
(49) C. E. Weir, J. Amer. Leather. Chem. Assoc. 44:108 (1949).

HYDROPHOBIC INTERACTIONS; THEIR ROLE IN THE ORGANIZATION OF SUPRAMOLECULAR SYSTEMS

THE INTRACAVITARY BASIS OF SOLUTE PARTITIONING IN DEXTRAN (SEPHADEX[R]) GELS AND THE ROLE OF VICINAL WATER

N.V.B. Marsden and Å.Ch. Haglund

Institute of Physiology and Medical Biophysics, University of Uppsala, Biomedical Center, Box 572, S-751 23 Uppsala, Sweden

ABSTRACT

The partitioning of hydrocarbons and some of their hydroxyl derivatives is described. Two types of solute can be distinguished, non-polar or weakly polar, and polar. The former have an affinity for the gel which appears to involve a hydrophobic interaction and the affinity increases with increasing number of methylene carbons. Polar solutes on the other hand exhibit a reversed type of behaviour, i.e. steric exclusion. The partitioning of all solutes in all the gels can, however, be generalized into a single equation yielding for each gel a generalized logarithmic distribution coefficient. This suggests that although there are different partitioning mechanisms for different solutes, a common structural property of the gel may underlie all partitioning. The structural property best correlated with the generalized distribution coefficient is the concentration of ether oxygens a property which reflects how the matrix changes with increased cross-linking. On a basis of a comparison with the Schardinger cycloamyloses which have similar partitioning properties it is suggested that a cavitary structure is also required and that further, matrix perturbed (vicinal) water may be responsible for the hydrophobic interaction.

INTRODUCTION

The dextran (Sephadex[R]) gels were first introduced for their molecular sieve property[1]. This property is exhibited by even the most highly cross-linked member whose steric selectivity ranges over only about a thousand daltons. Not unexpectedly, however, sorptive properties also become increasingly significant as the degree of cross-linking and the matrix concentration increases[2]. Whether steric hindrance or sorption dominates the partitioning of a solute depends very much on the polarity of the latter as is clearly evident from a study of saturated hydrocarbons and their hydroxylated derivatives.

In this account the properties of four dextran (Sephadex^R) gels, G-10, G-15, G-25 and G-50 together with a divinyl sulphone cross-linked dextran gel, DVS-10 and a polyacrylamide gel, Biogel P-2 with water as the sole solvent are discussed. The distribution coefficients (K_d) were determined at 25 ^oC by column chromatography and the columns were eluted sufficiently slowly, <2 cm h^{-1}, for quasi-equilibrial conditions to prevail.

SOLUTE PARTITIONING

The two classes of solute behaviour are exemplified by polar compounds such as the polyhydric alcohols, and non-polar species. Hydrocarbons are true representatives of the latter class but because of their poor aqueous solubility it is not possible to study a sufficient number of members of a series such as the linear paraffins in order to establish the relation between K_d and the number of carbon atoms in the molecule (n_c). However the polar hydroxyl group has only a short range influence and thus the ln K_d difference between two consecutive higher members should represent, to a first approximation at least, the contribution of a single methylene group. Under this condition the more soluble monohydroxy derivatives can be used as representatives for the non-polar hydrocarbons[3].

Towards polar solutes such as the perhydroxylated polyhydric alcohols, gels exhibit typical "gel filtration" behavior, functioning as molecular sieves. The distribution coefficients thus vary inversely with molecular weight or n_c. As shown in Fig. 1 there appears, in fact, to be a positive linear Gibbs free energy relation (LFER).

As Fig. 1 shows, in contrast to polar compounds, the K_d values of the less polar series all increase with increasing carbon number. Comparing the diols with the monoalcohols, the former have clearly less affinity and the 2,ω-1) members have clearly much less affinity than their corresponding 1,ω-isomers[4]. Fig. 1 suggests a close parallelism between higher members of different series and a tendency to a negative LFER. However, from a careful study[5] of the C-3 to C-11 linear 1-alcohols it was clear that there is not a LFER but instead a quadratic relation, the ln K_d values being related linearly and positively to n_c^2.

As is evident from Fig. 1 the increase of affinity with increasing n_c value which is a type of steric behaviour, albeit opposite in sign to steric exclusion, decreases as the proportion of OH groups on the hydrocarbon chain is increased and the molecule becomes more polar until finally in the perhydroxylated polyols steric exclusion predominates.

The affinity of apparently hydrophilic gels for non-polar solutes is perhaps their most striking property and almost certainly involves a hydrophobic interaction (HI) (3). The reduction of affinity due to the presence of a hydroxyl is consistent simply with suppression of the HI by this polar group (4).

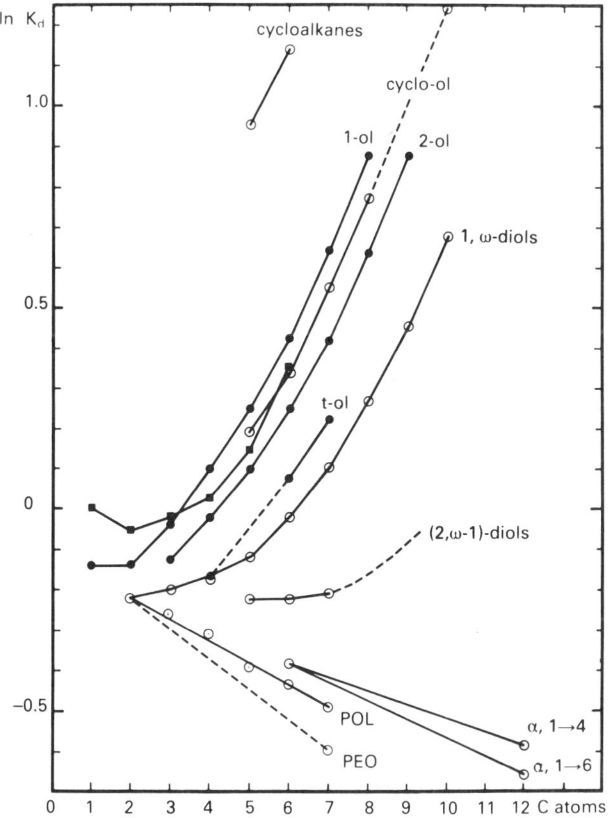

Fig. 1. Logarithms of distribution coefficients of Sephadex G-15 at 25 °C in relation to the number of C atoms in different aliphatic series. The alcohols are denoted by 1-ol etc and the amides by filled squares. POL are polyhydric alcohols and PEO is the slope connecting ethylene glycol and triethylene glycol; α, (1-4) and α, (1-6) are the slopes of lines connecting glucose to maltose and isomaltose respectively. They represent the slopes of the malto- and isomaltodextrins. (Redrawn from J. Polym. Sci., Polym. Lett. Ed., 18:271 (1980), and reproduced with permission.

Although the Sephadex gels have a high affinity for many aromatic compounds[6,7], aromatic hydrocarbons behave much as their aliphatic counterparts[4]. Hydroxylation, however, reveals a dramatic difference and unveils the well-documented aromatic affinity as shown in Fig. 2. Thus, in contrast to aliphatic compounds the presence of a hydroxyl on the benzene ring always increases the affinity.

A GENERALIZED LOGARITHMIC DISTRIBUTION COEFFICIENT

If the affinities of a less polar compound, such as a 1-alkanol or a

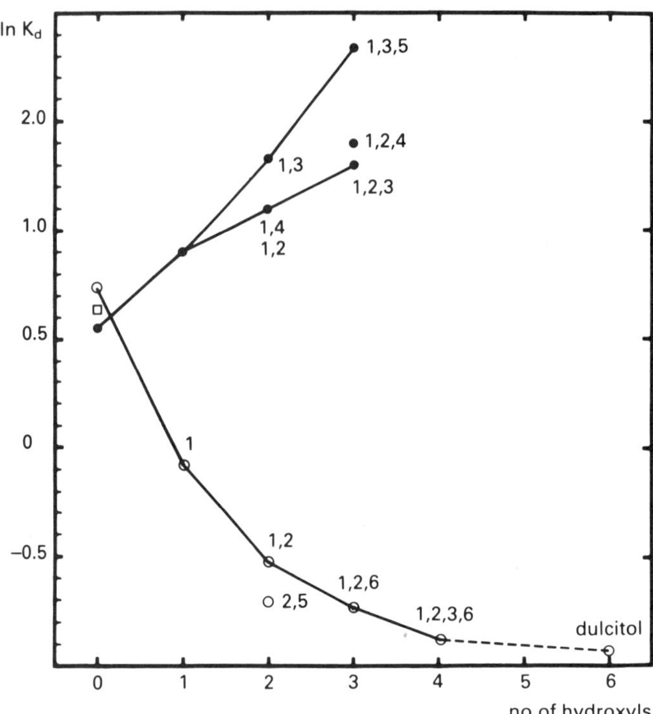

Fig. 2. Logarithms of distribution coefficients of hexane (O) and benzene (●) and some of their hydroxyl derivatives in Sephadex G-15 at 25 °C. The numerals beside the experimental points refer to the particular derivative. Although 1.2- and 1.4-benzenediols are shown as having the same values, the latter is slightly higher. The point for hexane was calculated assuming that ΔG^o(n-hexane) - ΔG^o(cyclohexane) = ΔG^o(1-hexanol) -ΔG^o(cyclohexanol). The point indicated by (□) refers to cyclohexanol. (Redrawn from J. Polymer. Sci., Polym. Lett. Ed., 18:271 (1980), and reproduced with permission.)

hydroxylated aromatic, for Sephadex gels of different degrees of cross-linking, and hence water regain, are compared, an interesting discontinuity is revealed as illustrated in Fig. 3. In the most tightly cross-linked gels G-15 and G-10 the affinity rises more steeply in relation to increasing matrix concentration. Yano and Janado[8] concluded that this relation represented a continuous function with the ln K_d value related to the fifth power of the matrix. It is, however, questionable whether a relation between the ln K_d values of a solute in different gels can be treated as a continous function and it should be noted that there appears to be a similar shaped relation, albeit with the opposite sign for excluded solutes[5]. Thus there is a similar discontinuity as evidenced by a disproportionately greater exclusion of some oligosaccharides in G-15 and G-10 than in gels of higher water content.

In Fig. 4 the ln K_d values of different gels are each compared with

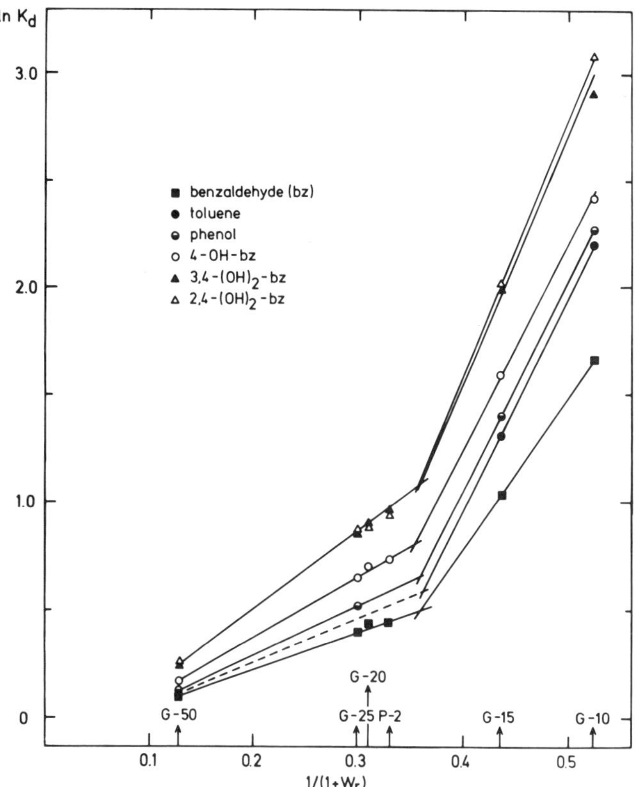

Fig. 3. Logarithms of the distribution coefficients of some aromatic solutes in relation to the matrix concentration in water-swollen Sephadex gels at 25 °C. The individual gels are denoted by arrows. G = Sephadex, P-2 = Bio-Gel polyacrylamide. Some of the experimental points are omitted in the "less steep" region for reasons of clarity. Reproduced with permission from Chromatography, E. Heftmann (Editor) Part A, Elsevier, Amsterdam, 1983, Ch. 8.

those of G-15 as reference. As is evident not only were the ln K_d graphs of all gels linear but those of the Sephadex gels had a virtually common intersection at K_d=1. In addition the graphs of the closely related divinyl sulphone cross-linked dextran gel (DVS-10) and the less closely related polyacrylamide Biogel^R P-2 intersected each other very near to the Sephadex point. The general equation of Fig. 4 is thus[5],

$$\ln K_d \text{ (gel i)} = A_i \cdot \ln K_d \text{ (G-15)} + B_i \qquad 1$$

By definition the slope (A) of G-15 is unity and the intercept (B) is zero.

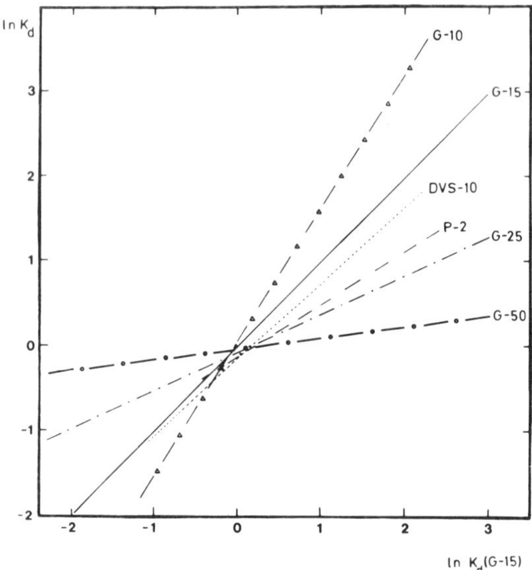

Fig. 4. ln K_d values at 25 °C in the different gels plotted against those in Sephadex G-15. Reproduced with permission from Acta Univ. Upsaliens, 423: (1982).

Eqn. 1 is valid for all solutes so far tested and may be regarded as generally representative of solute behaviour. A_i is thus a generalized logarithmic K_d value for a particular gel. Test species studied ranged widely in type between those whose partitioning is determined mainly by sorption (interactive) e.g. hydrocarbons, saturated alcohols, amides, phenols and halophenols and those for which steric exclusion is the prime determinant e.g. oligosaccharides and polyols. The fact that the A_i lines all have a virtually common intercept means that in all cases solute partitioning is proportional from gel to gel whatever the dominant mechanism of partitioning for a particular solute. This suggests that although solutes partition by different mechanisms all partitioning modes are determined by a common structural property of the gels.

THE COMMON BASIS OF PARTITIONING

As stated above partitioning behaviour in G-10 and G-15 seems to be more pronounced for both sorbed and excluded solutes. This indicates that as the matrix concentration rises above a certain limit both solute affinity and the degree of exclusion increase disproportionately. However, the term A_i includes both these modes of partitioning and if there is a common structural feature underlying partitioning this feature should exhibit a high degree of linear correlation with A_i. Matrix concentration is not apparently the significant variable since if A_i is plotted against it a graph similar to that for the aromatics of Fig. 3 is obtained. A_i is thus not correlated with the total water content of the gel. The best correlation[5] was obtained between A_i and the concentration of ether oxygens (including ring and glycosidic oxygens). This concentration is defined as the mol ratio between the ether oxygens and the water imbibed by the gel matrix. This is interesting because it suggests not surprisingly, that it is not only the cross-links but the whole matrix structure which contributes significantly to partitioning. The poor correlation between A_i and the matrix concentration itself underlines the significance of changes in the matrix as the degree of cross-linking increases.

Dextran gels are predominantly hydrophilic on the average and the presence of an increased concentration of ether oxygens should slightly increase the polarity[9]. Whereas it is true that the anhydroglucose residues (AGR) have non-polar regions it is unlikely that these can account for the high affinity for non-polar solutes unless they are arranged in a stable array[8]. This does not seem likely but a non-polar receptor site is, however, not a necessary condition for the sorption of non-polar solutes. In view of the present state of ignorance about the structural details of these gels it is instructive to consider the properties of the macrocyclic bacterial Schardinger cycloamyloses since these compounds exhibit close similarities to Sephadex as regards their affinities and whose properties suggest that an additional condition may be required for the dextran gels.

MODEL COMPOUNDS AND THE NECESSITY FOR A CAVITARY STRUCTURE

The cavity of the macrocyclic Schardinger cycloamyloses (CA) is lined with a row of the glycosidic oxygens each surrounded by a square array of four CH groups from adjacent AGR[10]. Notwithstanding this far from non-polar surface the CA can form inclusion complexes with non-polar and aromatic species[11] and like the dextran gels, HI and dispersion interactions are almost certainly involved. In both the CA and Sephadex, water is important for the affinity. If formamide is substituted for water the affinity of Sephadex G-10 for 1-alcohols vanished[12]: the order of elution is reversed and becomes that of molecular sieving. Although water is not actually essential for the formation of inclusion compounds with CA, these are much weaker in other solvents such as dimethyl sulphoxide, or dimethyl formamide[13]. A very interesting property of the CA is, however, that the formation of inclusion complexes apparently requires a macrocyclic ring structure[14]. As far as CA are concerned the necessity for a cavity is presumably due to the need for a water-poor or predominantly vicinal water region.

A MODEL FOR A HYDROPHOBIC INTERACTION

In the case of the affinity of CA for non-polar solutes in aqueous solution a plausible model for cyclohexoamylose has been described by Tabushi et al. (15); see also ref. 16. This model is essentially the same as that proposed earlier by Marsden[2] to explain the affinity of a dextran gel for weakly polar aliphatic alcohols. Its most prominent feature is the entropy gain on transfer of a non-polar solute into the water-poor CA cavity. Once inside the cavity the included solute can engage in dispersion interactions and the magnitude of these will depend on the closeness of approach. Thus it has been reported that more bulky branched alcohols have more negative ΔH^o values that their linear counterparts[17]. Although the possible thermodynamic differences between linear and more bulky species do not seem to have been studied in Sephadex it is clear that both HI and other, presumably dispersion interactions occur in these gels. Partitioning data, for example, strongly suggest that double bonded C atoms can interact with the gel.

In addition to the close similarities between the affinities of CA and Sephadex gels there are close structural resemblances. Thus, the interior of the CA behaves as though it were dioxane-like[18] and a dioxolane-like cross-linking structure has been reported for hydrolysed fragments of Sephadex G-25 (Holmberg, personal communication).

In both the Marsden and Tabushi models the HI results from an asymmetry between the states of water inside and outside the cavity. In the case of cyclohexaamylose the situation is almost extreme with only two water molecules in the otherwise empty cavity and the entering solute may displace these. Whereas the dimensions of the CA cavity are known with great precision this is not at all so in Sephadex. In fact calculations of the average centre-to-centre distances of the dextran chains are of seemingly little value. In the first place the average distance apart assuming randomly oriented dextran chains yields values for the most highly cross-linked gels which are impossibly small, being in fact less than the Van der Waals radii. However electron microscopic observations on such Sephadex gels do not support the existence of a random network. Instead there appears to be an open cellular mesh in the submicrometer range[19]. This picture of clustered chains very close to each other forming the walls of larger cells is similar to the microheterogeneous structure visualized by Flodin in his pioneering work on Sephadex[1].

The interior of the gel may therefore well consist of domains with a high matrix density and a very low water content alternating with more water-rich regions with a lower matrix density. The affinity for non-polar species might thus be due to their sequestration in a water-poor region and the partitioning of polar species to confinement in water-rich parts. However, attractive as this concept may appear, the partitioning pattern does not support such a dichotomy. Thus, since the more steeply rising affinity in the most highly cross-linked gels is paralleled by a greater degree of exclusion for polar solutes, it seems more probable that the same domains are involved in the partitioning of both polar and non-polar solutes.

THE POSSIBLE ROLE OF VICINAL WATER

Vicinal water is assumed to be perturbed by interaction with the matrix and the internal water of the gel may thus have a sandwich-like structure[20,21] with the matrix-perturbed water bounding regions of normal bulk structure sufficiently far from the surface.

Thus vicinal water may create together with the nearby matrix an environment favourable for the accumulation of non-polar solutes if the latter are unable to promote as much structuring in it as they do in bulk water (2). Whether vicinal water is more or less structured than in the bulk state is not clear; it presumably depends on the structure of the matrix. A cooperatively ordered more-structured state has been proposed for Sephadex[8]. On the other hand there seems to be some doubt as regards saccharide-water interactions[21,22] and the dioxolane nature of the cross-linking structure might be expected to have a disordering effect[23]. However, notwithstanding this uncertainty the only necessity for HI is that a low concentration of poorly water-soluble, non-polar nonelectrolytes does not increase significantly the established state of order of the vicinal water. Given this condition these solutes will accumulate and the distribution coefficient will depend on the hydrophobicity of the solute and the amount of vicinal water in the gel. We have used a simple model to calculate the latter as a function of the average pore size.

A VICINAL WATER MODEL

The gel is visualized as a regular cubic array of spherical cavities which contain all the internal water. The thickness of the layer of vicinal water lining these cavities was assumed to be equal to the distance over which ordering was observed in water i.e. 0.8 nm[24]. The graph relating the vicinal water (as a fraction of the total imbibed water) as a function of the distance apart of the matrix surfaces has a similar form to that of the aromatic compounds in Fig. 3 being considerably steeper when the spaces are smaller. Unfortunately it is not possible to fit the graph to that of Fig. 3 because of the lack of information about the internal dimensions of the gels.

ACKNOWLEDGEMENT

We thank the Swedish Natural Science Research Council (Grant No 2944) and Pharmacia Fine Chemicals for financial support .

REFERENCES

(1) P. Flodin, Dextran Gels and Their Applications in Gel Filtration, Pharmacia, Uppsala (1962).

(2) N.V.B. Marsden, Solute behavior in tightly cross-linked Dextran Gels, Ann. N.Y. Acad. Sci., 125:428 (1965).

(3) F. Franks, The hydrophobic interaction in " Water - A Comprehensive Treatise," Vol.4,"Aqueous Solutions of Amphiphiles and Macromolecules", F. Franks, ed., Plenum, New York (1975).

(4) Å.Ch. Haglund and N.V.B. Marsden, Hydrophobic and Polar Contributions to Solute Affinity for a highly cross-linked water-soluble (Sephadex) Gel, J. Polym. Sci., Polym. Lett. Ed., 18:271 (1980).

(5) Å.Ch. Haglund, "Solute Affinity in Dextran Gels," Acta Univ.Upsaliens., (Diss.), Uppsala, 432 (1982).

(6) B. Gelotte, Studies on Gel Filtration. Sorption Properties of the Bed Material Sephadex, J. Chromatogr., 3:330 (1960) .

(7) Å.Ch. Haglund, Adsorption of monosubstituted Phenols on Sephadex[R] G-15, J. Chromatogr., 156:317 (1978).
(8) Y. Yano and M. Janado, Hydrophobic interaction chromatography of unsubstituted Sephadex gels with high dextran concentrations, J. Chromatogr., 200:125 (1980).
(9) C. Hansch and A.J. Leo, Substituent Constants for Correlation Analysis in Chemistry and Biology, Wiley, New York, (1979).
(10) J.A. Thoma and L. Stewart, Cycloamyloses, in "Starch Chemistry and Technology", R.L. Whistler and E.F. Paschall, eds., Vol. 1, Academic Press (1965).
(11) D. French, The Schardinger Dextrins, Advan. Carbohydr. Chem., 12:189 (1957) .
(12) N.V.B. Marsden, Comment in "Hydrogen bonded solvent systems", A.K. Covington and P. Jones, eds., Taylor and Francis, London, p. 227 (1968).
(13) A. Schlenk, D.M. Sand and J.A. Tillotson, Stabilization of Autoxidizable Materials by Means of Inclusion, J. Amer. Chem. Soc., 77:3587 (1955).
(14) H. Schlenk and D.M. Sand, The Association of α- and β-Cyclodextrins with Organic Acids, J. Amer. Chem. Soc., 83:2313 (1961).
(15) I. Tabushi, Y. Kiyosuke, T. Sugimoto and K. Yamamora, Approach to the Aspects of Driving Force of Inclusion by α-Cyclodextrin, J. Amer. Chem. Soc., 100:916 (1978).
(16) F. Cramer and H. Hettler, Inclusion Compounds of Cyclodextrins, Naturwissenschaften, 54:625 (1967).
(17) Y. Matsui and K. Mochida, Binding Forces Contributing to the Association of Cyclodextrin with Alcohol in Aqueous Solution, Bull. Chem. Soc. Jap. 52:2808 (1979).
(18) R. L. van Ettan, J.F. Sebastian, G.A. Clowes and M.L. Bender, Acceleration of Phenyl Ester Cleavage by Cycloamyloses. A Model for Enzymatic Specificity, J. Amer. Chem. Soc., 89:3242 (1967).
(19) D. Hager, J. Theoretical Considerations of Molecular Sieve Effects, J. Chromatogr., 187:285 (1980).
(20) W. Drost-Hansen, Structure and Functional Aspects of Interfacial (Vicinal) Water as related to Membrane and Cellular Systems, Coll. Internat. du C.R.N.S., 246:177 (1976).
(21) M.V. Ramiah and D.A.I. Goring, The Thermal Expansion of Cellulose, Hemicellulose and Lignin, J. Polym. Sci., C 11:27 (1965).
(22) N.V.B. Marsden, The Role of Water in the Interaction of some Solutes in Aqueous Dextran Gel Systems, Acta Univ. Upsaliens. (Diss.), 123 (1972).
(23) H.S. Sage and S.J. Singer, The Properties of Bovine Pancreatic Ribonuclease in Ethylene Glycol Solution, Biochemistry, 1:305 (1962).
(24) A.H. Narten, M.D. Danford and H.A. Levy, X Ray Diffraction Study of Liquid Water in the Temperature Range 4-200 °C, Disc. Farad. Soc., 43:97 (1967) .

THE ROLE OF HYDROPHOBIC INTERACTION ON HYDRATION FORCES IN THIN AQUEOUS LAYERS

Gerhard Peschel, Mathilde M. Müller and
Manfred M. Müller

Institute of Physical and Theoretical Chemistry
University of Essen, 4300 Essen/FRG

INTRODUCTION

It is generally recognized that part of the water in biological systems has assumed a structure different from the bulk one [1,2]. This effect is attributed to the adsorptive interaction of water molecules with the different membranes and biomacromolecules. According to the diversity of cellular interfaces there will be a large number of regions of more or less restricted molecular mobility. By dealing with such a complex system the general question how an interface by its nature can carry its influence into the structure of water nearby might hardly be solved.

Nuclear magnetic resonance has become a valuable tool for investigating the behavior of water in cells[3,4]. The numerous studies revealed that the nmr relaxation times of the water nuclei are lower for structured than for bulk water. Nevertheless, a lot of difficulties are encountered. Thus focusing on transversal relaxation times small amounts of highly structured water and large quantities of less ordered water yield the same result. Therefore, it seems hard to get information about the extension of structured aqueous layers, otherwise a more sophisticated method is applied on a simple model system.

Recent work in this field is due to Israelachvili and Adams[5] and Pashley[6] who measured the surface forces between molecularly smooth mica sheets when being placed in aqueous alkali halide solutions. They observed short range repulsion forces extending to a separation distance of about 6 nm and exhibiting a decay length of about 1 nm which is only weakly dependent on electrolyte concentration.

Since surface hydration effects arose only along with the adsorption of hydrated cations onto the mica surfaces these experiments could hardly give information about the interface biomacromolecule/aqueous solution, because hydrogen bridging is assumed to contribute to a large extent to the interaction of water molecules with biological surfaces.

Related tests were carried out with an experimental device developed by us, particularly for the detection of surface hydration effects. A planar fused silica plate is fixed at the bottom of a small vessel which can be filled with the test solution. Opposite to the planar surface is located a spherically curved fused silica plate which is attached to a balance system and can thus be moved extremely slowly relative to the planar surface. Both silica surfaces are high-grade polished with a residual roughness of only about 1 nm; the moveable plate can be placed within $\pm$ 1 nm.

The movement of the balance is controlled by a specific electronic mechanism. The plate separation distance can be monitored via a strain gauge measuring bridge by a xy-recorder whose second coordinate refers to the force acting on the deflected balance beam. At sufficiently small plate separations excess forces come into play which allow an insight in the nature of surface forces. In accord with the findings of Israelachvili and Adams[5] and Pashley[6] our experiments by using hydroxylated fused silica plates indicate likewise the presence of structured water near interfaces.

In the present work we shall be concerned with surface forces within boundary layers of aqueous solutions of some selected alkali halides and urea and some of its derivatives. Just this problem might be of strong interest, since our knowledge about the influence of molecules, which are apt to interact by hydrophobing bonding, on the structure of immobilized surface layers is rather poor.

SURFACE FORCES IN THIN AQUEOUS BOUNDARY LAYERS

By approaching two plane surfaces within aqueous electrolyte solution overlap of the electrical double layers located at both the surfaces will occur at about 100 nm and smaller giving rise to a repulsion pressure (disjoining pressure) Π_{el}. According to the DLVO-theory[7] we can simply write

$$\Pi_{el} = 2\, n\, kT\, (\cosh u - 1). \tag{1}$$

n is the number of ions per cm^3 and $u = v\, e\, \psi_{h/2}/kT$; $\psi_{h/2}$ is the potential midway between the surfaces and k the Boltzmann constant.

The corresponding plate separation distance h is given by

$$h = - \frac{2}{\kappa} \int_{z}^{u} \frac{dy}{2\ (\cosh y - \cosh u)^{1/2}} \tag{2}$$

with $z = v\ e\ \psi_o/kT$ and $y = v\ e\ \psi/kT$. κ is the Debye-Hückel reciprocal length and ψ_o the surface potential. Eq. (2) can be evaluated by using tables[7].

For surface separations of h < 6 nm additional strong repulsion forces are often observed[5,6], which are attributed to the interpenetration of juxtaposed multimolecular hydration layers. For this structural disjoining pressure we had introduced the exponential expression

$$\Pi_s = C_s\ e^{-n_s h} \tag{3}$$

with C_s and n_s being parameters which can be determined via experimental data[8].

In the case of a spherical/planar plate system we have assuming a plate separation h for the electrical double layer force $K_{el,h}$

$$\Pi_{el,h} = \frac{K_{el,h} \cdot \kappa}{2\pi R} \tag{4}$$

and for the corresponding hydration force $K_{s,h}$ when h is sufficiently small

$$\Pi_{s,h} = \frac{K_{s,h} \cdot n_s}{2\pi R} \tag{5}$$

R is the curvature radius of the spherically formed plate. Specifically n_s^{-1} is the decay length of the hydration force arising between two approaching surfaces.

EXPERIMENTAL METHOD

The experimental method is outlined in its main features in the foregoing. Details will be found in a forthcoming paper[9]. A very important point is the purification of the fused silica plates and the chemicals, respectively. Before each run the silica plates were treated with hot chromic acid and thereafter with doubly-distilled water. Finally application of supersonic proved to be most effective in removing traces of contaminants. The temperature of the test solution was kept constant within ± 0.2 K during a run.

REPULSION EFFECTS IN A THIN LAYER OF AQUEOUS SODIUM CHLORIDE SOLUTION

The repulsion forces in thin layers of aqueous electrolyte solutions confined by two fused silica plates are commonly of electrostatic and structural origin, respectively. Fig. 1 demonstrates this effect for the case of a 10^{-4} M NaCl solution. For plate separations larger than about 6 nm a disjoining pressure is operative which is in excellent accord with the DLVO-theory, though the attractive contribution is apparently screened by the additional repulsion which dominates for $h < 6$ nm and gives evidence for vicinal structuring of water. The exact proof of the DLVO-theory, which was likewise obtained for LiCl and KCl solutions[8], supports the correctness of the results for the hydration forces, particularly for aqueous urea solutions as outlined in the following.

REPULSION EFFECTS IN THIN LAYERS OF AQUEOUS SOLUTIONS OF UREA AND SOME OF ITS DERIVATIVES

Urea is an ordinary constituent of blood and, therefore, of biological significance. It is believed to reduce the degree of water-water hydrogen bonding without interacting too strongly with water. Because of geometrical reasons urea is excluded from the tetrahedrally bonded water clusters and associated with the broken, more dense water species.

In Fig. 2 the decay length n_s^{-1} of the structured layer of some aqueous solutions of urea and a few of its alkylated derivatives is plotted against the concentration. At first glance it appears difficult to draw conclusions from the non-monotonous concentration dependence of n_s^{-1}. But recall that N-methylacetamide, which is closely related to urea, is known as a potent water structure breaker. Therefore, to have some relative guide, we studied at first the structural disjoining pressure in thin layers of aqueous solutions of this compound at different concentrations (Fig. 3). Two marked peaks at about 10^{-4} M and 10^{-1} M are evident. According to our experience with electrolyte solutions[8] the first peak originates from the superposition of two contrary effects. Adding a structure breaking solute to water greatly affects the tetrahedrally bonded structure of water, so that salting-in effects occur on the silica surface, i.e., the surface might be apt to decouple water structure more easily in favour of a specifically structured hydration layer. This layer, on the other hand, gets more and more disrupted by adsorption of the solute onto the silica surface.

In the concentration regime of the second peak the hydration envelopes of the solute molecules get more and more overlapped which in return by structure disrupting processes provides for a strengthened salting-in effect on the silica surface. The maximum value of n_s^{-1} lies at about 1.9 nm, i.e., 0.95 nm for one surface.

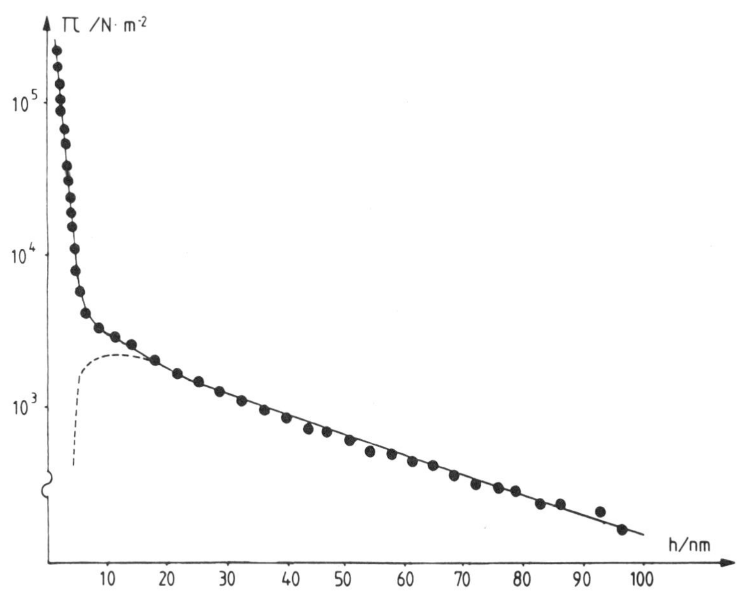

Fig. 1. Disjoining pressure in a thin layer of aqueous 10^{-4} M NaCl solution between two vitreous silica plates. T = 298 K; pH = 5.4. The smooth line was calculated according to the DLVO theory with ψ_o = 57.5 mV. The dotted line refers to van der Waals attraction. Reciprocal Debye length κ_{exp}^{-1} = 30.0 nm, κ_{theor}^{-1} = 30.7 nm.

Inspection of Fig. 2 shows that methylurea, which can be derived from N-methylacetamide by replacing a CH_3-group by a NH_2-group, displays a concentration dependence of n_s^{-1} analogous to that of N-methylacetamide. Only the peaks are shifted to higher concentrations. In view of the fact that the NH_2-group disrupts water structure more than the CH_3-group the peak shift at about 1 M might be reasonable since the scope of hydration is smaller for methylurea which implies an overlap of the hydration spheres

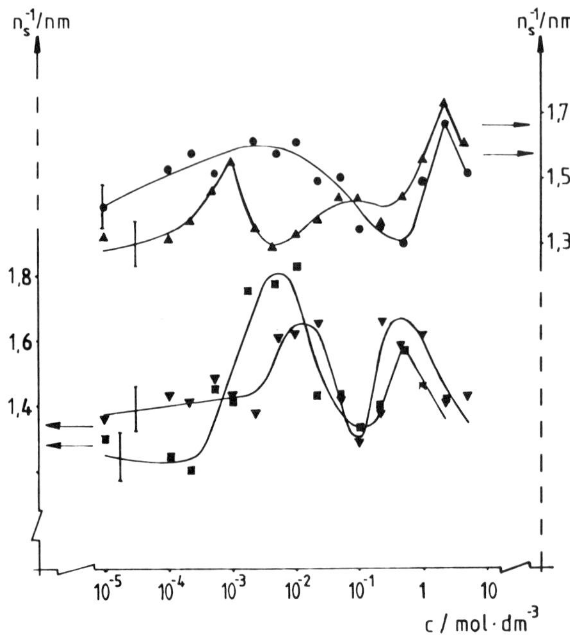

Fig. 2. Structural decay length n_S^{-1} for aqueous solutions of urea (▼), methylurea (▲), 1,3-dimethylurea (●), and tetramethylurea (■) in thin layers between vitreous silica plates for different concentrations. T = 298 K.

at higher concentrations. Tetramethylurea is equipped with four structure strenthening CH_3-groups; thus, the whole molecule might be the most potent structure maker in Fig. 2. Actually the salting-in effect on the high concentration side is smaller, on the low concentration side it is even missing. Instead, a strong hydration maximum arises at about 10^{-2} M, which was likewise found for electrolyte solutions. It is believed to originate from a matrix of hydrated solute molecules adsorbed on the silica surface up to a critical concentration where the hydration spheres of the adsorbed molecules start to get overlapped. This approach establishes the role of urea to be only a slight structure breaker since at about 10^{-2} M there actually exists a minor maximum, which in the case of a strong structure breaker like N-methylacetamide is degenerated to a shoulder only and even shifted to higher concentrations.

Adsorption of urea and its derivatives from aqueous solution onto charcoal lead to adsorption isotherms which particularly for

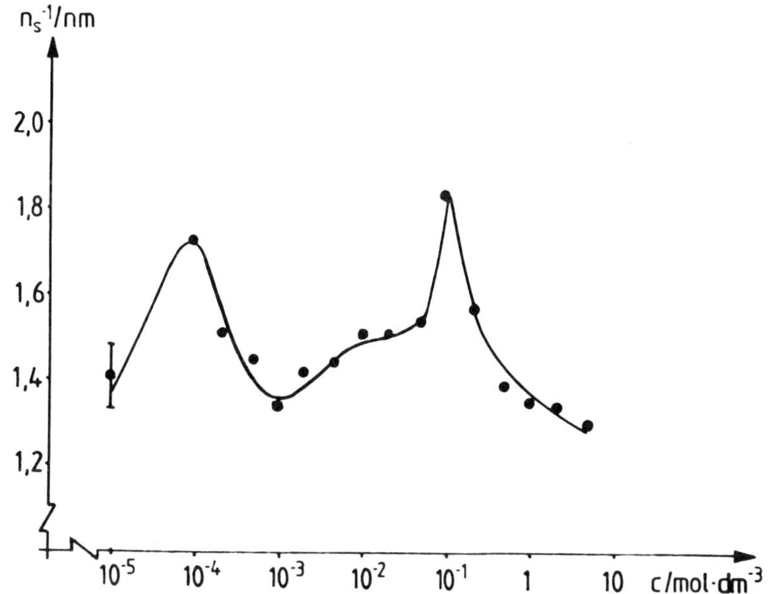

Fig. 3. Structural decay length n_s^{-1} for aqueous solutions of N-methylacetamide in thin layers between vitreous silica plates for different concentrations. T = 298 K.

the higher alkylated species showed a break in the concentration regime about 10^{-2} M[10] when, e.g., being plotted according to Freundlich[11]. From these studies which are supported by related adsorption tests conducted by Neitzel[12] it appears likely that some sort of hydrophobic interaction between the hydrocarbon groups of the adsorbed solute molecules contributes to the hydration maximum at about 10^{-2} M. Further work in this field is in progress.

Financial support by Deutsche Forschungsgemeinschaft and Fonds der Chemischen Industrie is gratefully acknowledged.

REFERENCES

1. W. Drost-Hansen, Structure and Properties of Water at Biological Interfaces, in: "Chemistry of the Cell Interface," H.D. Brown, ed., Academic Press, New York (1971).

2. C.F. Hazlewood, A View of the Significance and Understanding of the Physical Properties of Cell-Associated Water, in: "Cell-Associated Water," W. Drost-Hansen and J. Clegg, eds., Academic Press, New York (1979).

3. R. Mathur-De Vrê, The NMR Studies of Water in Biological Systems, Prog. Biophys. Mol. Biol. 35: 103 (1979).

4. S. Ratkovič, NMR Studies of Water in Biological Systems at Different Levels of their Organization, Scientia Yugoslavica 7: 19 (1981).

5. J.N. Israelachvili, Measurement of Forces between Two Mica Surfaces in Aqueous Electrolyte Solutions in the Range 0-100 nm, J. Chem. Soc. Faraday Trans. I 4: 975 (1978).

6. R.M. Pashley, DLVO and Hydration Forces between Mica Surfaces in Li^+, Na^+, K^+ and Cs^+ Electrolyte Solutions: A Correlation of Double-Layer and Hydration Forces with Surface Cation Exchange Properties, J. Colloid Interf. Sci. 83: 531 (1981).

7. E.J.W. Verwey and J.Th.G. Overbeek, "Theory of the Stability of Lyophobic Colloids," Elsevier Publishing Co. Amsterdam (1948).

8. G. Peschel, M.M. Müller, M.R. Müller, and R. König, The Interaction of Solid Surfaces in Aqueous Systems, Colloid & Polymer Sci. 260: 444 (1982).

9. M. Selig, M.M. Müller, M.R. Müller, R. König, and G. Peschel, An Improved Method for the Determination of DLVO and Hydration Forces, being prepared.

10. M.R. Müller and G. Peschel, being prepared.

11. Ch. Sheindorf, M. Rebhun, and M. Sheintuch, A Freundlich-Type Multicomponent Isotherm, J. Colloid Interf. Sci. 79: 136 (1981).

12. V. Neitzel, Experimental Investigation of the Adsorption of Phenol onto Different Carbon Surfaces from Aqueous and Nonaqueous Solutions, Diplomarbeit, University of Essen (1981).

ON THE INTERACTION OF LECITHIN VESICLES WITH CHAOTROPIC IONS

T. Pigor and R. Lawaczeck

Institute of Physical Chemistry
University of Würzburg
Marcusstr. 9/11 , 8700 Würzburg FRG

ABSTRACT

The intervesicular interaction between lecithin vesicles and chaotropic ions is described. The action of these ions leads to a precipitation of vesicles and at the same time to an increase of the turbidity. The tendency to induce this precipitation increases for monovalent anions in the following order: $Cl^- < CH_3COO^- < NO_3^- < N_3^- < Br^- < SCN^- < I^- < ClO_4^-$. The interpretation of the cations-effect is less straightforward. The ions-effect decreases with increasing temperature and does not show any significant change at the phase-transition temperature. A flocculation and/or aggregation of vesicles is discussed as origin for the observed ions-effect. Fusion processes seem to be slow and not predominant.

INTRODUCTION

The action of various ions on properties of water are widely discussed in terms of structure breaking and structure forming tendencies. The contributions by W.A.P. Luck et al. and G. Peschel et al. are examples along this line within these proceedings. With minor modifications the effects can be classified according to the Hofmeister series originally deduced from the precipitation of egg-white as function of the salts studied[1]. Though the Hofmeister series is known since about 100 years studies of the influence of structure breaking ions on the membrane stability are rather scarce [2-5].

Phospholipid bilayers which form the building blocks of biological membranes are stabilized by hydrophobic, electrostatic and steric interactions[6]. Changes of the aqueous milieu should therefore influence the physical parameters describing the balance between these parameters. The influences of pH, temperature, ionic strength and kind of ions on properties of lipid membranes are well documented and will not be discussed here. Instead we have focussed on the interaction between individual lipid vesicles. These inter-vesicular interactions were studied as function of various chaotropic ions. It will be shown that the structure breaking tendency of the ions parallels their behaviour to induce a precipitation of vesicles. The experiments suggest that the induced precipitation predominantly results from a flocculation or aggregation of vesicles. The rate of fusion seems to be low.

RESULTS AND DISCUSSION

Dimyristoyl- and Dipalmitoylphosphatidylcholine vesicles were prepared in aqueous 20 mM $CaCl_2$, 0.02% NaN_3 at neutral pH by sonication of the respective lipid dispersion above the crystalline to liquid-crystalline phase-transition temperature(T_c) for 10 min. The resultant vesicle solution was centrifuged to remove any larger fragments and then rapidly mixed with the respective ions in a stopped-flow apparatus. Either the intensity of the transmitted light or the light scattering perpendicular to the incident monochromatic beam were measured. Aggregation, fusion of vesicles or any increase in size by other mechanisms are manifested by an increase of the intensity of the scattered light of the vesicular solution with time[7]. For a quantitative interpretation of the light scattering data we have fitted the initial parts of the intensity-vs-time curves to a polynom of second order, i.e.

$$I_{l.sc.}(t) = a_o + a_1 t + a_2 t^2 \qquad (1)$$

in the time-interval $0 \leqslant t \leqslant t_1$; a_o , a_1 and a_2 are constants. From these functions the initial slope v_o is simply obtained:

$$v_o = (dI_{l.sc.}(t)/dt)_{t=0} = a_1 \ . \qquad (2)$$

The concentration dependence of the calculated initial slopes v_o allows an interpretation in terms of the underlying mechanism[7]. Any lipid-transfer shows a linear concentration dependence in v_o while collision-mediated processes like flocculation, aggregation or fusion cause a second order behaviour.

In Fig. 1 we present typical results. Vesicles derived from the same Dipalmitoylphosphatidylcholine (DPPC) stock-solution were successively mixed with the sodium salts of the respective anions while the intensity of the transmitted light at 400 nm was recorded. The final salt concentration of the mixtures was 200 mM. From Fig. 1

and similar experiments the action of anions was deduced and the following order was found:

$$Cl^- < CH_3COO^- < NO_3^- < N_3^- < Br^- < SCN^- < I^- < ClO_4^- .$$

The capability to induce a precipitation of vesicles or an increase in turbidity increases from left to right. The effect of 200 mM NaCl (at the left side of the series) is almost negligible and reveals that the respective vesicles do not act as osmometers in contrast to liposomes[8] and cellular membranes[9]. The ClO_4^--effect is most pronounced. The described tendency can favourably be compared with the Hofmeister series or with the structure breaking effect of the regarded anions[10]. The effect of cations is less pronounced and not as straightforward as the anions studied. For the zwitterionic lecithin vesicles the effect of alkaline cations decreases with increasing ion-radius where Li^+ forms an exception.

Studies on the dependence of the ClO_4^--concentration indicate that a threshold concentration exists. Only for concentrations above

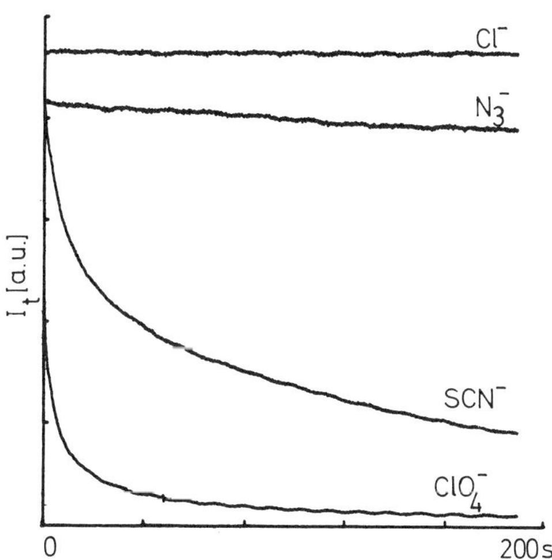

Fig. 1 Intensity of the transmitted light at 400 nm as function of time after mixing for the sodium salts indicated. The final salt concentration is 200 mM. DPPC vesicles, room temperature.

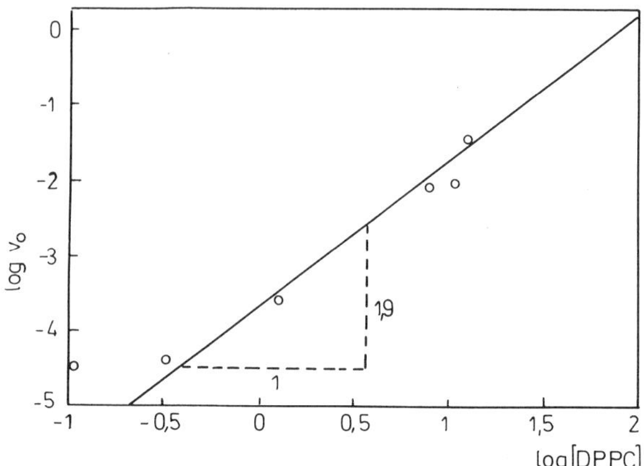

Fig. 2 Initial slope v_o of the intensity of the scattered light as function of the vesicle concentration on a double-logarithmic scale. Final $NaClO_4$-concentration 150 mM , DPPC vesicles, room temperature.

this threshold the action of ions is observable in the time-range studied (seconds up to a few minutes). For DPPC at concentrations of 1-2 mg/ml the threshold amount of ClO_4^- was 70 - 80 mM, however, these values might depend on the lipid to salt ratio.

The variation of the vesicle concentration reveals a second order process. In Fig. 2 we have plotted the initial slope v_o from the light scattering intensity versus the vesicle concentration on a double-logarithmic scale. The indicated slope of 1.9 obviously eliminates a lipid-transfer as possible mechanism for the observed effects. Indeed, the almost second order stresses a collision-mediated process as origin for the observed precipitation. Therefore the increase of the turbidity must be due to the flocculation, aggregation or fusion of vesicles.

Fig. 3 illustrates the temperature dependence of the ions-effect for ClO_4^-. The two symbols used reflect computer-fits to different initial parts of the experimental curves. The tendency to induce a vesicle flocculation, aggregation or fusion decreases with increasing temperature. Any marked breaks at the phase-transition temperature are not visible in these preliminary experiments. However, the experimental scatter becomes more pronounced in the region between the pre- and maintransition (35° and 41.5° C, respectively).

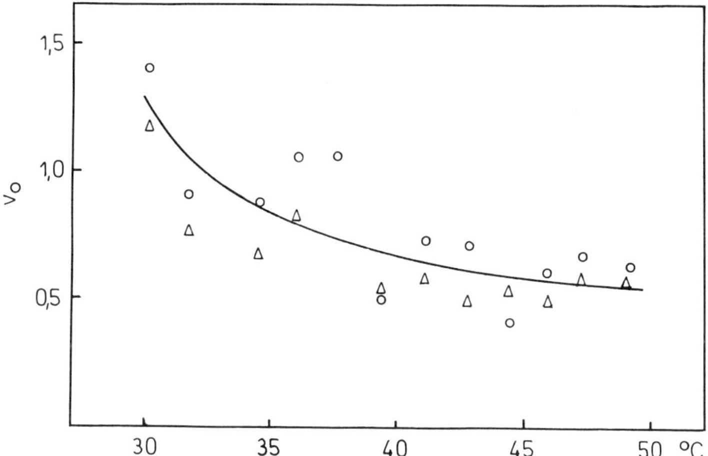

Fig. 3 Temperature-dependence. Initial slope v_o of the intensity of the transmitted light versus temperature. The two symbols correspond to computer-fits of 50 and 100 initial data-points, respectively. DPPC vesicles mixed with 300 mM $NaClO_4$.

The temperature-dependence points at an increasing back-reaction (redispersion of vesicles) with increasing temperature. As significant effects at T_c are absent most of the experiments were performed in the crystalline state.

With respect to the underlying mechanism we could show that intravesicularly encapsulated marker-molecules did not come into contact with the extravesicular milieu in the course of the vesicle precipitation. Thus the vesicles retain their individuality and/or large defects or pores are not generated during the action of the chaotropic ions studied. Dialysis experiments further confirmed that the observed ions-effect is almost reversible. For that purpose a vesicle solution was first exposed to e.g. 200 mM $NaClO_4$ for 1 h. After this period the transmission was read and subsequently the samples were dialysed at room-temperature against the salt-free buffer. After 26 h the transmission was read again. The values of table 1 indicate that under the chosen experimental conditions the action of ClO_4^--ions is almost reversible. These observations illustrate that the action of the studied chaotropic ions predominantly leads to a flocculation or aggregation of vesicles. From these assemblies fusion might proceed, however, the rate of fusion seems to be low under the chosen experimental conditions.

Table 1 Results of the dialysis experiments. DPPC vesicles in 20 mM $CaCl_2$, 0.02% NaN_3, neutral pH, room temperature.

	Transmission (%) at 400 nm 1 cm cuvette, room temperature.	
	after 1 h reaction time 200 mM salt	after 26 h dialysis against the salt-free buffer.
buffer (control)	93.3	87.9
NaCl	91	88.7
$NaClO_4$	23.7	74.1 +

+) after 5 min at 50° C and cooled down to room temperature a value of 83 was obtained.

It is generally accepted that fusion starts from aggregated vesicles where aggregation and also flocculation are reversible phenomena in contrast to fusion which is irreversible. At the present time we can not distinguish between aggregation and flocculation, though the temperature-dependence and the almost complete reversibility of the effect favour looser assemblies in the form of flocculations. The rates of aggregation and flocculation on one side and the rate of fusion on the other side might depend on the lipid to ion ratio.In the case of poly(ethylene glycol) (PEG) it is discussed that aggregates of vesicles occur at low and fusion at high PEG concentrations[11]. Whether similar conclusions hold for the action of the chaotropic ions studied is under current investigation.

Concerning any molecular mechanism one could speculate that the action of the structure breaking ions studied causes conformational changes at the polar head-group of the lipid-molecules. It is known from Raman studies[3] that structure breaking and structure forming anions have negligible influences on the packing of the hydrocarbon chains. Thus the membrane interior is not affected, however, the boundary layer at the lipid-water interphase and/or the conformation of the polar head-group might be changed due to the interaction with the chaotropic ions studied. Small changes at the polar head-group would be in agreement with early NMR results on the interaction of sodium halides with egg phosphatidylcholine vesicles[12]. Indeed, preliminary ^{31}P NMR spectra show differences between vesicles with and without small amounts of $NaClO_4$.

Support by the Fonds der Chemischen Industrie and the Deutsche Forschungsgemeinschaft (La 328/4-1) is highly acknowledged.

REFERENCES

1. F. Hofmeister, Zur Lehre von der Wirkung der Salze, Arch. Exptl. Pathol. Pharmakol. 24:247 (1888)
2. S. A. Simon, L. J. Lis, J. W. Kauffman and R. C. MacDonald, A calorimetric and monolayer investigation of the influence of ions on the thermodynamic properties of phosphatidylcholine, Biochim. Biophys. Acta 375:317 (1975)
3. L. J. Lis. J. W. Kauffman and D. F. Shriver, Effect of ions on phospholipid layer structure as indicated by Raman Spectroscopy, Biochim. Biophys. Acta 406:453 (1975)
4. M. K. Jain and N. M. Wu, Effect of small molecules on the Dipalmitoyl Lecithin liposomal bilayer: III phase transition in lipid bilayer, J. Membrane Biol. 34:157 (1977)
5. M. Gründel, R. Grupe, E. Preusser, H. Göring, Fluorimetrische Messungen zur Lipid-Ionenwechselwirkung an Single-Liposomen aus Eilecithin und Eilecithin-Cholesterin, studia biophysica 58:203 (1976)
6. J. N. Israelachvili, S. Marcelja and R. G. Horn, Physical principles of membrane organization, Quart. Reviews of Biophysics 13:121 (1980)
7. R. Lawaczeck, Intervesicular lipid transfer and direct fusion of phospholipid vesicles: a comparison an a kinetic basis, J. Colloid Interface Sci. 66:247 (1978)
8. A. D. Bangham, J. De Gier and G. D. Greville, Osmotic properties and water permeability of phospholipid liquid crystals, Chem. Phys. Lipids 1:225 (1967)
9. R. M. Blum and R. E. Forster, The water permeability of erythrocytes, Biochim. Biophys. Acta 203:410 (1970)
10. R. E. Verrall, Infrared spectroscopy of aqueous electrolyte solutions. Water a comprehensive treatise (F. Franks, Ed.)3 pp.211-264, Plenum Press New York 1973
11. L. T. Boni, T. P. Stewart, J. L. Alderfer and S. W. Hui, Lipid - Polyethylene Glycol Interactions:I. Induction of fusion between liposomes, J. Membrane Biol. 62:65 (1981)
12. G. L. Jendrasiak, Halide interaction with phospholipids: proton magnetic resonance studies, Chem. Phys. Lipids 9:133 (1972)

THE STRUCTURE OF SOME LECITHIN MONOLAYERS AT THE AIR/WATER INTERFACE

Maria Tomoaia-Cotişel, Emil Chifu and János Zsakó

Department of Physical Chemistry
"Babes-Bolyai" University, Arany J. 11
3400 Cluj-Napoca, Romania

INTRODUCTION

Generally, the monolayers are considered as halves of bilayers, and their physical properties can be compared to those of the biological membranes. Moreover, these systems fit to mimic biomembranes, providing considerable insight into the structural and functional role of their constituents, such as lecithins.

In our previous paper[1] the monolayer properties of some phospholipids, dipalmitoyl lecithin (DPL) and egg lecithin (EL), have been investigated by means of compression isotherms (surface pressure versus molecular area) at the air/water interface. In the present work this study is continued by investigating a similar phospholipid, namely distearoyl lecithin (DSL).

EXPERIMENTAL

The DSL used in our experiments was a synthetic commercial anhydrous L-α-distearoyl phosphatidylcholine (Sigma).

The monolayers were spread at the air/water interface from hexane solutions containing 2% - 10% absolute ethanol, in which DSL had been originally dissolved to obtain solutions of (4 to 8) . 10^{-4} M concentrations. The subphase was bidistilled water. Before compression the waiting time, ensuring evaporation of the solvent, was varied between 5 and 30 min.

The surface pressure versus area (π - A) curves were recorded discontinuously by using the Wilhelmy method, as previously described.[2] In our experiments compression speeds ranging between

0.02 and 0.08 nm^2/molecule.min were used, which allowed a curve to be recorded in about 50 - 15 min. No effect of waiting time and of compression speed on the (π — A) isotherms was observed. The measurements were made at room temperature (22 ± 2 °C). The accuracy of the surface pressure measurements was ± 0.5 mN/m and the (π - A) curves reproducible to within ± 0.02 nm^2/molecule. Eventually the (π - A) isotherm was selected from over 10 experiments.

RESULTS

The surface pressure versus area curve of DSL is given in Fig. 1 (curve a), together with the compression isotherms of DPL (curve b) and EL (curve c), the latters being taken from our previous paper.[1]

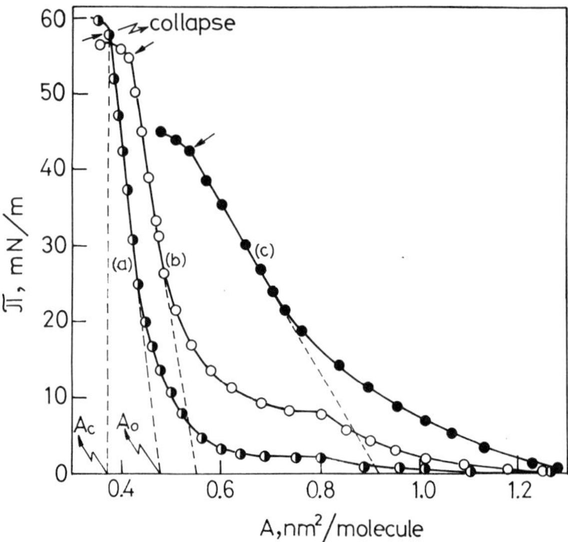

Fig. 1. Surface pressure - area curves of DSL (a), DPL (b) and EL (c) monolayers at the air/water interface.

As seen in this figure, the curves of DPL and DSL are of a similar shape and quite close to each other, and rather different from the compression isotherm of the egg lecithin film, the latter being on the one hand, more expanded than the former ones and on the other hand, it collapses at a much lower surface pressure than the monolayers of DPL and DSL.

On the compression isotherms of DPL and DSL a slope discontinuity is observed at approximately the same molecular area, of about 0.8 nm^2/molecule, but for EL no such discontinuity appears. This slope discontinuity will be referred to as "phase change". It is worth mentioning that a similar phase change in DPL monolayers has also been reported by other authors[3], and it was observed at about 0.76 nm^2/molecule in good agreement with our results. The same authors also give the compression isotherm of DSL,[3] but only for $A < 0.60$ nm^2/molecule and this is why the phase change has not been observed by them. In the literature also another surface pressure versus area curve is given for DSL[4] and even from as high molecular area values as A = 1.50 nm^2/molecule. These authors do not report a phase change, although their curve is very close to ours and its shape does not exclude the possibility of a phase change at about 0.90 nm^2/molecule.

Obviously, from the isotherms recorded by us, some characteristics of the monolayer could be derived. These surface properties: limiting molecular area A_o, collapse area A_c, collapse pressure π_c and surface compressibility C_s for surface pressures higher than 25 mN/m that can give information on the structure of lecithin monolayers are presented in Table 1. π_c and A_c values obtained for DSL are in good agreement with literature data[5].

DISCUSSIONS

Since the three lecithins (DSL, DPL and EL) have the same polar headgroup, the differences noted in the surface properties can be caused only by the length or the nature of the hydrocarbon chains of the substances. The unsaturated character of one of the

Table 1. Surface Characteristics of Lecithin Monolayers at the Air/Water Interface

Compound	A_o, nm^2	A_c, nm^2	π_c, mN/m	C_s, m/mN	Ref.
DSL	0.48	0.37	58	0.00389	this paper
DPL	0.55	0.42	55	0.00437	1
EL	0.92	0.54	42.5	0.00980	1

hydrocarbon chains of egg lecithin confers a partial rigidity to this chain, and in consequence this rigidity does not allow the hydrocarbon chain to attain a parallel orientation in the monolayers. This fact accounts for the high molecular areas A_o and A_c on the one hand, as well as for the higher compressibility of egg lecithin as compared to DPL or DSL, on the other. The absence of a parallel orientation makes the intermolecular van der Waals forces weaker, what leads to a lower collapse pressure.

Limiting molecular area A_o and collapse area A_c are close for the two saturated phospholipids, yet a little bit smaller for the values corresponding to DSL film. This fact illustrates the importance of the van der Waals forces between methylene groups of long hydrocarbon chains, what justifies the high values of the collapse surface pressure for DSL film. Considering the values of the surface compressibility, it may be noted that the phospholipids investigated display characteristic features of the typically condensed films[6] at surface pressures higher than 25 mN/m.

In order to find a correlation between the surface characteristics of the monolayers and the molecular structure of phospholipids, our rotating quasi-rigid prism model has been used. This model,[1] is based on the following assumptions:

1. The molecule is anchored in the water phase by means of its polar headgroup and the hydrocarbon chains are in the air phase;

2. The simple σ-bonds of the molecule allow almost free conformational transitions, but due to collision of the molecules, conformations with minimum space requirements will be preferred. At high surface pressures, this leads to a vertical orientation of the chain axes. This orientation will be favoured, and even stabilized by the intermolecular van der Waals forces. Thus the phospholipid molecules of the monolayer can be considered as quasi-rigid prisms;

3. The vertical molecular prisms have a random orientation, which changes permanently due to the collisions. The mean area requirements of a molecule will be the cross-sectional area of a cylinder, obtained by the rotation of the molecular prism around its vertical axis;

4. Near the collapse pressure, the rotation of the molecular prisms is presumed to be hindered and the area requirement of the molecules will be the cross-sectional area of the prism.

By presuming the vertical orientation of the hydrocarbon chains, the molecular area of DPL and DSL must be the same. This molecular area has been calculated for several extreme conformations of the

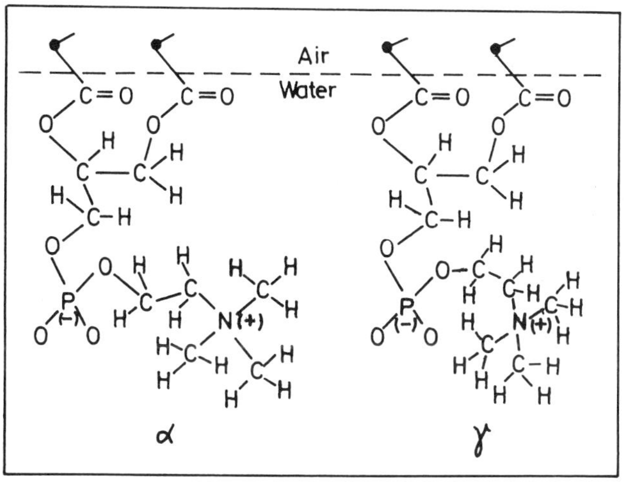

Fig. 2. Extended "zwitterion " (α) and folded "internal salt" (γ) conformations of the polar headgroup of lecithin molecules.

polar headgroups. Two of them, the extended "zwitterion" conformation, denoted by α , and the folded "internal salt" conformation, denoted by γ , are given in Fig. 2. In both conformations the dipole moment of the polar headgroup is presumed to be parallel with the interface, in agreement with literature data concerning the surface dipole moment values of DPL monolayers.[7]

Table 2. Molecular Areas of the Phosphatidylcholine Group in Different Conformations

Conformation	A_4, nm^2	A_6, nm^2	A_p, nm^2
α hydrated	1.25	1.09	0.81
α non-hydrated	0.88	0.77	0.40
γ	0.69	0.60	0.40

Three molecular area values have been calculated: A_4 represents the area requirement assuming a tetragonal close packing of the rotating prisms; A_6 represents the area requirement assuming a hexagonal close packing of the rotating prisms; A_p corresponds to the close packing of non-rotating prisms. These molecular area values, taken from our above mentioned paper,[1] are presented in Table 2 for conformations α, hydrated and non-hydrated, as well as for conformation γ.

It is worth mentioning that in both conformations " α non-hydrated" and γ, the molecule must not be presumed as being completely free of hydration water. Especially the PO_2^- moiety is very suited to fix water molecules. Due to the delocalized O - P - O π -bond an sp^2 hybridization can be presumed for both O - atoms. Thus, the longitudinal axes of the oxygen orbitals, containing the non-participant electron pairs, must lie in the same plane as the P and both O atoms. Since the negative charge of the PO_2^- moiety is localized in the O - atoms, the formation of H bridges with water molecules must be very favoured and one can presume the PO_2^- to fix four water molecules. The possible structure of this hydrated PO_2^- moiety is given in Fig. 3. In this figure the images of the oxygen orbitals containing the non-participant electron pairs are hatched. If the O···H – O hydrogen bridges are linear, all atoms situated in the area surrounded by a dashed line (Fig. 3) can be thought to have their nuclei in the same plane, which must be almost perpendicular to the interface. The water molecules linked to the PO_2^- moiety do not affect essentially the molecular area values calculated.

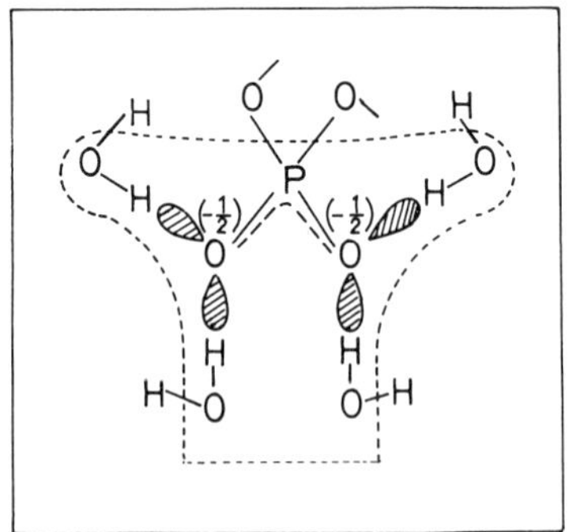

Fig. 3. Structure of hydrated PO_2^- moiety.

In the case of conformation "α hydrated" the molecular area values are increased mainly due to the presumed hydration of the quaternary ammonium moiety of the polar headgroup. Thus, our results and the above given considerations indicate only that the hydration shell of the quaternary ammonium moiety seems to contain no tightly bonded water molecules.

By comparing Tables 1 and 2, it is obvious that our experimental A_o data for DPL and DSL are quite close to A_6 value for conformation γ. Consequently, one may presume conformation γ to be preferred in the liquid condensed state. Since the A_o values, especially that of DSL, are even lower than A_6, the free rotation of the molecules, or their random orientation, must be partially hindered, which suggests the idea that micelles with close packed non-rotating molecules could appear at high surface pressure values. The experimental A_c values of DPL and DSL are practically identical to A_p calculated for both conformations, α non-hydrated and γ, pleading for the close packing of non-rotating molecules before collapse and excluding conformation α hydrated.

As far as the phase changes in DPL and DSL monolayers are concerned, we may assert that these can be attributed to a transition from the state of liquid expanded to that of liquid condensed. Since the phase change begins in both cases at a molecular area of about 0.8 nm^2, which is very close to the A_p value for the conformation "α hydrated" and also to the A_4 value for the conformation "α non-hydrated", one can presume the extended α conformation to be preferred at low surface pressures and the phase change to imply mainly an $\alpha \longrightarrow \gamma$ conformational transition. Investigations made on the behaviour of DPL monolayers, performed by means of an epifluorescence microscope,[8] revealed below 25 ^oC, a homogeneous fluorescence distribution even in the region of the "phase change" observed. This seems to be in agreement to our above hypothesis.

On the other hand, at such low surface pressures as 8 mN/m (DPL) and 2 mN/m (DSL), a close packing corresponding to an A_p value is not likely. Consequently, conformation "α hydrated" seems not to be realistic even at low surface pressures. The higher surface pressure at which the phase change of DPL begins, as compared to DSL, might be a consequence of the different length of the hydrocarbon chains of the two saturated phospholipids which leads to weaker van der Waals forces in the case of DPL as compared to DSL.

In reality, the phase changes observed could be complex phenomena: besides the conformational transition of the phosphorocholine group oriented parallel to the interface, other processes, such as partial elimination of the hydrate shell, formation of bidimensional micelles of oriented dipoles, fusion of such micelles, may also be present, the possibility of orientational transitions of the hydrocarbon chains cannot be excluded either.

The phospholipid conformation in a monolayer is a result of a balance of interaction between polar groups, between hydrocarbon chains, hydrocarbon chains with polar groups and polar groups with water as well as non-polar groups with water (hydrophobic hydration). Cooperativity of these effects in the monolayers may also mean that the changes in polar group interactions affect the interactions between the hydrocarbon chains and viceversa.

In conclusion, we consider that the compression isotherms, together with our molecular model can give valuable information concerning the conformation of the phospholipid molecules in monolayers, the structure of these films, and the processes occurring at phase changes.

REFERENCES

1. M. Tomoaia-Cotişel, J. Zsako and E. Chifu, Dipalmitoyl lecithin and egg lecithin monolayers at an air/water interface, Ann. Chim. (Rome), 71:189 (1981).
2. E. Chifu, M. Tomoaia and A. Ioanette, Behaviour of canthaxanthin at the benzene/water and air/water interfaces, Gazz. Chim. Ital., 105:1225 (1975).
3. M. C. Phillips and D. Chapman, Monolayer characteristics of saturated 1,2-diacyl phosphatidylcholines (lecithins) and phosphatidylethanol amines at the air/water interface, Biochim. Biophys. Acta, 163:301 (1968).
4. L. L. M. van Deenen, U. M. T. Houtsmuller, G. H. de Haas and E. Mulder, Monomolecular layers of synthetic phosphatides, J. Pharm. Pharmacol., 14:429 (1962).
5. P. Joos and R. A. Demel, The interaction energies of cholesterol and lecithin in spread mixed monolayers at the air/water interface, Biochim. Biophys. Acta, 183:447 (1969).
6. J. T. Davies and E. K. Rideal, "Interfacial Phenomena", p. 265, Acad. Press, New York and London (1963).
7. F. Vilallonga, Surface chemistry of L- α - dipalmitoyl lecithin at the air/water interface, Biochim. Biophys. Acta, 163:290 (1968).
8. V. von Tscharner and H. M. McConnell, An alternative view of phospholipid phase behaviour at the air/water interface, Microscope and film balance studies, Biophys. J., 36:409 (1981).

WATER AND IONS IN HETEROGENEOUS SYSTEMS

AN ESTIMATION OF THE DENSITY OF WATER IN *ARTEMIA* CELLS

James S. Clegg

Department of Biology
University of Miami
Coral Gables, FL. 33124 USA

INTRODUCTION

This paper is part of a long-range research program on the properties and roles of water in the cells of *Artemia* cysts[1,2]. Each cyst is a mass of about 4000 cells surrounded by a shell, most of which can be chemically removed[3]. This system is a useful model for such studies because the cells are capable of repeated cycles of dehydration-rehydration in a fully reversible way. Thus, one can measure a given property as a function of cell water content in *viable* animal cells. The cysts are available in large amounts, and details of their ultrastructure, biochemistry and biophysics are available[3]. In the present case I describe measurements of cyst density as a function of cell water content, and evaluate these data in terms of the density of intracellular water. A preliminary account has been published[4].

MATERIALS AND METHODS

Cysts from San Francisco Bay Brand, Newark, California, were washed, processed and stored at -20°C. All work was done on cysts whose outer shells had been removed by sodium hypochlorite treatment[4]. The thin inner cuticle that remains makes up less than 4% of the total volume and even less than that of the mass of cysts at maximum water content. Hence, results can be interpreted directly in terms of cellular properties. Cyst density (g/cc) was determined over the entire range of water content: $\approx$0 to 1.7 gH_2O/g dry mass (g/g) at 25°C[5].

Individual cyst densities were calculated from their sedimentation velocity through NaCl solutions in which the cysts were also hydrated to constant water content. Cysts are impermeable to ions. Sedimentation velocities were taken over 20 cm, and cyst density was calculated from the well known equation relating these two parameters[4]. That method is applicable only to cysts that have achieved sufficient water to become spherical (> 1.0 g/g). The density of cyst populations was measured by their displacement of n-heptane, using cysts of known mass and water content. A high precision 10.0 ml volumetric flask was weighed to the nearest 0.005 mg on a Mettler M5 microbalance. Cysts were added (between 200-300 mg) and the vessel reweighed to determine cyst mass. n-Heptane was added and the flask was weighed, giving the mass of heptane. The latter was converted to volume using the density of n-heptane, measured with 25 ml pycnometers. The average cyst density was then calculated.

The validity of this method, referred to as "mass-volume" (m-v) rests on the lack of interaction of n-heptane with the cysts, notably without changing their water content, and that the density and hydration of large masses of cysts accurately reflect those of individual cysts. Both of these conditions have been shown to apply[4].

RESULTS

The main part of Figure 1 shows average cyst density as a function of water content, using cyst populations (filled circles and triangles) and single cysts (open circles). Up to Wf of about 0.15, cyst density is not changed appreciably by water uptake. Above that point, the density decreases linearly. Vertical lines on the symbols in Fig. 1 are standard deviations: for m-v studies, n=5 and for single cysts, n >25. Symbols without error bars are single m-v deterinations. These errors are small, the standard deviation being about 0.5% of the mean cyst density. Note also that both methods produce the same results in the region of overlap (> 0.5 Wf).

The linear fit of these data is indicated by an $r^2 > 0.99$ (Fig. 1). It should be noted that the 4 data points below 0.15 Wf were not included in that statistical analysis. Coefficients of the regression equation at the top of Fig. 1 were obtained from 15 data points: 10 sedimentation-velocity and the 5 m-v measurements. Including data for single m-v measurements does not significantly alter the average values of the coefficients but does reduce r^2. Therefore, the relationship between cyst density and water content above 0.15 Wf is best described by this treatment.

The equation in the insert of Fig. 1 describes the relationship between the density of a solution and those of the solute and solvent as a function of weight fraction of solvent[6]. Although the hydrated cyst is hardly such a simple case I will apply the equation nonetheless, treating the hydrated cyst as the "solution" (subscript c), the non-aqueous components as the "solute" (s) and cyst water as the

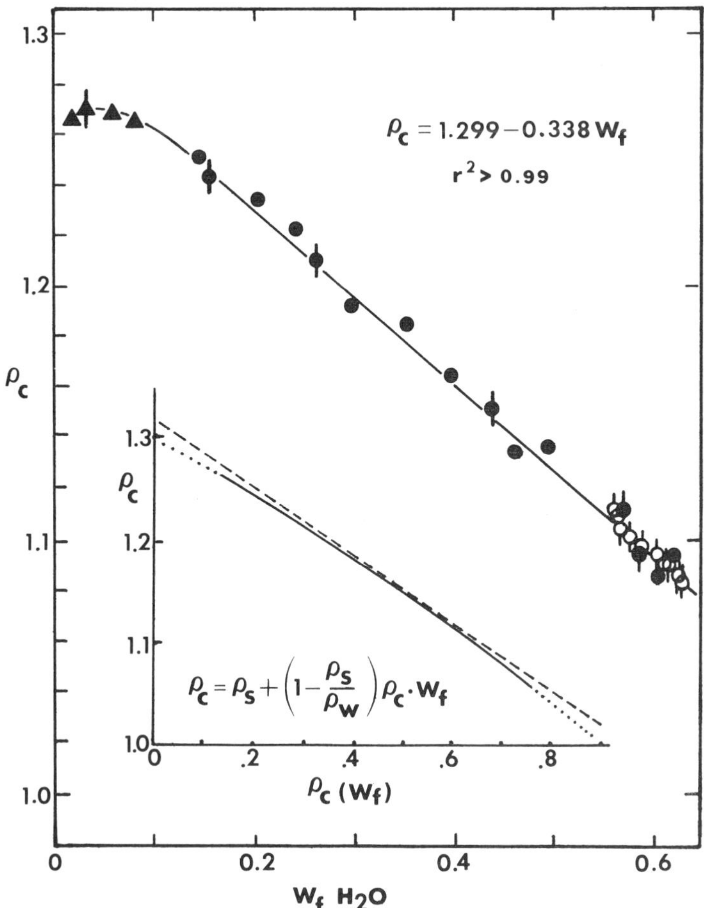

Fig. 1. Cyct dcnsity (ρ_c) as a function of the weight traction of water (W_f) and ρ_c·Wf (insert). The solid line in the insert represents data and the dashed line illustrates linearity. The dotted line is an extrapotation. See text for further description.

"solvent" (w). If the densities of the latter two were independent of each other as a function of Wf > 0.15 then the relationship shown in the insert should be linear with slope $(1- \rho_s / \rho_w)$. This is not the case for these cysts: the solid line represents cyst data and can be compared with the dashed line which has been added only to illustrate linearity. The slope changes from -0.28 at 0.15 Wf to -0.40 at 0.65 Wf, an increase of about 40%. Since the slope is determined by the ratio of the density of the nonaqueous components to that of cyst water it is evident that either one or, more likely both depend on cyst water content.

DISCUSSION

The first point to be made is that the "deshelled" cyst is effectively a mass of very tightly packed eucaryotic cells, with no detectable extracellular space[3]. Consequently, the data can be interpreted in terms of <u>intracellular</u> properties, including those of the water present.

Water added to almost completely dried cells causes very little change in density up to 0.15 Wf. A number of explanations are possible. Electrostriction of water molecules will occur in such severely dried cells, and that water should have a density greater than the pure water being added. Also, contractions in volume accompany the aqueous dissolution of certain solutes. Free glycerol is present in these cells at a concentration of about 5% by weight[5] and the addition of water to anhydrous glycerol results in a negative volume of mixing. Another contribution could arise from hydration-induced changes in the conformation of nonaqueous cell components, notably in macromolecules and ultrastructures such as membranes and organelles. Likely all of these are involved but, in the absence of detailed information about the behavior of all of these structures, little more can be said.

Cell density decreases when the water content exceeds 0.15 Wf and is a linear function of water content thereafter. What do these data tell us about the density of cell water? It seems impossible to arrive at a completely firm and unambiguous conclusion because of interactions between water and the nonaqueous components as more and more water is added to the cells. The constantly changing slope (insert in Figure 1) probably reflects these interactions. Some insight can be gained by evaluating the ratio of the density of nonaqueous cell components (ρs) to that of cell water (ρw). This ratio changes from 1.28 at 0.15 Wf to 1.40 at 0.65 Wf. The latter can be used graphically to obtain an estimate of the "partial density" of intracellular water for cells at 0.65 Wf. That estimate is 0.96, about 4% below the density of pure water at the same temperature and pressure. As expected, comparable results are obtained when these data are analyzed in terms of the specific volumes (V) of cells as a function of Wf. Using the method of slopes and intercepts[6] the partial specific volumes of the nonaqueous cell components (V_1) and cell water (V_2) are obtained. For cells at maximum water content V_2 is close to 1.035 cc/g compared to 1.003 for pure water. Cumulative experimental error is less than these differences; therefore, it appears that the density of water in these cells is significantly lower than that of pure water. Of course, the validity of this estimate is based on the method of extracting partial volumes and densities.

My observations are in contrast to those of previous attempts to estimate the density of water in amphibian oocytes[8] and in skeletal muscle[9]. In both cases cell water density was estimated to be 1-2% more dense than the pure liquid.

What is the basis for the apparent reduction in water density in *Artemia* cells? Reduced density requires that fewer nearest neighbors exist compared to pure water, perhaps due to enhanced tetrahedral bonding. I speculate that this could be caused by the enormous surface area that is presented to water by cell ultrastructure. Such surface area is vast, at least 10^4 μm^2 in a single rat liver cell for example[11], and that is clearly a gross underestimation[10]. Surfaces do indeed influence water over distances of at least 30 Å in biological systems[11,12] and non-biological ones[13]. The effective distance might be much larger, in excess of 500 Å according to Drost-Hansen[14] who coined the term "vicinal water" to refer to such surface-altered water.

Etzler has developed a theory for the structure of vicinal water which is consistent with the known properties of water near surfaces[15]. One of its predictions is that the density of vicinal water should be close to 0.97 g/cc. That prediction was confirmed[16] by studies on water-silica gel systems in which the water had a measured density of 0.966 $\pm$ 0.003 g/cc, remarkably (and likely fortuitously) close to the value obtained for the water in fully hydrated *Artemia* cells.

I note some additional predictions of Etzler's model for vicinal water: (1) Its permittivity should be lower than pure water, and we have presented microwave dielectic evidence for this in *Artemia* cells[2]; (2) The self-diffusion coefficient (D) of vicinal water should be reduced by a factor of about two, a prediction that is in fair agreement with results of pulsed NMR studies on D for *Artemia* cell water[1] and very close to the value of D obtained by quasi-electic neutron scattering[17]; (3) The partial specific heat capacity of *Artemia* cell water has been estimated to be about 20% higher than pure water, and that result[11] (although somewhat preliminary) is also in keeping with Etzler's theory.

ACKNOWLEDGMENTS

I thank the organizers of the Conference and their Romanian associates for the generous hospitality extended to me during and after the scientific session. The skilled assistance of Susan Dooley in the preparation of the manuscript is gratefully acknowledged. Research was supported by grant PCM 7925609 from the NSF (USA).

REFERENCES

1. P.K. Seitz, D.C. Chang, C.F. Hazlewood, H.E. Rorechach, and J.S. Clegg. The self-diffusion coefficient of water in Artemia cysts, Arch. Biochem. Biophys. 210:517 (1981).
2. J.S. Clegg, S. Szwarnowski, V.E.R. McClean, R.J. Sheppard, and E.H. Grant. Interrelationships between water and cell metabolism in Artemia cysts X. Microwave dielectric studies. Biochim. Biophys. Acta in Press (1982).
3. G. Persoone, P. Sorgeloos, O. Roels, and F. Jaspirs, eds. "The Brine Shrimp, Artemia" Vols. 1-3, Universa Press, Wettern Belgium (1980).
4. J.S. Clegg, and W. Drost-Hansen. On the density of intracellular water, J. Biol. Phys. 10:75 (1982).
5. J.S. Clegg. Interrelationships between water and cell metabolism in Artemia cysts VIII. Sorption isotherms and derived thermodynamic quantities. J. Cell Physiol. 94:123 (1978).
6. E.A. Moelwyn-Hughes, "Physical Chemistry", pp. 807-813, Pergamon Press, London (1961).
7. A.A. Newman, "Glycerol", Chemical Rubber Company Press, Cleveland (1978).
8. K. Hansson Mild, and S. Løvtrup. A note on the cell water density in amphibian eggs, J. Exp. Biol. 91:361 (1981).
9. S. Pócsik. Structure of water in muscle. Acta Biochem. Biophys. Acad. Sci. Hungary 4:395 (1969).
10. N.D. Gerehon, K.R. Porter, and B.L. Trus. The microtrabecular lattice and the cytoskeleton: Their volume, surface area and the diffusion of molecules through them, J. Cell Biol. 95:406 (1982).
11. J.S. Clegg. Metabolism and the intracellular environment: The vicinal water network model, in: "Cell-Associated Water", W. Drost-Hansen and J.S. Clegg, eds., Academic Press, New York (1979).
12. L.J. Lis, M. McAlister, N. Fuller, R.P. Rand, and V.A. Parsegian. Interactions between neutral phospholipid membranes, Biophys. J. 37:657 (1982).
13. J. Israelachvili. Double layer, van der Waals and hydration forces between surfaces in electrolyte solutions, in: "Biophysics of Water", F. Franks, ed., John Wiley and Sons, Inc., New York (1983).
14. W. Drost-Hansen. Water at biological interfaces-structural and functional aspects, Phys. Chem. Liq. 7:243 (1978).
15. F. Etzler. A statistical thermodynamic model for water near solid

interfaces, J. Coll. Interf. Sci. in Press (1983).
16. F.M. Etzler and D.M. Fagundus. The density of water and some other solvents in narrow pores, J. Coll. Interf. Sci. in Press (1983).
17. H.E. Rorschach, E.C. Trantham, D.B. Heidorn, C.F. Hazlewood, J.S. Clegg, R.M. Nicklow and N. Wakabayashi. The diffusion motion of protons in pure water, agarose gel and *Artemia* cysts by quasi-electic neutron scattering, in: "New Trends in the Study of Water and Ions in Biological Systems", V. Vasilescu, ed., Romanian Press, Bucharest, in Press (1983).

MICROWAVE CONDUCTIVITY OF FREE, "BULK" AND BOUND WATER

George Masszi and Ladislas Koszorus

Biophysical Institute
Medical University
Pécs, Hungary

Ultra-short radiowaves have been used in the research of biological structures since Höber's first investigation Höber (1910). Radiofrequency measurements in our Institute have shown the presence of bound K in muscle (Ernst, 1935; Masszi, 1958; Masszi, Tigyi-Sebes, 1962; Ernst, 1963).

The development of the microwave technique made it possible to use it for studying the structure of water (Hasted et al., 1948; Hasted, 1961) and the interrelation between the water structure and the biological organization (Cook, 1951; Grant, 1965; Masszi, Orkényi, 1967; Pennock, Schwan, 1969; Kaatze et al., 1975, Föster et al., 1979; Gascoyne et al., 1981; Koszorus, Masszi, 1982).

A schematic diagram of our dielectric microviscosimeter can be seen in Figure 1.

The conductivity (σ_o) and permittivity (λ_s) can be calculated from the measurement of standing wave-ratio and the voltage leaving the cell (Masszi, Orkényi, 1967; Koszorus, Masszi, 1982).

Figure 2 shows the microwave conductivity of free water over a frequency range of 2-4 GHz at different temperatures.

It can be seen from the figure that the conductivity of water is in linear connection with the square of frequency over this frequency range:

$$\sigma \simeq \frac{\Delta\varepsilon'}{54} \lambda_s f^2 + \sigma_o \; [\mathrm{mScm}^{-1}]$$

where λ_s is a "relaxation length for long wavelengths" (Debye, 1929; Budő, 1949).

$\Delta\varepsilon$ is the microwave dielectric decrement,

σ_0 is a conductivity, which is constant over the microwave range.

Figure 3 shows the microwave conductivity of:

i. Normal frog muscles (Rana esculenta)

ii. Muscle soaked in distilled water for 72 hours, and then in 15 per cent polyethylene glycol solution of a molecular weight of 20 000 daltons for 24 hours at a temperature +2 - +4 °C in a refrigerator. Thus we were able to perform comparative measurements on "de-ionized" samples, having same water content as the normal muscle.

iii. Muscle soaked in distilled water.

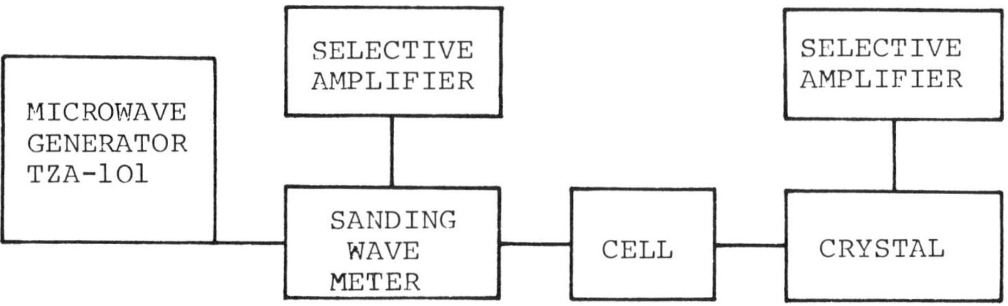

Fig. 1. Schematic diagram of the dielectric microviscosimeter.

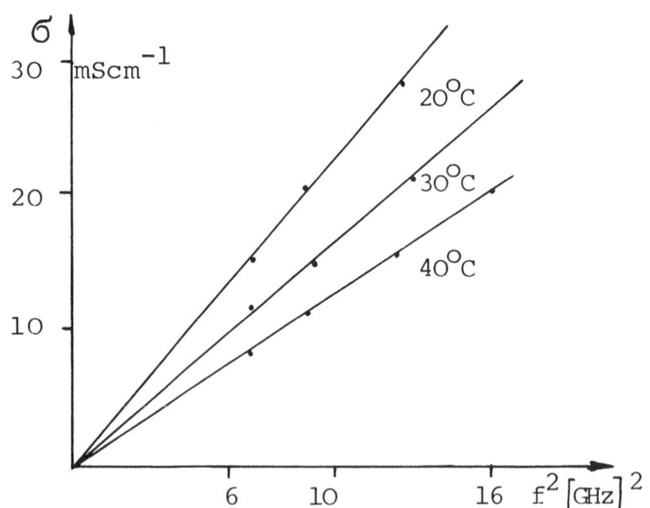

Fig. 2. Microwave conductivity of water.

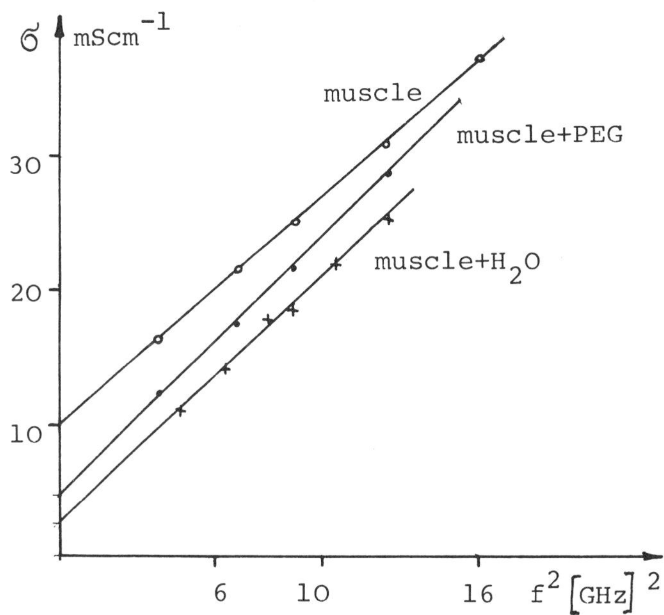

Fig. 3. Microwave conductivity of muscles.

Table 1. Relaxation Wavelength of Muscle Water (λ_s), and Microwave Conductivity of Muscle Extrapolated for 0 Frequency (σ_o)

N		H_2O		H_2O + PEG	
λ_s	σ_o	λ_s	σ_o	λ_s	σ_o
1.97	9.5	1.71	1.9	2.51	3.4
1.91	9.6	1.71	1.3	2.16	4.0
1.80	10.2	1.67	1.9	2.56	4.8
1.79	11.0	1.67	1.9	2.56	4.8
Average					
1.86	10.0	1.69	1.5	2.40	4.0

N: normal muscle
H_2O: muscle soaked in distilled water for 72 hours
H_2O + PEG: muscles soaked in distilled water for 72 hours, and then in 15 per cent PEG for 24 hours
λ_s : relaxation wavelength in cm
σ_o : specific conductivity in terms of $10^{-3}\Omega^{-1}cm^{-1}$ for distilled water;
λ_s = 1.53 cm, σ_o = 0.0

1. The relaxation wavelength of the "bulk" water can be calculated from the slope of the lines. It is longer than the relaxation wavelength of the free water, and so the microviscosity of "bulk" water is larger, too.

2. The axial sections of the straight lines represent a conductivity constant over the microwave range. This conductivity originates from the water molecules interacting directly with the proteins, having a relaxation wavelength of λ_S = 100 cm, and from the ionic conductivity of the samples, (Table 1).

There is a problem in the literature - whether the structure of the bulk water differs from that of the free water.

Table 2. There are some model-systems in the literature where the relaxation wave-length of the bulk water is larger than that of the pure water. (M: molecular weight in dalton c: concentration of solute).

Authors	Materials	
G. Masszi, 1972	Gelatine c = 0-50%	
U. Kaatze, 1975	Poly(vinylpirrolidone) M = 25 000 c $\geq$ 25%	
G. Masszi, Z. Szijártó, P. Gróf, 1976	Poly(ethyleneglycol) M = 20 000 c = 0-10%	
G. Masszi, G. Inzelt, P. Gróf, 1976	Poly(vinylalcohol) M = 110 000 c = 0-10%	
U. Kaatze, 1978	Poly(vinylpirrolidone) M = 12 000 - 350 000 c = 20-50%	
U. Kaatze, W. Y. Wen, 1978	Triethylenediamine in H_2O and D_2O M = 112 c = 0-35%	
C. Ballario, A. Bonincontro, C. Cametti, A. Di Biasio, 1980	Cellulose c = 30%	
C. Delbos, A. M. Bottreau, C. Marzat, J. L. Salefran, 1978	Dextran M = 11 000 c = 0-40%	
K. R. Foster, B. Epstein, P. C. Jenin, R. A. Mackay, 1982	O/W microemulsion (hexadecane, Tween 60 n pentanol C = 25-100%	A large quantity of interfacial water and bulk water

Table 3. There are some model-systems in the literature where the relaxation wave-length of the bulk water is shorter than that of the pure water.

Authors	Materials
U. Kaatze	C_{14}-Lecitin
H. Henze	C_{16}-Lecitin
A. Seegers	(intralamellar water)
R. Pottel 1975	λ_s 0.5 λ_{sH_2O}
J. P. Le Petit	water droplets
G. Delbos	(a few microns in
A. M. Bottreau	diameter) in oil.
J. Dutuit	$\lambda_s = 0.2\ \lambda_{sH_2O}$
C. Marzas	
R. Cabanas 1977	

Tables 2 and 3 show some model-systems where the relaxation wavelength of the bulk water is shorter and where it is longer than that of the pure water.

From these data it is difficult to explain why just the bulk water of the biological system is identical with the pure water.

The last table shows that the bound water is recently estimated to be two or three times larger than that which the former data showed.

Table 4. The bound or interfacial water, has been estimated with dielectric method in hemoglobin solution.

Schwan, H. P.	1957	0.3 - 0.4	gH_2O/gHb
Pennock, B. R. Schwan, H. P.	1969	0.2	-"-
Foster, K. R. et al.	1982	0.6	-"-

I think that the distribution of bound (or interfacial) and bulk water is an artificial distinction, and the structure of the water is more uniform in a system. The water-structure is like a fingerprint containing a considerable amount of information to characterize the subject to which it belongs.

REFERENCES

Ballario, C., Bonincontro, A., Cametti, C., Di Biasio, A., 1980, Dielectric properties of water-cellulose system at microwave frequencies, J.Coll.Interface Sci., 78:242.
Budó, A., 1949, Dielectric relaxation of molecules containing rotating polar groups, J.Chem.Phys., 17:686.
Cook, H. F., 1951, Dielectric behavior of human blood at microwave frequencies, Nature, 168:247.
Debye, P., 1929, in: "Polare Molekeln," S. Hirzel, Leipzig.
Delbos, G., Bottreau, A. M., Marzat, C., and Salefran, L. J., 1978, Microwave dielectric relaxation of aqueous solutions of dextran, J.Microwave Power 13:69.
Ernst, E., 1935, The Sechenov.J.Phys., USSR 21. Prog.XV Int. Physiol.Kongr.
Ernst, E., 1963, in: "Biophysics of the Striated Muscle," Akad.Kiadó Budapest.
Foster, K. R., Schepps, J. L., and Schwan, H. P., 1980, Microwave dielectric relaxation in muscle. A second look, Biophys.J., 29:271.
Foster, K. R., Epstein, B. R., Jenin, P. C., and Mackay, R. A., 1982, Dielectric studies on nonionic microemulsions, J.Colloid Interface Sci., 88:233.
Foster, K. R., Schepps, J. L., and Epstein, B R., 1982, Microwave dielectric studies on proteins, tissues, and heterogeneous suspensions, Bioelectromagnetics 3:29.
Gascoyne, P. R. C., Pethig, R., and Szent-Györgyi, A., 1981, Water structure-dependent charge transport in protein, Proc.Natl. Acad.Sci.USA., 78:261.
Grant, E. H., 1965, The structure of water neighboring proteins, peptides, and amino acids as deduced from dielectric measurements, Ann.New York Acad.Sci., 125:418.
Hasted, J. B., Ritson, D. M., and Collie, C. H., 1948, Dielectric properties of aqueous ionic solutions, J.Chem.Phys., 16:1.
Hasted, J. B., 1961, The dielectric properties of water, Prog.Dielectrics., 3:103.
Höber, R., 1910, Eine Methode die elektrische Leitfähigkeit im Innern von Zellen zu messen, Pfl.Arch., 133:237.
Kaatze, U. 1975, Dielectric relaxation in aqueous solutions of polyvinylpyrrolidone, Adv.Mol.Relax, Proc., 7:71.
Kaatze, U., Henze, R., Seegers, A., and Pottel, R., 1975, Dielectric relaxation in colloidal phospholipid aqueous solutions, Ber. Bunsenges.Phys.Chem., 79:42.
Kaatze, U., 1978, Dielectric relaxation in aqueous solutions of polymers, Progr.Colloid Polymer Sci., 65:162.
Kaatze, U., and Wen, W. Y., 1978, Molecular motion and structure of solutions of triethylenediamine in H_2O and D_2O as studied by dielectric relaxation measurements, J.Phys.Chem., 82:109.
Koszorus, L., and Masszi, G., 1982, Investigation of hydration of macromolecules. II. Study of ethyleneglycol and 1-3-dioxane

solutions by dielectric method. Acta Biochim.Biophys.Acad.Sci. Hung., 17:237.
Le Petit, J. P., Delbos, G., Bottreau, A. M., Dutuit, J., Marzat, C., and Cabanas, R., 1977, Dielectric relaxations of emulsions of saline aqueous solutions, J.Microwave Power, 12:335.
Masszi, G., 1958, Messung der Leitfähigkeit des Muskels mit Hochfrequenzstorm, Acta Physiol.Acad.Sci.Hung., 12:78.
Masszi, G., and Tigyi-Sebes, A., 1962, The state of potassium in muscle investigated by high frequency, Acta Physiol.Acad.Sci. Hung., 22:273.
Masszi, G., and Orkényi, J., 1967, Microwave investigation of biological substances II. Acta Biochim.Biophys.Acad.Sci.Hung., 2:69.
Masszi, G., 1972, Dielectric relaxation and water structure in gelatine solutions, Acta Biochim.Biophys.Acad.Sci.Hung., 7:349.
Masszi, G., Szijjártó, Z., and Gróf, P., 1976, Microwave investigation of polyethyleneglycol solutions, Acta Biochim.Biophys.Acad.Sci.Hung., 11:190.
Masszi, G., Inzelt, G., and Gróf P., 1976, Investigation of hydration of macromolecules, Acta Biochim.Biophys.Acad.Sci.Hung., 11:45.
Pennock, B. R. and Schwan, H. P., 1969, Further observations on the electrical properties of hemoglobin-bound water, J.Phys.Chem., 73:2600.
Schwan, H. P., 1957, Electrical properties of tissue and cell suspensions, in: "Biological and Medical Physics V. Acad. Press, New York.

BOUND WATER IN PLANTS

Stephen Pócsik and Ladislaus Koszorus

Biophysical Institute, Medical University
Pécs, Hungary

According to former experiments carried out in our Institute the vapor pressure, (Ernst et al., 1950) and the density, (Pócsik 1966; 1967) of the remaining water of the gradually dried muscle (gastrocnemius, sartorius of the frog etc.) gradually increases, which shows in our opinion the increasing boundedness of the remaining water.

In our present experiments we have examined the density of the remaining water of gradually dried plant sections (tomato, red beet, kohlrabi, turnip etc.). We cut from the vegetable sections of 20 x 10 x 1 mm^3 with the aid of metal patterns. We dried the segments gradually, step by step and at each step we measured their mass in air (M_i) and in a liquid of a known density (M_{bi}). The liquid we used was benzene of 23°C ($d = 0{,}8747 g\ cm^3$). The mass ($M_{13} = m$) and the volume ($V_{13} = v$) of the dry substance were obtained after drying at 105°C.

From these data the volume ($V_i = \frac{M_i - M_{bi}}{d_b}$) and the average density ($D_i = \frac{M_i}{V_i}$) of the vegetable segment and the average density of its water ($d_i = \frac{M_i - m}{V_i - v}$) can be calculated at each relative water content ($i = \frac{M_i - m}{m}$).

The results of the calculations are shown in columns 5-9 of Table 1. The experimental studies are also shown: in columns 1-4 of the table the phase of the experiment, the time (t) and temperature (T) of drying and the pressure (P) in the exsiccator.

Table 1. Experimental Data of a Plant Section

1	2		3	4	5	6	7	8	9
Phase	t		T	P			$i=\frac{M_i-m}{m}$	$D_i=\frac{M_i}{V_i}$	$d_i=\frac{M_i-m}{V_i-v}$
	hours	days	(C°)	Hgmm	M_i (mg)	M_{bi} (mg)		(g/cm³)	(g/cm³)
1	0		4-5	200	1031	199	3,71	1,084	1,024
2	2		4-5	200	896	183	3,09	1,099	1,030
3	18		4-5	200	588	143	1,69	1,156	1,051
4	29		4-5	200	480	129	1,19	1,196	1,074
5		2	4-5	200	404	117	0,85	1,231	1,089
6		2,5	4-5	200	336	106	0,53	1,278	1,114
7		3,5	4-5	200	259,4	90,0	0,19	1,359	1,138
8		4	4-5	10	245,6	87,2	0,12	1,356	1,161
9		5	23	10	237,3	85,7	0,08	1,369	1,210
10		21	23	10	232,1	84,4	0,06	1,375	1,22
11		22	23	10	231,7	84,3	0,06	1,375	1,23
12		56	23	10	230,0	84,0	0,05	1,378	1,26
13		69	105	10	218,9	80,6	-	1,384	-

In the columns of Table 1 are seen successively: the experimental phase, the time (t) and temperature (T) of drying, the pressure in the exsiccator (P), the mass of section measured in air (M_i) and in benzene (M_{bi}), the relative water content (i), the average density of the plant section (D_i), the average density of the remaining water of the plant section (d_i).

According to Table 1 the water density of the gradually dried vegetable segments gradually increases. We obtained similar results in the case of muscle (Pócsik 1967).

In the case mentioned above we assumed that the density of the dry substance (m/v) is constant during drying. This assumption can be disputed. Therefore we investigated the density calculations by other methods. Thus by the analytical method the above assumption did not emerge.

In greater detail we have given the volume of the vegetable slice (V) as a function of the water content ($\underline{m}$ = M - m)

$$V = f(\underline{m})$$

In Fig. 1 V was plotted against $\underline{m}$.

The curve fitting was best when we fitted a straight line at the points of starting phases and a curve of the second order was fitted in the remaining points.

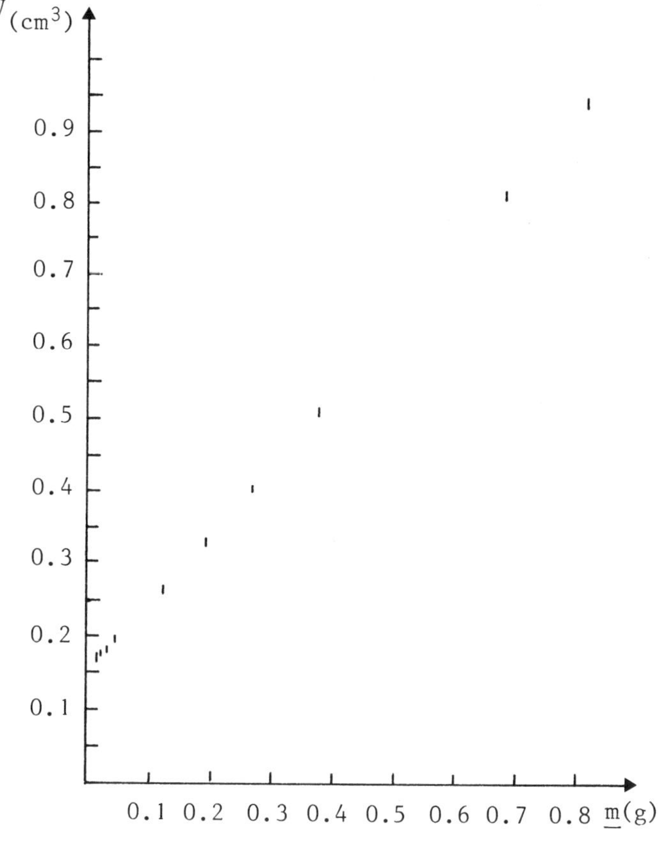

Fig. 1.

The equations of the curves are

$$V_{I} = 0{,}9917\ \underline{m} + 0{,}1443 \tag{1}$$

$$V_{II} = 2{,}1290\ \underline{m}^2 + 0{,}7968\ \underline{m} + 0{,}1580 \tag{2}$$

When differentiating these functions with respect to V we obtained the "differential" density of the plant water $(\frac{d\underline{m}}{dV})$

$$d_{I} = \frac{d\underline{m}}{dV_{I}} = \frac{1}{0{,}9917} = 1{,}008 = \text{const}$$

$$d_{II} = \frac{d\underline{m}}{dV_{II}} = \frac{1}{4{,}2580\ \underline{m} + 0{,}7968}$$

With the aid of the equations (1)(2) the average plant water density can also be calculated (Table 2).

Table 2. Values of the relative water content (i) the average density of water in plant section (d_i) the average density of plant section (D_i) at different phases of drying in plant (potatoes). (1) Supposing that the mass (m) and the volume (v) of the dry substance remain constant during drying. (2) Using analytical methods without supposing that m and v are constant

$$d_i = \frac{m_i}{V_i - v} \qquad D_i = \frac{m_i + m}{V_i}$$

Phase	i: g H_2O / g dry substance	d_i (g/cm^3) 1	d_i (g/cm^3) 2	D_i (g/cm^3) 1	D_i (g/cm^3) 2
1	3.71	1.024	1.026	1.084	1.086
2	3.09	1.030	1.029	1.099	1.098
3	1.69	1.051	1.048	1.156	1.152
4	1.19	1.074	1.065	1.196	1.190
5	0.85	1.089	1.090	1.231	1.232
6	0.53	1.114	1.143	1.278	1.290
7	0.19	1.138	1.132	1.359	1.339
8	0.12	1.161	1.171	1.356	1.358
9	0.08	1.210	1.196	1.369	1.369
10	0.06	1.22	1.21	1.375	1.374
11	0.06	1.23	1.21	1.375	1.375
12	0.05	1.26	1.22	1.378	1.376
13	0.00	-	-	1.384	1.385

The calculations have shown, that there is no significant difference between the results obtained by different methods.

Several authors have described that water density in different biological substances is higher than the density of common water (Pócsik, 1967; Bull and Breese, 1968; Bernhardt and Pauly, 1975; Hansson-Mild et al., 1979).

In our opinion the higher density of the biological water proves its higher boundedness.

REFERENCES

Bernhardt, J., and Pauly, H.J., 1975, Partial Specific Volumes in Highly Concentrated Protein Solutions. I. Water-Bovine Serum Albumin and Water-Bovine Hemoglobin, J.Phys.Chem., 79:584.

Bull, H. B., and Breese, K., 1968, Protein Hydration II. Specific Heat of Egg Albumin, Arch.Biochem.Biophys., 128:497.

Ernst, E., Tigyi, J., and Zahorcsek, A., 1950, Bindungszustand des Wassers und der Elektrolyte im Muskel, Acta Physiol.Acad. Sci.Hung., 1:5.
Hansson-Mild, K., Løvtrup, S., and Forslind, E., 1979, High Density Cell Water in Amphibian Eggs? J.Exp.Biol., 83:305.
Pócsik, S., 1966, Bound Water in Muscle, Acta Biochim.Biophys.Acad. Sci.Hung., 1:111.
Pócsik, S., 1967, Bound Water in Muscle, Acta Biochim.Biophys.Acad. Sci.Hung., 2:149.

THE ^{1}H NMR RELAXATION OF WATER IN AVIAN EGGS

C. Simaroj Thomas* and W. Derbyshire+

Department of Physics
University of Nottingham
Nottingham, NG7,2RD. U.K.

INTRODUCTION

Eggs were used as subjects during early developments of NMR imaging because they were of a size convenient for the existing magnets and because the development of the embryo in the fertilised egg provided an interesting system for study. As the image depended upon the relaxation properties of the different constituents parallel relaxation studies were undertaken to monitor changes in relaxation as functions of size and of age under different storage conditions. Eggs were collected, labelled, weighed and stored, at 6°C and low relative humidity, or at 36°C and 80% relative humidity to simulate incubation conditions. After an appropriate storage period the eggs were reweighed and the weight loss determined, broken open and four parts dissected out, the yolk, the thin and the thick white albumens and the chalazae. These were loaded into NMR sample tubes and various measurements undertaken. The reduction in weight during the ageing process is usually ascribed to a loss of water, mostly from the albumens, by evaporation through the shell. There may be an additional re-distribution of water between the albumen and yolk. As the NMR spectrometer was normally arranged to detect the slower relaxing ^{1}H components attributed to water, NMR relaxation should provide a reasonable mechanism for monitoring the water distribution and relaxation and implicitly motional properties. The rate of fractional weight loss was dependent upon storage condition and thermal history.

---

* Now at Department of Physics,Khan Kaen University,Khan Kaen, Thailand.
+ Now at Department of Physical Sciences,Sunderland Polytechnic.U.K.

THE 'H SPIN LATTICE RELAXATION

The 'H spin lattice relaxation of the three albumen samples, recorded over two orders of magnitude, were single component exponential within experimental error, whereas that of the yolk was markedly non-exponential and required two components for adequate characterisation. Reflecting the reduced water content the two spin lattice relaxation times were the order of 40 and 180 ms instead of the 1.1 to 1.2s recorded for the albumens. In previous work Fung et al[1] reported the occurrence of non-exponential 'H spin lattice relaxation of yolk samples, whereas James and Gillen[2] only observed single component relaxation in the range 58 to 72 ms consistent with the more rapidly relaxing component reported here. Measurements made on constituents from different eggs stored under equivalent conditions displayed a variation in relaxation times of the order of 10%. In spite of this, ageing effects and the influences of storage condition were apparent. Irrespective of the storage conditions the relaxation times of the three albumens, recorded at room temperature 20°C, showed an initial increase over a period of several days followed by a subsequent decrease, this is an initial increase in relaxation time in the presence of a weight loss assumed to be mostly of water from the albumen. A similar behaviour has been observed by Chang et al[3] in the early stages of rigor mortis in skeletal muscle. When the relaxation data were plotted against fractional weight loss instead of time, the scatter became significantly less and data for samples stored under different conditions appeared to be consistent, within experimental error and the sample variation referred to earlier. Several thermal regimes were studied, a few of which are shown in figure 2.
As the yolk relaxation was more complex than that of the albumens it was not surprising that the behaviour during storage was also correspondingly more complex. The relaxation time of the shorter component increased initially and then became constant, whereas that of the larger component increased to a maximum changing by a factor of two and then decreased. These changes were paralleled by changes in the relative populations of the two components, the relative amplitude of the short component displaying a maximum.

INTERPRETATION OF THE SPIN LATTICE RELAXATION DATA

Without absolute proof it has been convenient to assume that the recorded NMR signal can be attributed to water. However, it has not been established that all the water is recorded or that the recorded signal is entirely due to water. The data is interpreted in terms of the two phase model where it is assumed that a small fraction of the water is "bound" to more rigid constituents, and has an enhanced intrinsic relaxation rate due to a reduced mobility. It is further assumed that at temperatures above the bulk freezing point exchange between the more distant 'bulk' and the bound water is rapid and that averaged water relaxation rates

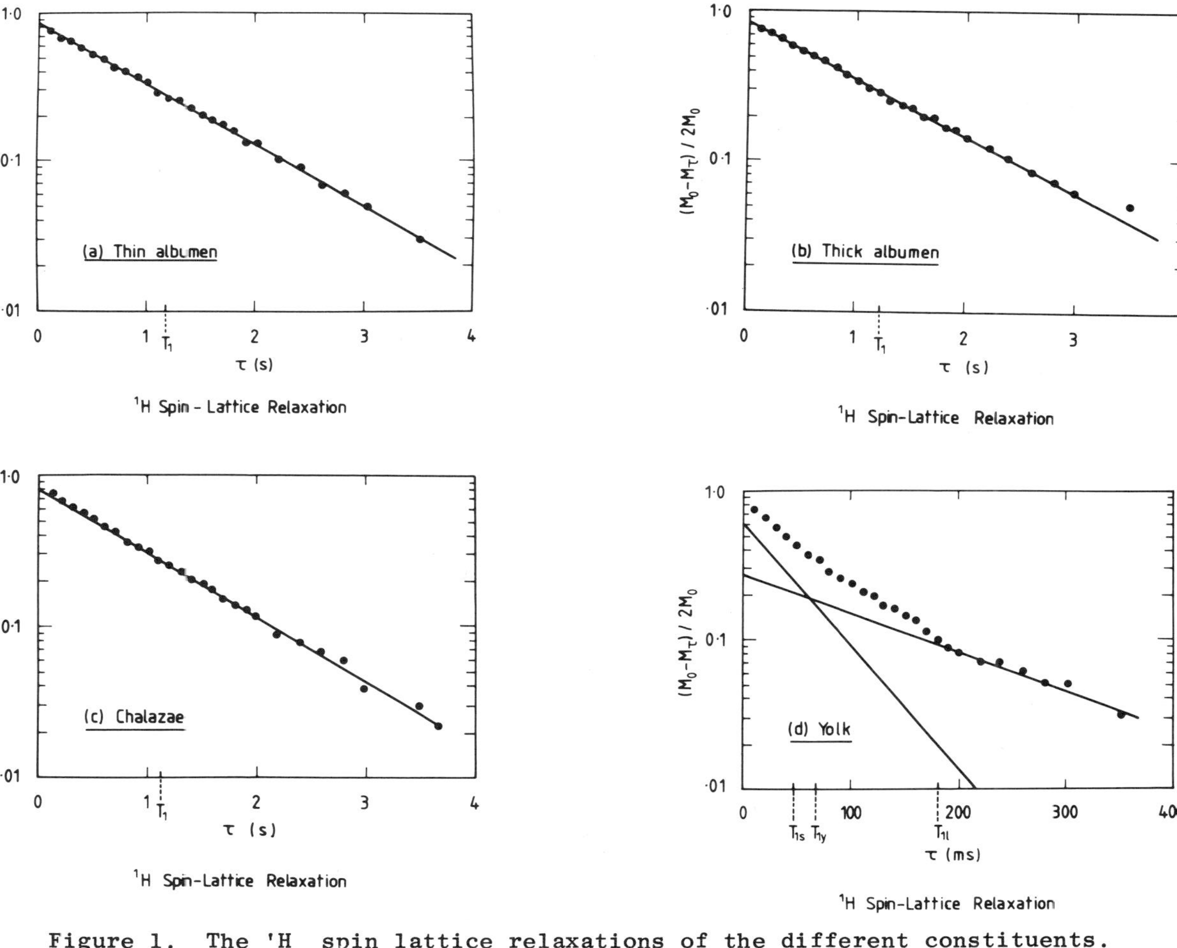

Figure 1. The 'H spin lattice relaxations of the different constituents.

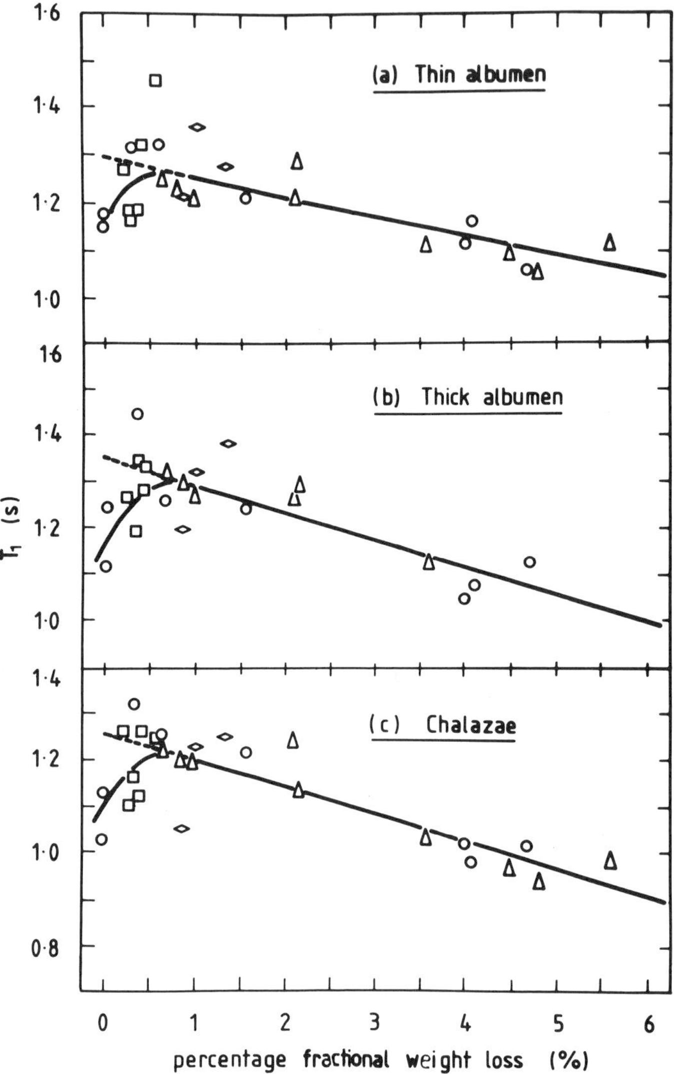

Figure 2. Graphs of the 'H spin lattice relaxation time against fractional weight loss.

o Eggs stored at 60°C.

□ Eggs stored at 35°C and a relative humidity of 80%

◇ Eggs stored at 35°C and a relative humidity of 80% for 9 days then at 6°C.

△ Eggs stored at 35°C and a relative humidity of 80% for 9 days, at 6°C for 4 days, and then at 35°C and a relative humidity of 80%.

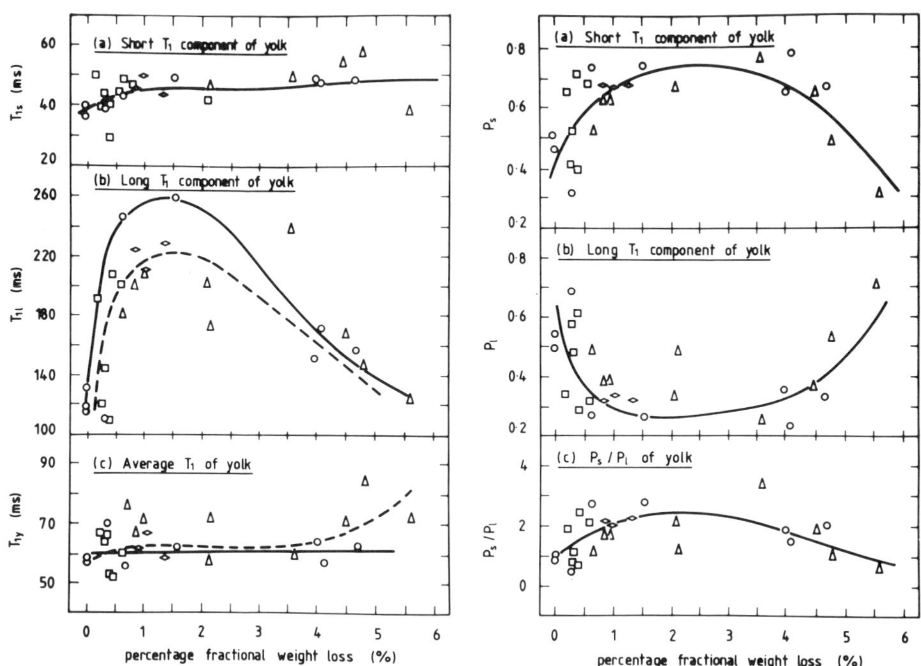

Figure 3. The dependences of the component 'H spin lattice relaxations of the yolks upon fractional weight loss (Ps and Pl represent the relative populations of the more rapidly and more slowly relaxing components).

are recorded. It is possible that spin exchange between the water and the host macromolecules and membrane protons might influence the results, affecting the numerical values of the relaxation times for the albumen and yolk samples, and possibly influencing, or even determining, the relative amplitudes of the two components of the yolk relaxation. Such spin exchange has been neglected in the treatments described below.

THE ALBUMENS

The reduction in the relaxation times of the albumen samples at longer storage times is generally consistent with a reduction in the free water content, a more careful analysis is required to establish if agreement is quantitative. However, the initial increase is not explicable in these terms. An increase in relaxation time in the presence of a reduction in water content implies a reduction in the amount of bound water,or in its intrinsic relaxation rate.

A convenient method of determining the bound water is to cool a sample to below the bulk freezing point. Under these conditions the bound water does not solidify,and as exchange between the bound water and the bulk ice phase is slow on an NMR timescale the NMR signals of the two "phases" are separately observable,and both the size and relaxation of the bound water can be obtained. Such measurements provided no evidence in support of the proposals of an initial decrease in the size of the bound water component or of its relaxation rate. If the water proton population of an albumen sample is P of which Pb is bound having a relaxation rate Rb and the remaining bulk water Pa is free having a relaxation rate Ra then at temperatures above the bulk freezing point where exchange between the bound and bulk phases is rapid

$$PR = P_a R_a + P_b R_b \text{ or } R = R_a + \frac{P_b}{P}(Rb-Ra)$$

It is assumed that there is a loss ΔP of free water during storage causing a reduction in the observed relaxation from the initial value R^o, and that the values R_b and R_a remain unchanged.

$$R = R_a + \frac{P_b}{(P^o - \Delta P)}(R_b - R_a)$$

$$\text{and } R^o = R_a + \frac{P_b}{P^o}(R_b - R_a)$$

$$\text{Hence } \frac{\Delta P}{P^o} = \frac{R - R^o}{R - R_a}$$

If W is the weight of the egg and α_j the weight fraction of the jth region, ρ_j the fraction of water in the jth region and β_j the fraction of the water loss ΔW due to the jth region then

$$\frac{\beta_j \Delta W}{\rho_j \alpha_j W} = \frac{R - R^o}{R - R_a}$$

Graphs of $(R - R^o)/(R - R_a)$ against the fractional weight loss are shown in figure 4 for samples of the thin and thick albumens and the chalazae. A choice of R^o has to be made, in figure 4a it is the original value, whereas in (b) it is the extrapolated value of the straight line portion of the graph. If the early portion of the graph representing the initial increase in relaxation time is ignored, the latter represents a better choice, with the least squares line fitted to the straight line portion passing through the origin. The graphs in figures 4c and d are also based upon an extrapolated value of R^o. Values of ρ and α have been estimated previously[4] for the various albumens. Hence values of β can be deduced from the gradient, these are 0.10 for the chalazae and 0.65 for the thin, and 2.05 for the thick albumens. The observation that the sum of the values of β exceeds unity implies an overestimate of the water loss from the albumens. It is conceivable that because of its low water content some water might have diffused into the yolk, and hence might not have contributed to the weight loss, but literature estimates are considerably lower than the 180% value necessary to account for these observations.

The treatment can be extended by assuming that the amount of bound water is proportional to the quantity of solid material present, equal to $h_j(1-\rho_j)\alpha_j W$ where h is a hydration value for each region

$$R = R_a + \frac{\left(P_b/P^o\right)}{\left(1 - \frac{\Delta P}{P^o}\right)}\left(R_b - R_a\right) = \frac{\frac{h(1 - \rho_j)}{\rho_j}\left(R_b - R_a\right)}{\left[1 - \frac{\beta_j}{\rho_j \alpha_j}\frac{\Delta W}{W}\right]}$$

Graphs of the observed relaxation rate R against $\left[1 - \frac{\beta_j}{\rho_j \alpha_j}\frac{\Delta W}{W}\right]^{-1}$

should be linear with gradients equal to $\frac{(1 - \rho_j)h}{\rho_j}(R_b - R_a)$

Values of $\beta_j/\rho_j\alpha_j$ can be obtained from the gradients of the previous graphs. The resulting graphs are linear if allowances are made for the inevitable scatter due to sample variation arising from the use of different eggs. The bound water relaxation rates can be deduced at 108, 117, 120 ms for the chalazae and thin and thick albumens, values generally consistent with but slightly larger than direct determinations of the bound water spin lattice relaxation.

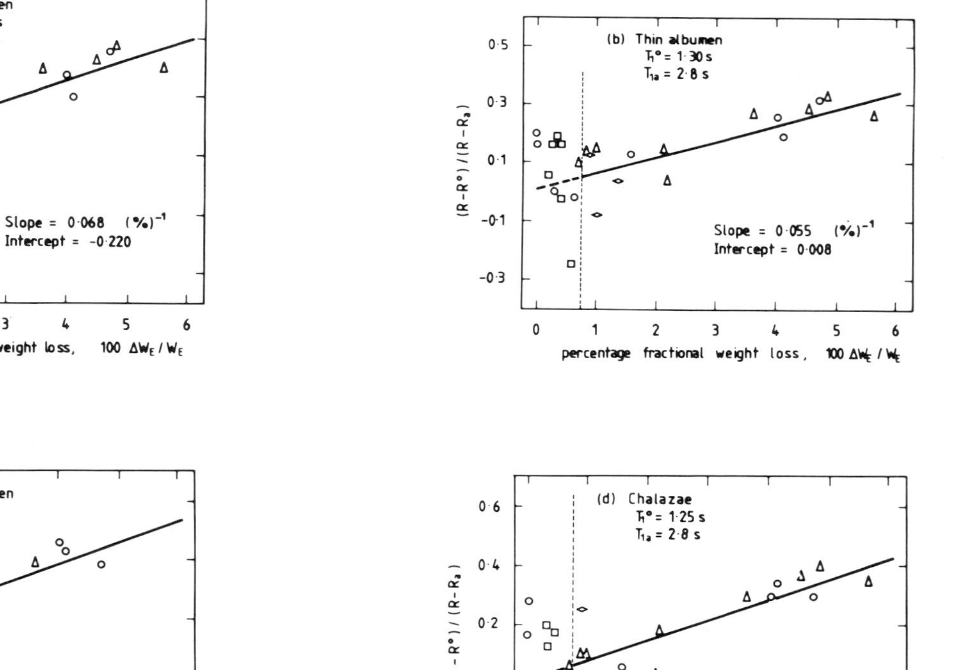

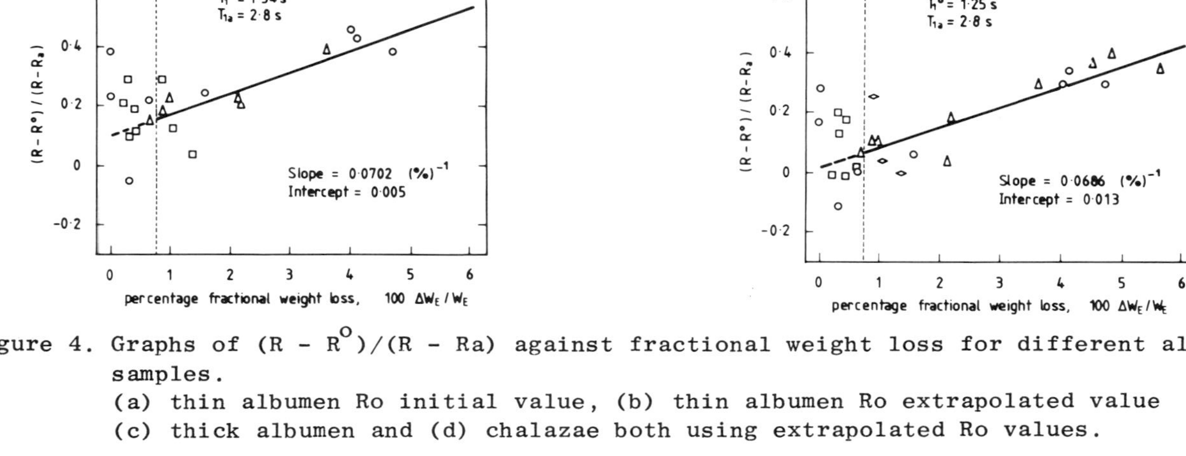

Figure 4. Graphs of $(R - R^O)/(R - Ra)$ against fractional weight loss for different albumen samples.
(a) thin albumen Ro initial value, (b) thin albumen Ro extrapolated value
(c) thick albumen and (d) chalazae both using extrapolated Ro values.

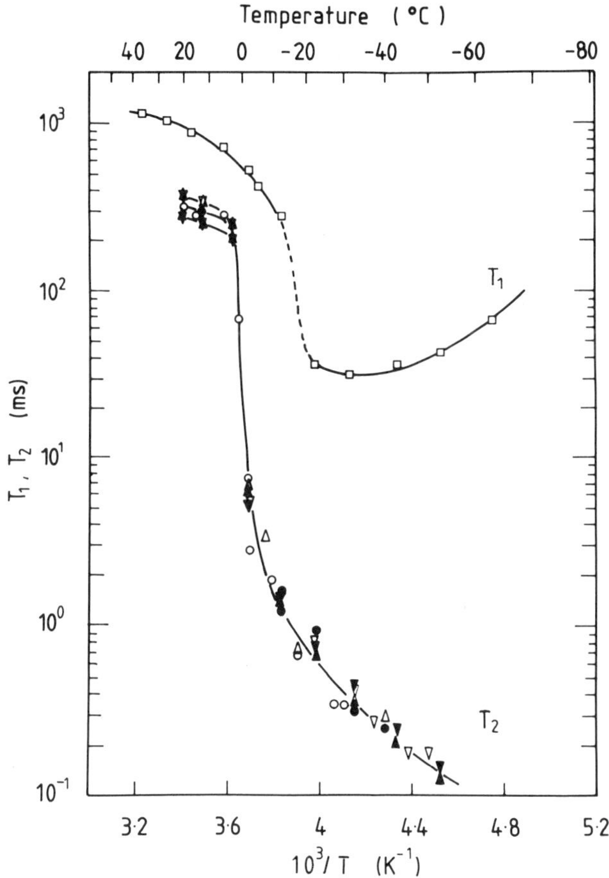

Figure 5. Graphs of the temperature dependencies of the 'H spin lattice and spin-spin relaxation of thick albumen samples above and below 0^oC.

THE YOLK

A simplified model is utilised to describe the two component spin lattice relaxation of the yolk. It is assumed that the yolk may be represented as two regions s and l between which exchange is slow on the NMR time scale giving rise to the two components observed. It is further assumed that each region can contain bound and free water, but that the original free water content of the s region is zero. The justification for this assumption is that the s relaxation is comparable to typical bound water relaxations. With time there might be a transfer of free water and solids together with their associated bound water between the regions and in and out of the yolk. These quantities can be deduced from the time dependences of the NMR relaxation times and of the relative populations of the slower and more rapidly relaxing components. There is some circumstantial evidence based upon the derived ratio of the solid matter in the ℓ and s regions 24:76 that s and ℓ may be associated with the continuous phase fluid and with the free floating lipoprotein droplets. As shown in figure 6 the model predicts a slight increase in the solids content of the s region and a slight increase in its free water content. The ℓ region displayed a near equivalent decrease in free water the overall content remaining reasonably constant. Certainly there was no evidence for the large influx ΔW suggested by the

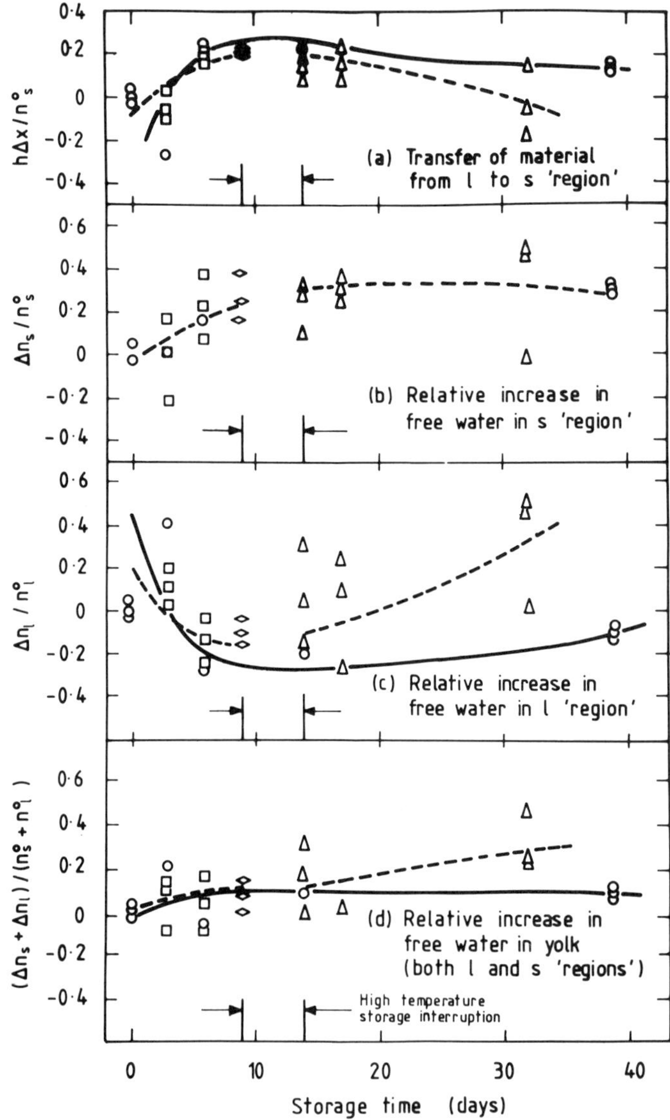

Figure 6. Movements in the yolk populations during ageing.

albumen relaxation data. With the yolk there was a dependence upon the storage conditions.

THE SPIN SPIN RELAXATION

The spin-spin relaxation was more complex requiring three and sometimes four components for adequate characterisation of the yolk and albumen samples, figure 7. In the case of the albumen samples it was found to be necessary to introduce some gaussian components. The observation of multi-component spin-spin relaxation for the albumens, and three and four component relaxation for the yolk samples indicates that the models developed for the interpretation of the spin lattice relaxation are over-simplistic. The spin-spin relaxation was determined by recording the amplitudes of 4000 echo signals produced by a Carr Purcell-Gill Meiboom pulse train with the pulse spacings selected to produce an adequate baseline. No effects on the observed relaxation due to diffusion or chemical exchange were detected. It is not clear if the derived components represent different phases of water, or merely serve to parameterise a continuous distribution of water molecule mobilities and relaxation times. Representation by stick diagrams, averages, second moments and skew parameters in units of log(relaxation time) again demonstrated the occurrence of ageing effects and size dependences. However, we feel that, because of sample variation detailed comparisons of the relaxations involving the extraction of many components is not warranted at this stage. For this it would be better to use a single instead of many eggs and to follow its development using some form of localised NMR technique. Instead we will apply the analysis applied to the spin lattice data to the fitted one and two-component spin-spin relaxation data.

Differences are immediately apparent, unlike the case for the spin lattice relaxation of the albumens there is a dependence of the single component spin-spin relaxation upon size and storage conditions, there was no initial rise in the observed relaxation time, instead for the thin albumen there was an initial decrease. This difference confirms the suspiscions aroused from the evaluation of the higher than expected water losses from albumen samples, that the single component spin lattice relaxation represents an over simplification and at best a first approximation. Application of the theory again gave higher than expected values of β. Deduced values of T_{2b} were in the range 20 to 50 ms somewhat higher than direct observation at temperatures below $0^{o}C$ although possibly consistent with extrapolated values.

The treatment of the fitted two component spin-spin relaxation of the yolk indicated no significant transfer of a solid material between the ℓ and s regions, the total free water content displayed a small increase, the free water component in the s state displayed a small increase and the ℓ component a broad maximum after some twenty days.

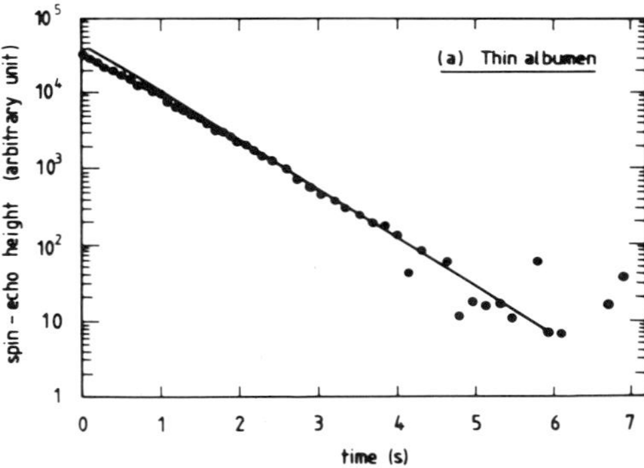

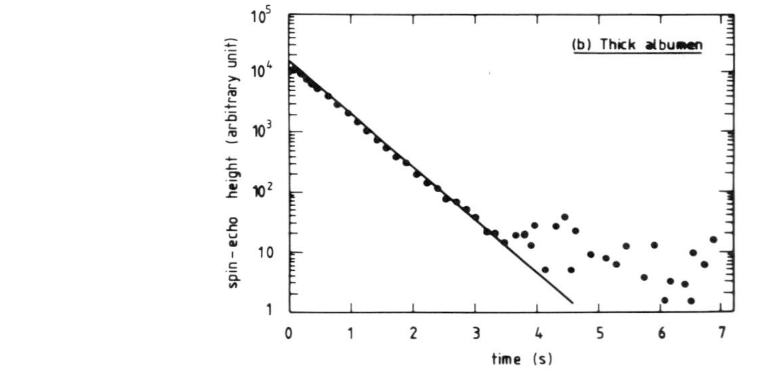

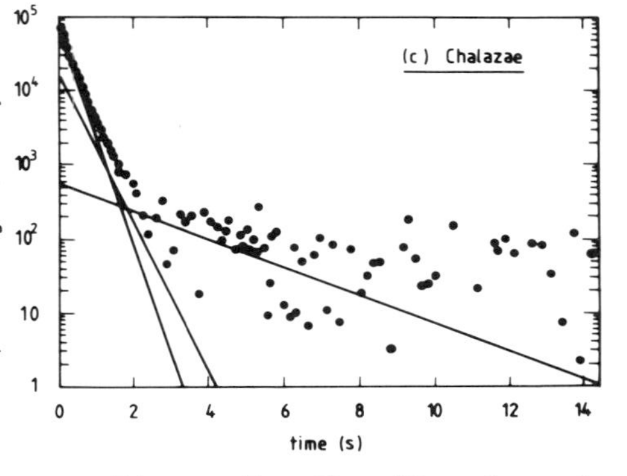

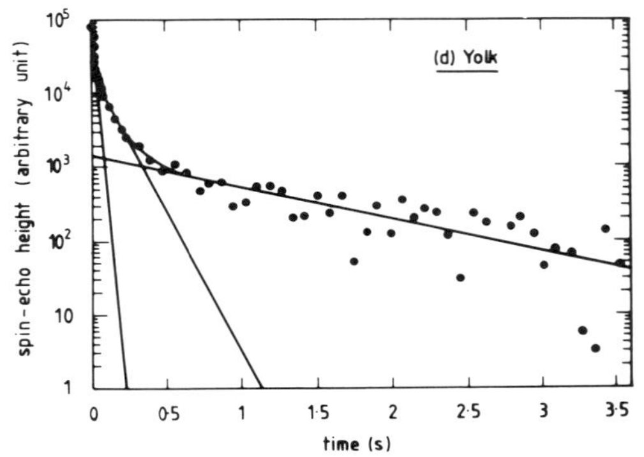

Figure 7. The 'H spin-spin relaxations of the different constituents.

CONCLUSION

The observation of ageing effects somewhat larger that variations between different individual eggs, on the 'H NMR relaxation of the egg components confirmed the promise of such techniques. The absence of a detailed consistency between the deductions made from the spin latice and spin-spin relaxations presumably reflect their dependences upon different motional properties.

REFERENCES

1. B.M. Fung, D.L. Durham and D.A. Wassil, The State of Water in Biological Systems as studied by Proton and Deuterium Relaxation. Biochim. Biophys, Acta. 399:191,(1975)

2. T.L. James and K.T. Gillen, N.M.R. Relaxation time and self diffusion constant of water in hen egg white and yolk. Biochim. Biophys. Acta. 286: 10, (1972).

3. D.C. Chang, C.F. Hazlewood and D.E. Woessner, The Spin Lattice Relaxation times of water associated with early post mortem changes in skeletal muscle. Biochim. Biophys. Acta. 437: 253 (1976).

4. A.L. Romanoff and A.J. Romanoff, The Avian Egg. Wiley, New York (1949).

ANOMALOUS VOLUME PROPERTIES OF VICINAL WATER AND SOME RECENT THERMODYNAMIC (DSC) MEASUREMENTS RELEVANT TO CELL-PHYSIOLOGY

W. Drost-Hansen

Laboratory for Water Research
Department of Chemistry, University of Miami
Coral Gables, Florida

I. INTRODUCTION

The properties of water and aqueous solutions adjacent to interfaces (particularly solid surfaces) differ from the corresponding properties of the bulk systems. The differences are ascribed to structural effects of the interface on the aqueous phase - the resulting interfacial water being referred to as vicinal water. Extensive reviews of the nature of vicinal water have been published (1, 2, 3, 4) over about a decade and will not be reviewed here; suffice it to mention that vicinal water appears to exist over notable distances from the interface namely of the order of 30 to 150 molecular diameters or in other words 0.01 to 0.05 microns (μm). Furthermore, vicinal water appears to differ energetically only slightly from bulk water (3, 5). Vicinal water undergoes abrupt changes in properties - and, hence, no doubt, in structure - near a number of discrete temperatures, namely near 15, 30, 45 and 60°C, referred to respectively as T_1, T_2, T_3, and T_4, or in general T_k. (An additional anomaly appears to occur near 74°C) (6). Previous measurements (7) of the thermal properties of vicinal water have suggested that the specific heat is approximately 25 percent larger than that of bulk water. It is of interest to note that in recent studies Angell and co-workers (8, 9) have observed notable increases in the specific heat of (bulk) water at very low temperatures (below say - 20°C) and at temperatures approaching the critical temperature.

II. VOLUME AND PRESSURE PROPERTIES

Etzler (10) has recently proposed a first order theory (EST) for vicinal water. The treatment is based on the Stanley and Teixeira (11) percolation model treatment of bulk water. Etzler takes as his basic input the enhanced value for the specific heat for vicinal water, namely 1.25 cal/°C. gram. In the EST theory the enhanced value for the specific heat capacity reflects the contribution to the specific heat in terms of fluctuations in volume and entropy. With the assumed value for the specific heat, the enhanced connectivity of the water lattice leads to a number of predictions including statements regarding the volume properties. Based on his calculations Etzler (10) predicts a value for the density of vicinal water of 0.97 g/ml (at room temperature). More recently Etzler and Fagundus (12) have made direct measurements of the apparent density of water in porous silica gel (with pore diameter of about 140 Å) assuming all the water in the pores is vicinal water. The result obtained was a value of 0.965 g/ml in remarkably good agreement with the predicted value. Further details regarding the calculations and the interpretation are discussed in a recent paper by Etzler and Drost-Hansen (5).

It is of interest to speculate on the observed (and calculated) density difference between vicinal water and bulk water. If indeed the specific volume of vicinal water is 3 percent larger than that of bulk, then, by Le Chatelier's principle, vicinal water should disappear as pressure is increased. Unfortunately there exists practically no data in the literature on which this prediction may be tested. Attention is called however to some high pressure measurements by Drickamer and co-workers on the diffusion of some electrolytes and non-electrolytes in aqueous solutions. The measurements have been analyzed in a paper on the effects of pressure on organisms by the present author. Drickamer and co-workers (13) proposed that the observed effects (namely, notable variations in the diffusion coefficient for instance of sodium sulphate or potassium sulphate as a function of pressure at various temperatures) reflected structural changes in the bulk aqueous phase. As no other evidence appears to exist for structural changes in bulk water at pressures of a few kilobars it was suggested by the present author (14) that these anomalies instead reflect changes in the vicinal water in the porous frits used in the diffusion cells. The phenomenon reported by Drickamer et al. often exhibit large effects: thus for pressures less than 200 atmospheres the diffusion coefficient of potassium sulphate increases almost by a factor of 2.5 (at 0°C) but decreases by a factor of 2 over the same pressure interval for sodium sulphate. Sharp maxima and minima in the diffusion coefficients are reported over pressure ranges from 1 atmospheres to 4 or 5 thousand atmospheres. The complexities of these extrema appear to increase with decreasing temperature and are particularly notable at 0°C.

On this basis it is proposed that the structure of vicinal water may indeed change more or less abruptly over narrow pressure intervals, similar to the changes in properties of vicinal water as a function of temperature. It is of interest also to note that a casual inspection of reported pressure effects on aquatic organisms often suggest abrupt changes at relatively modest pressures for instance near 70 to 150 atmospheres (corresponding to depths in the water column of about 700 to 1500 meters). These physiological effects likely reflect pressure induced changes due to the intracellular vicinal water. A number of volume properties of vicinal waters have been studied by the present author and his associates (5, 15). Thus it appears that the temperature of maximum density for vicinal water is notably lower than that of bulk water (data by Schufle and co-workers (16). Furthermore, the apparent thermal expansion coefficient of capillary held water appears to exceed that of bulk water. This observation is of interest in that the thermal expansion coefficient for bulk water is usually small compared to most "normal" liquids. In this respect, then, vicinal water appears to behave more like a "normal" liquid.

There exists presently no data on the isothermal compressibility of vicinal water. However in an earlier study (17) we have measured the ultrasonic velocity in polystyrene spheres suspensions in which a large fraction of the water present appears to be vicinal water. From these data very large values were obtained for the adiabatic compressibility. We are presently developing an automatic recording high precision piexometer for direct measurements of isothermal compressibility on large surface-to volume ratio systems.

III. THERMAL PROPERTIES; PARADOXICAL EFFECT

Cianci (18) has recently carried out an extensive differential scanning calorimetry (DSC) study of water in dispersed aqueous systems. Some of the results from this study have been discussed in previous papers from this laboratory. In general Cianci observes relatively large, often abrupt, fairly well defined endo or exothermic peaks in thermograms of water in systems as diverse as: water in porous quartz, polystyrene sphere suspensions and diamond powder in water. Table 1 shows the temperatures of these transitions for a variety of systems studied. It is seen that the temperatures of the transitions are, within the experimental error, the values reported by Peschel (19), Wiggins (20), and authors (1, 2, 21) for over 25 years. Most importantly, the temperatures are independent of the specific chemical characteristics of the surfaces studied. This constitutes a decisive proof of the "paradoxical effect" (3): vicinal water appears to exist regardless of the specific chemical details of the surface, owing its existence apparently merely to proximity to an interface. In this connection see also the papers by Clifford (22) and by Martini (23).

TABLE 1

AVERAGE TRANSITION TEMPERATURES (°C)

System	T_3	T_4	T_5
Silica Gel	45.9 ± 1.65	60.9 ± 1.95	73.2 ± 1.54
Diamond Powder	45.4 ± 1.58	59.8 ± 2.28	74.4 ± 1.36
Spheres (.176-.945)	45.2 ± 1.93	58.9 ± 2.75	74.6 ± 2.29
Spheres-Bulk (.176)	45.9 ± 2.37	57.8 ± 1.79	75.0 (*)
Spheres-Bulk (.945)	44.5 ± 1.50	-	75.0 (**)

(*) = Only one transition was observed in the region of T_5

(**) = Only two transitions (both at 75.0 °C) were observed in the region of T_5 .

TABLE 2

SALT SOLUTION / SILICAN GEL RUNS (°C)

Run Number	Solution	T_3	T_4	T_5
95	10^{-3}M LiCl	43	-	-
96	10^{-1}M LiCl	43	-	-
97	1.0 M LiCl	45	-	-
106	1.0 M LiCl	45	-	74
LiCl Totals :		44.0 ± 1.0		74.0
99	10^{-4}M NaCl	44	-	79
100	10^{-2}M NaCl	47	-	71
101	1.0 M NaCl	46	-	68
NaCl Totals :		45.7 ± 1.7	-	72.7 ± 6.3
92	10^{-3}M KCL	46	57	74
93	10^{-1}M KCL	47	-	71
94	1.0 M KCL	-	64	76
105	1.0 M KCL	44	57	72
KCL Totals :		45.7 ± 1.7	59.3 ± 4.7	73.3 ± 2.7
110	10^{-4}M RbCl	-	63	79
111	10^{-2}M RbCl	46	62	-
112	1.0 M RbCl	-	62	73
RbCl Totals :		46.0	62.3 ± 0.7	76.0 ± 3.0
107	10^{-4}M CsCl	47	-	-
108	10^{-2}M CsCl	49	64	72
109	1.0 M CsCl	47	-	-
CsCl Totals :		47.7 ± 1.3	64.0	72.0
TOTALS:		45.7 ± 1.67	61.3 ± 2.81	73.6 ± 3.18

Cianci (18) has also studied the effects of electrolytes, over wide concentration ranges, on the temperatures of the thermal transitions. The results are shown in part in table 2. An inspection of the data in this table shows that within the experimental error the temperatures of the thermal transitions are independent of the concentration of electrolyte from 10^{-4} Molar to 1 Molar. Furthermore the effect appears, at least to a first approximation, to be independent of the specific nature of the cation present: ranging from lithium ions to cesium ions. Again this is consistent with the paradoxical effect.

REFERENCES

1. Drost-Hansen, W., On the Structure of Water Near Solid Interfaces and the Possible Existence of Long-range Order, Industrial and Engineering Chemistry, 61, pp. 10-47.
2. Drost-Hansen, W., Structure and Properties of Water at Biological Interfaces, Chemistry of the Cell Interface, H. D. Brown, ed., Academic Press, N. Y., Chapter 6, pp. 1-184.
3. Drost-Hansen, W. Water at Biological Interfaces - Structural and Functional Aspects. Physics and Chemistry of Liquids, Gordon & Breach, Vol. 7, 1978, pp. 253-346.
4. Peschel, G., and Belouschek, P. "Cell-Associated Water," ed. W. Drost-Hansen and J. S. Clegg, Academic Press, N. Y., 1979, pp. 3-52.
5. Drost-Hansen, W., "The Occurrence and Extent of Vicinal Water," "Biophysics of Water," ed., F. Franks, Wiley & Sons, N. Y. 1982, pp. 163-169.
6. Cianci, J. J., "Thermal Anaomlies in Vicinal Water as Studied by DSC," Thesis for M.S. Degree, University of Miami, Coral Gables, Florida, 1981
7. Braun, Jr., Chester V. and Drost-Hansen, W., "A DSC study of Heat Capacity of Vicinal Water in Porous Materials," "Colloid and Interface Science," Vol. III, ed. M. Kerker, Academic Press, N. Y., 1976, pp. 533-541.
8. Angell, C. A., Oguni, M., and Sichina, W. J., Journal of Physical Chemistry, 86, 1982, pp. 998-1002.
9. Angell, C. A., and Tucker, J. C., Journal of Physical Chemistry, 84, 1980, pp. 268-272.
10. Etzler, F. M., "A Statistical Thermodynamic Model for Water Near Solid Interfaces," Journal of Colloid and Interfacial Science, 1983, in press.
11. Stanley, H. E. and Teixeira, J., Journal of Chemical Physics 73, 1980, p. 3404.
12. Etzler, F. M. and Fagundus, D., Journal of Colloid and Interfacial Science, 1983, in press.
13. Cuddeback, R. B., Koeller, R. C. and Drickamer, H. G., Journal of Chemical Physics, 21, 1953, pp. 589-597.

14. Drost-Hansen, W., in "The Effects of Pressure on Organisms," eds. M. A. Sleigh and A. G. MacDonald, SEB Symposium XXVI, 1972, pp. 61-101.
15. Etzler, F., and Drost-Hansen, W., Recent Thermodynamic data on Vicinal Water and a Model for Their Interpretation. Paper presented at VII International Summer Conference "Chemistry of Solid/Liquid Interfaces," Cavtat/Dubrovnik (Yugoslavia) June 1982. To be published in Croatia Acta Chemica, Vol. 56 (4), 1983.
16. Schufle, J. A. and Venugopalan, M., Journal of Geophysical Research, 72, 1967, 3271.
17. Bruum, S. G., Sorensen, P. G. and Drost-Hansen, W., "Studies of Vicinal Water Structuring in Suspensions II. Ultrasonic Velocity Measurements." Submitted for publication.
18. Cianci, J. J., "Thermal Anomalies in Vicinal Water as Studied by DSC," Thesis for M.S. degree, University of Miami, Coral Gables, Florida, 1981.(See Also Reference #5)
19. Peschel, G., and Aldfinger, K. H., Naturforsch, Z., 26a, 1971 707. See also Naturwissenschaften, 54, 1970, 614.
Wiggins, P., Clinical and Experimental Pharmacology and Physiology, 2, 1975, pp. 171-176.
21. Drost-Hansen, W., The Effects on Biological Systems of Higher-order Phase Transitions in Water. New York Academy of Science Annals, 125, pp. 471-501.
22. Clifford, J., "Water - A Comprehensive Treatise, " ed. F. Franks, Vol. 5, 1975, pp. 75-132.
23. Martini, G., Journal of Colloid and Interfacial Science, 80, 1981, pp. 39-48.

WATER AND SOLUTE TRANSPORT

VOLUME REGULATION IN CARP KIDNEY TISSUE AS INFLUENCED BY VARIOUS OSMOTIC AGENTS

Ivan Beneš, Karel Janáček
and Růžena Tauchová
Institute of Microbiology
Czechoslovak Academy of Sciences
142 20 Prague 4, Czechoslovakia

To what extent does the sodium pump (the plasma membrane Na,K-ATPase) contribute to volume regulation of animal cells? According to the original concept of Wilson (1954) and Leaf (1956) the sodium pump solves entirely the problem of volume maintenance in isotonic media, since "effectively impermeant" sodium ions (continually extruded by the pump) compensate for the Donnan excess of osmotic pressure, resulting mostly from the high concentration of diffusible counterions of macromolecules in the cytoplasm.

For Ling and his followers (see, e.g., Ling, 1982) such theoretical explanation is obviously unacceptable from the very start, since according to this school the sodium pump functionally does not exist and the exclusion of sodium ions from the cells is to be explained on the basis of the association-induction hypothesis. Accordingly, the isoosmotic volume regulation by human lymphocytes is being explained on the same basis (Negendank and Shaller, 1982).

However, at least when dealing with epithelial cells we are not prepared to give up the concept of the sodium pump and we hold that its possible role in cell volume maintenance and regulation should be discussed. The reality of transcellular active sodium transport across, e.g., the wall of frog urinary bladder cannot possibly be doubted and experiments from our laboratory showed that the extrusion of sodium ions from cells in the non-polar preparation of this organ is stimulated by the same hormones as the transcellular transport (Janáček and Rybová, 1970; Janáček et al., 1971). Hence the same pumping mechanism seems to be involved in both phenomena.

Still, as shown by abundant experimental evidence (see reviews by Kleinzeller, 1972; Macknight and Leaf, 1977) the plasma membrane Na,K-ATPase is not a complete solution to cell volume maintenance, mostly since the inhibition of the sodium pump by ouabain does not abolish entirely the control of the cell volume.

Surviving tissue slices of carp kidney are a convenient material to study the ouabain-insensitive regulation of cell volume. As a measure of the cell volume the water content of the tissue can be used, determined from the loss of weight by drying overnight at 95°C. The water content of tissue slices incubated in the Krebs-Henseleit medium remains stable for hours, it is not at all influenced by the presence of 0.1 mM ouabain but increases highly significantly in the presence of 0.1 mM 2,4-dinitrophenol (Beneš et al., 1982). Hence we concluded that the plasma membrane Na,K-ATPase is not responsible for the volume maintenance in isotonic media, the Donnan excess of osmotic pressure being there counteracted by another energy-dependent mechanism, and that it remains to be seen whether it manifests itself at all when the tissue is incubated in media of abnormal osmolarity and/or ionic composition. The media contained all the minor components of the Krebs-Henseleit saline (see, e.g., Beneš et al., 1982, 1983) including 28 mM glucose and their pH was 7.4 (Table 1).

Table 1. Content of ions in various media used (mM)

Medium	[Na^+]	[K^+]	[Cl^-]	milliosmols
Normal Krebs-Henseleit	145	6	128	337
Hypertonic with mannitol	145	6	128	639
Ions two-fold diluted	72	3	64	183
Ions two-fold concentrated x/	264	11	256	621
Hypertonic with potassium chloride	145	130	252	586
Hypertonic with Tris chloride	145	6	282	645
Hypertonic with Na^+ benzenesulph.	299	6	128	645

x/ with the exception of $NaHCO_3$

The water content of tissue after 2-h incubation is shown in Table 2.

Table 2. Water content in carp kidney tissue

Medium	kg H_2O/kg dry solids controls		0.1 mM ouabain		0.1 mM DNP
Normal Krebs-Henseleit	4.00	~	4.05	<<	4.50
Hypertonic with mannitol	2.69	~	2.71	<<	3.37
Ions two-fold diluted	4.85	<<	5.53	~	5.63
Ions two-fold concentrated	2.84	<	3.04	<<	3.37
Hypertonic with potassium chloride	3.59	<<	3.78	<<	4.16
Hypertonic with Tris chloride	2.61	<<	2.88	~	2.96
Hypertonic with Na^+ benzenesulphonate	3.15	<<	3.40	<	3.53

~ not significant
< significant ($p < 0.05$)
<< highly significant ($p < 0.01$)

It may be seen that neither in the normal Krebs-Henseleit medium, nor in the medium with osmolarity almost doubled with mannitol does 0.1 mM ouabain bring about swelling, as compared with the respective controls; thus Na,K-ATPase is not the mechanism maintaining the cell volume under these conditions. On the other hand, the water content of the tissue is highly significantly increased with 2,4-dinitrophenol, suggesting an ATP-dependent character of the volume-maintaining mechanism.

In media of abnormal ionic composition significant swelling is brought about by ouabain; obviously, Na,K-ATPase is important (in the medium hypertonic with Tris chloride possibly all-important) in the cell volume regulation when both the osmolarity and the ionic composition of the medium are changed. Still, in most cases the presence of 2,4-dinitrophenol brings about an additional swelling, demonstrating the presence of an additional energy-dependent regulating device.

Some information concerning the nature of the 2,4-dinitrophenol-sensitive mechanism can be obtained from changes of ion contents, accompanying the additional swelling brought about by this inhibitory agent. Thus the sodium content of the tissue in the medium hypertonic with sodium benzenesulphonate is not increased by the additional dinitrophenol-induced swelling, being (in mmol/kg dry solids ± S.E.M.) 902 ± 42 /n=8/ with ouabain and 900 ± 18 /n=7/ with dinitrophenol. In the medium hypertonic with Tris chloride the sodium content is even highly significantly reduced with 2,4-DNP (by 59 ± 11 mmol/kg dry solids, $p < 0.001$) as compared with ouabain-treated tissue. Thus whatever the character of the second ouabain-insensitive cell volume regulating mechanism may be, it is not correct to call it "the second sodium pump". On the whole, it may be shown that carp kidney tissue behaves as an osmometer gaining or losing sodium salts and/or potassium salts in media of various osmolarities and with inhibitors; the tissue water content is linearly dependent on the reciprocal value of osmolarity corrected for the observed shifts of salts (Beneš et al., 1983).

REFERENCES

Beneš, I., Tauchová, R., and Janáček, K., 1982, Ouabain-insensitive mechanism of cell volume maintenance in the carp kidney, Physiol. bohemoslov.31:237.

Beneš, I., Janáček, K., and Tauchová, R., 1983, Cell water content in carp kidney tissue slices as influenced by various osmotic agents, Physiol. bohemoslov., in press.

Janáček, K., and Rybová, R., 1970, Nonpolarized frog bladder preparation - the effects of oxytocin, Pflügers Arch., 318:294.

Janáček, K., Rybová, R., and Slavíková, M., 1971, Independent stimulation of sodium entry and sodium extrusion in frog urinary bladder by aldosterone, Pflügers Arch., 326:316.

Kleinzeller, A., 1972, Cellular transport of water, in: "Metabolic Pathways: Metabolic Transport," L. E. Hokin, ed.,Vol. 6, Academic Press, New York (1972).

Leaf, A., 1956, On the mechanism of fluid exchange of tissues in vitro, Biochem. J., 62:241.

Ling, G. N., 1982, The physical state of K^+ and water in the living cell according to the association-induction hypothesis and experimental evidence, in: Second International Conference on Water and Ions in Biological Systems, Abstracts, Bucharest.

Macknight, A. D. C., and Leaf, A., 1974, Regulation of cellular volume, *Physiol. Rev.*, 57:510.
Negendank, W., and Shaller, C., 1982, Isoosmotic volume regulation by human lymphocytes, *in*: Second International Conference on Water and Ions in Biological Systems, Abstracts, Bucharest.
Wilson, T.H., 1954, Ionic permeability and osmotic swelling of cells, *Science*, 120:104.

MODIFICATIONS OF HUMAN ERYTHROCYTE MEMBRANES AND THEIR EFFECT ON WATER PERMEABILITY STUDIED BY A NUCLEAR MAGNETIC RESONANCE TECHNIQUE

Gheorghe Benga[1], Octavian Popescu[1], Victor I. Pop[1], Ross P. Holmes[3], Tudor Pavel[2] and Mihai Ionescu[4]

Departments of Cell Biology[1] and Physiology[2], Faculty of Medicine, Medical and Pharmaceutical Institute Cluj-Napoca, Roumania, Burnsides Research Laboratory[3], University of Illinois, Urbana-Champaign, USA and Institute of Physics and Nuclear Engineering Bucharest-Magurele[4], Roumania

INTRODUCTION

Although the process of water transport across biological membranes is of considerable importance for many physiological processes and various hypotheses have been proposed and investigated, the mechanism controlling water movement has not been fully elucidated. Because of its simple structure, lacking internal membranes, the red blood cell is ideally suited for studying water permeability.

The NMR[1] method (Conlon and Outhred, 1972; Fabry and Eisenstadt, 1975; Morariu and Benga, 1977, Pirkle et al., 1979) has been shown to be a valuable technique for the study of the physiology and pathology of erythrocyte diffusional water permeability (Benga and Morariu, 1977; Benga et al., 1981). However, there is limited NMR data related to changes in water permeability following exposure of erythrocytes to inhibitors of this transport process and to the effects of proteolytic digestion (Ashley and Goldstein, 1981, Benga et al., 1982, Benga et al., 1983). The suggestion that band 3 is involved in

---

[1]Abbreviations: DTNB= 5,5'-dithiobis-2-nitrobenzoate: FMA= fluoresceinmercuric acetate; H_2DIDS=dihydro-4,4'-diisocyanostilbene-2,2'-disulfonate;IAM= iodoacetamide; NEM= N-ethylmaleimide; NMR= nuclear magnetic resonance; PCMB= p-chloromercuribenzoate; PCMBS= p-chloromercuribenzene sulfonate; SDS= sodium dodecyl-sulphate.

water transport (Brown et al., 1975; Sha'fi and Feinstein, 1977) was based on labelling experiments with DTNB and mercurials and on studies examining the inhibitory effects of such compounds on osmotic permeability of erythrocytes.

The aim of our paper is to present NMR studies on changes in water diffusion through erythrocyte membranes following exposure to various inhibitors and proteolytic enzymes and to correlate the observed effects with the binding of a radioactive labelled inhibitor to membrane proteins. Significant differences were found regarding the effects of DTNB and mersalyl on water diffusion compared with data of other authors on osmotic permeability. No changes in water diffusion were noticed after exposure of intact erythrocytes to ^{203}Hg-PCMBS. The findings are discussed in relation to the molecular mechanisms proposed for water transport.

MATERIALS AND METHODS

Blood Sample Preparations

Human blood was obtained by venipuncture in heparinized tubes and used within 4 hours. The donors were healthy male subjects, 20-40 years old. The erythrocytes were isolated by centrifugation, and washed twice in 150 mM NaCl, 5 mM Hepes (pH 7.4).

The incubation of erythrocytes with various reagents was performed as indicated in the legends of figures and tables. After incubation, reagents were removed by two washings of the erythrocytes in 150 mM Nacl, 5 mM Hepes, each followed by a centrifugation.

NMR Measurements

The erythrocytes were suspended in a wash solution containing 0.1% glucose and 1% defatted bovine serum albumin, at a hematocrit of 25%. Samples for NMR measurements were prepared by carefully mixing 0.2 ml cell suspension and 0.1 ml doping solution (40 mM $MnCl_2$, 100 mM NaCl).

The water proton relaxation time of the erythrocytes (T'_{2a}) was measured by the spin-echo method as previously described (Morariu and Benga, 1977; Morariu et al., 1981). The T'_{2a} is dominated by the exchange process through the membrane and the water exchange time through erythrocyte membranes is inversely related to the water permeability of erythrocytes as shown by Conlon and Outhred (1972). The inhibition of water diffusion across human red blood cell membranes was calculated assuming that the permeability coefficient is inversely related to T'_{2a}, according to the formula:

$$\% \text{ Inhibition} = \frac{\frac{1}{T'_{2a}\text{ (control)}} - \frac{1}{T'_{2a}\text{ (sample)}}}{\frac{1}{T'_{2a}\text{ (control)}}}$$

The measurements were performed with and Aremi-78 spectrometer (Institute of Physics and Nuclear Engineering, Bucharest-Mågurele, Roumania). The temperature was controlled to 37± 0.2°C by air flow over an electrical resistance using the variable temperature unit attached to the spectrometer.

Preparation of Erythrocyte Ghosts and Polyacrylamide Gel Electrophoresis

Ghosts were prepared as described by Dodge et al., (1963) and modified by Fairbanks et al., (1971). Membrane peptides were separated using the discontinuous SDS polyacrylamide gel system designed by Laemmli (1970), as previously described (Benga et al., 1983). The protein concentration was estimated according to Lowry et al., (1951).

Measurement of Radioactivity

In order to determine the location of radioactively labelled peptides on gels, two procedures were used. Gel slabs were cut into 2 mm wide slices and the radioactivity was measured with a well-type scintillation counter (Procedure A). In another series of experiments, the gel slabs were stained, destained and dried; the estimation of radioactivity was performed by densitometry of films and this was compared with densitometry of stained gels (Procedure B).

RESULTS AND DISCUSSIONS

A wide variety of reagents and chemical treatments of erythrocyte membranes have been tested to examine their effect on the water exchange time. As shown in Table 1, mercurials, including mercalyl, were the only sulfhydryl reacting reagents that were efficient inhibitors. When the optimal conditions of inhibition were found (time and temperature of incubation, concentration of inhibitor) all mercurials produced the same degree of inhibition, around 45%. This is in agreement with the results of other authors (Macey et al., 1972; Fabry and Eisenstadt, 1975; Ashley and Goldstein, 1981) who also have reported similar degrees of inhibition of diffusional permeability induced by mercurials. However, an important difference in the behavior of FMA compared with other mercurials became apparent when the reversibility of inhibition by cysteine was investigated (Table 1). FMA was the only mercurial which appeared to inhibit irreversibly water diffusion through erythrocyte membranes. The significance of this finding for the mechanisms of water transport has been previously discussed (Benga et al., 1982).

Other SH reagents, aside from mercurials, did not appear to inhibit diffusional permeability significantly (Table 2). It should be emphasized that this also applies to DTNB, for which a degree of inhibition of only 10% was noticed, in contrast to a 60% inhibition of osmotic water transport reported by Naccache and Sha'afi (1974). From Table 2 it can also be seen that SH reagents which do not contain mercury (DTNB, IAM and NEM) neither prevented nor potentiated the inhibitory effect of a mercurial. This suggests that the SH group involved in water transport does not react with any of these SH reagents.

Table 3 lists the degree of inhibition noticed with a variety of reagents. None of these markedly increased water exchange time. This included H_2DIDS, a specific inhibitor of the anion transport system in erythrocyte membranes. Phloretin which inhibits anion transport and other facilitated transport processes in the red blood cell membrane also did not affect the water permeability. Other compounds that react with amino groups, including N-succinimidyl-3-(2-pyridylthio) propionate (Carlsson et al., 1978), had no significant inhibitory effect on water exchange. Water diffusion appears not to be energy-dependent since blocking glycolysis with NaF did not

Table 1. Effect of Mercury Containing Reagents on Water Diffusion Across Human Red Blood Cell Membranes.

Compound [a]	% Inhibition
0.1 mM HgCl	42.0
1 mM Mersalyl	35.2
1 mM PCMB	44.1
1 mM PCMB, followed by 10 mM Cysteine	5.2
1 mM PCMBS	43.6
2 mM PCMBS, followed by 10 mM Cysteine	5.0
0.25 mM FMA	46.0
0.25 mM FMA, followed by 10 mM Cysteine	43.2

[a] All incubations were performed for 60 min at 37°C, except for mersalyl and cysteine, where incubations of 15 min at 37°C were used. The results are means of 2-22 determinations. Inhibition was calculated as described in "Materials and Methods."

Table 2. Effect of Some Sulfhydryl Reacting Reagents on Water Diffusion Across Human Red Blood Cell Membranes.

Compound[a]	% Inhibition
1 mM DTNB	10.1
1 mM DTNB, plus 1 mM PCMB	48.0
1 mM IAM	6.9
1 mM IAM, followed by 1 mM PCMB	39.7
1 mM NEM	6.5
1 mM NEM, followed by 1 mM PCMBS	46.0
1 mM IAM, plus 1 mM NEM followed by 1 mM Mersalyl	50.2
1 mM IAM, plus 1 mM NEM, followed by 1 mM PCMBS	46.0
1 mM IAM, plus 1 mM NEM, followed by 1 mM PCMBS and 0.25 mM FMA	45.5

[a] All incubations were performed for 60 min at 37°C, except for mersalyl where an incubation of 15 min at 37°C was used. The are means of 2-12 determinations.

affect the water exchange time. No changes in water diffusion were noticed after exposure of erythrocytes to tryspin and chymotrypsin. The enzymatic treatment of membranes also did not prevent the inhibition induced by mercurials. In contrast, the effect of mercurials appeared to be slightly potentiated by the enzymatic treatment.

Using polyacrylamide gel electrophoresis Sha'afi and coworkers (Brown et al., 1975, Sha'afi and Feinstein, 1977) have found that the band 3 protein in human red cell membranes can be selectively labelled by ^{14}C-DTNB or ^{14}C-PCMB, both inhibitors of osmotic water transport. However, these experiments must be reevaluated, because: (a) DTNB does not inhibit diffusional water permeability (Table 2), and (b) the selective labelling of band 3 by ^{14}C-PCMB was observed after preincubations of erythrocytes with 1 mM each of IAM, NEM and mersalyl, compounds which were thought not to inhibit water transport (Sha'afi and Feinstein, 1977). As shown in Table 1, mersalyl is a strong inhibitor of diffusional water permeability.

Considering these factors, we performed labelling experiments with radioactive PCMBS, with and without preincubation with IAM, NEM and FMA.

Table 3. Effect of Various Compounds on the Water Diffusion Across Human Red Blood Cell Membranes

Compound[a]	% Inhibition
10 μM DIDS	10.8
10 μM DIDS, followed by 1 mM PCMBS	39.5
1 mM Phloretin	11.2
40 mM NAF	9.2
0.5 M N-succinimidyl-3 (2-pyridylthic) propionate	6.6
Trypsin	13.5
Trypsin, followed by 1 mM PCMBS	51.2
Trypsin, followed by 250 μM FMA	46.2
Chymotrypsin	9.7
Chymotrypsin, followed by 1 mM PCMBS	54.4
Chymotrypsin plus trypsin	13.1

[a] All incubations were performed for 60 min at 37°C, except for FMA, where an incubation of 30 min was used. The results are means of 2-14 determinations.

Preincubation of erythrocytes with IAM and NEM were considered to block SH groups not related to water transport (Sha'afi and Feinstein, 1977). However, we found that preincubation with IAM and NEM results in an enhanced labelling of all peptides in erythrocyte membrane (Figure 1A). If 0.25 mM FMA was added together with IAM and NEM in the preincubation medium the labelling of all peptides is reduced, with a slightly enhanced reduction of labelling of a band 3 (Figure 1B).

When the preincubation with IAM and NEM was omitted and the labelling with radioactive PCMBS and the preparation of ghosts were performed in the presence of FMA, very little radioactivity could be detected in the membrane peptides (Figure 2A). If FMA was present only in the preincubation and was omitted in the washing media the radioactivity occurred in all membrane peptides, being relatively smaller in band 3 (Figure 2B).

When ^{203}Hg-PCMBS-labelled erythrocytes were exposed to trypsin no alteration in the peptide pattern was observed (Figure 3C). However, the band 3 was digested by chymotrypsin (Figure 3B). In this latter case the radioactivity in band 3 was greatly reduced, while a corresponding increase in both peptide concentration and radioactivity occurred in the region of band 4.5, where one of the two peptides resulting from the action of chymotrypsin is located (Passow et al., 1977).

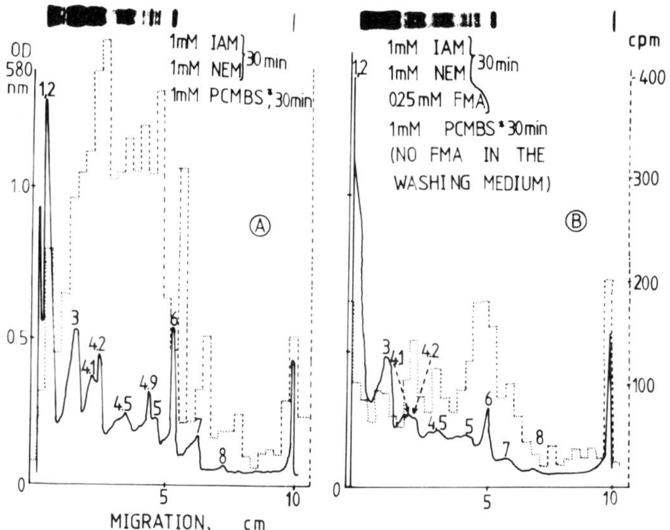

Fig. 1. SDS- polyacrylamide gel electrophoresis of erythrocyte membrane peptides isolated from ^{203}Hg-PCMBS-labelled intact red cells. The washed erythrocytes were incubated for 30 min at 21°C with 1 mM IAM and 1 MM NEM (Figure 1A) and 0.25 mM FMA (Figure 1B), followed by an incubation for 30 min at 21°C with 1 mM radioactive PCMBS. Ghosts were analysed for distribution of radioactivity by Procedure A described in "Materials and Methods."

The above labelling experiments indicated that, despite the various conditions we used, selective labelling of band 3 protein could not be obtained. Therefore, these results do not allow us to conclude whether a particular peptide of erythrocyte membranes is associated with the water channel. As proteolytic digestion did not affect water diffusion of the inhibitory action of mercurials, this suggests that if band 3 is associated with glycophorin in the formation of the water channel, this association must occur in portions of these proteins not accessible to proteolytic digestion. Similarly, treatment with trypsin or chymotrypsin has no inbibitory effect on anion transport (Cabantchik and Rothstein, 1974; Passow et al., 1977). However, the two types of transport are inhibited by two different classes of agents: disulfonic stilbene derivatives (e.g. H_2DIDS) inhibit anion transport and mercurials inhibit water transport.

It therefore appears difficult to determine whether the anion or water pathways are located in the same channel as suggested by Solomon (personal communication) or two channels (Sha'afi, 1981).

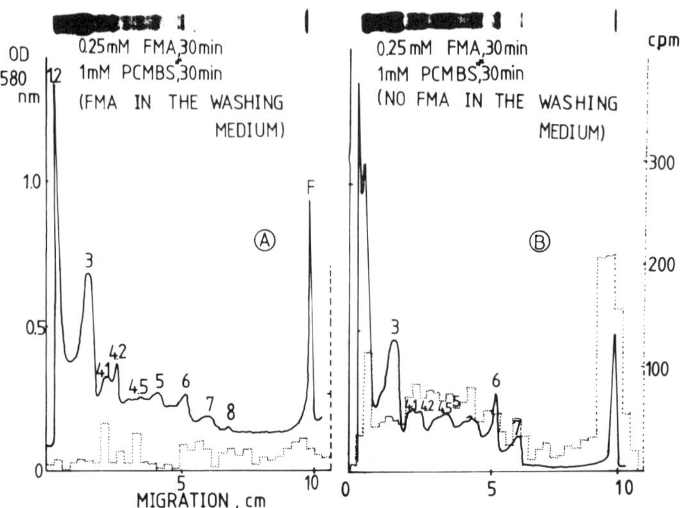

Fig. 2. SDS-polyacrylamide gel electrophoresis of erythrocyte membrane peptides isolated from ^{203}Hg-PCMBS-labelled intact red cells. The washed erythrocytes were incubated for 30 min at 21°C with 0.25 mM FMA, followed by an incubation of 30 min at 21°C with 1 mM radioactive PCMBS. Ghosts were analysed for radioactivity distribution by Procedure A described in "Materials and Methods".

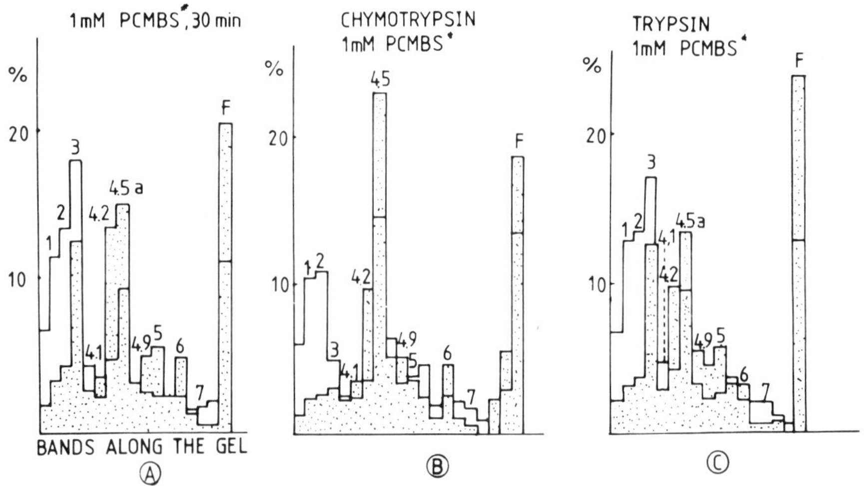

Fig. 3. Integrated areas of electropherograms and autoradiograms of erythrocyte membrane peptides isolated from ^{203}Hg-PCMBS labelled intact red cells (A), after preincubation for 60 min at 37°C with 0.6 mg.ml chymotrypsin (B) or trypsin (C), followed by an incubation of 30 min at 37°C with 1 MM radioactive PCMBS. The results are expressed as percentages of total area. The empty columns correspond to stained peptide profiles and the dotted columns correspond to the binding of radioactive PCMBS.

Nevertheless we may conclude that the NMR method is a useful tool for studying changes in water diffusion in erythrocyte membranes following various chemical manipulations of the membranes aimed at locating the water channel.

The financial support of the Academy of Medical Sciences of Roumania, of the Wellcome Trust, U.K., and of the Burnsides Research Fund, University of Illinois, USA, is gratefully acknowledged. N-succinimidyl-3-(2-pryridylthio) propionate was a gift from Pharmacia (Uppsala, Sweden).

REFERENCES

Ashley, D. L., and Goldstein, J. H., 1981, Time dependence of the effect of p-chloromercuribenzoate on erythrocyte water permeability: a pulsed nuclear magnetic resonance study, J. Membrane Biol., 61:199.

Benga, G. H., and Morariu V. V., 1977, Membrane defect affecting water permeability in human epilepsy, Nature, 265:636.

Benga, G. H., Pop, V.I., Ionescu, M., Holmes, R. P., and Popescu, O., 1982, Irreversible inhibition of water diffusion through erythrocyte membranes by fluoresceinmercuric acetate, Cell Biol.Int.Rep., 6:775.

Benga, G. H., Pop, V. I., Popescu, O., Ionescu, M., and Mihele, V., 1983, Water exchange through erythrocyte membranes, III. Nuclear magnetic resonance studies on the effects of inhibitors and of chemical modification of human membranes, J. Membrane Biol., 76:129.

Brown, P. A., Feinstein, M. B., and Sha'afi R. I., 1975, Membrane proteins related to water transport in human erythrocytes, Nature, 254:553.

Cabantchik, Z. I., and Rothstein, A., 1974, Membrane proteins related to anion permeability of human red blood cells, II. Effect of proteolytic enzymes on disulfonic stilbene sites of surface proteins, J.Membrane Biol., 14:227.

Carlsson, J., Drevin, H., and Axen, R., 1978, Protein thiolation and reversible protein-protein conjugation. N-succinimidyl-3-(2-pyridylthio) propionate a new heterobifunctional reagent, J.Biochem., 173:723.

Conlon, T., and Outhred, R., 1972, Water diffusion permeability of erythrocytes using an NMR technique, Biochem Biophys Acta., 288:354.

Dodge, J. T., Mitchell, C., and Hanahan, D. J., 1963, The preparation and chemical characteristics of hemoglobin free ghosts of human erythrocytes, Arch.Biochem. Biophys., 100:119.

Fabry, M. E., and Eisenstadt, M., 1975, Water exchange between red cells and plasma. Measurement by nuclear magnetic relaxation, Biophys J., 15:1101.

Fairbanks, G., Steck, T. L., and Wallach, D. F. H., 1971, Electrophoretic analysis of major polypeptides of the human-erythrocyte membrane, Biochemistry., 10:2606.

Jennings, J. L., and Passow, H., 1979, Anion transport across the erythrocyte membrane in situ proteolysis of band 3 protein, and cross-linking of proteolytic fragments by 4,4'-diisothiocyano dihydrostilbene-2,2'-disulfonate, Biochim Biophys Acta., 554:498.

Laemmli, U. K., 1970, Cleavage of structural proteins during the assembly of the hand of bacteriophage, T_4, Nature, 227:690.

Lowry, O. H., Rosebrough, H. J., Farr, A. L., and Randall, R. J., 1951, Protein measurement with the Folin phenol reagent, J.Biol.Chem., 193:265.

Macey, R. I., Karan, D. M., and Farmer, R. E. L., 1972, Properties of water Channels in Human Red Cells, in: "Biomembranes," F. Kreuzer, J. F. G. Slegers, eds., Plenum Press, New York.

Morariu, V. V., and Benga, G. H., 1977, Evaluation of a nuclear magnetic resonance technique for the study of water exchange through erythrocyte membranes in normal and pathological subjects, Biochim Biophys Acta., 469:301.

Morariu, V. V., Pop, I. V., Popescu, O., and Benga, G. H., 1981, Effects of temperature and pH on the water exchange through erythrocyte membranes: evidence for state transitions, J. Membrane Biol., 62:1.

Naccache, P., and Sha'afi, R. I., 1974, Effect of PCMBS on water transfer across biological membranes, J.Cell Physiol., 83:449.

Passow, H., Fasold, H., Lepke, S., Pring, M., and Schuhmann, B., 1977, Chemical and ensymatic modification of membrane proteins and anion transport in human red blood cells, in: "Membrane Toxicity," M. W. Miller, A. E. Shamoo, eds., Plenum Press, New York.

Pirkle, J. L., Ashley, D. L., and Goldstein, J., 1979, Pulse nuclear magnetic resonance measurements of water exchange across the erythrocyte membranes employing a low Mn concentration, Biophys J., 25:389.

Sha'afi, R. I., 1981, Permeability for water and other polar molecules, in: "Membrane Transport," S. L. Bonting, H. H. M., Du Pont, eds., Elsevier, Amsterdam.

Sha'afi, R. I., and Feinstein, M. B., 1977, Membrane water channels and SH-groups, in: "Membrane Toxicity," M. W. Miller and A. E. Shamoo, eds., Plenum Press, New York.

CATION TRANSPORT DURING THE CELL CYCLE OF NEUROBLASTOMA CELLS[1]

J.Boonstra, C.L.Mummery, E.J.J.van Zoelen, P.T.van der Saag and S.W.de Laat

Hubrecht Laboratory
Uppsalalaan 8
3584 CT Utrecht
The Netherlands

[1]This work has been supported by the Netherlands Cancer Foundation (Koningin Wilhelmina Fonds) and Shell Internationale Research Maatschappij B.V.

INTRODUCTION

The plasma membrane of mammalian cells is considered as having a major role in the regulation of cell growth. Acting as a barrier between the extra- and intracellular environment, the plasma membrane determines the compositional differences between them by selective permeability and by specific membrane-bound transport systems for ions and nutrients. In addition, the plasma membrane as the first cell organelle interacting with external factors, functions as the transducer of all extracellular signals which may affect the cellular status. Modulation of its compositional and structural properties might in turn be involved in the ability to perform this function.

The association of changes in growth behaviour with changes in cation transport, have suggested an important mechanism by which the plasma membrane properties might mediate its role in the regulation of cell growth (for reviews see 1-3). We have studied the properties and regulation of cation transport, in particular of Na^+ and K^+, during the cell cycle of neuroblastoma cells, clone Neuro-2A, and their possible role in control of its progression. Clone Neuro-2A is particularly suitable for cell cycle studies, because it has a relatively short generation time (8-10 hrs; 4,5) and the cells can be synchronized by mitotic shake-off (4).

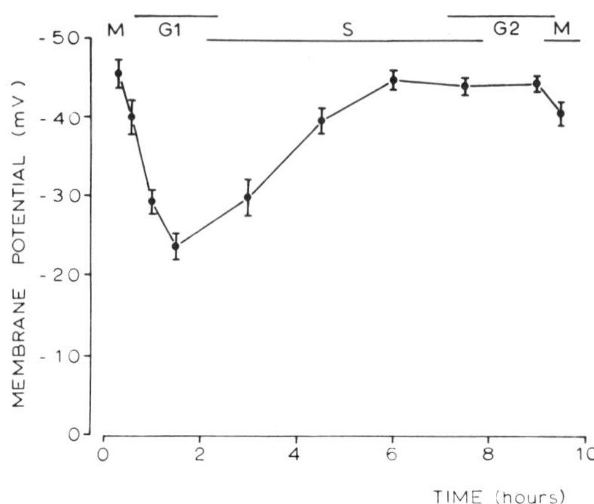

Fig. 1. Membrane potential during the cell cycle of neuroblastoma cells. The membrane potential was determined using a conventional 3M KCl-filled micro-electrode. Data presented as Mean $\pm$ S.E.M. From: Boonstra et al. 1981. J.Cell. Physiol. 107: 75-83

This method of synchronization has the advantage above other procedures in that it does not induce a time lag and has no additional side effects.

Membrane Potential During the Cell Cycle

The membrane potential (E_m) across the plasma membrane develops as a result of the transmembrane K^+ and Na^+ gradients and their respective permeabilities, and as such, is a convenient parameter for assessing net Na^+ and/or K^+ transport. In Neuro-2A cells, E_m can be described according the modified Goldman-Hodgkin-Katz equation with K^+ and Na^+ as the permeable ions and Cl^- in thermodynamic equilibrium with E_m (6). E_m was measured using a conventional 3 M KCl-filled microelectrode in synchronized Neuro-2A cells in various phases of the cell cycle (7). As shown in Fig.1, E_m was maximally hyperpolarized in mitosis (-45 mV), but rapidly depolarized to -23 mV upon progression to G_1 phase. This was followed by a gradual repolarization, reaching a value of -44 mV in mid-S phase. This level was maintained throughout the remainder of the cell cycle until the next mitosis. These results clearly indicate cell cycle phase specific modulations in K^+ and/or Na^+ gradients, and/or changes in the plasma membrane permeability properties towards these ions.

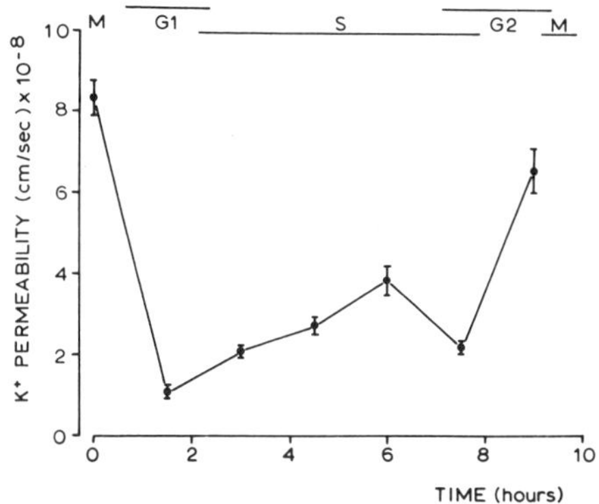

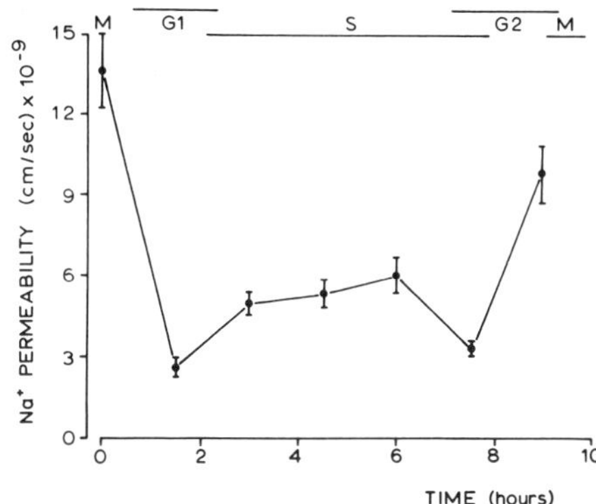

Fig. 2. K^+ and Na^+ permeabilities during the cell cycle of neuroblastoma cells. The K^+ and Na^+ permeability have been calculated from electrophysiological and tracer-flux data, as described in detail in Ref. 6 and 7. Data presented as Mean ± S.E.M. From: Boonstra et al. 1981. J.Cell.Physiol. 107: 75-83

Cation Permeability During the Cell Cycle

The relative membrane permeability to Na^+ and K^+ can be determined electrophysiologically from the dependence of E_m on external K^+ concentrations under conditions of equimolar replacement of Na^+ by K^+(6). During the cell cycle of Neuro-2A cells, the permeability ratio, P_{Na} / P_K, increases from 0.16 in mitotic cells to 0.26 in G_1 phase, followed by a gradual decrease to 0.16 in mid-S phase, with no further variations during the remainder of the cell cycle (7). The membrane permeability to Na^+ and to K^+ were distinguished by the study of K^+ efflux from the cells, since K^+ efflux is mainly of electrodiffusional origin. K^+ efflux (pmoles K^+ / sec / cm^2) was found to decrease from 6.5 to 0.9 pmoles K^+ / sec / cm^2 during progression from mitosis to G_1 phase, followed by a slow increase in S phase to 2.9 pmoles K^+ / sec / cm^2. A second decrease was observed in G_2 phase and a large increase just before the next mitosis (7). From these data the K^+ permeability was calculated (6) and found to decrease markedly as cells pass from mitosis to G_1 phase, followed by a small increase during S-phase. A second decrease was observed during G_2 phase and a large increase before the next mitosis as shown in Fig.2^A (7). The Na^+ permeability (P_{Na}) was calculated using P_K (Fig.2^A) and the permeability ratio, P_{Na} / P_K, and was found to be modulated during the cell cycle qualitatively in a similar way as P_K (Fig.2^B).

The changes of the permeability characteristics of the plasma membrane during the cell cycle might well be explained by variations of the compositional and structural properties of the membrane. It has been shown recently in Neuro-2A cells that supplementation of the growth medium with unsaturated fatty acids results in a marked modulation of K^+ and Na^+ permeability (8). Furthermore, evidence is available that changes in membrane composition and structure occur mainly around mitosis and early G_1 phase (1-3,9). In Neuro-2A cells the numerical and lateral distribution of intramembraneous particles also show a cell cycle dependence (10). Further, dynamic features of the plasma membrane, such as membrane fluidity and lateral mobility of lipids, are modulated during the cell cycle, being minimal in mitosis and increasing markedly in G_1 phase (4,11).

Changes in membrane permeability during the cell cycle of neuroblastoma cells are parallelled by variations in a number of compositional, structural and dynamic membrane properties, and it seems quite probable that the modulations of membrane permeability during the cell cycle is a reflection of the latter membrane properties.

K^+ Influx During the Cell Cycle

The changes in membrane permeability of K^+ (Fig.2A) during the cell cycle alone are not sufficient to explain the observed modulations of E_m (Fig.1), indicating that there must be an additional mechanism operating in its control. This is illustrated by measurements of the intracellular K^+ activity using a K^+-selective micro-electrode (7). Maximal values were observed during mitosis (127 mM), followed by a rapid decrease to 79 mM as cells progressed to G_1 phase. A gradual increase was observed during S-phase to 118 mM, after which no further variations were observed. Thus an additional mechanism involved in the control of E_m was sought among K^+ transport process, in particular the Na^+,K^+-pump. The pumping activity of this enzyme can be measured in whole cells from the accumulation of K^+, or Rb^+ as a convenient experimental analogue (6,12), in the absence and presence of a specific inhibitor, the cardiac glycoside ouabain. The ouabain-sensitive, or active K^+ influx, was measured from a 5 minute pulse labelling with $^{42}K^+$ as a radioactive tracer. As shown in Fig.3, the active K^+ influx decreases rapidly by more than 4-fold as cells pass from mitosis to G_1 phase. These low values are maintained through G_1 phase. On entry into S phase a transient increase is observed, after which a gradual increase occurs with progression through S phase. Immediately prior to the next mitosis, in G_2 phase, a small but significant decrease was observed (12). Using the data available on K^+ efflux and influx and on the intracellular K^+ content during the cell cycle, the net K^+ flux was calculated and correlates well with the modulations in K^+ content. Thus on passing from mitosis to G_1 phase there is net efflux, while on entry into S phase net flux

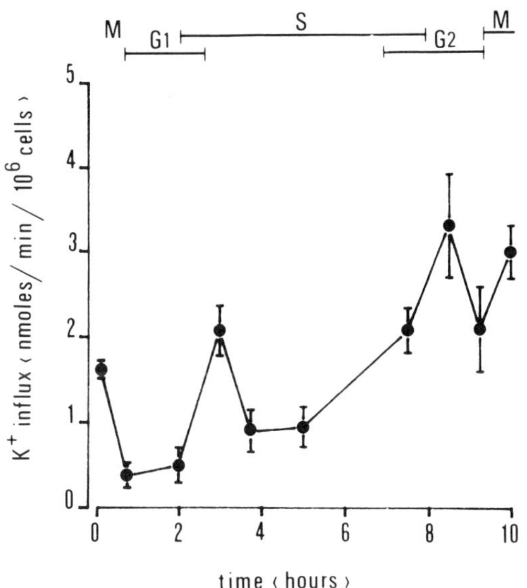

Fig. 3. Na^+, K^+-pump mediated K^+ influx during the cell cycle of neuroblastoma cells. The Na^+, K^+-pump mediated K^+ influx was taken as the difference between total and ouabain-insensitive influxes, using a 5 minute pulse labelling time with $^{42}K^+$ as radioactive tracer. Ouabain was used at a 5 mM final concentration. Data presented as Mean $\pm$ S.E.M. From: Mummery et al. 1981. J.Cell.Physiol. 107: 1-9

becomes inwardly directed and although virtually zero during mid-S phase, increases inwardly in a particularly pronounced way in late S phase (12). The net flux is of the correct order of magnitude to account for the changes in intracellular K^+ content during the cell cycle. The data described are sufficient to account for the modulations of E_m during the cell cycle.

Regulation of Na^+, K^+ Pump Activity During the Cell Cycle

Having characterized the modulations in membrane permeability and K^+ gradient during the cell cycle, the question arises as to the mechanism by which these modulations are realised. Since modulations in the K^+ gradient are mostly the result of the activities of the Na^+, K^+-pump, understanding the regulation of the Na^+, K^+ pump activity may lead to the answer of this question. Three possible regulatory mechanisms may, in principal, be involved: a) variations in substrate availability; b) changes in the physico-chemical properties of the membrane resulting in conformational and/or activity changes of the pump; c) variations in the number of active pump sites.

a) Substrate availability. Determination of the involvement of substrates (K^+, Na^+, Mg^{2+}, ATP and P_i) in the regulation of the Na^+, K^+-pump activity can be obtained from the comparison of the activity of the enzyme under optimal substrate conditions with the activity under physiological conditions. The optimal activity has been determined from ATP hydrolysis measurements in cell homogenates (12). During the cell cycle of Neuro-2A cells this activity decreases rapidly 3-fold during progression from mitosis to G_1 phase and rises linearly to the next mitosis (Fig.4). Comparison of this optimal activity with the functional activity, the active K^+ influx, reveals a major difference around the G_1/S phase transition. The absence of the transient increase in optimal Na^+, K^+-pump activity compared with the functional activity indicates a particular involvement of the enzyme substrates in the regulation of pump activity in this period. A number of studies have shown that the intracellular Na^+ is likely to be the most important substrate in the regulation of Na^+, K^+-pump activity in whole cells. (13-16). Therefore Na^+ was measured during the cell cycle of Neuro-2A cells (17), using $^{22}Na^+$ or $^{24}Na^+$ as radioactive tracer.

On passing from mitosis to G_1 phase, Na^+ influx decreased almost 2-fold, but increased rapidly and transiently upon entry into S-phase, returning to a steady level during the remainder of the cell cycle (Fig.5). In particular the transient increase in Na^+

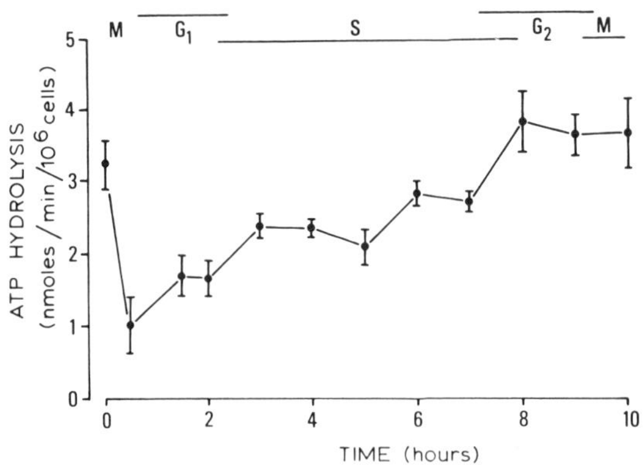

Fig. 4. (Na^+, K^+)ATPase activity during the cell cycle of neuroblastoma cells. The (Na^+, K^+)ATPase activity was determined from ATP hydrolysis in cell homogenates of cells from different cell cycle phases under optimal substrate conditions. Data presented as Mean $\pm$ S.E.M. From: Mummery et al. 1981. J.Cell.Physiol. 107: 1-9

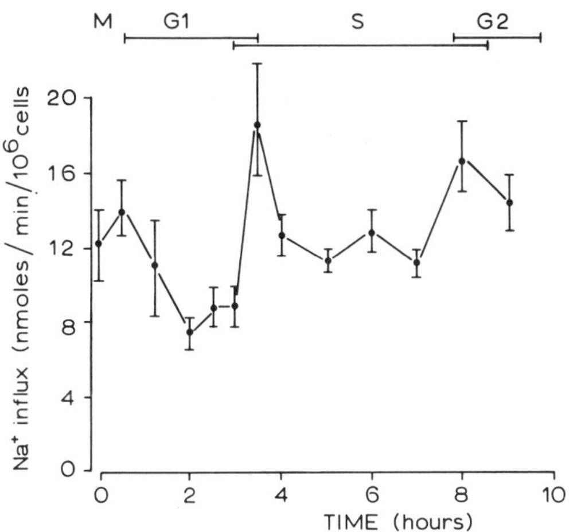

Fig. 5. Na^+ influx during the cell cycle of neuroblastoma cells. Na^+ influx was determined for a 3 minute labelling with $^{22}Na^+$ as radioactive tracer in the presence of 5 mM ouabain to prevent back flux of the tracer via the Na^+, K^+-pump. Data presented as Mean ± S.E.M. From: Mummery et al. 1982. J.Cell.Physiol. 112: 27-34

influx at the G_1 phase transition occurred simultaneously with the transient increase in active K^+ influx, the functional Na^+, K^+--pump activity, as was shown in a double labelling experiment (17). In addition amiloride, an inhibitor of Na^+/H^+ exchange in a number of cell systems (18,14), exhibited a strikingly cell cycle-dependent ability to inhibit Na^+ and active K^+ influx. No significant effect was observed on the uptake rates in early G_1 and S phases, while there was complete inhibition of the transient activation of both transport processes at the G_1/S phase transition (17). Taking into account a) the substrate dependency of Na^+, K^+-pump activity, b) the amiloride-sensitivity and c) the absence of significant changes in membrane potential during this cell cycle phase, it is suggested that the transient activation of the Na^+, K^+-pump at the G_1/S phase transition is due to an increased electroneutral, amiloride-sensitive Na^+ influx system. By analogy with other studies but in particular with serum stimulation studies this may also be a Na^+, H^+--exchange system (18-20). Preliminary experiments have confirmed this suggestion, showing that a Na^+-dependent amiloride-sensitive H^+ extrusion was principally active during the G_1/S phase transition, while hardly active in the other phases of the cell cycle (Mummery et al., unpublished results).

b) Number of active sites of the Na^+, K^+-pump. Another possible mechanism for regulation of Na^+, K^+-pump activity per cell, is by an alteration in the number of active pump sites. During the cell cycle the number of enzyme copies must double so that sufficient copies are available for each daughter cell. Usually the number of copies of the (Na^+, K^+)-ATPase is determined from ouabain-binding studies (16). Upon progression from mitosis to G_1 phase, the number of ouabain-binding sites increases, remains almost constant during late G_1 and early S phase, and increases rapidly in late S and G_2 phase (21), almost doubling as expected, and parallels to a large extent the increase in cell surface area. These observations suggest that the density of (Na^+, K^+)-ATPase molecules per unit of surface area is virtually constant during the cell cycle (21).

c) Optimal activity per pump site. The optimal activity per pump site can be determined in whole cells, provided the number of active pump sites and the pump activity under optimal substrate conditions is known. The optimal pumping activity has been determined in the presence of extracellular ATP, which enhanced the plasma membrane permeability for Na^+ and increased the affinity of the pump for binding of ouabain (16). Comparing the optimal pumping activity of the Na^+, K^+-pump and the number of ouabain-binding sites during the cell cycle demonstrated that the optimal pumping activity per (Na^+, K^+)ATPase molecule is modulated such that there is a high activity during mitosis, followed by a 2-fold decrease on passing to G_1 phase. During the remainder of the cell cycle the activity per copy gradually increased (21). For such modulations changes in membrane environment of the (Na^+, K^+)ATPase are necessary during the cell cycle, which may be reflected in modulation in physico-chemical membrane properties.

From the data presented above, it appears that the Na^+, K^+-pump activity during the cell cycle is regulated by a) substrate availability, in particular around the G_1/S transition, b) the number of active sites, in turn related to the cell surface area and c) the optimal activity per enzyme copy, in a most pronounced way between mitosis and G_1 phase.

Role of Cation Transport in Regulation of the Cell Cycle

Evaluation of any specific role of K^+ and Na^+ transport in cell progression presents a number of difficulties, due to the complex network of inter-relationships between K^+ and Na^+ on the one hand, and a wide variety of cellular metabolites and processes on the other. In this respect it is at present impossible to determine whether changed growth behaviour is exclusively due to changed K^+/Na^+ concentrations and/or transport or to other parameters which are also coupled to the cation transport systems. Taking these considerations into account, a variety of reports have, however, indicated the involvement of cation transport in growth regulation; increased malignancy of a variety of tumors coincides with an increased Na^+/K^+ ratio (22,23), while increased K^+ influx has been observed upon growth stimulation of quiescent cells by serum or purified growth factors, in all cases preceded by increased Na^+ influx (13,14,24,25). During the cell cycle of neuroblastoma cells, the Na^+/K^+ ratio is also modulated (17), during the G_1 phase the relative loss of Na^+ from the cells being significantly higher than the loss of K^+, the resulting Na^+/K^+ ratio being similar to that reported for untransformed 3T3 cells (22). A rapid recovery occurs upon entry into S phase which is maintained during the remainder of the cell cycle, the Na^+/K^+ ratio then reaching the value reported for transformed 3T3 cells (22). These observations clearly show a specific Na^+/K^+ ratio in various phases of the cell cycle.

Addition of low concentrations of amiloride has been shown to inhibit the transient increase of Na^+ influx during the G_1/S phase transition and, as a consequence, the transient activation of the active K^+ influx, while no effects were observed during the other phases of the cell cycle (17). Amiloride increased the cell generation time and decreased the probability of a transition from G_1 to S-phase. Similar results were obtained by inhibition of the transient activation of the active K^+ influx by low concentrations of ouabain (12). Thus inhibition of the ionic events at the G_1/S phase transition by amiloride or ouabain resulted in a marked effect on cell cycle growth kinetics and suggest a causal relationship. However, as mentioned above, Na^+ and K^+ transport, as well as their intracellular concentrations, are linked closely to a variety of other ions (Ca^{2+}, H^+) and nutrients (amino acids, sugars) and complete evaluation of the effect of K^+ and Na^+ on cell cycle progression will be only possible when the properties of the related ions and nutrients have been characterized during the cell cycle.

Concluding Remarks

The results described above, demonstrate that during the cell cycle the cation transport systems are modulated in such a way that a relationship between these properties and the compositional, structural and dynamic plasma membrane properties is evident (1,2). Large modulations occur mainly in the G_1 phase and during mitosis, a cell cycle phase in which membrane growth is most prominent. During these cell cycle phases it might be expected that large modulations occur in the composition, the ultrastructure and dynamic properties of the membrane, as has indeed been shown (4,10,11).

Furthermore, there appears to be a cell cycle specific response to external growth factors, thus in G_1 phase there may be specific receptor expression for growth factors, such as has been described for NGF in neuroblastoma (26) so that the cells then initiate a sequence of ionic events triggering entry into S-phase, analogous to those occurring when serum or purified growth factors are added to quiescent cells in a constant state of receptor expression.

References

1. S. W. de Laat and P. T. van der Saag, The plasma membrane as a regulatory site in growth and differentiation of neuroblastoma cells, Int. Rev. Cytol. 74:1-54 (1982).
2. S. W. de Laat, J. Boonstra, W. H. Moolenaar, C. L. Mummery, P. T. van der Saag, and E. J. J. van Zoelen, Cation transport and growth control in neuroblastoma cells in culture, in: "Membranes in Growth and Development," G. Giebisch and J. Hofman, eds., Alan R. Liss Inc., New York 211-236 (1982).
3. J. Boonstra, C. L. Mummery, E. J. J. van Zoelen, P. T. van der Saag and S. W. de Laat, Monovalent cation transport during the cell cycle, Anticancer Res. 2:265-274 (1982).
4. S. W. de Laat, P. T. van der Saag and M. Shinitzky, Microviscosity modulation during the cell cycle of neuroblastoma cells, Proc. Natl. Acad. Sci. USA. 74:4458-4461 (1977).
5. E. J. J. van Zoelen, P. T. van der Saag and S. W. de Laat, Family tree analysis of a transformed cell line and the transition probability model for the cell cycle, Exp. Cell Res. 131:395-406, (1981).
6. J. Boonstra, C. L. Mummery, L. G. J. Tertoolen, P. T. van der Saag and S. W. de Laat, Characterization of $^{42}K^{+}$ and $^{86}Rb^{+}$ transport and electrical membrane properties in exponentially growing neuroblastoma cells, Biochim. Biophys. Acta 643: 89-100 (1981).
7. J.Boonstra, C. L. Mummery, L. G. J. Tertoolen, P. T. van der Saag and S. W. de Laat, Cation transport and growth regulation in neuroblastoma cells. Modulations of K^{+} transport and electrical membrane properties during the cell cycle, J. Cell. Physiol., 107:75-83 (1981).
8. J. Boonstra, S. A. Nelemans, A. Feijen, A. Bierman, E. J. J. van Zoelen, P. T. van der Saag and S. W. de Laat, Effect of fatty acids on plasma membrane lipid dynamics and cation permeability in neuroblastoma cells, Biochim. Biophys. Acta 692:321-329 (1982).
9. J. G. Bluemink and S. W. de Laat, Plasma membrane assembly as related to cell division, in: "The synthesis, assembly and turnover of cell surface components," G. Poste and G. L. Nicolson, eds., Elsevier North Holland Biomedical Press, Amsterdam, pp.403-461 (1977).
10. S. W. de Laat, L. G. J. Tertoolen, P. T. van der Saag and J. G. Bluemink, Quantitative analysis of modulation in numerical and lateral distribution of intramembrane particles during the cell cycle of neuroblastoma cells, J. Cell. Biol. In press, (1982).
11. S. W. de Laat, P. T. van der Saag, E. L. Elson and J. Schlessinger, Lateral diffusion of membrane lipids and proteins during the cell cycle of neuroblastoma cells, Proc. Natl. Acad. Sci. USA. 77:1526-1528 (1980).
12. C. L. Mummery, J. Boonstra, P. T. van der Saag and S.W. de Laat,

Modulation of functional and optimal $(Na^{+}-K^{+})$ATPase activity during the cell cycle of neuroblastoma cells, J.Cell.Physiol. 107:1-9 (1981).
13. J. B. Smith and E. Rozengurt, Serum stimulation of the Na^{+},K^{+} pump in quiescent fibroblasts by increasing Na^{+} entry, Proc. Natl. Acad. Sci. USA. 75:5560-5564 (1978).
14. W. H. Moolenaar, C. L. Mummery, P. T. van der Saag and S. W. de Laat, Rapid ionic events and the initiation of growth in serum-stimulated neuroblastoma cells, Cell 23:789-798 (1981).
15. J. Boonstra, P. T. van der Saag, W. H. Moolenaar and S. W. de Laat, Rapid effects of nerve growth factor on the Na^{+},K^{+}-pump in rat pheochromocytoma cells, Exp.Cell Res.131:452-455 (1981)
16. E. J. J. van Zoelen, L. G. J. Tertoolen, J. Boonstra, P. T. van der Saag and S. W. de Laat, Effect of external ATP on the plasma membrane permeability and $(Na^{+}-K^{+})$ATPase activity of mouse neuroblastoma cells, Biochim.Biophys.Acta 720:223-234 (1982).
17. C. L. Mummery, J. Boonstra, P. T. van der Saag and S.W.de Laat, Modulations of Na^{+} transport during the cell cycle of neuroblastoma cells, J. Cell. Physiol. 112:27-34 (1982).
18. W. H. Moolenaar, J. Boonstra, P. T. van der Saag and S.W.de Laat, Sodium/proton exchange in mouse neuroblastoma cells, J. Biol. Chem. 256:12883-12887 (1981).
19. R. D. Moore, Stimulation of $Na^{+}:H^{+}$ exchange by insulin, Biophys. J. 33:203-210 (1981).
20. M. J. Rindler and M. H. Saier, Evidence for Na^{+}/H^{+} antiport in cultured Dog Kidney Cells (MDCK), J. Biol. Chem. 256:10820-10825 (1981).
21. E. J. J. van Zoelen, C. L. Mummery, J. Boonstra, P. T. van der Saag and S. W. de Laat, Membrane regulation of the Na^{+},K^{+} -ATPase during the neuroblastoma cell cycle. Correlation with protein lateral mobility, submitted
22. M. Lubin, Control of growth by intracellular potassium and sodium concentrations is relaxed in transformed 3T3 cells, Biochem. Biophys. Res. Comm. 97:1060-1067 (1980).
23. L. Zs-Nagy, G.Lustyik, V.Zs-Nagy, B.Zarandi and C.Freddari--Bertoni, Intracellular $Na^{+}:K^{+}$ ratios in human cancer cells as revealed by energy dispersive K-ray microanalysis, J. Cell Biol. 90:769-777 (1981).
24. E. Rozengurt and L.A. Heppel, Serum rapidly stimulates ouabain sensitive $^{86}Rb^{+}$ influx in quiescent 3T3 cells, Proc.Natl. Acad. Sci. USA. 72:4492-4495 (1975).
25. S. A. Mendoza, N. H. Wigglesworth, P. Polijanpelto and E. Rozengurt, Na entry and Na-K pump activity in murine, hamster and human cells, J. Cell. Physiol. 103:17-27 (1980).
26. R. Revoltella, L. Bertolini and M. Pediconi, Unmasking of nerve growth factor membrane-specific binding sites in synchronized murine C1300 neuroblastoma cells, Exp. Cell Res. 85:89-94 (1974).

SELECTIVITY RULES FOR THE ION TRANSPORT IN NARROW CHANNELS THROUGH BIOLOGICAL MEMBRANES FROM GEOMETRICAL AND DYNAMICAL PROPERTIES

Jürgen Brickmann and Walter Fischer

Institut für Physikalische Chemie
Technische Hochschule Darmstadt
Petersenstr. 20, D-6100 Darmstadt, FRG

ABSTRACT

The influence of geometrical and dynamical properties of narrow transmembrane channels is studied within two model approaches both analytically and numerically. The parameters for the models are chosen in close relation to the Gramicidin A channel. It is demonstrated that anomalous diffusion behaviour (in comparison with the diffusion in aqueous solution) of different cations can be related to differences in the activation entropy resulting from geometrical properties of the channel, and to dynamical effects which are strongly correlated to the softness of the vibration of the pore forming molecular groups.

INTRODUCTION

The migration of cations through biological membrane channels often takes place via transport mechanisms which are totally different from those of ion diffusion in aqueous solution. It is well known from a variety of experimental [1-3] and theoretical investigations [4-6] that this transport can be explained empirically within rate theoretical models. In these models, the overall transport rate is the product of three different contributions: the association (dehydration) of the ion with the channel, its transport from one end of the channel to the other (migration) and the dissociation (hydration) of the ion from the channel to the liquid. Since the geometrical parameters of ion channels in biomembranes are of the order of magnitude of the diameter of small molecules there is no a priori justification for using concepts appropriate to a free liquid to describe transport phenomena. Each of the three main steps (dehydration, migration, hydration) has to be studied from a microscopic point of view. Each of them can substantially contribute to ion specific transport properties. The subject of this paper is related to the migration in the channel interior. On the one hand the influence of structural inhomogeneities of the channel on the migration rate of different ions is studied analytically within a simple geometrical model, on the other hand the dynamics of the ion transport is analyzed as a function of the vibrational frequencies of polar groups of model pores interacting predominantly with the migrating particles. The latter is treated within the molecular dynamics simulation technique scheme.

In general, the details of the structure of the channels and also the details of the channel-ion interaction are not known. The situation is a little bit better for the synthetic channel formed by a dimer of the Gramicidin A molecule. For this channel the structure and also the single file diffusion rates for some ions are well known from experiments [1,2,7,8]. We used model systems with parameters in close relation to the Gramicidin A channel in the analytical and numerical treatments.

GEOMETRICAL EFFECTS

It is one aim of this paper, to find a possible geometrical explanation for the fact that heavier (and larger) univalent cations may migrate faster through a transmembrane channel than lighter (and smaller) ones. In order to treat the ion transport kinetics analytically within a rate theoretical concept, a model is used which is able to describe the major contributions of the channel-ion interaction. A detailed discussion of this model will be presented elsewhere [14,15]. Its major properties can be summarized as follows:

Model Approach

(1) The pore is modelled as a collection of polar groups along a cylinder of infinite length.

(2) The ions can interact with these groups by electrostatic and van der Waals type forces.

(3) The electrostatic interaction between the migrating ion and the channel is reduced to a sum of point charge interactions between polar groups (carbonyl groups) near the channel surface and the ionic charge (see fig. 1a). The positive pole of the polar group is fixed on a cylinder with radius R_+ (outer channel radius) while the negative charges are located inside the channel (distance $R_- < R_+$ from the axis).

(4) The polar groups are assumed to exist in pairs of two dipoles, with one negative pole pointing in the forward and one in the backward direction of the channel. This model assumption is related to the Gramicidin A molecule where the carbonyl groups form

an alternating sequence along the helix chain (see fig.1a).

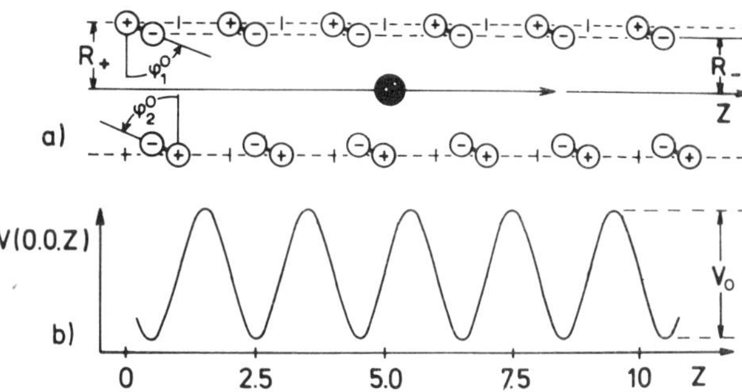

Fig. 1. (a) Arrangements of the carbonyl groups in the channel (schematically) and (b) potential energy along the channel axis z.

(5) The carbonyl groups are placed opposite one another in the channel. This simplifying arrangement was chosen to avoid complicated geometrical effects which are not important for the understanding of the transport mechanism. It was shown earlier [9] that the alternating arrangement of the polar groups creates a potential energy function V(z) for the migrating ion along the axis (diffusion coordinate z) which is a periodic function of z with minima (quasi equilibrium sites) at positions with maximal coulomb attraction between ions and polar groups and maxima (transition state sites) at positions with maximal coulomb repulsion (see fig.1b).

(6) The motion of the ion is confined to the channel by introducing a repulsive interaction between the migrating particle and the cylinder surface (radius R_+). The repulsive potential is modelled by a d^{-12}-term where d is the distance between the wall and the particle.

(7) The non-ionic interactions between the polar groups and the ion is modelled by the Lennard-Jones (12,6) potential[10]:

$$V_{LJ}^{12,6} = 4\varepsilon \left[\left(\frac{\sigma}{r}\right)^{12} - \left(\frac{\sigma}{r}\right)^{6} \right] \qquad (1)$$

This potential is the most common analytical potential function for non-bounding interactions of spherical particles. It has only two parameters, the range σ and the dissociation energy ε.

can be roughly identified with the radius of the sphere in a hard sphere model. The LJ parameters for dislike partners can be obtained from those for two particles of the same kind by applying combination rules [10]

$$\sigma_{12} = (\sigma_1 + \sigma_2)/2 \qquad \varepsilon_{12} = (\varepsilon_1 \cdot \varepsilon_2)^{1/2} \tag{2}$$

The potential function V(x,y,z) for the migrating ions is the sum of the contributions described above. Perpendicular to the channel axis this function has quite different forms in the x-direction (containing the polar groups) for the quasiequilibrium position (see fig. 2a) and the transition state position (see fig. 2b), while

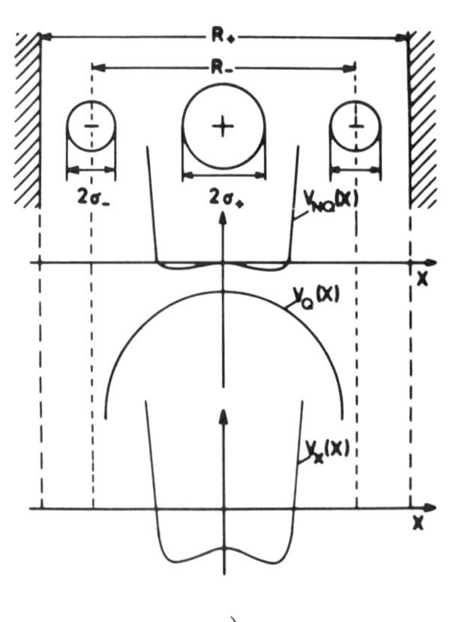

a)

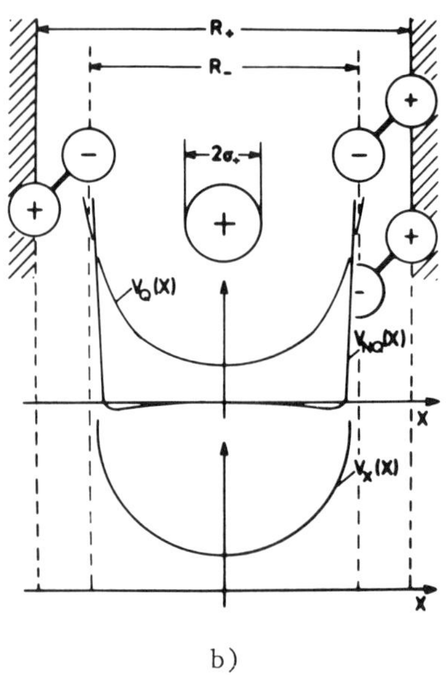

b)

Fig.2. (a) Nonionic potential $V_{NQ}(x)$, ionic contribution $V_Q(x)$ and total potential function $V_x(x)$ for the ion perpendicular to the diffusion coordinate in the model system at the position of the quasiequilibrium site (schematically) (b) Potentials $V_{NQ}(x)$, $V_Q(x)$, and $V_x(x)$ as in (a) at the position of the transition state

in the y-direction (perpendicular to the x-and z-coordinate), the two potentials are similar. We have to focus our attention to the x-coordinate, since for this coordinate strong ion specific differences between equilibrium and transition states are expected.

The potential function V(x) along the x-coordinate contains two parts, the coulomb interaction $V_Q(x)$ and the non-ionic contribution $V_{NQ}(x)$ (see fig. 2).

At the equilibrium position (fig. 2a) the ion "sees" an attractive coulomb force originating in the negative charges in the polar groups, i.e., the potential $V_Q(x)$ has a flat maximum value along the diffusion coordinate (x=0) and decreases if the ion moves toward the channel wall along x. This motion, however, is bounded by the Lennard-Jones repulsion term of the non-ionic potential contribution $V_{NQ}(x)$. The total potential $V_x(x)$ is consequently a slight modification of $V_{NQ}(x)$, i.e., the ion can move along the x-coordinate between $\pm x_o = \pm (R_- - \sigma_+)$ with relatively small variations of the potential in this region.

At the transition site position the potential cut $V_x(x)$ is not very much effected by the repulsive part of the LJ potential because the main repulsion is related to the coulomb repulsion. Consequently, this part of the potential is nearly the same for all univalent ions. The ion specific differences of $V_x(x)$ for the quasiequilibrium position and the transition state are the origins for different activation entropies for the ions, as is demonstrated in the next section [14].

Relative Ion Diffusion Rates from Rate Theory

It is well known from experimental observations that most hopping diffusion processes in ordered material have an Arrhenius temperature dependence [11]. The diffusion constant can be formally given as

$$D = D_{\infty}(T) \exp(-E_a/kT) \tag{3}$$

where k is Boltzmann's constant and E_a is the activation energy. The preexponential factor $D_{\infty}(t)$ is in general smoothly dependent on the absolute temperature T. For the mean jump length a between two quasi equilibrium sites the diffusion coefficient D is related to the site to site migration rate by the relation $D = a^2 \nu/2$. In Eyring's formulation of the absolute rate theory [12,13] the rate constant of a thermally activated rate process is given by the expression

$$\nu = \frac{M^{\ddagger}}{M} (kT/h) \Omega^{\ddagger} \Omega^{-1} \exp(-V_o/kT) \tag{4}$$

where $M^{\ddagger}/M$ is the ratio of the number of activated states to the number of sites, k and h are Boltzmann's and Planck's constants respectively, $\Omega^{\ddagger}$ is the partition functions of the transition state not including the reaction coordinate (diffusion coordinate), and Ω is the total partition function of the equilibrium state. The value V_o is the difference of the potential energy between transition state and quasiequilibrium well (see fig. 1). As has been demonstrated recently by the present authors [14] the partition functions Ω and $\Omega^{\ddagger}$ can be estimated for the simple model presented above leading to a rate expression

$$\nu = (f_z kT/2\pi f_x^{\ddagger} m) L_x^{-1} \exp\left[-E_a/kT\right]$$
$$= \nu_z e^{-1/2} \exp(\Delta\bar{S}/R) \exp(-E_a/kT) \qquad (5)$$

where f_z and $f_x^{\ddagger}$ are the force constant b for the ion vibration at a quasiequilibrium site along z and at the transition state along x, respectively. $L_x = 2R_- - \sigma_+ - \sigma_-$ is the width of potential wells along x, where σ_+ and σ_- are the LJ parameters for the cation and the anions of the carbonyl groups respectively. Different ions have different activation entropies $\Delta\bar{S}$ according to eq. (5). The pre-exponential factor shows the usual behaviour with respect to the mass m, i.e., $\nu \sim m^{-1/2}$ but there is strong dependence on the size of the ions via the parameter L_x, $\nu \sim (2R_- - \sigma_+ - \sigma_-)^{-1}$ This factor may cause the small ions to migrate more slowly than larger ions. From eq.(5), the ratio R_{12} of the diffusion rates of two univalent ions with masses m_1 and m_2 and "diameters" σ_{1+} and σ_{2+} can be roughly estimated as

$$R_{12} = \left(\frac{m_2}{m_1}\right)^{1/2} \frac{2R_- - \sigma_{2+} - \sigma_-}{2R_- - \sigma_{1+} - \sigma_-} \qquad (6)$$

This expression was used to calculate the relative rates listed in table I.

Table 1. Ratios $R_{12} = \nu_{M^+} / \nu_{Na^+}$ of the Ionic Diffusion Rates ν in Gramicidin A Like Channels for M = Li, K, Rb

M	exp. (ref.2)	R_{12} from eq (6) [a]	R_{12} from MD simulation [b] $\tilde{\nu} = 150 cm^{-1}$	$\tilde{\nu} = 500 cm^{-1}$
Li	0.29 ± 0.04	1.50 ± 0.03	1.37	0.50
K	2.09 ± 0.20	1.85 ± 0.35	0.74	1.76
Rb	3.04 ± 0.15	2.20 ± 0.54	0.53	1.33

[a] see ref. 14 [b] see ref. 15

The details of this geometrical approach to the ion diffusion rates are given in Ref. 14.

The numerical test of our analytical expression show that not only experimental data but also the results of numerical computer simulation (see next section) on the diffusion of different ions through the extensively studied Gramicidin A channel are well fitted by our formulas.

We found that the inversion of the normal mass effect on the diffusion rates of different ions is mainly caused by strong entropy effects. The activation entropies can be explicitely estimated from microscopic data. However, for small ions and (or) wide channels, the entropy effect is accompanied by a change of the actual activation energy comparing different ions.

DYNAMICAL EFFECTS

The analysis of the ion transport within a rate theoretical concept on the basis of particular properties of a one particle potential cannot reflect dynamical properties of the channel and its specific dynamical interaction with the ion because the normal rate theory is a mean field theory in which all the dynamical details are removed by averaging procedures. Nevertheless, the dynamical details of the ion-channel system may also become important as has been demonstrated recently [15]. In this section some molecular dynamics results are presented related to the question of how the flexibility of the polar groups in the channel may cause drastic changes in the ion transport process. The model assumptions are similar to those in the last section, and represent more closely the Gramicidin A channel.

Model Approach

(i) The channel is now more realisticly modelled by a hexagonal chain of fixed carbon atoms arranged on a helix. The geometrical parameters are those of Gramicidin A.

(ii) Each carbon atom is a member of a carbonyl group. The corresponding oxygen atoms are displaced towards the axis of the channel with fixed distance but flexible angles φ with respect to it.

(iii) The carbonyl groups are allowed to perform bending vibrations around equilibrium angles of 20 degrees with respect to the axis (alternating in forward and backward direction).

(iv) The carbonyl groups interact via dipole-dipole forces.

(v) The interaction between the migrating ion and the channel is simulated by a sum of electrostatic and van der Waals interactions. The electrostatic part is reduced to point charge interactions between the polar groups and the ionic charge. The van der Waals part is modelled as before by Lennard-Jones (12,6) potentials.

(vi) The motion of the ion is confined to the channel by introducing repulsive interactions between the migrating particle and the cylinder surface of the pore.

ION MIGRATION TRAJECTORIES FROM MOLECULAR DYNAMICS SIMULATION

In the molecular dynamics technique the classical equations of motion $\dot{q}_i = \partial H/\partial p_i$ $\dot{p}_i = -\partial H/\partial q_i$ for components q_i and p_i for the position and the momenta of the different particles respectively, are solved simultaneously by a numerical integration scheme. $H = T + V$ is the Hamiltonian of the system which contains the kinetic energy T and the potential energy V. A Runge-Kutta integration scheme with a time step of $\Delta t = 2.5 \cdot 10^{-15}$s was used and 30 000 - 75 000 integration steps were carried out in a single calculation corresponding to an integration time of 10-25 h on a PDP 11/60 mini Computer. Periodic boundary conditions were used in all calculations with a unit cell of 30 dipoles, i.e. the ligand motion is periodic after five turns of the pore helix. The system of dipoles was "thermalized" to a temperature T by the choice of proper initial conditions [9,15]. The total energy of the system (channel dipoles plus ion) is then stable during each calculation series, i.e. the total system represents a microcanonical ensemble.

As has been demonstrated before [9,15], the ion migration rate and so the diffusion coefficient can be obtained directly from the ion trajectories. These trajectories are represented by the local coordinates $\vec{r}(t) = (x(t), y(x), z(t))$ as functions of time. The analysis of the z-coordinate of the trajectory shows that the ion carries out discrete jumps from one binding site into another. One can count the total number N_j of jumps from one site to a neighbouring one in the time period τ of the computer simulation. The jump rate is then simply given by $k' = N_j/2\tau$ and the diffusion coefficient D can be obtained from k' and the distance a of adjacent sites (jump length) as $D = a^2 \cdot k'$. The temperature T of the oscillating polar groups can be calculated as a time average of the kinetic energies $E_{kin}{}^i$ of the dipoles as[15]

$$T = \frac{2}{k} \sum_{i=1}^{f} \langle E^i_{kin} \rangle^{\Delta t'} \qquad (7)$$

where f is the number of degrees of freedom and $\Delta t'$ a finite time interval of the simulation series. It was shown earlier [9,15] that the jump rate as a function of temperature, as calculated with the MD scheme, shows Arrhenius type behaviour predicted by the normal rate theory, and also that the activation energy E_a is not necessarily equal to the height of the potential barrier seperating two wells in the total energy surface. All calculations reported in this paper are carried out at only one temperature T = 1200 K. This unrealistic temperature was chosen in order to obtain a sufficiently high number of jumps per single simulation series for a statistical interpretation of the results. It has already been shown that the results can be extrapolated to lower temperatures[15].

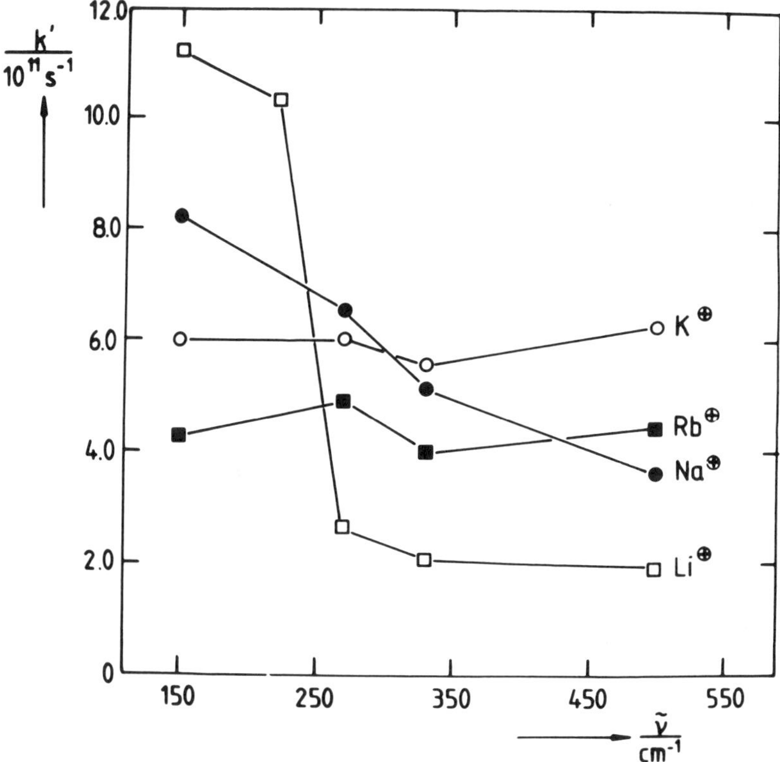

Fig. 3. Site to site transition rate for different alkali cations as a function of softness of the channel (represented by the CO binding mode $\tilde{\nu}$).

As has been pointed out, the geometrical parameters of the present model channel are taken from the Gramicidin A molecule. This is not possible for most of the dynamical parameters.

The most important one of this type is the force constant γ (or corresponding wavenumber $\tilde{\nu}$) of the restoring force for the dipole vibration. In the undistored case $\gamma \to 0$, the dipoles are freely rotating while for $\gamma \to \infty$ the dipoles become rigid. Consequently, the force constant can be viewed as a "softness" parameter of the model channel and it is directly related to the nuclear polarizibility of the helix.

In our calculations the wavenumber was varied in the range $100cm^{-1}$ - $600cm^{-1}$. The results are drawn in fig.3 and listed in table 1. It can be seen that there is a drastic change in the relative hopping rate for most ions as a function of the channel "softness" parameter $\tilde{\nu}$. For soft channels (low values of $\tilde{\nu}$) the "normal" mass dependence of the jump rates results,i.e. the lighter univalent ions migrate faster that the heavier ones. This sequence is nearly converted for rigid channels (high values $\tilde{\nu}$). For these the geometrical arrangements of the polar groups become increasingly important and the normal mass effects of the rates are overcompensated by a geometrical entropy effect as described above.

REFERENCES

1. S. B. Hladky and D. A. Haydon, Biochim. Biophys. Acta 274:294 (1972).
2. E. Neher, J. Sandblom, and G. Eisenman,J. Membrane Biol. 40:97 (1978).
3. G. Eisenman and J. P. Sandblom, in:"Physical Chemistry of Transmembrane Ion Motions", ed.C.Troyanowski, Elsevier, Paris (1982).
4. P. Läuger, Biochim. Biophys. Acta 311:423 (1973).
5. P. Läuger, Biophys. Chem. 15:89 (1982).
6. H. Schröder, "Rate Theoretical Analysis of Ion Transport in Membrane Channels with Elastically Bound Ligands", J. Chem. Phys., in press.
7. D. W. Urry, Proc. Nat. Acad. Sci. USA 68:672 (1971).
8. D. W. Urry, M. C. Goodall, J. D. Glickson, and D. F. Mayers, Proc. Nat. Acad. Sci. USA 68:1907 (1971).
9. W. Fischer, J. Brickmann, and P. Läuger, Biophys. Chem. 13:105 (1981).
10. W. Fischer and J. Brickmann, Ber. Bunsenges. Phys. Chem. 86:650 (1982).
11. See, for example, Topics in:"Applied Physics", Vol. 28, 29, "Hydrogen in Metals I, II" ed. G. Alefeld and J. Völkl, Springer, Berlin (1978).
12. B. J. Zwolinsky, H. Eyring, and C. E. Reese, J. Phys. Chem. 53:1426 (1949).
13. T. L. Hill, "Statistical Thermodynamics", Addison-Wesly, Reading (1960).
14. J. Brickmann and W. Fischer, "Entropy Effects on the Ion Diffusion Rate in Transmembrane Protein Channels",Biophys.Chem.,(1983).
15. W. Fischer and J. Brickmann, "Ion Specific Diffusion Rates through Transmembrane Protein Channels: A Molecular Dynamics Study", (to be published).

IONIC REGULATION OF GROWTH IN NORMAL AND TUMOR CELLS[1]

Ivan L. Cameron and Nancy K.R. Smith

Department of Anatomy
The University of Texas
Health Science Center at San Antonio
San Antonio, Texas 78284

Abstract

That cellular growth is regulated by intracellular ions such as Na, K, Mg, Ca and H stems from several lines of evidence. Electron probe X-ray microanalysis or EDS gives us the potential to study the relationship between ionic content and growth in cells with carefully characterized growth kinetics. The EDS technique can measure the content of several ions simultaneously at a subcellular level. EDS can be applied both to cells in culture or *in vivo*. This technique, as with other techniques does not give data on the free or the bound state of the ions measured. EDS has been used to measure elemental concentration differences between rapidly and slowly dividing cell populations and to measure differences between tumor and their non-tumor counterpart cells. We find that increased intracellular Na is related to rapid cell proliferation but show that increased Na is more dramatically related to oncogenesis than to mitogenesis. Our recent results from the injection of amiloride (a drug which blocks the passive influx of Na into mitotically activated mammalian cells) into tumorous mice, indicates that the intranuclear concentration of Na, but not Mg or K is correlated to the proliferative activity of rapidly dividing duodenal crypt enterocytes and also to rapidly proliferating H6 hepatoma cells in mice. That Na but no other ion measured changed in parallel with cell proliferative activity implicates Na in the regulation of cell replication. This finding does not rule out H or other unmeasured ions in the ionic regulation of cell proliferation.

[1] Supported by Grant #PCM80484 from the U.S. National Science Foundation and by Grant #CA30956 from The National Cancer Institute.

Introduction

Several ions have been implicated as growth regulators in mammalian cells. Among these are Na, K, Mg, and Ca. It is likely that there exist more than one kind of ionically mediated excitation-response coupling mechanism in growth regulation. A definitive explanation of these regulatory mechanisms has remained elusive because technical problems make critical experiments difficult to perform (see 4 for a review).

Recent studies have implicated Na as playing a key role in both mitogenesis and oncogenesis (1-14, 16-21). For example, rapidly proliferating normal cells have higher intracellular concentrations of Na than slowly proliferating normal cells, and transformed cells have much higher concentrations of Na than do rapidly proliferating normal cells (2-5, 14, 19-20). It has been speculated that the high concentrations of Na in normal and tumor cells is mitogenic (6,9-10,14,17,19).

The following report is a compilation and a summary of recent reports from this laboratory (5,17) and deals with our attempt to determine the role of high intracellular concentration of Na in normal and in transformed cells. The effects of amiloride, which is reported to inhibit passive sodium influx and proliferation of normal mammalian cells (9,10,13,16-18,21) was tested on cell proliferation in a rapidly dividing normal cell population and on cell proliferation in a rapidly dividing tumor cell population. The intranuclear elemental content of this normal cell and this tumor cell was also measured to determine any positive correlation between ion content and cell proliferation.

Materials and Methods

The H6 hepatoma was propagated in male A/J mice by the subcutaneous injection of a 1.5 ml suspension of cells in lactated Ringer's solution on the flank. The amiloride used in this study was supplied by Dr. E. Cragoe of Merck, Sharp and Dohme (West Point, PA.). To determine how amiloride treatment influenced the intranuclear content of Na and other elements and the cell proliferative activity, the following experiment was performed. Mice harboring a 1 cm diameter H6 hepatoma 13 days after inoculation were placed into 3 groups. The tumor bearing (TB) mice in the first group were given three injections of amiloride made up in lactated Ringer's solution at a concentration of 1.0 μg/g body weight. The drug was made up fresh for i.p. injections and was given at eight hour intervals. The second group of TB mice were given the same series of injections of lactated Ringer's but without amiloride. Non-tumor bearing mice (NTB) in a third group were given a similar series of three injections of lactated Ringer's without amiloride. All mice were given an injection of tritiated thymidine at 1 μ Ci/gm body weight, (sp. ac. 6.0 Ci/mM) three hours after the last injection of amiloride or Ringer's solution and were killed by decapitation one hour later at noon.

Segments of the tumor, the liver and the duodenum were then processed either for electron probe X-ray microanalysis as described elsewhere (5,15,17) or for autoradiography.

To measure the proliferative activity in the various tissues, pieces of the tissue were fixed in 10% neutral buffered formalin, processed for histological sectioning by embedment in paraffin and were sectioned at 4 μm thickness. These histological sections were mounted on glass slides for autoradiography. The histological slides were deparaffinized and were processed for autoradiography by dipping into liquid NTB-2 emulsion (Kodak). Autoradiographic exposure time was eight weeks and then the autoradiographs were developed in Microdol X (Kodak). They were then stained through the emulsion with hematoxylin and eosin. The percentage of cells within 25 μm of a blood vessel in the tumor and the number of labeled cells per crypt were scored. A total of 20 such crypts, the lumen of which was visible from the base to the mouth, were scored to obtain the average in each mouse.

Results and Discussion

The data obtained from determining how the tumor and how amiloride treatment influenced intranuclear content of Na in the tumor and in three tissue cell types are summarized in tables I & II. Table I indicates that the three injections of amiloride at a dose of 1.0 μg/g body weight into mice bearing H6 hepatomas resulted in a significant decrease in the intranuclear content of Na but not the content of Mg, Cl or K as measured by electron probe x-ray microanalysis in the H6 hepatoma cells. Table II gives the results of a two-way analysis of variance for Na. Two independent experimental variables were arranged in a 3x3 factorial design. The first

Table 1. Effect of three injections of amiloride at 1.0μg/g body weight each on intranuclear content of Na, Cl, K and Mg in H6 hepatoma cells Table modified from (17)

	Content in mMol/kg dry weight (means ± s.e.m.)	
Element	Ringer's* (4)	Amiloride (1.0μg/g)* (3)
Na	499±60	275±26**
Cl	328±25	234±26+
K	556±51	664±66***
Mg	54±3	67±8***

* Values represent means from 3 animals in the amiloride group and 4 animals in the Ringer's group. Ten X-ray spectra were collected from each animal.

* P-value < .05, significantly different from Ringer's control values

\+ P-value < .10

*** Not significantly different from Ringer's control value

independent variable, type of tissue, consisted of the three tissue cell types listed in table II. The second independent variable consisted of treatment such as: (1) mice without tumors, (2) mice with tumors and (3) mice with tumors that were given the three injections of amiloride. Two-way analysis of variance was used to determine significant effects in element content due to (1) tissue cell type, (2) treatments (tumor presence or absence and amiloride treatment) and (3) possible interactions between tissue cell type and the treatments.

Analysis of variance showed a significant main effect of tissue cell type for Na, K and Cl. Briefly, this shows that significant differences exist for Na, K, and Cl concentration in the three tissue cell populations. Because this study was not concerned with differences between tissue cell types this significant main effect will not be further pursued here. The bottom of Table II indicates that there was a significant main effect of treatment without interaction only in the case of intranuclear Na concentration. Neither K, Cl nor Mg (not

Table 2. Effect of tumor presence and amiloride injections on intracellular Na concentration (mM/kg dry weight) of host tissue cells (mean ± S.E.M.)

	Tissue (T)			
Treatment (TR)	fibroblasts	liver	duodenum	Row means
Untreated controls	257±30	170±21	245±15	217
Tumor bearing	314±33	287±16	243±17	283
Tumor bearing + amiloride	251±40	164±22	226±27	214
Column means	277	206	238	

Results of two-way analysis of variance

	F Ratio	p value
TR	4.07	<0.05
T	3.46	<0.05
TRxT	1.06	NS

Results of multiple range tests (p value)

	liver	duodenum
fibroblasts	<0.05	NS
duodenum	NS	

	tumor bearing amiloride	tumor bearing (no amiloride)
control	NS	<0.05
tumor bearing	<0.05	

* Table modified from (5)

illustrated) showed a significant effect of treatment. A multiple comparison test procedure is therefore appropriate to determine which treatment (tumor absence or presence or tumor presence plus amiloride treatment) were significantly different from one another across the three tissue cell types in the case of Na. The Student-Newman-Keuls multiple comparison test showed that the intranuclear Na concentration was significantly higher ($p < 0.05$) in the tissue cell types of the TB mice than in the tissue cell types of the TB mice treated with amiloride and in the NTB mice. The latter two groups of mice were not shown to be significantly different in Na concentration.

The tumor caused a significant overall increase in intranuclear Na concentration over that of NTB mice in the host cell types studied, while amiloride treatment of tumor bearing mice countered the effect of the tumor and reduced the intranuclear Na concentration to the levels of that in the NTB mice.

Amiloride at dosages of 1.0 μg/g body weight per injection significantly inhibited tumor cell proliferation as measured by the tritiated thymidine autoradiography labeling index (table III). Cell proliferation activity in the duodenal crypts as measured by autoradiography of the number of labeled cells per crypt showed that the mice without tumors and the mice with tumors were not significantly different in number of labeled cells per crypt whereas the mice with tumors that were treated with amiloride has significantly fewer labeled cells per crypt. Thus, amiloride lowered the proliferative activity in the duodenal crypts of the mice with a tumor.

Table 3. Effect of three injections of amiloride (1.0 μg/g body weight at eight hour intervals) on cell proloferation activity (autoradiographic labeling index[b]) of H6 hepatoma cells and number of labeled cells duodenal crypt section in vivo (mean ± S.E.M.)

Treatment group	Number of mice	Percent labeled cells within 25 μg of a blood vessel	Number of labeled cells per duodenal crypt
With tumor			
Ringer's	5	44.3±1.6	12.52±0.885
Amiloride in Ringer's	5	36.6±1.9[c]	8.40±.0.81[c]
Without tumor			
Ringer's	5		12.050±0.60

[a] Table from (4) with permission of the publisher, Academic Press, Inc., N.Y.
[b] The mice were injected with tritiated thymidine 1 hr before sacrifice.
[c] Significantly lower ($p < 0.025$) than the Ringer's treatment alone.

The intracellular and intranuclear concentration of Na has been shown to be involved in both oncogenesis and mitogenesis (2-5,9,10). The high level of Na in tumor cells appears to be related to the high proliferative activity of tumor cells. The present report, based on other recent reports from our laboratory (4,5), shows that injections of TB mice with the diuretic amiloride (which is reported to inhibit sodium influx in rapidly dividing mammalian cells), lowered the intranuclear Na but not the concentration of Mg, P or K and also lowered the proliferation rate in duodenal enterocytes and in tumor cells _in vivo_ (4,5).

Thus, a correlation was found between intranuclear Na and cell proliferation rate. Sodium, therefore, seems to be strongly implicated in the ionic regulation of cell proliferation in normal and tumor cells. A change in intracellular Na concentration is therefore directly linked to regulation of mitogenesis and oncogenesis.

REFERENCES

1. Cameron, I.L., N.K.R. Smith and T.B. Pool. 1979. J. Cell Biol. 80: 444.
2. Cameron, I.L., N.K.R. Smith, T.B. Pool, B.G. Grubbs, R.L. Sparks and J.R. Jeter, Jr. 1980. In: Nuclear-Cytoplasmic Interactions in the Cell Cycle, G.L. Whitson (ed.), Academic Press, Inc., New York, N.Y. 249.
3. Cameron, I.L., N.K.R. Smith, T.B. Pool and R.L. Sparks. 1980. Caner Res. 40: 1493.
4. Cameron, I.L.,N.K.R. Smith and P. Skehan. 1983. In: Ions, Cell Proliferation and Cancer, W.L. McKeenan and A. Boyton (eds.) (in press).
5. Cameron, I.L. and K.E. Hunter. 1983. Cancer Res.(in press)
6. Cone, C.D., Jr. 1971. J. Theor. Biol. 30: 151.
7. Frantz, C.N., D.G. Nathan and C.D. Scher. 1981. J. Cell Biol. 88: 51-56.
8. Kaplan, J.G. 1978. Ann. Rev. Physiol. 40: 19.
9. Koch, K.S. and H.L. Leffert. 1979. Cell 18: 153.

10. Leffert, H.L. and K.S. Koch. 1980. Ann. N.Y. Acad. Sci. 339: 201.
11. Mendoza, S.A., N.M. Wigglesworth, P. Pohjanpelto and E. Rozengurt. 1980. J. Cell Physiol. 103: 17.
12. Mendoza, S.A., N.M. Wigglesworth and E. Rozengurt. 1980b. J. Cell. Physiol. 105: 153.
13. Moolenaar, W.H., C.L. Mummery, P.T. van der Saag and S.W. de Laat. 1981. Cell 23: 789.
14. Pool, T.B., I.L. Cameron, N.K.R. Smith and R.L. Sparks. 1981. In: The Transformed Cell, I.L. Cameron and T.B. Pool (eds.), Academic Press, Inc., New York, N.Y. 398.
15. Pool, T.B., N.K.R. Smith, K.H. Doyle and I.L. Cameron. 1980. Cytobios 28: 17.
16. Rozengurt, E. and S. Mendoza. 1980. Ann. N.Y. Acad. Sci. 339: 175.
17. Sparks, R.L., T.B. Pool, N.K.R. Smith and I.L. Cameron. 1983 Cancer Res. (in press).
18. Smith, J.B. and E. Rozengurt. 1978. Proc. Natl. Acad. Sci. USA. 75: 5560.
19. Smith, N.R., R.L. Sparks, T.B. Pool and I.L. Cameron. 1978. Cancer Res. 38: 1952.
20. Smith, N.K.R., S.B. Stabler, I.L. Cameron and D. Medina. 1981. Cancer Res. 41: 3877.
21. Villeral, M. 1981. J. Cell. Physiol. 107: 359.

EFFECTS OF pH AND OF TEMPERATURE ON SATURABLE TRANSPORT PROCESSES

Arnošt Kotyk and Jaroslav Horák

Institute of Microbiology
Czechoslovak Academy of Sciences
142 20 Praha - Krč, Czechoslovakia

INTRODUCTION

In their molecular nature, specific membrane transport processes resemble enzyme reactions taking place in a continuous phase: They involve binding of a "substrate" molecule to a specific receptor site, the translocation proper corresponding to the substrate → product conversion, and dissociation of the "product", in the transport case simply release of the bound molecule or ion to another aqueous phase, separated from the starting aqueous phase by the membrane. The amount of evidence supporting the above sequence of events is vast and need not be reiterated here. Kinetically, all such transports are characterized by a half-saturation constant which is formally identical with the Michaelis constant of enzyme kinetics and is a similarly complicated function of various rate constants comprised in the mechanism, and by a maximum rate of transport, involving the total amount of carrier protein present in a given amount of cells and a combination of first-order rate constants, again depending on the complexity of the system under consideration.

Like enzyme reactions, carrier transport reactions go to equilibrium which may represent an equal distribution of the transported solute at both membrane sides (this situation is unique for transport since the chemical potentials of an uncharged solute before transport and after transport are the same in all physiological solutions) or an unequal distribution if an input of free energy is required for the reaction to proceed (such "unequal" distribution always occurs in enzyme kinetics because the actual chemical potential of the product is generally different from that of the substrate).

It appears almost trivial to speak of effects of pH and of temperature on the rates of enzyme reactions and of the information that can be gleaned from such studies (cf. Laidler and Bunting 1973) but, the similarity notwithstanding, very few quantitative data have been reported so far on such effects in transport. The situation is even less satisfactory with respect to such effects on transport equilibria. We intend to provide here examples, based on nonelectrolyte transport in yeast, of such quantitative information and to point out some of the pitfalls that might be encountered while obtaining such information.

pH AND RATES OF TRANSPORT REACTIONS

Plotting the initial rate of a transport reaction against pH generally yields a bell-shaped curve like that in Fig. 1.

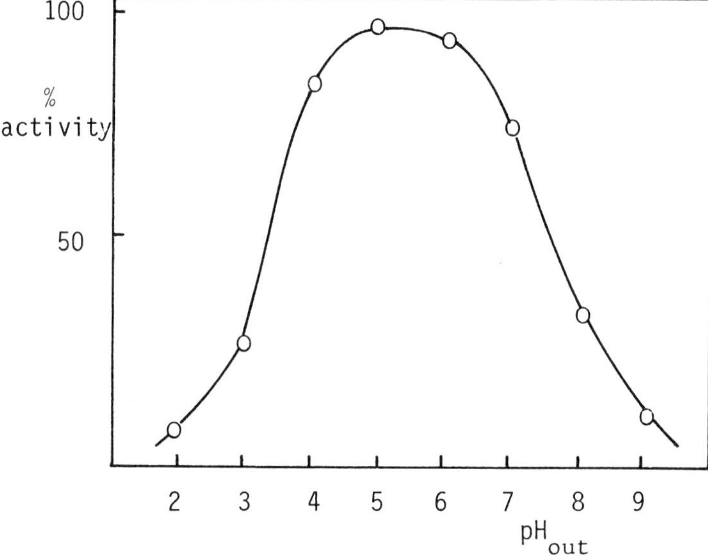

Fig. 1. Effect of pH on the initial rate of uptake of 1 mM glycine by cycloheximide-treated baker's yeast.

The information one gains from such a figure is that the transport proceeds most rapidly (and hence the carrier functions optimally) at, say, pH 5. Proceeding from the only reasonable hypothesis of pH effects presently available, this means that the "active site" of the carrier requires a certain degree of protonation to be functional. To determine this degree and to estimate the functional groups involved we need to determine the pH dependence of the maximum rate of transport and of the half-saturation constant and plot them as shown in Fig. 2. However, it should be noted that the crossing points of the linear extrapolations would represent the negative logarithms of the dissociation constants of the active site only if the carrier were taken to be exposed to the same pH at both membrane sides. This is probably not the case, the intracellular pH of most cells being buffered. In yeast cells there are two plateaus of pH_{in}: at external pH below 6 the pH_{in} lies at 5.6 - 5.9, depending on the species while at external pH above 7 the pH_{in} is 7.2 - 7.8 (cf. Kotyk 1963; Slavík 1982). A calculation based on the assumption that only the properly protonated carrier can move across the membrane yields the following expressions for the kinetic parameters:

$$J_{max} = \frac{c_t k_{CH} k_{CHS}}{k_{CH}(1 + H^+_I/K_{CHSH} + K_{CSH}/H^+_I) + k_{CHS}(1 + H^+_{II}/K_{CHSH} + K_{CH}/H^+_{II})}$$

$$K_T = \frac{k_{CH} K_{CHS}(2 + H^+_I/K_{CHH} + K_{CH}/H^+_I + H^+_{II}/K_{CHH} + K_{CH}/H^+_{II})}{k_{CH}(1 + H^+_I/K_{CHSH} + K_{CSH}/H^+_I) + k_{CHS}(1 + H^+_{II}/K_{CHSH} + K_{CS}/H^+_{II})}$$

where K_{CH}, K_{CHH}, K_{CSH}, K_{CHS}, and K_{CHSH} are the dissociation constants of complexes CH, CHH, CSH, CHS, and CHSH with respect to H^+, H^+, H^+, S, and H^+, respectively; k_{CH} and k_{CHS} are the translocation rate constants of CH and CHS, respectively; c_t is the total carrier concentration. C stands for carrier, S for the translocated solute, H for the hydrogen ion. It will be seen from the formulae that omission of terms containing H^+_{II} yields the enzyme formula for pH effects on rates.

The introduction of terms containing H^+_{II} has the consequence of decreasing the maximum rate and, in the graphic presentation, of spreading farther out the crossing points in the plots of all parameters, the shift increasing with higher pH_{in} (Fig. 2, right-hand panel). As the intracellular pH shifts from near 6 to near 8 the logarithmic plots of the parameters become asymmetric, with a more gentle slope at the alkaline side.

Fig. 3 shows two actual cases from different yeast species. L-Proline uptake by baker's yeast shows a fairly regular pattern where pK_1 is at 6.8, pK_2 at 4.9, pK_{a1} at 7.5 and pK_{a2} at 3.8. Considering the above-mentioned shift due to intracellular pH one can ascribe the protonations to histidine imidazolium group, γ-carboxyl of glutamic acid, histidine in a different milieu, and β-carboxyl of aspartic acid. The pattern of wuinovose transport in *Rhodotorula glutinis* is not so regular. First of all, the H^+-dissociation constants of the free carrier are close to those of the substrate-binding one, this causing a lower slope of the straight line. Secondly, there is apparently an effect of changing intracellular pH and, thirdly, as the transport in question is very likely a H^+ symport, there will be an effect of acid pH to decrease the K_T - hence the increase of K_T toward acid pH caused by overprotonation will be more gentle. However, here again, histidine imidazolium and γ-carboxyl of glutamic acid appear to be involved.

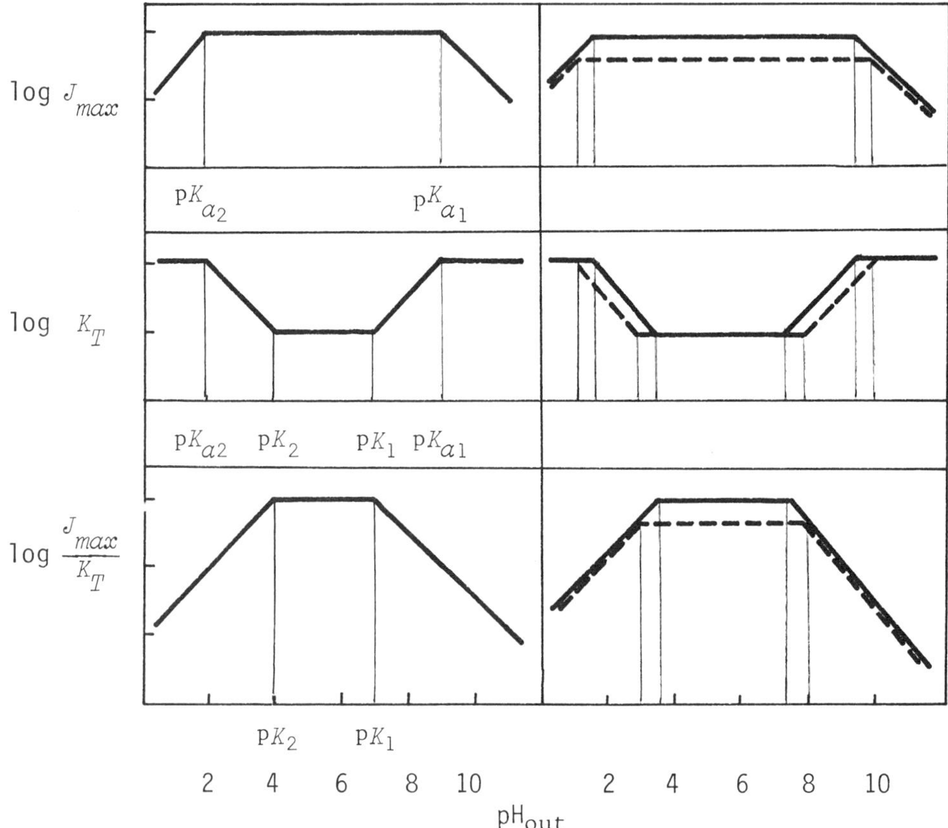

Fig. 2. Determination of dissociation constants of a specific carrier mechanism, based on the pH dependence of the maximum rate J_{max}, the half-saturation constant K_T, and of their ratio. Left: independent of pH_{in}; right: for pH_{in} = 6, full lines, and for pH_{in} = 8, dashed lines.

It goes without saying that such information as is described in connection with Fig. 2 and Fig. 3 is relevant only if no parallel transport system operates in the membrane, say, a distinct carrier for the same substrate with a different pH dependence, or simple diffusion (such transport should exhibit no pH dependence).

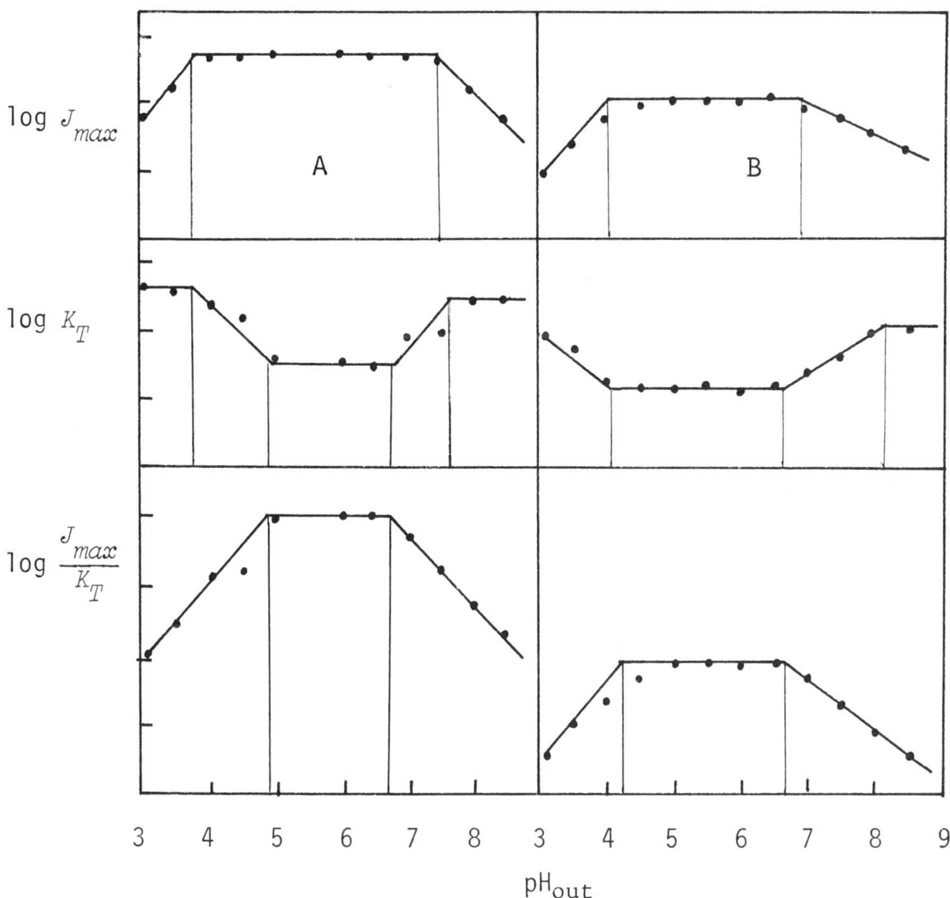

Fig. 3. Logarithmic plot of the kinetic parameters of transport of L-proline in *Saccharomyces cerevisiae* (A) and of quinovose (6-deoxy-D-glucose) in *Rhodotorula glutinis* (B).

pH AND TRANSPORT EQUILIBRIUM

Unlike the initial rate, the equilibrium position of a transport reaction of a nonelectrolyte is not sensitive to pH. It would only be so if the proton itself took part, directly or indirectly, in the transport reaction. One such situation obtains for the transport of dissociable solutes, either anionic (acids) or cationic (bases). If only the uncharged form (e.g., propionic acid, or ammonia) permeates or is carried across the membrane it will distribute itself according to the local pH. This unequal distribution is made use of in the estimation of intracellular pH, using the formula

$$pH_{in} = pH_{out} + \log [(c_{in}/c_{out})(1 + 10^{pK_a - pH_{out}}) - 10^{pK_a - pH_{out}}]$$

for weak acids, and

$$pH_{in} = pH_{out} - \log [(c_{in}/c_{out})(1 + 10^{pH_{out} + pK_b - 14}) - 10^{pH_{out} + pK_b - 14}]$$

for weak bases (Kotyk 1963).

A completely different situation arises if the proton serves as the driving ion in the so-called symport mechanisms where the carrier binds the solute and the hydrogen ion which endows it with either greater mobility or greater affinity for its substrate, or both. Since the driving force in such a symport is compounded from the electrical transmembrane potential $\Delta\phi$ and from a hydrogen ion concentration term, $(RT/F)\ln(H^+_{in}/H^+_{out})$, we may predict, on the assumption of obligatory coupling with the H^+ gradient and from the knowledge of the membrane potential and the difference between intracellular and extracellular pH, what the maximum accumulation ratio would be. Concretely speaking, if the energy stored in the electrochemical potential gradient of H^+ ($\Delta\tilde{\mu}_{H^+} = -\Delta\phi F - RT \ln(H^+_{in}/H^+_{out})$) is fully utilized for accumulating a symported solute S so that $\Delta\tilde{\mu}_{H^+} = -\Delta G^o = RT(\ln S_{in}/S_{out})$ we will have

$$S_{in}/S_{out} = (H^+_{out}/H^+_{in})e^{-\Delta\phi F/RT}$$

This accumulation ratio may then be used for diagnosing proton symports. If a proton symport has been established, say, by a pH_{out} increase on adding the putatively symported solute and by depolarizing the membrane (whose potential is known to be due to an electrogenic H^+ pump) we may expect it to obey the above equation. If, like in Fig. 4 (cf. Kotyk *et al.* 1982) this is not the case we are forced to look for alternative explanations, such as (1) inaccuracy of our $\Delta\phi$ and pH determination, (2) a mechanism taking up protons for the symport directly in the membrane rather than from the bulk solution, (3) existence of an energy source different from $\Delta\tilde{\mu}_{H^+}$, available simultaneously. We favor the second explanation, as depicted in Fig. 5.

One of its basic tenets, i.e. the intrinsic pH dependence of the proton pump, is in agreement with data obtained in yeasts. Most of the transports presumably driven by a proton gradient show an optimum accumulation at pH 5.5 - 6.5 (where the protonmotive force is quite low), which is the range of maximum activity of the recently isolated H^+-ejecting ATPase (5.7 - 6.1; Goffeau and Slayman

1981). (A more detailed analysis of the H^+ ion effects on equilibria can be found in Kotyk, 1983.)

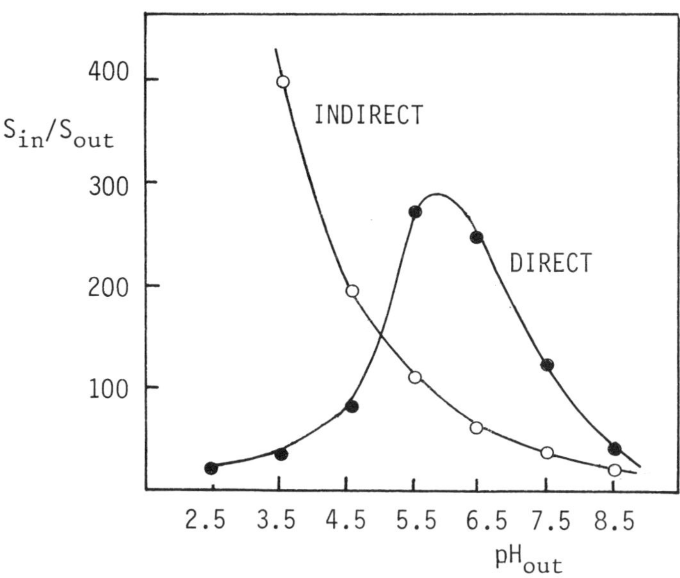

Fig. 4. pH dependence of quinovose accumulation by *Rhodotorula glutinis*. INDIRECT: Values based on independent measurement of $\Delta\phi$ and ΔpH as components of the driving force for quinovose accumulation. DIRECT: Values obtained directly from the distribution of ^3H-labeled quinovose.

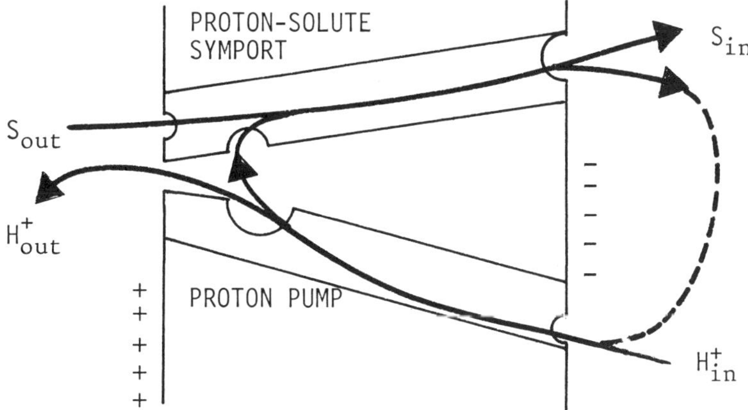

Fig. 5. Schematic representation of the assumed topological coupling of a proton-ejecting pump (below) with a proton-symporting solute carrier (above). The carrier "sees" the proton concentration in the membrane not in the bulk solution. Overflow of protons may be predominant if few H^+-symport carriers are functioning.

TEMPERATURE AND RATES OF TRANSPORT REACTIONS

Every first-order rate constant is in principle temperature-dependent, physical processes displaying a linear dependence on absolute temperature - thus, diffusion in continuous phase is, to a good approximation, unaffected by temperature in the physiological range (an increase by 3% from 20 to 30 °C). However, chemical reactions and those catalyzed by enzymes or carriers, show a different temperature dependence, based on the assumption of an activated state of the reaction catalyst where the rate-determining step of translational movement from the activated state to the enzyme-plus-product state is equivalent to an equilibrium situation between the "substrate" and the catalyst (Glasstone *et al.* 1941). This then follows the empirical Arrhenius equation, formally identical with van't Hoff's isobaric equation, where

$$\ln(k_2/k_1) = (E_a/R)(1/T_1 - 1/T_2)$$

The rate constants (or rates, if measured at identical substrate concentrations) k correspond to two different temperatures T; R is the gas constant (8.314 J $mol^{-1}K^{-1}$) and E_a the so-called activation energy of the reaction. Plotting ln k versus $1/T$ should yield a straight line with slope equal to $-E_a/R$.

Because of the complexity of the fundamental kinetic parameters it is generally recommended in enzyme kinetics to use the maximum rate of reaction in such plots as it may under certain conditions contain a single rate constant. The same caution applies to transport reactions although here, in most cases, the rate-limiting step is the translocation constant which thus governs the reaction rate at all concentrations. Still, the example in Fig. 6 indicates that the activation energy may shift from 30 kJ mol^{-1} at 0.7 μM sugar to 58 kJ mol^{-1} at 7 mM sugar.

There is in transport a different complication to temperature dependence studies. The activation energy of translocation is apparently influenced by the state of the membrane lipids, the energy being lower when all lipids are in the liquid crystalline phase and greater when crystalline domains separate out, i.e. below the so-called transition temperature. Somewhat unexpectedly, the activation energy may decrease again as all lipids solidify. It may be conjectured that a different partial step of the transport reaction becomes rate-limiting (Fig. 7).

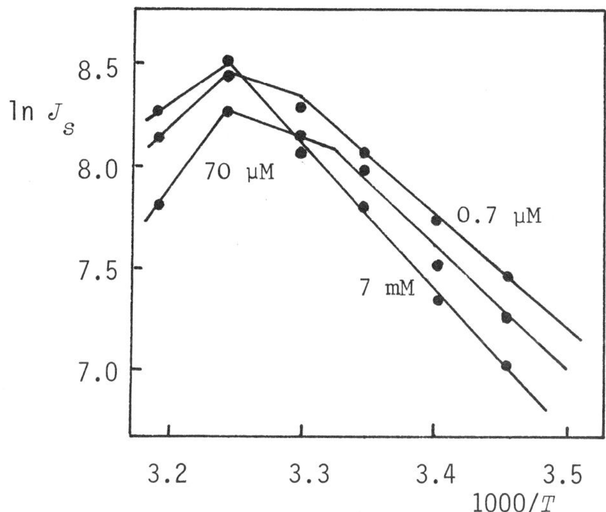

Fig. 6. Temperature dependence of the initial rate of uptake of different concentrations of quinovose by *Rhodotorula glutinis*.

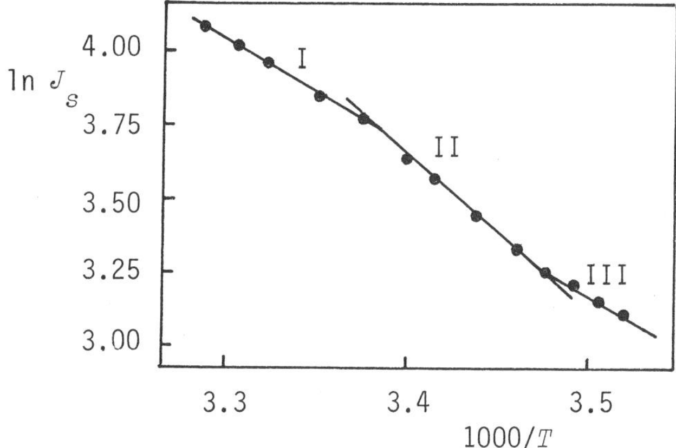

Fig. 7. Temperature dependence of the initial rate of uptake of L-proline in *Saccharomyces cerevisiae*. The activation energies measured in the various temperature ranges were 71, 105, and 71 kJ mol^{-1}.

TEMPERATURE AND TRANSPORT EQUILIBRIUM

The least explored area of quantitative transport studies is the effect temperature exerts on the final distribution ratio of an active transport. For reasons mentioned in the introduction, temperature will only affect the equilibrium of transports requiring energy i.e. "active", very often simply going uphill irrespective of whether this energy is provided by the splitting of ATP or by the dissipation of a H^+ or Na^+ electrochemical potential gradient. The effect will be described quantitatively by

$$\ln(K_{eq_2}/K_{eq_1}) = (\Delta H/R)(1/T_1 - 1/T_2)$$

from basic thermodynamics, the K_{eq} standing for the equilibrium constant at different temperatures (in our case simply the final ratio S_{in}/S_{out}), the ΔH being the change in enthalpy attendant on such a reaction. For $\Delta H > 0$, the reaction is endothermic, for $\Delta H < 0$ it is exothermic. Fig. 8 shows a slightly endothermic accumulation by one yeast species (15 kJ mol^{-1}) and an exothermic situation in another (-45 kJ mol^{-1}). Strangely enough, $\Delta\tilde{\mu}_{H^+}$, the assumed driving energy for solute transport, is established endothermically in *Rhodotorula glutinis*. The $\Delta\phi$ component has a ΔH of 32 kJ mol^{-1}, the ΔpH component one of 34 kJ mol^{-1}.

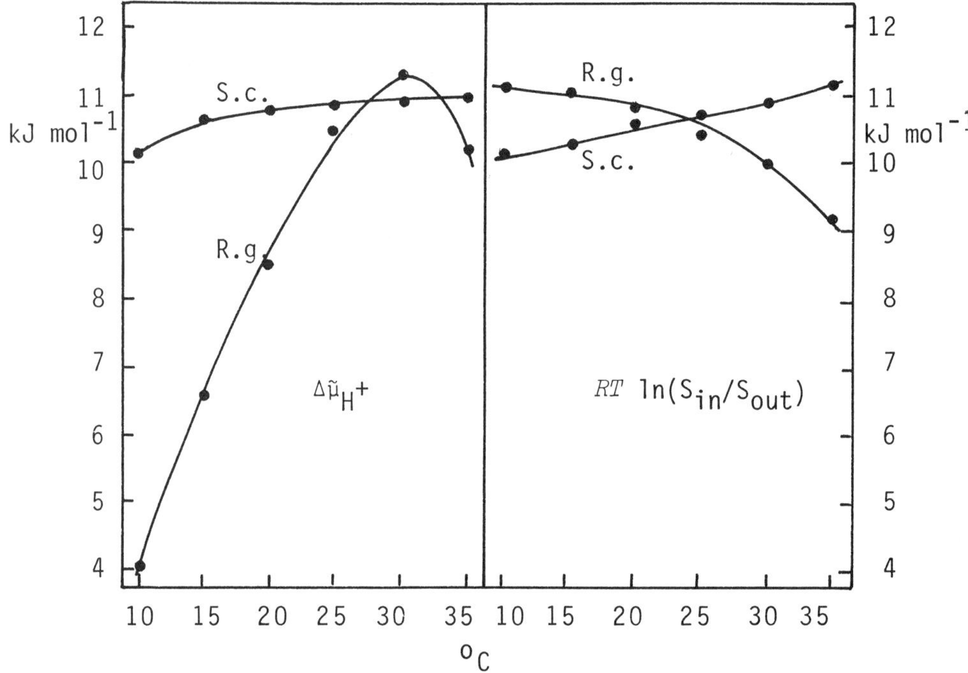

Fig. 8. Temperature dependence of the electrochemical potential gradient (left) and of solute accumulation (right) by two yeast species, *Saccharomyces cerevisiae*, S.c., *Rhodotorula glutinis*, R.g. The solute examined in *S.c.* was L-proline, that in *R.g.* quinovose, both at 0.1 mM concentrations.

Finally, making use of another fundamental equation of thermodynamics

$$\Delta G^o = \Delta H^o - T\Delta S^o$$

we may calculate that the accumulation of quinovose in *Rhodotorula glutinis* is associated with a decrease of entropy, amounting to 110 J $mol^{-1}T^{-1}$ while the establishment of ΔpH is accompanied by an increase of entropy, amounting to 138 J $mol^{-1}K^{-1}$. It remains to be seen what, if any, relationship these entropy changes may have to changes in order or structurality of the system.

It should be observed that here, just as in the case of the pH dependence of substrate accumulation, the changes of measured accumulation ratios do not always correspond to changes in the driving energy with changing temperature. Although this evidence is rather meager it again suggests that the bulk parameters of $\Delta\phi$ as well as ΔpH may not be the ones which determine the actually membrane-felt electrochemical potential difference of protons.

The wealth of biochemical data on specific transport processes is only then able to provide the picture of a functioning carrier if both kinetic and thermodynamic information is included. This information is not technically difficult to obtain and more effort should be expended in this direction.

REFERENCES

Glasstone, S., Laidler, K.J., and Eyring, H., 1941, "The Theory of Rate Processes", McGraw-Hill, New York.

Goffeau, A., and Slayman, C.W., 1981, The proton-trasnlocating ATPase of the fungal plasma membrane, Biochim. Biophys. Acta, 639:197.

Kotyk, A., 1963, Intracellular pH of baker´s yeast, Folia Microbiol., 8:27.

Kotyk, A., Stružinský, R., and Slavík, J., 1982, Electrochemical potential of protons and active transport in several yeast species, Studia Biophys., 90:17.

Kotyk, A., 1983, Coupling of $\Delta\tilde{\mu}_{H^+}$ with secondary active transport, J. Bioenerget. Biomembr., in press.

Laidler, K.J., and Bunting, P.S., 1973, "The Chemical Kinetics pf Enzyme Action", Clarendon Press, Oxford.

Slavík, J., 1982, Intracellular pH of yeast cells measured with fluorescent probes, FEBS Lett., 140:22.

TRANSPORT OF BINARY AND TERNARY ELECTROLYTES THROUGH CHARGED MEMBRANES WITH ALLOWANCE FOR INTERACTIONS

F. Ludwików, M. Bartoszkiewicz and S. Miekisz

Department of Biophysics
Medical School
Wrocław, Poland

INTRODUCTION

One of the fundamental problems in theoretical description of substance transport through membranes treated as a separate bulk phase is to decide on the character and way of description of interactions between the permeating components and the membrane as well as among the components. Specification of the assumptions on the character of the interactions determines transport properties of the membrane-permeating solution system and is fundamental for any membrane transport model. Allowance for interactions, particularly electrostatic and non-electrostatic in the case of electrolytes, is connected with many difficulties if we want to go through the difficult road from the potentials of molecular interactions to averaged quantities and macroscopic equations of transport. Such an approach has been followed by many authors[1-5]. As it often occurs in this type of models, the starting point is quite general. The final result, however, if given in an analytic form, is restricted by a series of additional simplifying assumptions and is thus much less general.

In view of that, wanting to avoid these drawbacks, we have made an attempt to allow both qualitatively and quantitatively for the interaction between the permeating solution and the membrane, sticking to the idea that the membrane is an external agent with respect to the permeating solution. The material presence of the membrane causes certain changes in the thermodynamic state of the permeating solution resulting at least in a perturbation of its ground state. The idea was recurrently used in models which treated the membrane either as an external field described by certain potential barriers, their number and shape depending on particular model[6-9] or as a factor which modified in a definite way the activities of the permeating components[10-12].

In the present work aimed at specification of the membrane as an external factor we assume from the beginning that the bulk transport and equilibrium properties of sufficiently diluted electrolyte solutions are determined by the ionic strength independently of the type of ion. This fact, according to a hypothesis put forward by one of the present authors[13], has been extended to electrolyte solutions which are constrained by the presence of electrically charged membrane of certain fixed charge density, owing to the assumption that the electrostatic (and not only that) membrane-solution interaction can be considered by using the so-called effective ionic strength which is calculated with allowance for all ions present, both those of the solution and the fixed membrane charges. The above assumption enables one to write explicite formulae which express the constitution of equilibrium and transport physico-chemical variables describing the thermodynamic properties of non-free electrolyte solution, provided such formulae are known for free solution. Such a procedure leads to a broad class of transport models which satisfy the concentration limit law for electrolytes and are constructed according to a definite scheme. This makes it possible to compare and discuss the influence of various factors on transport properties of membranes for the whole class of models of tranport which were thus generated.

BASIC ASSUMPTIONS OF THE MODEL

The membrane together with the permeating solution is a homogeneous material medium. The membrane material contains groups which can dissociate into ions. In general, there may be several types of such groups. Here, in detailed considerations, we assume for the sake of simplicity that there is only one type of fixed groups. We assume also that the fixed membrane ions can form pairs with counterions of the permeating solution, thus changing or neutralizing the membrane charge. In other words we assume that ions of the permeating solution may react with membrane ion groups, kinetic equations for these reactions being of the mass action law type. At this point we indicate two possible cases. In the first case the reaction rate is comparable with the rate of diffusion. Then, describing nonstationary processes, we have to consider the nonstationary chemical kinetics in its entirety. Of course, this involves considerable mathematical complications. In the second case the rate of ion pair formation is substantially larger than the rate of diffusion. This allows to apply a simplifying assumption that during a nonstationary transport the chemical reaction is in equilibrium. Thus the differential equation of chemical kinetics becomes an algebraic equation known as the Guldberg-Waage mass action law.

Let M denote an ion of valency z_M created by dissociation of a membrane substance group and let A , B , C (or 1, 2, 3 in equivalent denotation of concentration) denote ions of the solution that can penetrate the membrane, their valencies being z_1 , z_2 , z_3 respectively. According to the assumptions, ion M may form pairs with counterions, i.e. with one or two among ions A , B , C , this occuring with a definite stoichiometry determined not only by electric properties of ion M but also by the ion steric packing within the membrane, i.e. topology and other forces that may favour such pair formation. In general the problem reduces to a number of particular cases arising from the fact that some variants of the pair-formation reaction schemes are possible; their diversity being, for reasons mentioned earlier, substantially more numerous, as the formed ion groups may not be completely electroneutral.

For instance we give the following schemes of association:

$$w_{MA} M + \bar{w}_A A \rightleftarrows M_{w_{MA}} A_{\bar{w}_A} \tag{1}$$

$$M + w_A A \rightleftarrows MA_{w_A} \tag{2}$$

$$w_{MB} M + \bar{w}_B B \rightleftarrows M_{w_{MB}} B_{\bar{w}_B} \tag{3}$$

$$M + w_B B \rightleftarrows MB_{w_B} \tag{4}$$

$$MA_{w_A} + w_{AB} B \rightleftarrows MA_{w_A} B_{w_{AB}} \tag{5}$$

$$MB_{w_B} + w_{BA} A \rightleftarrows MB_{w_B} A_{w_{BA}} \tag{6}$$

Scheme (1) (also scheme (3)) corresponds to a situation in which for sterical reasons it is possible that two or a number of ionic membrane groups associate with ions of the permeating solution. In order to illustrate the notion of effective ionic strength, we will consider henceforth scheme (4). In this case

$$I^{eff} = \frac{1}{2} \left(z_1^2 c_1 + z_3^2 c_3 + z_2^2 c_2^f + z_M^2 c_M^f + \delta_M^2 c_M^b \right) \tag{7}$$

where c_2^f is the concentration of free ions of type B , c_M^f is the concentration of free ion group of the membrane, $\delta_M = w_B z_2 + z_M$ is the resultant charge of an ion pair, if it is not completely electroneutral (for neutral pairs $\delta_B = 0$).

The mass action law for scheme (4) has in general the form:

$$K = \frac{\left[M\, B_{w_B}\right]}{[M]\,[B]^{w_B}} \tag{8}$$

where [ ] denotes the activity of the component and K is the chemical equilibrium constant. In the case of electrolytes concentrations cannot be substituted for activities without drastic assumptions about dilution. It is generally known that activity coefficients essentially depend on ionic strength. Thus, having in view (7) and (8), we may say that in general the effective ionic strength can be expressed by a complicated, implicit equation. It is no problem, however, if we decide to expand in series the terms which contain the ionic strength with respect to the parameter $\sqrt{I^{eff}}$, preserving such number of the terms of the expansion which is necessary for solving a particular transport model. Let us now be more specific.

According to scheme (4) we have the following relationships between free ions (index f) and ions associated in ionic pairs (index b):

$$c_M = c_M^b + c_M^f \tag{9}$$

$$c_2^b = w_B\, c_M^b \tag{10}$$

$$c_2 = c_2^f + c_2^b \tag{11}$$

where $c_M = \nu_M \xi_M / \bar{v}_M$, ν_M is the fraction of molar concentration of the membrane substance that transforms, as a multiplier, the membrane substance concentration into concentration of its ionized groups (the fraction may depend on pH and I^{eff}), ξ_M is the membrane volume fraction and $\bar{v}_M$ is the partial volume of the membrane substance.

The mass action law (8), using molar activity coefficients y and assuming that the activity of bound ion pairs is equal to their concentration, has the form:

$$K \cdot c_M^f \cdot y_m \cdot (c_2^f)^{w_B} \cdot (y_2)^{w_B} = c_M^b \quad . \tag{12}$$

Using (9), (10) and (11) we express c_M^b and c_2^f in (12) and (7) by c_2, c_M and c_M, thus obtaining:

$$K \cdot c_M^f \cdot y_M \cdot \left[c_2 - w_B \left(\nu_M \frac{\xi_M}{\bar{v}_M} - c_M^f \right) \right]^{w_B} (y_2)^{w_B} = \nu_M \frac{\xi_M}{\bar{v}_M} - c_M^f \tag{13}$$

and

$$c_M^f = \left[z_2^2 w_B + z_M^2 - \delta_B^2 \right]^{-1} \cdot \{ 2 I^{eff} - z_1^2 c_1 - z_3^2 c_3 - z_2^2 (c_2 - w_B \nu_M \xi_M / \bar{v}_M) - \delta_B^2 \nu_M \xi_M / \bar{v}_M \} \quad . \tag{14}$$

From equations (13) and (14) we obtain the following implicit equation:

$$K \, y_M(I^{eff}) \cdot \left[z_2^2 w_B + z_M^2 - \delta_B^2 \right]^{-1} \cdot \{ 2 I^{eff} - z_1^2 c_1 - z_3^2 c_3 - z_2^2 \left[c_2 - w_B \nu_M (pH, I^{eff}) \, \xi_M / \bar{v}_M \right] - \delta_B^2 \nu_M (pH, I^{eff}) \, \xi_M / \bar{v}_M \} \cdot \{ c_2 - w_B [\![\nu_M (pH, I^{eff}) \xi_M / \bar{v}_M - \left[z_2^2 w_B + z_M^2 - \delta_B^2 \right]^{-1} \{ 2 I^{eff} - z_1^2 c_1 - z_3^2 c_3 - z_2^2 \left[c_2 - w_B \nu_M (pH, I^{eff}) \xi_M / \bar{v}_M \right] - \delta_B^2 \nu_M (pH, I^{eff}) \xi_M / \bar{v}_M \}]\!] \}^{w_B} \times \{ y_2 (I^{eff}) \}^{w_B}$$

$$= \nu_M (pH, I^{eff}) \xi_M / \bar{v}_M - \left[z_2^2 w_B + z_M^2 - \delta_M^2 \right]^{-1} \cdot \{ 2 I^{eff} - \left[z_1^2 c_1 - z_3^2 c_3 - z_2^2 c_2 - w_B \nu_M (pH, I^{eff}) \xi_M / \bar{v}_M \right] - \delta_B^2 \nu_M (pH, I^{eff}) \xi_M / \bar{v}_M \} \tag{15}$$

Because we chose as independent the following variables:

$$I^{eff} \; ; \; \xi_W \; ; \; \bar{I} = \frac{1}{2} (z_1^2 c_1 + z_3^2 c_3) \; ; \; (\xi_M = \text{const}) \tag{16}$$

equation (15) has to be supplemented by two additional ones:

$$\bar{v}_1 (I^{eff}) c_1 + \bar{v}_2 (I^{eff}) c_2 + \bar{v}_3 (I^{eff}) c_3 + \xi_M + \xi_W = 1 \tag{17}$$

$$\bar{I} = \frac{1}{2} (z_1^2 c_1 + z_3^2 c_3) \tag{18}$$

(the round brackets by quantities y_M, y_2 and $\bar{v}_i$ in Eqs. (15) and (17) denote functional dependence on the effective ionic strength).

Equations (15), (17) and (18) enable one to obtain, in the form of a power series with respect to I^{eff}, a transformation of concentrations c_1, c_2 and c_3 to new variables. By using them it is possible to calculate for each particular model of electrolyte all the necessary derivatives of concentrations c_1 and c_2 with respect to I^{eff}, though the equations are implicit (see e.g. Korn[14]).

The above considerations illustrate sufficiently the way of including the scheme of ion pair formation reaction into description of electrostatic interactions under the assumption that the rate of ion pair formation reaction is substantially larger than the rate of diffusion. When the rates of the two processes are comparable, the problem complicates still further. For the sake of simplicity we do not consider this case here. Detailed considerations on this point are given in 15.

The derivation of the transformation equations is the first step towards the formation of a transport model. The next step is to consider a particular model of bulk electrolyte which would describe both equilibrium and transport properties as a function of various physico-chemical parameters, in particular ionic strength and concentration. Then in such a model we make a modification introducing in place of ionic strength the effective ionic strength according to formula (7) or to the transformation formulae.

This prescription allows to obtain the constitution of the basic physico-chemical quantities of an electrolyte within a membrane and thus a local equation of transport within the membrane. If $\phi_i(I)$ is a functional dependence of a physico-chemical quantity of an electrolyte solution on the ionic strength, then according to the above prescription the same functional dependence $\phi_i(I^{eff})$, possibly with altered structural parameters, describes quantities ϕ_i inside the membrane. This can be shown in the scheme:

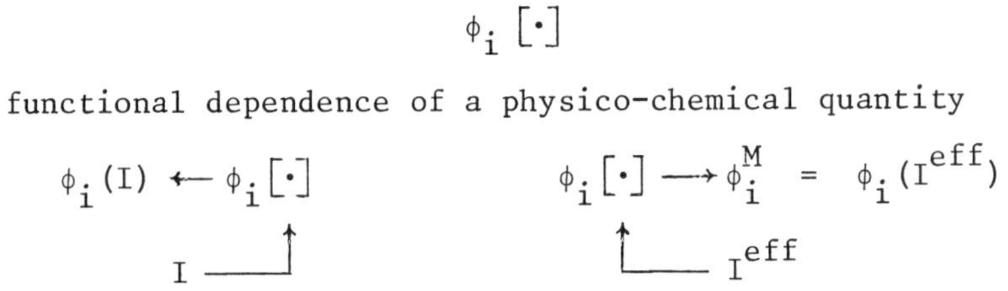

$$\phi_i\,[\cdot]$$

functional dependence of a physico-chemical quantity

$$\phi_i(I) \longleftarrow \phi_i[\cdot] \qquad\qquad \phi_i[\cdot] \longrightarrow \phi_i^M = \phi_i(I^{eff})$$

$$I \uparrow \qquad\qquad\qquad\qquad \uparrow I^{eff}$$

Fig. 1. Scheme of the procedure for obtaining constitution of physico-chemical quantities of an electrolyte solution in a membrane.

One of the problems with putting the procedure into practice is the lack of consistent formulae relevant to the situation in the membrane which would connect the Onsager phenomenological coefficients L_{ij} with measured transport quantities such as: self-diffusion coefficients, transference numbers, hydration numbers, electric conductivity.

In the scientific literature there are many reports on the ionic-strength dependence of activity coefficients for many electrolytes in broad range of ionic strength. Instead, transport properties of electrolytes in the range of concentrations suitable for membrane transport description are reported only fragmentary in conjunction with various theories, this making it the more difficult to bring them to a common point of reference. Various attempts at a systematic description of electrolyte transport phenomena have been made[16-20]. They are of equal usefulness for implementing the above present concept of electrolyte-membrane interaction. However, in order to transfer the results of those papers to the field of membrane transport theories one has to take into account the specific character of membrane transport as well as to analyse critically the assumptions made in the derivations of relationships between the various types of transport quantities.

Owing to the availability of experimental data on self-diffusion coefficients[21-23], we have adopted the viewpoint of paper 16. This paper gives for binary electrolytes a relationship between the Onsagerian coefficients of transport and measurable transport quantities such as: electric conductivity, self-diffusion coefficients, transference and hydration numbers. In the derivation of these relationships some assumptions were made which may not be satisfied in the case of membrane transport. A critical analysis of them has been made[24] and the previously mentioned relationships between the Onsagerian coefficients and measurable quantities for binary and ternary electrolytes have been derived anew, putting much attention to their usefulness in solution-membrane interaction models.

In the present paper we build on the results of the mentioned paper. We assume a linear transport equation in the form:

$$J_i = \sum_k L_{ik} X_k \tag{19}$$

where J_i is the flux of i-th component in the laboratory frame of reference (fixed to the membrane), $L_{ik} = L_{ki}$ are the Onsagerian transport coefficients, whereas

$$X_k = -(\mathrm{grad}\ \mu_k)_T - e_k\ \mathrm{grad}\,\varphi \tag{20}$$

are the thermodynamic forces. Index k = 0 refers to the solvent, assuming that $e_0 = 0$. For other (ionic) components $e_i = z_i F$.

Next, the following relations are satisfied[24]:

$$L_{ii} = D_i^*(x_i/RT) \qquad i = 0, 1, 2 \tag{21}$$

$$L_{01} = L_{10} = S_1\left[(\kappa/2e_1^2) + D_1^*(x_1/2RT) - D_2^*(x_2/2RT)(e_2/e_1)^2\right] \tag{22}$$

$$L_{02} = L_{20} = S_2\left[(\kappa/2e_2^2) + D_2^*(x_2/2RT) - D_1^*(x_1/2RT)(e_1/e_2)^2\right] \quad (23)$$

$$L_{12} = L_{21} = (\kappa/2e_1e_2) - D_1^*(x_1/2RT)(e_1/e_2) - D_2^*(x_2/2RT)(e_2/e_1) \quad (24)$$

for binary electrolytes, and for ternary electrolytes:

$$L_{ii} = D_i^*(x_i/RT) \qquad i = 0, 1, 2, 3 \quad (25)$$

$$L_{01} = L_{10} = S_1\, \kappa t_1/e_1^2 \quad (26)$$

$$L_{02} = L_{20} = S_2\, \kappa t_2/e_2^2 \quad (27)$$

$$L_{03} = L_{30} = S_3(1 - t_1 - t_2)\, \kappa/e_3^2 \quad (28)$$

$$L_{12} = L_{21} = -\tfrac{1}{2}(\kappa/e_1e_2)\{1 - 2t_1 - 2t_2 + D_1^*(e_1^2/\kappa)(x_1/RT) + D_2^*(e_2^2/\kappa)(x_2/RT) - (e_3^2/\kappa)(x_3/RT)\, D_3^*\} \quad (29)$$

$$L_{13} = L_{31} = -\tfrac{1}{2}(\kappa/e_1e_3)\{2t_2 - 1 + D_1^*(e_1^2/\kappa)(x_1/RT) - D_2^*(e_2^2/\kappa)(x_2/RT) + (e_3^2/\kappa)(x_3/RT)\, D_3^*\} \quad (30)$$

$$L_{23} = L_{32} = -\tfrac{1}{2}(\kappa/e_2e_3)\{2t_1 - 1 - (e_1^2/\kappa)(x_1/RT)\, D_1^* + D_2^*(e_2^2/\kappa)(x_2/RT) + D_3^*(e_3^2/\kappa)(x_3/RT)\} \quad (31)$$

In the quoted formulae the symbols have the following meanings:

$$\kappa = -\left(\sum_i e_i J_i / \operatorname{grad} \varphi\right)_{(\operatorname{grad} \mu_k)_T = 0}$$

is the conductivity of the electrolyte,

$$t_i = e_i \left(J_i / \sum_k e_k J_k\right)_{(\operatorname{grad} \mu_j)_T = 0}$$

is the transference number of i-th ion, D_i^* is the self-diffusion coefficient of i-th component[16,21-23], x_i is the molar fraction of i-th component, μ_k is the chemical potential of i-th component, and S is the hydration number of i-th ion determined by the relation: $J_0 = \sum_i S_i J_i$ for $(\operatorname{grad} \mu_k)_T = 0$, $i = 1, \ldots, N$.

In each consistent model of electrolyte solution are known explicit relations between quantities y_i , κ , t_i , S_i , D_i^* (or other equivalent quantities[16,17-19,25]) and structural parameters (e.g. dielectric constant, viscosity of the solvent, ionic radii etc.) as well as ionic strength or concentration. The above-mentioned quantities correspond to the symbol ϕ_i in the scheme shown in Fig. 1. The last step towards final formulation of the model of interactions consist in the use of relations (21) - (31) and the prescription of Fig. 1 for writing the constitutions of the physico-chemical quantities as functions of the effective ionic strength and, possibly, the modified structural constants of the solution.

The present model can be adapted to the case of a membrane under the action of strong electric fields. Such a case occurs with biological membranes where a potential difference of 100 mV exists across a membrane of 100 Å thickness, this making an electric field strength of 10^7 V/m. This aspect of membrane transport has already been considered[26] and is known in the literature as the Wien effect[27-29]. As applied to membrane-permeating solution interaction the Wien effect has also been accounted for[15]. In the present paper we merely signal such a possibility by indicating that in the mass action law (8) the equilibrium constant depends on electric field, though the dependence is rather complicated. Introduction of this dependence has an essential influence on the basic transformations of physico--chemical quantities to new variables, e.g. the effective ionic strength. Moreover, the remaining physico-chemical quantities also undergo modification under the action of strong electric field. New problems also arise in connection with the linear equations of transport for systems in strong electric fields, as nonlinear effects may appear that differ from those which are due to the Wien effect.

One more remark concerning equation (8). Activity of ionic groups of the membrane substance appears in this equation. This quantity can be assumed on the basis of the theory of electrolytes if the number of these groups is not great, or on the basis of the polyelectrolytes theory if the groups are numerous. This is also important for the final transformation of variables.

SPECIFIC CASE - BINARY ELECTROLYTE, NO REACTIONS

In this case the effective ionic strength can be expressed as follows:

$$I^{eff} = \frac{1}{2}\,(z_1^2 c_1 + z_2^2 c_2 + z_M^2 c_M) \tag{32}$$

We supplement this formula with the equation:

$$c_1 V_1 + c_2 V_2 + \xi_W + \xi_M = 1 \tag{33}$$

where V_1 and V_2 are the apparent molar volumes of ions 1 and 2 respectively and ξ_W and ξ_M are the volume fractions of water and membrane substance respectively.

Using these equations we get the following transformations of concentrations:

$$c_1 = (z_1^2 V_2 - z_2^2 V_1)^{-1} [(2I^{eff} - z_M^2 c_M) V_2 - z_2^2 (1 - \xi_M - \xi_W)] \quad (34)$$

$$c_2 = (z_2^2 V_1 - z_1^2 V_2)^{-1} [(2I^{eff} - z_M^2 c_M) V_1 - z_1^2 (1 - \xi_M - \xi_W)] \quad (35)$$

Transformations (34) and (35) enable one to write linear equation (19) in the form:

$$J_i = - [\ell_{i1} (\text{grad } p)_T + \ell_{i2} (\text{grad } I^{eff})_T + \ell_{i3} (\text{grad } \xi_W)_T + \ell_{i4} \text{ grad } \varphi] \quad (36)$$

where:

$$\ell_{i1} = \sum_{k=0}^{2} L_{ik} \bar{v}_k \quad (37)$$

$$\ell_{i2} = \sum_{k=0}^{2} L_{ik} RT \left[\left(\frac{\delta \ln f_k}{\delta I^{eff}} \right) + \frac{1}{x_k} \left(\frac{\delta x_k}{\delta I^{eff}} \right) \right] \quad (38)$$

$$\ell_{i3} = \sum_{k=0}^{2} L_{ik} RT \frac{1}{x_k} \left(\frac{\delta x_k}{\delta \xi_W} \right) \quad (39)$$

$$\ell_{i4} = \sum_{k=1}^{2} L_{ik} e_k \quad (40)$$

where $\bar{v}_k$ is the partial molar volume of component k.

Assuming that $\Gamma = \sqrt{I^{eff}}$ is small, one can express the coefficients L_{ij} in the form of the series:

$$\ell_{ij} \simeq \ell_{ij}^{(0)} + \ell_{ij}^{(1)} \Gamma + \ell_{ij}^{(2)} \Gamma^2 + \ldots \quad (41)$$

Detailed calculations for an extended Debye-Hückel theory are given in paper 30. Transport equations (16) have also been used[31] for describing nonstationary transport of binary electrolytes across charged membranes under the following additional assumptions:

(i) the volume fraction of water inside the membrane is constant and the membrane is homogeneous and does not "swell", (ii) the electric field strength is constant (i.e. it does not depend on space coordinates), (iii) the membrane is as thin as lipid bilayer membranes and the electric field strength is sufficiently large, (iv) the dielectric constant is a function of the ionic strength and it does not depend on electric field, (v) the Wien effect is negligible. Then a possibility of multiple steady states in the transporting system has been shown. The number of the states depends on water flux which in turn depends on boundary condictions (the pressure difference in particular). The paper concludes that water flux determines the stability area and the multiplicity of stationary states of the transporting system.

ACKNOWLEDGEMENT

This work was partly supported in the realm of a Department Problem of Basic Research of MNSzWiT, R.1.9: Evolution of Biological Control Systems.

REFERENCES

1. J. Schröter and G. West, Annalen der Physik, 7 Folge, 32:19 (1975).
2. J. Schröter and G. West, Annalen der Physik, 7 Folge, 32:33 (1975).
3. J. Schröter and G. West, Annalen der Physik, 7 Folge, 32:227 (1975).
4. G. West, A. Grauel and J. Schröter, Annalen der Physik, 7 Folge, 33:125 (1976).
5. R. J. Bearman and J. G. Kirkwood, J. Chem. Phys. 28:136 (1958).
6. G. S. Manning, J. Chem. Phys. 49:2668 (1968).
7. G. S. Manning, J. Phys. Chem. 76:393 (1972).
8. G. S. Manning, in: "Biophysics of Membrane Transport School Proceedings, Part II", S. Miękisz and W. Gomułkiewicz, eds., Wrocław (1974).
9. Yu. A. Chizmadjev and S. K. Aityan, in: "Biophysics of Membrane Transport, Part III", S. Miękisz and W. Gomułkiewicz, eds., Wrocław (1974).
10. Y. Kobatake, J. Chem. Phys. 28:146 (1958).
11. Y. Kobatake, N. Takeguchi, Y. Toyoshima and H. Fujita, J. Phys. Chem. 69:3981 (1965).
12. Y. Kobatake, Y. Toyoshima and N. Takeguchi, J. Phys. Chem. 70:1187 (1966).
13. F. Ludwików, "Transport of binary and ternary electrolytes through charged membranes with allowance for interactions", unpublished (1976).

14. G. A. Korn and T. M. Korn, "Mathematical Handbook for Scientists and Engineers", Mc Graw-Hill, New York (1976).
15. F. Ludwików, "Transport of binary and ternary electrolytes through charged membranes with allowance for interactions. II. General Model", in preparation for publication.
16. B. Baranowski and A. S. Cukrowski, Z. phys. Chem. 228:292 (1965).
17. D. G. Miller, J. Phys. Chem. 70:2639 (1966).
18. D. G. Miller, J. Phys. Chem. 71:616 (1967).
19. D. G. Miller, J. Phys. Chem. 71:3558 (1967).
20. D. D. Fitts, "Nonequilibrium Thermodynamics", Mc Graw-Hill, New York (1962).
21. E. Endom, H. Hertz, B. Thül and M. Zeidler, Ber. Bunsenges. Phys. Chem. 71:1008 (1967).
22. H. Hertz, Ber. Bunsenges. Phys. Chem. 75:183,572 (1971).
23. H. Hertz, K. Harris, R. Mills and L. Wolf, Ber. Bunsenges. Phys. Chem. 81:664 (1977).
24. F. Ludwików, "Transport of binary and ternary electrolytes through charged membranes with allowance for interactions. I. Relationship between the Onsager transport coefficients and the self-diffusion coefficients, transport and hydration numbers", in preparation for publication.
25. R. A. Robinson and R. H. Stokes, "Electrolyte Solutions", Butterworths Scientific Publications, London (1959).
26. B. Neumcke, D. Walz and P. Läuger, Biophys. J. 10:172 (1970).
27. L. Onsager, J. Chem. Phys. 2:599 (1934).
28. L. Onsager and S. K. Kim, J. Phys. Chem. 61:198 (1957).
29. L. Onsager and S. K. Kim, J. Phys. Chem. 61:215 (1957).
30. F. Ludwików, A. Albiniak and A. Marszałek, "Transport of binary electrolytes through charged membranes with allowance for interactions. II. Case of no chemical reaction", in preparation for publication.
31. F. Ludwików, M. Wierzchaczewski and S. Miękisz, "Stationary State Analysis of Equations for Transport of Binary Electrolytes across charged Membranes", Energetics and Regulation of Membrane Transport, Praha-Zvikovske Podhradi, 1 - 4.10.1979.

EFFECT OF PROTEIN CROSS-LINKING ON EXCITABLE AND NON-EXCITABLE MEMBRANES

Doru-Georg Mărgineanu

Laboratory of Biophysics
Faculty of Biology, University of Bucharest
Bucharest 76201, Romania

INTRODUCTION

In the last decade the interest of molecular membranologists gradually shifted towards the study of membrane proteins[1], recognized to play the prominent roles in all the main membrane activities. Within a whole gamut of procedures devised for characterising membrane proteins, the use of cross-linking agents has the particular feature of revealing the "near neighbors".

A wealth of recent studies deal with cross-linking membrane proteins by means of small bifunctional organic molecules reactive towards the functional groups on the side chains of proteins, especially -SH and $-NH_2$ groups. Apart the analytical purposes[2,3] some of the cross-linking reagents, especially the glutaraldehyde, are widely used for immobilizing enzymes, organelles and even whole cells within artificial macroscopic membranes. On the other hand, a rather minor attention was paid to the effects of these reagents on membrane activities and properties. Among such attempts are our recent studies on the nerve[4] and on epithelia[5,6].

Here I summarize the kinetic investigations of impulse blocking in frog sciatic nerve by a representative selection of aldehydes, the effects of glutaraldehyde on water and ionic permeabilities of frog epithelia and its effects on the osmotic behaviour of erythrocytes.

Frog sciatic nerves, mounted in a thermostated stimulation chamber, were stimulated with square pulses (40 Hz and 0.04 ms duration) whose amplitude was taken so that to obtain a maximal compound action potential (AP), displayed on the screen of a Tektronix 564B storage oscilloscope. The nerves were initially immersed into conventional Ringer solution which was afterwards replaced by aldehyde solutions. These were prepared by diluting into Ringer the commercial formaldehyde (FA), crotonaldehyde (CrA), butyraldehyde (BA), glutaraldehyde (GA) and cinnamaldehyde (CinA).

All these aldehydes gradually decrease the amplitude of the AP, up to the complete block, the rate of this effect depending on the nature and concentration (C) of aldehyde and on temperature (T). In fig.1, the time evolution of AP amplitude, consecutive to nerve immersion in GA-Ringer is plotted, for the same concentration at three temperatures.

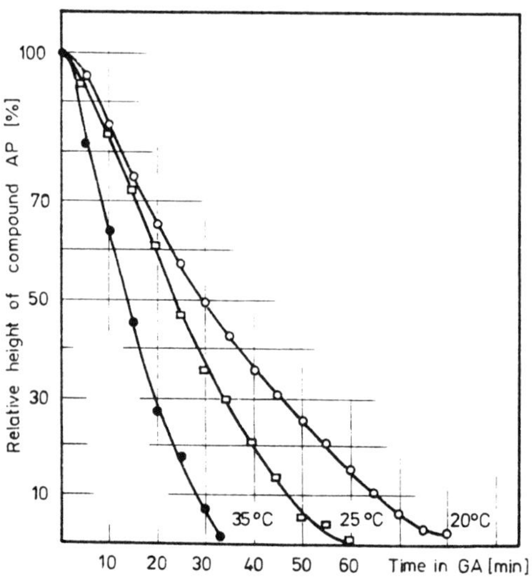

Fig.1. The relative height of the compound AP in frog sciatic nerve following exposure to GA 0.5% (w/v)[4].

The effect shows an identical curve for all the five aldehydes, irrespective of T and C, these parameters influencing only its rate.

The sigmoid dependence of AP amplitude on the immersion time (t), as represented in Fig.1, is fairly described by the equation:

$$y = 1 - \left[1 - \exp(-t/\tau)\right]^n \qquad (1)$$

where y is the normalized (relative) AP amplitude. The computer fitting of the experimental points with eq.(1) revealed that the parameter n does not vary but with the type of aldehyde. If now, for a given aldehyde, n is set constant, irrespective of C and T, one easily obtains that the blocking time τ_b for which the amplitude falls below the 1% sensitivity of the equipment, i.e. y = 0.01, is simply the time constant τ appearing in eq.(1), multiplied by a constant:

$$\tau_b = \left[- \ln(1 - 0.99^{1/n})\right] \cdot \tau \qquad (2)$$

Equation (2) shows that τ depends on C and T exactly as τ_b experimentally obtained, so that I shall further refer to only τ_b.

The blocking time in GA-Ringer solution is plotted in Fig. 2 as function of temperature. It clearly appears the linear dependence of τ_b on T at each aldehyde concentration. If the solid lines which fit the experimental points are extrapolated, they intercept the abscissa in the same point. This appears as a critical temperature (T_c) characterizing the effect of GA on frog sciatic nerve. Such a significant regularity holds in the case of other aldehydes too, each having a specifical T_c as listed in Table 1.

The reduction in AP amplitude up to the complete block is irreversible for GA, FA, BA and CrA even at the smallest concentrations. Only in the case of CinA, nerve excitability is restored almost completely upon reimmersion in normal Ringer, so that the effect of a second treatment can be studied. The second blocking time is much shorter than the first one (about half of it) at the same T and C, but - when plotted together - they all reveal the same T_c.

The assembly of these data show that the dependence of τ_b on temperature and aldehyde concentration has in all cases the form:

$$\tau_b = (A/C^{\alpha}) \cdot (T_c - T) \qquad (3)$$

where the constants A and α, as well as T_c, depend only on the type of aldehyde. Thus, the effects of these five aldehydes on impulse propagation in the sciatic nerve of the frog are sufficiently regular to be accomodated within a simple empirical equation. The values of T_c, n and α are given in Table 1.

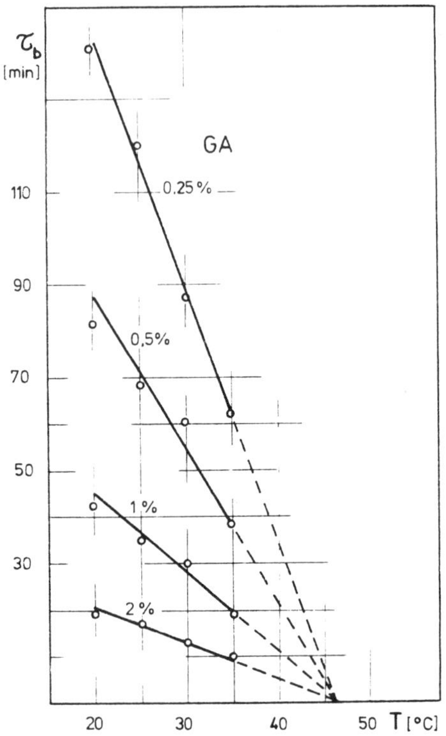

Fig.2. The blocking time of the nerve impulse as a function of temperature, at four GA concentrations[4].

Table 1. Empirical Parameters describing Impulse Blocking by Aldehydes in Frog Nerve[4]

	FA	CrA	BA	GA	CinA
T_c (°C)	49.5	43.0	57.5	46.5	48.0
n	2.98	1.32	1.38	1.92	1.84
α	0.23	0.78	0.53	0.95	0.54

The reduction in AP amplitude by aldehydes is a gross consequence of the block of ionic channels, in agreement with data on single axons[7].

The main sites of aldehyde attack on proteins are the free amino groups[8] so that one can infer that the ionic channels actually bear such free groups. T_c has the meaning of temperature at which nerve impulse blocking by a given aldehyde would by instantaneous, no matter of concentration. Its values are specific for each aldehyde, so that it characterizes the chemical changes induced in the ionic channels. It is noteworthy that all T_c's lie in the domain $(43 - 58)^{o}C$ where thermal denaturation of proteins occurs, this being suggestive for the nature of molecular changes produced by aldehydes.

The order of the blocking potency of aldehydes is: CrA > CinA > BA > FA > GA , without any obvious correlation with the molecular weight of aldehydes, so that it does not reflect mere differences in diffusion rate, but the actual chemical interaction. Notice that the only aldehyde (in the group investigated) whose blocking effect on nerve impulse is reversible is that one having a benzene ring.

GLUTARALDEHYDE EFFECTS ON TRANSPORT THROUGH EPITHELIA

A convenient model for the study of the effect of cross-linking agents on transport phenomena through biological membranes are the isolated frog epithelia whose permeability properties result from a series/paralel summation of the properties of cell membranes.

High GA concentrations (2.5% w/v) produce a marked increase (+40%) in the diffusional permeability of water, probably due to the stabilization of membrane hydrophilic domains[5]. Both the osmotic and the diffusional permeabilities for sodium and potassium are increased at increasing GA concentrations (from 0.5% to 2.5%), but this effect is reversed at temperatures above $20^{o}C$. As a purely tentative explanation we assume that the thermal activation of the chemical cross-linking (leading to spatial obstructions for the passage of ions) and of the diffusion itself results in such a mixed global effect.

As concerns the effect of GA on the transepithelial active transport of sodium across isolated frog skin, our preliminary results show that the treatment with small aldehyde concentrations (0.01-0.1%) results in exponential decays of the active pumping, indicating the gradual block of (Na+K)ATPases.

GLUTARALDEHYDE EFFECT ON THE OSMOTIC HAEMOLYSIS

The osmotic fragility of human red blood cells was studied by the classical method of Drabkin. The treatment consists in keeping the washed erythrocytes into glutaraldehyde-isotonic phosphate buffer for 5 min, after which they are washed three times in the buffer (pH 7.4). As reference (100%) haemolysis is taken that one produced by Triton X-100. The rate of haemolysis was measured by introducing 1.7 ml of 50 mM NaCl buffered solution over 50 μl stock suspension in the thermostated cuvette of a spectrophotometer, and recording the time course of the decrease in absorbance at 690 nm. At this wavelength, haemoglobin does not absorb, but the intact erythrocytes scatter the light. The rate of haemolysis is expressed as % decrease in absorbance/sec, calculated in the interval 3-6 sec from the outset of recording.

Increasing concentrations of GA gradually diminish the level of maximal haemolysis as it appears from Fig. 3. Apart from this, two other aspects are equally seen: a) the salt concentrations producing the maximal haemolysis are almost the same, no matter what the GA concentration; b) the aldehyde causes a kind of "ground level" haemolysis at normally non-haemolising salt concentration, the extent of which increases with the concentration of GA. The same characteristics appear at other temperatures too, in the domain (22-42)^{o}C, the differences being merely quantitative.

Fig.4 shows, in a semi-logarithmic representation, the time dependence of the relative absorbance of the erythrocyte suspensions at 690 nm, following the treatment with GA at 37°C. The absorbance of an equally dense suspension into isotonic buffer is the reference (100%).

Increasing GA concentrations gradually diminish the rate of haemolysis at all the temperatures, but this effect cannot be accomodated into any simple formula.

The fact that GA is known to cross-link between amino-groups and some sulfhydryl groups[8] suggests that its antihaemolytic activity is mainly due to an increased mechanical resistance of the cytoskeleton and of a newly created network of integral proteins. The data of Eitan et al.[9] further suggest that cross-kinking of the spectrin network is less important, as GA has rather similar effects on both maximal haemolysis and the rate of haemolysis, this latter being relatively independent on spectrin network[10]. Thus, I propose that GA converts the sparsely distributed membrane integral proteins into a kind of "membrane skeleton" able to support the osmotic stretching. The fact that some of the treated cells are no longer able to tolerate even minor, normally nonhaemolysing stretching, suggests that GA treatment is formally equivalent with an artificial "ageing" process, attenuating the differences between the most resistant ("young") and the most fragile ("aged") erythrocytes. The marked decrease in the osmotic fragility of GA-treated cells at

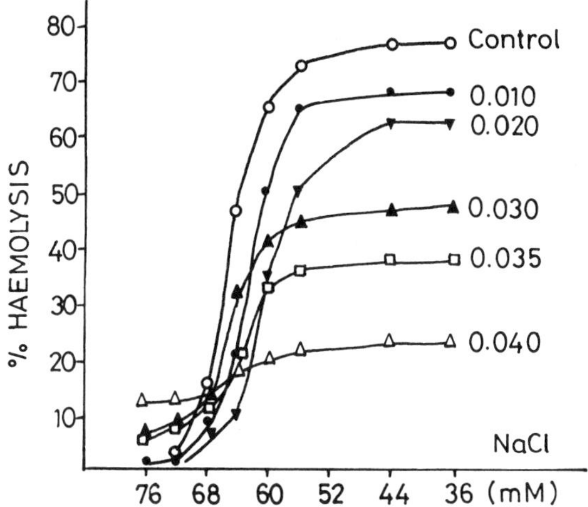

Fig. 3. Osmotic fragility curves of human erythrocytes at 37°C after the treatment with GA in the concentrations given (in % w/v) near each curve.

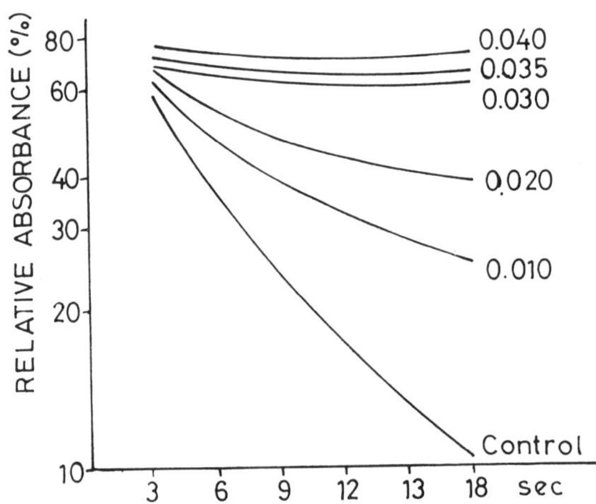

Fig. 4. Optical absorbance at 690 nm in suspensions of human erythrocytes treated with GA (% concentrations indicated near each curve) in 50 mM NaCl-phosphate buffer at 37°C.

higher temperatures could be attributed to the cummulative effects of a thermally increased fluidity of of the lipid bilayer, it making more available lipid for cell swelling, and to the appearance of even more cross-links as the reaction of GA with the protein side groups is thermally activated.

CONCLUSION

The cross-linking aldehydes, particularly the glutaraldehyde appear to be useful tools for investigating the effects of _in situ_ chemical modification of the proteins on transport phenomena through both excitable and non-excitable biological membranes.

REFERENCES

1. A. Rothstein, The cell membrane - A short historical perspective, _Curr.Topics Membr.Transp_. 11:1 (1978).
2. K. Peters and F. M. Richards, Chemical cross-linking reagents and problems in studies of membrane structure, _Ann.Rev. Biochem_. 46: 523 (1977).
3. R. B. Freedman, Cross-linking reagents and membrane organization, _Trends Biochem.Sci._ 3:193 (1979).
4. D. G. Mărgineanu, E. Katona and J. Popa, Kinetics of nerve impulse blocking by protein cross-linking aldehydes, _Biochim. Biophys.Acta_ 649:581 (1981).
5. D. G. Mărgineanu and M. L. Flonta, Effects of cross-linking reagents on water and ionic fluxes through epithelia, _Studia Biophys_. 84:49 (1981).
6. C. Rucăreanu, D. Popescu and D. G. Mărgineanu, Glutaraldehyde effect on diffusional permeabilities of frog urinary bladder, _Rev.Roum.Biochim._ 19:57 (1982).
7. R. Horn, M. S. Brodwick and D. C. Eaton, Effect of protein cross-linking reagents on membrane currents of squid axon, _Am.J.Physiol_. 238:C127 (1980).
8. K. Ziegler, J. Schmitz and H. Zahn, Introduction of new cross-links into proteins, _Adv.Exp.Med.Biol_. 86A:345 (1977).
9. A. Eitan, B. Aloni and A. Livne, The erythrocyte membrane site for the effect of temperature on osmotic fragility, _Biochim.Biophys.Acta_ 465:46 (1977).
10. K. Araki and J. M. Rifkind, The rate of osmotic haemolysis, _Biochim.Biophys.Acta_ 645:81 (1981).

ISOSMOTIC VOLUME REGULATION BY HUMAN LYMPHOCYTES

William Negendank and Calvin Shaller

Hematology-Oncology Section, Department of Medicine
Hospital of the University of Pennsylvania
3400 Spruce Street, Philadelphia, PA USA 19104

INTRODUCTION

The conventional assumptions in cell physiology, and specifically in the membrane-osmotic pump theory, are that cellular water and ions exist primarily in a physical state like that of a dilute aqueous solution, that the macroscopic concentration gradients of ions mirror true chemical or electrochemical activity gradients, that fixed charges on intracellular macromolecules require balancing by counter-ions that tend to follow a Donnan equilibrium, and that the outwardly-directed Na pump is required to maintain normal cellular volume against the forces in the Donnan equilibrium that tend to produce an excessive osmotic pressure. These concepts have been crystallized in terms of the "pump-leak, double-Donnan" formulations of the maintenance of cellular volume.

In human lymphocytes, the plasma membrane Na,K-ATPase is assumed to be a pump responsible for the normal net extrusion of Na, coupled to the net accumulation of K. Ways to inhibit this putative Na pump would include depletion of ATP, incubation at low temperature, treatment with ouabain, and incubation in medium that contains very low concentrations of K. In fact, each of these manipulations causes lymphocytes to lose their normal K and to gain Na. Na reaches a level that is slightly greater than that in the external medium, and by this criterion pumping of Na outward is fully inhibited. However, these cells fail to swell in the manner expected in the pump-leak, double-Donnan concept. In this presentation, we report these results and we suggest a different theory to explain them.

Table 1. Water and Ions in Lymphocytes in 145 mM Na, 5.5 mM K, 136 mM Cl

	Water (% wet wt)	Na (mM)	K (mM)	Cl (mM)
Control	77.8 ± 0.6	37.2 ± 4.7	179 ± 8.0	94.7 ± 7.8
ATP-Depleted	77.7 ± 2.4	223 ± 11	13.1 ± 1.6	203 ± 22
0°	77.5 ± 3.0	165 ± 3.4	46.3 ± 2.2	104 ± 3.7
Ouabain	78.6 ± 0.3	177 ± 11	20.0 ± 2.4	150 ± 17
0.1 mM K_{ex}	77.7 ± 2.5	188 ± 9.5	17.2 ± 6.4	-

Data are means ± SEM of 4-14 separate determinations. Water and ions were time-independent and measured between 24 and 48 hours. Ouabain was 5×10^{-5}M. Results at 0° are from reference 3 (confirmed in 4), and at 0.1 mM K_{ex} from reference 1. The remainder of data, and all chlorides, are described in a paper recently submitted.

METHODS

Human lymphocytes prepared from peripheral blood were incubated in isosmotic medium and time courses were followed for 48 hours to ensure full equilibration of ions. Medium was essentially that of Hanks and contained 145 mM Na, 5-6 mM K, 136 mM Cl, 1.3 mM Ca, 0.9 mM Mg, 14.3 mM HCO_3, 0.34 mM HPO_4, 1.3 mM H_2PO_4 and 5.6 mM glucose, pH 7.4, and 10% autologous serum[1]. Experimental conditions included controls at 37° C, cells at 0° C, cells in 10^{-5}M ouabain, cells in nearly 0 mM K, and cells metabolically inhibited by treatment with IAA,N_2. In 1 mM IAA, lymphocytes lose ATP within 5 hours[2]. Cell water was determined by drying, Na and K by atomic absorption spectroscopy of 0.1 NHCl or HNO_3 extracts, Cl by an Orion Cl-sensitive electrode in a HNO_3 extract, and trapped medium in cell pellets by ^{14}C-polyethylene glycol[1].

RESULTS

Control cells in 145 mM Na, 5.5 mM K, 136 mM Cl at 37° C have a water content of 78% of wet weight. This and their ionic concentrations are shown in the top line of Table 1.

Cells were treated with iodoacetate (IAA) and nitrogen to deplete ATP, incubated at 0°, incubated in medium that contained 0 K, or incubated in 10^{-5} M ouabain. In each case, they lost K and

Table 2. Water and Ions in Lymphocytes in 22 mM Na, 5.5 mM K, 20 mM Cl

	Water (% wet wt)	Na (mM)	K (mM)	Cl (mM)
Control	69.8 ± 1.7	29.0 ± 3.2	113 ± 10	33.0 ± 4.0
0°	69.2 ± 4.1	110 ± 5.0	53.0 ± 8.9	37.2 ± 7.7
Ouabain	75.8 ± 1.5	43.0 ± 3.3	20.0 ± 3.7	21.1 ± 2.3

Data are means ± SEM of 3-5 separate experiments. Water, Na and K at 0° are from reference 5; the remainder of data are described in a paper recently submitted.

gained Na as expected were any outwardly-directed Na pumps fully inhibited (Table 1). However, the cells did not swell under any of these conditions. Moreover, this failure to swell occurred in the absence of metabolism, required neither a net exclusion of Na nor a net exclusion of Cl, and clearly was not due to impermeability or another restriction on ion equilibration. These points are also emphasized by the results obtained when lymphocytes are incubated in medium that is isotonic but contains lower concentrations of Na and Cl (Table 2). In fact, at 0° or in ouabain the cells actually gain Na to levels higher than in the medium.

The failure of "inhibited" lymphocytes to swell is not due to a hydrostatic pressure exerted by a rigid membrane or matrix, since these cells swell readily when immersed in a hypotonic medium. The results also are not explained by an altered colloid-osmotic pressure, an altered Donnan charge ratio, a subcellular compartmentation of water or ions, or a variable amount of "nonsolvent water". The reasons for this are given in detail elsewhere[6].

If we assume that the greater concentration of cellular Na under these conditions mirrors the tendency of Na to approach a Donnan equilibrium in a system bearing fixed and/or impermeant negative charges, then, in the absence of swelling, the following equations define the relation between r, the Donnan ratio of cell/medium ion concentrations, and R^-, the value (expressed in mmoles/l cell water) of the intracellular fixed and impermeant negative charges.

$$r = \frac{[M^+]_{in}}{[M^+]_{ex}} = \frac{[A^-]_{ex}}{[A^-]_{in}} = \frac{[M^{++}]_{in}}{[M^{++}]_{ex}}^{\frac{1}{2}} = e^{\Psi F/RT}, \quad (1)$$

Table 3. Apparent Values of the Donnan Ratio, r, Derived From Results in Tables 1 and 2

	Derived from: Na	K	Cl
In 145 Na, 5.5 K, 136 Cl			
ATP-Depleted	1.5	2.4	0.7
Ouabain	1.2	3.7	0.9
0°	1.1	8.4	1.3
In 150 Na, 0.1 K, 136 Cl			
0.1 K_{ex}	1.3	170	-
In 22 Na, 5.5 K, 20 Cl			
0°	5.0	9.6	0.5
Ouabain	2.0	3.6	1.0

where the ions M^+, M^{++} and A^- are freely-exchangeable and have activity coefficients in cellular water that are similar to those in medium water, and all of cellular water is a solvent for these ions. Ψ is the Donnan potential. Neglecting polyvalent anions in the medium (which are only about 2 mM),

$$[M^+]_{ex} + [2M^{++}]_{ex} = [A^-]_{ex} . \quad (2)$$

In the absence of a significant number of freely-dissolved and freely-exchanging or permeant polyvalent anions within the cell, electroneutrality requires that:

$$[M^+]_{in} + [2M^{++}]_{in} = [A^-]_{in} + R^-, \quad (3)$$

where R^- includes fixed charges and impermeant anions, such as organic phosphates.

From equations 1-3,

$$R = r[M^+]_{ex} + 2r^2 [M^{++}]_{ex} - \frac{[A-]_{ex}}{r} . \quad (4)$$

We then calculate values of r and from equation 4 the corresponding values of R^- for cells in the two different kinds of external medium used and the results are shown in Table 3. It is evident that the concentrations of Na in cells in 145-150 mM Na in the presence of ouabain, in low external K, or at 0° would be compatible with a Donnan equilibrium in which there is a Donnan ratio, r, of between 1.1 - 1.3, and a corresponding fixed or impermeant charge, R^-, of between 40-80 mmoles/l cell water. Cellular Cl at 0° would be compatible with r=1.3. There are, however, three major problems in fitting the remainder of the data in Table 3 to a Donnan equilibrium.

First, the cellular K levels and their r values are too high (Tables 1 and 2). One could suppose, of course, that a fraction of the K either is "bound" or is contained within a subpopulation of cells that is not sensitive to these various experimental manipulations. Second, in cells depleted of ATP the concentration of Na is high, 223 mM (Table 1). This could occur if there were an increase in R^- from its usual value of 40-80 mM to about 120 mM. However, cellular Cl deviates markedly from its expected value; instead of decreasing it rises. Third, in cells incubated in medium containing low concentrations of Na and Cl (Table 2), the cellular Na level at 48 hours at 0° is much higher than expected. This could have occurred if the concentrations of fixed (or inexchangeable) charges, R^-, were to have increased from the usual value of 40-80 mM to around 230 mM. It is difficult to imagine the source of such a high concentration of fixed or impermeant charges. Moreover, cellular Cl, which should have decreased markedly, remains much higher than expected, both at 0° and in ouabain.

DISCUSSION

The results of this study indicate the following: (1) Lymphocytes depleted of ATP, incubated at 0°, treated with ouabain, or incubated at low external K gain Na and lose K as expected were the putative Na pump, the Na,K-ATPase, suppressed under each of these conditions. However, contrary to expectations based on the pump-leak, double-Donnan concept, they do not swell. (2) The maintenance of normal volume in isosmotic medium is not due simply to blockage of passive Na influx or K efflux, does not necessarily require active metabolism, and does not necessarily require net extrusion of Na or Cl. (3) Cells depleted of ATP accumulate Na and Cl to apparently hypertonic levels (439 mM) without significant swelling or shrinkage. (4) Ions in "inhibited" lymphocytes do not consistently approach a Donnan equilibrium. In cells depleted of ATP or incubated in low external Na and Cl, the cellular Na levels would require an increase in intracellular fixed negative charge, but the Cl levels, which then would be expected to fall, either increase or remain inappropriately high. (5) Under conditions in which there is a reduction in total Na + K, there often is some shrinkage of the cells, and the lost

Na + K usually is balanced by a loss of Cl.

We suggest that in lymphocytes the supposed Donnan forces expected to cause swelling do not exist, first, because cellular water is itself sufficiently ordered that it excludes solutes to equilibrium levels lower than in the external medium, and second, because the majority of cellular ions are not free in solution in cellular water but are closely associated with or adsorbed onto fixed charges on intracellular macromolecules. When Na replaces K under the influence of ATP depletion, low temperature, ouabain, or low external concentrations of K, the gained Na also is adsorbed onto macromolecules.

In this view, which is based on the association-induction hypothesis of Ling[7,8], the primary force that determines the normal water content of the cell is the polarization of water in multilayers by alternating electronegative, electropositive regions of extended polypeptide chains, while the osmotic pressure exerted by the relatively small amounts of ions dissolved within this water is of only secondary importance. The primary restraining force that limits cellular swelling is the presence of reversible salt-linkages between fixed negative and fixed positive charges on macromolecules within the cell. The equilibrium water content relative to dry weight, hence, the volume of lymphocytes mirrors the net resultant of four forces: equilibration of chemical activity of water, polarization of water in multilayers, exclusion or entry of solutes, and cohesive forces due in part to salt-linkage formation. Some of the forces resulting from these processes are interdependent, and they may oppose or potentiate one another.

These concepts explain the following results in Tables 1 and 2: (1) A normal cellular volume is maintained in isosmotic medium in the face of a variation of equilibrium levels of total cell K plus Na plus Cl that range from 84 to 439 mM. (2) Normal cellular volume is maintained in the absence of active metabolism. (3) In cells depleted of ATP, there is a marked increase in concentrations of Na and Cl (223 and 203 mM, respectively) in the absence of a significant change in volume or water content; this is due to two different parameters that are influenced by ATP, and in its absence (a) Na replaces K on adsorption sites and (b) cellular water is depolarized allowing it to dissolve more solutes relative to their concentrations in the external medium.

These concepts also explain results obtained when cells are incubated in medium that is isosmotic but contains low concentrations of both Na and Cl (Table 2). These include: (1) the ability of cells in this medium when at 0° or treated with ouabain to accumulate Na, and to retain K and Cl, all at concentrations higher than those in the external medium (this is because the accumulated ions are adsorbed onto fixed negative or positive charges); and (2) the mild shrinkage and loss of total K plus Na in these cells (this is because the decreased abundances of Na and K, as well as of Cl, permit fixed positive and negative charges to form salt-linkages between or within macromolecules). The salt-linkage formation reduces the equilibrium volume of water associated with the macromolecules.

Finally, we suggest that the phenomenon of reversible macromolecular salt-linkage formation underlies the ability of lymphocytes, lymphoblasts, and certain red cells to regulate their volume when immersed in a hypotonic medium.

These, and other aspects of ion and water physiology of human lymphocytes are summarized in more detail in a recent review[6].

ACKNOWLEDGEMENTS

This work was supported by the Office of Naval Research Biophysics Program, by the Veterans Administration Medical Research Service, and by a donation in memory of James E. Casey.

REFERENCES

1. W. Negendank and C. Shaller, Potassium-sodium distribution in human lymphocytes: Description by the association-induction hypothesis. J. Cell. Physiol. 98:95 (1979).
2. W. Negendank and C. Shaller, The effect of metabolic inhibition on ion contents and sodium exchange in human lymphocytes. J. Cell. Physiol. 110:291 (1982).
3. W. Negendank and C. Shaller, A critical temperature transition of K-Na exchange in human lymphocytes. J. Cell. Physiol. 103:87 (1980) (errata 105:108).
4. W. Negendank and C. Shaller, The effects of temperature and ouabain on steady-state Na and K exchanges in human lymphocytes. J. Cell. Physiol. 113:440 (1982).
5. W. Negendank and C. Shaller, Simultaneous net accumulation of both K and Na by lymphocytes at 0° C. Biochim. Biophys. Acta 640:368 (1981).
6. W. Negendank, Studies of ions and water in human lymphocytes Biochim. Biophys. Acta.694:123 (1982).
7. G.N. Ling, "A physical theory of the living state: The association-induction hypothesis," Blaisdell, Waltham, Mass. (1962).
8. G.N. Ling, M.M. Ochsenfeld, C. Walton, and T.J. Bersinger, Mechanism of solute exclusion from cells: The role of protein-water interaction. Physiol. Chem. & Phys. 12:3 (1980).

BAND 3 PROTEIN-MEDIATED ANION TRANSPORT ACROSS THE RED CELL MEMBRANE: THE SITE OF ACTION OF THE INHIBITORS 4,4′-DI ISOTHIOCYANO DIHYDROSTILBENE-2,2′-DISULFONATE (H_2DIDS) AND 1-FLUORO-2,4-DINITROBENZENE (N_2ph-F)

H. Passow, L. Kampmann, and S. Lepke

Max-Planck-Institut für Biophysik
Frankfurt am Main
Fed. Rep. of Germany

Like all other cells, the red blood cells are surrounded by a hydrophobic lipid bilayer that is virtually impermeable for hydrophilic anions such as HCO_3^- and Cl^-. Nevertheless, the HCO_3^--Cl^- exchange that constitutes an important step in CO_2 transport by the blood, is completed within the short time that the red cells pass through the capillary bed of the peripheral tissues or the lung (less than one second). Hence there must exist a transport system that facilitates the anion exchange across the bilayer. This transport system consists of one of the most abundant integral membrane proteins of the red blood cell membrane, the so-called band 3 protein (Cabantchik and Rothstein, 1974; Passow, Fasold, Zaki, Schuhmann and Lepke, 1975).

The name of this protein is derived from its location on SDS polyacrylamide gel electropherograms of the red cell membrane: it is the third major band from the top. This location indicates a molecular weight of about 96,000 Daltons, which has been confirmed by analytical ultracentrifugation (Dorst and Schubert, 1979) and amino acid analysis (Steck, Koziarz, Singh, Reddy and Köhler, 1978). In the red cell membrane, the protein exists in the form of monomers, dimers and tetramers (Steck, 1978; Margaritis, Elgsaeter and Branton, 1977). When dissolved in suitable non-ionic detergents, reversible association equilibria between these forms have been demonstrated to exist (Pappert and Schubert, 1982). It remains to be seen whether or not the associations observed in the red blood cell membrane also represent true reversible equilibria, or if interactions with the cytoskeleton at the inner surface of the membrane impose constraints that affect or even prevent the establishment of such equilibria. Interactions with the cytoskeleton are obvious, since in the intact membrane band 3 is nearly indiffusable

(Peters, 1974), while after reversible detachment, or irreversible destruction of the cytoskeleton, lateral diffusion may occur (for review, see Peters, 1981). Based on evidence supplied by Jennings and Passow, 1979 and Macara and Cantley, 1981, it may be assumed that the monomers of band 3 are independently operating units, even when they form constituents of dimers or tetramers.

Today there exist many methods for the isolation and purification of the band 3 protein (e.g. Herbst and Rudloff, 1982; Schubert, 1978; Drickamer, 1980; Steck, Koziarz, Singh, Reddy and Köhler, 1978; Grinstein, Ship and Rothstein, 1978; Jenkins and Tanner, 1977). Attempts have been made to insert the purified protein into liposomes and to reconstitute transport. Most of the earlier results remained ambiguous. However, up to the present time there appeared at least a few publications that report a successful reconstitution of some of the properties that the transport protein shows in the intact red blood cell membrane (Cabantchik, Volsky, Ginsburg and Loyter, 1980; Köhne, Haest and Deuticke, 1981).

Many attempts have been made to use the easy supply with pure band 3 for the study of the primary structure (Drickamer, 1980; Rao and Reithmeier, 1979; Steck et al., 1976, 1978; Rothstein 1982; Rothstein, Ramjeesingh, Grinstead, Knauf, 1980; Tanner, Williams, Jenkins, 1980). The application of the conventional methods of protein sequence analysis encountered, however, considerable difficulties and thus only a short piece of the sequence has been elucidated (Mawby and Findlay, 1982). Further attempts at sequencing are under way (e.g. by Tanner, Marchesi, personal communications), but the results have not yet been published. In view of the difficulties, several laboratories now try to derive the sequence from sequencing the nucleic acids that contain the genetic information about the structure of the band 3 protein.

In spite of the difficulties described, the combined efforts of many laboratories have provided a reasonably well established picture of the arrangement of the peptide chain in the lipid bilayer (Drickamer, 1980; Rothstein, 1982; Tanner et al., 1980; Rothstein, et al., 1980). This picture is important for the subsequent discussion of the relationship between protein structure and anion transport and needs to be considered in some more detail.

Fig. 1 shows that the peptide chain passes several times across the lipid bilayer. The N terminus resides on the plasmatic surface in a large hydrophilic piece of 42,000 Daltons that seems to protrude into the cell interior. The location of the C-terminus is not yet established. However, from the point of view of this talk, it is important to note that near the C-terminus carbohydrates are attached to the peptide chain; altogether they amount to about 7% by weight of the protein (for review, see Steck, 1978). These carbohydrates project into the external medium. Thus, two major portions

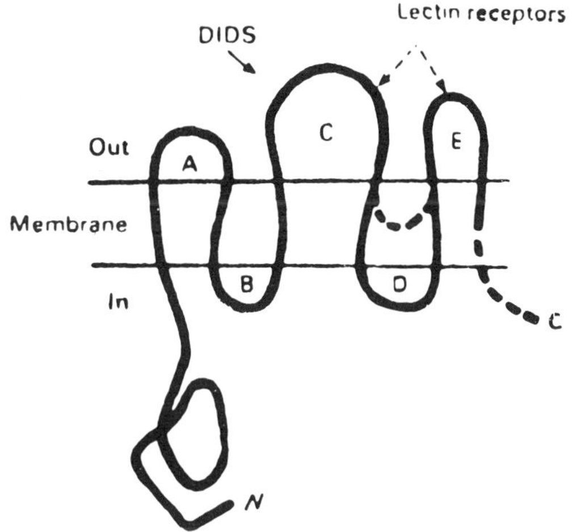

Figure 1

The organisation of the band 3 protein in the lipid bilayer, according to Tanner et al., (1980).

SDS polyacrylamide gel electropherograms of the H_2DIDS treated red cell membrane. (1) Exposure to external chymotrypsin (1 mg/ml cell suspension for one hour at 37°C, pH 7.4). (2) Chymotrypsinised red cells exposed to H_2DIDS at pH 9.5 for one hour at 37°C. (3) Control: neither treated with chymotrypsin nor H_2DIDS (from ref. 8).

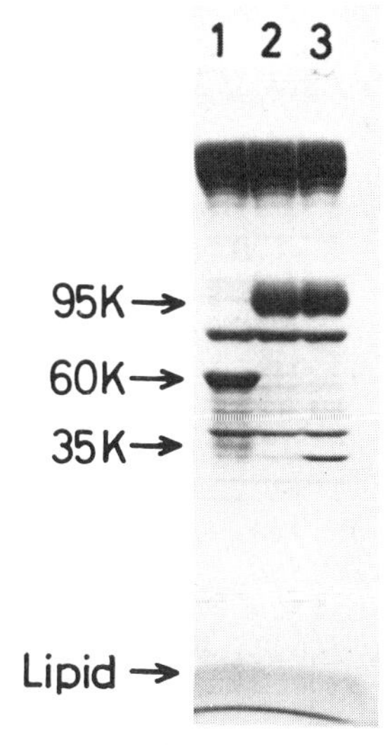

Figure 2

of the transport molecule are hydrophilic. Since they are located on opposite surfaces of the membrane, they prevent the molecule from rotating across the lipid bilayer. This implies that the anion transport cannot take place by rotational diffusion of the transport molecule around an axis parallel to the surface of the bilayer (Jennings and Passow, 1979). Thus, it is most likely that the anions are transported across the interspaces between different segments of the transport protein. These interspaces do not form a simple aqueous pore. This conclusion is based on the following considerations:

Each penetrating anion carries one negative charge with it. From the easily measurable rate of anion equilibrium exchange across the membrane, it is possible to determine how many electrical charges are transferred across the membrane per unit time. From this number, one can calculate an expected electrical conductance of the membrane. Estimates of the conductance by several independent techniques have shown, however, that the electrical conductance of the human red cell membrane is many thousand times smaller than the conductance calculated from the isotope exchange experiments (Scarpa, Cecchetto and Azzone, 1970; Harris and Pressman, 1967; Hunter, 1971). As a consequence, one has to assume that the rate-limiting step of anion transport involves anion binding, one or several subsequent conformational changes of the transport protein, and a dissociation of the anion at the other membrane surface.

For the understanding of what follows, it is useful to note that the intracellular 40,000 Dalton fragment can be cleaved off by intracellularly applied trypsin (Steck, Ramos and Strapson, 1976), without inhibition of anion transport (Lepke and Passow, 1976; Grinstein, Ship and Rothstein, 1978). Thus, this segment does not seem to be involved in transport. The protein can be split into two fragments by external chymotrypsin (Cabantchik and Rothstein, 1974; Passow, Fasold, Lepke, Pring and Schuhmann, 1977). One fragment has a molecular weight of 60,000 Daltons, the other a molecular weight of 35,000 Daltons. The cleavage does not affect the anion transport (Cabantchik and Rothstein, 1974; Passow, Fasold, Lepke, Pring and Schuhmann, 1977). Apparently, the lateral pressure of the lipid bilayer and/or the tertiary and quatenary forces within the protein molecule preserve its structure in the functional state. When external papain is applied, the 35,000 Dalton segment is digested, and inhibition ensues (Jennings and Passow, 1979). This suggests that the 35,000 Dalton segment participates in transport. Chemical modification of a specific lysine residue on this segment confirmed this suggestion (Jennings, 1982). Experiments with several covalently binding inhibitors have shown that the 60,000 Dalton segment is also required for transport (for review, see Rothstein, 1982).

Currently many laboratories around the world - in Denmark,

France, Poland, Germany, England, Canada, the USA, Japan, etc. -are engaged in elucidating the details of the structure of band 3 and of the reaction sequence that leads to anion transfer. Most of the publications that result from this effort would deserve to be discussed in detail; but within the short length of time available for this talk, a fair appraisal of them will not be feasible. It may suffice to mention that the present state of knowledge has been amply reviewed, e.g. by Drickamer, 1980; Tanner, Williams and Jenkins, 1980; Cabantchik, Knauf and Rothstein, 1978; Rothstein, 1982; Knauf, 1979; Knauf, 1982; Passow, 1982; Macara and Cantley, 1982). We shall confine ourselves, therefore, to a discussion of some work from our laboratory, much of which was done jointly with Doctors Fritzsch and Fasold. This work is concerned with the exploration of the reactions of H_2DIDS, a selective inhibitor of anion transport, with the anion transport protein and with the inferences we drew from the binding studies with respect to the transport mechanism. The inhibitor is related to 4,4'-diisothiocyanate-stilbene-2,2'-disulfonate (DIDS), a compound that had first been used in the study of anion transport by Cabantchik and Rothstein in 1974 (Cabantchik and Rothstein, 1974). This compound and many of its covalently or non-covalently reacting derivatives have played an important role in the study of anion transport. The dihydroderivative H_2DIDS is a symetrical molecule that carries on a diphenylethane backbone two negatively charged SO_3^- groups and two NCS groups.

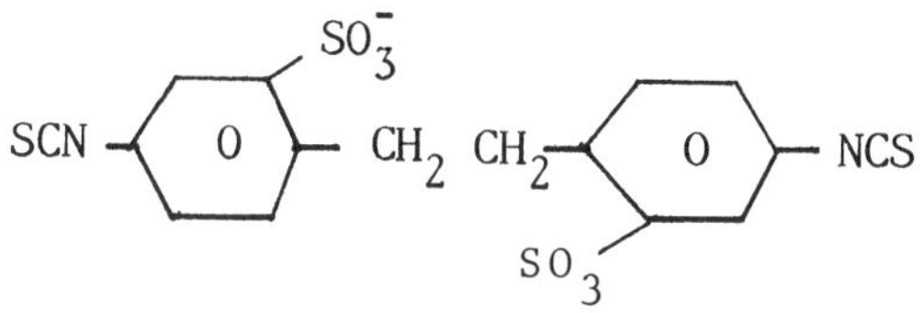

$$SCN-\langle O \rangle(SO_3^-)-CH_2\,CH_2-\langle O \rangle(SO_3^-)-NCS$$

The binding of the H_2DIDS molecule to the band 3 protein shows a 1:1 stoichiometry (Lepke, Fasold, Pring and Passow, 1976; Ship, Shami, Breuer and Rothstein, 1977). It takes place in two steps. The first is reversible. It is complete in less than one second and associated with inhibition. This binding seems to be competitive with the substrate Cl^- (Shami, Rothstein, Knauf and McCulloch, 1978; Fröhlich, 1982), suggesting that substrate and H_2DIDS bind to overlapping sites, or that substrate and H_2DIDS bind to sites that are allosterically linked to each other. After reversible binding, slow covalent bond formation takes place: the isothiocyanate groups react with lysine residues to form thiourea bonds:

$$R_1\text{-NCS} + H_2N\text{-}R_2 \longrightarrow R_1\text{-NH-CS-NH-}R_2.$$

The two isothiocyanate groups of H_2DIDS can react with two lysine residues called lys _a_ and lys _b_, which reside on the 60,000 Dalton and the 35,000 Dalton chymotryptic fragments, respectively. They thus form an intramolecular cross-link. This was shown by

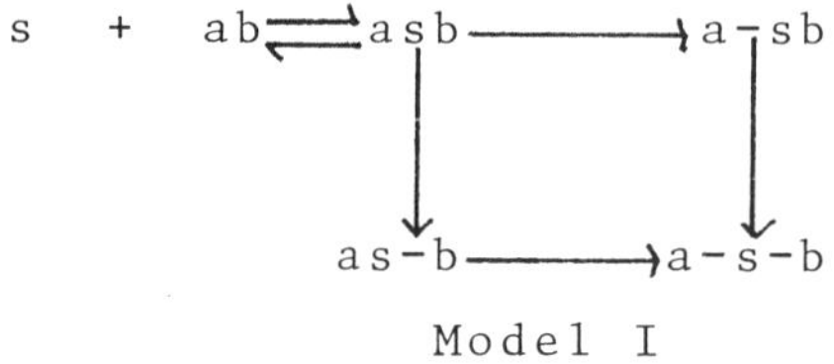

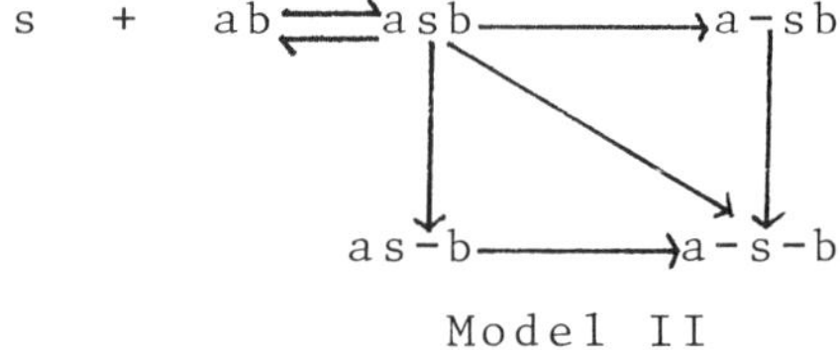

In Model II the transition asb⟶a-s-b would probably require an intermediate transition state.

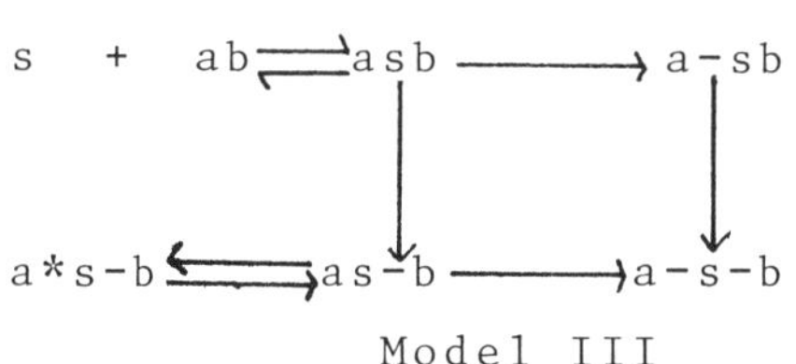

Figure 3

Model of the simplest possible cross-linking reaction (Model I) and of 2 extensions of the model (Models II and III) that yield better fits to the data: a = lysine residue a on the 60,000 Dalton segment; b = lysine residue b on the 35,000 Dalton segment; s = H_2DIDS. The equations describing the time course of the covalent reaction of H_2DIDS with a and b have been published by Kampmann, Lepke, Fritzsch, Fasold and Passow, 1982. asb represents non-covalent H_2DIDS binding in the immediate vicinity of a and b. The hyphens indicate covalent bond formation.
a-sb, as-b and a-s-b refer to covalent H_2DIDS binding to lys a, lys b or both lys a and lys b, respectively. In a*s-b, lys a is non-reactive;

Jennings and Passow, (1979) as follows (Fig. 2):

Red cells were exposed to chymotrypsin until all band 3 molecules were split into 60,000 and 35,000 Dalton fragments. Subsequently, H_2DIDS was added. At low pH, the H_2DIDS reacts predominantly with one of its SCN groups with lys a. When the pH is increased, the other isothiocyanate group reacts with lys b. The two chymotryptic fragments are joined together, and band 3 reappears on SDS polyacrylamide gel electropherograms (Fig. 2).

We have now followed the kinetics of the cross-linking reaction in more detail. Fig. 3 (Model I) shows that the H_2DIDS molecule (denoted by s) that is non-covalently bound to band 3 in the vicinity of lys a and lys b (denoted by asb) has two options: either to react covalently first with lys a or with lys b. Thus, as time goes on, the numbers of 60,000 Dalton and 35,000 Dalton segments that carry a covalently bound H_2DIDS molecule (i.e. a-sb and as-b, respectively; the hyphens indicate the covalent bonds) should increase. Since those H_2DIDS molecules that first reacted with lys a will further react with lys b, and those that reacted first with lys b will further react with lys a, the number of cross-linked molecules a-s-b will increase at the expense of a-sb and as-b. These latter forms will pass through a maximum, while the cross-linked form will increase monotonically, until all fragments will be cross-linked. These predictions can be formulated mathematically and tested experimentally (Passow, Fasold, Jennings and Lepke, 1982). Fig. 4 shows that the predictions can be verified qualitatively. H_2DIDS binding to the 35,000 Dalton and 60,000 Dalton fragments pass in fact through maxima and the formation of the reconstituted 95,000 Dalton peptide continues monotonically. When one tries, however, to fit the data to the equations that describe the expected behaviour quantitatively, deviations become apparent (Fig. 4).

Amongst a number of extensions of the reaction scheme that we thought could account for these deviations of the cross-linking reaction from Model I, we found two that can be fitted reasonably well to the data (Kampmann, Lepke, Fritzsch, Fasold and Passow, 1982). The first (Model II in Fig. 3) postulates that in addition to the two reaction pathways: (1) asb $\longrightarrow$ a-sb $\longrightarrow$ a-s-b and (2) asb $\longrightarrow$ as-b $\longrightarrow$ a-s-b there exists a third pathway where a simultaneous reaction between lys a, lys b and the reversibly bound s takes place: (3) asb $\longrightarrow$ a-s-b. The other extension (Model III in Fig. 3) stipulates that the lysine residue a may exist in two states: one in which it is reactive (a) and another in which it is non-reactive (a*) and that a slow transition between the two states affects the rate of the reaction.

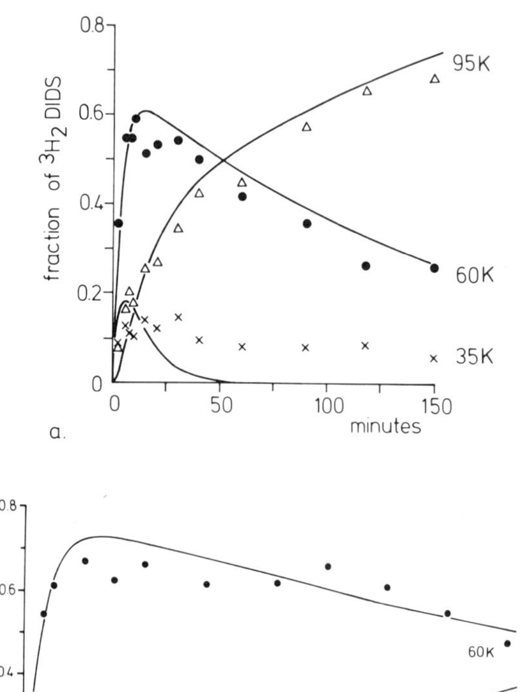

Figure 4

Time course of cross-linking reaction at pH 8.01 (Fig. 4a) or pH 7.36 (Fig. 4b), 37°C, in the presence of a large excess of H_2DIDS in the supernatant (i.e. under conditions where the equilibrium s+ab $\rightleftharpoons$ asb is far to the right: see the reaction schemes in Fig. 3). The curves represent non-linear least squares computer fits to Model I (from Kampmann et al., 1982).

$$asb \longrightarrow as\text{-}b \longrightarrow a\text{-}s\text{-}b$$
$$as\text{-}b \rightleftarrows a^{*}s\text{-}b$$

Evaluation of the data in terms of both Model II and Model III yielded good fits with nearly identical sums for the least squares of the errors (Kampmann et al., 1982). Examples of the quality of the fits to Model III are represented in Fig. 5. This model is supported by independent observations about the reactivity of lys a with 1-fluoro-2,4-dinitrobenzene (N_2ph-F). These observations will be presented later.

Within the scheme of the cross-linking reaction, it is experimentally feasible to isolate one partial reaction: the reaction a-sb ⟶ a-s-b. This is due to the fact that at low pH H_2DIDS reacts much faster with lys a than with lys b. Therefore, after exposure of the red cells to H_2DIDS at low pH for a short time, most of the inhibitor is covalently bound in the form a-sb. After removal of the H_2DIDS in the medium and of the reversibly bound H_2DIDS by washing with bovine serum albumin, most of the H_2DIDS that remains attached to the membrane exists in the form a-sb. The transition of this form into a-s-b can now be followed at any desired pH without the superimposition of the other reactions described above. The isolated reaction a-sb ⟶ a-s-b is of the first order (Fig. 6). The observed rate constants agree quite well with the corresponding rate constants derived from the computer evaluation of the time course of reaction sequence as measured in the presence of an unlimited supply of H_2DIDS in the medium and as exemplified in Fig. 5. Interestingly enough, the agreement between these rate constants was independent of the specific model used for the evaluation (Table I). This indicates that one of the 2 branches of the reaction sequence is correctly described by the reactions: asb ⟶ a-sb ⟶ a-s-b, while the description of the other branch of the reaction sequence (asb ⟶ as-b ⟶ a-s-b) represents an oversimplification. Additional reactions such as those included in Models II and III in Fig. 3 seem to be involved.

The isothiocyanate groups of H_2DIDS can react only with the deprotonated form of the amino group of a lysine residue

$$-NH_2 + SCN\text{-}R \longrightarrow -NH\text{-}CS\text{-}NH\text{-}R,$$

but not with the protonated form NH_3^+. The pH dependence of the reaction rate should, therefore, vary parallel to changes of the dissociation equilibrium:

$$NH_3^+ \rightleftarrows NH_2 + H^+.$$

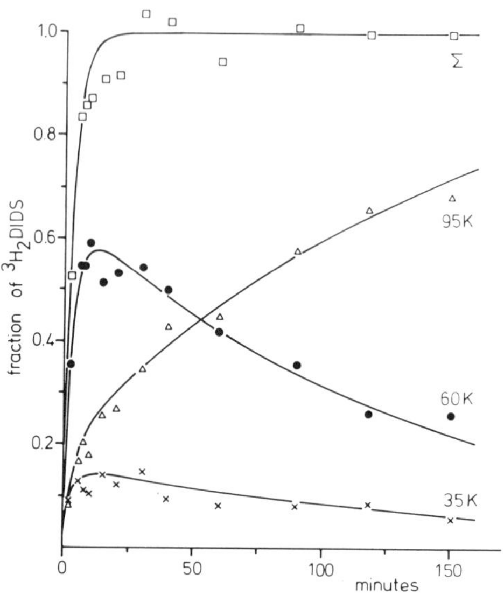

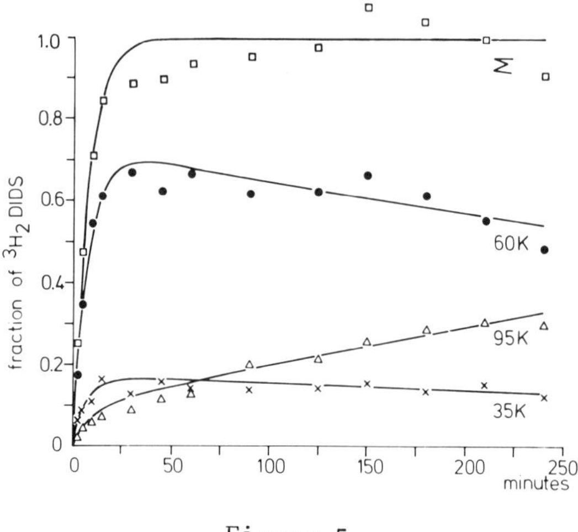

Figure 5

Same data as in Fig. 4. The fitted curves pertain to the model III represented in Fig. 3. Fits to Model II in Fig. 3 are virtually indistinguishable from those to Model III (from Kampmann, et al., 1982).

TABLE I

pH	Rate Constants for the Transition a-sb ⟶ a-s-b			Direct Estimate
	Model I	Model II	Model III	
7.25	0.0018	0.0012	0.0012	0.0014
7.36	0.0018	0.0013	0.0013	0.0017
7.40	0.0027	0.0029	0.0025	0.0019
7.90	0.0082	0.0065	0.0065	0.0060
8.01	0.0063	0.0071	0.0072	0.0077
8.40	0.020	0.022	0.0215	0.0187
8.70	0.036	0.034	0.038	0.036
8.76	0.037	0.039	0.039	0.041

Rate constants calculated for the transition a-sb ⟶ a-s-b, in min^{-1}. The figures in the first 3 columns were obtained by computer evaluations of experiments of the type represented if Fig. 5, in which the time course of H_2DIDS binding was followed while the cells were incubated in the presence of a large excess of H_2DIDS in the medium. The values in the last column were calculated, using a pK of 9.95 and a maximal rate of thiocyanylation of 0.678 min^{-1}. These parameter values had been derived from a non-linear least squares fit to experimentally determined data points of the isolated reaction a-sb ⟶ a-s-b, as demonstrated in Fig. 4b of the paper of Kampmann et al., 1982. All data are based on work with intact red cells.

Thus, the pH dependence of the rate constants of the reaction a-sb ⟶ a-s-b should be governed by the pK value of the dissociation equilibrium for lys *b*.

The rate constants measured at the various pH values can be fitted by a non-linear least squares method to an S-shaped curve that pertains to a single pK value of 10.0 (Fig. 7). This is close to the expected value for the ε amino group of a lysine residue at high dielectric constant in an electrically neutral environment (pK ≈10.5, Tanford, 1962).

Thus the work on the kinetics of covalent H_2DIDS binding has provided two results:

(1) The time course can be predicted only approximately by the simplest possible cross-linking reaction (Fig. 3, Model I). One way of obtaining better predictions consists of stipulating that one of the two lysine residues, lys *a*, may exist in two states: reactive and non-reactive; and that the rate limiting step for the establishment of the cross-link is the conversion of the non-reactive form of lys *a* into the reactive form (Fig. 3, Model III). (2) The pK value for lys *b* on the 35K segment is about 10.0.

It is useful to discuss the implications of these results with respect to the structure of the H_2DIDS binding site using Fig. 8. This figure shows an H_2DIDS molecule that has established thiourea bonds between lys *a* and lys *b*, thereby bridging a distance of about 20 Å.

We shall first consider the finding that the pK of lys *b* was close to 10.0. Although this value agrees with the pK value of a lysine residue in an electrically neutral environment, we are not entitled to believe that the dissociating amino group is indeed located in such an environment. The reason is the following: The H_2DIDS molecule possesses two SO_3^- groups that are introduced into the immediate vicinity of lys *b*. These negatively charged groups should shift the pK value of lys *b* to values far above the observed value of 10. Since this is not the case, we may suspect that the two negative charges of H_2DIDS are compensated for by at least two positive charges near lys *b*. Recent work in our laboratory (Zaki, 1981; Zaki, 1982) and in Copenhagen (Wieth, Bjerrum, Borders, 1982; Wieth and Bjerrum, 1982) suggests that arginine residues are involved in anion binding and transport by band 3. The charged groups that we postulate to exist near lys *a* and lys *b* may be identical with these arginine residues. Thus, within a distance similar to the size of an H_2DIDS molecule there seems to be located a cluster of at least four, perhaps even five, charged groups, two lysine residues and two or three additional residues that are possibly arginine residues. At least the lysine residues reside on dif-

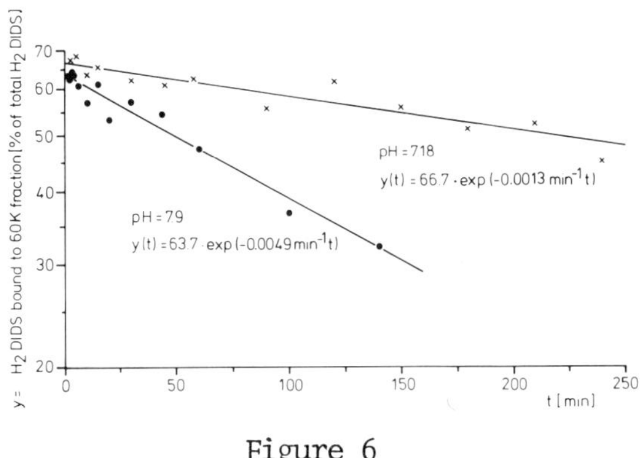

Figure 6

Semilog plot of the time course of the isolated reaction a-sub ⟶a-s-b. The sigure shows the decrease with time of the number of H_2DIDS labeled chymotryptic 60,000 Dalton fragments (a-sb). The straight lines represent non-linear least squares computer fits to the data, from which the numerical values were derived, that are indicated in the equations in the figure (from Kampmann et al., 1982).

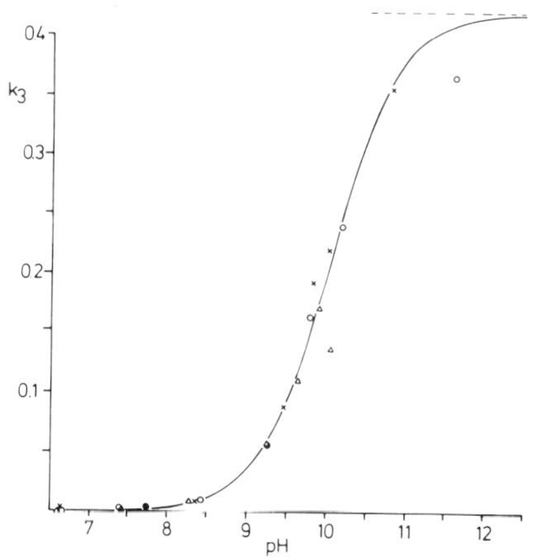

Figure 7

Rate constant k_3 for the transition a-sb⟶a-s-b as a function of pH. A drawn line is the result of a non-linear curve-fitting procedure assuming a single pK value of 10.0. The calculated pK value does not change significantly when the two points at the hightest pH value are omitted in the computer fitting procedure. Experiments with resealed red cell ghosts (from Kampmann et al., 1982).

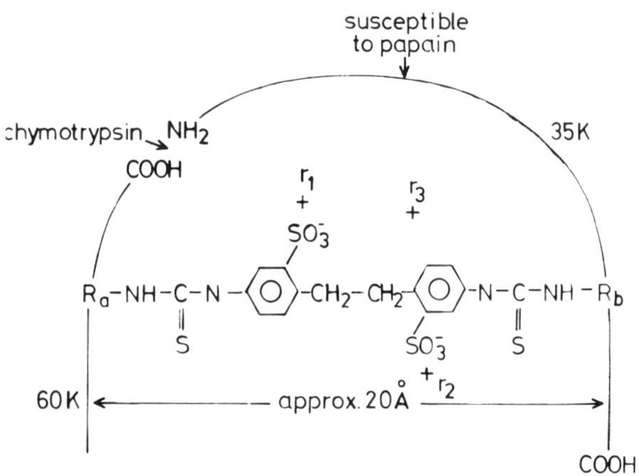

Figure 8

Schematic representation of the H_2DIDS binding site. r_1, r_2 and r_3 represent positively charged groups, possibly arginine residues whose existence has been postulated on the basis of the discussion presented in the text (from Passow et al., 1982).

ferent segments of the peptide chains, but in close juxtaposition.

We are now turning to lys a. To make our kinetic data fit the model, we introduced the assumption that the H_2DIDS binding site may exist in two states with different reactivities of lys a. Such assumption appeared plausible to us for the following reasons:

As has already been mentioned above, transport involves anion binding to the substrate binding site at one surface, conformational changes of the complex between substrate site and anion, and a release of the anion at the other surface. Thus, the anion transport protein should be able to exist in at least two different conformational states, depending on whether the substrate binding site faces inward or outward. Since the amino acid residues involved in the covalent binding of H_2DIDS should be allosterically linked to the substrate binding site, one could expect that they change their reactivity in response to changes of the occupancy and location of the substrate binding site. Limitations of time do not permit me to provide a detailed description of the experimental evidence that supports this idea. However, the following pieces of evidence may lend some plausibility to this stipulation.

Lys a is capable of reacting with another amino reagent, 1-fluoro-2,4-dinitrobenzene (N_2ph-F), which has only one reactive group, and hence cannot form a cross-link. This reagent inhibits both anion transport and the capacity of band 3 to combine covalently with H_2DIDS (Passow, 1979). The effect on transport is roughly proportional to the effect on covalent H_2DIDS binding, suggesting a stoichiometrical relationship (Fig. 9).

Although the band 3 molecule contains many amino acid residues that, in principle at least, should be capable of reacting with N_2ph-F (e.g. 28 lysine residues, see Steck, Koziarz, Singh, Reddy and Köhler, 1978), at 70-75% inhibition of anion transport, no more than about four N_2ph-F residues are bound per band 3 molecule. Of these residues, only one resides on the so-called 17,000 Dalton fragment, which represents the transmembrane portion of the 60,000 Dalton chymotryptic fragment. Another is found on the 35,000 Dalton fragment (Fig. 10a). When the protein is treated with H_2DIDS prior to its exposure to N_2ph-F, N_2ph-F binding is only reduced on the 17,000 Dalton fragment, but not on the 35,000 Dalton fragment (Fig. 10b). Conversely, N_2ph-F binding to the band 3 protein prior to its exposure to H_2DIDS, prevents the binding of the latter (not shown). Thus, on the 17,000 Dalton fragment, there is only one amino acid residue that, under conditions where transport is inhibited, combines with either N_2ph-F or H_2DIDS. Amino acid analysis of the ^{14}C N_2ph-F-labeled, isolated 17,000 Dalton segment showed that all of the N_2ph-F is bound to a lysine, just like the H_2DIDS (Lepke and Passow, 1982). We concluded, therefore, that binding of H_2DIDS and N_2ph-F take place at the same lysine residue (or on allosterically-

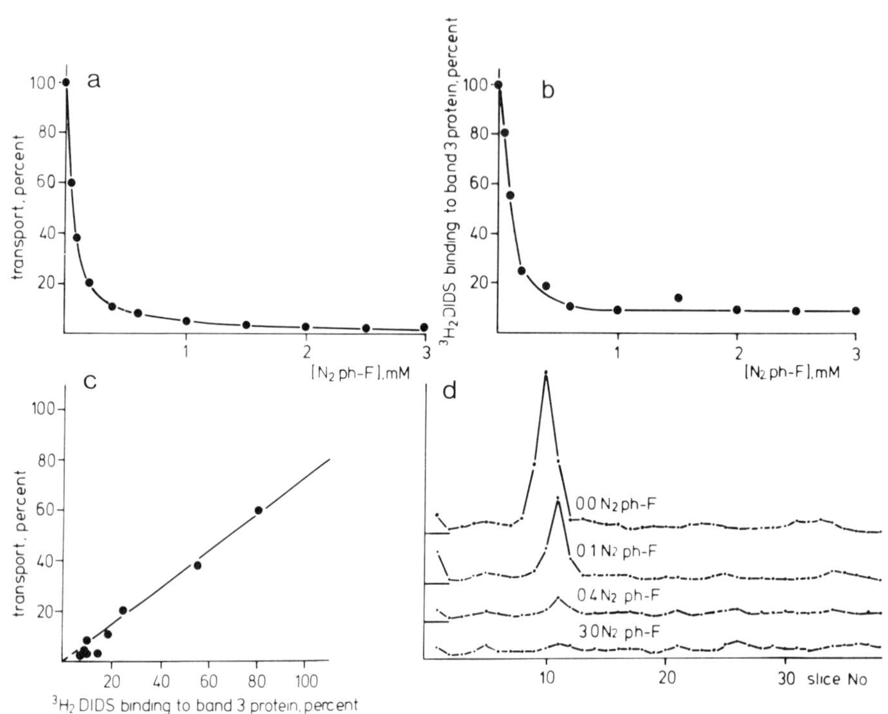

Figure 9

Effect of dinitrophenylation of the red blood cell membrane on anion transport (9a) and covalent H_2DIDS binding to band 3 (9b). Reduction of transport is proportionate to the reduction of the capacity of the dinitrophenylated cells to bind H_2DIDS (9c). The labeling profiles of SDS polyacrylamide gel electropherograms, from which the effects of H_2DIDS binding were inferred, are represented in Fig. 9d. Red cell ghosts had been exposed to the N_2ph-F concentrations indicated in the figure for 0.5 hrs. at 37°C, pH 7.4. After the removal of excess N_2ph-F, each batch of cells was divided into two; one was used for the determination of covalent $^3H_2$DIDS binding, the other for the measurement of sulfate equilibrium exchange (pH 7.4, 37°C, from Passow, 1979). Non-covalent binding of H_2DIDS that precedes the covalent bond formation is much less affected by dinitrophenylation than the subsequent covalent binding. This indicates that lys a plays at best a minor role in the non-covalent H_2DIDS binding (Passow, 1979; Passow et al., 1980a).

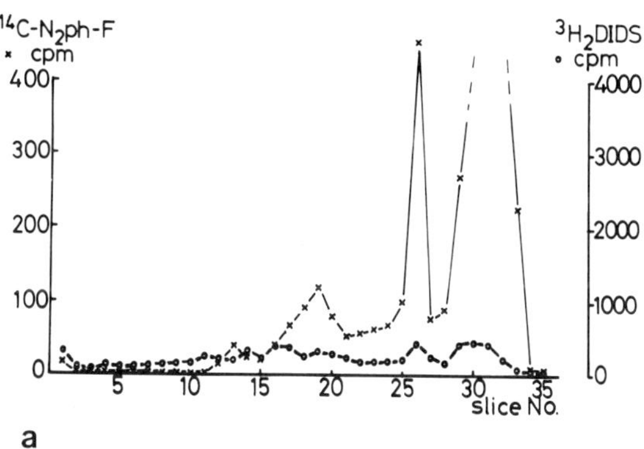

(a) labeling profiles of red cell membranes that were obtained from resealed ghosts after treatment with external chymotrypsin and internal trypsin and subsequent stripping with PCMBS. The labeling was performed at a N_2ph-F concentration of 175 µM (30 min. at pH 7.0, 37°C), which produces 71-75% inhibition of anion transport. o represents labeling with H_2DIDS that was performed after treatment with N_2ph-F. x represents binding of ^{14}C N_2ph-F prior to exposure to $^3H_2$DIDS. (b) same as Fig. 10(a), but H_2DIDS was added prior to ^{14}C and N_2ph-F (from Lepke and Passow, (1982).

Experimental procedure: Prepare resealed ghosts. Treat with external chymotrypsis to split band 3 into 60,000 and 35,000 Dalton segments, which leaves transport unaltered. Expose to N_2ph-F and H_2DIDS, as indicated under (a) and (b). Disrupt the membrane by washing with saponin-containing medium. Treat with trypsin to remove the 42,000 Dalton fragment from the chymotryptic 60,000 Dalton fragment and to remove other membrane proteins. Strip with PCMBS. Dissolve in SDS for electrophoresis.

Figure 10

linked lysine residues) on the 17,000 Dalton fragment of the band 3 protein, i.e. on lys a. Thus, we are entitled to believe that the effects of N_2ph-F are due to a reaction with lys a.

If it were true that lys a is allosterically coupled to the substrate binding site, then we would expect that saturation of that site with a substrate, e.g. chloride, should change the accessibility of lys a. This should influence the rate of dinitrophenylation. We observe indeed that saturating the transport system with Cl^- enhances both the rate of dinitrophenylation of lys a and the inhibition of anion transport by dinitrophenylation (Fig. 11). Thus, we conclude that lys a may indeed exist in two states, accessible and inaccessible for a reaction with N_2ph-F and that the ratio between the two states is a function of anion binding. When Cl^- is bound, lys a is reactive, when no Cl^- is bound, it is non-reactive. It may be added in passing that the inhibition of anion transport by arginine-selective reagents is reduced when the exposure to the agents is performed in the presence of Cl^- (Zaki, 1982). Thus, when lys a becomes more reactive, an anion transport-controlling arginine residue becomes less reactive.

To sum up the experimental findings presented today, one may state that the H_2DIDS binding site consists of a cluster of positively charged groups: two amino groups, and two or three as yet unidentified groups, possibly arginine residues. At least one of the amino groups changes its reactivity in response to substrate binding.

These observations may be integrated into a very speculative picture of the mechanism of the band 3 mediated transport (Figs. 12 and 13. Passow, 1982; Passow, Kampmann, Fasold, Jennings and Lepke, 1980):

We assume that the anions migrate between more or less parallel segments of the peptide chain. These segments traverse the lipid bilayer. Perhaps they form helices in which the hydrophobic amino acid residues in consecutive windings are lying on top of one another and face outwards towards the lipids; the hydrophilic amino acid residues may also lie on top of each other, but facing inward. Thus, three or more segments of the peptide chain could form a hydrophilic channel, with a hydrophobic surface that faces the lipids (Fig. 12). Since the transport does not contribute to the conductance, one would need to postulate the existence of a gate that combines with the substrate. This gate should reside near the outer surface, since the hydrophilic H_2DIDS does not penetrate (Cabantchik and Rothstein, 1974). Lys a, lys b, r_1 and r_2 may be arranged as suggested in Figs. 12 and 13. When an anion approaches, it will electrostatically attract r_1 and r_2. This leads to the formation of a complex. Under these conditions the reactivities of r_1 and r_2 with arginine-selective reagents will be reduced, the reactivity of lys b will be enhanced. A movement can occur only

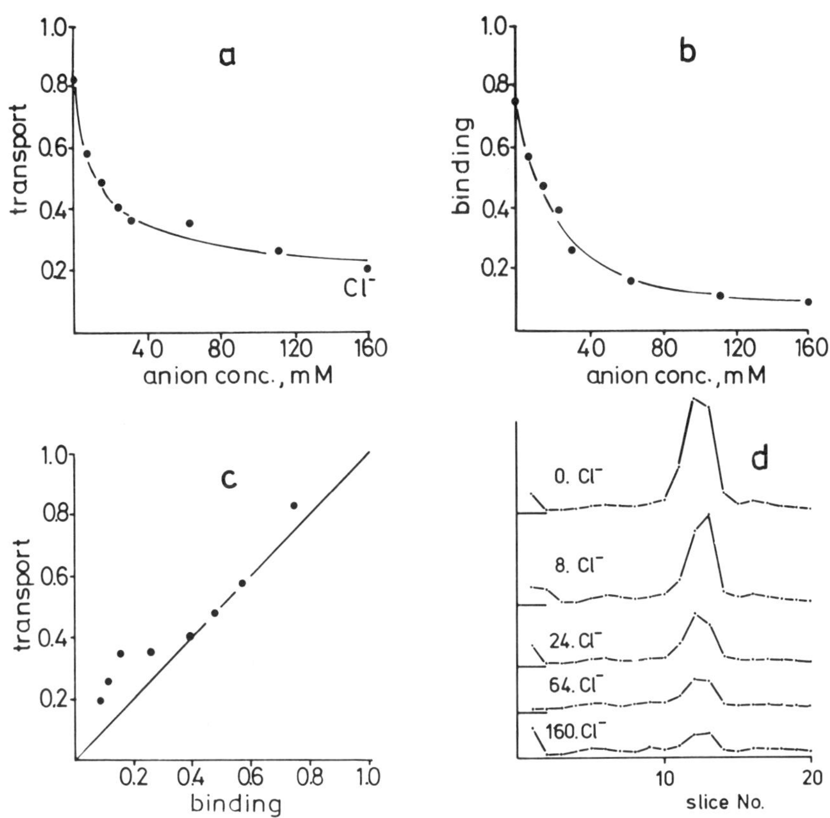

Figure 11

Effect of chloride on dinitrophenylation of the transport protein at a constant concentration of N_2ph-F of 125 μM. (a) Effect on anion transport; (b) effect on capacity of band 3 to bind H_2DIDS; (c) proportionality of effects on transport and H_2DIDS binding; (d) $^3H_2$DIDS labeling profiles of SDS polyacrylamide electropherograms, on which the data in panel b are based. Red cell ghosts had been exposed to a fixed concentration of N_2ph-F for 30 minutes at the chloride concentrations indicated in the figure, pH 7.4, 37°C. After removal of excess N_2ph-F, each batch was subdivided into 2, and $^3H_2$DIDS binding and sulfate equilibrium exchange were measured (from Passow et al, 1980).

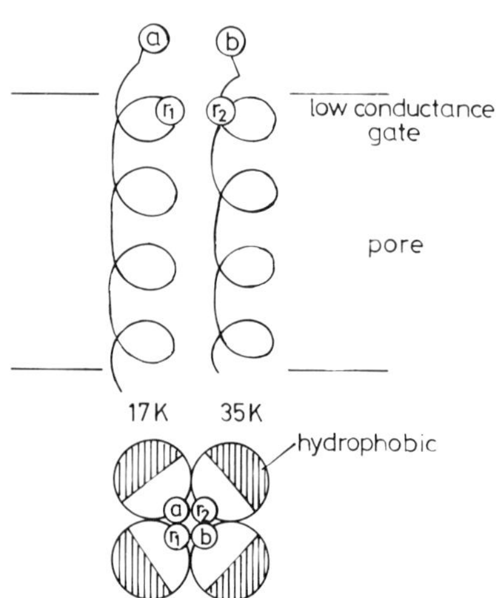

Figure 12

Hypothetical structure of the anion channel. An anion gate is located at the external opening of a channel that is formed by several different intramembrane segments of the band 3 protein. Shaded aread in the cross sectional view represent the location of predominantly hydrophobic amino acid residues. The areas that are not shaded represent the location of predominantly hydrophilic amino acid residues. 17K and 35K refer to 2 different segments of the peptide chain. The former carries lys a (denoted a), the latter lys b (denoted b). The anion binding amino acid residues r_1 and r_2 may be identical with r_1 and r_2 in Fig. 8, but evidence to this effect is still inadequate.

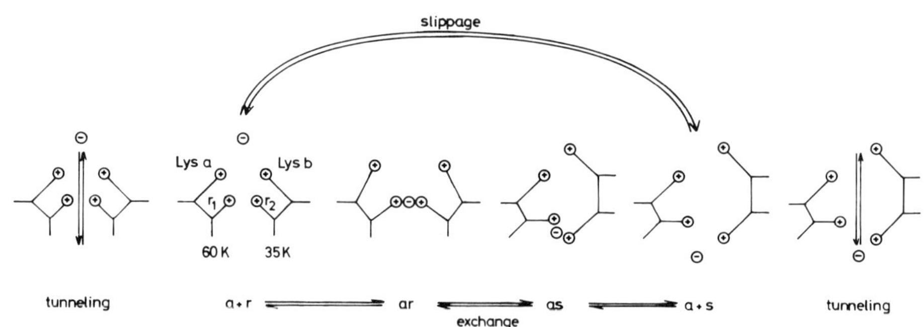

Figure 13

The low conductance gate (from Passow et al., 1980b).

when the gate plus the enclosed anion are moved together by a conformational change. When this change is accomplished, the anion may be released. The gate is now ready to accept an anion for the journey in the opposite direction. The rearrangement of the protein without an entrapped anion ("slippage") seems to be a rare event . Possibly, it is energetically difficult for the two positively charged groups, r_1 and r_2 to reorient themselves without a negatively charged anion present. If this idea were correct, one could say that the anion catalyses the cis-trans conformational change of the transport protein (Fig. 13).

I should like to add that the model also explains, apparently, a recently observed H_2DIDS-sensitive flow component that contributes to the conductance of the membrane (Kaplan, Pring and Passow, 1980; Knauf and Law, 1980). Occasionally, an anion may pass through the gate before the gate finds time to lock the anion in ("tunnelling", a term suggested by Gunn and Fröhlich, personal communication). On these occasions, the anion transport protein acts like an ordinary pore; but as has been emphasised above, this is a very rare event.

It is evident that the model described above is based more on imagination than fact. Especially the nature of the complex-forming process at the gate is still debatable. Instead of the mechanism described above, it has been suggested, for example, that the gate consists of salt bridges between COO^- and arginine residues that are opened by the approaching anion (Wieth, Bjerrum, Brahm and Andersen, 1982; Wieth et al, 1982; Cantley and Macara, 1982). Nevertheless, whatever the final truth, many people in our field would agree that transport takes place by diffusion through a channel that is formed by several more or less parallel segments of the peptide chain and that the diffusion process is interrupted by anion binding to a gate and subsequent conformational changes of the complex between anion and gate. The more specific model presented above may add some color to the general statements. In addition, it may help to visualise the various flux components of anion transport and of a number of related results that affect the transport process.

REFERENCES

Cabantchik, Z.I., Volsky, D.J., Ginsburg, H., Loyter, A. (1980) Ann. New York Acad. Sci. 341: 444-454

Cabantchik, Z.I., Knauf, P.A., Rothstein, A. (1978) Biochim. Biophys. Acta 515: 239-302

Cabantchik, Z.I., Rothstein, A. (1974) J. Membrane Biol. 15: 207-226

Cantley, L.C. and Macara, I.G. (1982) submitted for publication

Dorst, H.-J. and Schubert, D. (1979) Hoppe Seylers Z. Physiol. Chem. 360: 1605-1618

Drickamer, K. (1980) Ann. New York Acad. Sci. 341: 419-432
Fröhlich, O. (1982), J. Membrane Biol. 65: 111-123
Grinstein, S., Ship, S., Rothstein, A. (1978) Biochim. Biophys. Acta, 507: 294-304
Harris, E.J. and Pressman, B.C. (1967), Nature 216: 918-919
Herbst, F. and Rudloff, V. (1982) "Protides of the Biological Fluids," 113-116, Pergamon Press, Oxford, H. Peters, ed.
Hunter, M.J. (1971) J. Physiol. 218: 49P-50P
Jenkins, R.E. and Tanner, M.J.A. (1977) Biochem. J. 161: 139-147
Jennings, M.L. and Passow, H. (1979) Biochim. Biophys. Acta 554: 498-519
Jennings, M.L. (1982) J. Biol. Chem. 257: 7554-7559
Kampmann, L., Lepke, S., Fritzsch, G., Fasold, H., Passow, H. (1982) J. Membrane Biol. In the press.
Kaplan, J.H., Pring, M., Passow, H. (1980) in: "Membrane Transport in Erythrocytes," Alfred Benzon Symp. 14: 494-497 Munksgaard, Copenhagen, U.V. Lassen, H.H. Ussing, J.O. Wieth, eds.
Knauf, P.A. (1979) Current Topics in Membrane Transport, 12: 248-263
Knauf, P.A. (1982) in "Membranes and Transport," Vol. 2, A.N. Martonosi, ed. Plenum Press, New York and London, pp.441-449
Knauf, P.A., Law, F.Y. (1980) in: "Membrane Transport in Erythrocytes," Alfred Benzon Symp. 14: 488-493, U.V. Lassen H.H. Ussing, J. O. Wieth, eds., Munksgaard
Köhne, W., Haest,C.W.M. and Deuticke, B. (1981) Biochim. Biophys. Acta, 664: 108-120
Lepke, S. and Passow, H. (1976) Biochim. Biophys. Acta, 455: 353-370
Lepke, S., Fasold, H., Pring, M., Passow, H. (1976), J. Membrane Biol., 29: 147-177
Lepke, S. and Passow, H. (1982) to be submitted for publication.
Macara, I.G. and Cantley, L.C. (1981) Biochemistry, 20: 5095-5105
Margaritis, L.H., Elgsaeter, A. and Branton, D. (1977), J. Cell Biol., 72: 47-56
Mawby, W.J. and Findlay, (1982) Biochem. J. 205: 465-475
Pappert, G. and Schubert, D. (1982) in: "Protides of the Biological Fluids," H. Peeters, ed., 29: 117-120
Passow, H., Fasold, H., Gärtner, E.M., Legrum, B., Ruffing, W., Zaki, L. (1980a) Ann. New York Acad. Sci., 341: 361-383
Passow, H. (1982) in: "Membranes and Transport," Vol. 2, A.N. Martonosi, ed. Plenum Press, New York and London, pp. 451-460
Passow, H., Fasold, H., Jennings, M.L., Lepke, S. (1982) in: "Chloride Transport in Biological Membranes," J.A. Zadunaisky, ed. Academic Press, New York, pp. 1-31.

Passow, H. (1979) in: "Cell Membrane Receptors for Drugs and Hormones," R.W. Straub and L. Bolis, eds. Raven Press, New York, pp. 203-218

Passow, H., Fasold, H., Zaki, L., Schuhmann, b., Lepke, S. (1975) in: "Biomembranes: Structure and Function," G. Gárdos, I. Szász, eds. pp. 197-214, North Holland, Amsterdam and the Publishing House of the Hungarian Academy of Sciences.

Passow, H., Fasold, H., Lepke, S., Pring, H. and Schumann, B. (1977) in: "Membrane Toxicity," M.W Miller and A. Shamoo, eds., pp. 353-377, Plenum Press, New York

Passow, H., Kampmann, L., Fasold, H., Jennings, M., Lepke, S. (1980b) in: "Membrane Transport in Erythrocytes," Alfred Benzon Symp. 14, 345-367, U.V. Lassen, H.H. Ussing, J.O. Wieth, eds. Munksgaard, Copenhagen. pp.345-367

Peters, R., Peters, J., Tews, K.H., Böhr, W. (1974) Biochim. Biophys. Acta 367: 282-294

Peters, R., (1981) Cell Biol. International Reports, 5: 733-760

Rao, A. and Reithmeier, A.F. (1979) J. Biol. Chem., 254: 6144-6150

Rothstein, A. (1982) in: "Membrane Transport," Vol. II, A. Martonosi, ed. 435-440, Plenum, New York and London

Rothstein, A., Ramjeesingh, Grinstein, S., Knauf, P.A. (1980) Ann. New York Acad. Sci., 341: 433-443

Scarpa, A., Cecchetto, A., Azzone, G.F. (1970) Biochim. Biophys. Acta 219: 179-188

Shami, Y., Rothstein, A., Knauf, P.A., McCulloch, L. (1978) Biochim. Biophys. Acta, 508: 357-363

Schubert, D. and Domning, B. (1978) Hoppe Seyler's J. Physiol. Chem., 359: 507-515

Ship, S., Shami, Y., Breuer, W., Rothstein, A. (1977) J. Membrane Biol., 33: 311-323

Steck, T.L. (1978) J. Supramol. Structure 8: 311-324

Steck, T.L. Ramos, B., Strapazon, E. (1976) Biochemistry, 14: 1154-1161,

Steck, T.L. Koziarz, J.J., Singh, M.K., Reddy, G., Köhler, H (1978) Biochemistry 27: 1216-1222

Tanford, C. (1962) Advances in Protein Chemistry, 17: 69-165, C.B. Anfinsen and J.T. Edsall, eds., Academic Press, New York

Tanner, M.J.A., Williams, D.G., Jenkins, R.E. (1980) Ann. N.Y. Acad. Sci., 341: 455-464

Wieth, J.O., Bjerrum, P.J. and Borders, C.L., (1982) J. Gen.Physiol., 79:283-312 Wieth, J.O. and Bjerrum, P.J. (1982) J. Gen. Physiol. 79: 253-282

Wieth, J.O., Bjerrum, P.J., Brahm, J. and Andersen, O.S. (1982). Tokai J. Exp. Clin. Med. In the press

Zaki, L. (1981) Biochim. Biophys. Res. Comm., 99: 243-251

Zaki, L. (1982) in: "Protides of the Biological Fluids, H. Peeters, ed. Pergamon Press, Oxford, pp. 279-282

WATER PERMEABILITY OF THE LIPID BILAYER:

EXPERIMENTAL STUDIES OF THE STRUCTURE OF SOME SIMPLE BILAYERS

Daniel C. Petersen

Physics Department
Harvey Mudd College
Claremont, California, U.S.A.

INTRODUCTION

The past few years I have been studying the water permeability of some simple lipid bilayers with the goal of finding out to what extent the lipid bilayer looks like an appropriately chosen bulk hydrocarbon liquid to the water molecules passing through (Petersen, 1980). Let me point out right off that the membranes I am studying have no special water channels--I am talking about water permeation through the lipid matrix.

The basic experimental approach is to (1) measure the water permeability of the bulk hydrocarbon liquid; (2) use the liquid hydrocarbon assumption to predict the water permeability of the lipid bilayer; (3) measure the water permeability of the lipid bilayer (or use the data of others); (4) compare the results of (2) and (3) and draw conclusions on the validity of the liquid hydrocarbon assumption.

The measurements on the lipid bilayer have been done on a planar lipid bilayer membrane (or black lipid membrane) which is formed on a hole in a hydrophobic septum. This type of bilayer requires some sort of "solvent" molecule in addition to the lipid of interest in order to satisfy the boundary conditions between the membrane and the hydrophobic support (White et al., 1976). The systems I have chosen to study consist of monoolein (glyceryl-1-monoolein) plus a solvent of either n-hexadecane or triolein. These systems were chosen because (1) the monoolein/n-hexadecane bilayer has been well-characterized: its thickness and the hexadecane solvent fraction (both needed for the predictions just mentioned) are well known as a function of temperature (White, 1976; White, 1977); its water permeability coefficient has already been measured for various temperatures by Fettiplace (1978); (2) capacitance measurements of the monoolein/triolein bilayer have shown that there is less than 3% by volume of triolein in the bilayer (Waldbillig and Szabo, 1979); and the hydrocarbon tails of monoolein and triolein are identical, so the presence of any triolein makes no difference in the predictions of the liquid hydrocarbon model. The bulk liquid hydrocarbon analogs in these two cases are (1) 8-heptadecene/n-hexadecane (in the correct volume fractions) and (2) 8-heptadecene, respectively.

EXPERIMENTAL APPROACH

The rate of water permeation through the bulk hydrocarbon liquid is given by Fick's first law of diffusion:

$$dm/dt = - DA\ dc/dx$$

where dm/dt is the mass transport rate of water, D is the diffusion coefficient of water in the hydrocarbon, A is the cross-sectional area of the hydrocarbon layer, and dc/dx is the concentration gradient of water in the hydrocarbon. The experiment is performed as shown in Fig. 1, following the procedure of Schatzberg (1965). Water is put into a carefully made hydrophobic cup and then a layer of known thickness of the appropriate hydrocarbon liquid is floated on top. This cup is put on the weighing pan of a recording microbalance which is in a zero-relative-humidity environment. The water slowly diffuses through the oil and evaporates; the rate of weight loss of the cup is dm/dt, which in this steady state situation is given by

$$dm/dt = -P_h A c_w^w$$

where P_h is the permeability coefficient for water of the hydrocarbon liquid and c_w^w is the concentration of water in the aqueous phase below the hydrocarbon layer (1.00 g/ml). Here the permeability coefficient P_h is given by

$$P_h = DK/d_h$$

where K is the partition coefficient of the water in the hydrocarbon and d_h is the thickness of the hydrocarbon layer.

Using the liquid hydrocarbon assumption one can then predict the water permeability coefficient P_m for the bilayer membrane of thickness d_m:

$$P_m = (d_h/d_m)P_h$$

since D and K are assumed to be the same in bulk and bilayer. The thickness d_m is obtained from capacitance measurements:

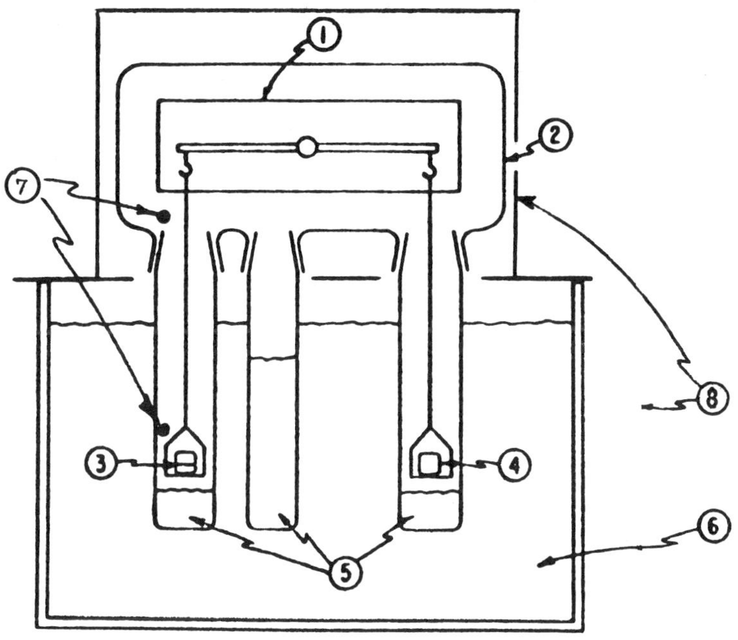

Fig. 1. Apparatus for the water permeability coefficient of hydrocarbon liquids. (1) recording microbalance; (2) closed glass system with hangdown tubes; (3) sample cup (hydrocarbon liquid over water); (4) tare cup (hydrocarbon liquid alone); (5) dessicant; (6) controlled-temperature bath; (7) thermistors; (8) insulating enclosures. Cups were precisely machined from stress-relieved polychlorotrifluoroethylene (KEL-F). The temperature at the cup was held constant to 0.02° C; the top of the hangdown tube was kept approximately 0.1° C warmer than the cup to minimize convection. (Adapted from Schatzberg (1965))

$$d_m = \varepsilon_0 \varepsilon / C_m$$

where ε_0 is the permittivity of the vacuum (8.854×10^{-12} F/m), ε is the dielectric coefficient of the membrane interior (2.1-2.2) (Requena and Haydon, 1975), and C_m is the capacitance per unit area of the membrane.

The water permeability of the lipid bilayer is obtained from the flow of water induced by an osmotic pressure difference across the membrane (review: Fettiplace and Haydon, 1980). The membrane is formed over a 1.5 mm diameter hole in a teflon septum, thereby producing a closed inner aqueous volume. The NaCl concentration in the outer chamber is then increased, leading to a flow of water out of the inner chamber and through the membrane. This causes the membrane to bow inward. This bowing is observed via a microscope (which is also used to measure the membrane area), and a plunger attached to a micrometer drive in the inner chamber is adjusted to keep the membrane flat. The micrometer reading then gives the water volume flux, and hence the osmotic permeability coefficient P_f:

$$J_v = -P_f \bar{V}_w \nu (g_2 c_2 - g_1 c_1)$$

where J_v is the volume flux of water through the bilayer, $\bar{V}_w$ is the partial molar volume of water, $\nu = 2$ when NaCl is used to establish the osmotic pressure difference across the bilayer, c_i is the concentration of NaCl on side i of the bilayer, and g_i is the rational osmotic coefficient associated with the concentration c_i. <u>If the liquid hydrocarbon model is correct, then $P_f = P_m$.</u>

RESULTS AND DISCUSSION

The results of the measurements on the <u>monoolein/hexadecane</u> system are shown in Fig. 2. The important thing to note here is that these two Arrhenius plots have significantly different slopes. That is, <u>the activation energy for the water permeation through the lipid bilayer is significantly greater than for water permeation through the bulk hydrocarbon liquid</u>. The liquid hydrocarbon model is not precisely correct. However in the case of the <u>monoolein/triolein</u> system (Fig. 3) the two slopes are essentially the same. <u>The activation energy predicted by the liquid hydrocarbon model agrees well with the measured value</u>. The numerical results (Table I) show that the two bulk hydrocarbon systems and the monoolein/triolein bilayer have essentially the same activation energy, whereas the monoolein/hexadecane bilayer (normalized to constant thickness) has an activation energy about 50% larger.

Before discussing these results further, I should comment on the discrepancy between the predicted and measured permeability coefficients for monoolein/triolein. The predicted value is 40% above the measured value. Why the discrepancy? Well, in the first place, since $d_h/d_m = 10^6$, a factor of 1.4 is not bad! Secondly, there may be an unstirred water layer near the membrane which reduces the osmotic gradient, hence the permeation rate. A simple calculation shows this would need to be 10 μm thick to account for the effect. That's possible. Or maybe there is some effect of an altered water structure near the membrane, or there could be headgroup contributions. In any case the error is small, and the good agreement with activation energy supports the liquid hydrocarbon model, at least in comparison with the monoolein/hexadecane system.

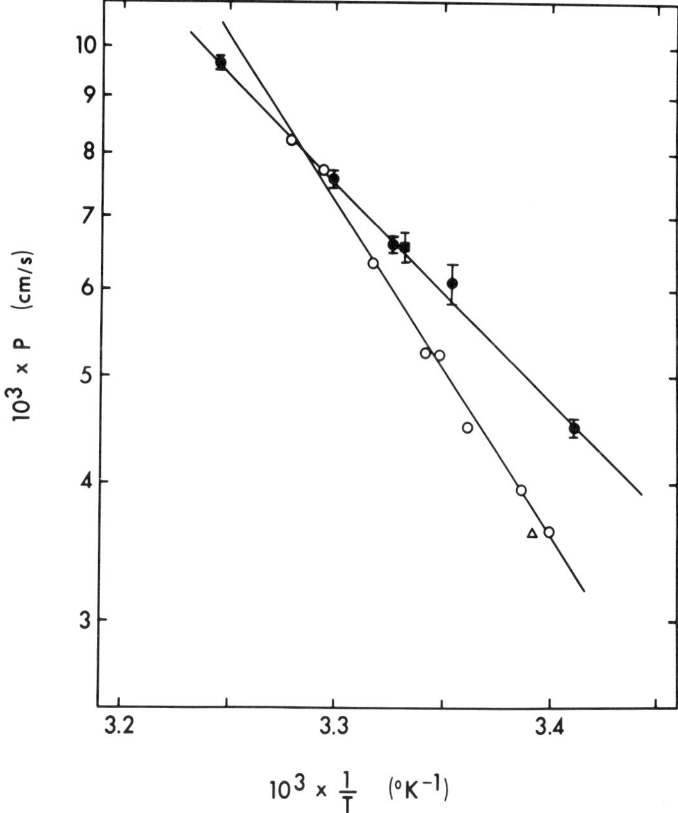

Fig. 2. Arrhenius plots of measured and predicted water permeability coefficients for monoolein/n-hexadecane bilayers. Open symbols: measured osmotic permeability coefficients; o, data from Fettiplace (1978); Δ, datum from Petersen (1980). Filled symbols: water permeability coefficients predicted by the liquid hydrocarbon model from measurements on bulk 8-heptadecene/n-hexadecane mixtures. ●, d_h = 2.0 mm; ■, d_h = 3.0 mm. Error bars ± 1 S.D. Volume fractions chosen for n-hexadecane in 8-heptadecene (increasing from 0.200 to 0.275 for temperatures from 20° to 35° C, respectively) were based on the measurements of White (1976), as were the values of the membrane thickness d_m needed for the predictions (d_m increased from 3.1 nm to 3.5 nm for 20° to 35° C, respectively) (Adapted from Petersen (1980)).

Why does monoolein/triolein seem to be quite well represented by the liquid hydrocarbon model whereas monoolein/hexadecane does not? One might at first glance assume that, since hexadecane is a

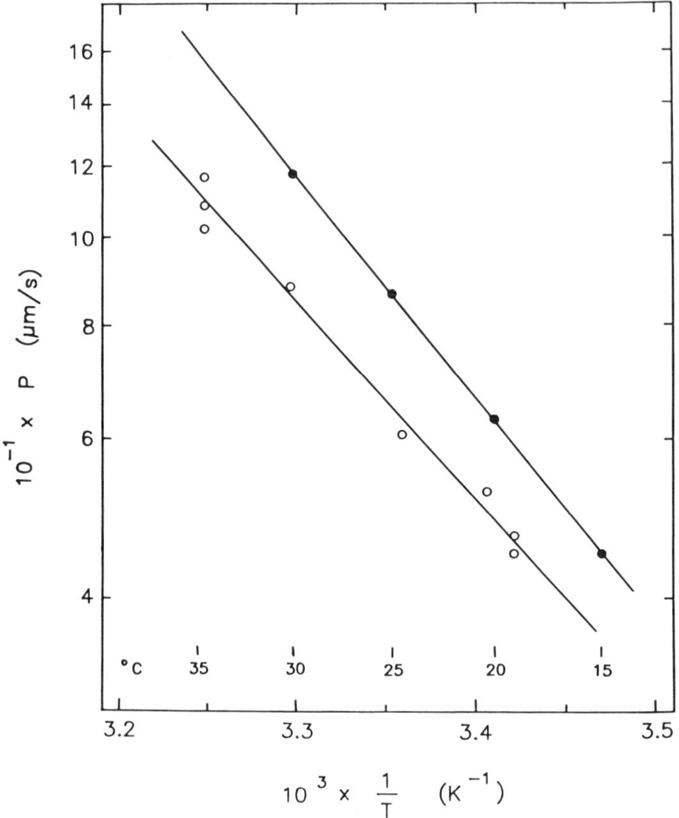

Fig. 3. Arrhenius plots of measured and predicted water permeability coefficients for monoolein/triolein bilayers. o, measured osmotic permeability coefficients; •, water permeability coefficients predicted by the liquid hydrocarbon model from the measured water permeability coefficients of bulk 8-heptadecene. Liquid hydrocarbon thickness d_h = 2.0 mm; bilayer thickness (from capacitance measurements) d_m = 2.45 nm.

pure hydrocarbon with no polar headgroup, the presence of hexadecane would make the membrane more like a bulk hydrocarbon liquid--but that is just the opposite of what is observed. The thermodynamics determining the molecular configuration must be more complex than that simple assumption. The molecular organization appears to be determined by several thermodynamical constraints. First of all, the volume of the monoolein acyl chain in the bilayer is fixed at essentially its bulk value, 0.475 nm^3 (Requena and Haydon, 1975; Nagle and Wilkinson, 1978). Secondly, the bilayer surface area per monoolein molecule remains at approximately 0.39 nm^2 for a wide range of alkane

Table I. Activation Energies $E^{\dagger}$ for Water Permeation

SYSTEM	$E^{\dagger}$(kcal/mole)
8-heptadecene	11.1 ± 0.2
8-heptadecene/n-hexadecane[a]	10.5 ± 0.3
monoolein/triolein	9.8 ± 0.7
monoolein/n-hexadecane[b]	15.6 ± 1.1

[a]Petersen (1980); [b]Fettiplace (1978), normalized to constant membrane thickness (see Fig. 2 caption). The activation energy is defined by $P = A\exp(E^{\dagger}/RT)$, where P is the water permeability coefficient, A is a constant, R is the universal gas constant and T is the absolute temperature.

solvents (Fettiplace et al., 1971). Finally, for small volume fractions of solvent (which is the case for both systems used here) the monoolein acyl chains extend on the average half way across the bilayer, while essentially never extending much beyond the midplane of the bilayer (White, 1977; Gruen and Haydon, 1981).

The combination of all these constraints means that the hexadecane molecules, which constitute approximately 25% of the monoolein/hexadecane bilayer volume (Requena et al., 1975; White, 1976), interdigitate between the monoolein acyl chains without increasing the surface area per monoolein molecule. This means that the acyl chains must be somewhat extended perpendicular to the membrane surface to accommodate the hexadecane. A measure of this extension can be obtained from the ratio of the average length $\langle L \rangle$ of the monoolein acyl chain normal to the membrane surface divided by the length L_o of the fully extended monoolein chain, 2.1-2.2 nm (Seelig and Seelig, 1977; White, 1977). The average length $\langle L \rangle$ is given, in turn, by $d_m/2$, where d_m comes from capacitance measurements. For the monoolein/hexadecane bilayer, d_m = 3.3 nm (Requena et al., 1975; White, 1976), giving $\langle L \rangle / L_o = 0.77$. In contrast, the monoolein/triolein bilayer thickness is only 2.45 nm (my measurement), giving $\langle L \rangle / L_o = 0.57$.

Thus the hexadecane acts to order the monoolein acyl chains perpendicular to the membrane surface. Comparison of the monoolein acyl chain for the monoolein/triolein and monoolein/hexadecane systems (Fig. 4) shows the marked increase in order induced by the presence of hexadecane. This greater order is reflected in the higher activation energy for water permeation in the monoolein/hexadecane bilayer. In the monoolein/triolein bilayer, on the other

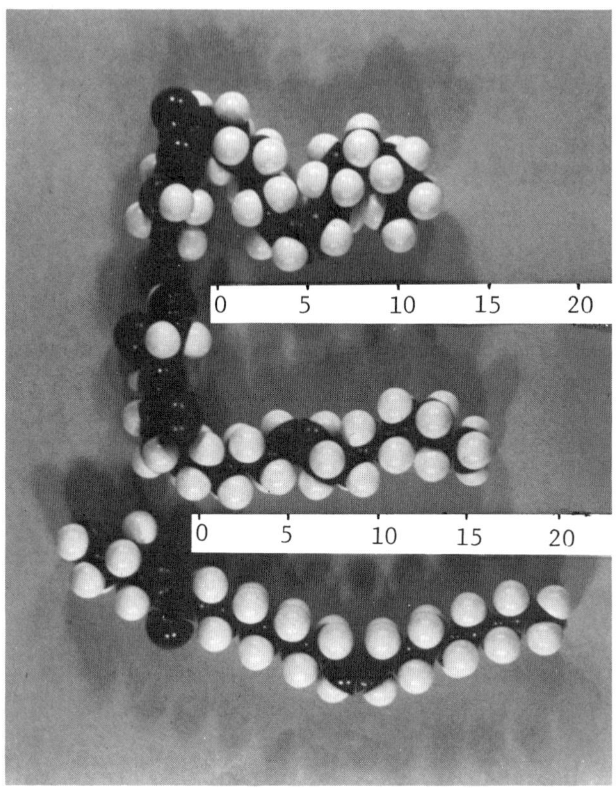

Fig. 4. Molecular models of monoolein depicting the average extension of the acyl chain in two lipid bilayer membranes. (Top), monoolein/triolein bilayer; (middle), monoolein/n-hexadecane bilayer; (bottom), monoolein in fully extended configuration (note cis double bond). Scales are in Angstrom units. Glycerol backbone is to the left in all cases. The long axis of the acyl chain is assumed to be perpendicular to the membrane surface. With the glycerol backbone oriented roughly parallel to the membrane surface (top two configurations) the surface area per monoolein molecule is approximately 0.39 nm^2 (see text). The acyl chain in the monoolein/triolein bilayer (top) fills in the space "below" the glycerol backbone, whereas in the monoolein/hexadecane bilayer a space remains, allowing room for an occasional hexadecane molecule. Note: While these models are consistent with the measured bilayer thicknesses and solvent concentrations, no precise information is yet available on actual acyl chain configurations. Many other configurations could satisfy these criteria also, but it is difficult to find configurations in which the acyl chain is <u>more</u> ordered in triolein than hexadecane.

hand, the acyl chains are considerably less ordered and similar enough to the bulk liquid hydrocarbon to give reasonable agreement between the measured water permeability coefficient and the prediction based on the liquid hydrocarbon assumption.

REFERENCES

Fettiplace, R., 1978, The influence of the lipid on the water permeability of artificial membranes, Biochim. Biophys. Acta, 513:1.

Fettiplace, R., Andrews, D. M., and Haydo , D. A., 1971, The thickness, composition and structure of some lipid bilayers and natural membranes, J. Membrane Biol., 5:277.

Fettiplace, R., and Haydon, D. A., 1980, Water permeability of lipid membranes, Physiol. Revs., 60:510.

Gruen, D. W. R., and Haydon, D. A., 1981, A mean-field model of the alkane-saturated lipid bilayer above its phase transition, II, Biophys. J., 33:167.

Nagle, J. F., and Wilkinson, D. A., 1978, Lecithin bilayers: density measurements and molecular interactions, Biophys. J., 23:159.

Petersen, D. C., 1980, Water permeation through the lipid bilayer membrane: test of the liquid hydrocarbon model, Biochim. Biophys. Acta, 600:666.

Requena, J., Billett, D. F., Haydon, D. A., 1975, Van der Waals forces in oil-water systems from the study of thin lipid films, I, Proc. R. Soc. Lond. A, 347:141.

Requena, J., and Haydon, D. A., 1975, Van der Waals forces in oil-water systems from the study of thin lipid films, II, Proc. R. Soc. Lond. A, 347:161.

Schatzberg, P., 1965, Diffusion of water through hydrocarbon liquids, J. Polymer Sci. C, 10:87.

Seelig, A., and Seelig, J., 1977, Effect of a single cis double bond on the structure of a phospholipid bilayer, Biochem., 16:45.

Waldbillig, R. C., and Szabo, G., 1979, Planar bilayer membranes from pure lipids, Biochim. Biophys. Acta, 557:295.

White, S. H., 1976, The lipid bilayer as a "solvent" for small hydrophobic molecules, Nature, 262:421.

White, S. H., 1977, Studies of the physical chemistry of planar bilayer membranes using high-precision measurements of specific capacitance, Ann. N. Y. Acad. Sci., 303:243.

White, S. H., Petersen, D. C., Simon, S., and Yafuso, M., 1976, Formation of planar bilayer membranes from lipid monolayers: a critique, Biophys. J., 16:481.

ACTIVE MONOVALENT CATION TRANSPORT IN CANINE CARDIAC TISSUES

H. M. Rhee

Oral Roberts University
School of Medicine
Tulsa, OK 74171, USA

INTRODUCTION

Previous investigations suggest that functionally different cardiac tissues have different electrochemical and physiological properties[1-3]. The intracellular ionic distribution of the sino-atrial node or other specialized cardiac conducting fibers differs from the ionic concentrations of contractile ventricular muscle[4-7]. Electrophysiological sensitivity to cardiac steroids such as ouabain in canine Purkinje fiber exceeds the sensitivity to ouabain in the ventricle[8]. K-strophanthin decreases conduction velocity in canine Purkinje fibers at a concentration which did not significantly alter the parameter in the ventricular muscle[9]. A greater sensitivity to ouabain in Purkinje fibers than the ventricular fibers was confirmed by other differences such as K^+ fluxes[10-12]. However, recent studies indicate that partially purified Na^+,K^+-ATPase prepared from the bovine false-tendon had no greater sensitivity to ouabain than the comparable Na^+,K^+-ATPase preparation from the papillary muscle of the same animal[13]. Although there may be no significant difference in ouabain sensitivity of Na^+,K^+-ATPase activity *in vitro*, the exact biochemical basis of such electrophysiological differences between the conducting fibers and the contractile muscle has not yet been documented.

Therefore, the purpose of this investigation was to examine the magnitude of the active transport of monovalent cations and Na^+,K^+-ATPase activity in functionally different cardiac Tissues. The possible differential sensitivity to ouabain in the active transport of $^{86}Rb^+$ between the Purkinje fiber and the ventricular muscle was also investigated.

METHOD

Experiments were conducted with healthy mongrel dogs of either sex, weighing between 15 to 25 kg body weight. A mid-line sternotomy was performed under pentobarbital (30 mg/kg i.v.) anesthesia and the heart was quickly excised and left beating in ice-chilled saline to wash out the blood in the chambers. In each heart seven different cardiac tissues were identified and removed for the study of cation active transport.

Tissue Identification and Isolation

The sino-atrial node (SAN) was identified by the sinus nodal artery, the first branch of the right coronary artery as described [14]. The SAN lies along the sulcus terminalis at the junction of the superior vena cava and the right atrium[15] in dogs. The atrio-ventricular node (AVN) is in the lower interatrial septum below the fossa ovalis, anterior to the ostium of the coronary sinus. Since it is impossible to dissect the SAN or the AVN in pure form, in the initial study the contamination of other tissues was routinely examined by a light microscope. The nodal tissues consist of many P fiber clusters which distinctly differ from the contractile elements in the atria.

Strands of free-running Purkinje fiber (false-tendon) from both the ventricles were easily identified. The papillary muscle of the right ventricle and a piece of the right atrium and the ascending aorta were also removed. A piece of the left ventricle was used after removal of the epicardial and the endocardial layers. The entire procedure was carried out at 4°C, and it usually took 5 to 7 minutes after removal of the heart as reported[14]. Upon isolation of these tissues, each tissue was immediately incubated in K^+-free cold Krebs-Henseleit (K-H) solution[16], pre-equilibrated with 95% O_2 and 5% CO_2 (pH 7.4) at 4°C. Tissue slices (0.5 mm thickness) were prepared by a McIlwain tissue chopper (Brinkman Instrument, Westbury, NY) after a brief trimming of fat and connective tissues.

Determination of $^{86}Rb^+$ Transport and Preparation of Na^+,K^+-ATPase

The modified method of Ku et al.[17] originally described by Bernstein and Israel[18] was used to determine the active transport of Rb^+ as described[19]. The detail procedures for the preparation of Na^+,K^+-ATPase were reported previously[20]. Either the ventricular muscle or the specialized nodal tissues (usually under 50 mg) was minced into small pieces, placed in 1 ml of 1 M KCl and homogenized with Polytron PT-10/ST (Brinkmann Instruments, Inc., Westbury, NY) for 30 sec at setting 4 of the instrument. The homogenate was passed through a double layer of cheese cloth and centrifuged at 1000 x g for 10 min. The sediment was suspended in 8 ml of a solution containing 50 mM KCl and 50 mM Tris-HCl (pH 7.4) and centrifuged again as indicated above. This washing procedure was repeated twice with 50 mM Tris-HCl (pH 7.4). The final pellet was suspended in 0.5 ml of 1 mM Tris-EDTA (pH 7.4) with the aid of a hand homogenizer and used as the source of enzyme. All the above procedures were done at 0-4°C.

3-O-Methylfluorescein Phosphate (MFPase) assay

Since it is not possible to measure Na^+,K^+-ATPase activity in the crude cardiac homogenate of dogs because of high Mg^{++}-dependent ATPase activity, K^+-activated partial reaction of Na^+,K^+-ATPase has been used. Utilizing a fluorescent artificial substrate of 3-O-methylfluorescein phosphate (MFP), hydrolysis of phosphate group from MFP was measured by a continuous recording of fluorescence emitted from the product of methylfluorescein in a colorimeter as reported [20]. That is, assays were conducted at 37°C in a mixture containing 20 μM 3-O-methylfluorescein phosphate, 4 mM $MgCl_2$, 1 mM EDTA, 10 mM KCl, 80 mM Tris-HCl (pH 7.4) and 50 μl of the KCl-enzyme in a final volume of 1 ml. The ligand concentrations used are optimal for enzyme activity at the indicated substrate concentration. The K^+-independent portion of the activity was determined either by the omission of K^+ from the mixture or by the addition of 1.5 mM ouabain to the complete mixture. Protein was assayed by Lowry procedure[21].

RESULTS

Tissue Isolation and Histological Studies

The dispersion of tissue weight of the specialized cardiac tissues obtained from seven dogs is summarized (Table 1). The SAN showed the least variability in tissue weight, 46.2 ± 3.2 (s.e.) mg per dog. This small variation is largely due to the fact that the SAN is easily identifiable along the sulcus terminalis and the superior vena cava. The Purkinje fiber (33.5 mg, wet weight) presents in most dogs the smallest amount among the three specialized tissues. But the Purkinje fiber from each dog did provide at least 3 to 4 determinations of Rb^+ uptake. Since the dimension of the AVN is bigger than the SAN[22], a bit more tissue was expected from the AVN.

$^{86}Rb^+$ Uptake Studies

The characteristics of Rb^+ uptake in the left ventricular tissue of dogs is shown in Fig. 1. Total Rb^+ uptake (O) assayed in the absence of ouabain was dependent upon time of incubation. Non-specific Rb^+ accumulation (▲) was assayed in the presence of 10^{-4} ouabain. The specific Rb^+ uptake (X) obtained from the difference

Table 1. Sampling of Specialized Canine Cardiac Tissues.

Tissue	N[a]	Tissue weight (mg, wet weight)	
		Range	Mean ± S.E.M.
Sino-atrial node (SAN)	7	34.4 - 55.4	46.2 ± 3.2
Atrio-ventricular node (AVN)	7	36.2 - 70.5	55.2 ± 5.1
Purkinje fiber	7	12.7 - 54.7	33.5 ± 5.1

[a] N indicates number of animals used
[b] Animal bodyweight was 22 ± 0.5 kg

between the total and the nonspecific Rb^+ accumulation was directly linear to the incubation time up to 60 minutes. In order to test whether the Rb^+ uptake system is temperature dependent, Rb^+ uptake

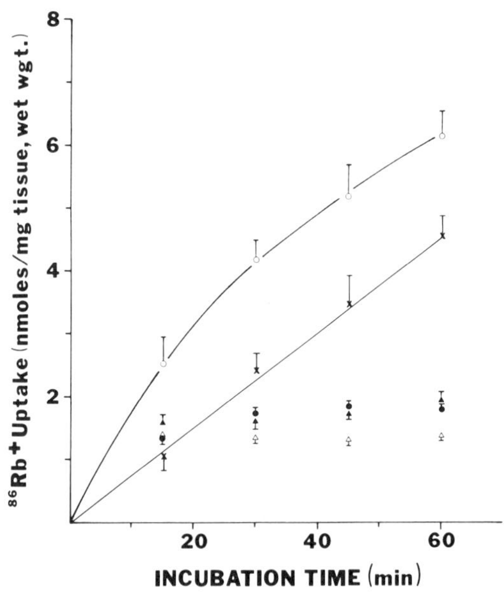

Fig. 1. $^{86}Rb^+$ Uptake in Cancine Heart. Rb^+ uptake was conducted in the absence of 10^{-4} M ouabain as the control (O). The difference of the Rb^+ uptake assayed in the presence (▲) and in the absence of 10^{-4} M ouabain was considered the active transport (X) of Rb^+. The incubation was also conducted at 0°C (●) or with 1 mM iodoacetic acid (△). (From Rhee[19] by permission of Naunyn-Schmiedeberg's Arch. Pharmacol).

Table 2. $^{86}Rb^+$ Uptake in Seven Canine Cardiac Tissues[a].

	$^{86}Rb^+$ Uptake (nanomoles/mg tissue, wet weight)					
VM[b]	RA	Pap	PF	SAN	AVN	A_o
2.89[c] ±0.26	2.98 ±0.31	3.28 ±0.29	8.37 ±0.72	3.69 ±0.35	3.26 ±0.39	0.71 ±0.08

[a] Total 21 determinations from 9 dogs were used in this experiment.
[b] VM indicates ventricular muscle, RA for right atrium, Pap for papillary muscle, PF for Purkinje fiber, SAN for sino-atrial node, AVN for atrio-ventricular node, and A_o for aorta, respectively.
[c] Each value represents means ± S.E.
[d] Modified from Rhee,[14] by permission of Brit.J.Pharmacol.

also was carried out at 0°C (●). An addition of 1 mM iodoacetic acid (△) to the complete incubation medium decreased the Rb^+ uptake to the level of the nonspecific accumulation which was assayed in the presence of 10^{-4} M ouabain (▲).

A comparison of the specific uptake of Rb^+ in functionally different tissues is shown in Table 2. The nonspecific accumulation of Rb^+, assayed in the presence of 0.1 mM ouabain, was similar in all tissues except the aorta; this similarity indicates that membrane leakness to Rb^+ is not significantly different in these tissues. The specific uptake of Rb^+ in the right atrium was not significantly different ($P > 0.05$) from that either in the left ventricle or the papillary muscle of the right ventricle. The specific Rb^+ uptake in the papillary muscle appeared significantly ($P > 0.05$) greater than the Rb^+ uptake in the left ventricle.

Differential Rb^+ Transport in Purkinje Fiber and Ventricle

In order to compare the possible differential sensitivity to ouabain between these two tissues in the uptake of Rb^+, experiments were performed as described under "Methods" in the presence of various concentrations of ouabain. In 19 tissues obtained from 11 dogs, the control Rb^+ uptake was 9861 ± 683 pmoles/mg (wet weight) for the Purkinje fiber and 4990 ± 389 for the left ventricle. The values are highly significantly ($P < 0.001$) different from each other. An exposure to ouabain in the concentration of 10^{-7} M decreased Rb^+ uptake approximately 30% in the Purkinje fiber and 20% in the left ventricle respectively. Ouabain (10^{-6} M) inhibited Rb^+ uptake as much 60% in the Purkinje fiber while only 43.5% was inhibited by the same dose of ouabain in the left ventricle. Almost complete inhibition of Rb^+ uptake was achieved in both of the Purkinje fiber and the left ventricle with 10^{-5} M ouabain since there was no statistically significant further inhibition in the uptake of Rb^+ in these two tissues with a higher concentration of ouabain.

This greater inhibition does not provide evidence that the Purkinje fiber has a greater sensitivity to ouabain than does the ventricular muscle in the uptake of Rb^+. Therefore, the active transport of Rb^+ was expressed as percent of the control active transport assayed in the absence of ouabain (Fig. 2). An inspection of the two ouabain dose and inhibition curves indicates that, at any concentrations of ouabain tested, there must be a similar degree inhibition of active Rb^+ transport in the Purkinje fiber and the ventricular muscle. Concentration of ouabain required for half-maximal inhibition I_{50} was 2×10^{-7} M for the Purkinje fiber and 5×10^{-7} M for the ventricular muscle, which are not significantly different.

MFPase Activity in Purkinje Fibers

Na^+,K^+-ATPase is considered the molecular machinery of monovalent cation transport system in mammalian cells[23-25]. There-

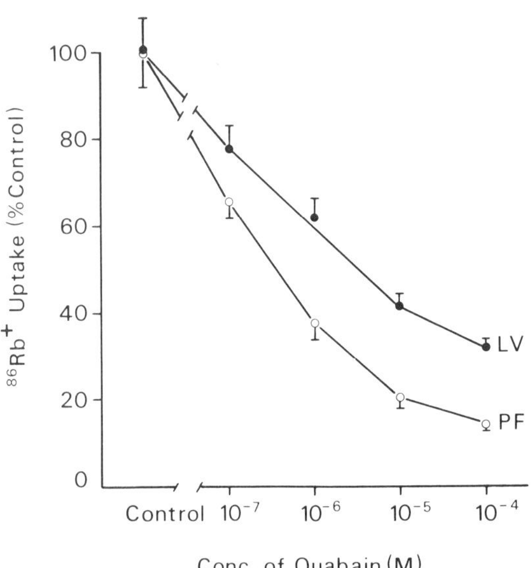

Fig. 2. Sensitivity of active Rb^+ transport to ouabain in Purkinje fiber and left ventricle. The nonspecific uptake of Rb^+ was 1.6 nmol/mg tissue in the ventricle (●) and 1.8 nmol in the Purkinje fibers (0) respectively. (From Rhee[19] by permission of Naunyn-Schiedeberg's Arch. Pharmacol.)

fore, in order to test the possibility that the different K^+-activated MFPase may account for the differential accumulation of $^{86}Rb^+$ in functionally different cardiac tissues, K^+-MFPase was assayed as an index of Na^+,K^+-ATPase activity. As shown in Fig. 3, MFPase was very sensitive to ouabain, and MFPase was inhibited by ouabain dose dependently in both Purkinje fiber and the left ventricular preparation. Concentrations of ouabain required for half-maximal inhibition of K^+-MFPase were somewhat lower as shown in Fig. 2. Also, there was no significant difference in I_{50} for the left ventricle (8×10^{-8} M) from the Purkinje fiber (6×10^{-8} M) in K^+-MFPase activity.

DISCUSSION

The heart consists of several functionally and morphologically different cell groups; impulse generating nodal tissues, impulse

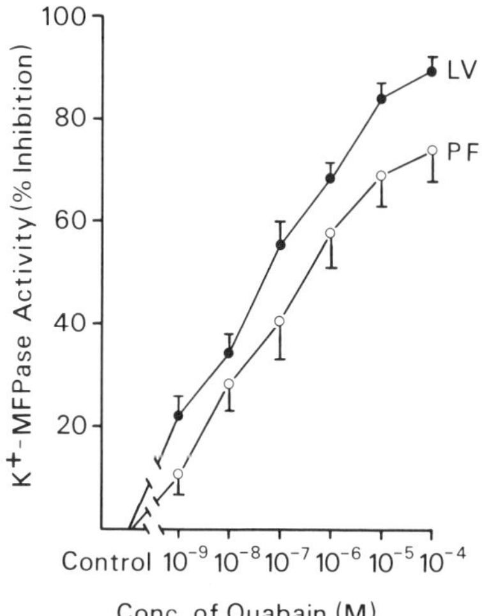

Fig. 3. Inhibition of K^+-MFPase by Ouabain in the Purkinje Fiber (PF) and the Left Ventricle Muscle (LV). K^+-MFPase activity was calculated from the difference between the total MFPase and nonspecific MFPase activity assayed in the absence of 10 mM K^+. Each point represents the mean of 10 determinations with S.E.M.

conducting Purkinje fibers, and mechanically contracting ventricular muscle (see Introduction). The heterogeneous cardiac cell groups have a specific functional role in the myocardial excitation-contraction cycle. One of the most striking differences between impulse-forming pacemaker cells and cardiac ventricular muscle is their concentrations of intracellular Na^+ and K^+. The intracellular Na^+ concentration in nodal tissue (150 mmol/Kg) is similar to that in extracellular space[7]. The intracellular Na^+ concentration in the left ventricle is very low, while the intracellular K^+ concentration in the ventricle is quite high as in the skeletal muscle. This unique differential distribution of monovalent cations may be responsible for the differential role of each specific tissue.

Since Na^+,K^+-ATPase is the single molecular entity which is responsible for the interaction with ouabain[24], it is reasonable to investigate the inhibitory action of ouabain in the specialized cardiac tissues. However, a direct chemical assay of Na^+,K^+-ATPase by conventional means in the crude homogenate of cardiac tissues has been a problem because the cardiac tissue is a very poor source of the enzyme. Moreover, since a considerable purification is a prerequisit for the enzyme assay, a relatively large mass of tissue is required. This has prohibited the assay of the enzyme from a small specialized cardiac tissue such as canine Purkinje fiber or nodal tissue. As a consequence, ouabain sensitive $^{86}Rb^+$ uptake in myocardial tissue slices or in the myocardial biopsy samples has been used as a valid index of monovalent cation pumping activity[17,26].

The $^{86}Rb^+$ uptake system in this present study was active transport process because it was sensitive to incubation temperature and was inhibited by ouabain specificially (Fig. 1). The uptake system was also dependent upon metabolic energy since iodoacetic acid (1 mM IAA) inhibited the uptake of Rb^+. Nonspecific accumulation of Rb^+ (determined in the presence of 10^{-4} M ouabain) was quite similar in seven different tissues tested in the study. However, specific uptake of Rb^+ was remarkably different from a tissue to another. The most outstanding fact is that the active uptake of Rb^+ in Purkinje fiber was the highest among the seven functionally different cardiac tissues tested. This high Rb^+ uptake in Purkinje fiber is at least two to three times greater than any other cardiac tissues (Table 2) from the same animal.

In order to study the differential toxic effect of cardiac glycoside, Polimeni and Vassalle[11] compared K^+ fluxes in canine Purkinje fiber and papillary muscle of cats. An electrical stimulation of the tissues increases the uptake of $^{42}K^+$ as much as 39% in Purkinje fiber, and only 6% of the papillary muscle. Under this condition, almost three times higher concentration of ouabain was required for the papillary muscle to produce a comparable inhibition of $^{42}K^+$ uptake achieved for the Purkinje fiber. Thus, this study confirms that the Purkinje fiber has a greater capacity of the active transport of monovalent cations than the ventricular or the papillary muscle does (Table 2). Since active ion pumping capacity in the Purkinje fiber is greater than any other functionally different tissues, it is understandable that cardiac glycosides inhibit the pump severely in the Purkinje tissue. The onset of the inhibitory effect of cardiac steroids on ion pump may occur early in the Purkinje fiber than in other tissues. Therefore, the orderly fluxes of ions which are essential for the rhythmic discharge of impulse may be disturbed primary in the Purkinje fiber.

Although this present study confirms the greater active Rb^+ uptake in the Purkinje fiber, this study disproves quite clearly the greater sensitivity to ouabain in the Purkinje fiber in comparison to the ventricular muscle. The well-known differential electrophysiological sensitivity to the conducting and contractile tissue may not be attributed to the difference in sensitivity to ouabain in active transport of monovalent cations. This present investigation may be in agreement with the recent report of Palfi et al.[13]. They reported that sensitivity to ouabain in Na^+,K^+-ATPase activity of the partially purified enzyme from canine Purkinje fiber was not greater than the sensitivity to ouabain of comparable Na^+,K^+-ATPase preparation from the papillary muscle. The specific Na^+,K^+-ATPase activity from the Purkinje fiber was less than 25% of the specific activity of the ventricular muscle. Na^+,K^+-ATPase activity per mg tissue (wet weight) was only under 6% of the comparable value from the left ventricle. Two to threefold higher K_i values for strophanthin, digoxin, and ouabain in calf His-bundle Na^+,K^+-ATPase relative to calf left ventricular Na^+,K^+-ATPase have also been reported[27], although it may not be much meaningful to compare quantitatively the active Rb^+ uptake in intact tissue slices with an *in vitro* assay of Na^+,K^+-ATPase in detergent-treated enzyme.

An effort has also been directed to test the potential possibility that differential uptake of ouabain by the functionally different cardiac tissues may be related to the differential sensitivity to cardiac glycosides in the tissues[14,28]. Tissue accumulation of ouabain per gram of the Purkinje fiber was significantly lower than the tissue content of ouabain is most of contractile muscles. Therefore, this study also does not support the contention that a greater electrophysiological sensitivity of the conducting system to cardiac glycosides is responsible for the greater uptake of cardiac steroids as studied in the sheep heart[28].

However, the specialized conducting fibers have a greater water content than the contractile muscles[7]. This difference was recently confirmed by a new method derived from voltage and cell volume relationship[29]. As much as 51% of the Purkinje fiber accounted for the extracellular space while only 23% of the contractile papillary muscle accounted for the extracellular space. Thus, the dry weight of the Purkinje fibers may be far less than the dry weight of the ventricle. If this inference is correct, the specific binding of ouabain per unit mass of dry weight would be greater in the Purkinje fibers than the ventricular muscle. Accordingly, the picture of the active monovalent cation uptake in these two tissues may be changed, if one expresses this parameter in terms of dry tissue weight basis.

SUMMARY

Ouabain inhibitable, active uptake of $^{86}Rb^+$ was compared in functionally different seven tissues of canine heart. The highest active transport of Rb^+ was noted in the Purkinje fiber (8.4 nmoles/mg tissue, wet weight) and the least in the ascending aorta (0.8 nmoles/mg). The capacity of Rb^+ active transport in impulse conducting Purkinje fiber was significantly ($P < 0.001$) greater than the active Rb^+ uptake in the contractile ventricular muscle. This observation is consistent with the fact that I_{50} for ouabain in the Purkinje fiber is significantly lower than the I_{50} value obtained from the ventricular muscle. However, there was a greater inhibition of transport adenosinetriphosphatase (E.C. 3.6.1.3) prepared from the ventricular muscle in vitro in comparison with the enzymatic activity obtained from the Purkinje fiber from the same animal. Therefore, it can be concluded that the well-known differential electrophysiological sensitivity to ouabain in these two functionally different tissues cannot account totally for the difference in the sensitivity of active transport of monovalent cation in this species.

Acknowledgements

This research was supported in part by grants from NIH, Heart Lung Blood Institute (HL-21196), and ORU Institutional Research Fund. Travel grant for the presentation of this research from the Office of Naval Research is greatly appreciated.

REFERENCES

1. P. F. Cranefield, "The Conduction of the Cardiac Impulse," Futura Publishing Co., Mount Kisco, New York (1975).
2. L. Sherf and T. N. James, Amer.J.Cardiol., 44:345-370 (1979).
3. M. Vassalle and C. I. Lin, Am.J.Physiol., 236(5):H689-H697 (1979).
4. F. Davies, R. E. Davies, E. T. B. Francis, and R. Whittam, J.Physiol., 118:278-281 (1952).
5. P. I. Polimeni, Amer.J. Physiol., 227:676-683 (1974).
6. P. A. Poole-Wilson and I. R. Cameron, Amer.J.Physiol., 229:1229-1304 (1975).

7. R. L. Vick, D. C. Chang, B. L. Nichols, C. F. Hazlewood, and M. C. Harvey, Annals of the New York Acad.Sci., 204:575-606 (1973).
8. M. Vassalle, J. Karis, and B. F. Hoffman, Am.J.Physiol., 203: 433-439 (1962).
9. G. K. Moe and R. Méndez, Cir., 4:729-734 (1951).
10. P. B. Hollander, J.Clin.Pharmacol., 15:560 (1975).
11. P. I. Polimeni and M. Vassalle, Am.J.Physiol., 218(5):1381-1388 (1970).
12. P. I. Polimeni and M. Vassalle, Am.J.Cardiol., 27:622-629 (1971).
13. F. J. Palfi, H. R. Besch, Jr., and A. M. Watanabe, J.Mol.Cell. Cardiol., 10:1149-1155 (1978).
14. H. M. Rhee, Brit.J.Pharmacol., 73:81-86 (1981).
15. T. N. James, Anat.Rec., 143:251-256 (1962).
16. S. Winegard and A. M. Shanes, J.Gen.Physiol., 45:371-394 (1962).
17. D. Ku, T. Akera,C. L. Pew, and T. M. Brody, Naunyn-Schmideberg's Arch.Pharmacol., 258:185-200 (1974).
18. J. C. Bernstein and Y. Israel, J.Pharmacol.Exp.Ther., 174:323-329 (1979).
19. H. M. Rhee, Naunyn-Schmiedeberg's Arch.Pharmacol., 318:344-348 (1982).
20. W. Huang, H. M. Rhee, T. H. Chiu, and A. Askari, J.Pharmacol. Exp.Ther., 211:571-582 (1979).
21. O. H. Lowry, N. J. Rosenbrough, A. L. Farr, and R. J. Randall, J.Biol.Chem., 193:265-275 (1951).
22. R. C. Truex, "Cardiac Arrhythmias," L.S. Dreifus and W. Likoff, eds., Grune and Stratton, New York, pp. 1-12 (1973).
23. S. L. Bonting, K. A. Simon, and N. M. Hawkins, Arch.Biochem. Biophys., 95:416-423 (1961).
24. A. Schwartz, G. E. Lindenmayer, and J. C. Allen, Pharmacol.Rev., 27:1-134 (1975).
25. J. C. Skou, Biochim.Biophys.Acta., 42:6-23 (1960).
26. T. J. Hougen and T. W. Smith, Circ.Res., 42(6):856-863 (1978).
27. W. Kübler and P. von Smekal, Acta Cardiol.Suppl., 17:103-113 (1975).
28. H. Hammerman, A. Hernandez, and D. Goldring, J.Lab.Clin.Med., 78:799 (1971).
29. S. R. Houser and A. R. Freeman, Amer.J.Physiol., 236(3):H519-H524 (1979).

TRANSPORT OF ANIONS IN THE FRESH-WATER ALGA *Hydrodictyon reticulatum*

Renata Rybová, Karel Janáček
and Ludmila Nešpůrková

Institute of Microbiology
Czechoslovak Academy of Sciences
142 20 Prague 4, Czechoslovakia

The present paper compares the transport of three anionic species in the alga Hydrodictyon reticulatum; chloride ions represent non-metabolizable species, whereas bicarbonate and sulphate anions play the role of basic nutrients. The possible transport of hydroxyl anions is also considered.

The experiments with the transport of chloride anions (Rybová et al., 1972) were performed in artificial pond water containing 1.0 mM KCl, 0.1 mM NaCl and 0.1 mM $CaCl_2$, the overall concentration of the chloride anion being 1.3 mM. Since the intracellular concentration of this ion is about 75 mM, the Nernst--Donnan potential for chloride anions is about + 100 mV. The membrane potential, recorded with a glass microelectrode, however, is close to - 100 mV, inside negative. The active character of chloride ion uptake is thus obvious. Influx of this ion is remarkably dependent on light conditions, being high in the light and very low in the dark. The inhibitor DCMU (3´- (3,4-dichlorphenyl) 1,1´-dimethylurea) reduces the influx in the light to low values close to those in darkness. Since DCMU is an inhibitor of the second photosystem (it stops electron flow near the water splitting end of the photosynthetic chain), it may be that the active transport of chloride anions is energized by the reducing equivalents. However, not only the uphill transport of chloride ions, but also their outflux in the direction of simple physico-chemical forces is substantially reduced in the dark, so that there is no wasteful loss of chlorides under the

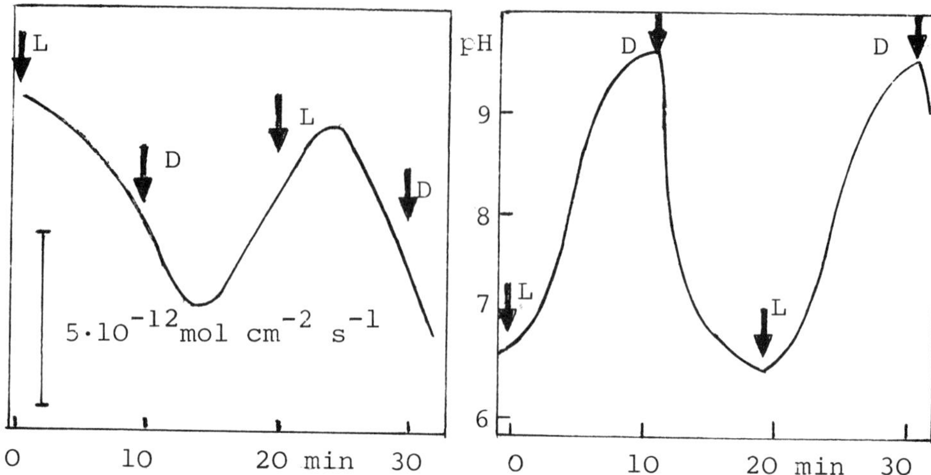

Fig. 1. Net outflow of Cl^- from the medium into the alga in the light and in darkness

Fig. 2. Effects of light conditions on pH of the algal suspension

conditions of a limited energy supply. Accordingly, there is an increase of the plasmalemma ohmic resistance in darkness. As shown in Fig. 1. the response of the chloride transporting system to changes in light conditions is rather rapid; the time lag does not exceed several minutes.

The response of the pH in algal suspensions to changes in light conditions is still more rapid, reflecting, as it seems, rapid changes in bicarbonate anion transport rate. In terrestrial and most aquatic plants it is carbon dioxide which enters the cells as the carbon source for photosynthetic fixation. In the genus Hydrodictyon, however, the bicarbonate anion appears to be the carbon source used by the alga (Raven, 1968). Indeed, when transferring a thick suspension of algal cells from darkness to the light the pH in the suspension increases from originally slightly acidic value to pH above 9 or 10 (Fig. 2). At such alkaline pH practically no carbon dioxide is available in the medium. Since ultimately it is carbon dioxide entering photosynthetic reaction, it is reasonable to assume that one hydroxyl ion is released at the membrane level for each bicarbonate ion taken up by the alga, thus accounting for the alkalization of the medium.

The pH change in time is exponential and the

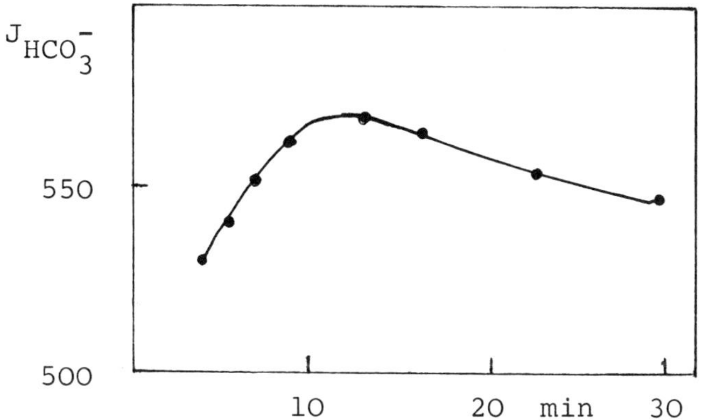

Fig. 3. Time dependence of bicarbonate anion inflow into the cells of H. ret.

apparent proton outflow from the medium in pH-stat experiments is linear, the coefficients of determination in the two cases being extremely high. Applying various conservation and equilibrium relations to these regression equations, time courses of individual concentrations and fluxes can be calculated, otherwise not available experimentally (Rybová and Janáček, 1982). The survival value of elaborate alkalization mechanism seems to consist in promoting the diffusion of carbon dioxide from distant water surface to the alkaline surface of algae. Typical steady-stae flux under experimental conditions amounted to 550 nmol m^{-2} s^{-1} (Fig. 3). The process of hydroxyl efflux and HCO_3 ion influx is very sensitive to DCMU, suggesting its dependence on the second photosystem. Moreover, alkalization is impaired by inhibitors of carbonic anhydrase.

The transport of sulphate anions reveals properties which are in many respects different. Responses of the transport in light conditions are relatively slow (Fig. 4); whereas with other anions at most minutes, with sulphate at least one hour is required. Also the action of inhibitors is different when compared with their effects on transport of chloride or bicarbonate anions. Table 1 shows the effect of some of the inhibitors used on sulphate uptake in the light after a 5-hour incubation:

Table 1. Inhibitors of sulphate uptake

controls	$5 \cdot 10^{-7}$CCCP	$5 \cdot 10^{-7}$DCMU	10^{-5}DNP	$5 \cdot 10^{-4}$DIDS
100	77	104	5	10

It may be seen that the transport of sulphate anions

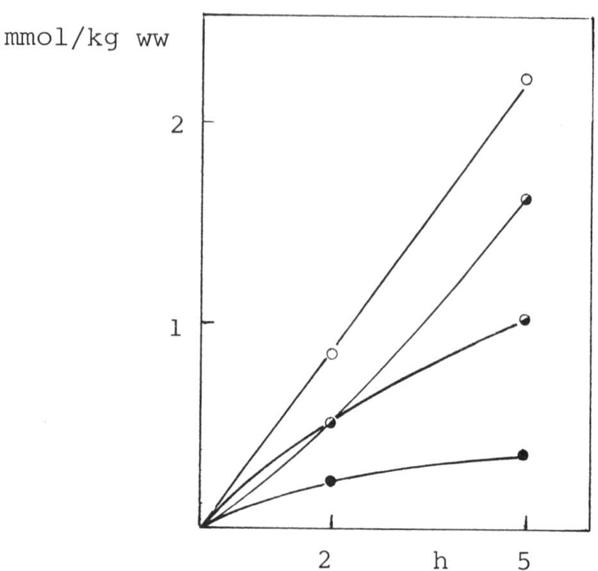

Fig. 4. Sulphate uptake in the alga Hydrodictyon reticulatum by cells kept permanently in the light (o) and darkness (•) as compared with cells pretreated for 12 h in the opposite light conditions (◐)

is sensitive to the action of uncouplers, CCCP and DNP, whereas DCMU has no effect. The inhibitory action of the anionic channel inhibitor DIDS (diisothiocyanostilbene-2,2′-disulphonic acid) is still more pronounced than in the case of chloride anion transport.

Sulphate anion is accumulated in the cells from the concentration 0.5 mM in the medium to almost 30 mM. Its efflux is practically negligible, but after permeabilization of the plasmalemma with nystatin it is rapidly released.

The concentration dependence of the initial rate of the sulphate anion uptake suggests that at low concentrations two mechanisms of active transport function in parallel. One of the mechanisms has a high affinity, and can be interpreted as a transport proportional to the external sulphate concentration, proceeding, however, through sites the number of which decreases exponentially with the sulphate concentration. At concentrations higher than 0.2 mM this energetically demanding mechanism is switched off and the transport appears to proceed by a conventional Michaelis-Menten saturable mechanism (Rybová et al., 1982).

REFERENCES

Raven, J. A., 1968, The mechanism of photosynthetic use of bicarbonate by Hydrodictyon africanum, J. Exp. Bot., 19:153.

Rybová, R., Janáček, K., and Slavíková, M., 1972, Ionic relations of the alga Hydrodictyon reticulatum. The effects of light conditions and inhibitors, Z. Pflanzenphysiol., 66:420.

Rybová, R., and Janáček, K., 1982, Mathematical model of the alkalization of suspension of the alga Hydrodictyon reticulatum in light, Bioelectrochem. Bioenerg., 9:509.

Rybová, R., Nešpůrková, L., Janáček, K., and Stružinský, R., 1982, Sulphate-uptake isotherm of the alga Hydrodictyon reticulatum, Studia Biophys., in press.

IONIC CONDUCTIVITIES IN GASTRIC MUCOSA UNDER OSMOTIC GRADIENTS

Leopoldo Villegas

Centro de Biofísica y Bioquímica,
Instituto Venezolano de Investigaciones Científicas
Apartado 1827, Caracas 1010A, Venezuela

ABSTRACT

The effect of osmotic gradients on the transmucosal ionic conductance, intracellular potential and the serosal-to-mucosal conductivity ratios of the ephitelial and oxyntic cells were explored. Transmucosal and intracellular potentials as well as the changes produced by transmucosal 0.1 mA/cm^2, 1 sec, current pulses were measured. In 10^{-4} M histamine-stimulated mucosae the transmucosal conductance is significantly ($P < 0.001$) reduced from 3.73 ± 0.09 mmohs/cm^2 obtained with isosmotic (231 mOsm/Kg of water) solutions at both surfaces to 3.17 ± 0.05 mmohs/cm^2 by using hyperosmotic (508 mOsm/Kg of water) solution at the mucosal surface. Simultaneously, the transmucosal electrical potential difference referred to the serosal surface significantly increases ($P < 0.001$) from -23.3 ± 0.5 mV to -26.6 ± 0.5 mV by effect of hyperosmotic solution at the mucosal surface. No significant changes ($P > 0.05$) in the short circuit current, the serosal-to-mucosal conductivity ratios of ephitelial and oxyntic cells and the intracellular potentials were produced by the use of hyperosmotic solution at the mucosal surface. These results suggest the possibility that hypertonicity at the mucosal surface affects mainly the paracellular pathway and in a less extent the transcellular pathway for diffusion.

The author is also professor at the Instituto Internacional de Estudios Avanzados, Caracas, Venezuela.

INTRODUCTION

The present paper is concerned with the mechanism utilized in the regulation of ion flux across the gastric mucosa. An *in vitro* preparation of frog gastric mucosa obtained by blunt disectio from the stomach of *Rana pipiens* was used. This preparation includes the mucosa, the *lamina propria*, the muscularis mucosa and part of the submucosa. The mucosa itself is heterogeneous. The surface and the pits regions are covered by the ephitelial cells in contact with the bulk solution at the mucosal surface. The tubular region is covered by oxyntic cells (Ito, 1967). The mucosal surface of oxyntic cells is in contact with the content of the gastric gland, and indirectly with the bulk solution. The junctional complexes are between the apical cells borders. Behind the junctional complexes are the intracellular spaces opened to the serosal surface (Sedar and Forte, 1964). These intracellular spaces and the cell membrane in contact with the basement membrane are separated from the solution at the serosal surface by the lamina propria, the muscularis mucosa and the part of submucosa remaining from the disection. So that, the acid secretion produced by the gastric mucosa occurs primarily from the extracellular serosal space to the extracellular mucosal space. Two main roles of these spaces on the regulations of ionic fluxes have been proposed: First, as restricted compartments with composition different from the bulk solutions, and second as a paracellular pathway for ion flux across the mucosa (Hill,1980; Durbin 1981).

In this preparation bathed by isosmotic solutions in open circuit conditions, the transmucosal net ionic movement is mainly made up of two fractions: the acidic fraction formed by the hydrogen ion secretion and the acidic chloride output and the non acidic fraction formed by the chloride transported in excess of the acidic fraction, which has at its upper limit the short circuit current (Hogben, 1955). To a large extent, the cations move passively along the electrochemical gradient created by the non acidic chloride transported. The use of hyperosmotic solutions at both surfaces reduces the acidic fraction of the ionic secretion without a significant effect on the acidic fraction (Villegas, 1975). Simultaneously the net water flux from serosal-to-mucosal surface is reduced, without significant changes in the unidirectional water diffusion fluxes (Villegas, 1982). No significant changes are obtained by changing the solution tonicity at the serosal surface. On the contrary, changes in the solution tonicity at the mucosal surface alone reduce both water diffusion fluxes from the serosal-to-mucosal surface and from thc mucosal-to-serosal surface.

The effect of hyperosmotic solution at the mucosal surface with isosmotic solution at the serosal surface in the non acidic fraction of the gastric secretion was investigated. Electrophysiological parameters were measured during incubation with isosmotic solution at both surfaces and in the same mucosae using isosmotic solution at the serosal surface and hyperosmotic solution at the mucosal surface.

METHODS

The assembly and physiological solution used in the experiments were those previously described (Villegas, 1962). The mucosa was mounted between two lucite chambers, with its serosal surface facing the lower one. This lower chamber had two openings through which the physiological solution was continuously circulating. The upper chamber was opened to permit the measurements with microelectrodes. Both chambers were filled with physiological buffered solution. In the initial period isotonic solution was used in both chambers. The composition of this isotonic solution, in mmoles/1 was: NaCl, 84.6; KCl, 3.2; CaCl2, 1.5; KH_2PO_4, 0.8; $MgSO_4$, 0.8; $NaHCO_3$, 17.8; and glucose 22.0. The measured osmolality of this solution was 231 mOsm/Kg water. In the second period of measurements hypertonic solution, prepared by adding glucose to isotonic solution was used. The measured osmolality of the hyperosmotic solution was 508 mOsm/Kg water. During all the experiments, the solutions were bubbled with O_2 - CO_2, 95-5. Some of the experiments were performed with spontaneously-secreting mucosae and other with histamine-stimulated mucosae. For stimulation, 10^{-4}M histamine diphosphate was added to the solution in both chambers.

Two external calomel electrodes, connected to the chambers by 3% agar-physiological solution bridges were used to register the transmucosal electrical potential difference. Two Ag-AgCl electrodes located at each chamber entered them and were used to send 1 sec, 0.1 mA/cm^2 current pulses for the transmucosal and intracellular conductance measurements. The intracellular potentials were measured with microelectrodes of suitable shape, filled with 3M KCl, connected by 3% agar-3M KCl to a calomel electrode. The tip resistance of these microelectrodes was always between 5 and 10 megohms, and the tip potential difference between 0 and -3 mV.

RESULTS

The effect of hypertonicity of the solution in contact with

the mucosal surface on the transmucosal electrical potential difference referred to the solution at the serosal surface, measured across the two extracellular calomel electrodes is presented in Table I.

TABLE I

EFFECT OF THE MUCOSAL SOLUTION HYPERTONICITY ON THE TRANSMUCOSAL POTENTIAL DIFFERENCE REFERRED TO SEROSAL SOLUTION

SOLUTIONS TONICITIES mOsm/Kg water		POTENTIAL DIFFERENCE* mV	
SEROSAL	MUCOSAL	NON-STIMULATED	HISTAMINE-STIMULATED
231	231	-24.3 ± 0.5	-23.3 ± 0.5
231	508	-30.2 ± 0.4	-26.6 ± 0.5
DIFFERENCES			
0	277	- 5.9 ± 0.8	- 3.3 ± 0.7

* Each potential difference value is the mean from 10 experiments ± standar error of the mean.

In both groups of non-stimulated and histamine-stimulated mucosae, the use of hyperosmotic solution at the mucosal surface induced a significant increment of the transmucosal electrical potential difference ($P < 0.001$).

The effect of hypertonicity of the solution at the mucosal surface on the transmucosal conductance measured in the same two groups of mucosae is presented in Table II.

TABLE II

EFFECT OF MUCOSAL SOLUTION HYPERTONICITY ON THE TRANSMUCOSAL CONDUCTANCE

SOLUTIONS TONICITIES mOsm/Kg water		TRANSMUCOSAL CONDUCTANCE* mmohs/cm^2	
SEROSAL	MUCOSAL	NON-STIMULATED	HISTAMINE-STIMULATED
231	231	3.83 ± 0.07	3.73 ± 0.09
231	508	3.41 ± 0.05	3.17 ± 0.05
DIFFERENCES			
0	277	-0.42 ± 0.09	-0.56 ± 0.10

* Each transmucosal conductance value is the mean from 10 experiments ± standard error of the mean.

The conductances were calculated from the potential changes induced by square pulses of 0.1 mA/cm^2 and 1 sec duration. In both, spontaneously-secreting and histamine-stimulated mucosae, the transmucosal electrical conductance is significantly reduced ($P < 0.001$) by the use of hyperosmotic solution at the mucosal surface with isosmotic solution at the serosal surface.

The intracellular potential referred to the extracellular electrode located in the solution immersed in the serosal surface is presented in Table III.

TABLE III

EFFECT OF THE MUCOSAL SOLUTION HYPERTONICITY ON THE INTRACELLULAR POTENTIAL REFERRED TO SEROSAL SOLUTION

SOLUTIONS TONICITIES mOsm/Kg water		INTRACELLULAR POTENTIALS (mV)*			
		NON-STIMULATED		HISTAMINE STIMULATED	
SEROSAL	MUCOSAL	EPHITELIAL	OXYNTIC	EPHITELIAL	OXYNTIC
231	231	-55.9 ± 1.1	-18.9±0.9	-51.4± 1.1	-14.8± 0.9
231	508	-53.4 ± 0.9	-19.9±1.1	-51.1± 1.0	-17.7± 0.8
DIFFERENCES					
0	277	2.5 ± 1.4	- 1.0±1.1	0.3± 1.5	- 2.9± 1.2

* Each intracellular potential value is the mean from 10 experiments ± the standard error of the mean.

Two populations of measurements were obtained in each mucosa. All cells are negative with respect to the electrode immersed in the solution in contact with the serosal surface. The cells that are negative with respect to the extracellular electrodes immersed in the solution at the mucosal surface are the ephitelial cells. The oxyntic cells are positive with respect to the extracellular electrode immersed in the solution in contact with the mucosal surface (Villegas, 1962). The electrode impaled one cell for each measurement. Consequently, the measurements obtained with isosmotic and with hyperosmotic solution at the mucosal surface correspond to the same mucosae but not to the same cells. The differences shown in the last line of table, are between the means of the populations. In this condition no significant effect of hyperosmotic solution at the mucosal surface is observed in the oxyntic and ephitelial cells of spontaneously-secreting or histamine-stimulated mucosae.

The conductivity ratios calculated from the changes induced in each cell surface by application of transmucosal current flow are presented in Table IV. In both non-stimulated and histamine-stimulated mucosae the asymmetry on the response of the oxyntic cells is larger than those of ephitelial cells ($P < 0.001$) (Villegas, Michelangeli and Sananes, 1970). However, no significant effect of hypertonic solution at the mucosal surface is observed, either in the spontaneously-secreting or the histamine-stimulated mucosae.

TABLE IV

EFFECT OF THE MUCOSAL SOLUTION HYPERTONICITY ON THE RATIO $K_{SEROSAL}/K_{MUCOSAL}$ IN THE EPHITELIAL AND THE OXYNTIC CELLS

SOLUTIONS TONICITIES mOsm/Kg water		$K_{SEROSAL}/K_{MUCOSAL}$ * NON-STIMULATED		HISTAMINE-STIMULATED	
SEROSAL	MUCOSAL	EPHITELIAL	OXYNTIC	EPHITELIAL	OXYNTIC
231	231	3.55 ± 0.20	6.11 ± 0.32	2.79 ± 0.16	5.69 ± 0.28
231	508	3.36 ± 0.23	5.97 ± 0.30	2.80 ± 0.15	5.49 ± 0.23
DIFFERENCES					
0	277	-0.39 ± 0.30	-0.14 ± 0.44	0.01 ± 0.22	-0.20 ± 0.36

* Each value is the mean from 10 experiments ± standard error of the mean.

DISCUSSION AND CONCLUSION

Estimation of the secretory surface per square centimeter of chamber ranges from 44 to 189 cm^2 in the non-stimulated mucosae and from 129 to 557 cm^2 in the histamine-stimulated mucosa (Hellander, 1977; Hellander and Durbin, 1977). Even considering that these calculations are partially based on approximations and generous assumptions and the authors emphasized that there may be errors of considerable magnitude, the ratio stimulated to spontaneously secreting area in both papers ranges between 2.9 and 3.0 times.

Considering these differences in areas, the possible effect on transmucosal conductance due to hyperosmolality of the solution at the mucosal surface must affect the ratio Ks/Km in a different extent on spontaneously-secreting and histamine-stimulated mucosae. This expected differential effect is not observed in the results shown in Table III and Table IV. However, there are significant effects of hyperosmolality of the solution at the mucosal surface on the transmucosal potential differences and conductances shown in Table I and Table II. Hyperosmolality at the mucosal surface may affect a paracellular pathway for ion diffusion. A sweeping effect of the osmotic flux, inducing changes in the extracellular serosal or mucosal spaces must produce an asymmetry in the response to positive and negative current fluxes. This asymmetry was not observed in the experiments reported. It is reasonable to conclude that the effect of hyperosmolality at the mucosal surface is mainly on the restriction offered by the paracellular pathway for ion diffusion.

REFERENCES

Durbin, R.P., 1981. Osmosis in Ephitelial Membranes, J. Membrane Biol.,61: 141.

Hellander, H.F., 1977. An attempt to correlate functional and morphological data for the gastric parietal cells, Gastroenterology, 73: 956-957.

Hellander, H.F., and R.P. Durbin, 1977. Secretory surface area and phosphatase activity of frog gastric mucosa. Am.J. Physiol., 232: E48-E52.

Hill, A., 1980. Salt-Water Coupling in Leaky Ephitelia, J. Membrane Biol., 56: 177-182.

Hogben, C.A.M., 1955. Active transport of chloride by isolated frog gastric ephitelium. Origin of the gastric mucosal potential, Am.J. Physiol. 180: 641-649.

Ito, S., 1967. Anatomic structure of the gastric mucosa. Handbook of Physiology, Ed.: C.F. Code. American Physiological Society, Washington Section 6, Vol. 2, p. 705-741.

Sedar, A.W., and J.Q. Forte, 1964. Effects of calcium depletion on the junctional complex between oxyntic cells of gastric glands. J.Cell. Biol., 22: 173-188.

Villegas, L., 1962. Cellular location of the electrical potential difference in frog gastric mucosa. Biochim. Biophys. Acta, 64: 359-367.

Villegas, L., 1975. Response of active transport of ions and spontaneous water flux to osmotic gradients in gastric mucosa. Am. J. Physiol., 228: 738-741.

Villegas, L., 1982. Water diffusion under osmotic gradients in frog gastric mucosa. Biochim. Biophys. Acta, 685: 249-252.

Villegas, L., F. Michelangeli, and L. Sananes, 1970. Asymmetrical response of oxyntic cells to direct current in non-stimulated frog gastric mucosa, Biochim. Biophys. Acta, 219: 518-520.

WATER TRANSPORT IN ROOT SYSTEMS AND THE NATURE OF ROOT PRESSURE

Vladimir N. Zholkevich

K.A.Timiriazev Institute of Plant Physiology
Academy of Sciences of the USSR
Moscow, USSR

So far some authors have reduced the water transport mechanism in plants to simple ultra-filtration and mass flow under the pressure generated by a water potential gradient in a soil-plant-atmosphere system. From such a point of view a plant behaves like a canal between soil and atmosphere. Sometimes attempts are even made to draw a parallel between uprising water flow in plant and water movement along a strip of blotting paper whose lower end is immersed into a glass of water. Although water is transported in plants for functioning living cells, it follows, that these cells do not promote water flow; on the contrary, they only resist it. However, a series of facts contradicts such simplified schemes and evidences for an active part of the parenchyma cells in water movement along the plant. We shall refer to only some of these facts concerning water transport in root systems.

According to the osmotic conception, the value of root pressure is expressed by the equation

$$SP = k(OP_i - OP_e) \quad (1)$$

where SP is the root pressure, OP_i - the osmotic pressure of the exudate, OP_e - the osmotic pressure of the ambient solution, k - the coefficient, dependent on the hydraulic conductivity of the root.

If equation (1) is true, exudation is impossible in hypertonic solutions; it must stop when OP_i equals OP_e. However, exudation continues in hypertonic solutions, obeying the equation

$$SP = k(OP_i - OP_e) + NOC \quad (2)$$

where NOC is the non-osmotic component of the root pressure. It was called so by those investigators who revealed an inconsistency of the osmotic exudation conception (van Overbeek, 1942; Broyer, 1951; Mozhaeva and Pil'shchikova, 1972).

In our laboratory it was shown that NOC was highly labile, its fluctuations under the effect of various chemical agents ranging from 0 to 80% of the root pressure value; the latter was measured by means of the compensation method. Applying this method, we supposed that the root pressure value equals the osmotic pressure of the ambient solution which stops exudation. NOC is energy-dependent and decreases strongly when the cell membrane structure is disturbed; the exudation rate is positively correlated mainly with the NOC value, but not with the osmotic gradient (Zholkevich et al., 1979, 1981; Zholkevich, 1981).

60 years ago Ursprung and Blum (1921) revealed a so-called endodermal jump of the water potential. Even till now this jump seems to be a mysterious phenomenon, as it may evidence for the radial water flow via the parenchyma cells against the water potential gradient. Recently the endodermal jump existence has been supported (Borisova and Zholkevich, 1982): when moving in the radial direction from the epidermis to the last layer of the cortex cells, the water potential of each subsequent cell is lower than that of the previous one. However in endodermal cells the water potential increases sharply, and when moving further towards xylem vessels, the water potential of each subsequent cell is always higher than that of the previous one (Table 1).

Table 1. Water potential (ψ) distribution in root cells of Vicia faba and Zea mays (Average standard error is 1%)

Cell location	Ψ, MPa	
	Vicia faba	Zea mays
Epidermis	-0,02	-0,02
Cortex, layer 1	-0,15	-
- " - - " - 2	-0,18	-
- " - - " - 3	-0,19	-
- " - - " - 4	-0,21	-
- " - - " - 5	-0,22	-
- " - - " - 6	-0,23	-0,15
Endodermis	-0,08	-0,08
Pericycle	-0,06	-0,05
Parenchyma cells adjacent to xylem vessels	-0.04	-0,02

Table 2. Effect of chemical agents on the endodermal jump of the water potential in Zea mays roots (Average standard error is 1%)

Cell location	ψ, MPa						
	Control	2,4-di-nitro-phenol $2,5 \times 10^{-4}$M, 1 hr	Pipol-phene 4×10^{-4}M, 20 min,	Pipol-phene 4×10^{-4}M, 20 min, then $CaCl_2$ 1×10^{-2}M, 20 min	$CaCl_2$ 1×10^{-3}M, 2 hr	Cyto-chala-zin B 1×10^{-6}M, 2 hr	Colchi-cine 1×10^{-3}M, 2 hr
Cortex, layer 6	-0,15	-0,08	0	-0,15	-0,19	0	0
Endodermis	-0,08	-0,08	-0,08	-0,08	-0,08	0	0

We studied the effect of various chemical agents on the endodermal jump value (Borisova, Lazareva and Zholkevich, 1982). Similar to NOC, the endodermal jump appeared to be highly labile. It is dependent on the cell energy supply and membrane structure intactness. The endodermal jump disappears under the action of 2,4-dinitrophenol (applied in the concentration inducing an uncoupling effect on oxidation and phosphorylation) or pipolphene (this anesthetic violates the cell membrane structure at the expense of calcium ions extrusion). But if the roots are immersed into $CaCl_2$ (which stabilizes the membrane structure) following pipolphene treatment, the endodermal jump is restored. Under the treatment of $CaCl_2$ alone the endodermal jump becomes higher. The endodermal jump disappears under the action of the contractile protein inhibitors, such as cytochalasin B (which destroys microfilaments, containing actomyosin-like proteins) and colchicine (which destroys microtubules, containing tubulin) (Table 2).

When endodermal jump disappears, exudation considerably decelerates and its temperature coefficient Q_{10} decreases approximately down to unity (Table 3).

As Q_{10} gives us information on the nature of the phenomenon under consideration, one may suppose that when the endodermal jump is absent the root pumps water by means of more simple, mainly physical, processes. Thus, the disappearance of the endodermal jump is associated with the elimination of an energy-dependent metabolic root pressure constituent. It is quite possible that this constituent and the non-osmotic component of root pressure are one and the same.

Table 3. Effect of chemical agents on the endodermal jump of the water potential, rate and Q_{10} of the exudation of Zea mays roots

	Ψ, MPa			
	Cortex, layer 6	Endodermis	J_v, µl/hr	Q_{10}
Control	-0,15	-0,08	1,3 ± 0,1	3,05
2,4-dinitrophenol $2,5\times10^{-4}$M, 1 hr	-0,08	-0,08	0,7 ± 0,2	1,02
Cytochalasin B 1×10^{-6}M, 2 hr	0	0	0,8 ± 0,1	1,07
Colchicine 1×10^{-3}M, 2 hr	0	0	0,3 ± 0,1	1,17

The existence of NOC and the endodermal jump contradicts the osmotic exudation conception. Probably, exudation is not of a pure osmotic but of a more complicated nature. More that 50 years ago Bose (1927) advanced a general thesis postulating the participation of cell pulsations in sap movement along the plants. So far this thesis seems almost fantastic. However, in recent years a series of data (Mozhaeva and Pil'shchikova, 1972; Petrov, 1974; Mozhaeva et al., 1975; Kubichek, 1976; Zialalov, 1981; Ushakov and Koltunova, 1982) has supported Bose's hypothesis. These data are concerned with water transport in roots, xylem transport and transpiration, i.e. they are concerned with the whole uprising water flow in plants. Root pressure studies established the following. Exudation and water uptake by a decapitated root system display an impulsive rhythmicity, the rhythms of exudation being alternated with those of water uptake: the maximum of exudation coincides with the minimum of water uptake (Mozhaeva and Pil'shchikova, 1972). Rhythmic alternations of water uptake and exudation are accompanied by rhythmic microoscillations of the cortex, stele and whole root cross-section areas; the maximum of exudation coincides with a contraction of the root; root contraction is energy-dependent and sensitive to n-chloromecuri benzoate, which may inhibit an actomysin-like protein activity (Mozhaeva et al., 1975). As was noted above, exudation is also sensitive to such contractile protein inhibitors, as cytochalasin B and colchicine. An actomyosin-like protein was really detected in roots (Mozhaeva and Bulycheva, 1971; Zholkevich et al., 1979; Abutalybov et al., 1980, 1981). So the contractile proteins can take part in water pumping by roots.

The data obtained on the rhythmic alternations of water uptake and exudation and on the cell root contraction at the moment of exudate extrusion allow us to suppose that water uptake by root and subsequent water extrusion during exudation represent a two-phase process: at first, the root cells uptake the water and increase their own volume (phase one) and then they extrude the water towards the xylem vessels (phase two). So, exudation does not simply result from a passive water ultra-filtration via the root tissues; on the contrary, the root tissues take an active part in water pumping. Data, obtained by Petrov (1974), also evidence for the active part of living cells in water pumping. According to these data, root pressure is dependent on the amount of cells participating in root pressure build-up: the longer the root the higher is the root pressure. If the root pressure is of a pure osmotic nature, such a phemomenon will not take place.

If the water transport mechanism is connected with cell pulsations, water must be translocated in the radial direction via living cells, but not only via free space. In fact, experimental data and corresponding calculations show that in decapitated root system water moves mainly by a symplasm and vacuolar pathways (Newman, 1976; Burkina and Guseinova, 1981, 1982).

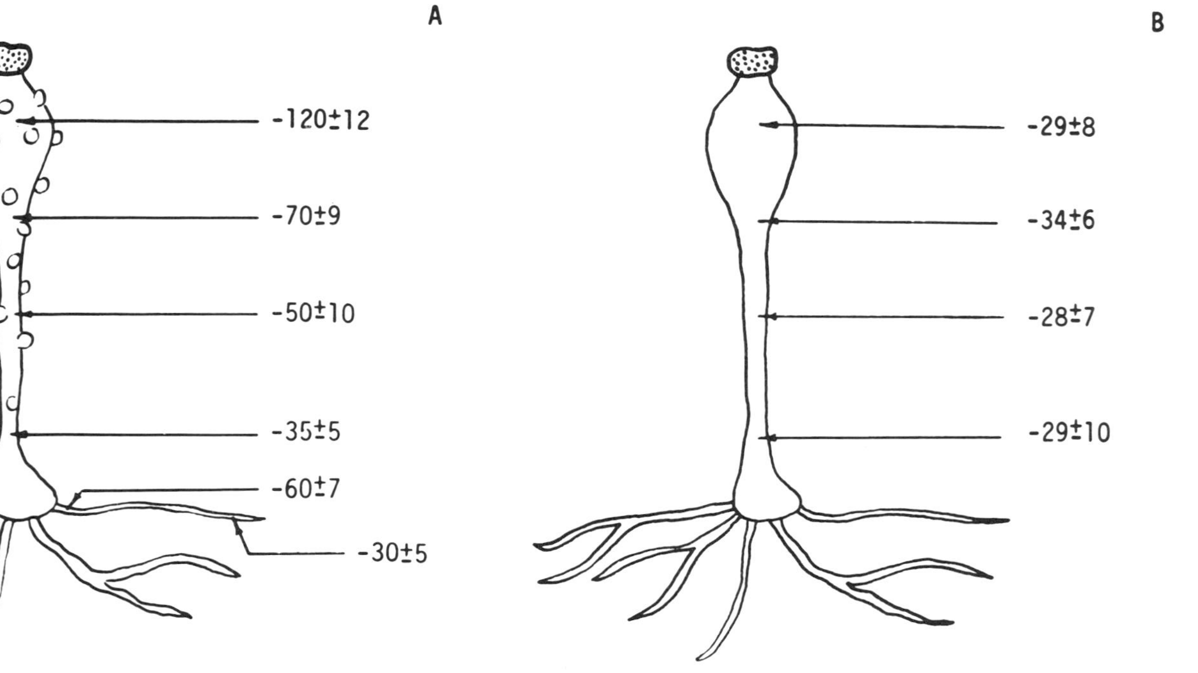

Fig. 1. Interrelationship between guttation and membrane potential (in mv) gradient of *Pilobolus* cell. Control (A), effect of sulphuric ether or chloroform, 10 min (B).

If pumping of water by the root system is connected with cell pulsations, a cell polarization must be an indispensable condition for unidirectional water flow. In fact, experiments with fungus Pilobolus (Tarakanova et al., 1982) demonstarted a certain interrelationship between electric polarization of the cell and unidirectional water flow. Pilobolus is well-known thanks to its abundant guttation. The Pilobolus body represents a gigantic cell; its lower parts uptake water and its upper parts secrete water in the form of numerous drops. The measurements by means of the microelectrode technique showed that the water moved towards the increasing membrane potential (Fig. 1). When anesthetic substances (sulphuric ether or chloroform) are applied, the membrane potential gradient disappears and the guttation ceases.

The above-mentioned facts reveal that water transport in plants is a rather complicated physiological process, and a great deal of work is still needed to elucidate its mechanisms.

Abutalybov, V. F., Shushanashvili V. J., and Zholkevich, V. N., 1980, Isolation of actin-like protein from sunflower roots, Doklady Akad. Nauk SSSR, 252:1023.

Abutalybov V. F., and Zholkevich,V. N., 1981, Effect of Ca^{2+}, Mg^{2+}, Na^{+} on the activity of actomyosine-like ATPase isolated from sunflower roots, Phyziologia rastenii, 28:442.

Borisova, T. A., Lazareva, N. P., and Zholkevich, V.N., 1982, Effect of chemical agents on the endodermal jump of water potential and on Zea mays L. root exudation, Doklady Acad. Nauk SSSR, 267:766.

Borisova, T.A., and Zholkevich, V.N., 1982, On the endodermal jump of water potential in root system, Physiologia rastenii, 29:737.

Bose, J. C., 1927, "Plants autographs and their revelations", Longmans, Green and Co, London.

Broyer, T. C., 1951, Exudation studies on the water relations of plants, Amer. J. Bot., 38:157.

Burkina, Z. S., and Guseinova, G.M., 1981, Application of H_2O^{18} for studying water exchange in maize roots in relation to energy supply and membrane state, Studia biophysica, 85:21.

Burkina, Z. S., and Guseinova, G. M., 1982, Effect of temperature on heavy oxigen water H_2O^{18} exchange, in "Water relations regulation in plants", Kiev, USSR (in press, in Russian).

Kubichek, S. A., 1976, Secretion of liquid drops by guard cells, Phyziologia rastenii, 23:403.

Mozhaeva, L. V., and Bulycheva, E.M., 1971, Properties of a contractile protein isolated from pumpkin roots, Izvestiya Timiriazevskoi selskochozaiatvennoi akademii, 2:3.
Mozhaeva, L. V., and Pil'shchikova, N. V., 1972, On the nature of pumping water process by plant roots, Izvestiya Timiriazevskoi selskochozaistvennoi akademii, 3:3.
Mozhaeva, L. V., Pil'shchikova, N.V. and Zaitseva, N. V., 1975, The study on the contractile properties of root cells in relation to rhythmicity of plant exudation, Izvestiya Timiriazevskoi selskochozaistvennoi akademii, 1:3.
Newman, E. J., 1976, Water movement through root systems, Philos.Trans. Roy. Soc. London, 273:463.
Overbeek van J., 1942, Water uptake by excised root systems of tomato due to non-osmotic forces, Amer. J. Bot., 29:677.
Petrov, A. P., 1974, "Bioenergetic aspects of water relations and drought-resistance of plants", Kazan, USSR (in Russian).
Tarakanova, G. A., Panichkin, L. A., and Zholkevich, V. N., 1982, Unidirectional water flow and electric polarization of Pilobolus cell, in "Substance transport and bioelectrogenesis in plants", Gorki, USSR (in press, in Russian).
Ursprung, A., und Blum, G., 1921, Zur Kenntnis der Saugkaft. IV. Die Absorptionszone der Wurzel. Der Endodermisspung, Ber. dtsch. bot. Ges., 39:70.
Ushakov, V. Yu., and Koltunova, I.R., 1982, On the pulsatory nature of transpiration and water uptake into leaves of plants, Doklady Akad. Nauk SSSR, 266:766.
Zholkevich, V. N., 1981, On the nature of root pressure, in "Structure and function of plant roots", R.Brouwer et al., eds., Martinus Nijhoff/ Dr.W. Junk Publishers, The Hague/Boston/London.
Zholkevich, V. N., Sinitsina, Z. A., and Peisakhzon, B. I., 1981, On physiological regulation of water transport in root system, Studia biophysica, 85:17.
Zholkevich, V. N., Sinitsina, Z. A., Peisakhzon, B. I., Abutalybov, V. F., and D'yachenko, I. V., 1979, On the nature of root pressure, Phyziologia rastenii, 26:978.
Zialalov, A. A., 1981, Water state and the autocorrelation analysis of subcrown internode three-dimentional oscillations, Doklady Akad. Nauk SSSR, 256:247.

ROLE OF WATER AND IONS IN MEMBRANE STRUCTURE AND FUNCTION WATER AND IONS IN EVOLUTION OF BIOLOGICAL SYSTEMS

FREEZING: A TOOL FOR THE PRESERVATION AND SEPARATION OF ISLETS OF LANGERHANS

Harvey L. Bank

Department of Pathology
Medical University of South Carolina
Charleston, S.C. 29425

Many forms of diabetes can be successfully treated by oral hypoglycemic drugs or by subcutaneous administration of insulin or insulin analogs. A major disadvantage of such treatment is the difficulty in adjusting the insulin dose (hence blood glucose levels) in response to food consumption. The ideal therapy would utilize a continuous feedback network which permits minute by minute surveillance of the blood glucose level with appropriate alteration of hormonal output, just as the alpha and beta cells of the islets respond in the normal individual. One approach is to reimplant islets or beta cells in patients whose islets of Langerhans are not responding at an adequate level (Lacy, 1967, 1980; Mattas, 1976, 1977). The probability of a successful transplant would be greatly enhanced if large numbers of tissue-typed islets could be cryogenically stored until an immunologically suitable diabetic recipient is available. Since viability of frozen tissue stored in liquid nitrogen is essentially independent of the storage time, the current constraints involved in obtaining good tissue cross-matches can be overcome. Reimplantation of nonimmunogenic islets of Langerhans will provide the minute by minute regulation of blood glucose levels necessary to prevent the peripheral vascular complications of insulin dependent diabetes, including diabetic retinopathy, clotting disorders and glomerular nephritis.

The technology for the isolation and short-term culture of isolated islets of Langerhans from a number of species is well established (Scharp, 1973, 1980). Most investigators have reported a time dependent loss in the ability of isolated islets to secrete insulin in response to a glucose challenge (Andersson, 1976; Buitrago, 1975; Kostianovsky, 1972; Lacy, 1967). Since isolated islets of Langerhans have potential for therapeutic transplantation into Juvenile Onset Diabetics and since the mechanics of such transplants are far simpler than for transplantation of intact pancreases, we have attempted to develop methods for long-term storage of isolated islets. Practical long-term preservation will require freezing the islets in a fashion which is not deleterious to either the morphology or the physiological responsiveness of the islets. The development of techniques for long-term storage without loss of either morphological integrity or physiological capacity will be a fundamental step in the development of a practical method for transplantation of islets between noninbred individuals. Long-term storage of isolated islets would allow sufficient time for tissue typing and relax the temporal constraints in matching histocompatible donors and recipients. Potentially, such technology may permit pooling of islets from several donors for a single transplant. These techniques may also be useful in long-term *in vitro* experiments where it is desirable to have an "unchanged" control islet to compare to experimentally treated tissue.

Several investigators have utilized continuous cooling protocols for the preservation of islets of Langerhans at cryogenic temperatures (Bank, 1979; Ferguson, 1976; Rajotte, 1977). Others have attempted to prolong the survival of islets *in vitro* by 1) liquid storage at 4-8°C (Ferguson, 1976; Frankel, 1976, 1978; Knight, 1973), 2) tissue culture at 24-37°C (Andersson, 1972; Ferguson, 1976; Frankel, 1976; Kostianovsky, 1972; Lacy, 1979; Nielsen, 1979; Ono, 1979; Weber, 1977), 3) using fetal pancreases (Agren, 1980; Brown, 1980; Hellerstrom, 1979; Kemp, 1978; Lazarow, 1978; Mazur, 1976).

The objective of our cryobiological studies is to devise methods for the long term preservation of islets of Langerhans. In order to accomplish this broad objective, it is essential to 1) develop an understanding of the events which occur preceding, during and subsequent to the cryobiological preservation of islets of Langerhans. Including: a) type(s) of cryoprotective agent(s), b) temperature of exposure to the agent(s), c) length of exposure to the agent(s) prior to freezing, d) continuous cooling rates, e) temperature and holding time for two-step cooling, f) rate of warming, g) dilution protocol following thawing. 2) Evaluation of frozen and thawed islets with respect to their ability to synthesize and secrete insulin, glucagon and somatostatin *in vitro* and to maintain morphological integrity after freezing. 3) Confirming the physiological competence of optimally preserved islets by the ability of transplanted islets to reverse streptozocine-induced diabetes.

We have operationally defined viability as the net insulin

release, i.e. the gross insulin release by a given tube of islets when exposed to high glucose concentrations minus the amount of insulin released when the same tube of islets are exposed to low concentrations of glucose. Net insulin release was determined both before and after freezing so that the percent change in viability can be determined. To assess the efficacy of the freezing procedures, all islets were placed in organ culture for a minimum of 18 hours prior to the glucose challenge.

Our studies showed that isolated rat islets of Langerhans can be frozen and thawed with little alteration in their physiology or morphology. The critical freezing parameters were found to be the type of cryoprotective agent used, the time and temperature of exposure to Me_2SO, the rate of cooling, the temperature at which dilution from the freezing medium occurs and the presence of serum during the dilution procedures. When the freezing parameters are optimized, between 75 and 95% of the islets survived as measured by net glucose sensitive insulin release.

Our approach differs in many respects from the approach used by other investigators attempting cryogenic preservation of isolated islets (Ferguson, 1976; Kemp, 1978; Mazur, 1976; Rajotte, 1977). Other investigators have fully equilibrated islets in a balanced salt solution containing Me_2SO or Me_2SO and 10% fetal calf serum as a cryoprotective agent, and have used cooling rates of 0.5 to 3°C per minute, followed by rapid warming. The assumption made by these investigators in attempting to devise strategies for the cryogenic preservation of islets of Langerhans has been that the islet is the osmotically reacting unit which must be protected from freezing damage (Bank, 1979). However, the capsule of the islet does not pose an osmotic barrier to the flow of water. This can be demonstrated by placing isolated islets in hypertonic solutions and observing that no change in islet diameter occurs. Therefore, the osmotically reacting unit must be the individual islet cell.

We utilized a relatively short exposure [6 min.] to 1M Me_2SO which may allow sufficient time for most of the osmotically active water to be removed from the cell but insufficient time for the Me_2SO to penetrate the cells and restore them to their isotonic volume. This partial osmotic dehydration apparently renders the cells less susceptible to subsequent freezing damage. As the extracellular solution is frozen, the cells dehydrate further in response to the osmotic pressure gradient. The lack of intracellular metastable ice crystals is indicated by the observation that rapid thawing is not necessary or even desirable (Bank, 1979). We found that maximum functional survival was obtained when the islets were thawed at 7.5°C per minute. If the islets contained intracellular ice, such a slow thawing rate would allow sufficient time for the ice crystals to grow and mechanically damage the cells. The relatively slow thawing rate coupled with the stepwise dilution procedures should minimize osmotic damage to the islets during warming and subsequent "dilution shock" (Farrant, 1977).

Previously we reported that most islets frozen under conditions which showed relatively high functional viability appeared morphologically intact after thawing. Functional viability increased approximately three-fold as the cooling rates were increased from 1°/min. to 75°/min. but decreased at faster cooling rates. The functional viability of cells frozen at 75°/min. was highest when the islets were warmed at rates >3.5°/min. For islets frozen with these optimized conditions, net insulin released into the culture medium during a glucose challenge was ~75% of the amount released by the same islets prior to freezing.

Having established an "optimal" cooling velocity in the presence of 1M Me_2SO, a number of other variables were tested including time of exposure to Me_2SO and dilution procedures. For such rapid cooling rates, maximal functional viability was obtained for islets exposed to 1M Me_2SO at 4° for 5 min. prior to freezing. Exposure times shorter than 1 min. or longer than 80 min. resulted in poor functional viability after freezing. The islets frozen in the presence of the culture medium containing Me_2SO released 1½ times more insulin than did islets frozen in phosphate buffered saline containing Me_2SO. Provided there was serum present in the dilution medium, the presence of serum CMRL had little or no effect on glucose dependent insulin release. When serum was present in the dilution medium, we found a five-fold increase in total glucose-dependent insulin release compared to islets diluted at 37° in the same medium with serum. Long term storage experiments showed no alteration in functional viability for islets stored at -196° for up to 1 1/2 years.

Although maximal functional viability for isolated islets was obtained after rapid cooling at ~75°C/min., it is difficult to reproducibly obtain such cooling rates with larger sample sizes. Thus, it became desirable to devise simpler, more reproducible methods. We have recently developed a two step cooling technique to preserve isolated islets in dimethyl sulfoxide [Me_2SO] (Bank, 1979). The two-step cooling procedure is a reproducible technique which requires a minimum of specialized apparatus (Farrant, 1976, 1977; McGann, 1976a,b). Less technical expertise is necessary for achieving satisfactory preservation because there is no need for precise control of the kinetics of cooling. After exposure to the Me_2SO, the islets are simply transferred to a specific subzero temperature for a specific time interval and then quenched in liquid nitrogen.

The functional survival of rat islets of Langerhans subjected to two-step freezing was dependent on both the temperature of the holding bath and the length of exposure at that temperature. In general, the lower the temperature of the holding bath, the shorter the time necessary to protect the islets from further cooling to -196°C. Relatively high functional survival was obtained when the islets were held at -25, -30 and -35°C for 15 minutes. Islets held in a -20°C bath showed a time dependent increase in their ability to withstand further cooling to liquid nitrogen, while those islets held in a -40°C bath had a high functional survival after 10 minutes but functional survival decreased with increasing holding time.

Since the islets of Langerhans comprise only approximately 1% of the mass of the total pancreas, much emphasis has been placed on devising techniques which would enable the isolation of large numbers of islets for morphologic, physiologic and studies on transplantation. Most investigators are currently using modifications of the technique of Lacy (1967). Basically, the technique consists of exposing the pancreas in a test animal, clamping off the pancreatic ducts, cannulating the proximal end of the duct and inflating it with Hanks Balanced Salt Solution. The pancreas is then excised, washed and repeatedly minced until pieces approximately 1 mm a side are obtained. The tissue is then digested in a buffered media containing crude collagenase. This digestion process is relatively inefficient since it relies upon the differential sensitivity of the acinar cells vs. the islets to crude collagenase. Presumably, the few islets which survive the digestion procedure are those which were located in the center of the minced tissue prior to the onset of digestion. During this prolonged digestion period, the islets are not only exposed to the crude collagenase, but the contaminating trypsin and chymotryptic enzymes, plus enzymes released from the zymogen granules in the disintegrating acinar tissue and the proteases released by the resident blood cells and macrophages. After digestion, the islets are usually isolated either by handpicking or by separation on an ascending Ficoll gradient.

We recently devised a new method of isolating islets of Langerhans which appears to be approximately in order of magnitude more efficient in the separation of this tissue (Bank, 1981). The method relies upon the differential sensitivity of islets of Langerhans and acinar tissue to the effect of freezing. Basically, the technique involves a rapid mincing of the excised pancreatic tissue at 4°C followed by a very brief collagenase digest (approximately 1/6 the normal digestion time). Once the extracellular matrix has been somewhat loosened by the action of the collagenase, the tissue is washed, cooled to 4°C and placed in a solution containing 1-1.5 M cryoprotective agent for approximately 2 min. Glycerol or sucrose are used as extracellular protective agents, since there is little or no permeation of these compounds at 4°C in intact cells. The tissue is then transferred directly to a bath at -28 to -30°C, held for 2 min. and immediately thawed. During this brief cooling step, ice crystals apparently form in most of the acinar cells which causes the mechanical disruption of the cells and tissue organization. The islet cells, being much smaller, are able to dehydrate more rapidly in the presence of the extracellular cryoprotective agent. Thus, most of the islet cells do not form intracellular ice crystals, nor are they damaged by the brief exposure to the hypertonic solution which forms as water is converted into ice crystals, leaving behind a hypertonic salt solution. After thawing and gradual dilution of the tissue from the extracellular cryoprotective agent, the tissue is washed and digested in a solution containing purified collagenase, dispase, DNAse and heparin. This combination of enzymes was chosen since its action is almost entirely on extracellular constituents. Islets with intact membranes are relatively uneffected by the treatment. Two to three 10 minute digestions with this enzyme solution causes the dissociation of virtually all of the acinar tissue. The released islets can then be washed free of the enzyme solution and isolated either by handpicking or on a Ficoll gradient if desired. Typical yields from a rat pancreas are 2,500-4,000 islets, compared to approximately 250 islets obtainable using the conventional collagenase digestion procedure. Of these isolated islets, approximately 75% appear to be morphologically intact and as judged by their appearance in phase contrast and Hoffman modulation contrast, light microscopy and approximately half are morphologically intact or judged by their reaction to fluorochromes such as ethidium bromide, propidium iodide and fluorescein diacetate and acridine orange. Currently more definitive viability assays are being performed including the ability of the islets to release insulin and glucagon in response to a glucose challenge.

As the techniques for the preservation of isolated human islets of Langerhans are perfected, islet transplantation should become an increasingly important procedure in the treatment of insulin dependent diabetic patients (Morris, 1980; Sutherland, 1980). The use of this islet isolation technique may increase the number of islets obtainable from a single donor sufficiently to overcome much of the variability currently observed in diabetic animals which have received islet transplantations.

REFERENCES

Agren, A., Andersson, A., Bjorken, C., Groth, C. G., Gunnarsson, R., Hellerstrom, C., Lindmark, G., Lundqvist, G., Petersson, B., and Swenne, I., 1980, Human fetal pancreas, culture and function in vitro, Diabetes, 29:64.

Andersson, A., Borg, H., Growth, C. G., Gunnarsson, R., Hellerstrom, C., Lundgren, G., Westman, J., and Ostman, J., 1976, Survival of isolated human islets of Langerhans maintained in tissue culture, J. Clin. Invest., 57:1295.

Bank, H. L., Davis, R. F., and Emerson, D., 1979, Cryogenic preservation of isolated rat Islets of Langerhans: Effect of cooling and warming rates, Diabetologia, 16:195.

Bank, H. L., and Reichard, L., 1981, Cryogenic preservation of isolated islets of Langerhans: Two-step cooling, Cryobiology, 18:489.

Bank, H. L., In press, Isolation of rat islets of Langerhans: a high yield method using differential sensitivity to freezing, Cryobiology.

Brown, J., Kemp, J. A., Hurt, S., and Clark, W. R., 1980, Cryopreservation of human pancreas, Diabetes, 29:70.

Buitrago, A., Gylfe, E., Hellman, B., Idahl, L. A., and Johansson, M., 1975, Function of microdissected pancreatic islets cultured in a chemically defined medium. I. Insulin content and release, Diabetologia, 11:535.

Farrant, J., Walter, C. A., and Knight, C. S., 1976, Cryopreservation and selection of cells, Cryoimmunologie/Cryoimmunology, Inserm., 62:61.

Farrant, J., Walter, C. A., Lee, H., and McGann, L. E., 1977, Use of two-step cooling procedures to examine factors influencing cell survival following freezing and thawing, Cryobiology, 14:273.

Ferguson, J., Allsopp, R. H., Taylor, R. M. R., and Johnston, I. D. A., 1976, Isolation long term preservation of pancreatic islets from mouse, rat and guinea pig, Diabetologia, 12:115.

Ferguson, J., Allsopp, R. H., Taylor, R. M. R., and Johnson, I. D. A., 1976, Isolation and preservation of islets from the mouse, rat, guinea-pig and human pancreas. Brit. J. Surg., 63:767.

Frankel, B. J., Gylfe, E., Hellman, B., and Idahl, L. -A., 1976, Maintenance of insulin release from pancreatic islets stored in the cold for up to 5 weeks, J. Clin. Invest., 57:47.

Frankel, B. J., Gylfe, E., Hellman, B., Idahl, L. -A., Landstrom, U., Lovtrup, S., and Sehlin, J., 1978, Metabolism of cold-stored pancreatic islets, Diabetologia, 15:187.

Hellerstrom, C., Lewis, N. J., Borg, H., Johnson, R., and Freinkel, N., 1979, Method for large-scale isolation of pancreatic islets by tissue culture of fetal rat pancreas, Diabetes, 28:769.

Knight, M. J., Scharp, D. W., Kemp, C. B., Ballinger, W. F., and Lacy, P. E., 1973, Effects of cold storage on the function of isolated pancreatic islets, Cryobiology, 10:89.

Kemp, J. A., Mullen, Y., Weissman, H., Heininger, D., Brown, J., and Clark, W. R., 1978, Reversal of diabetes in rat using fetal pancreases stored at -196°C, Transplantation, 26:260.

Kostianovsky, M., Lacy, P. E., Greider, M. H., and Still, M. F., 1972, Long term (15 days) incubation of islets of Langerhans isolated from adult rats and mice, Lab. Invest., 27:53.

Lacy, P. E., Davie, J. M., and Finke, E. H., 1979, Prolongation of islet allograft survival following in vitro culture (24°C) and a single injection of ALS, Science, 204:312.

Lacy, P. E., and Kostianovsky, M., 1967, Method for isolation of intact islets of Langerhans from the rat pancreas, Diabetes, 16:35.

Lazarow, A., Wells, L. J., Carpenter, A. M., Hegre, O. D., Leonard, R. J., and McEvoy, R. D., 1973, Islet differentiation, organ culture and transplantation, Diabetes, 22:877.

Matas, A. J., Sutherland, D. E. R., and Najarian, J. S., 1976, Current status of islet and pancreas transplantation in diabetes, Diabetes, 25:785.

Matas, A. J., Sutherland, D. E. R., Steffes, M. W., and Najarian, J. S., 1977, Islet transplantation, Surg. Gynecol. Obstet., 145:757.

Mazur, P., Kemp, J. A., and Miller, R. H., 1976, Survival of fetal rat pancreases frozen to-78 and -196°C, Proc. Natl. Acad. Sci. USA 73:4105.

McGann, L. E., and Farrant, J., 1976a, Survival of tissue culture cells frozen by a two-step procedure to -196°C. I. Holding temperature and time, Cryobiology, 13:261.

McGann, L. E., and Farrant, J., 1976b, Survival of tissue culture cells frozen by a two-step procedure to -196°C. II. Warming rate and concentration of dimethyl sulphoxide, Cryobiology, 13:269.

McGann, L. E., 1978, Differing actions of penetrating and nonpenetrating cryoprotective agents, Cryobiology, 15:382.

Nielsen, J. H., Brunstedt, J., Andersson, A., and Frimodt-Moller, C., 1979, Preservation of beta cell function in adult human pancreatic islets for several months in vitro. Diabetologia, 16:97.

Ono, J., Takaki, R., Okano, H., and Fukuma, M., 1979, Long-term culture of pancreatic islet cells with special reference to the Beta-cell function, In Vitro, 15:95.

Rajotte, R. V., Stewart, H. L., Voss, W. A. G., Shnitka, T. K., and Dossetor, J. B., 1977, Viability studies on frozen-thawed rat islets of Langerhans, Cryobiology, 14:116.

Weber, C. J., Hardy, M. A., Lerner, R. L., and Reemtsma, K., 1977, Tissue culture isolation and preservation of human cadaveric pancreatic islets, Surgery, 81:270.

HYDRATION DEPENDENT LIPID PHASE TRANSITIONS IN A BIOLOGICAL MEMBRANE

John H. Crowe and Lois M. Crowe

Department of Zoology
University of California
Davis, California 95616, U. S. A.

INTRODUCTION

Phospholipids spontaneously form ordered structures in water, usually bilayers with the hydrocarbon chains in a fluid state, depending on the temperature, length of hydrocarbon chains, and the nature of the headgroup(1,2). When these same phospholipids are dehydrated they undergo phase transitions to the gel state (again depending on the temperature) or to other crystalline states such as the well-known hexagonal II (H_{II}) phase, in which the lipids are reorganized into tubes with the polar head groups oriented into the aqueous phase in the center of the tube. These phase transitions are well documented in pure phospholipid systems, but they have been poorly studied in complex mixtures of phospholipids or in intact biological membranes, even though there is some suggestion that such mixtures of phospholipids may exhibit different phase behavior from that of pure phospholipids (3). We have undertaken such a study, using isolated sarcoplasmic reticulum (SR) as a model membrane, and we report some of our progress in this regard in the present paper.

RESULTS AND DISCUSSION

1. Physical State of Water Near Membranes

Using differential thermal analysis and differential scanning calorimetry, Chapman, et al. (1) showed that about 0.3g H_2O/g phospholipid (g/g) is tightly bound to the phospholipid and behaves differently from the bulk water in that it does not freeze, even at liquid nitrogen temperatures. Since that time a variety of other physicochemical techniques have been applied to this problem (4), and most agree that phospholipids strongly bind about 0.3 g/g, which amounts to about 10-12 H_2O molecules/ phospholipid. In addition, proteins appear to bind a similar amount of H_2O (5). Following this lead, we have measured the freezable H_2O content of SR vesicles as described previously (6). The results show clearly that as the water content of the SR is decreased, there is a decline in the equilibrium freezing point until a limiting water content of 0.28-0.32 g/g is reached, below which freezable H_2O was not detected. Similarly, we have measured the non-freezable water content of SR vesicles using calorimetric methods, as previously described (6). These measurements again showed that the membranes contain about 0.3 g non-freezable H_2O/g membranes.

There is some suggestion in the literature that this unusual H_2O may be associated with head groups in phospholipids, while others have suggested that H2O exists in an ice-like form near the hydrocarbon chains (7). However, for our present purposes, no matter where the water resides in the membrane, its removal might be expected to result in upheavals in membrane structure and function. This is indeed the case, as shown in the following paragraphs.

2. Thermal Analysis During Dehydration

We have developed a simple procedure for measuring thermal activity in a membrane sample during dehydration that permits simultaneous measurement of the water content. Typical results with this procedure (Fig. 1) show that in the region of 0.3 g/g, a series of exotherms are found, suggesting that some components of the membranes are undergoing crystallization.

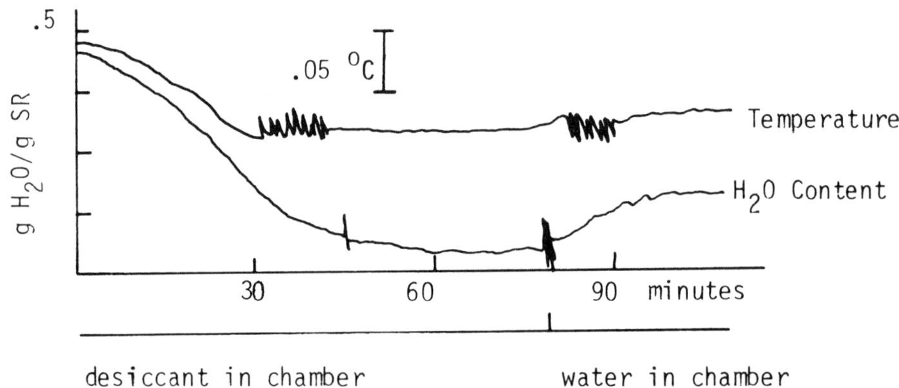

Fig.1. Thermogram and H_2O content of SR vesicles undergoing dehydration in the presence of a desiccant and rehydration over H_2O vapor. Note the series of exotherms during dehydration and endotherms during rehydration, both at water contents in the region of 0.3g H_2O/g SR.

When the membranes that had been dried were rehydrated over water vapor, they adsorbed water and again in the region of 0.3 g/g a series of endotherms was seen (Fig. 1). We suspect that these endotherms represent a return of the membrane components to a liquid-crystalline state.

3. NMR Studies

Several workers have shown that ^{31}P NMR can be used to detect phase transitions from the bilayer to other phases (3,8). While interpretation of the NMR spectra is not unequivocal (8,9), NMR is nevertheless a useful, non-invasive technique from which one can make suggestions about the phase behavior of phospholipids in a dynamic fasion. The hydrated vesicles show a line shape and chemical shift that has been identified with the bilayer phase, while the dehydrated membranes show spectra similar to what has been identified with H_{II} (Fig. 2). Curiously, this H_{II} signal predominates the spectra in the dry membranes even though the freeze fracture appears to show that much of the membrane is in a bilayer. When the dry samples are cooled in the NMR tube from 25°C to -10°C, the H_{II} signal gradually undergoes a chemical shift and change in line shape to one typical of bilayer. Based on this finding, we suggest that the H_{II} siganl may be due to phase separated phosphatidylethanolamine (PE) which is known to exist in H_{II} in pure preparations of the lipid at room temperature and to shift to bilayer phase when cooled in bulk solution.

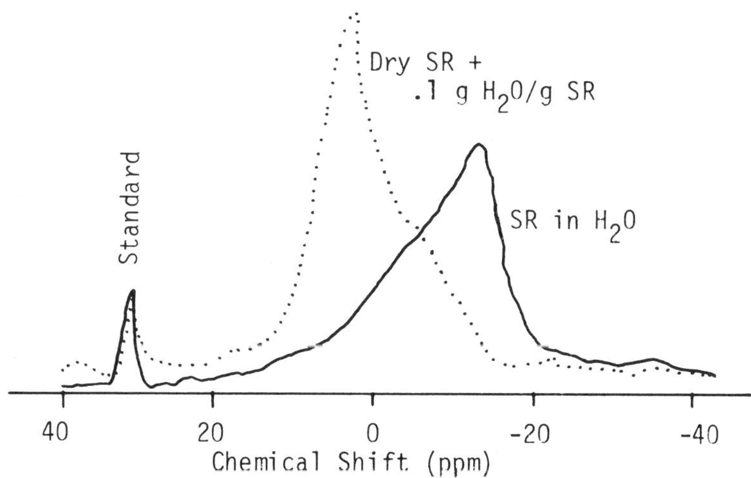

Fig. 2. ^{31}P NMR spectra of SR in bulk H_2O and partially hydrated SR that had previously been dried. The spectrum of the sample in bulk H_2O is characteristic of the bilayer phase, while that of the "dry" sample is characteristic of hexagonal II phase (but see references 8,9 for precautions).

4. <u>Freeze Fracture Studies</u>

We have confirmed that membrane phospholipids do undergo phase transitions leading to complex crystals during dehydration, using freeze fracture. The results show that the fresh vesicles are of approximately uniform size and possess intramembrane particles, known to represent the Ca-ATPase, mainly on the PF face of the membrane (Fig. 3). When the vesicles are dehydrated, massive fusion occurs, accompanied by phase separation of components like cholesterol (which forms crystals, Fig. 4), phospholipids, and intramembrane particles. Further dehydration results in formation of H_{II} crystals (Fig. 5) and more complex crystals, the structure of which is only poorly understood (Fig. 6).. When the dry membranes are rehydrated, large vesicles are obtained which possess redistributed intramembrane particles (Fig. 7). We have previously shown that these rehydrated vesicles have lost the capacity for Ca transport (10,11).

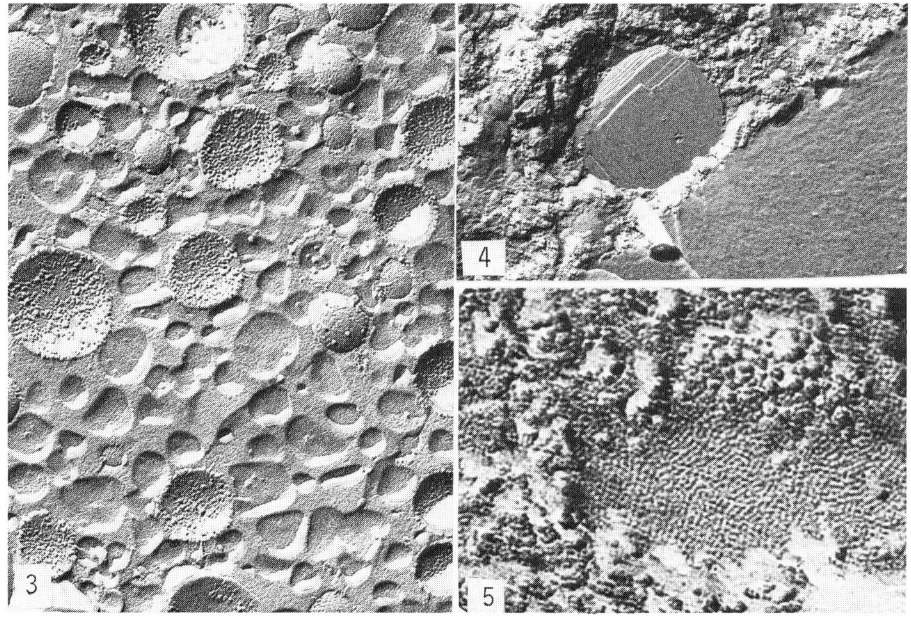

Figs. 3-5. Freeze fractures of SR vesicles in bulk water (Fig. 3), a cholesterol crystal in dehydrated SR (Fig. 4), and a cross-fracture through H_{II} phase rods in dehydrated SR (Fig. 5).

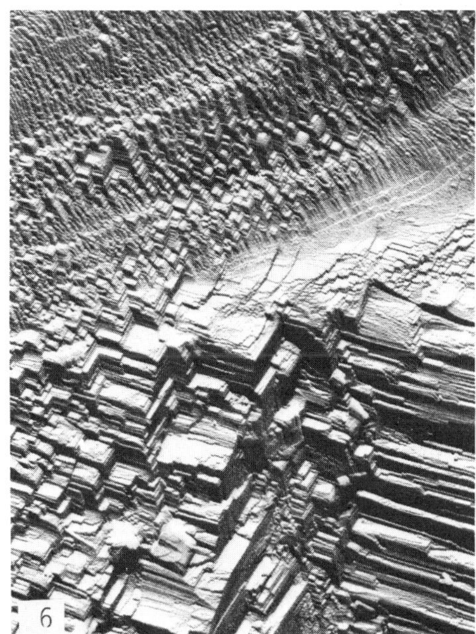

Figs. 6 and 7. Freeze fractures of SR. A complex crystal in dry SR (Fig. 6), rehydrated SR vesicles (Fig. 7). Note the distribution of intramembrane particles (known to represent the Ca-ATPase in these membranes) in Figure 7 and compare to Figure 3.

CONCLUSIONS

Intact biological membranes have physical properties and hydration dependent phase behavior similar to those of isolated phospholipids and proteins. As we have shown here and elsewhere, removal of the final 0.3 g H_2O/g membrane results in massive and deleterious changes in the membrane. Nevertheless, numerous organisms are capable of surviving such dehydration (10,11). The mechanisms by which they do so constitute the subject of our present studies (12).

REFERENCES

1. D. Chapman, R.M. Williams, and B.D. Ladbrooke. Physical studies of phospholipids. VI. Thermotropic and lyotropic mesomorphism of some 1,2 diacyl-phosphatidylcholines (lecithins). Chem. Phys. Lipids 1: 445 (1967)

2. J. Stumpel, E. Hansjorg, and A. Nicksch. X-ray analysis and calorimetry on phosphatidylcholine model membranes. The influence of length and position of acyl chains upon structure and phase behavior. Biochim. Biophys. Acta 727: 246 (1983).

3. P.R. Cullis, G.W.M. Van Dijck, B. DeKruijff, and J. De Gier. Effect of cholesterol on the properties of equimolar mixtures of synthetic phosphatidylethanolamine and phosphatidylcholine. Biochim. Biophys. Acta 513: 21 (1978).

4. H. Hauser. Water/phospholipid interactions. In: "Water Relations of Foods", R.B. Duckworth, ed., Academic Press, London (1975).

5. M. Ruegg, U. Moor, A. Lukesch, and B. Blanc. Hydration and thermal stability of α-lactalbumin. A calorimetric study. In: "Application of Calorimetry in Life Science", A. Lamprecht and B. Schaarschmidt, eds., De Gruyter, Berlin (1977).

6. J.H. Crowe, L.M. Crowe, and S.J. O'Dell. Ice formation during freezing of Artemia cysts of variable water contents. Mol. Physiol. 1: 145 (1982).

7. C.-H. Chen, D.S. Berns, and A.S. Berns. Thermodynamics of carbohydrate-lipid interactions. Biophys. J. 36: 359 (1981).

8. J. Seelig. ^{31}P nuclear magnetic resonance and the head group structure of phospholipids in membranes. Biochim. Biophys. Acta 515: 105 (1978).

9. J.H. Noggle, J.F. Marecek, S.B. Mandal, R. Van Venetie, J. Rogers, M.K. Jain, and F. Ramirez. Bilayers of phosphatidylglycerol and phosphatidylcholesterol give ^{31}P NMR spectra characteristic for hexagonal and isotropic phases. Biochim. Biophys. Acta 691: 240 (1982).

10. J.H. Crowe, and L.M. Crowe. Induction of anhydrobiosis: membrane changes during drying. Cryobiology 19: 317 (1982).

11. J.H. Crowe and J.S. Clegg, eds. "Dry Biological Systems", Academic Press, New York (1978).

12. J.H. Crowe, L.M. Crowe, and S.A. Jackson. Preservation of structural and functional activity in lyophilized sarcoplasmic reticulum. Arch. Biochem. Biophys. 220 (in press).

MECHANISMS OF PROTON-HYDROXIDE FLUX ACROSS MEMBRANES

David W. Deamer

Department of Zoology
University of California
Davis, CA 95616

J. Wylie Nichols

Department of Physiology
Emory Univ. Medical School
Atlanta, GA

ABSTRACT

In both model and biological membranes, measurements of proton-hydroxide flux produce permeability coefficients that are orders of magnitude greater than those obtained with other monovalent ions. As a working hypothesis, we suggest that this large variance can be explained by the flux of protons along hydrogen-bonded strands of associated water or protein in the hydrophobic plane of the membrane.

INTRODUCTION

The barrier properties of biological membranes and model membranes (lipid bilayers) have been studied for a number of years, and there is reasonable agreement about relative permeabilities to various ions and molecules. For instance, in liposome systems typical values for cations like sodium and potassium are in the range of 10^{-10} to 10^{-14} cm/s (1,2) and anions like chloride are in the range of 10^{-11} cm/s (3). In the past few years, several laboratories have attempted to add proton and hydroxide permeabilities to this list. Surprisingly, the reported values are typically much higher than expected from knowledge of other ion permeabilities. Furthermore, the results are remarkably diverse, and the extreme values disagree by at least five orders of magnitude. Most of these measurements have been made by measuring decay rates of pH gradients established across liposome membranes, and some examples of reported values are given in Table 1. As can be seen, measured values range from 10^{-3} to 10^{-9} cm/s. Since it has been widely assumed that membranes, and particularly the lipid

Table 1. Proton-hydroxide permeability coefficients measured in lipid bilayer systems.

System	Measurement	Permeability coefficient cm/s	Reference
LUV, PC:PA 98:2	9-AA fluoresence	$P_{net} = 1.4 \times 10^{-4}$	Nichols et al. (4)
LUV, PC:PA 90:10	Glass electrode	$P_{net} = 3.2 \times 10^{-4}$	Nichols and Deamer (8)
LUV, POPC	Glass electrode	$P_{net} = 0.7 \times 10^{-4}$	Deamer and Nichols (9)
LUV, mixed PL	Pyranine fluorescence	$P_{net} = 1.7 \times 10^{-4}$	Rossignol et al. (5)
SUV, PC	3P-UBF fluorescence	$P_{OH} = 1.4 \times 10^{-4}$	Pohl (6)
SUV, DMPA	Cresol red absorbance	$P_{net} \sim 10^{-3} - 10^{-5}$	Elamrani and Blume (10)
LUV, PC	Glass electrode	$P_{H} = 3 \times 10^{-9}$	Nozaki and Tanford (2)
SUV, PC	Diffusion potential	$P_{H} \sim 10^{-8} - 10^{-9}$	Cafiso and Hubbel (12)
PLM, PC	Conductance	$P_{H} = 3 \times 10^{-9}$	Gutknecht and Walter (11)
PLM, PC-cholesterol 1:1	Conductance	$P_{OH} = 1.8 \times 10^{-9}$	Gutknecht and Walter (25)

Abbreviations: LUV, large unilamellar vesicles (liposomes); SUV, small unilamellar vesicles; PLM, planar lipid membranes; PC, phosphatidylcholine; PA, phosphatidic acid; POPC, synthetic 1-palmitoyl, 2-oleoyl phosphatidylcholine; DMPA, dimyristoyl phosphatidic acid; 9-AA, 9-aminoacridine; 3P-UBF, 3-palmitoyl-7-oxyumbilliferone.

bilayer moiety, are essentially impermeable to protons unless specific carriers or channels are present, it is important to resolve this discrepancy in order to provide some convincing values for proton-hydroxide permeability.

Definitions of proton flux and permeability coefficient

Before continuing, it will be useful to summarize some of the mechanisms by which proton equivalents can cross membranes, and to define relevant terms. In general, it is not possible to distinguish between proton and hydroxide flux, particularly in neutral pH ranges. Therefore some investigators (4,5) have simply reported values as net proton-hydroxide flux according to the relationships below:

1) J = measured flux (moles $cm^{-2}s^{-1}$) and $J = P \Delta [H^+]$

2) $J_{net} = P_H([H^+]_o - P_H [H^+]_i) + P_{OH}([OH^-]_i - [OH^-]_o)$

3) $P_{net} = P_H + P_{OH}$

Other investigators have worked at relatively low or high pH ranges, and assumed that the flux contributing to decay of a pH gradient was composed of the predominant species present: protons at low pH and hydroxide at high pH. To distinguish between these possibilities, we will use P_{net} when referring to the net proton-hydroxide permeability coefficient and P_H and P_{OH} for specific proton and hydroxide permeability coefficients. The most general term we will use is proton equivalent, which simply refers to the measured decay of a pH gradient without specifying the mechanism.

Some possible ways by which proton equivalents might cross membranes are shown in Figure 1. Discrete protons or hydroxide ions may cross either as bare ions or as hydrated species (1 A,B). A second possibility to be described in detail later is that proton flux occurs along hydrogen-bonded strands of water or protein (1C).

Proton equivalents may also be carried by weak acids, bases or protonophores (1 D,E,F) but this does not reflect intrinsic proton permeability and must be excluded by the experimental conditions. We must also exclude artifactual flux through membrane defects produced by contaminants such as oxidation damage, hydrolysis products, or traces of solvents and detergents remaining from the preparation method.

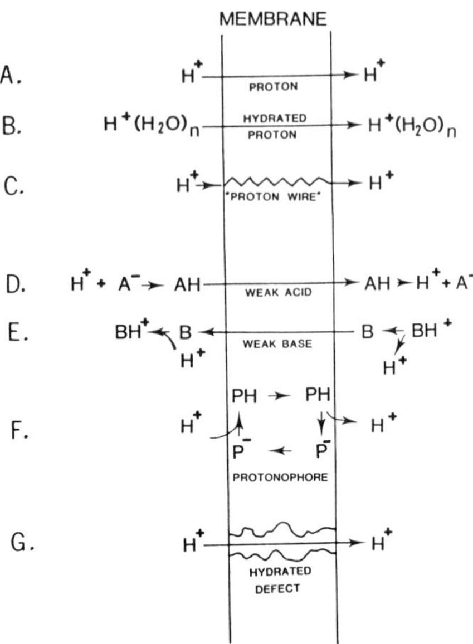

Figure 1. Mechanisms for proton flux across membranes. (See text for details).

Comparison of liposome systems

As a first step toward understanding how the extraordinary range of reported values for proton-hydroxide permeability could have arisen, we will ask what experimental variables were present in the different liposome systems that have been studied to date. The most obvious include differences related to lipid composition, such as surface charge and acyl chain properties. A second variable is the size of the vesicles, which range from 25-50 nm in small unilamellar preparations prepared by sonication, and 0.1-0.2 μm in large unilamellar preparations. However, these variables do not appear to be an important factor, since no pattern emerges when proton-hydroxide permeabilities from different liposome preparations are compared with respect to lipid composition and vesicle diameter.

A third possible variable is that artifactual proton transport processes may be occurring. As noted earlier, the presence of lipid hydrolysis products, oxidation damage, and solvent or detergent contamination could all conceivably enhance proton flux. This possibility is more difficult to rule out, since no systematic attempts have been made to compare proton flux under all the different conditions that might give rise to such artifacts.

However, since the artifactual flux would always be relatively high, we may again compare the experimental systems to see if certain conditions always produced high values for proton-hydroxide permeability. High values have been measured in liposomes prepared by sonication (6,7) solvent vaporization (4,8) detergent dialysis (2,9) simple dispersion by agitation (10), and liposome compositions included natural lipids (4,9) synthetic lipids (7,10) and lipid mixtures (5). The only factor common to all of these experiments is exposure to organic solvents. However, this was also true of those studies that determined relatively low permeabilities, so one difference might be that certain experimental conditions permit significant traces of solvents to remain. To be finally convincing, a careful study of solvent effects on proton-hydroxide permeability is required, and until this is resolved, the possibility that trace solvents enhance proton flux must remain open.

A fourth variable is the magnitude of the pH gradient used, since all of the experimental procedures involved measurement of proton flux down pH gradients. Protons are charged species, and if it is assumed that no other ionic species are able to cross the membrane, proton flux down large pH gradients will develop a significant diffusion potential that ultimately will reach electrochemical equilibrium with the driving force. As a result, the net flux of protons will approach zero. Of course, under experimental conditions there will always be a slow leak of other ions, either as counter- or co-ions, so that the final measured decay rate will simply reflect the leak of the other ions driven by the proton diffusion potential.

Several investigators have tested this possibility. For instance, Biegel and Gould (7) found two rates at which large pH gradients decayed in a liposome system, one with a halftime of about 300 milliseconds, the other with a halftime measured in minutes. The authors proposed that the first rate reflected the intrinsic rate of proton flux before a significant diffusion potential developed, and that the second rate was limited by the proton diffusion potential. This could be checked by addition of valinomycin and potassium, which would be expected to provide a counterion current of potassium that would release the diffusion potential and permit proton flux to increase. When this was tested experimentally, the second rate greatly increased, and the authors concluded that a proton potential had in fact developed and was limiting proton-hydroxide flux.

This result is relevant to the study by Nozaki and Tanford (2) who reported the lowest measured value for proton permeability of liposomes. In this paper, relatively large pH gradients (≃ 3 pH units) were permitted to decay, and it was assumed that other ions present in the system provided sufficient flux so that proton diffusion potentials did not develop. In a typical experiment, decay rates were measured over periods up to 40 hours, much longer than the rates observed by other workers, and reported values for proton permeability were in the range of 10^{-9} cm/s. The authors also assumed that most of the proton flux occurred as neutral protonated species such as chloride and nitrate. When the assumed flux was subtracted from the measured flux, a final calculated value of 10^{-12} cm/s was obtained for proton permeability.

These two assumptions -- lack of a diffusion potential and proton transport as a neutral species -- could markedly affect the final estimate of intrinsic proton permeability. We have therefore repeated the experiments and tested the assumptions in two ways (9). The first was tested by adding valinomycin so that a potassium counterion current could occur if a diffusion potential were limiting proton flux. The second was tested by carrying out the experiments in media containing sulfate, an anion that has been shown to be unable to carry protons as a neutral species (11). Some results are shown in Figure 2, which is redrawn from reference 9. First, note that proton efflux is in fact very slow in the absence of counterion current. The apparent permeability coefficient calculated from the decay between 1 and 2 hours is 9 x 10^{-9}, in the same range as that reported by Nozaki and Tanford. However, when valinomycin was added, the flux increased markedly. This suggested that efflux of protons was limited by a diffusion potential. Since these fluxes were measured in sulfate media, we could also conclude that proton flux did not necessarily occur as a neutral species. We went on to repeat the experiments under conditions that permitted only a small pH gradient to decay, and found that valinomycin had no effect. The measured proton permeability was in the range of 10^{-4} cm/s, similar to that reported by ourselves and other investigators (see Table 1).

Proton-Hydroxide Permeability of Planar Lipid Membranes

We may now turn our attention to proton-hydroxide permeability and conductance values measured for planar lipid membranes. In past studies, PLM conductance was measured at very low pH ranges, for instance, 0.1 M HCl, and the specific proton conductance was found to be in the same range as that measured for sodium or potassium (11). However, the comparison of proton-hydroxide permeability values obtained at low pH to those obtained in the neutral pH range normally used in liposome studies may be misleading. For example, in the study by Hopfer et al. (12) diffusion potentials resulting from equal concentrations of HCl on both sides of a phosphatidylglycerol bilayer indicated $P_H/P_{Cl} \cong 19$. However, when proton gradients in the pH range of 5.1 were established across the same membrane, proton potentials as high as 22 mV/pH unit were observed. If the Goldman equation is used to calculate relative permeabilities, the maximum $P_H P_{Cl} = 4.5 \times 10^4$.

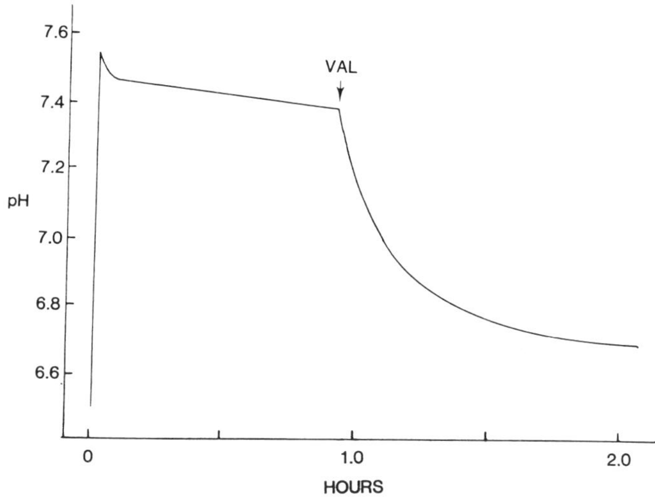

Figure 2. Effect of counterion current on proton flux. Liposomes were prepared in 40 mM potassium aspartate, pH 4.2, in 0.25 M potassium sulfate at a concentration of 10 mg phosphatidylcholine/ml. 0.5 ml of the liposome preparation was added to 0.5 ml of the same buffer in the chamber of a recording pH meter, and sufficient NaOH was added to bring the external pH to near 7.5. This produced a pH gradient of 3.3 pH units across the liposome membranes. Since aspartate is a poor buffer in the neutral pH range, the change in pH that follows reflects proton efflux from the liposomes. Upon addition of 1.0 μM valinomycin (VAL) the flux rate markedly increased, suggesting that lack of counterion current was limiting efflux of protons during the initial phase. If the same experiment was done in sodium sulfate so that potassium was excluded from the medium, valinomycin had little effect on efflux rate.

The fact that the ratio P_H/P_{Cl} varies by over three orders of magnitude between pH 1 and pH 5.1 illustrates the difficulty in comparing permeability measurements made at different pH ranges even in the same experimental system.

The fact that proton-hydroxide diffusion potentials can be established at neutral pH ranges suggests that the bilayer conductances measured in distilled water or in low salt may be due to proton-hydroxide flux. As the concentration of NaCl is reduced below 10 mM the conductance of phosphatidylcholine bilayers approaches that measured in distilled water -- in the range of

3×10^{-8} S-cm^2 (13). If we assume that this conductance is due to proton-hydroxide flux, the calculated permeability for net proton-hydroxide at pH 7.4 is 0.4×10^{-4} cm/sec, very similar to the net proton permeability calculated for liposomes in this pH range. Further experiments designed specifically to measure proton-hydroxide permeability of PLM's in the neutral pH range are required, but the evidence to date indicates that the proton-hydroxide permeability of liposomes and PLM's is similar when measured in the same pH range.

This leads to a more interesting question. Perhaps proton and hydroxide ions do not flux across phospholipid bilayers independently of each other. If so, the usual equations for calculating electrical conductances and permeability coefficients would not be appropriate. Earlier results from our laboratory (8) support this possibility, since the measured net proton-hydroxide flux for small pH gradients centered around pH 6, 7 and 8 could not be predicted by a model based on independent proton-hydroxide diffusion across the membrane. Instead, the results indicated that the apparent permeability coefficients varied inversely with the proton or hydroxide concentrations. In order to account for this surprising result, we have been led to consider the possibility that protons do not enter one side of the bilayer and leave the other, but instead may interact with hydroxide entering from the transbilayer surface to form water before net transfer can occur. This model would account for the pH independence of proton-hydroxide flux described above for liposomes since for a given pH gradient the product of proton activity and hydroxide activity on opposite sides of the membrane would be a constant at any pH. This model would also account for the widely divergent values for proton permeability calculated at different pH ranges because the flux would remain constant regardless of the proton activity.

Relative permeabilities of lipid bilayers

We may now consider an interesting observation: in the liposome systems, the measured values for proton permeability are all at least several orders of magnitude greater than equivalent values for other monovalent ions. In past studies (4,8) we have proposed that this difference could reflect an interaction between protons and water in the hydrophobic phase of the membrane. There is no doubt that a certain amount of water exists in this phase, since water is relatively permeable, with a permeability coefficient of 10^{-3} cm/s across lipid bilayers (15). It has also been shown that bulk phase water does not extend very far into the hydrophobic phase (16) and this has led to the general conclusion that water crosses the membrane by diffusion as monomers (17). However, we may speculate that some fraction of the membrane water exists in association with other water molecules. This would provide a unique mechanism for proton flux within the bilayer phase, since proton transfer along hydrogen bonds is a well-known phenomenon (18).

Elamrani and Blume (10) have tested this hypothesis in an interesting manner. It is known that certain lipids pass through temperature dependent phase transitions, and ion and water permeability have both been measured as a function of this transition (19,20). The general finding is that ion permeability reaches a maximum just at the phase transition, then decreases again, while water permeability simply increases by two orders of magnitude. It follows that if protons do have a mechanism depending on water content of the membrane, their permeability change during a lipid phase transition should reflect that of water, rather than being similar to other ions. To test this, Elamrani and Blume formed liposomes from dimyristoylphosphatidic acid (DMPA) taking advantage of the fact that this synthetic lipid could be obtained in a highly purified state and has a phase transition at 53°C. Also, because of its negative charge, DMPA forms small unilamellar liposomes by simple dispersion. It was found that proton permeability increased from 10^{-5} to 10^{-3} cm/s as the lipid passed through the phase transition. Furthermore, the activation energy of the increase was very high, in the range of 20 kcal, which also is near that of water (21). These investigators concluded that their results supported the hypothesis that proton flux across liposome membranes was somehow related to water in the hydrophobic phase. Rossignol et al. (5) had earlier reached similar conclusions based primarily on the high measured activation energy of proton flux.

Relation to biological membranes

We may now ask how these results relate to proton flux across biological membranes. To set this question in perspective, there is a basic problem related to proton permeability that must be solved by a variety of different membranes. Mitochondria, chloroplasts, bacteria, lysosomes, and chromaffin granules all have membrane transport systems that pump proton equivalents so that electrochemical gradients of protons are produced. It follows that for such membranes to function, they must be sufficiently impermeable to proton current so that the gradients do not decay as fast as they are formed. Obviously, the lipid bilayer must play a role as the primary barrier to free proton flux. From the results described earlier, it appears that the bilayer may be much more permeable to protons than expected from comparison with other ions. Does this present any difficulties for membranes that maintain proton gradients?

Table 2 shows a compilation of the published estimates of proton-hydroxide permeability in several biological membranes. It is enteresting that the estimate for mitochondria is an order of

Table 2. Proton permeabilities of biological membranes

Membrane	Permeability coefficient, cm/s	Reference
Mitochondria	$P_{net} \sim 10^{-3}$	Mitchell and Moyle (26) Nichols and Deamer (8)
Frog muscle sarcolemma	$P_{net} \sim 10^{-3}$	Itzuzu (27)
Sarcoplasmic reticulum	$P_{net} \sim 10^{-3}$	Meissner (28)
Erythrocyte membrane	$P_{OH} = 2 \times 10^{-4}$	Crandall et al. (29)
Root plasma membrane	$P_{H} = 0.6 \times 10^{-5}$	Pitman et al. (30)

magnitude greater than even the highest values obtained for model membranes. We may conclude that although the lipid bilayer may be more permeable than expected, it still offers a sufficient barrier to proton current. A moment's reflection suggests how this apparently paradoxical situation arises. Even though proton permeability may be much greater than that of other cations, concentrations of protons are in the micromolar range, while the concentrations of other ions are around 0.1 M. It follows that proton permeability may be relatively high, but the actual current of protons driven by micromolar concentrations is negligible with respect to the function of the organelles involved.

Mechanisms of proton flux

We may finally inquire into the mechanisms that produce a relatively high proton permeability in biological membranes. The obvious possibility is that there exist transmembrane proteins that provide either specific or non-specific channels for proton flux. It should be noted that proton permeability is again several orders of magnitude greater than that of other monovalent cations. For instance, in the axon membrane, which has specific channels for potassium, the potassium permeability coefficient for the resting membrane is in the range of 10^{-7} cm/s (22) while typical values for proton permeability are in the range of 10^{-3} cm/s for biological membranes.

We are therefore led to the conclusion that proton flux in biological membranes may also occur along a unique pathway. One intriguing mechanism has been published which considers that proton flux may occur along organized hydrogen bonds. This was originally described by Nagle and Morowitz as a "proton wire" (23) and consists of transmembrane protein chains with amino acid groups arranged in such a manner that protons can be transported as shown in Figure 3.

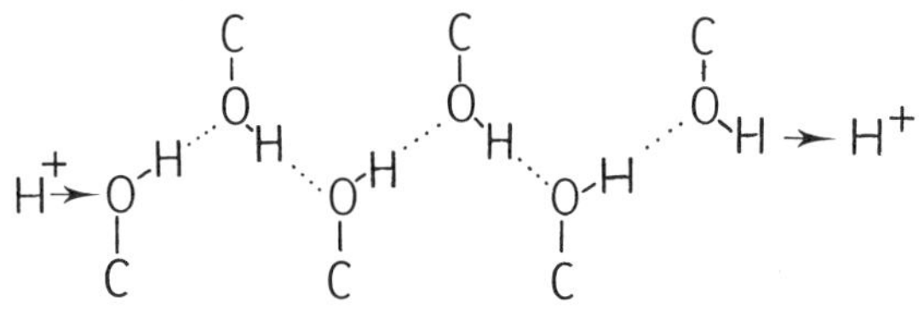

Figure 3. A "proton" wire." In this illustration, adapted from Nagle and Morowitz (23) serine hydroxyl groups are arranged to form a hydrogen-bonded strand. A proton entering one end of the strand forms an ionic defect that transports the charge to the right, where it leaves the strand. Rotation of the hydroxyl groups around the C-O bond must then occur to return the "wire" to its original state before a second proton can be transported. (See review by Scheiner (31) for a useful quantum mechanical discussion of this type of proton transport.)

A second model was suggested by Edmonds (24) who proposed that water could be organized in hydrogen bonded structures as shown in Figure 4. This was originally suggested as a means for hydrated

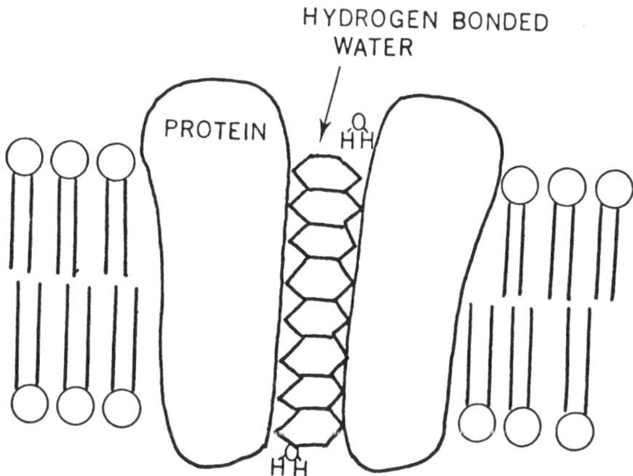

Figure 4. Edmonds (24) has proposed that transmembrane channel proteins could cause water to fall into a hydrogen bonded structure. The water takes the form of multiple dodecahedra with a central pore about 1 nm in diameter. The aqueous pore provides a low energy mechanism by which ions can cross membranes without losing water of hydration.

ions to find a low energy pathway through specific protein channels, but the hydrogen bonded structure would clearly act as a "proton wire" as well.

A third possibility follows from the fact that all biological membranes so far studied have high proton permeabilities, even though they may not be specialized for proton transport. Any defect in a lipid bilayer that is caused by an integral protein very likely will permit water to enter along with it, either to fill the voids caused by the presence of the protein strand, or carried with the polar groups composing the transmembrane peptide. In either case, it seems likely that the water will find favorable hydrogen bonding structures and provide the kind of "proton wire" envisaged by Nagle and Morowitz.

References

1. Hauser, H., Oldani, D. and Phillips, M.C. Biochemistry 12: 4507 (1973).

2. Nozaki, Y. and Tanford, C. Proc. Natl. Acad. Sci. USA 78: 4324 (1981).

3. Papahadjopoulos, D., Nir, S. and Ohki, S. Biochim. Biophys. Acta 266:561 (1972).

4. Nichols, J.W., Hill, M.W., Bangham, A.D. and Deamer, D.W. Biochim. Biophys. Acta 596:393 (1980).

5. Rossignol, M., Thomas, P. and Grignon, C. Biochim. Biophys. Acta 684:195 (1982).

6. Pohl, W.G. Z. Naturforsch 37c:120 (1982).

7. Biegel, C.M. and Gould, J.M. Biochemistry 20:3474 (1981).

8. Nichols, J.W. and Deamer, D.W. Proc. Natl. Acad. Sci. USA 77:2038 (1980).

9. Deamer, D.W. and Nichols, J.W. Proc. Natl. Acad. Sci. USA 80:1983 (in press).

10. Elamrani, K. and Blume A. Biochim. Biophys. Acta (in press).

11. Gutknecht, J. and Walter, A. Biochim. Biophys. Acta 641:183 (1981).

12. Hopfer, V., Lehninger, A.L. and Lennarz, W.J. J. Membrane Biol. 2:41 (1970).

13. Papahadjopoulos, D. and Ohki, S. In Surface Chemistry of Biological Systems. Plenum Press, 1970. pp. 155-173.

14. Cafiso, D. and Hubbel, W. Biophys. J. 33:114a (1981).

15. Bangham, A.D., DeGier, J. and Greville, G.D. Chem. Phys. Lipids 1:225 (1967).

16. Griffith, O.H., Dehlinger, P.J. and Van, S.P. J. Membrane Biol. 15:159 (1974).

17. Cass, A. and Finkelstein, A. J. Gen. Physiol. 50:1756 (1784).

18. Eigen, M. and DeMaeyer, L. Proc. R. Soc. London Ser A 247: 505 (1958).

19. Kanehisa, I.M. and Tsong, T.Y. J. Am. Chem. Soc. 100:424 (1978).

20. Papahadjopoulos, D., Jacobson, K., Nir, S. and Isac, T. Biochim. Biophys. Acta 311:330 (1973).

21. Peterson, D.C. Biochim. Biophys. Acta 600:666 (1980).

22. Katz, B. In Nerve, Muscle and Synapse. McGraw-Hill, p. 61. (1966).

23. Nagle, J.F. and Morowitz, H.J. Proc. Natl. Acad. Sci. USA 75:298 (1978).

24. Edmonds, D. Biochem. Soc. Symp. 46:91 (1980). London: The Biochemical Society.

25. Gutknecht, J. and Walter A. Biochim. Biophys. Acta 645:161 (1981).

26. Mitchell, P. and Moyle, J. Biochem. J. 104:588 (1967).

27. Itzuzu, K.T. J. Physiol. (Lond) 221:15 (1972).

28. Meissner, G. and Young, R.C. J. Biol. Chem. 255:6814 (1980).

29. Crandall, E.D., Klocke, R.A. and Forster, R.E. J. Gen. Physiol. 57:664 (1971).

30. Pitman, M.G., Anderson, W.P. and Schaefer, N. In Regulation of Cell Membrane Activities in Plants. Marre E. and Ciferri O. eds. pp. 147-160. North Holland, Amsterdam (1977).

31. Scheiner, S. Ann. New York Acad. Sci. 367:493 (1981).

ION PERMEABILITY AND MEMBRANE POTENTIAL IN VIRAL DISEASE

C.A. Pasternak, C.L. Bashfor, M.A. Gray
and K.J. Micklem*

Department of Biochemistry, St George's Hospital Medical School, Cranmer Terrace, London SW17 ORE, U.K.
*Present address: Department of Biochemistry, University of Oxford, South Parks Road, Oxford

INTRODUCTION

The ionic content of cells is an important determinant of their function. A raised intracellular Na^+ in lens cells has been shown to be a direct cause of cataract (Piatigorsky, 1980) and changes in intracellular monovalent cations, as well as in Ca^{2+} have been implicated in diseases varying from hepatitis (Alam et al., 1978) to hypertension (Reid, 1982; Swales, 1982). Indeed, in many diseases in which the surface membrane of cells is primarily affected, it is an altered cation content that appears to be at fault (Griffin and Pasternak, 1982). Following our observation that certain viruses make cells permeable to low molecular weight compounds and ions without lysis (Pasternak and Micklem, 1973, 1974), we have been studying the effects of viruses on surface membrane function: our aim has been to determine to what extent such changes underly the processes of viral infection and disease. In this report we concentrate on changes in ion permeability and membrane potential.

Our studies received impetus from the suggestion (Carrasco and Smith, 1976; Carrasco, 1977) that virally-mediated changes in membrane permeability, leading to an increase in intracellular Na^+, are part of the mechanism by which virally-infected cells switch from synthesising host proteins to the synthesis of viral proteins. In fact we found that the induction of non-specific permeability changes is limited to haemolytic paramyxoviruses (Poste and Pasternak, 1978), and that these act as toxins rather than as infectious agents (Pasternak and Micklem, 1983): thus the permeability changes result directly from the fusion (Knutton, 1978) of an inherently or potentially permeable viral membrane with the cell surface, and are not dependent on the introduction of infectious nucleic acid into cells (Wyke et al., 1980). Although, therefore, limited in clinical significance (Pasternak, 1983), the permeabilisation of cells by haemolytic paramyxoviruses has proved useful as a tool to study biological processes such as the Ca^{2+}-stimulated secretion of histamine from mast cells (Gomperts et al., 1981, 1983) or of ACTH secretion from anterior pituitary cells (Gillies et al., 1981; Pasternak, 1983). In the first part of this report we present data on the use of haemolytic paramyxovirus to study the surface membrane potential of cells in culture.

The second part is devoted to a study of precisely what changes in ion permeability and membrane potential do occur in cells infected with a number of different viruses.

RESULTS

The toxin-like action of haemolytic paramyxoviruses: effect on membrane potential

Haemolytic paramyxoviruses such as Sendai virus induce a non-specific permeability change in susceptible cells: monovalent cations (Pasternak and Micklem, 1974; Poste and Pasternak, 1978) and Ca^{2+} (Impraim et al., 1980) leak down their respective gradients, presumably because the cellular pumps, - (Na^+/K^+) ATPase and (Ca^{2+}) ATPase(s), - are not able to maintain an asymmetric distribution; at present, there is no evidence that the pumps themselves are affected (Poste and Pasternak, 1978). These changes, which are reversible and which can be controlled by extracellular Ca^{2+} (Pasternak and Micklem, 1973; Pasternak et al., 1976; Micklem and Pasternak, 1977; Impraim et al., 1980; Masuda and Goshima, 1980), lead to functional alterations in physiologically-competent cells: neuronal cells, for example, lose excitability, and heart cells stop beating; both types of cell recover their normal function within minutes (Forda et al., 1982). The clinical implications of such changes have been discussed elsewhere (Pasternak, 1983). The mechanism by which they occur is presumably the result of an increased conductance and a collapsed membrane potential: certainly Sendai virus has just this latter effect on human amnion (Okada et al., 1975), HeLa (Fuchs et al., 1978) and Lettre (Impraim et al., 1980) cells. But while it is well-documented that the membrane potential of excitable cells is indeed mainly due to the asymmetric distribution of K^+ across the cell membrane, several reports have recently suggested that this might not be so for non-excitable cells (Heinz et al., 1981; Hacking and Eddy, 1981; Pietrzyk et al., 1978). We have now used Sendai virus to explore this point. In order to measure membrane potential, we have used the dye oxonol-V (Smith et al., 1976; Bashford et al., 1979); because this dye is negatively-charged, it is much easier to handle than positively-charged dyes such as the cyanines, which report on mitochondrial, as well as surface, membrane potential (Philo and Eddy, 1978). First, we found that the membrane potential of Lettre cells (a line of Ehrlich ascites cells grown in mice) is relatively unaffected by alteration of extracellular K^+ (Fig. 1); the potential depolarises only when the K^+ ionophore valinomycin is added in the presence of external K^+ ions.

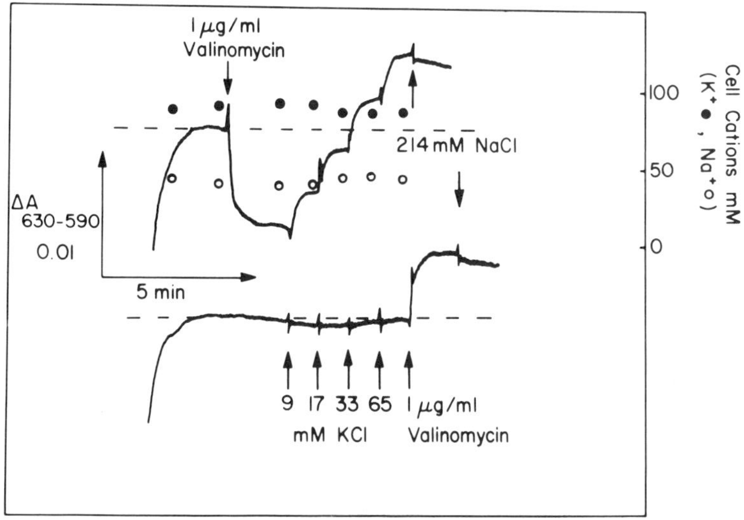

Fig. 1 Effect of extracellular K^+ on membrane potential of Lettre cells

Lettre cells ($5x10^6$/ml) were suspended with stirring in a medium containing 150mM NaCl, 5mM KCl, 5mM Na-Hepes, 5mM glucose, 2mM $CaCl_2$, 1mM $MgCl_2$, 1mM Na-pyruvate, 2μM oxonol-V, pH 7.4, at 37°C, and the difference between the absorbance at 630 and 590nm measured continuously. Samples were removed at intervals and intracellular Na^+ and K^+ measured after spinning through oil (Impraim et al., 1979).

Both traces show the effect of increasing external K^+ by the amounts (final concentration) indicated by the arrows.

In the upper trace addition of valinomycin is seen to hyperpolarize the cells (indicated by a decrease in absorbance); subsequent addition of K^+ depolarizes the cells. Note that cellular Na^+ and K^+ are unaltered throughout the experiment. In the lower trace addition of K^+ is seen to cause no change until valinomycin is added.

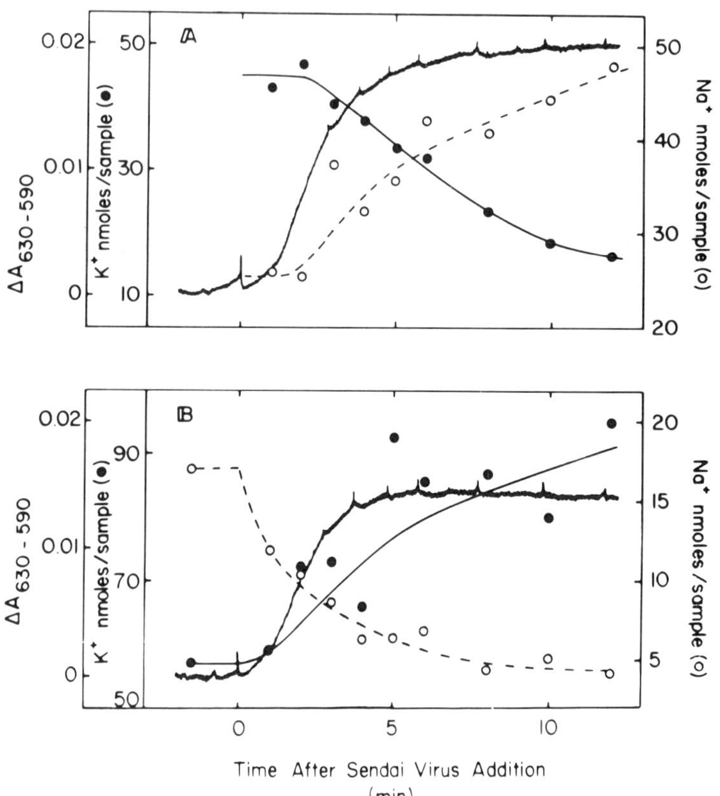

Fig. 2 Effect of Sendai virus on membrane potential and cation content of Lettre cells

Lettre cells ($5x10^6$/ml) were suspended with stirring in a medium containing 5mM Na-Hepes, 1mM $MgCl_2$, 2μM oxonol-V, pH 7.4, and either 150mM NaCl, 5mM KCl (A) or 155mM KCl (B), at 33°C, and the difference between the absorbance at 630 and 590nm measured continuously. At 0 time Sendai virus (2 HAU/ml final concentration) was added, and samples removed and analysed as described for Fig. 1. In each case Sendai virus causes a depolarization (indicated by an increase in absorbance); in A, intracellular K^+ decreases (while intracellular Na^+ increases), whereas in B intracellular K^+ increases (while intracellular Na^+ decreases).

This is in direct contrast to excitable cells, which are depolarized by increasing extracellular K^+. The insensitivity of Lettre cells to K^+ allowed us to suspend them in high K^+ (Na^+-free) medium without abolishing their membrane potential (Fig. 2). We then compared the effect of Sendai virus on cells suspended in low and in high external K^+. In each case membrane potential is abolished: that this is due to the 'permeabilising' effect of Sendai virus as previously described (Pasternak and Micklem, 1973; Poste and Pasternak, 1978; Impraim et al., 1980) is shown by the fact that depolarization is halted and partially reversed, when Ca^{2+} is added (Impraim et al., 1980). The decrease of membrane potential is accompanied by a loss of intracellular K^+ from cells suspended in low K^+, and by a gain of intracellular K^+ in cells suspended in high K^+. These results are clearly incompatible with the notion that membrane potential in Lettre cells is set by the K^+ gradient. Other experiments with inhibitors have confirmed our conclusion that, under the conditions of our experiments, membrane potential in Lettre cells is set predominantly by cellular pumps, not transmembrane ion gradients.

Ion and membrane potential changes in virally-infected cells

The hypothesis has been advanced that during infection, cells switch from synthesis of host to viral proteins as a result of an increase in intracellular Na^+ brought about by a general increase in membrane leakiness (Carrasco and Smith, 1976, 1980; Carrasco, 1977). We have measured various parameters of membrane function, including intracellular cation content, in cells infected with (non-haemolytic) paramyxoviruses, orthomyxoviruses, rhabdoviruses and togaviruses, and have found no such change in cells that were actively synthesising viral proteins (Pasternak et al., 1982 and unpublished experiments with G. Agnarsdottir, C.L. Lancashire and F. Brown). Although in one instance, namely in Semliki Forest virus (SFV)-infected cells, intracellular Na^+ did increase, this was subsequently shown to be more likely a consequence and not a cause of the initiation of viral protein synthesis (unpublished experiments of M.A. Gray, K.J. Micklem and C.A. Pasternak). However some membrane changes do occur in virally-infected cells (reviewed by Pasternak and Micklem, 1981). Membrane potential, as measured by the K' null-point method with various dyes (Bashford, 1981; Bashford et al., 1981), is one such change.

Figure 3 shows the membrane potential of BHK cells infected with SFV (a toga virus) or vesicular stomatitis virus (VSV, a rhabdovirus). Note that intracellular K^+ did not change under these conditions; since the membrane potential of these BHK cells is only partly set by the transmembrane K^+ gradient, - a situation intermediate between that of an excitable cell and that of the Lettre cells mentioned above, - it is perhaps not surprising that a decrease in potential is not accompanied by a decrease in intracellular K^+. The physiological significance of such changes has not yet been evaluated, though it may well be related to alterations in the physiological function of neuronal (Fukada and Kurata, 1981) and heart (Batra et al., 1976) cells infected with Herpes simplex virus in vitro, or to alterations of neuronal function in animals infected with Herpes viruses in vivo (Dolivo, 1980).

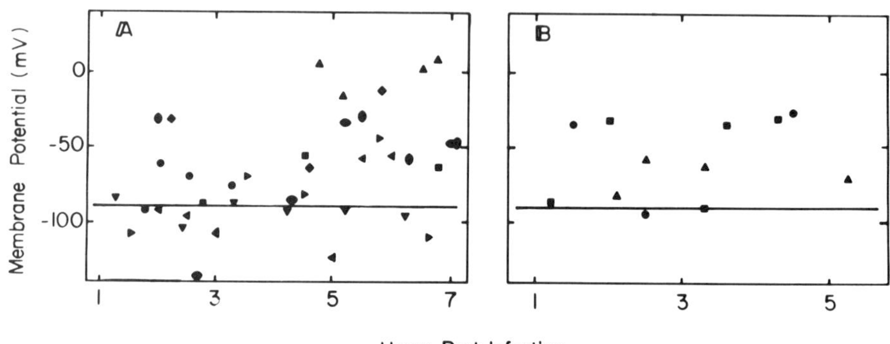

Fig. 3 Membrane potential of virally-infected cells
The membrane potential of BHK cells in monolayer, infected with SFV (A) or VSV (B), was measured with optical indicators as described by Bashford (1981); intracellular K^+ remained constant, at approximately 150mM, under all conditions. The membrane potential of mock-infected cells, shown as a solid line, did not vary over the course of the experiment. Different symbols refer to separate experiments.

CONCLUSION

The surface membrane of cells is affected by viruses in two ways. (1) During entry of haemolytic paramyxoviruses, a non-specific increase in membrane permeability to ions and a collapse of the membrane potential occur; this change has not been observed with other viruses, and hence its clinical importance is likely to be limited. (2) During the infectious cycle of a number of viruses in cultured cells, the membrane potential is decreased without extensive changes in intracellular cations. The clinical relevance of such alterations remains to be elucidated.

ACKNOWLEDGEMENTS

We are grateful to the SERC, Roche Products Ltd., Central University of London Research Fund and the Cell Surface Research Fund for financial assistance.

REFERENCES

Alam, A.N., Poston, L., Wilkinson, S.P., Golindano, C.G. and Williams, R., 1978, Clin.Sci.Mol.Med., 55:355.
Bashford, C.L., 1981, Biosci.Rep., 1:183.
Bashford, C.L., Chance, B., Smith, J.C. and Yoshiba, T., 1979, Biophys.J., 25:63.
Bashford, C.L., Foster, K.A., Micklem, K.J. and Pasternak, C.A., 1981, Biochem.Soc.Trans., 9:80.
Batra, G.K., Nahmias, A.J. and De Haan, R.L., 1976, Nature, 259:677.
Carrasco, L., 1977, FEBS Lett., 76:11.
Carrasco, L. and Smith, A.E., 1976, Nature, 264:807.
Carrasco, L. and Smith, A.E., 1980, Pharmac.Ther., 9:311.
Dolivo, M., 1980, Trends in Neuro Sciences, 3:149.
Forda, S.R., Gillies, G., Kelly, J.S., Micklem, K.J. and Pasternak, C.A., 1982, Neurosci.Lett., 29:237.
Fuchs, P., Spiegelstein, M., Haimsohn, M., Gitelman, J. and Kohn, A., 1978, J.Cell.Physiol., 95:223.
Fukada, J. and Kurata, T., 1981, Brain Res., 211:235.
Gillies, G., Micklem, K.J. and Pasternak, C.A., 1981, Cell Biol. Int.Rep., Suppl. A, 5:20.
Gomperts, B., Micklem, K.J. and Pasternak, C.A., 1981, Cell Biol. Int.Rep., Suppl. A, 5:20.
Gomperts, B.D., Baldwin, J.M. and Micklem, K.J., 1983, Biochem.J., in press.
Griffin, G.E. and Pasternak, C.A., 1982, Clin.Sci., 63:1.
Hacking, C. and Eddy, A.A., 1981, Biochem.J., 194:415.
Heinz, A., Sachs, G. amd Schafer, J.A., 1981, J.Membr.Biol., 61:143.
Impraim, C.C., Micklem, K.J. and Pasternak, C.A., 1979, Biochem. Pharmacol., 28:1963.
Impraim, C.C., Foster, K.A., Micklem, K.J. and Pasternak, C.A., 1980, Biochem.J., 186:847.
Knutton, S., 1978, Micron, 9:133.
Masuda, A. and Goshima, K., 1980, Biochim.Biophys.Acta, 599:596.
Micklem, K.J. and Pasternak, C.A., 1977, Biochem.J., 162:405.
Okada, Y., Koseki, I., Kim, J., Maeda, Y., Hashimoto, T., Kanno, Y. and Matsui, Y., 1975, Exp.Cell Res., 93:368.
Pasternak, C.A., 1983, Virally-mediated changes in cellular permeability. In: Membrane Processes: Molecular Biological Aspects and Medical Applications, G. Benga, H. Baum and F. Kummerow, eds., Springer Verlag, New York, in press.
Pasternak, C.A. and Micklem, K.J., 1973, J.Membr.Biol., 14:293.
Pasternak, C.A. and Micklem, K.J., 1974, Biochem.J., 144:593.
Pasternak, C.A. and Micklem, K.J., 1983, Altered function of excitable cells caused by paramyxoviruses. In: Viruses and demyelinating diseases, C.A. Mims, ed., Academic Press, London, in press.
Pasternak, C.A., Strachen, E., Micklem, K.J. and Duncan, J., 1976, New approaches for studying surface membranes. In: Cell surfaces and malignancy, P.T. Mora, ed., U.S. Government

Printing Office, Washington DC, Fogarty Int.Center Proc.28.
Philo, R.D. and Eddy, A.A., 1978, Biochem.J., 174:801.
Piatigorsky, J., 1980, Current Topics in Eye Research, 3:1.
Pietrzyk, C., Geck, P. and Heinz, E., 1978, Biochim.Biophys.Acta, 513:89.
Poste, G. and Pasternak, C.A., 1978, Cell Surface Revs., 5:306.
Reid, J.L., ed., 1982, Proc.Ninth Meeting Int.Soc.Hypertension, Clin.Sci., 63: Suppl. 8.
Smith, J.C., Russ, P., Cooperman, B.S. and Chance, B., 1976, Biochemistry, 15:5094.
Swales, J.D., 1982, Biosci.Rep., 2:967.
Wyke, A.M., Impraim, C.C., Knutton, S. and Pasternak, C.A., 1980, Biochem.J., 190:625.

MOLECULAR ASYMMETRIES IN BIOLOGICAL EVOLUTION AND THE ROLE OF WATER AND IONS

C. Portelli

Faculty of Medicine
Department of Biophysics
8, Dr. Petru Groza Blvd., Bucharest, Romania

In a previous paper[1], a model was presented according to which the living organisms manifests a selectivity for L-amine acids and beta-D-pentose molecules because of the asymmetric nitrogen atom included in the structure of the nucleotide bases and at the level of the peptide bonds of protein molecules. This asymmetric nitrogen atom presents a first covalent bond $(C^{+} \rightarrow N^{-})$, which has a major electric polarization, a second covalent bond $(C^{+} \rightarrow N^{-})$, which has a minor electric polarization, and a third covalent bond with a hydrogen atom. The hydrogen atom, covalently bound to the asymmetric nitrogen atom, is preferentially directed to the side of the molecular plane of the nucleotide base in a position from where an observer should see the sense of the major electric polarization as being in the counter clockwise direction of the molecular ring.

Since the relationship existing between the electric polarization of the covalent bonds and the preferential orientation of some hydrogen bonds appears to be an invariant of evolution, it is probably that it is also present at the level of some other molecules. In this paper, I will analyse from this point of view the connection of the ATP molecule with the bound water molecule and the role of fixed ions.

In a solution with the pH 7, ATP molecule has three fully ionized oxygen atoms, and a fourth oxygen atom which is ionized in a proportion of 50 per cent[2]. Cytosol is a solution which has the pH 6-6.5, and it is rich in potassium and magnesium ions[3]. When an ATP molecule is present in cytosol, the oxygen O(1) and O(2) atoms are ionized and they fix a magnesium ion, the oxygen O(3) atom is ionized and it can fix a potassium ion, while the oxygen 0(4) atom is ionized in a proportion lesser than 50 per cent. Owing to the electric asymmetry induced in the framework of the ATP molecule by the ionization of the oxygen 0(3) atom and the non-ionization of the oxygen 0(4) atom, the electric polarization of the covalent bond ($P^{+}_{III} \rightarrow O^{-}_{II}$) is greater than the electric polarization of the covalent bond ($P^{+}_{II} \rightarrow O^{-}_{II}$), (figure 1,a).

On the other hand, the interstitial fluid is a solution which has the pH 7.2-7.4, and it is rich in sodium and calcium ions[2]. If an ATP molecule is present in the interstitial fluid, the oxygen 0(1) and 0(2) atoms are ionized and they can fix a calcium ion, the oxygen 0(3) atom is also ionized, while the oxygen 0(4) atom is ionized in a proportion higher than 50 per cent. The ionized oxygen 0(3) and 0(4) atoms can fix two sodium ions (figure 1,b). Due to the fact that the distance existing between the ionized oxygen 0(3) atom and the phosphorus P_{II} atom is greater than the distance between the ionized oxygen 0(4) atom and the phosphorus P_{III} atom, the electric polarization of the covalent bond $P^{+}_{II} \rightarrow O^{-}_{II}$) is greater than the electric polarization of the covalent bond ($P^{+}_{III} \rightarrow O^{-}_{II}$) bond (figure 1,b).

The available data show that a water molecule must be linked at the level of the high energy bond of the ATP molecule before to its hydrolysis in (ADP + Pi).[4] In a first approximation we can consider that the connection of the water molecule with the ATP molecule can be produced as well at the covalent bond ($P^{+}_{III} \rightarrow O^{-}_{II}$) as at the covalent bond ($P^{+}_{II} \rightarrow O^{-}_{II}$). But certainly the water molecule will "prefer" for its connection the bond having the major electric polarization. Subsequently, the water molecule will be attached to the covalent bond ($P^{+}_{III} \rightarrow O^{-}_{II}$) in the case of the ATP^{3-} molecule, and it will be linked to the covalent bond ($P^{+}_{II} \rightarrow O^{-}_{II}$) in the case of the ATP^{4-} molecule, (figure 1,a,b). In both situations, one hydrogen atom of the water molecule will form a hydrogen bond with the oxygen O_{II} atom of the ATP molecule, whereas the oxygen atom of the water molecule will be linked by van der Waals forces at the phosphorus P_{II} or P_{III} atoms (figure 1,a,b).

According to the relationship existing between the electric polarization of the covalent bonds and the orientation of some hydrogen atoms, the second hydrogen atom of the water molecule will be preferentially directed to a certain side of the intermolecular ring. It will be in the position from where an observer will see the polarization of the covalent bond of the water molecule, which is included in the molecular ring, as being directed in the counter clockwise sense of the respective ring (figure 1,a,b).

Experiments performed with H_2O^{18} molecules have proved that the hydrolysis of the ATP molecule in (ADP + Pi) always develops in

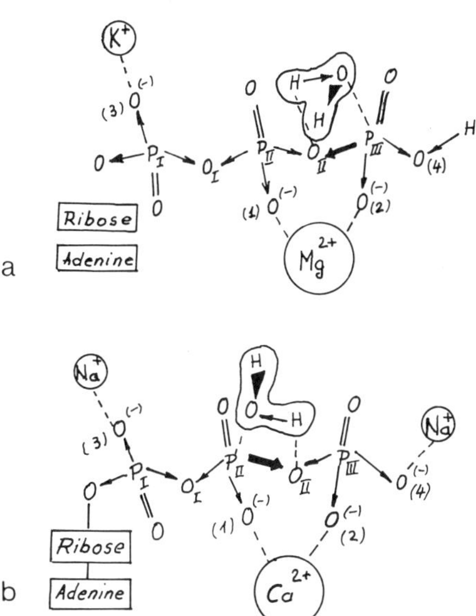

Fig. 1: a) Electric polarization of the covalent bonds of ATP^{3-} molecule and the positions of bound water molecule and ions
b) Electric polarization of the covalent bonds of ATP^{4-} molecule and the positions of bound water molecule and ions

such a way that the hydroxyl ($O^{18}H$) group is connected to the delivered phosphate[4]. Therefore, the water molecule must be previously fixed at the covalent bond ($\overset{+}{P}_{III} \longrightarrow \overset{-}{O}_{II}$) of the ATP molecule, while the another alternative is not suitable to produce the hydrolysis. The explanation would be the following: when the water molecule is fixed at the covalent bond ($\overset{+}{P}_{III} \longrightarrow \overset{-}{O}_{II}$), its second hydrogen atom has an adequate orientation to achieve a hydrogen bond with an oxygen atom of the ATP, this second bond is not possible if the water molecule is linked to the covalent bond ($\overset{+}{P}_{II} \longrightarrow \overset{-}{O}_{II}$). The hydrolysis of the ATP molecule requires the connection of the water molecule in three points at the level of ATP molecule.

In conclusion, a close relationship appears to be between the ionic composition of a liquid and the electric polarization of the covalent bonds of the immersed biological molecules. The electric polarization of the covalent bonds determines the orientation of some hydrogen atoms and hydrogen bonds, and also the position of the bound water molecules.

REFERENCES

1. C. Portelli, J. theor. Biol., 93, (93-96), 1981
2. A. Lehninger, Biochemistry, Worth Publishers, New York, 1975
3. E.D.P. De Robertis, E.M.P. De Robertis, Cell and Molecular Biology, 7-th ed. Saunders, Philadelphia, 1980
4. M. Serban, Dita Cotariu, The biochemistry of the muscle contraction, Acad. R.S.R., Bucharest, Romania

ELECTRICAL PROPERTIES OF THE PHOSPHOLIPID/WATER INTERFACE

Saša Svetina, Alenka Luzar and Boštjan Žekš

Institute of Biophysics, Medical Faculty and "J.Stefan" Institute, E.Kardelj University, Ljubljana, Yugoslavia

The subject of this article is a theoretical study of the behavior of solution ions close to the phospholipid/water interface of charged and uncharged phospholipids. Representative examples from the existing experimental data indicating adsorption of monovalent and divalent cations to phospholipids are presented first. Then the theoretical evidence for spontaneous polarization of water in the layer adjacent to the phospholipid/water interface is given. Finally, a simple model is introduced describing the effect of water polarization on the distribution of ions in the solid/ionic solution interfacial region.

An indication of the adsorption of monovalent cations to bilayer membranes containing negative phospholipids came from measurements of electrophoretic mobilities of phospholipid vesicles (Eisenberg et al., 1979). The zeta potentials obtained suggested a specific interaction between the monovalent cations and phosphatidylserine (PS) membrane where the interaction of these ions with PS decreased in the sequence $Li^+ > Na^+ > NH_4^+ > K^+ > Rb^+ > Cs^+ > (Et)_4N^+ > (Me)_4N^+$. Similar studies in the presence of divalent cations have shown that upon addition of divalent cations, the initially negative zeta potential increases towards zero and may, at sufficiently high concentrations, even become positive (McLaughlin et al., 1981). The interaction of divalent cations with PS is also cation specific and decreases in the sequence $Ni^{2+} > Mn^{2+} > Ca^{2+} > Mg^{2+}$. A complementary set of experimental data has been obtained by the determination of the effect of divalent cations on the surface potential of the PS monolayers (Ohki and Kurland, 1980). Divalent cations interact also with neutral phospholipids. A positive zeta potential has been determined on phosphatidylcholine (PC) vesicles in the presence of divalent cations (McLaughlin, Grathwohl and McLaughlin, 1978).

The phenomena described were satisfactorily accounted for by the Stern equation, a combination of the Gouy-Chapman equation from the theory of the diffuse double layer, the Boltzmann relation, and the Langmuir adsorption isotherm. However, the molecular mechanism for the adsorption of cations to phospholipid membranes is not yet clear. In this paper an attempt is made to interpret the adsorption of cations on the basis of modifications of the water structure at the interface.

A non-specific factor which may possibly affect cation behavior at the interfaces is spontaneous water polarization in the interfacial region. We have recently shown by theoretical analysis (Luzar, Svetina and Žekš, in preparation) that such polarization is due to the restrictions in orientations of water molecules with their oxygen atom located within a layer of thickness equal to the length of the O-H bond. The calculated average dipole moment pointing to the liquid phase is shown in Fig. 1 as a function of the distance from the membrane/water contact plane. The resulting electrical potential profile calculated by assuming that the dielectric constant retains its bulk value in the layer containing polarized water is also shown in Fig. 1. The detailed procedures which lead to the result presented in Fig. 1 will be presented elsewhere: therefore only the basic ideas employed are given here. Firstly, a simple model of water was developed based on the following assumptions. Each water molecule may participate in the formation of at most four hydrogen bonds. The hydrogen bonds arising from a given oxygen atom are oriented in the direction of the corners of the tetrahedron with the circumradius of the length of the water hydrogen bond ($\sim$0.3 nm). Water is treated as an ensemble of a large number of O-H••O pairs, the hydrogen bond of a pair being either formed or broken. Parameters of the model are the bonded fraction of the phase space pertaining to the O-H••O pair and the energy of the hydrogen bond. Their values are chosen in such a way that the predicted temperature dependence of the fraction of hydrogen bonded O-H groups fits that measured. Secondly, the spontaneous polarization in the layer of the thickness of the length of the O-H bond was calculated by making the following assumptions. Density of water in the interfacial region is equal to the bulk value. In the interfacial layer of the thickness of the length of the O-H••O bond the number of neighbouring water molecules available for the formation of hydrogen bonds is diminished. Furthermore, in the layer of thickness equal to the length of the O-H bond not all orientations of water molecules are allowed because of the steric restrictions.

For the determination of the effect of water polarization in the interfacial layer on the behavior of ions a simple model is introduced, schematically depicted in Fig. 2. The effect of water polarization is here simulated by two equally but oppositely charged infinite planes having the surface charge density σ_d, separated by the distance a. A suitable value for this distance estimated from Fig. 1 is 0.08 nm whereas the value for σ_d is 0.1 As/m^2. It is assumed that the fixed membrane charges are uniformly smeared on the surface with the surface charge density σ (0.23 As/m^2 for PS and 0 for PC; here it is assumed that each lipid molecule occupies an area of 0.7 nm^2 and that there is one unit charge per lipid molecule in PS). It is furthermore assumed that the dielectric constant in the polarized layer (ε_2) differs from the bulk value (ε_1), and that different cations have different values for their distance of closest approach (b(i) for the i-th ion) (Valleau and Torrie, 1982). The distance of closest approach of the anion is taken to be equal to zero. The values given are the values employed in our calculations unless otherwise stated.

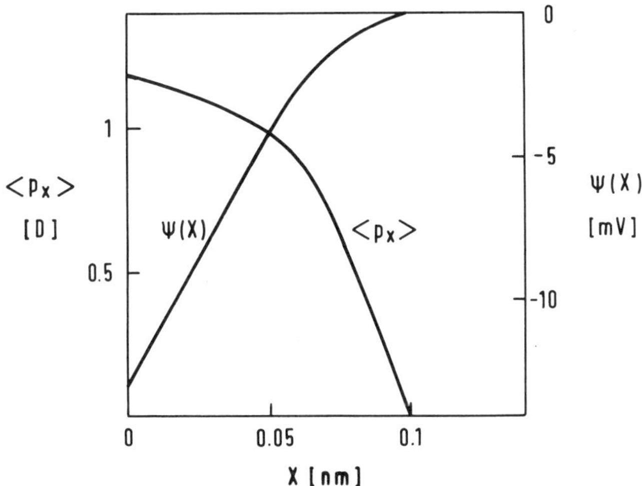

Fig. 1. Calculated statistical average of the component of the water dipole moment perpendicular to the unpenetrable solid surface, $\langle p_x \rangle$, and the electrical potential calculated by assuming that the dielectric constant also retains its bulk value in the layer of polarized water, $\psi(x)$, as a function of the distance from the membrane/water contact plane.

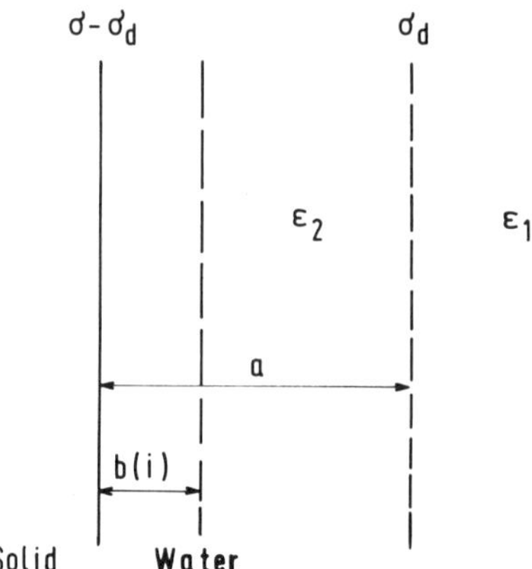

Fig. 2. Schematic representation of the model described in the text.

The electrical potential profile resulting from the charge distribution given by the above model can be obtained by using the standard procedures of the Gouy-Chapman diffuse double layer theory (Aveyard and Haydon, 1973). The Poisson-Boltzmann equation is solved by taking into consideration the additional boundary condition imposed by the charged plane σ_d. The model presented is a possible extension of the diffuse double layer theory and it is therefore instructive to compare its predictions with the result of the ordinary Gouy-Chapman theory. In Fig. 3 are shown the electrical potential profiles calculated for two values of dipolar charges, one value being zero, and for two values of the polarized layer dielectric constant. The polarized water layer causes the increase of the membrane potential for the negatively charged membranes. However, at the distances from the membrane where the hydrodynamic plane of shear is expected to lie in the electrophoretic mobility experiments, the electrical potential obtained by the model is smaller than the one predicted by the Gouy-Chapman theory. The effect is enhanced at smaller values of the dielectric constant of the polarized layer.

Fig. 4 shows, for two values of dipolar charge, the calculated potential at the outer boundary of the polarized layer, $\psi(a)$, as a function of the polarized layer dielectric constant and the cation distance of closest approach for the 0.1 molar concentration of the monovalent salt. The closer a cation approaches the interface, the less negative is the potential $\psi(a)$. This result can be related to the observation (Eisenberg et al., 1979) that smaller monovalent cations exhibit less negative zeta potentials. Therefore it is tempting to interpret the specificity of cation adsorption by differences in their distances of closest approach.

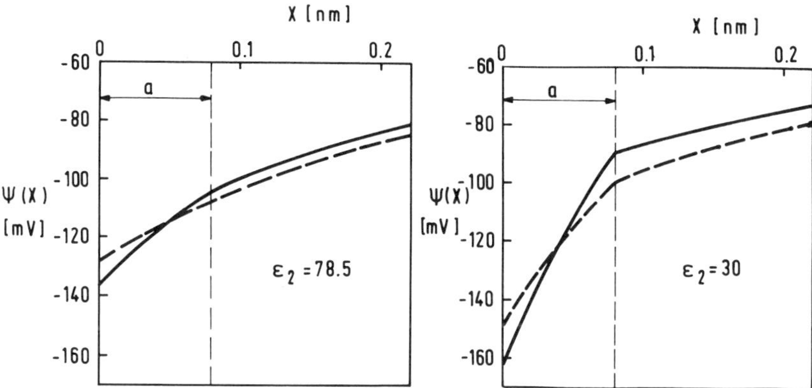

Fig. 3. Comparison of the electrical potential profiles obtained by the Gouy-Chapman diffuse double layer theory and the present model for two different values of the dielectric constant and two different values of the dipolar surface charge densities: σ_d = 0.1 As/m^2 (solid lines) and σ_d = 0 (broken lines).

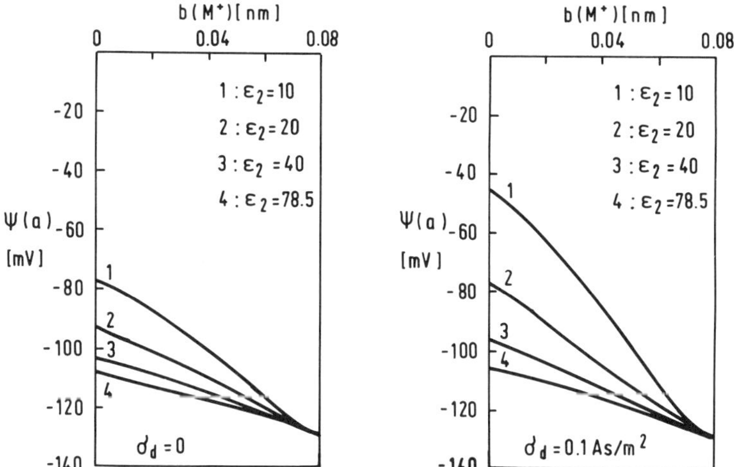

Fig. 4. Dependence of the electrical potential at the outer boundary of the polarized layer, $\psi(a)$, on the dielectric constant of the polarized layer and the distance of closest approach of the monovalent cation $b(M^+)$.

The results shown in Fig. 4 indicate that in order to describe the existing data within a reasonable range of model parameters it is necessary to assume the existence of all three factors appearing in the model, the polarized layer, its lower dielectric constant, as well as the effect of different distances of closest approach. A further confirmation of this notion can be obtained from the analysis of data with the divalent ions present. The effect that in the presence of divalent cations the zeta potential may attain a zero value is considered first because the model parameters which predict $\psi(a) = 0$ predict also zero zeta potential provided that the hydrodynamic plane of shear lies at a distance from the interface larger than the distance a. The zeta potential has zero value for instance in the presence of 0.1 molar NaCl and 0.08 molar $CaCl_2$. In Fig. 5 are shown values of parameters ε_2 and σ_d which give $\psi(a) = 0$ for three different distances of closest approach for Ca^{2+}. Higher values of σ_d require a less diminished value of the polarized layer dielectric constant. It is interesting to note (this is not explicitly shown) that the divalent ion concentration at which $\psi(a) = 0$ is larger at higher values of the divalent cation distance of closest approach, provided that the values of all other model parameters are kept constant. By noting that the zeta potential is zero in 0.02 molar $NiCl_2$ (McLaughlin et al., 1981), and assuming $b(Ni^{2+}) = 0$, the minimum value for $b(Ca^{2+})$ is 0.02 nm. This value is used in later calculations.

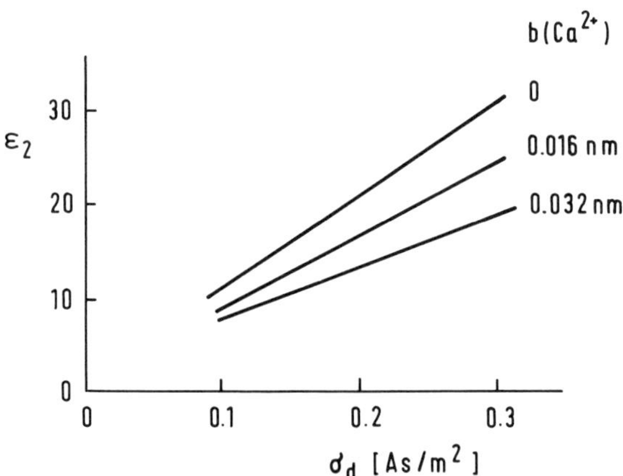

Fig. 5. Values of the polarized layer dielectric constant and the dipolar surface charge density which give zero zeta potential at 0.08 molar divalent ion concentration and 0.1 molar monovalent ion concentration. Distance of closest approach for the monovalent cation is taken to be zero.

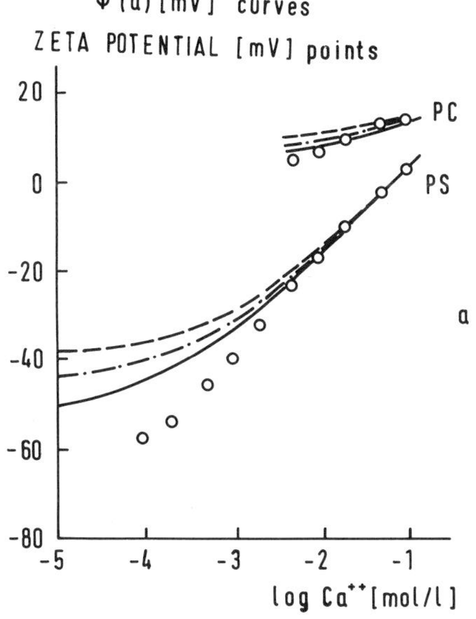

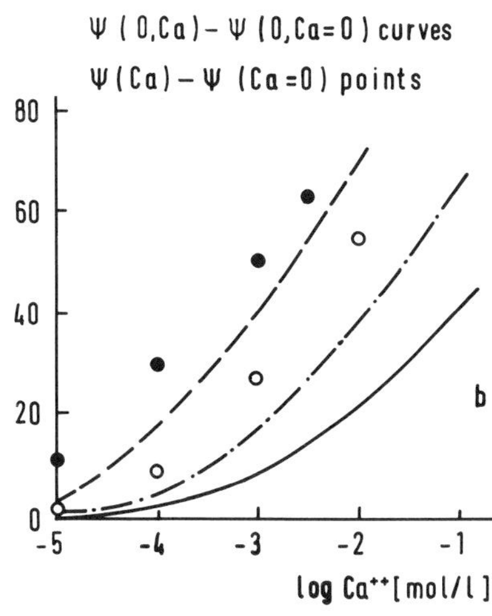

Fig. 6. (a) Dependence of the calculated electrical potential at the outer boundary of the polarized layer, $\psi(a)$, and the dependence of the zeta potential on the concentration of Ca^{2+} ions in aqueous solution containing 0.1 M NaCl. Points represent the experimental zeta potentials obtained by McLaughlin et al. (1981).

(b) Dependence of the difference between the membrane potentials in the presence and absence of divalent cation, $\Delta\psi(0)$, on the concentration of Ca^{2+} ion in the aqueous solution containing monovalent ions. Points represent the experimental data obtained by Ohki and Kurland (1981): o in 0.1 M NaCl, • in 0.1 M $(Me)_4NCl$.

Lines in (a) and (b) represent the theoretical curves, calculated according to our model with three different distances of closest approach for the monovalent cation, $b(M^+)$: 0 (——), 0.032 nm (----), 0.016 nm (-•--•-). ε_2 is determined from Fig. 5 in order to fit the zero zeta potential concentration of Ca^{2+}. The values of other parameters used in calculations are as given in the text.

The zeta potential attains zero value at ionic concentrations where the ionic strength of the electrolyte solution is mostly determined by the divalent cation concentration. The mutual effect of monovalent and divalent cations on the electrical potential profile is expected to be more pronounced at lower concentrations of divalent cations. This is illustrated in Fig. 6a where the calculated electrical potential $\psi(a)$ is shown as a function of the Ca^{2+} for different values of the distance of closest approach for the monovalent ion. From the comparison of calculated and measured differences in the membrane potential due to the presence of divalent cations it is possible to make an estimate of distances of closest approach for some monovalent cations. In Fig. 6b are given calculated membrane potentials for three values of the monovalent cation distance of closest approach and also the existing experimental values for Na^+ and $(Me)_4N^+$ (Ohki and Kurland, 1981). The distance of closest approach of these ions obtained by interpolation are $b(Na^+) = 0.023$ nm and $b((Me)_4N^+) = 0.038$ nm. The larger distance obtained for $(Me)_4N^+$ is in accord with the higher value observed for the zeta potential in the presence of this ion.

The zeta potentials obtained experimentally with PS vesicles in the presence of Ca^{2+} are more negative than the potentials $\psi(a)$ calculated with the model parameters, which are otherwise consistent with other data discussed (Fig. 6a). This may mean that the hydrodynamic plane of shear might lie slightly closer to the interface than the outer boundary of the polarized layer. A near coincidence of these planes would suggest that the hydrodynamic plane of shear is defined by the same restrictions of orientations of water molecules which cause the spontaneous polarization treated above.

The present model also accounts for the appearance of positive zeta potentials on PC in the presence of divalent cations. The same set of model parameters which were used to describe the interaction of cations with PS also consistently describes the data on PC (Fig. 6a).

SUMMARY

The tendency for the number of hydrogen bonds to be maximized, as well as steric hindrance, lead to water polarization in the solid/water interface. The existence of this water polarization points to a possible interpretation of cation adsorption to charged and uncharged phospholipid membranes. Ion specificity can be attributed to differences in distances of closest approach for different cations. The simple model derived on the above premises describes a set of experimental data on cation binding to phosphatidylserine and phosphatidylcholine membranes.

REFERENCES

Aveyard, R., and Haydon, D.A., 1973, An Introduction to the Principles of Surface Chemistry, Cambridge University Press, London.

Eisenberg, M., Gresalfi, T., Riccio, T., and McLaughlin, S., 1979, Adsorption of Monovalent Cations to Bilayer Membranes Containing Negative Phospholipids, Biochemistry, 18: 5213.

McLaughlin, A., Grathwohl, C., and McLaughlin, S., 1978, The Adsorption of Divalent Cations to Phosphatidylcholine Bilayer Membranes, Biochim. Biophys. Acta, 513: 338.

McLaughlin, S., Mulrine, N., Gresalfi, T., Vaio G., and McLaughlin, A., 1981, Adsorption of Divalent Cations to Bilayer Membranes Containing Phosphatidylserine, J. Gen. Physiol., 77: 445.

Ohki, S., and Kurland, R., 1981, Surface Potential of Phosphatidylserine Monolayers. II. Divalent and Monovalent Ion Binding, Biochim. Biophys. Acta, 645: 170.

Valleau, J.P., and Torrie, G.M., 1982, The Electrical Double Layer. III. Modified Gouy-Chapman Theory with Unequal Ion Sizes, J. Chem. Phys., 76: 4623.

DSC-STUDY ON THE INTERATION BETWEEN MONOVALENT CATIONS AND DPPC

F. Tölgyesi, S. Györgyi and M. Szögyi

Institute of Biophysics
Semmelweis Medical University
PO Box 263, 1444 Budapest

While Na^+ and K^+ ions are very important participants of the cellular processes, the other three alkali cations (Li^+, Rb^+ and Cs^+) play a negligible role in biological systems. At the same time the study and the comparison of the biophysical behavior of these five cations provide important data on the molecular mechanism of ion transport across membranes.

In our previous work the alkali ion transport and selectivity of red blood cells and that of bimolecular lipid membrane (BLM) were studied by a radioactive tracer method and electric conductivity measurements[1,2,3].

Taking into account the observations that the lipid bilayer plays a significant role in the regulation of passive and active transport of alkali cations one has to deal with the interaction between alkali cations and phospholipid bilayers in order to shed more light on the transport processes.

In the last years many papers have been devoted to this problem[4-26], but most of them concentrate on the interaction between divalent cations (mainly Ca^{++}) and charged phospholipids. The monovalent cations and the more frequent zwitterionic phospholipids have gained much less publicity in the literature so far and the results obtained on these systems are more or less contradictory.

In our recent experiments the interaction between the alkali cations and one phospholipid, dipalmitoyl-phosphatidylcholine (DPPC) has been studied using microcalorimetry and the results were compared with data obtained in a parallel study on the permeability of DPPC-liposomes.

MATERIALS AND METHODS

Synthetic DPPC was purchased from Sigma Chemical Co. and was used without further purification. (The purity of lipid was checked by thin-layer and gas-liquid chromatography).

The samples were prepared by mixing dry lipid with an appropriate amount of twice distilled water or that of various salt-solutions (LiCl, NaCl, KCl, RbCl, CsCl) in a glass vessel. The mixture was kept at 50°C, above the phase transition temperature of lipid, for 30 minutes at the same time dispersing by a vigorous vortex-mixer. The lipid solution weight ratio was 1:4. Ten μl of the mixture was sealed hermetically in an Al-pan.

The measurements were carried out on a Dupont 990 Thermal Analyzer at a heating rate of 5°C/min in the sensitivity range of 160-400 μW/cm.

Onset temperatures were taken as the transition temperatures to reach better reproducibility.

RESULTS AND DISCUSSION

The results obtained for pre- and main transition temperatures and enthalpies of pure DPPC-water dispersion (T_p = 35.5°C, T_m = 41.5°C, ΔH_p = 7.9 J/g, ΔH_m = 56.9 J/g) are in good agreement with the values generally accepted in the literature.

In the presence of various monovalent cations at various concentrations the shape of the DSC curves did not change remarkably but (1) both the pre- and main transition shifted on the temperature-scale, and (2) the enthalpies belonging to these transitions changed.

The effect of Rb^+ at concentrations of 0.3, 0.5, 1, 2, 3M on the transition temperatures (T_p, T_m) and enthalpies (ΔH_p, ΔH_m) is demonstrated in Figs. 1, 2 and 3.

At lower concentrations the effect is small and ambiguous, but at higher concentrations the relation between T_m, T_p and the concentration is more expressed and nearly linear. The correlation coefficients for T_p- and T_m-lines are 0.997 and 0.530 respectively. Gratsheva has obtained similar results on the effect of RbCl on the phase transition[27].

Such strict correlation concerning the enthalpies could not be found because of their relatively great error, but the tendency of the values is a decrease with increasing salt concentration.

From the interpretation of these data the next comparison can be made.

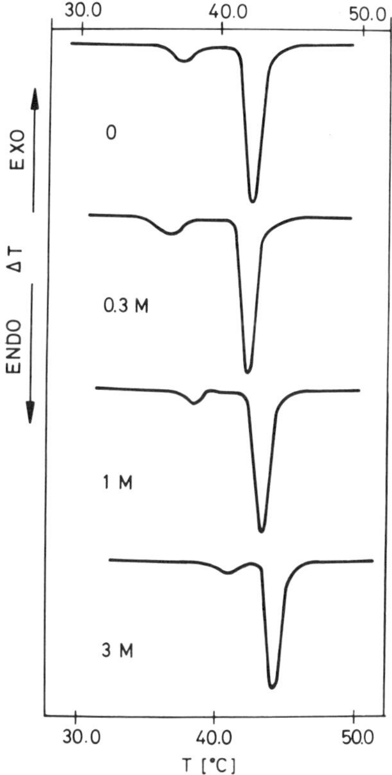

Fig. 1. DSC-curve of DPPC-water-RbCl dispersion with increasing RbCl-concentration. Heating rate 5°C/min.

Examining some nonionic surfactants it has been found that the surfactant molecules added to DPPC-liposomes increase the ^{42}K-permeability of the liposomes, and accordingly lower the transition temperature and enthalpy. From these and other results it was concluded that the surfactant molecules are partly embedded into the hydrophobic region of the bilayer[28].

In the presence of monovalent cations the effect of the surfactants on the permeability and on the phase transitions is less dominant. The effectiveness of the surfactant decreases nearly linearly with increasing salt concentration. So the structure of the lipid bilayer alters under the influence of the monovalent cations in the way that the packing of the lipid molecules will be closer hindering the penetration of the surfactant molecules between the hydrocarbon chains.

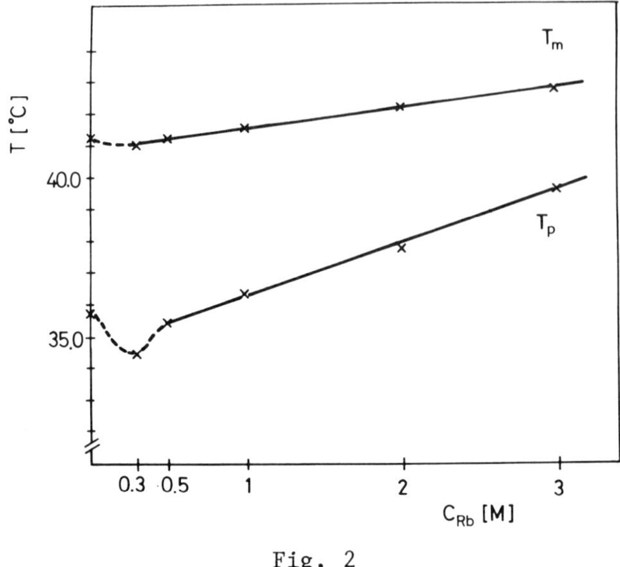

Fig. 2

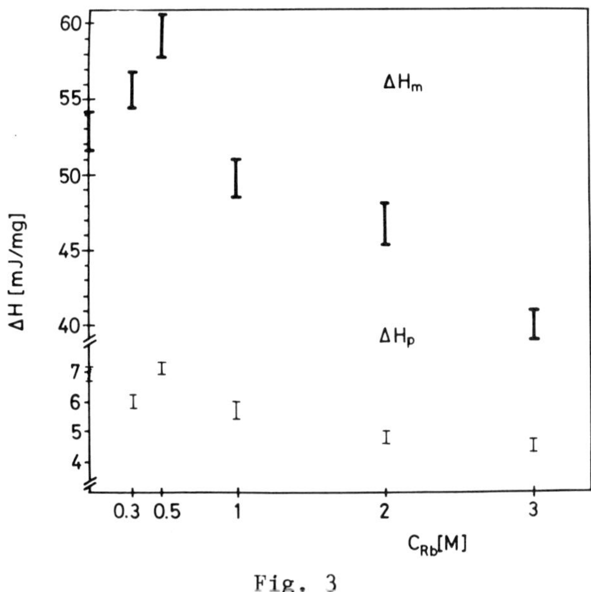

Fig. 3

Figs. 2 and 3. The pre- and main transition temperatures (T_p, T_m) and enthalpies (ΔH_p, ΔH_m) of DPPC-water-RbCl dispersion vs RbCl-concentration.

Table 1. The Effect of Monovalent Cations on the Phase Transition Parameters of DPPC + Water Mixture.

	T_p (°C)	T_m (°C)	ΔH_p ($\frac{J}{g}$)	ΔH_m ($\frac{J}{g}$)
DPPC	35.5	41.5	7.9	56.9
+ 3M LiCl	-	43.9	-	69.7
+ 3M NaCl	36.6	42.2	7.9	62.7
+ 3M KCl	40.2	43.5	3.8	48.6
+ 3M RbCl	41.8	43.9	6.4	50.4
+ 3M CsCl	41.9	44.1	5.5	47.4

The experimental errors are ±0.2°C in the temperature and 5-10% in the enthalpy-values. (The lower errors belong to ΔH_m, the higher ones to ΔH_p).

The effect of the five alkali cations on the phase transition parameters was investigated at the concentration of 3M. The results are summarized in Table 1.

Comparing these data with the results of Simon and coworkers [11], apart from some differences (e.g. the effect of LiCl), an agreement which seems to be important can be mentioned. In our measurements, the monovalent cations of biological interest, K^+ and Na^+, showed different behavior. While K^+ increased both the pre- and main transition temperature and decreased the enthalpies, Na^+ had practically no effect on the transition temperature but raised the transition enthalpy slightly. The different behavior of K^+ and Na^+ is probably due to the difference in the state of water in their primary hydration shells[29].

The interpretation of the data summarized in Table 1 is rather difficult. However, on the basis of these results the ions can be divided into the usual two groups of alkali cations: Li^+, Na^+ and K^+, Rn^+, Cs^+ proving that this method can supply further information on the ion-membrane interaction.

Additional contribution to the explanation could be obtained from measurements on the effect of alkali cations on the phase behavior of phospholipids with different polar head groups (e.g. serine, ethanolamine). Some thin-layer chromatographic data on the interaction between alkali cations and phosphatidylserine, phosphatidylcholine are already available[30] proving the role of the head group polarity and the ionic hydration in these interactions.

REFERENCES

1. S. Györgyi and B. Kanyár, Acta Biochim.Biophys.Acad.Sci.Hung., 7:359 (1972).

2. S. Györgyi and K. Blaskó, Acta Biochim.Biophys.Acad.Sci.Hung., 9:97 (1974).
3. K. Blaskó, S. Györgyi, and I. Horváth, J.Antibiotics, 32:408 (1979).
4. H. Träuble and H. Eibl, Proc.Natl.Acad.Sci.USA, 71:214 (1974).
5. J. F. Tocanne, P. H. J. T. Ververgaert, A. J. Verkleij, and L. L. M. van Deenen, Chem.Phys.Lipids, 12:201 (1974).
6. A. J. Verkleij, B. de Kruijff, P. H. J. T. Ververgaert, J. F. Tocanne, and L. L. M. van Deenen, Biochim.Biophys.Acta, 339:432 (1974).
7. P. H. J. T. Ververgaert, B. de Kruijff, A. J. Verkleij, J. F. Tocanne, and L. L. M. van Deenen, Chem.Phys.Lipids, 14:97 (1975).
8. K. Jacobson and D. Papahadjopoulos, Biochemistry, 14:152 (1975).
9. H. J. Galla and E. Sackmann, J.Am.Chem.Soc., 97:4114 (1975).
10. R. C. MacDonald, S. A. Simon, and E. Baer, Biochemistry, 15:885 (1976).
11. S. A. Simon, L. J. Lis, J. W. Kauffman, and R. C. MacDonald, Biochim.Biophys.Acta, 375:317 (1975).
12. K. Harlos , J. Stümpel, and H. Eibl, Biochim.Biophys.Acta, 555:409 (1979).
13. S. K. Hark and J. T. Ho, Biochem.Biophys.Res.Comm., 91:665 (1979).
14. M. J. Hope and P. R. Cullis, Biochem.Biophys.Res.Comm., 92:846 (1980).
15. K. Harlos and H. Eibl, Biochim.Biophys.Acta, 601:113 (1980).
16. K. Harlos and H. Eibl, Biochemistry, 19:895 (1980).
17. K. Arnold, R. Pausch, J. Frenzel, and E. Winkler, Studia Biophys., 51:81 (1975).
18. Y. Inoko, T. Yamaguchi, K. Furuya, and T. Mitsui, Biochim. Biophys.Acta, 413:24 (1975).
19. I. S. Puskin, Membrane Biol., 35:39 (1977).
20. I. Haller and M. I. Freiser, Biochim.Biophys.Acta, 455:739 (1976).
21. M. Gründel, R. Grupe, E. Freusser, and H. Göring, Studia Biophys., 58:203 (1976).
22. D. Papahadjopoulos, W. I. Vail, K. Jacobson, and G. Poste, Biochim.Biophys.Acta, 394:483 (1975).
23. M. M. Sacre and I. F. Tocanne, Chem.Phys.Lipids, 18:334 (1977).
24. K. Toko and K. Yamafuji, Chem.Phys.Lipids, 26:79 (1980).
25. L. J. Lis and J. W. Kaufman, Biochim.Biophys.Acta, 406:453 (1975).
26. B. Karvaly and E. Loshchilova, Biochim.Biophys.Acta, 470:492 (1977).
27. O. Gratscheva, personal communication.
28. S. Györgyi, F. Tölgyesi, M. Szógyi, and K. Blaskó, in: "Abstracts of the 4.Int.Liq.Cryst.Conf. of Socialist Countries," Tbilisi, p. 156 (1981).
29. B. E. Conway, "Ionic Hydration in Chemistry and Biophysics," Elsevier Sci.Publ.Co., Amsterdam, p. 672 (1981).
30. T. Cserháti and M. Szógyi, Chem.Phys.Lipids, in press.

CHLORIDE FLUXES IN THE ERYTHROBLASTIC LEUKEMIC CELL: AN EXAMPLE OF MEMBRANE DIFFERENTIATION

H.G. Hempling and S. White

Department of Physiology
Medical University of South Carolina
Charleston, S.C.

INTRODUCTION

The erythroblastic leukemic cell from the Long-Evans rat has been used in our laboratory as a model of the immature component of the erythroid line (Wise, '74; Hempling and Wise,'75; Wise,'76). We have used this cell line to determine which membrane functions found in the mature erythrocyte are already present in the immature stage and which membrane functions present in the immature stage disappear when the cells lose their nuclei and mature. It was found (Hempling and Wise,'75) that the rapid permeability of the membrane to water and to urea, so distinctive for the membrane of the mature erythrocyte, was characteristic of the erythroblastic leukemic cell also. In contrast, Wise found (Wise,'76) that the membrane of the immature erythroblastic leukemic cell possessed three mediated pathways for the transport of neutral amino acids which were dependent on external Na^+. On the other hand, the rat reticulocyte lost one mediated pathway, while the mature erythrocyte lost all three.

We have considered as a first hypothesis that the basic membrane pattern which underlies the passive process for the transport of water and non-electrolytes is laid down early and remains after maturation and the loss of genetic material. In contrast, those specialized transport systems which are responsible for cumulative processes, like amino acid transport, may be extensive in the immature cell, but less so as genetic information is lost. Since electrolyte transport is another major mediated transport system, it seemed appropriate to assess what its status was in this cell line and what this might imply about membrane maturation.

Toward this end we have made an extensive study of the water and electrolyte content of these cells. Further we have used a number of inhibitors of electrolyte transport which include furosemide, dihydro DIDS, and phloretin. Steady state kinetics of Cl^- revealed two cellular compartments. From differential responses of the cell populations to inhibitors and to the substitution of Cl^- with $SO_4^=$ a model of two separate cell populations has been proposed. Two transport systems for Cl^- have also been proposed. One system mediates Cl^- - Cl^- exchange. The other system mediates Cl^- - $SO_4^=$ The cell populations may differ in their representation of each system. Different populations may represent different stages in the maturation of Cl^- transport as the populations differentiate.

METHODS

Methods for initiating and maintaining the tumor line have been described in previous papers (Hempling and Wise,'75; Wise,'74). Na^+ and K^+ were measured with a flame photometer, Model 343 (Instrumentation Laboratory Inc., Lexington, Mass.), calibrated with standard solutions of $Na^+/K^+/Li^+$ ranging from 1.0 mEq/l to 0.05 mEq/l Na^+/K^+ in 1% acetic acid and 15 mEq/l $LiNO_3$. Chloride ions were measured with an Aminco-Cotlove chloride titrator (American Instrument Co., Silver Spring, Md.). The same standard solutions were used for calibration. Phloretin was prepared in 100% ethyl alcohol at a concentration of 1 mg/ml. Then cells were exposed to a final concentration of 0.25 mM when 0.05 ml of this solution was added to every 8 ml cell suspension. Controls received only 0.05 ml of 100% ethyl alcohol for every 8.0 ml cell suspension. Dihydro-DIDS was prepared in K^+-Na^+ Ringers without glucose at concentrations between 0.009 and 0.05 mg/ml. Cells were exposed to these concentrations and allowed to equilibrate for 15 minutes at room temperature. A stock solution of 20 mM furosemide was prepared by dissolving 66 mg in 10 ml of K^+-Na^+ Ringer solution adjusted to a pH of 9 with NaOH. Once dissolved, the pH was readjusted to 7.4 with 1N HCl. To prepare a 1 mM solution, 0.1 ml of stock solution was used with every 2 ml of cell suspension.

The chemicals used in the experiments were obtained as follows: Phloretin (K & K Laboratories, Plainview, N.Y. 11803). Dihydro-DIDS (courtesy of: Dr. C. Levinson, Dr. A. Rothstein). Furosemide was a gift from Hoechst-Roussel Pharmaceuticals, Somerville, N.J.

RESULTS

Thirteen experiments were carried out under steady-state conditions at 25°C to obtain values for Na^+, K^+, and Cl^-. Electrolyte values are summarized in Table 1. Electrolyte data have been expressed per liter of osmotically active water. When changes in content are emphasized, the data may be expressed per 10^7 cells. All the data necessary to make the conversions are available. Chloride concentration is as high in these cells as it is in the mature rat erythrocyte. A theoretical potential difference (PD) was calculated from the distribution ratio of internal to external chloride ion and averaged - 18.9±0.7 mV.

Chloride fluxes: steady-state

Ten experiments were carried out to measure steady-state fluxes of Cl^-, using Cl^{36}. On the average, chloride fluxes were characterized by the kinetics of two cell compartments and the rate coefficients, fluxes, and compartment sizes appear in Table 1. Several experiments, however, provided excellent data with kinetics for a single cell compartment. Although these findings were disconcerting at first, in retrospect they have served to support a hypothesis which we will begin to develop that the two cell compartments represent two populations of cells which possess different mechanisms for chloride transport. The apportionment of these different mechanisms in these populations is evidenced by the relative sizes of the two cell compartments and their respective rate coefficients. Whether we see one cell compartment or two may be a function of the stage of maturation of chloride transport in the erythroblastic leukemic cells when we isolate them from the rat liver. We may have one population which possesses one constellation of mechanisms of chloride transport or a population whose chloride transport is the resultant of a different constellation. In such a circumstance, the kinetics would be that of a single cell compartment. On the other hand, if both populations were present, then we could expect the kinetics of two compartments. We could also anticipate a transition stage in which each population carried both constellations of mechanisms in a variety of proportions, in which case the rate coefficients would reflect the mix of each constellation of mechanisms.

Chloride fluxes: the effect of 4,4' diisothiocyano 1,2-diphenylethane 2,2' disulfonate (dihydro DIDS)

Dihydro DIDS is considered by some as a non-penetrating agent that binds to the external surface of the plasma membrane (Cabantchik and Rothstein, 1972) and has been used as a specific inhibitor of mediated anion transport across the membrane of the mature human erythrocyte (Rothstein et al., 1976). Binding to the membrane may be reversible or irreversible and inhibit both chloride and sulfate transport. Inhibition of transport in the human erythrocyte is linearly related to the number of molecules bound to Band 3 proteins (Rothstein et al., 1976; Lepke et al., 1976; Ship et al., 1977). Levinson (1976) demonstrated that dihydro DIDS when present in the medium inhibits both chloride and sulfate transport in the Ehrlich mouse ascites tumor cell. However, if dihydro DIDS is removed from the environment and the cells washed, only sulfate transport is inhibited by the dihydro DIDS remaining with the cells.

Therefore we chose dihydro DIDS to help us to distinguish between the parallel and the series model of chloride flux.

Table 1. Osmotic Properties of Rat Erythroblastic Leukemic Cells

MEAN CORPUSCULAR VOLUME	VOLUME OF OSMOTICALLY INACTIVE MATERIAL	VOLUME OF OSMOTICALLY ACTIVE WATER	% WATER
677 ± 16	196 ± 13	458 ± 23	71±0.02
(13)	(13)	(13)	

ELECTROLYTE CONCENTRATIONS*

POTASSIUM	SODIUM	CHLORIDE
112 ± 2	37 ± 0.02	83 ± 2
(95)	(95)	(95)

ALL VOLUMES IN CUBIC MICRA *MEQ/LITER OSMOTICALLY ACTIVE WATER

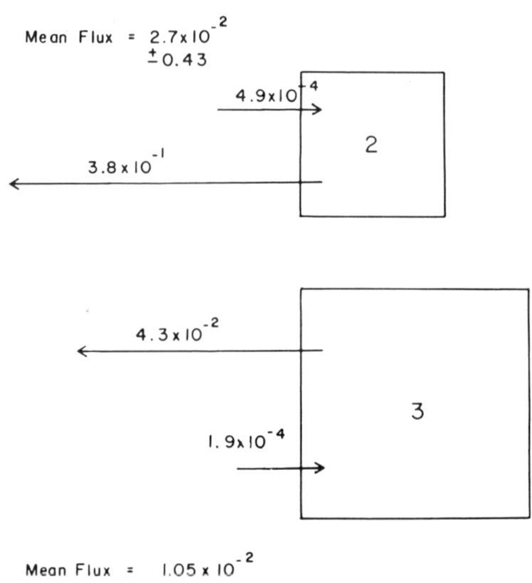

Fig. 1. Parallel model of steady-state Cl^- fluxes at 25°C

We reasoned that if dihydro DIDS obliterated one flux but not the other, then this demonstration would support a parallel model. A summary of our findings is in Table 2.

Dihydro DIDS at 9.5 micrograms/ml was tested on a population of 2.5 to 3 x 10^7 cells/ml which was exhibiting the kinetics of two cell compartments. Dihydro DIDS produced the kinetics of one cellular compartment. The rate constants of this compartment were characteristic of the control values from the larger, slow compartment even though the fluxes had increased by 43%. One may conclude from this experiment that the population of cells represented by the small fast compartment contained at least two mechanisms: (A) one which was susceptible to low concentrations of dihydro DIDS and (B) one which was not sensitive to dihydro DIDS at that concentration. In addition the rate constants of mechanism A were greater than those for mechanism B. The rate coefficients of the control were a function of the contribution of the two mechanisms. In the presence of the low concentrations of dihydro DIDS, the rate coefficients of the sensitive component were reduced to 0 and the rate coefficients of the insensitive component were equal to the rate coefficient of the population of cells which made up the large, slow compartment. Therefore the kinetics took on the appearance of one compartment with the kinetic parameters of the insensitive mechanism(s). The fluxes of the insensitive mechanism were higher because both populations were contributing to this mechanism and the compartment size consequently was larger.

In a second experiment, only 82% of the cell chloride was exchangeable under control conditions and the exchangeable chloride existed in two compartments. When the cells were exposed to 79 micrograms/ml of dihydro DIDS, 61% of the flux of the small, fast A compartment was inhibited and 88% of the flux of the large, slow B compartment was inhibited. At a lower concentration of dihydro DIDS of 39 micrograms/ml, one compartment was eliminated and the one that remained had the kinetic characteristics of the large, slow B compartment, much like the first experiment. Since the total exchangeable chloride was 79% of the total cell chloride and close in value to control, it would appear that the dihydro DIDS had converted the chloride fluxes from one mechanism to another in both populations which would exchange chloride. At even lower concentrations of 20 micrograms/ml dihydro DIDS, more of the chloride transport was blocked completely because the exchangeable compartment was reduced significantly to 73% of total cell chloride. The single compartment which was left had a flux which was 58% of the small, fast A compartment of the control; but now was 158% of the large B compartment of the control. Models which deal with these observations are considered in the Discussion section.

Steady-state chloride fluxes: the effect of sulfate ion

In two experiments, external chloride was replaced with sulfate. In each experiment, cells lost chloride accompanied by a loss in K^+

Table 2. The Effect of Dihydro-DIDS on Steady-State Chloride Fluxes at 25°C.

PARALLEL MODEL

	SMALL, FAST COMPARTMENT				LARGE, SLOW COMPARTMENT			
EXPT	SIZE	K12	K21	FLUX	SIZE	K13	K31	FLUX
CONT.	0.21	5.2	4.6	2.4	0.79	3.0	5.7	1.4
9.5*					1.00	3.2	5.2	2.1
CONT	0.25	7.6	9.8	4.6	0.75	2.8	12.0	1.7
20					1.00	3.5	15.5	2.7
39					1.00	3.1	13.0	1.6
79	0.43	2.8	2.2	1.8	0.57	0.3	1.9	0.2

RATE CONSTANTS: MIN-1

K12=VALUEX10-4 K21=VALUEX10-1 K13=VALUEX10-4 K31=VALUEX10-2

*MICROGRAMS DIHYDRO-DIDS/ML FLUX= VALUEX10-2 MICROMOLES/10 7 CELLS MIN.

Table 3. The Effect of Replacement of External Chloride by Sulfate on the Steady-State Fluxes of Chloride.

PARALLEL MODEL

	SMALL, FAST COMPARTMENT				LARGE, SLOW COMPARTMENT			
EXPT	SIZE	K12	K21	FLUX	SIZE	K13	K31	FLUX
154*	0.44	7.0	4.9	5.8	0.56	0.2	1.0	0.14
20	0.42	4.0	1.7	0.5				
151	0.15	8.8	9.1	4.1	0.85	3.6	6.5	1.7
82	0.60	9.3	1.6	2.3				
47	0.54	7.8	1.4	1.1				
14	0.40	4.2	0.9	0.2				

RATE CONSTANTS: MIN-1

K12=VALUEX10-4 K21=VALUEX10-1 K13=VALUEX10-4 K31=VALUEX10-2

*EXTERNAL CHLORIDE: MEQ/L FLUX= VALUEX10-2 MICROMOLES/10 7 CELLS MIN.

and water. No significant losses in cell Na^+ occurred.

Table 3 summarizes the kinetic results. The obvious effect was the elimination of one cell compartment. It was evident at 82 mEq/liter of external Cl^-. The second obvious effect was a progressive decrease in the exchangeable chloride. Note that the compartment sizes refer to the exchangeable compartments. Actually 28% of the total cell chloride was not exchangeable. Since both the chloride content and the percent exchangeable were decreasing, the total exchangeable pool declined. These results are plotted in Figure 2 upper as a function of external chloride. In addition, the non-exchangeable content also decreased as more and more of the external chloride was replaced with sulfate (figure 1 lower). The non-exchangeable chloride content may represent that population of cells with membranes which can only exchange their chloride with sulfate and cannot use a chloride-chloride exchange mechanism (Villereal and Levinson,'77).

Chloride fluxes: the effect of phloretin

The effect of phloretin on steady-state chloride fluxes is summarized in Table 4. The effect was a decrease in all rate coefficients and fluxes. At a concentration of 2.5×10^{-4} M phloretin did not block mechanism A or mechanism B exclusively. It was more effective on the slow, compartment B, blocking up to 80% of the flux while the rapid A compartment was blocked by 60%. Since phloretin is not specific in its inhibitory action and can inhibit urea permeability in these cells (Hempling and Wise, 1975), these findings are not surprising.

In all experiments, total exposure to phloretin did not exceed fifteen minutes through the course of the experiment. During this time, intracellular K^+ and Cl^- content did not change. However, in all three experiments, the internal content of Na^+ although constant during the experiment, was always an average of 30% higher in the presence of phloretin. This gain in Na^+ content must have occurred within the first two minutes of exposure to phloretin.

Steady state chloride fluxes: chloride fluxes: the effect of furosemide

Erythroblastic leukemic cells were exposed to 0.1 mM, 1 mM, or 5 mM furosemide for 15 minutes at 25°C after which steady-state chloride fluxes were measured with Cl^{36} for an additional sixty minutes. Intracellular chloride concentrations varied from experiment to experiment in a range between 30 to 110 mEq/liter of osmotically active water, but in each experiment furosemide had no significant effect on intracellular chloride concentration.

Table 4. The Effect of Phloretin on Chloride Fluxes in the Steady-State at 25°C.

PARALLEL MODEL

	SMALL, FAST COMPARTMENT				LARGE, SLOW COMPARTMENT			
EXPT	SIZE	K12	K21	FLUX	SIZE	K13	K31	FLUX
CONT	0.17	6.4	6.8	3.2	0.83	2.6	5.6	1.30
250	0.19	2.2	2.0	1.4	0.80	0.4	0.8	0.25
%INHIB		66	71	56		85	86	81
CONT	0.15	5.0	4.0	2.2	0.85	1.9	2.7	0.83
250	0.18	1.9	1.4	0.86	0.81	0.36	0.61	0.16
%INHIB		62	65	61		81	77	81

RATE CONSTANTS: MIN-1

K12=VALUEX10-4 K21=VALUEX10-1 K13=VALUEX10-4 K31=VALUEX10-2

*PHLORETIN: MICROMOLES/L FLUX= VALUEX10-2 MICROMOLES/10 7 CELLS MIN.

Table 5. The Effect of Furosemide on Steady-State Chloride Fluxes at 25°C.

Parallel Model

	Fast Compartment				Slow Compartment			
Expt.	Size	K12	K21	Flux	Size	K13	K31	Flux
Control	0.79	7.6	1.8	3.9	0.21	0.003	0.3	0.017
1 mM	0.38	2.0	1.0	1.0	0.62	0.20	0.61	0.11
Control	0.58	4.7	2.7	2.8	0.42	0.067	0.54	0.039
5 mM	0.35	0.87	1.1	0.52	0.65	0.027	0.19	0.016
Control	0.49	14.8	5.3	6.8	0.5	0.31	1.0	0.14
0.1 mM	0.30	5.2	5.9	2.2	0.70	1.1	5.4	0.47
1 mM	0.41	5.6	4.4	2.7	0.59	0.56	3.0	0.27
Control	0.69	8.8	2.2	5.2	0	0	0	0
0.1 mM	0.63	5.3	1.4	3.2	0.06	0.004	0.16	0.003

Rate constants: Min^{-1}

K12 = value x 10^{-4} K21 = value x 10^{-1} K13 = value x 10^{-4} K31 = value x 10^{-2}

Flux = value x 10^{-2} micromoles/10^{7} cells min

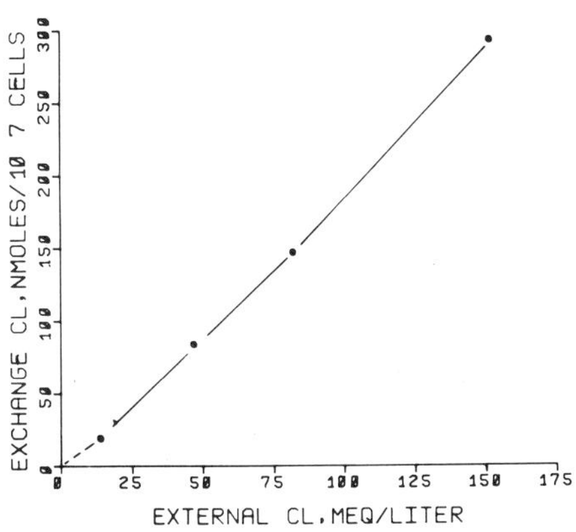

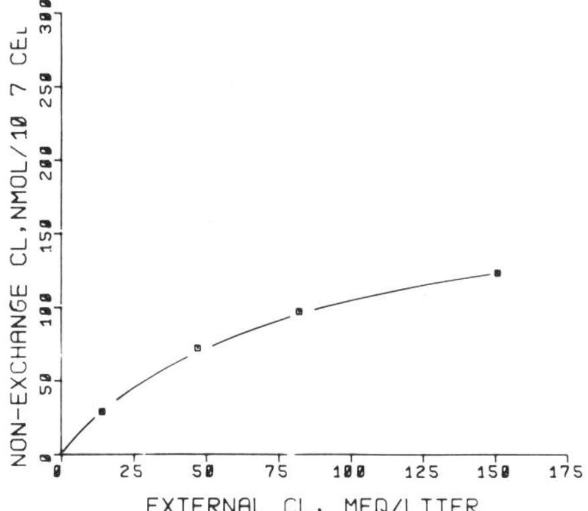

Fig. 2. The effect of sulfate on the exchangeable chloride compartment in rat erythroblastic leukemic cells. Upper: The response of the exchangeable compartment to the replacement of external chloride by isosmotic equivalents of sulfate. Lower: The response of the non-exchangeable compartment to the replacement of external chloride by isosmotic equivalents of sulfate.

Furosemide at 0.1 mM or 1 mM had two effects on steady-state chloride fluxes (Table 5): 1. Furosemide inhibited the fluxes across the fast A compartment by as much as 74% with both influx coefficient and efflux coefficient being affected. The compartment size was reduced as well. 2. Furosemide increased the steady-state fluxes across the slow B compartment almost six fold. The compartment size was increased as well.

At the higher concentration of 5 mM, the overall effect for both compartments was inhibition by as much as 81%.

DISCUSSION

We propose two populations of erythroblastic leukemic cells which differ in those membrane functions which regulate Cl^- transport. These two populations may represent extremes in the maturation of these particular membrane functions. We can anticipate equally well a range of intermediate populations which represent stages in the maturation process. The evidence for two cell populations is illustrated most easily with the replacement experiments. Replacement of chloride with sulfate eliminated the slow compartment completely (Table 3). This compartment may represent a population of cells which possesses a system for Cl^- - Cl^- exchange. The fact that a non-exchangeable cell compartment remains which diminishes as external sulfate increases implies a chloride transport system which will only exchange with the sulfate anion.

The effect of dihydro DIDS also supports a hypothesis of two cell populations because one could convert two compartments into one compartment whose kinetics were similar to one of the two control compartments. Taken alone this finding does not rule out a series model, but when included with the sulfate results, it is consistent with two populations with varying contributions from Cl^- - Cl^- exchange and from Cl^--$SO_4^=$ exchange. The stilbene inhibitors such as dihydro DIDS inhibit sulfate transport by 90% or more in Ehrlich mouse ascites tumor cells (Levinson, '78; Villereal and Levinson,'76).

Cl^- - Cl^- exchange was inhibited by no more than 37%. In our work with the erythroblastic leukemic cell, it was the slow compartment which was least sensitive to dihydro DIDS and responded only after the concentrations were increased, while the fast compartment was virtually two populations of cells, one with a preponderance of Cl^- - Cl^- exchange (the slow compartment) and one with a preponderance of Cl^--$SO_4^=$ exchange (the fast compartment). Obviously, studies on $S^{35}O_4$ exchange are in order.

Applying the same model to the phloretin results, in the Ehrlich mouse ascites tumor cell phloretin is not selective in its inhibition of the Cl^- - Cl^- exchange and the Cl^--$SO_4^=$ exchange. As much as 80% to 90% of chloride transport can be eliminated with phloretin. In our studies (Table 4), up to 80% of the slow compartment (Cl^- - Cl^- exchange) was blocked and 60% of the fast compartment.

The response of the two hypothesized populations to furosemide has its counterpart in the mature human erythrocyte. As analyzed by Brazy and Gunn ('75) the mature erythrocyte has a catalytic site and a transport site which subserve chloride fluxes and a modifier site which when occupied by Cl^- inhibits chloride fluxes. In their analysis, furosemide has a dual effect on chloride fluxes. It is capable of inhibiting the exchange fluxes, but when as an anion, it replaces the Cl^- at the modifier site, it may release the self-inhibition by Cl^- and stimulate exchange fluxes. We observed a similar pattern in the erythroblastic leukemic cell. From Table 5, furosemide at 0.1 mM inhibited the fluxes across the fast compartment (the Cl^--$SO_4^=$) exchange and stimulated the fluxes across the slow compartment (the Cl^- - Cl^- exchange). It is reasonable to hypothesize that the slow compartment represents Cl^- - Cl^- exchange with a high degree of self regulation.

In summary, our general approach from this kinetic analysis has been to argue for two cell populations. The best test of this hypothesis will require the isolation of the two populations and a demonstration that the kinetics of the two populations analyzed separately predict the kinetics of the two populations when combined.

REFERENCES

Brazy, P.C. and Gunn, R.B., 1975, Furosemide inhibition of chloride transport in human red blood cells, J. Gen. Physiol., 68:538-599.

Cabantchik, Z.I. and Rothstein, A., 1972, The nature of the membrane sites controlling anion permeability of human red blood cells as determined by studies with disulfonic stilbene derivatives, J. Membr. Biol., 10:311-330.

Hempling, H.G. and Wise, W.C., 1975, Maturation of membrane function: the permeability of the rat erythroblastic leukemic cell to water and to non-electrolytes, J. Cell. Physiol., 85:195-207.

Lepke, S., Fasold, F., Pring, M., and Passow, H., 1976, A study of the relationship between inhibition of anion exchange and binding to the red cell membrane of 4,4'-diisothiocyanostilbene-2,2' disulfonic acid and its dihydroderivative (H_2DIDS), J. Membr. Biol., 29:147.

Levinson, C., 1978, Chloride and sulfate transport in Ehrlich ascites tumor cells: evidence for a common mechanism, J. Cell. Physiol., 35:23-32.

Levinson, C. and Villereal, M.L., 1976, The transport of chloride in Ehrlich ascites tumor cells, J. Cell. Physiol., 88:181-192.

Rothstein, A., Cabantchik, Z.I. and Knauf, P., 1976, Mechanism of anion transport in red blood cells: role of membrane proteins, Fed. Proc., 35:3-10.

Ship, S., Shami, Y., Breuer, W. and Rothstein, A., 1977, Synthesis of tritiated 4,4'-diisothiocyano-2,2'stilbene-disulfonic acid ([^{3}H]DIDS) and its covalent reaction with sites related to anion transport in human red blood cells, J. Membr. Biol., 33:311-323.
Villereal, M. and Levinson, C., 1976, Inhibition of sulfate transport in Ehrlich ascites tumor cells by SITS, J. Cell. Physiol., 89: 303-312.
Villereal, M.L. and Levinson, C., 1977, Chloride-stimulated sulfate efflux in Ehrlich ascites tumor cells: evidence for 1 : 1 coupling, J. Cell. Physiol., 90:553-564.
Wise, W.C., 1974, A transplantable erythroblastic stem cell leukemia, Cancer Research, 52:611-612.
Wise, W.C., 1976, Maturation of membrane function: transport of amino acid by rat erythroid cells, J. Cell. Physiol., 87:199-212.

ROLE OF VICINAL WATER IN CELLULAR EVOLUTION

W. Drost-Hansen

Laboratory for Water Research
Department of Chemistry, University of Miami
Coral Gables, Florida

I. INTRODUCTION

Water is crucial to all living systems; indeed, life undoubtedly developed in an aqueous environment and life depends on, and reflects, the unusual properties of water. Phrased differently, life is intimately tied to the nature and extent of hydrogen bonding, not only through the role of hydrogen bonding in determining the structure and properties of water, but in addition through the role of H-bonding in determining the stability and functioning of proteins and nucleic acids. Hydrogen bonding - and hence its manifestations in the form of the nature of water - is truly a time invariant. Thus, whatever adaptations cellular systems may have had to conform to and take advantage of H-bonding, those features must have been the same since life began and to the present time. Specifically, therefore, whatever information we may now be able to deduce about living systems, reflecting the structural properties of water, must be generally true from protocells to the most complex of present day organisms.

It is generally recognized that water is unique amongst liquids. Unfortunately, the structure of bulk water and bulk aqueous solutions remain to be definitely understood. Some of the structural effects of water are exceedingly subtle, yet of critical importance. However, in the present paper, attention is primarily focused on interfacial water, the structure and properties of which have been modified by proximity to a surface. Such water is referred to as vicinal water (vw); a brief introduction to vicinal water is presented in the second section of this paper. It is important to recognize that vicinal water - just as bulk water - must have been an invariant over geological time. Thus, the experiences now attainable regarding the role of vicinal water in living systems must apply equally well to the origin of life and all subsequent evolution.

II. VICINAL WATER

Water adjacent to an interface (particularly a solid interface) differs in its properties from bulk water. The characteristics of vicinal water have been presented in various papers(1,2,3)and are summarized below in a condensed version.

A. Vicinal water appears to exist adjacent to most (or all) solid interfaces and extending into the aqueous phase for a distance of about one hundred Å (and possibly for as much as five hundred Å) - in other words distances ranging from 0.01 μm up to 0.05 μm (this corresponds roughly to 30 to 150 molecular diameters.)

B. Within certain limits, vicinal water appears to occur relatively independently of the specific details of the chemical nature of the solid surface and relatively independent of the nature and concentration of the solutes present(4,5).This phenomenon is referred to as the "paradoxical effect."

C. The properties of vicinal water undergo more or less abrupt changes over narrow temperature intervals - transitions resembling first and/or higher order phase transitions(1,2,6).The transition temperatures are near 15, 30, 45 and 60°C; to be referred to, respectively, as T_1, T_2, T_3, and T_4, or in general T_k. Conversely, when thermal anomalies are observed over narrow temperature intervals at or near T_k, the presence of the anomalies may be taken to be indicative of the presence of vicinal water. (Note: in general, and particularly in more complicated systems especially of biochemical or biological interest, changes, such as protein denaturation or lipid phase transitions (in lipid/ aqueous systems) do of course occur, independently of vicinal water. Note also that we are not concerned with local, high energy interactions such as local ion-hydration, or dipole-water interactions.)

D. Macromolecules in aqueous solutions appear to posess vicinal water ("vicinal hydration")(5). It seems that a critical size of solute exists below which no vicinal hydration occurs and above which vicinal hydration appears to occur frequently (or invariably). The critical size range appears to be between one thousand and a few thousand Daltons.

E. Vicinal water exists independently of electrical double layers(2,7).

F. Vicinal water does not owe its origin to "hydrophobic

hydration" (in the traditional sense) nor to ion-water or dipole - water interactions (such as, for instance, involved in the hydration of ionic sites on macromolecules).

G. The effects of vicinal water are super-imposed on those effects which are normally discussed in surface and colloid chemistry (i.e. London-van der Walls forces, electrical double layer effects, etc.)

H. By virtue of the paradoxical effect it is concluded that vicinal water must also exist in all cellular systems - consistent with the observation that many thermal anomalies have been reported in cellular functioning at the temperature ranges where vicinal water in physicochemically well defined system exists such anomalies(1, 2,8,9).

For purposes of the present paper a few specific properties of vicinal water merit mentioning. It appears that the density of vicinal water is less than that of bulk water and thus, by the La Chatelier's principle must be expected to be influenced by application of pressure. This must have played a role in the process of evolution in the development of deep sea organisms. The specific heat of vicinal water is approximately twenty to twenty five percent larger than that of bulk water(4,5)and the thermal expansion coefficient for vicinal water appears to be larger than that of bulk water(3).Various other properties appear unusual such as dielectric properties(11,12),adibatic compressibility(13),viscosity(14),and energies of activation for ionic conduction in vicinal water(15).

Finally special mention should be made of the unusual ion selective properties of vicinal water. Wiggins(16),has demonstrated the propensity of vicinal water to enhance the concentration of structure breaking ions over those of structure making ions. The effect is quite notable and exhibits highly characteristic thermal behavior, namely sharp minima and maxima as a function of temperature with peaks of ion enhancement of structure breaking ions near or at 15, 30 and 45° C - in other words exactly at the temperatures of T_k.

III. PSEUDO PHASE DIAGRAMS OF LIFE

In the process of evolution plants and animals have evolved from some prebiotic forms of single cells (or protocells) to multicellular organisms of enormous complexity. Changes and adaptations of incredible variety have occurred. Part of these changes have resulted from intra and inter-species interactions while other changes have resulted from changing physical environmental factors. The purpose of the present paper is to stress and briefly discuss some <u>invariant</u> aspects of evolution related to physico-chemical properties, reflecting the time-invariance of the structure and properties of bulk and vicinal water and aqueous solutions.

Temperature (T) and pressure (P) are the two most fundamental thermodynamic parameters which determine the ranges possible for living organisms. In Figure 1 is a sketch of a "pseudo phase-diagram" in which is plotted the approximate ranges of life as a function of logarithm of T and P.

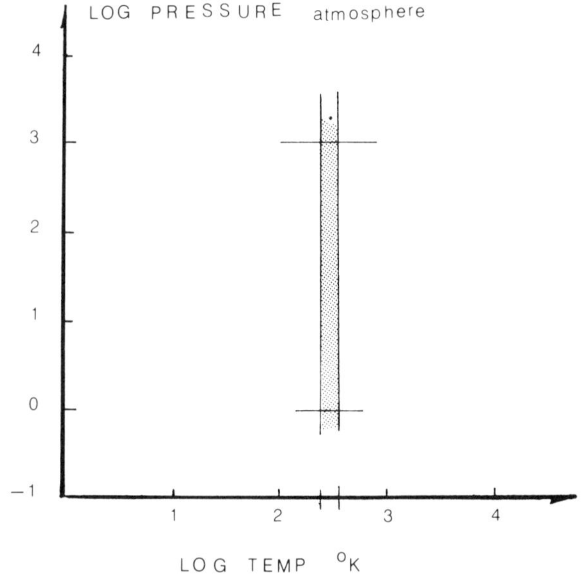

Fig. 1. Approximate domain, in a phase-diagram, of conditions under which life is likely to exist.

Obviously, the domain of living systems is bounded (primarily) by the range in which water exists as a liquid. At 1 atmosphere pressure this corresponds to a very narrow interval (273 to 373°K) and although some high-temperature barophilic bacteria exist (up to 383°K or 110°C) there are no reports of higher organisms existing much above approximately 50 or 60°C. On the other hand, life exists in a range of pressure from less than one atmosphere and up to approximately 1000 atmospheres (such as found, for instance in deep ocean trenches) and possibly higher.

It is of interest to attempt to expand the notion of a "life phase - diagram" (even though it obviously has no counterpart in thermodynamic reality). Thus, Figure 2 shows such an attempt. Here the approximate species diversity is plotted as a function of T and P (in a highly qualitative manner.) (Note the change in ranges of T and P shown.) Finally, a graph of the total estimated biomass (M) as the independent variable in the "pseudo phase-diagram" is likely to resemble qualitatively the graph of number of species.

IV. FUNCTIONAL AND EVOLUTIONARY ASPECTS OF VICINAL WATER

The P-T behavior sketched above refers solely to the bulk attributes of water. This implies primarily the role of water as an external ecological factor. However, more importantly, water plays a dominant role as the internal medium in which life takes place. By way of an example, Tracey (17) points out that, for instance, a muscle is 40 Molar in water, 0.15 M in fat, 0.2 M in NaCl and very little else - apart from a miniscule 0.002 M in protein. One way of looking at "the rest", the 0.002 M protein, is to suggest that the presence of this material serves to provide the necessary "interfaces" which transforms much (if indeed, not all) of the cellular water to vicinal water. In this light, life is merely complex biochemistry in an extensive multivarious colloid matrix. Thus it is hardly surprising that cell functioning reflects many aspects of interfacially modified water, - the vicinal water.

As an example of the likely role of vicinal water in determining cell functioning consider the size of cells. A minimum cell size appears to exist corresponding to an equivalent diameter of about 0.2 μm. Traditionally, the existence of a minimum cell size has been related to the intrinsic volumes of the macromolecules required in living cells. If indeed, this is so - if this consideration is the main or only consideration - minimum cell size must have been an invariant in time over the span in which the present day known macromolecules were present in earlier forms of cells.

It is suggested in this connection that the minimum cell size may also - or instead - reflect the need for the existence of vicinal water in the cell. In view of the suspected minimum geometric extent vicinal water, in the range from 100 - 500 Å, it is hardly surprising that a minimum cell size exists. This becomes particularly likely if the detailed functioning of the metabolic processes in the cell requires the presence of both vicinal and

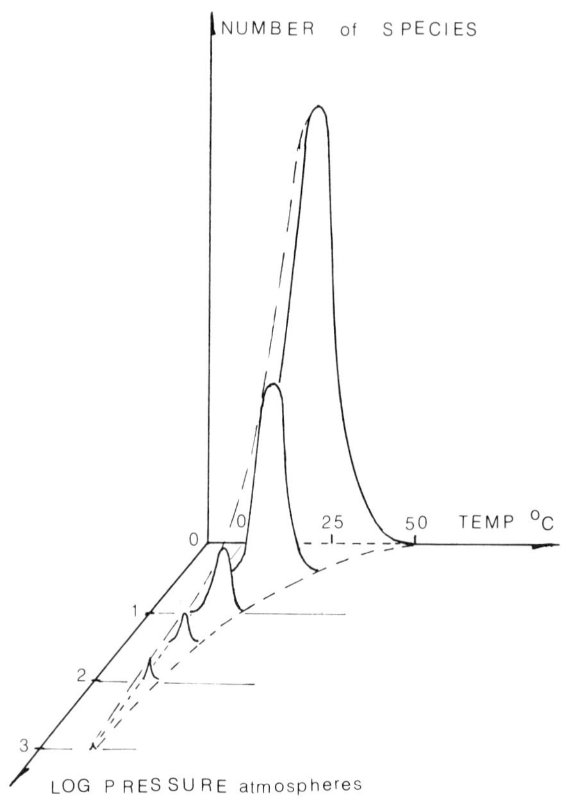

Fig. 2. A (pseudo) Phase-diagram of estimated number of species as function of pressure and temperature (highly schematized).

bulk water. This notion has been discussed by Clegg (18) regarding internal cell compartmentization. In Clegg's model a bulk-like phase for the diffusion of electrolytes and small metabolites is required with the majority of the macromolecules, specifically, the enzymes, to be located near interfaces (in the vicinal aqueous phase).

In order to have a compartment with essentially bulk-like characteristics, bulk water filled "void space" must be present in which no confining surface (membrane, micro-structural element or macromolecule) is present over a distance of about two times 100 (to 500) Å; in other words 200 to 1000 Å, or 0.02 to 0.1 μm. this is at least within an order of magnitude of the known minimal cell size.

As discussed earlier, vicinal water exhibits a certain amount of ion selectivity: thus there exists a tendency to concentrate structure breaking solutes over structure making solutes (Wiggins (16);Hurtado and Drost-Hansen(19). Hence, the interior milieu of the first proto-cell may likely have been characterized by a relatively higher K^+ - to - Na^+ concentration ratio than in the surrounding "sea". This enhanced ratio constituted the "normal" interior environment and it is interesting to speculate that active transmembrane transport developed as a means of insuring and/or enhancing the unequal ion distribution originally present.

V. THERMAL ZONATION

Vicinal water undergoes notable changes at the transition temperatures, T_k. These changes influence cellular functioning, sometimes dramatically (1). The influence may be mediated, for instance by the effects of the water structure changes on membrane functioning (20) or the hydration of the biomacromolecules (21,22). Recall that the first three transition temperatures are near 15, 30 and 45°C. - in other words in the range of most physiological phenomena.

Andrewartha and Birch (23) have discussed the most likely range of temperatures of organisms. They observed that no individual species is known which can thrive over a range from 0 - 50°C. It is of interest to quote these authors "It will be noticed...that all the animals with narrow ranges (16°C or less) were aquatic, living in places where the difference between minimal and maximal temperature is small, and that most of those with wide ranges (20°C or more) were terestrial or amphibious, living in places where the temperature is more variable." Andrewartha and Birch go on to mention a study by Moore who tabulated the range of occurence of 41 aquatic animals including Crustacea, Echinodermata, Mollusca, Tunicata and Pisces and found that 90% of them were restricted to a range of 16°C or less while for 42% the range was even more narrow namely, 14°. Amphibians and some insects were known to exist over temperature ranges exceeding 20°. It is proposed, here, as discussed previously by the present author, that a range of survival of just about 15° may be no coincidence but rather the manifestation of upper and lower thermal limits, determined by the thermal temperatures (T_k) imposed by the vicinal water transitions.

Thermal boundaries - both upper and lower - are sometimes exceedingly abrupt, extending over no more than one or two degrees (see in particular the papers by Thorhaug and the present author 24,25). Furthermore, the mechanism of thermal death is poorly understood (see for instance Fry (26)); and even less obvious is the sharpness of the thermal boundaries. It appears unlikely that such sharp boundaries can be explained by a combination of detrimental simultaneous and/or consequtive reactions: combinations of van't Hoff and/or Arrhenius equations do not allow for extremely abrupt changes (21). Instead criticality can be explained only by the existance of highly cooperative, collective phenomena. Such cooperativity, however, is exactly what characterizes vicinal water. Thus, if one or more reactions which may be "critical" (important) to the functioning of a cell depends on or involves vicinal water, a change of one or two degrees (near, for instance, 30°C) may have catastrophic results. Finally, recall that the characteristics of water and vicinal are time invariants. Thus, the criticality now observed for vicinal water must have been present since the first occurrence of liquid water on earth (or in space). From the origin of the first proto-cell to the evolution of the most complicated organism, life must have been subject to and hence reflect the constraints imposed by vicinal water. Adaptations have obviously occured: however, it is unlikely the temperature ranges near T_k could have been transgressed with impunity.

In a previous paper (1), it has been suggested that the existence of the thermal transitions - with an associated decrease in biological activity at the temperatures of T_k - may have resulted in a general tendency for microorganisms to fall into one of several categories. Thus, the standard classification of psychrophiles, mesophiles, and thermophiles may reflect this tendency to thermal stratification. Indeed, once again, the existence of the thermal boundaries must have been invariant in geological time. No doubt adaptations may have occured but, as discussed in our previous papers on multiple growth optima (27), most microorganisms so far tested, do indeed exhibit multiple growth optima when cultured on minimal media (25) suggesting that, at least, a decrease in activity may always have occured at the temperature of the thermal transitions.

Many micro-organisms, particularly unicellular organisms, have developed the ability to exist over temperature ranges wider than 15°. Thus for some organisms growth is observed from, for instance,

5° up to near 45 or even 55^oC. Some years ago Oppenheimer and the present author (27) proposed that the organisms do so by invoking different metabolic pathways in the different temperature intervals between consecutive thermal transition temperatures. Growing many of these organisms on minimal media tends to bring out bi- or multimodal patterns; We have described several such examples over the past 25 years (22). In each case distinct growth minima are observed at or near the temperatures of the thermal transitions and occasionally no less than three distinct, separate growth optima have been observed.

VI. EVOLUTION OF HOMEOTHERMY

Homeothermy offers a mechanism for protecting the organism from the capriciousness of climate. At the same time it must be imperative to avoid operating the system close to anyone of the thermal transition temperatures, T_k(1,2). In fact it must have been desirable to be as far removed from any of these transitions as possible - in other words it would be preferable to operate a homeothermic system near the midpoint between two consecutive thermal transition temperatures. The ranges then for optimum body temperature would be near 23°, near 37° or near 53^oC. The latter temperature range obviously would correspond to an excessive temperature differential between the core of the organism and the environment under the normal climatic conditions which have prevailed in the geological time period in which homeothermy developed. Conversely, a body temperature of 23° would involve the risk of notably higher environmental temperatures during summer (especially in sub-tropical and tropical climates) with a resultant difficulty in heat dissipation. Thus, as the least undesirable alternative, it would appear the range around 37° has been selected for the body temperature of mammals, and to a first approximation, for birds.

An inspection of the body temperatures of about 160 mammals suggets that the temperatures are distibuted in a remarkably narrow peak around 37 to 38° with a half-width of approximately 2 to 3^o(1). Consistent with this notion is also the observation that no mammal (or bird) is able to withstand body temperatures in excess of 45 degrees. Profound physiological changes also occur around 30° (also in those mammals which are able to withstand cooling - the hibernators - below 30 degrees.)

The average body temperature of birds appear to be notably higher than that of mammals, namely around 41 to 42°. This elevated body temperature is interpreted as a concession to the requirement for rapid energy delivery in order to sustain flight which is one of the most energetically demanding physiological activities. It is of interest in this connection to note that such non-flying birds as the Kiwi, Ostrich, and Penguin, etc. have lower body temperatures, namely near 37 to 39° - in other words in the range of the mammals. Thus when the need for rapid energy delivery for flight disappears it is more favorable to "reset" the homeostatic thermostat at a lower set-point. Note, however, that for birds as well as for mammals an upper thermal limit appears to be 45°. This means that the range available to birds for coping with infectious diseases is correspondingly smaller than that of mammals (excepting the non flying birds mentioned above). On the other hand, the higher body temperature may provide an internal environment which is more unfavorable for malignancies. Thus it would be of interest to know from avian pathology if birds tend to suffer less from malignant tumors than mammals and specifically to know if there is significant difference in the occurrence of malignancy between the flying and non-flying birds. If indeed birds have less tendency to cancer, this finding would lend credence to the proposed benefits of hyperthermia treatment of cancer.

VII. CONCLUSIONS

Life originated in an aqueous environment. This external environment also became the internal environment; hence, any living organism carries with it the benefits and restrains imposed by the nature of water - particularly vicinal water. The legacy of the structural attributes of water and vicinal water constitute time invariant aspects of molecular thermobiology. Specifically, the restrains imposed by vicinal water may prove to offer a central, unifying theme in our attempts to understand cellular functioning. The most obvious of the restraints imposed by vicinal water are clearly delineated by the thermal response of living systems.

The structure of bulk water and bulk aqueous solutions continue to escape an "einwandfrei" description. The complexity is apparently enormous. Even less is known with certainty about the structure and properties of vicinal water. Yet, in spite of these shortcomings it seems evident that an understanding of water is imperative to understanding cellular physiology on the molecular level. However, the ultimate understanding of the structural properties of water and vicinal water may provide a unifying foundation for unraveling part of the enormous complexity of living systems. However incomplete our understanding may be, any theory which can be based on molecular arguments is likely to be far more valuable than merely descriptive empirical observations.

Tracey(17) has put the problem of water in biological systems in proper perspective by quoting Herman Boerhaave who apparently held the chairs of Chemistry, Botany and Medicine at Leyden University in the eighteenth century. Thus, from a book on chemistry by Boerhaave; "As water is common to be met with and comes to be used on every occasion, men are apt to imagine they thoroughly understand its nature; but those who have carefully applied themselves to the examination thereof, find it is one of the most difficult subjects in all natural philosophy to be acquainted with."

REFERENCES

1. Drost-Hansen, W., Structure and properties of water at biological interfaces, Chemistry of the Cell Interface, H.D. Brown, ed., Academic Press, New York, 1971, Chapter 6, pp. 1-184.
2. Drost-Hansen, W., Water at biological interfaces - structural and functional aspects, Physics and Chemistry of Liquids Volume 7, Gordon & Breach, 1978, pp. 253-346.
3. Drost-Hansen, W., "The Occurrence and extent of vicinal water," "Biophysics of Water," F. Franks, Wiley & Sons, New York, 1982, pp. 163-169.
4. Braun, C., Drost-Hansen, W., "A DSC study of heat capacity of vicinal water in porous materials," "Colloid and Interface Science", Volume III ,M. Kerker, ed., Academic Press, New York, 1976, pp. 533-541.
5. Etzler, F., Drost-Hansen, W., Recent thermodynamic data on vicinal water and a model for their interpretation. Paper presented at VII International Summer Conference "Chemistry of Solid/Liquid Interfaces", Volume 56 (4), Cavtat/Dubrovnik (Yugoslavia) 1983. To be published in Croatia Acta Chemica.
6. Drost-Hansen, W., On the structure of water near solid interfaces and the possible existence of long-range order. Industrial and Engineering Chemistry,1961, pp. 10-47.
7. Drost-Hansen, W., Effects of vicinal water on colloidal stability and sedimentation processes. J. Colloid Interface Science, Volume 58 (2), 1976, pp. 251-262.
8. Drost-Hansen, W., Structure and functional aspects of interfacial (vicinal) water as related to membranes and cellular systems. Colloq. Internationaux du C.N.R.S. as "L'eau et les Systemes Biologiques", No. 246, 1976, pp. 177-186.
9. Drost-Hansen, W., Phase transitions in biological systems - manifestations of cooperative processes in vicinal water. Ann. New York Academy of Science, 204, 1973, pp. 100-112.
10. Etzler, F., Fagundus, O., Journal of Colloid and Interfacial Science, in press, 1983.
11. Ballario, C., Bonicontro, A., Cametti, C., DiBiasio, A., Journal of Colloid and Interfacial Science,78, 1980 p. 242.
12. Henry, F., These de Doctorat d'Etat et Sciences, l'Universite Pierre et Marie Curie, Paris VI, 1982.
13. Bruun, S.G., Sorensen, P.G., Drost-Hansen, W., Studies of vicinal water structuring is suspensions. II. Ultrasonic velocity measurements. 1978, Submitted for publication.
14. Peschel, G., Adlfinger, Naturwissenschaften, 56, 1969, pp. 558-559.

15. Schufle, V.A., Huang, C.T., Drost-Hansen, W., Surface conductance and vicinal water. Journal of Colloid and Interfacial Science, 54, 1976 pp. 184-202.
16. Wiggins, P.M., Clinical and Experimental Pharmacology and Physiology, 2, 1975, pp. 171-176.
17. Tracey, M.V., "Water Relations of Foods", ed. R.B. Duckworth, Academic Press, New York, 1975, pp. 659-662.
18. Clegg, J.S., "Cell-Associated Water," ed. W. Drost-Hansen and J.S. Clegg, Academic Press, New York, 1979 pp. 363-413. See also Collective Phenomena, 3, 1981, pp. 289-311.
19. Hurtado, R.M., Dorst-Hansen, W., Ion selectivity of vicinal water in the pores of a silica gel. Monograph, Academic Press, 1978, pp. 115-123 in "Cell-Associated Water."
20. Thorhaug, A., Drost-Hansen, W., Temperature effects in membrane phenomena. Nature, 215, 1967, pp. 506-508.
21. Etzler, F.M., A role for water in biological rate processes Monograph, Academic Press, New York, 1978, pp. 125-164 in "Cell-Associater Water."
22. Etzler, F.M., Drost-Hansen, W., A role for vicinal water in growth, metabolism, anc cellular organization. Advances in Chemistry Series 188, 1980, pp. 485-497.
23. Andrewartha, H.G., Birch, L.C., "The Distribution and Abundance of Animals", University of Chicago Press, Chicago, Illinois 1954
24. Drost-Hansen, W., Biologically allowable thermal pollution limits. Final Report to Enviromental Protection Agency, Office of Research and Monitoring. AS EPA #660/3-74-003.
25. Drost-Hansen, W., Gradient device for studying effects of temperature of biological systems. Journal of the Washington Academy of Science, 71 (4), pp. 187-201.
26. Fry, F.E., "Thermobiology", ed. A.H. Rose, Academic Press, New York, 1967, pp. 375-409.
27. Oppenheimer, C.H., Drost-Hansen, W., A relationship between multiple temperature optima for biological systems and the properties of water. Journal of Bacteriology, 80, pp. 21-24.

PROTON TRANSLOCATION AND BIOENERGETIC PHENOMENA

APPLICATION OF CHEMICAL MODIFICATION AND SPIN LABELING TECHNIQUES TO THE STUDY OF ENERGY CONVERSION BY BACTERIORHODOPSIN

L. Packer, A.T. Quintanilha, and R.J. Mehlhorn

Dept. of Physiology/Anatomy; Membrane Bioenergetics Group
Lawrence Berkeley Laboratory
University of California, Berkeley, CA 94720

INTRODUCTION

Light generates a pH gradient and an electrical potential across the purple membrane of Halobacterium halobium. We are investigating the time resolved changes in protonation of the side chains of specific amino acid residues and the correlation of these changes with photon absorption and the ensuing photoreaction cycle. We seek to determine the precise molecular description of the photocycle and of the time dependent steps in the uptake, translocation, and release of protons by the retinal proton catalyst in this membrane, bacteriorhodopsin (BR).

The photocycle is initiated by absorption of visible radiation (Abs. max at 570 nm) by the all-trans retinal chromophore. The initial steps of the photocycle involve the isomerization of the chromophore to the 13-cis retinal. A deprotonated Schiff base forms with a $t_{1/2}$ of about 50 μsec. Proton release occurs in this time range (illustrated in the upper half of Fig. 1). Later, the deprotonated species decays ($t_{1/2}$ about 3-5 msec), restoring the original all-trans form of the chromophore. During this time, protons are taken up from the cytoplasmic surface of the protein. This process is shown in the bottom half of Fig. 1. The deprotonated photointermediate is referred to as the M_{412} species.

To assess significant structural features of the lipid and the protein important for proton movement, we have used chemical modification of amino acid side chains, spin labeling and deuterium isotope effects on purple membranes. Purple membranes have also been reconstituted into liposomes before and after such treatment

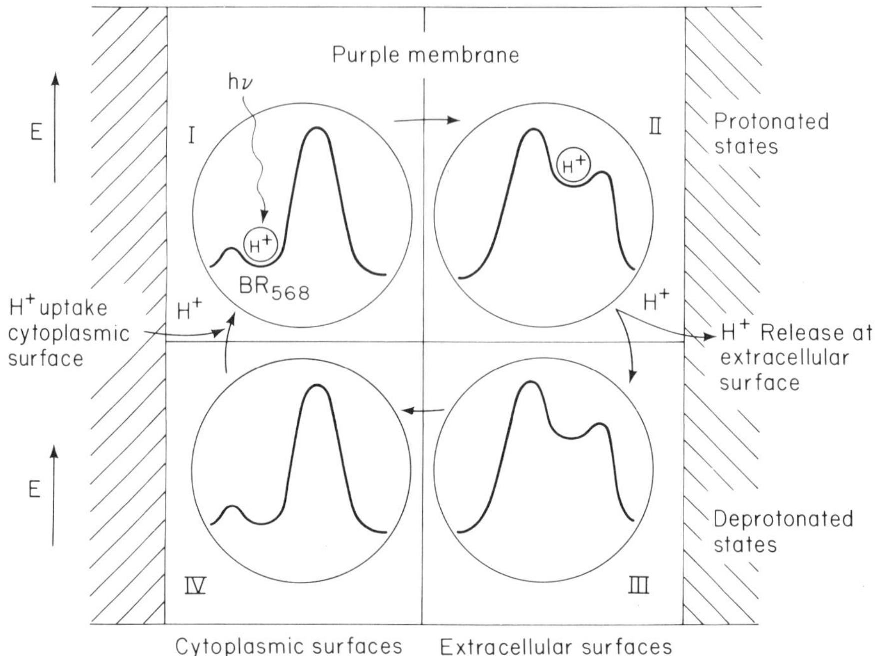

Fig. 1. Protein Conformation States of Bacteriorhodopsin Associated with Proton Translocation across the Purple Membrane

in order to derive information relevant to the occurrence of proton movement at the cytoplasmic surface of the purple membrane. It is hoped that this information will contribute to an understanding of the importance of the electrochemical potential developed across the closed vesicle membrane in the regulation of light energy conversion.

CHEMICAL MODIFICATION STUDIES

Modification of Positively Charged Amino Acids

Lysine. Modification of up to 80% of the ε-amino groups of lysine (5 of the 6 available lysine residues[1]) with imidoesters (methyl acetimindate, ethyl acetimidate and methyl butyrimidate) which introduce progressively more bulky groups onto the lysyl residues but leave the charge substantially unaltered, showed little effect on the rate of M_{412} formation and decay.[2] Similar results were obtained when the charge of the amino group was changed by the use of succinic or acetic anhydride, or pyridoxyl phosphate. These studies demonstrate that the ε-amino groups of lysine are probably not involved in a coordinated pathway of proton translocation by the protein. However, lysine groups do play a structural role essential for activity. Using either dimethyl adipimidate (DMA) (8.3 Å) or glutaraldehyde (7.5 Å), bifunctional reagents with chemically crosslinked lysyl residues, we found a very strong inhibition of M_{412} decay[3] and a decreased proton uptake in reconstituted liposomes. Crosslinking with longer chain bifunctional reagents (like dimethyl suberimidate [11.3 Å]) of individual BR molecules to each other had no effect on activity. These effects suggest that intramolecular conformational changes and not protein-protein interactions in the membrane are essential for activity.

Arginine. With both 2,3-butanedione (BD) and phenylglyoxyl (PGO) treatments, similar effects were obtained for the modified purple membranes.[4] Modification of 3 or the 7 arginine residues resulted in a 33- and 18-fold inhibition of the $t_{1/2}$ of the first and second phases of M_{412} decay, respectively, and a 19-fold increase in the amount of M_{412} stationary state assayed at pH 8. These results indicate that the positive charges of the guanidinium groups of the unmodified residues are essential for photocycling activity.

Modification of Negatively Charged Amino Acids

Glutamate and Aspartate. Water soluble carbodiimides cause the carboxyl groups of these amino acids to become either neutral or positive. These modifications slowed down the M_{412} decay, had no effect on its formation, and caused a large increase in its photostationary state.

In the presence of the added nucleophile, glycine methyl ester (GME), 1-ethyl-3-(3-dimethylaminopropyl) carbodiimide (EDC) treatment is expected to yield the amide adduct. With increasing concentrations of GME (1-250 mM), and in the presence of EDC, the degree of inhibition of M_{412} decay was partially decreased, suggesting that GME may be competing with a nucleophilic group on the protein. Evidence supporting this has come from double modification experiments. EDC treated samples previously modified by ethyl acetimidate (EA) showed M_{412} decay kinetics similar to those observed in samples modified by EDC in the presence of high GME concentrations.

The similarity suggests that EA modified lysines have become unavailable for reaction with EDC, preventing the formation of intramolecular crosslinks. Since EDC can only crosslink closely lying groups, we suspect that the ε-amino groups of lysine must be very close to the carboxyl groups.

Other important results of the carbodiimide modification are the disappearance of one of the photointermediates (O_{640}) in the photocycle and the inability to generate the acid stable form of the protein (Abs. max. at 605 nm). These results suggest that the protonation of one (or more) carboxyl residues is necessary for the generation of that photointermediate as well as for the acid stabilization of the protein.

Tyrosine. By investigating the time course of lactoperoxidase catalyzed iodination of purple membranes, we have been able to obtain evidence of an involvement of at least one tyrosyl residue in proton release and one or several tyrosyl residues involved in reprotonation.[5,6] Modification of one type of tyrosyl residue (localized on the cytoplasmic surface, according to modification studies done on intact cell envelope vesicles) markedly slows M_{412} decay but does not alter the optical and Raman spectral characteristics of the chromophore. Iodination of another tyrosyl residue alters the optical and Raman spectra of the chromophore and markedly stimulates M_{412} formation.

The modifications exhibit a strong effect on the pH dependence of the photocycle, shifting the curve to lower pH values. This suggests that the introduction of bulky groups onto the tyrosyl residues alters the pKs and sterically effects their ability to take part in proton movements. What follows provides evidence for this.

Deuterium Isotope Effects

In agreement with previous reports,[7] we find that the rise time of M_{412} is slowed by a factor of 4.7 in D_2O whereas the decay time is slowed by a factor of only 2.1. We have measured similar isotope effects for a number of chemically modified purple membranes. The similarity indicates that the basic mechanism remains similar in the modified samples (see Table 1). The activation parameters for the rates of rise and decay of M_{412} were determined from their temperature dependence.[8] For tyrosine modified samples, for example, the activation parameters are consistent with a decreased pK for a tyrosine residue (lowered E_a), an increased steric effect caused by iodine substitution (negative S^+), and an overall decrease in the rate of reprotonation (increased G^+) of the Schiff base. If the tyrosine residue is directly involved in the reprotonation of the Schiff base, the effect of its modification on the activation parameters is not surprising.

Table 1. Kinetic Parameters and Isotope Ratios for Rise and Decay of M_{412}

Sample[c]	Rise[a]			Decay[a]		
	H $t_{1/2}$	D $t_{1/2}$	k_1^H/k_1^D	H $t_{1/2}$	D $t_{1/2}$	k_2^H/k_2^D
Control	0.069	0.33	4.76	3.65	7.7	2.11
EAC (Carboxyl)	0.099	0.42	4.24	90[b]	131[b]	1.45
I_2 (Tyrosine)	0.026	0.11	4.15	420[b]	1042	2.48
NBS (Tryptophan)	0.043	0.147	3.40	19.2	67.3[b]	3.5
DMA (Lysine)	0.072	0.35	4.86	33.0	46.2[b]	1.4
EA (Lysine)	0.062	0.32	4.9	10.8	27.4[b]	2.5
BD (Arginine)	0.096	0.49	5.1	359[b]	1055	2.9
PGO(Arginine)	0.142	0.71	5.0	355[b]	724[b]	2.0

[a]Half lives in milliseconds.
[b]Decay kinetics are biphasic, only slow phase given.
[c]Assay conditions: protein 0.15-0.30 mg/ml in 100 mM KCl and 5 mM phosphate; pH 7.5-8.0

SPIN LABELING STUDIES

Nitroxide spin probe techniques have been developed for measuring membrane bioenergetic functions.[9,10,11] Electrical surface potentials at outward-facing membrane-water interfaces are measured with impermeable amphiphilic spin probes which give rise to distinct spectral lines from aqueous and membrane environments.[9] Transmembrane pH gradients are measured with spin-labeled amines and carboxylic acids by quantitating intra- and extracellular concentrations of the nitroxides. Impermeable paramagnetic line-broadening agents, e.g., ferricyanide, are used to quench extracellular signals for these quantitations.[10] Cell volumes are determined similarly, using a freely permeable spin probe, which does not respond to electrochemical potentials (TEMPONE). Transmembrane electrical potentials are estimated from the distributions of spin-labeled phosphorium ions which slowly permeate membranes and hence respond to equilibrium gradients.[11] Unfortunately, rate limitations rule out applications of these nitroxides to kinetic measurements.

PARAMETER	PROBE	MEMBRANE PREPATION	BASIS OF MEASUREMENT	ASSAY CONDITIONS
ΔpH^{*}	TA TC	Sealed vesicles or cells	Increase (TA) or decrease (TC) of line height with H^{+} uptake	Impermeable paramagnetic broadening agent outside
$\Delta\psi^{*}$	T P	"	Increase of line height as interior becomes more negative than exterior	
Volume	TK	"	Line height	
$\Delta\psi_s^{**}$	CAT_n	Vesicles,cells or membrane sheets	Relative intensities of aqueous and membrane bound signals	Low ionic strength

* Absolute gradients can be determined from simultaneous measurements with ^{14}N and ^{15}N probes E.G., TK and TϕP or TA and TC.

** Absolute surface potentials can be determined relative to the binding of unchanged or oppositely charged probes.

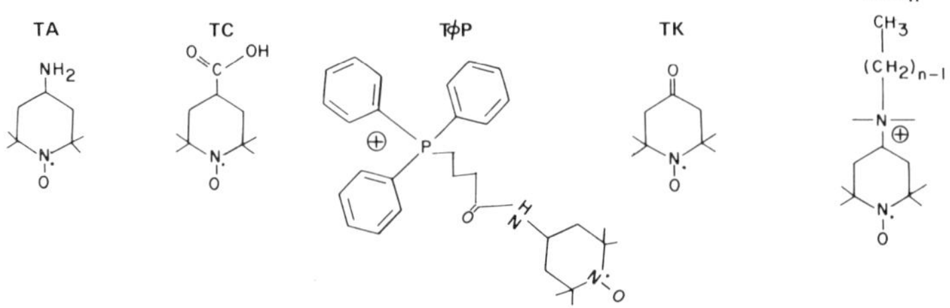

Fig. 2. Some ESR Probes of Membrane Electro-Chemical Potentials and Gradients

Light-Induced Surface Potential Changes in Purple Membranes

The positively charged paramagnetic amphiphile 4-(dodecyl dimethyl ammonium)-1-oxyl-2,2,6,6-tetramethyl piperidine bromide (CAT_{12}) partitions between the membrane and the aqueous phase. Therefore, CAT_{12} was useful in monitoring changes in surface potential associated with the release and uptake of protons from the purple membrane.[12] Using laser flash EPR and photolysis, we showed that the uptake and release of the CAT_{12} probe closely follows the kinetics of the rise and decay of the M_{412} intermediate.[13] This correlation indicated that the probe senses exposed negative charges at the surface from which protons are released. The number of light-induced surface charges per M_{412} in the photostationary state was calculated to be approximately 1. The value is more than 50% decreased for carboxyl and guanidinium group modification and for heavily modified tyrosine samples, suggesting that some of these groups may be located at the surface of the protein and be involved in the uptake and release of protons from the aqueous environment.

Control of the Photocycle by the Transmembrane Potential Gradient

When reconstituted into liposomes, BR will generate light-induced electrochemical gradients across the lipid bilayer. Our reconstituted systems generate pH gradients of 0.6 - 1.0 and transmembrane electrical potentials of 60-80 mV.[14] The photostationary state of M_{412} at the same light intensity and protein concentration was much higher in the reconstituted system than in the isolated purple membranes. Valinomycin (0.5 μM) decreased the photostationary state of M_{412} by one order of magnitude in the reconstituted system but had no effect in isolated purple membranes. Nigericin (0.5 μM) had no effect on either system. Because we observed that at the concentration used, valinomycin collapses the transmembrane potential and nigericin collapses the pH gradient, we assume that the electrical gradient was probably the major controlling factor in the kinetics of the photocycle.

Steady state light-induced electrochemical gradients across BR-containing liposomes showed a substantial slowdown of the M_{412} decay kinetics with virtually no effect on the rise kinetics. Again, valinomycin, but not nigericin, brought the decay kinetics of M_{412} to those measured in the absence of steady state illumination.

These responses suggest that within the protein light may generate strong electric dipoles that could be very sensitive to the transmembrane electric potential gradients, which can be established across the bacterial membrane.

DISCUSSION

The sharp pH dependence seen in controls and the downward shift of the dependence in the rise and decay of the M_{412} species in tyrosine modified purple membrane preparations suggests a direct involvement of reversible tyrosine protonation in the pathway of proton movement through BR (Figure 3). Reversible dissociation of protons at both the outer and inner purple membrane surface would be expected to be pH and pK dependent, hence regulated by the electrochemical gradient. Since the pH gradients are small (<1), pK

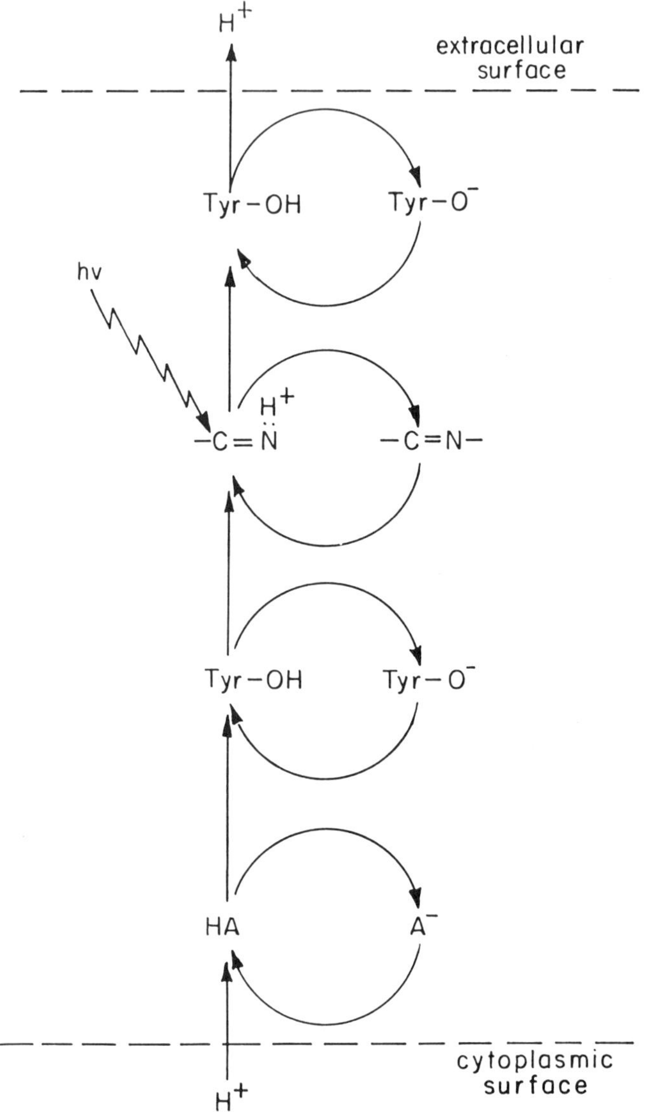

Fig. 3. A Proposed Pathway for Uptake and Release of Protons by Bacteriorhodopsin. (We include on the pathway the reversible deprotonation of the tyrosine residues before and after the Schiff base.)

effects on the photocycle are not to be expected, and indeed, we find that the main control parameter is the transmembrane potential.

Proton Uptake. Several structural requirements appear to be important for reprotonation of BR. Our results with modification of carboxyl and guanidinium groups suggest that a charge transfer complex essential for reprotonation may exist between the two types of groups. Moreover, the large changes in the entropy of activation for the M_{412} decay provides strong supporting evidence that interaction between these two types of groups is an essential structural feature involved in proton uptake and the reprotonation phase of the photocycle. Furthermore, as judged by double modification experiments, an interaction between carboxyl groups and at least one amino group of lysine also appears to be important for reprotonation. The above structural requirements together with the inhibitory effects of bifunctional crosslinking reagents which restrain motion of intramolecular amino groups of lysine located within 8.3 Å of one another suggests that protein conformational changes are essential to proton uptake and M_{412} reprotonation. Tyrosine may also act as a donor for the Schiff base reprotonation.

Proton Release. The specific groups involved in release of protons at the outer surface of the purple membrane probably include tyrosine residues. This conclusion is drawn from the observed effects of iodination and nitration[15] that shift the pK of the dissociation of the phenolic hydroxyl group of tyrosine residues. The pK shift may accelerate proton movement from this group on the tyrosine to a nearby basic group which may be water, OH^-, a group on the lipid of the purple membrane, or a buffer anion. The acceleration would be expected to occur in the range where proton release is pH dependent. That the tyrosine residues are involved in the early stages of the photocycle, associated with proton release, can also be argued by the pronounced deuterium isotope effect on proton release as seen before and after iodination of tyrosine.

Carboxyl groups may also play a role in the movement of protons. Modification of these groups has inhibited the generation of one of the intermediates in the photocycle and the formation of an acid state species.

Light-induced changes in the surface potential of the purple membrane suggest that protons are released from protonated groups at the surface because these surface potential changes are closely coupled to the formation of the M_{412} species. Our experiments provide evidence that the groups involved may be carboxyl and tyrosyl residues.

REFERENCES

1. W. Stoeckenius and R. Lozier, Light Energy Conversion in Halobacterium halobium, J. Supramol. Struct. 2:769-774 (1974).
2. T. Konishi, S. Tristram and L. Packer, The Effect of Cross-linking on the Photocycing Activity of Bacteriorhodopsin, Photochem. Photobiol. 29:353-358 (1979).
3. T. Konishi and L. Packer, Light-Dark Conformational States in Bacteriorhodopsin, Biochem. Biophys. Res. Commun. 72:1437-1441 (1976).
4. L. Packer, S. Tristram, J. Herz, C. Russell and C. Borders, Chemical Modification of Purple Membranes: Role of Arginine and Carboxylic Acid Residues in Bacteriorhodopsin, FEBS Letts. 108:243-248 (1979).
5. P. Scherrer, C. Carmeli, P. Shieh and L. Packer, The Effect of Chemical Modification of Tyrosyl Residues on Proton Conductivity Mediated by Bacteriorhodopsin, Biophys. J. 25:207a (1979).
6. P. Scherrer, L. Packer, S. Seltzer and M. Lin, Involvement of Tyrosine Residues in Bacteriorhodopsin Activity, (submitted for publication).
7. R. Korenstein, W. Sherman and R. Caplan, Kinetic Isotope Effects in the Photochemical Cycle of Bacteriorhodopsin, Biophys. Struct. Mech. 2:267-276 (1976).
8. P. Sullivan, A. Quintanilha, S. Tristram and L. Packer, Isotope Effects and Activation Parameters for Chemically Modified Bacteriorhodopsin, FEBS Letts. 117:359-362 (1980).
9. R.J. Mehlhorn and L. Packer, Membrane Surface Potential Measurements with Amphiphilic Spin Labels, in: "Methods in Enzymology," S. Fleischer and L. Packer, eds., Academic Press, New York, Vol. 56, pp. 515-527, (1979).
10. R.J. Mehlhorn, T. Racanelli and L. Packer, Simultaneous Measurements of Proton Movement and Membrane Potential Changes in the Wild Type and Mutant Halobacterium halobium Vesicles, in: "Methods in Enzymology," L. Packer, ed., Academic Press, New York, Vol. 88, pp. 334-344 (1982).
11. R.J. Mehlhorn, P. Candau and L. Packer, Measurements of Electrochemical Gradients with Spin Probes, in: "Methods in Enzymology," L. Packer, ed., Academic Press, NY, Vol. 88, pp. 761-762, (1982).
12. C. Carmeli, A. Quintanilha, and L. Packer, Light-Induced Surface Potential Changes in Purple Membranes and Bacteriorhodopsin Liposomes, in: "Membrane Bioenergetics," C.P. Lee, G. Schatz, L. Ernster, eds., Addison Wesley Publishing Co., Reading, MA (1980).
13. C. Carmeli, A. Quintanilha, and L. Packer, Surface Charge Changes in Purple Membranes and the Photoreaction Cycle of Bacteriorhodopsin, PNAS (USA), 77:4707-4011 (1980).
14. A. Quintanilha, Control of the Photocycle in Bacteriorhodopsin by Electrochemical Gradients, FEBS Letts. 117:8-12 (1980).
15. M. Camps-Cavieres, T. Moore and R. Perham, Effect of Modification of the Tyrosine Residues of Bacteriorhodopsin with Tetranitromethane, Biochem. J. 179:233-238 (1979).

ACKNOWLEDGEMENTS

This work has been supported by the Office of Biological Energy Research, Divison of Basic Energy Sciences, U.S. Department of Energy, under contract No. W-7405-ENG-48.

MECHANISM OF PROTON TRANSLOCATION BY THE $b-c_1$ COMPLEX OF MITOCHONDRIAL RESPIRATORY CHAIN

S. Papa, F. Guerrieri, M. Lorusso and D. Boffoli

Institute of Biological Chemistry, Faculty of Medicine and Centre for the Study of Mitochondria and Energy Metabolism, C.N.R., University of Bari, Italy

The energy conserving segments of redox chains directly convert chemical energy into transmembrane electrochemical proton gradient[1,2].

According to the chemiosmotic rationale[3] coupling in redox enzymes between chemical catalysis and vectorial proton translocation is an immediate consequence of diffusion of the same substrates and/or chemical groups along specifically oriented conduction pathways through the catalytic center. Mitchell has proposed[4] a quinone cycle for the cytochrome system, which explains the effective translocation of two protons from the negative (N) to the positive (P) aqueous domain per electron flowing along the chain, as the consequence of conduction of hydrogen atoms from the N to the P side of the membrane by quinones, acting in a cycle on both sides of b cytochromes, and transmembrane electron conduction from P to N by b cytochromes and cytochrome oxidase. This model doesn't attribute much of a role to the polypeptides, but passive involvement as scaffolds holding the metal centers in the proper orientation in the membrane and possibly facilitating specific mobilities of quinones. There is, however, overwhelming evidence that the apoproteins feel the transfer of electrons at the metal center in a much more active way[5]. The occurrence of cooperative linkage between redox transitions of the metal and protolytic events elsewhere in the proteins, is shown for the hemes of cytochrome oxidase[6], the b cytochromes[7] and the Fe-S protein[8] of the $b-c_1$ complex by pH dependence of the midpoint redox potential. These linkage phenomena are denominated redox Bohr effects by analogy to those displayed by hemoglobin[2,5,9].

When considering the involvement of redox Bohr effects in the function of membrane enzymes their topographical arrangement is of particular relevance[10].

The Bohr effect could, in fact, result in proton pumping if the redox linked protolytic group is exposed at one side of the membrane, the negative, when electrons arrive at the metal the pK is consequently enhanced and protons are taken up at this side. Then the protonated group, or another accepting protons from the first, attains exposure at the opposite side of the membrane, upon oxidation of the metal the pK is lowered and protons are released from this side[10]. However, a vectorial Bohr mechanism based exclusively on H^+ conduction by ionizable group in hemoproteins, is likely to produce variable stoicheiometries depending on the actual values of the pK's of these groups with respect to proton activity in the N and P domains. There is also an intermediate possibility, where redox-linked protolytic groups can serve to facilitate proton transfer through channels in the protein from one aqueous phase to an hydrogen carrier (ubiquinone) in the membrane and from this to the opposite aqueous phase (Fig. 1). Such model incorporates in series both the concept of vectorial protonmotive catalysis and cooperative proton transfer by polypeptides.

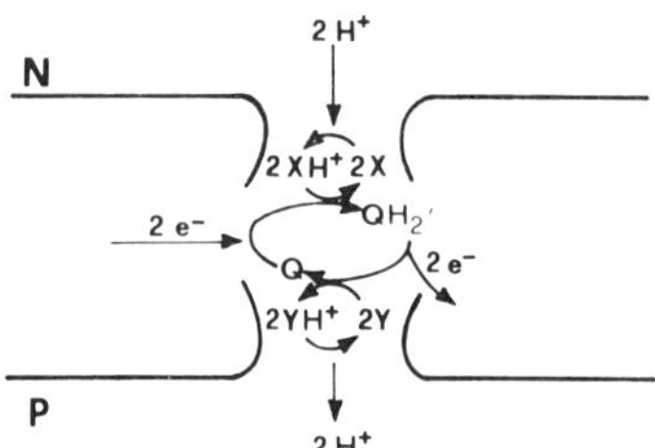

Fig.1 - Model of proton pump based on combined ligand conduction and Bohr effect

Redox Bohr effects in the cytochrome system in the mitochondrial membrane and isolated redox complexes have been directly determined[11,12].

Oxidation of electron carriers by oxygen, under conditions where vectorial processes are abolished, has to result in equal consumption of protons for protonation of reduced oxygen to water. However if the oxidation of the metal is accompanied by release of protons from the proteins, this acid reaction results in a smaller proton consumption with respect to the metals oxidized (Fig. 2).

The difference reveals the occurrence of Bohr effects in the cytochrome chain with an H^+/e^- coupling number of 0.7 in the pH range 6.8-9, which means that as much as 70% of the oxido-reductions of electron carriers are linked to H^+ transfer by ionizable groups.

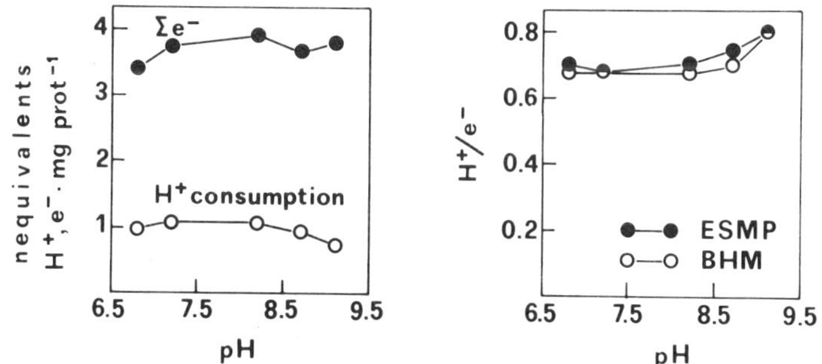

Fig.2 - pH dependence of oxidation of electron carrier and related scalar proton transfer reactions induced by oxygen pulses of anaerobic beef-heart mitochondria (BHM) and submitochondrial particles (ESMP). For experimental conditions and procedure see ref. 11.

In Table 1 measurements of redox Bohr effects in the isolated $b-c_1$ complex are summarized. The aerobic oxidation of redox centers of the $b-c_1$ complex, effected through traces of purified cytochrome oxidase and cytochrome c, amounting to 7 electron equivalents resulted in the consumption of only two proton equivalents. Thus five proton equivalents are released from the enzyme corresponding to 1.2 H^+ released per b cytochromes plus Fe-S protein oxidized. It should be recalled that c cytochromes are pH independent. Correction for the protons eventually released in the oxidation of bound quinol to semiquinone leaves with an H^+/electron ratio of 0.8. In the presence of an amount of antimycin which prevented net oxidation of b cytochromes, the H^+ electron coupling number for the Fe-S protein was also 0.8. On the right measurement of redox Bohr effects in cytochrome oxidase is presented. Also in this case proton consumption was smaller than the sum of redox carriers oxidized from which an H^+/electron ratio of 0.9 could be measured.

Table 1. Redox Bohr-effects in isolated cytochrome b-c_1 complex and in isolated cytochrome c oxidase from BHM. The values are expressed in ngion or nmol/mg protein of complex. pH 7.2. For explanations and other experimental details see ref.12.

	b-c_1 complex				Cytochrome c oxidase
			+Ant.A		
H^+ uptake	2.06		1.60	H^+ uptake	11.59
cyt.b	1.92		—	a+a_3	10.55
Fe-S, g=1.90	2.38		2.19	copper	10.55
cyt.c+c_1	2.94		2.75	cyt.c_1	1.23
				PMS	0.49
Σe^-	7.24	Σe^-	4.94	Σe^-	22.82
ΔH^+ uptake	5.18		3.34	ΔH^+	11.23
$\frac{\Delta H^+}{\text{cyt.b,Fe-S}}$	1.20	$\frac{\Delta H^+}{\text{Fe-S}}$	1.52	$\frac{\Delta H^+}{a,a_3}$	1.06
$\frac{\text{Bohr } H^+}{\text{cyt.b,Fe-S}}$	0.82	$\frac{\text{Bohr } H^+}{\text{Fe-S}}$	0.84	$\frac{\text{Bohr } H^+}{a,a_3}$	0.89

Papa et al. have proposed that redox Bohr-effects can be directly involved in the protonmotive activity of the quinone-cytochrome c segment of the mitochondrial respiratory chain[10,13]. Later Wikström has proposed that redox Bohr effects or similar cooperative phenomena can result in proton pumping also by cytochrome oxidase[14]. A close scrutiny of the results available on the overall proton electron stoicheiometry for the cytochrome chain of mitochondria[15] and the protonmotive events associated to operation of cytochrome oxidase[16] suggests, however, that acceptance of a proton pump in cytochrome oxidase requires more solid experimental evidences than that at present available (see ref. 9 for review).

The proton pumping activity of the b-c_1 complex is on the contrary a clearly established process. To study this activity, experiments on isolated b-c_1 complex reconstituted in liposomes have been performed. In the experiment shown in Fig. 3 reconstituted b-c_1 vesicles were supplemented with an excess of duroquinol and traces of soluble cytochrome oxidase. Electron flow was initiated by adding ferricytochrome c. This system allows measurements of pure vectorial proton transport since an amount of protons equivalent to those deriving from the oxidation of quinol to quinone is consumed in the external medium for reduction of oxygen to water by the soluble oxidase. Electron flow, measured potentiometrically as oxygen consumption, resulted in a fully reversible proton release which was completely abolished by FCCP. The initial rates of proton translocation and oxygen uptake give an H^+/electron ratio for net H^+ transport of 1 at pH 7, corresponding to an overall H^+/electron ratio of 2.

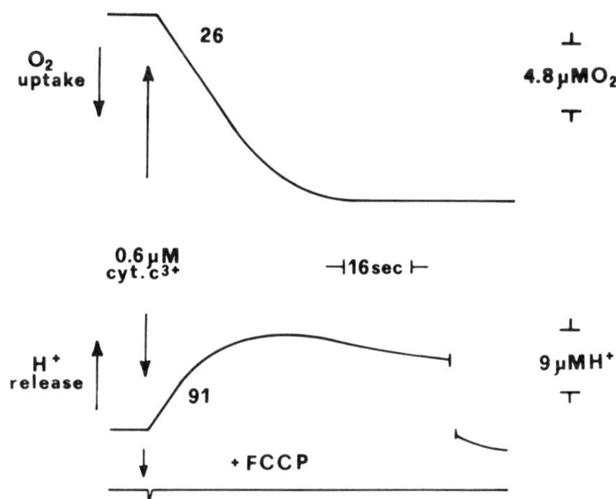

Fig.3 - Active proton transport in b-c_1 vesicles. b-c_1 vesicles (1μM cyt.c_1), suspended in 150 mM KCl and 5 mM $MgCl_2$, were supplemented with 1μg valinomycin, 0.25μM cytochrome oxidase and 23μM DQH_2. pH 7.0. The reaction was initiated by adding ferricytochrome c. Where indicated 4μM FCCP was present. Figures on the traces represent initial rates of H^+ release and O_2 consumption.

The pH dependence of the H^+/electron stoicheiometry in the b-c_1 vesicles (Fig. 4) shows a ratio of 2 at acidic and neutral pH values, then this ratio declined to 1.5 as the external pH was increased to 8.3. Further pH increase resulted in a small but significant enhancement of the H^+/electron ratio. The decline of the H^+/electron ratio in the pH range 7 to 8 was not due to an enhanced rate of proton back-flow. In fact the rate of passive proton equilibration induced by a potassium diffusion potential, imposed by valinomycin, was negligible when compared to proton equilibration by FCCP and even decreased at alkaline pH's[10]. It has been already

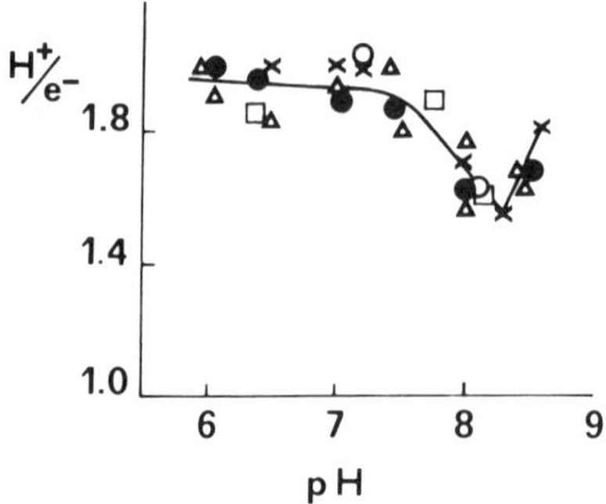

Fig.4 - pH dependence of H^+/e^- stoicheiometry in $b-c_1$ vesicles. The experimental conditions are those described in the legend to Fig.3. Symbols refer to different experiments.

pointed out that a pH dependence of the H^+/electron stoicheiometry is not explained by the Q cycle[5,9,10].

If the proton pumping activity of the complex resulted directly from H^+/electron coupling in b cytochromes, as it has been proposed[17], one would expect the stoicheiometry of the pump to follow the pH profile of effective hydrogen conduction by b cytochromes. Thus the protons transported would be expected to raise from 0 to a maximum value going with pH from 6 to 8. The actual H^+/electron stoicheiometry, on the contrary, decreases in the pH range 7 to 8. Thus transmembrane proton translocation by the complex cannot be a direct consequence of redox Bohr effects of b cytochromes.

Our group has developed a mechanism for proton pumping by the $b-c_1$ complex, different from the Q cycle[5,10], which incorporates a primary role of a specific quinone system identified in the $b-c_1$ complex and redox Bohr effects (see Fig. 5). It is proposed that quinol of the pool is oxidized at the N side of the membrane to semiquinone by an antimycin insensitive reaction. The semiquinone is oxidized at the P side of the membrane by b_{566} which then transfers electrons to b_{562} at the N side and from this to bound semiquinone through an antimycin sensitive reaction. The two electrons reunite at the level of the bound quinone and from this pass to Fe-S protein and to cytochrome c_1. The cytochrome b shunt explains the antimycin promoted oxidant-induced reduction of b cytochromes, as well as the drop of the H^+/e^- ratio from 2 to 1.5 from pH 7 to 8.5, where b cytochromes change from electron to hydrogen carriers.

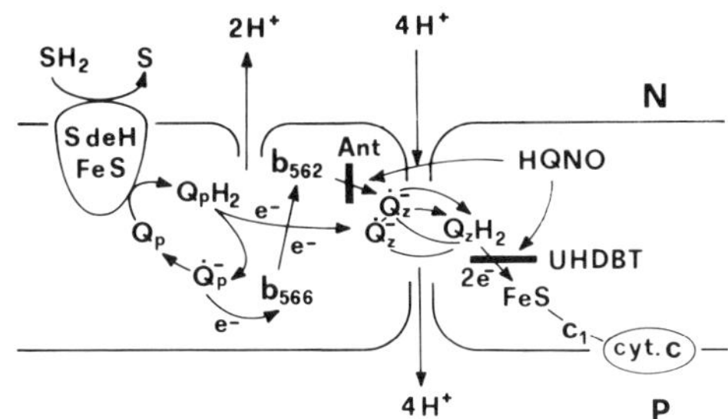

Fig.5 - Model of proton pumping by $b-c_1$ complex. Q_p indicates the ubiquinol pool; Q_z special bound ubiquinone. Sdh=succinate dehydrogenase.

Protonation of the UQ/UQH_2 couple from the N side of the membrane upon reduction and proton release, upon oxidation, at the P side can result in transmembrane translocation of 2 to $4H^+$ per $2e^-$ traversing the systems. The actual stoicheiometry will depend on the pK of the semiquinone and the actual pH at the N and P domains. Since the data available would assign to semiquinone a pK around 6, this can transfer $4H^+$ per $2e^-$ at physiological pH's. Two of the protons removed from the N space are replaced here by $2H^+$ from substrate oxidation and two positive charges will be compensated by net transfer of $2e^-$ from the dehydrogenase at the N side to cytochrome c at the P side.

pK shifts of ionizable groups in apoproteins of the complex, linked to oxido reduction of the quinone center, can favour H^+ transfer in the N half of the channel of the H^+ required for protonation of the reduced quinone and release of the H^+, produced in the oxidation of quinol to semiquinone, to ionizable groups in the P half of the channel. A general feature of this model is that Bohr effects in redox proteins, probably arising from electrostatic pairing of protolytic positive and negative residue, cooperate with protonmotive covalent bond exchange (Fig. 2).

Table 2 shows that treatments of the $b-c_1$ complex with DCCD that is a specific reagent for carboxyl residue or with TNM, specific for tyrosil residue, inhibit the proton pumping activity of the $b-c_1$ vesicles. It can, however, be noted that whereas TNM inhibits to the same extent both the rate of proton translocation and electron flow, so that, the H^+/e^- ratio is unaffected, DCCD at the con-

Table 2. Effect of DCCD and TNM (tetranitromethane) on active proton transport in b-c_1 vesicles.
3 mg protein of b-c_1 complex were treated with an amount of DCCD of. TNM corresponding to 100 nmoles of reagent per mole of cytochrome c_1. After 20 min incubation the complex was reconstituted in liposomes. Redox and proton transport activity were activated by adding to the vesicles, supplemented with 1 μg of valinomycin and 10μM ferricytochrome c, 11μM durohydroquinone. Figures represent the initial rates of proton translocation and e^- flow expressed as $\mu M \cdot min^{-1}$.

	Rate H^+ release	Rate e^- flow	H^+/e^-
Control	204.7	105	1.95
+ TNM	26	13.9	1.87
+ DCCD	127	96	1.32

centrations used in this experiment inhibited more profoundly proton transport than e^- flow with a consequent decrease of the H^+/e^- stoicheiometry.

Clearly further experiments are needed to better characterize the reaction of the modifiers used and their effect on the functional parameters. However the observations reported here and elsewhere[18,19] provide evidence for involvement of aminoacid residues sequence of apoproteins in the redox linked protonmotive activity of the b-c_1 complex.

REFERENCES

1. P. Mitchell, Compartmentation and communication in living systems. Ligand conduction: a general catalytic principle in chemical, osmotic and chemiosmotic reaction systems, Eur.J.Biochem. 95:1 (1979).
2. S. Papa, Proton translocation reaction in the respiratory chain, Biochim.Biophys.Acta 456:39 (1976).
3. P. Mitchell "Chemiosmotic Coupling in Oxidative Phosphorylation", Glynn Research, Bodmin, England (1966).
4. P. Mitchell, Possible molecular mechanism of the protonmotive function of cytochrome system, J. Teor.Biol. 62:327 (1976).

5. S. Papa, Molecular mechanism of proton translocation by the cytochrome system and the ATPase of mitochondria. Role of proteins, J.Bioenerg.Biomembr. 14:69 (1982).
6. V.Y. Artzabanov, A.A. Kostantinov and V.P. Skulachev, Involvemement of intramitochondrial protons in redox reactions of cytochrome a, FEBS Lett. 87:180 (1978).
7. P.F. Urban and M. Klingenberg, On the redox potential of ubiquinone and cytochrome b in the respiratory chain, Eur.J.Biochem. 9:519 (1969).
8. R.C. Prince and P.L. Dutton, Further studies on the Rieske iron-sulfur center in mitochondrial and photosynthetic systems: a pK on the oxidized form, FEBS Lett. 65:117 (1976).
9. S. Papa, Mechanism of active proton translocation by cytochrome systems, in: "Membrane and Transport", A.N. Martonosi ed., Plenum Press Vol. 1, pp.363-368 (1982).
10. S. Papa, M. Lorusso and F. Guerrieri, Energy transfer by redox proteins in mitochondria, in: "Cell and Differentiation", G. Akoyunoglou et al. eds.,Alan R.Liss,Inc.,New York, pp.423-437 (1982).
11. S. Papa, F. Guerrieri and G. Izzo, Redox Bohr-effects in the cytochrome system of mitochondria, FEBS Lett. 105:213 (1979).
12. F. Guerrieri, G. Izzo, I. Maida and S. Papa, Redox Bohr effects in isolated cytochrome $b-c_1$ complex and cytochrome c oxidase from beef-heart mitochondria, FEBS Lett. 125:261 (1981).
13. S. Papa, F. Guerrieri, M. Lorusso and S. Simone, Proton translocation and energy transduction in mitochondria, Biochemie 55:703 (1973).
14. M.K.F. Wikström and K. Krab, Proton pumping cytochrome c oxidase, Biochim.Biophys.Acta 549:177 (1979).
15. S. Papa, F. Guerrieri, M. Lorusso, G. Izzo, D. Boffoli, F. Capuano, N. Capitanio and N. Altamura, The H^+/e^- stoicheiometry of respiration-linked proton translocation in the cytochrome system of mitochondria, Biochem. J. 192:203 (1980).
16. M. Lorusso, F. Capuano, D. Boffoli, R. Stefanelli and S. Papa, The mechanism of transmembrane $\Delta\tilde{\mu}H^+$ generation in mitochondria by cytochrome c oxidase, Biochem. J. 182:133 (1979).
17. G. von Jagow and W.D. Engel, A model for the cytochrome b dimer of the ubiquinol cytochrome c oxidoreductase as a proton translocator, FEBS Lett. 111:1 (1980).
18. B.D. Prince and M.D. Brand, Proton translocation by the mitochondrial $b-c_1$ complex is inhibited by NN'-dicyclohexylcarbodiimide, Biochem. J. 206:419 (1982).
19. M. Degli Esposti, J.B. Sans, J. Timoneda, E. Bertoli and G. Lenaz, The inhibition of proton translocation in the mitochondrial $b-c_1$ region by dicyclohexylcarbodiimide, FEBS Lett. 147:101 (1982).

THE STRUCTURAL AND FUNCTIONAL ORGANIZATION OF WATER CLEAVAGE IN PHOTOSYNTHESIS

G. Renger, H.-J. Eckert, W. Weiss and G. Dohnt

Max-Volmer-Institut für Biophysikalische und
Physikalische Chemie, Technische Universität Berlin
Strasse des 17. Juni 135, 1000 Berlin 12, Germany

1. The overall organization scheme of photosynthetic water cleavage

Photosynthetic water cleavage is the "heart" of the biological solution of solar energy exploitation of the biosphere. The process takes place within the thylakoid membrane and is supported by two photosystems which are connected by a plastoquinone pool (for recent review see ref. 1). The light-induced reaction sequence in system II leads to water cleavage into molecular oxygen and hydrogen bound to plastoquinone which is of moderate reducing power. In system I the electronic energy of metabolically bound hydrogen is further enhanced through the light-driven transfer to $NADP^+$. Free energy is additionally stored as a transmembrane electrochemical potential gradient which is used for ATP synthesis. This organizational scheme indicates that the essential steps for water cleavage are realized by system II.

Regardless of the molecular mechanism, photolytic water cleavage into molecular oxygen and plastohydroquinone requires the realization of three operationally distinct processes:

A. The transformation of an electronically excited state, generated at a specific chromophor, into a ground state radical pair of sufficient oxidizing and reducing power and of a stability assuring a high efficiency of the subsequent reactions with water.

B. The cooperation of four oxidizing redox equivalents, ⊕, giving rise to formation of molecular oxygen.

C. The cooperation of two reducing equivalents, ⊖ , leading to plastohydroquinone.

Biological systems are able to perform chemical reactions in a well-defined manner by the use of a unique material for embedding the functional groups which participate in a reaction. This material are proteins. Their great variability allows a perfect adaptation of the reaction coordinates so that almost any degree of specificity, efficiency and regulatory control required for a biological function has been achieved during the evolutionary selection process. Accordingly, in order to understand the functional and structural organization of photolytic water cleavage, we have to unravel the functional groups and their reactivity in their special native protein matrices. At least three protein complexes were found to perform the water cleavage in system II: the reaction center complex, the water splitting enzyme system Y, and the plastoquinone-containing protein.

2. The reaction center complex of system II

The formation of the oxidizing redox equivalents occurs by exciton dissociation at a special chlorophyll a, referred to as

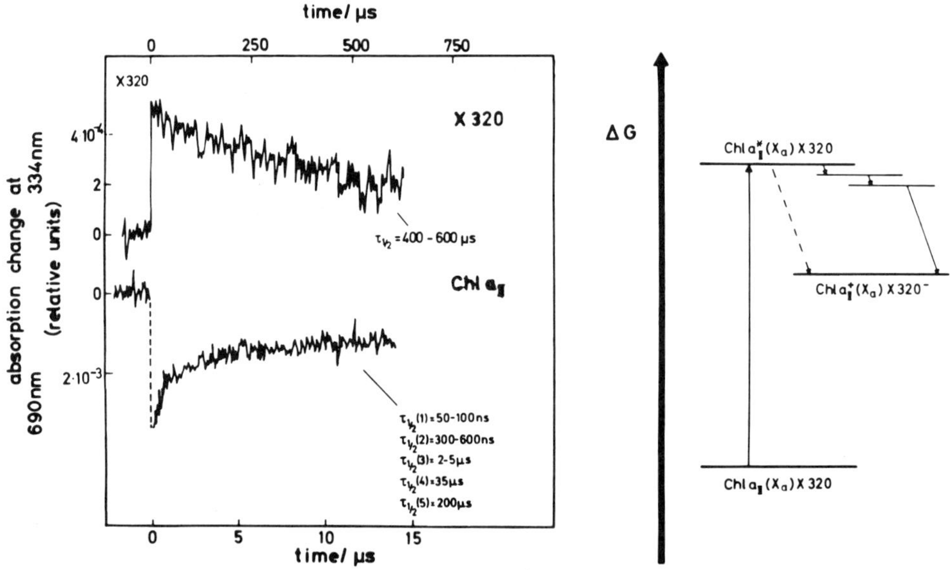

Fig. 1. Absorption changes at 334 nm and 690 nm as a function of time (5, 6) and energetic scheme of charge separation via "direct jump"-(broken line) and "staircase"-(solid line) mechanism. (X_a) represents intermediary redox groups.

$Chla_{II}$ (2). From the excited $^1Chla^*_{II}$ an electron is ejected which is transferred within 1 ns (3) to a special plastoquinone molecule (4) (designated as X320) under formation of a plastosemiquinone anion radical. Typical flash-induced absorption changes reflecting the turnover of $Chla_{II}$ and X320 are shown in Fig. 1. The flash-induced rise at 334 nm, limited by the experimental time resolution, reflects the formation of the plastosemiquinone anion radical of X320, the relaxation predominantly the subsequent reoxidation by another plastoquinone, B (or R)(7,8).

The negative absorption change at 690 nm indicates the photo-oxidative bleaching of $Chla_{II}$, generating a π-cation radical, $Chla^+_{II}$ which becomes reduced by a secondary donor, D_1 (vide infra), via multiphasic kinetics in the ns- and µs-time range (for a recent discussion see ref. 6).

The electron transfer from $^1Chl^*_{II}$ to X320 is a vectorial process coupled with the formation of an electric potential difference, $\Delta\psi$, across the thylakoid membrane. Different lines of evidence led to the conclusion that $Chla_{II}$ and X320 are located at the inner and outer side, respectively, of the thylakoid membrane, i.e. the electron transfer pathway within the reaction center is of the order of 2-3 nm. Accordingly, with respect to the mechanism of charge separation, generally two different types have to be considered (see Fig. 1, right): a) A direct electron jump from $^1Chla^*_{II}$ to X320 or b) participation of further components acting as functional redox components, thereby mediating a staircase sequence.

Independent lines of evidence favor a staircase transfer mechanism (9,10) resembling that of the charge transfer in bacterial reaction centers (11). Pheophytin was inferred to act as intermediary redox carrier (9,11). Latest EPR-measurements suggest the existence of a further redox component, mediating the electron transfer between $^1Chla^*_{II}$ and pheophytin (12). All redox groups involved in the staircase electron transfer from $^1Chla^*_{II}$ to X320 will be summarized by the symbol X_a. X320 as a plastoquinone was surprisingly found to act as 1-electron acceptor (13)(see section 5). As $Chla_{II}$ can be photooxidized, while X320 stays reduced, it remained to be clarified if the photoreaction causes a transmembrane electric potential difference and whether the oxidizing redox equivalents of $Chla^+_{II}$ are able to support water oxidation. Repetitive double flash group experiments (14) confirmed that $Chla_{II}$ can be photooxidized even if X320 is still reduced. However, neither O_2 evolution nor a significant "stable" $\Delta\psi$-formation can be observed under these circumstances. Therefore, X320 is required for a sufficient stabilization of the charge separation to assure water cleavage in system II.

In order to achieve the high quantum efficiency (close to one, see ref.15) of the overall charge separation within the reaction center complexes, a definite balance of the rate constants for the forward and back reaction of each individual electron transfer step is required. This is obviously realized by selection of appropriate redox groups with definite mutual electronic coupling established by a special protein matrix. Interestingly enough, according to our present state of knowledge all redox groups in the reaction centers of system II and photosynthesizing bacteria are cyclic π-electron systems involving heteroatoms (chlorophylls, pheophytins and quinones). Only few data are known about the orientation of these groups (e.g. the plane of pheophytin in system II was inferred to be arranged perpendicular to the thylakoid membrane, see ref. 16). The redox steps of the sequence are assumed to occur via vibronically coupled electron tunneling (for review see ref. 17). The reaction coordinate of the overall process and the mutual orientation of the active redox groups as well as their reactivities are determined by the protein matrix, which accordingly, has to be considered as the apoprotein of the reaction center complex. Unfortunately, information about the structure of the protein matrix of system II reaction centers is completely lacking. Likewise, the determination of the redox properties of the couple, $Chla^{+}_{II}/Chla_{II}$, by the protein matrix also remains a completely unresolved problem.

3. The functional coupling of the reaction center and the water splitting enzyme system Y

Based upon the classical experiments of Joliot and Kok (18,19), each system Y was inferred to be coupled with one reaction center complex. Accordingly, $Chla^{+}_{II}$ oxidizes system Y via a sequence of univalent redox reactions until, after the accumulation of four redox equivalents, molecular oxygen is evolved. As the reduction of $Chla^{+}_{II}$ occurs via multiphasic kinetics in the ns- and µs-range (see Fig. 1), while system Y is oxidized stepwise with half times of 100-1000 µs (20) at least one redox component, referred to as D_1, has to exist, which connects both operational units. The reduction kinetics of $Chla^{+}_{II}$ were inferred to depend on the redox state of system Y (21) and the pH of the inner space of the thylakoids (22). The $Chla^{+}_{II}$-reduction by D_1 becomes significantly retarded, if chloroplasts are selectively deprived of their oxygen evolving capacity (23). Under these conditions the shape of the EPR-spectrum which reflects the turnover of D_1, remains unaffected, while relaxation kinetics and saturation behavior are changed (24). The functional redox group of D_1 in tris-washed chloroplasts binds a proton in the reduced state ($pK \gtrsim 8$) which is released upon oxidation by $Chla^{+}_{II}$ (25). Furthermore, the active redox group of D_1 was inferred to be incorporated into a protein which appears to be rather resistent to mild trypsinization either from the outer (26) or the inner side (Renger and Völker, in preparation). The chemical nature of D_1 is completely unknown. It has not yet even been clarified if D_1 is an integral part of system Y.

4. The water splitting enzyme system Y

The molecular mechanism of photosynthetic water oxidation still withstands clarification. A reasonable starting point for attacking this problem appear to be theoretical considerations based on the attempt to correlate the intermediary redox states of system Y with the four-step univalent reaction sequence leading from H_2O to O_2. Obviously, the free redox intermediates of this pathway, i.e. OH radical, H_2O_2 and superoxide radical, cannot coexist with biological material. Accordingly, the basic problem of biological water oxidation is the energetic and kinetic control of these intermediary species in system Y. Transition metal-protein complexes with defined ligand coordination shell appear to be especially suited to fulfill this function. There is now ample evidence that manganese plays a key role in photosynthetic water oxidation (for recent review see ref. 27). Based upon these experimental findings, system Y was assumed to be a manganoprotein (28). The above-mentioned consideration about the reactivity together with the data of Joliot and Kok (18,19) and the kinetic EPR data of Babcok et al. (20), led to the postulation of the following molecular model for the reaction sequence of photosynthetic water oxidation (29):

$$\left[(H_2O)_2^*\right]M \xrightarrow[<100\ \mu s]{\oplus} \left[(H_2O)_2^*\right]M^+ \tag{1}$$

$$\left[(H_2O)_2^*\right]M^+ \xrightarrow[100\ \mu s]{\oplus} \left[(H_2O)^*(OH)^*\right]M^+ + H^+ \tag{2}$$

$$\left[(H_2O)^*(OH)^*\right]M^+ \xrightarrow[400\ \mu s]{\oplus} \left[(H_2O_2)^{**}\right]M^+ + H^+ \tag{3}$$

$$\left[(H_2O_2)^{**}\right]M^+ \xrightarrow[1\ ms]{\oplus} \left[(O_2^- \ldots .H^+)^{**} + H^+\right]M^+ \xrightarrow{k_b} \left[(O_2)^{**}\right]M + 2H^+ \tag{4}$$

$$\left[(O_2)^{**}\right]M + 2H_2O \xrightarrow{k_{O_2}} \left[(H_2O)_2^*\right]M + O_2 \tag{5}$$

The model involves three basic assumptions:

a. There exists a redox component, M, of rather moderate oxidizing power in state M^+ ($E_{m,7}$ of the couple M/M^+ is in the range of 200-400 mV), which is stable in the dark and, therefore, accounts for the exceptionally long life time of one oxidizing redox equivalent stored in system Y (19). This component is not necessarily a transition metal group.

b. The intermediary redox states formally corresponding to OH, H_2O_2 and HO_2 are transition metal groups containing a special ligand shell involving water. These states are referred to as "crypto-hydroxyl", "crypto-hydrogen peroxide" and "crypto-superoxide" and are symbolized by $(OH)^*$, $(H_2O_2)^{**}$ and $(HO_2)^{**}$ [1)], respectively. They are thermodynamically stabilized so that the exchange reactions $H_2O + C \longrightarrow (H_2O)^* + F$ are highly endergonic (C = complexed intermediate, F = free intermediate) and, therefore, free intermediates do not exist. The stabilization energy is estimated to be of the order of 80-100 kJ/Mol. One or two asterisks indicate mono- or binuclear complexation, respectively (see footnote).

c. The dioxygen bond is formed at the redox level of peroxide, which is binuclearly complexed at two transition metal centers. Abstraction of two electrons from this state (vide infra) leads to oxygen complexed either mono- or binuclearly which becomes released via an exergonic exchange reaction.

These postulations (a)-(c) imply that the thermodynamic pattern of water oxidation via four sequential univalent steps in photosynthesis significantly differs from that in aqueous solutions as is discussed in ref. 29.

Fig. 2 depicts the essential features of crypto-hydrogen peroxide with Me representing transition metal ions attaining the formal oxidation state $(n + \delta)$. The apoprotein (symbolized by hatched area) is assumed to contribute to the ligand shell, thereby significantly affecting the redox state of the group. L_b is a bridging ligand which could couple the two metal centers analogously to what is known in cytochromoxidase (30). It has been argued that two manganeses - sufficiently close together to allow formation of an oxygen-oxygen bond - function as centers of these complexes (29,31).

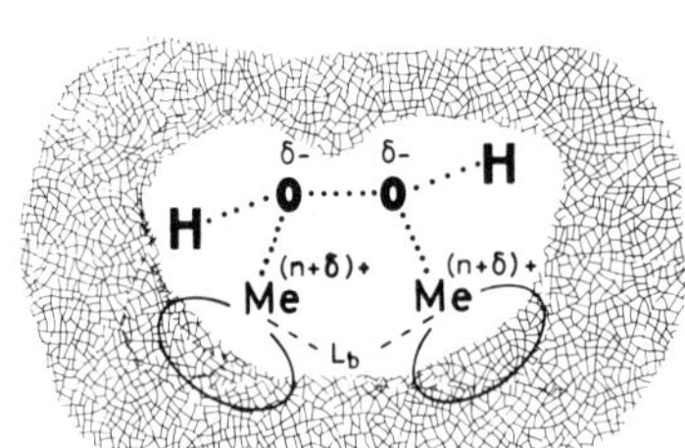

Fig. 2. Schematic representation of "crypto-hydrogen peroxide"

1) Cryptosuperoxide and dioxygen could be complexed mono- or binuclearly as will be discussed later.

Their substantiation as manganese is not of pivotal importance for the molecular model described by eqs. (1) - (5), but there is now ample evidence to support this idea (27,32). Furthermore, EXAFS studies indicate that the first coordination shell of manganese very likely contains oxygen (33). The valence state of manganese is not yet clarified. Latest EPR studies (32) favor the idea that manganese is predominantly in the state Mn(III), while X-ray absorption edge studies place the average oxidation state between +2 and +3 (34).

The thermodynamic stabilization of the intermediates is assumed to be due to electronic delocalization. This assumption is in line with previous findings (35) on peroxide binuclearly complexed in cytochrom-oxidase at an iron and copper center, giving rise to asymmetric charge delocalization.

In contrast, a recent model proposed by Dismukes et al. (32) which is basically quite similar to ours (29,36), assumes that the manganese centers are in definite redox states. The state corresponding to "crypto-hydrogen peroxide" is described by an equilibrium existing between hydrogen peroxide complexed in a tetramer which contains four Mn(III) and two hydroxyl ions complexed to an array of two Mn(III) and two Mn(IV).

The water splitting enzyme system is very fragile. Thus, an isolation and purification has not yet been achieved.

Further information about the intermediary redox states should be obtainable by spectral analysis in the visible and UV-range. Previous data suggest that components absorbing in the UV are involved in photosynthetic water oxidation (37,38). However, a detailed spectral characterization of the redox states of system Y could not be achieved mainly because of the interfering binary oscillation of the acceptor side of system II (see Section 5). The latter effect can be eliminated by mild trypsinization of chloroplasts (39). Under these conditions absorption changes have been detected, exhibiting relaxation kinetics characterized by half time of 1 ms which coincides in their extent with the oscillation pattern of the oxygen yield (39,40). Accordingly, these absorption changes are inferred to reflect a precursor state of photosynthetically evolved oxygen. The difference spectrum of these absorption changes is depicted in Fig. 3. It reveals a pronounced band peaking around 320 nm and a smaller band at 265 nm whose origin has not yet been determined.

Based on the molecular model described by eqs. (1)-(5) it is concluded that the spectrum of fig. 3 reflects the absorption difference between $[(H_2O_2)^{**}]M^+$ and $[(H_2O)_2^*]M$ (40).

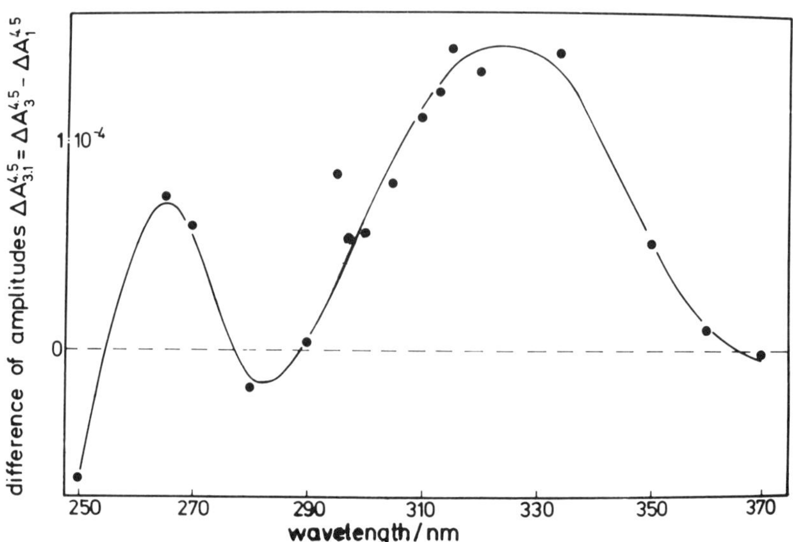

Fig. 3. Difference spectrum of the precursor state of photosynthetically evolved oxygen (for details see ref. 40)

The positive peak around 320 nm might suggest participation of plastosemiquinones in the reaction sequence of system Y. This is in line with experimental findings that plastoquinone is required for restoration of the oxygen-evolving capacity in plastoquinone extracted thylakoids (41,42). The possibility of the identification of the M/M^+-redox couple with the PQH_2/PQH-redox system has been outlined in ref. 40.

Another possible function of plastoquinone acting as a bridging ligand (see Fig. 2) was also discussed (32).

For the time being a possible role of plastoquinone in system Y has not been clarified.

For further discussion of the mechanism of photosynthetic water oxidation it appears reasonable to consider the properties of other closely related enzyme systems. Especially interesting in this respect is the cytochrome oxidase which has been developed by higher heterotrophic organisms as a unique system for catalysis of the reverse of water oxidation, i.e. the O_2-reduction to H_2O. The primary step in the reduction sequence leading from O_2 to H_2O is probably the formation of a loosely bound mononuclear complex between O_2 and the reduced Fe^{2+} center of the cytochrome a_3-moiety which closely resembles oxyhemoglobin (30). It therefore seems reasonable to assume that complexed O_2 and also cryptosuperoxide might be bound mononuclearly rather than binuclearly (see footnote, p. 564).

Concerning the oxidative sequence leading from "cryptohydrogen peroxide" to complexed oxygen (see eq.(4)), it is interesting to note that the reverse process in cytochromoxidase is realized by an almost simultaneous electron transfer from Fe^{2+} of cytochrome a_3 and from $Cu^+_{a_3}$. Accordingly, the reaction sequence described by eq.(4) could take place via an analogous mechanism with M^+ having the same function as oxidative group in system II, which is fulfilled by $Cu^+_{a_3}$ as reductive group in cytochrome oxidase. The overall process of eq. (4) would be triggered by electron abstraction from the manganese center of "crypto-hydrogen peroxide" by $Chla^+_{II}$ (via D_1 and/or further intermediary redox groups).

The molecular model described by eqs. (1)-(5) also defines a pattern for proton release coupled with water oxidation. It depends on the protonization properties of the M/M^+ couple. Previously, the experimentally detected release pattern was found to be in fair correspondence with the theoretical model (29). However, the experimentally detected proton release pattern does not necessarily provide information about the intrinsic proton release pattern of the water oxidation in system Y (for further discussion see ref. 25,36).

The considerations presented in this section indicate that some progress has been achieved during the last years. However, the crucial mechanistic problems of photosynthetic water oxidation still remain to be solved.

5. The plastoquinone containing protein

Analogously to water oxidation via a four-step univalent reaction sequence, plastohydroquinone formation occurs via a two-step univalent reductive pathway with a special plastoquinone B (or R) acting as functional group for the intermediary storage of an electron (7,8). B becomes reduced sequentially by the primary quinone acceptor, X320, in situ acting as 1-electron carrier (13). Accordingly, it appeared logical to assume that the special reactive behavior of the plastoquinone molecules X320 and B is determined by a protein and, therefore, to ask for its properties and mode of action. Based on the results obtained in mildly trypsinized chloroplasts it was inferred that X320 and B are incorporated into a functional protein matrix (see ref.46) referred to as X320-B-apoprotein (47) which makes the following activities possible: (a) It regulates the electron transport from X320 to B; (b) it functions as a barrier to H^+-transport and highly retards the electron transfer from $X320^-$ to exogeneous redox agents; (c) it contains the binding sites for DCMU-type inhibitors which are postulated to act via an allosteric mechanism.

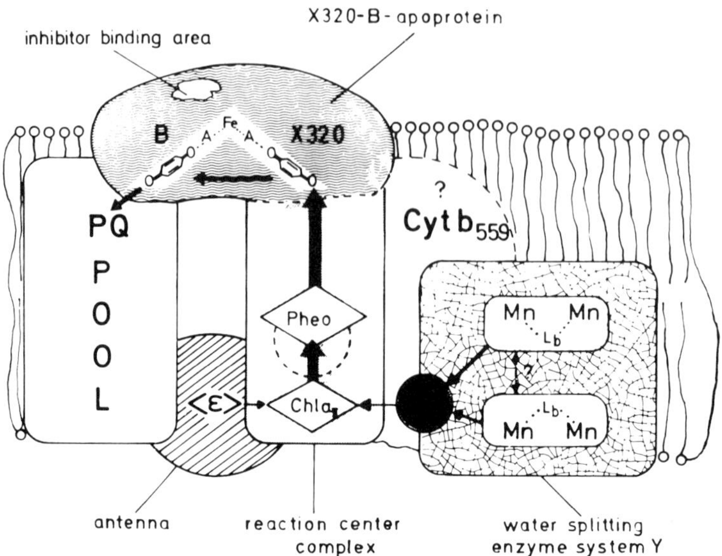

Fig. 4. Schematic representation of the functional and structural organization of system II (for the sake of simplicity, a possible heterogeneity of the acceptor side is not shown. For the same reason, the functional connection with system I via the cytochrome b_6/f-complex (49) has been omitted).

These basic ideas have been experimentally confirmed, especially the B-binding polypeptide is now well characterized (for latest data see ref. 48). Unfortunately, limited space does not allow to describe further details of the X320-B-apoprotein and of the functional coupling between both plastoquinone molecules X320 and B.

6. Summary

The consideration of the present communication led to a functional and structural organization scheme which is depicted in Fig. 4. It shows that three functional proteins (protein complexes) are coupled in order to realize water cleavage by visible light. However, it has to be emphasized that the scheme of Fig.4 provides only a tentative, static model because essential mechanistic and structural aspects of this fundamental bioenergetic process are still unresolved questions. Interestingly enough, latest findings of the effect of a single site nuclear mutation on the polypeptide pattern indicate, that system II has to be considered as a physiological unit whose assembling in the membrane occurs independently of the other major complexes, such as photosystem I or the cytochrome b_6-f-complex (50).

Acknowledgements

The authors would like to thank Dipl.-Chem.R.Hagemann and Dipl.-Chem.M.Völker for their contributions to this paper, I.Jürgens, S. Hohm-Veit and C.Behm for skillful technical assistance. The financial support provided by Deutsche Forschungsgemeinschaft, Fonds der Chemischen Industrie and Bundesminister für Forschung and Technologie is gratefully acknowledged.

References

1. Renger, G. (1982) in: Biophysik (W. Hoppe, W.Lohmann, H.Markl and H. Ziegler, eds.) Springer Verlag, Berlin-Heidelberg, pp. 532-561
2. Döring,G., Renger,G., Vater,J. and Witt,H.T. (1969) Z.Naturforsch. 24b, 1139-1143
3. Trissl, H.W. and Gräber,P. (1980) Biochim.Biophys.Acta 595, 96-108
4. Stiehl, H.H. and Witt, H.T. (1969) Z.Naturforsch.24b, 1588-1598
5. Renger,G. and Wolff, Ch. (1976) Biochim. Biophys. Acta 423, 610-614
6. Eckert,H.-J., Brettel,K., Renger,G., and Witt,H.T. (1983) FEBS-Letters (submitted)
7. Bouges-Bocquet, B. (1973) Biochim. Biophys.Acta 314, 250-256
8. Velthuys, B.R. and Amesz,J. (1974) Biochim. Biophys. Acta 333, 85-94
9. Shuvalov, V.A., Klimov, V.V., Dolan, E.P. and Ke, B. (1980) FEBS-Letters 118, 279-289
10. Eckert, H.-J., and Renger, G. (1980) Photochem. Photobiol. 31, 501-511
11. Fajer,J., Davis,M.S., Forman, A., Klimov, V.V., Dolan E. and Ke, B. (1980), J.Amer. Chem. Soc. 102, 7143-7145
12. Rutherford, A.W. (1981) Biochem.Biophys.Res.Commun. 102,1065-1070
13. Witt, K. (1973) FEBS-Letters 38, 116-118
14. Renger, G. and Eckert, H.J. (1980) Bioelectrochemistry and Bioenergetics 7, 101-124
15. Sun, A.S.K. and Sauer,K. (1971) Biochim.Biophys.Acta 239,399-414
16. Ganago, J.B., Klimov,V.V., Ganago,A.O., Shuvalov,V.A. and Erokin, Y.E. (1982) FEBS-Letters 140, 127-130
17. DeVault, D. (1980) Quart. Rev. 13, 387-564
18. Joliot, P., Joliot,A., Bouges, B. and Barbieri,G. (1971) Photochem. Photobiol. 11, 287-305
19. Kok, B., Forbush, B. and McGloin, M. (1970) Photochem. Photobiol. 10, 457-475
20. Babcock, G.T., Blankenship,R.E. and Sauer, K. (1976) FEBS-Letters 61, 286-289
21. Gläser,M., Wolff, Ch. and Renger, G. (1976) Z.Naturforsch. 31c, 712-721
22. Renger, G., Gläser, M., and Buchwald,H.-E. (1977) Biochim. Biophys.Acta 461, 392-402

23. Conjeaud, H. and Mathis, P. (1980) Biochim. Biophys.Acta 590, 353-359
24. Blankenship,R.E., Babcock, G.T., Warden, J.T. and Sauer,K. (1975), FEBS-Letters 51, 287-293
25. Renger, G. and Völker,M. (1982) FEBS-Letters, 149, 203-207
26. Renger, G. and Eckert,H.-J. (1981) Biochim. Biophys. Acta 638, 161-171
27. Livorness,J. and Smith, T.D. (1982) in: Structure and Bonding Vol. 48, Springer Verlag, Berlin, pp. 1-44
28. Renger, G. (1970) Z. Naturforsch. 25b, 966-970
29. Renger, G. (1977) FEBS-Letters, 81, 223-228
30. Powers, L., Chance, B., Ching, V. and Angiolillo, P. (1981) Biophys.J. 34, 465-498
31. Renger, G. (1977) in: Topics in Current Chemistry, Vol. 69 (F.L.Boschke, ed.) Springer Verlag, Berlin-Heidelberg-New York, pp. 39-90
32. Dismukes, G.C., Ferris,K. and Watnick,P. (1982) Photobiochem. Photobiophys. 3, 243-246
33. Kirby, J.A., Robertson, A.S., Smith, J.P., Thompson, A.C., Cooper,S.R. and Klein,M.P. (1982) J.Am.Chem.Soc. 103,5529-5537
34. Kirby, J.A., Goodin, D.B., Wydrzynski, T., Robertson, A.S. and Klein, M.P. (1981) J. Am.Chem.Soc. 103, 5537-5542
35. Denis, M. and Clare, G.M. (1981) Plant Physiol. 68, 229-235
36. Renger, D. (1978) in: Photosynthetic Water Oxidation (H. Metzner ed.) Academic Press, London, pp. 229-248
37. Mathis, P. and Havemann, J. (1977) Biochim. Biophys.Acta 461, 167-181)
38. Velthuys, B.R. (1981) in: Proc.5th Int.Congr.Photosynthesis (Akoyunoglu,G. ed.) Balaban Intl.Sci.Serv., Vol.II, pp.75-85, Pennsylvania
39. Renger, G. and Weiss. W. (1982) FEBS Letters 137, 217-221
40. Renger, G. and Weiss. W. (1983) Biochim.Biophys.Acta 722,1-11
41. Sadewasser, D.A. and Dilley, R.A. (1978) Biochim.Biophys.Acta 501, 208-216
42. Goldfield, M.B., Blumenfeld, L.A., Dimitrovski, L.G. and Mikoyan, V.D.(1981) Molekul.Biol.14, 635-641 (Engl.transl.)
43. Rich, P.R. (1981) Biochim. Biophys. Acta 637, 28-33
44. Renger, G. Erixon,K., Döring, G. and Wolff, Ch. (1976) Biochim. Biophys.Acta 440, 278-286
45. Renger, G. (1976) FEBS Letters 69, 225-230
46. Renger, G. (1976) Biochim. Biophys. Acta 440, 279-289
47. Renger, G., Hagemann, R. and Dohnt, G. (1981) Biochim. Biophys. Acta 636, 17-26
48. Reisfeld, A., Mattoo, A.K. and Edelmann, M. (1982) Eur.J. Biochem. 124, 125-129
49. Hurt, E. and Hauska, G. (1981) Eur.J.Biochem. 117, 591-599
50. Metz, J.G. and Miles, D. (1982) Biochim. Biophys. Acta 681, 95-102

ANALYSIS OF PROTON TRANSLOCATION THROUGH HYDROGEN-BONDED CHAINS USING MOLECULAR ORBITAL METHODS

Steve Scheiner and Eric A. Hillenbrand

Department of Chemistry and Biochemistry
Southern Illinois University
Carbondale, IL 62901 U.S.A.

The chemiosmotic hypothesis has come to be one of the most widely accepted and useful concepts in biochemistry [1-5]. A basic tenet of this theory is that energy may be transduced across a biomembrane via a protonmotive force which is capable of pushing protons against a pH gradient and/or electric field. Bacteriorhodopsin, for example, is known to function as a "proton pump" when it is energized by light of the proper frequency [6-9]. The energy stored in a pH gradient may be harnessed for the purpose of biological work when protons are allowed to be transported across the membrane in much the same way that discharge of an electric capacitor is a source of energy. The synthesis of ATP by H^+-ATPase is thought to be driven by the passage of protons through the F_0 segment of the transmembrane protein [5,10-13].

There is evidence to confirm the reasonable presumption that in passage across the membrane, protons avoid the very hydrophobic lipid bilayer and instead make their way through integral protein molecules that completely traverse the membrane [5-13]. However, little is known at this point about the specific mechanism by which the protons are conducted through the protein. In bacteriorhodopsin, the membrane protein for which a good deal of structural data exists, it is possible to rule out the existence of an aqueous pore through the protein or any "carrier" group [14-16].

An attractive alternative mechanism, and one which is consistent with available structural information, involves a "hopping" of protons from one group to the next along a chain of hydrogen-bonded residues within the protein [17-19]. A schematic illustration of this type of chain and its proposed functioning is presented in Fig. 1. The chain must extend from one aqueous face

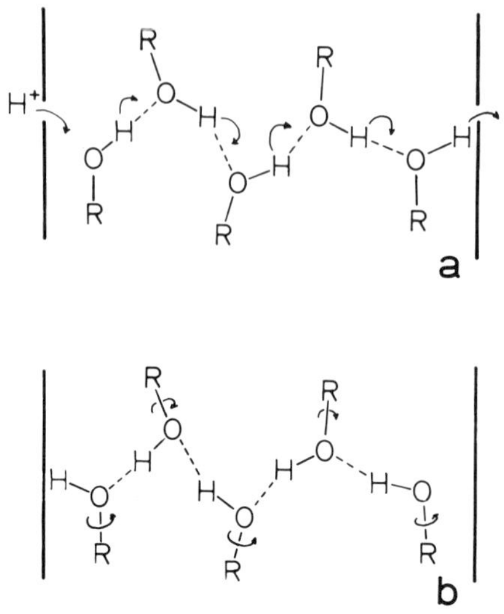

Fig. 1. Proton conduction mechanism through chain of hydrogen-bonded residues. Thick vertical lines represent interfaces of membrane with aqueous regions. Hops of protons between oxygen atoms are indicated by curved arrows in a. Arrows in b represent rotations of hydroxyl groups about R-O bonds.

of the membrane to the other, represented in the figure by the two vertical lines. In general, at least 20 groups would be needed for a chain of the proper length; however, for purposes of clarity, only 5 residues are included in Fig. 1. Another simplification contained in the figure is the representation of all groups by hydroxyls. While this group, present in the side chains of Ser, Thr, and Tyr, as well as water of hydration, is expected to play a major role in the chain, a number of other groups will generally be involved as well. These groups include amide of peptide bonds and Gln and Asn side chains, amino of Lys and Arg, and COOH of Asp and Glu. The R symbols in Fig. 1 represent the groups to which each OH is directly connected. Although the present lack of any high-resolution diffraction data of membrane proteins precludes the positive identification of any appropriate hydrogen-bonding chain, the known features of the structure of bacteriorhodopsin [9] hint at the presence of such a chain.

Initially, the chain is in the OH--OH--OH configuration wherein all hydroxyl groups "point to the right", as in Fig. 1a. A proton from the external medium approaches and binds to the leftmost oxygen, leaving this atom with an excess proton and formal positive charge. As shown by the curved arrow in Fig. 1a, the proton initially bound covalently to this oxygen and H-bonded to the second oxygen of the chain, then transfers across to the latter atom. This hop is followed by similar transfers of each hydrogen atom from one oxygen to the next along the chain. As indicated on the right of Fig. 1a, the hopping terminates with the expulsion of the last hydrogen into the medium external to the membrane. The entire process to this point has resulted in two net effects. A single proton has been transported from one side of the membrane to the other and the chain has been changed from its original OH--OH--OH configuration to HO--HO--HO, as illustrated in Fig. 1b. Transformation of the chain to its original configuration and ready to transport a second proton is accomplished by rotation of each hydroxyl group about its respective R-O bond.

The viability of the process described in Fig. 1 for transporting protons in a membrane protein depends on the kinetics of both the proton hopping and bond rotation steps. The research described here centers about the former process involving the transfer of protons from one residue of the protein to the next. Fig. 1 has been drawn to emphasize the fact that the geometries of the hydrogen bonds along the chain are expected to be quite variable. That is, the OO bondlengths exhibited by the chain will occur over a range of values and many of the bonds will deviate significantly from their optimal linear configurations. These stretches and bends of the H-bonds result from the large number of geometrical constraints imposed on the chain by the overall structure of the protein molecule. It should be remembered that the R groups of Fig. 1 are in fact all interconnected through the same polypeptide backbone of the protein.

On chemical grounds, one would expect the particular geometry of the H-bond to have a strong influence on the proton transfer process within it. A general potential energy function for the motion of the proton from one O atom to another is presented in Fig. 2. The left and righthand minima respectively correspond to the situation wherein the proton is bound to the left (OH--O) or right (O--HO) atom. In order to transfer between the two oxygens, the proton must pass through a region of high energy. Whether the proton transfers via classical thermal "climbing over the barrier" or by the process of quantum mechanical tunneling, the rate of the transfer will be extremely sensitive to the height of the energy barrier, $E^{\dagger}$, which, in turn, will have some functional dependence upon the geometry of the H-bond. In order to analyze the dynamics of the proposed proton transport mechanism, it is therefore essential to have at hand information about the barrier height for a wide range of different H-bond geometries.

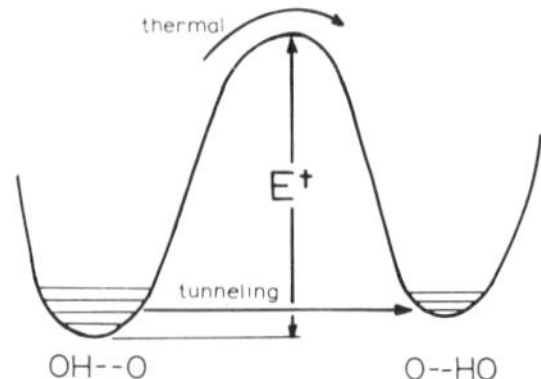

Fig. 2. Potential energy curve for proton transfer from left to right. Vibrational energy levels are indicated by the horizontal lines in each well.

Although rather inaccessible by current experimental techniques, the required data may be obtained using modern ab initio molecular orbital methods [20,21]. The relative orientations of the molecules involved in the H-bond may be chosen in a systematic fashion and known with great precision. The proton transfer may be "frozen" at any point in order to study the evolution of various factors during the process. Moreover, judicious choice of theoretical method leads to results of sufficient reliability.

It is of course impossible to perform the appropriate molecular orbital calculations on a molecule as large as a protein. It is therefore necessary to simulate the biological situation with a suitable small model system. The assumption is made that the business part of the transport mechanism is the array of hydroxyl (or other H-bonding) groups and that the actual character of each R group (see Fig. 1) provides only fine tuning of the properties of the chain. We therefore model the system to a first approximation by an array of water molecules; i.e., each R group is replaced by a hydrogen atom. In fact, it is quite likely that water molecules themselves will serve as links in the chain since their mobility allows them to fill in gaps in the chain of protein residues. The validity of treating other side chain groups as H atoms will be tested below.

METHODS AND RESULTS

Calculations were performed using the ab initio Hartree-Fock procedure wherein molecular orbitals are constructed as a linear combination of atomic orbitals (LCAO) centered on each nucleus

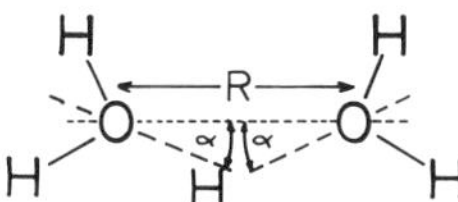

Fig. 3. Parameters used to specify geometry of hydrogen bond in $(O_2H_5)^+$. Dashed lines bisect the HOH angle of each water molecule where θ(HOH) = 115°; r(OH) = 0.95 Å.

[21]. The 4-31G basis set [22] used in the calculations is termed a "split-valence" set since each valence orbital; e.g. 1s for H, 2s,2p for O, is represented by two different functions for increased flexibility of the basis set. Previous calculations have provided evidence that treatment of proton transfer processes with this basis set yields results in excellent agreement with much more sophisticated and time-consuming theoretical methods [23-27]. All computations were performed with the GAUSSIAN-70 [28] and GAUSSIAN-80 [29] packages of computer codes.

We begin our investigation with the transfer of a proton between two water molecules. The geometry of the $(H_2OHOH_2)^+$ system is illustrated in Fig. 3 where R is defined as the distance between the two oxygen atoms [23]. The deviations from linearity of the hydrogen bond are described in terms of the angles α between the O--O axis and the HOH bisector of each water molecule. In the optimal situation, α would be equal to 0° and the central proton would lie directly along the O--O axis. However, for nonzero values of α the central H will generally prefer to lie somewhat off this axis, as indicated in Fig. 3. There are of course alternate modes of bending that could be considered other than that in which both water molecules are rotated by equal amounts α. Some of these modes have in fact been treated in previous publications [23,25,30] but we will focus our attention here on the above mode for purposes of illustration.

For each configuration of the H-bond, described by the values of R and α, the energy was calculated as the proton was allowed to translate between the two O atoms. The energy barrier to proton transfer, $E^†$, was evaluated as the difference in energy between the bottom of the lefthand well and the top of the barrier (see Fig. 2). These barriers are presented in Fig. 4 as a function of both R and α. It is immediately clear that both stretches and bends of the H-bond lead to rapid increase in the height of the energy barrier. The sensititivy of $E^†$ to both of these parameters and not just the interoxygen distance is emphasized. For example, stretching the bondlength from 2.55 Å to 2.75 Å in a

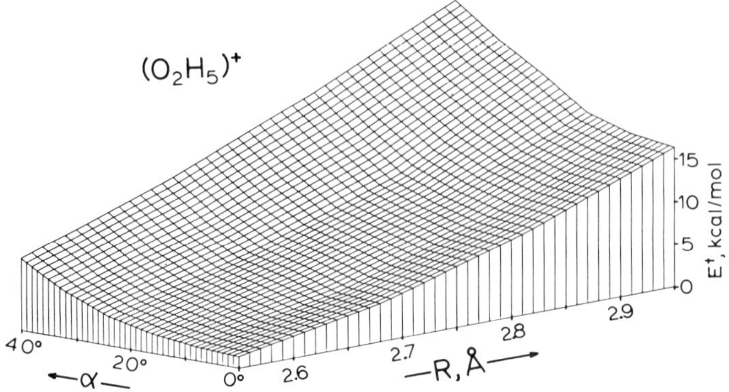

Fig. 4. Height of energy barrier to proton transfer in $(O_2H_5)^+$ as a function of both R and α.

linear fashion increases the barrier from 1.4 to 7.6 kcal/mol while rotating the two water molecules by 40°, with the R(OO) distance held fixed at 2.55 Å results in a greater increase to 8.7 kcal/mol.

pH Dependence

As described above, the hydrogen-bonded chain of protein residues provides an intramembrane connection between media of different pH. It is therefore of some importance to determine the manner in which changes in pH may affect the functioning of the proposed proton transport mechanism. Each water molecule in the $(H_2OHOH_2)^+$ cation serves as a model of a neutral hydroxyl group.[2] To simulate a change in pH, which has the effect of

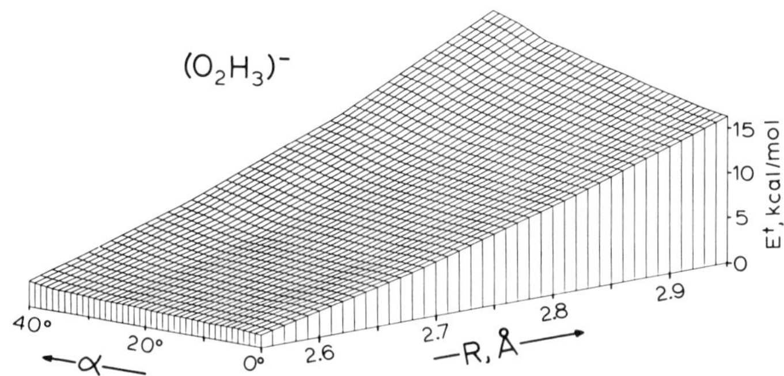

Fig. 5. Energy barriers in $(O_2H_3)^-$. R again refers to the R(OO) distance while α represents the bending of each terminal hydrogen from its optimal θ(OOH) angle of 108.7° towards greater angles.

altering the state of ionization of many protein residues, each water molecule was replaced by a hydroxide anion [31]. The transfer of a proton between two such hydroxides was studied using the $(HOHOH)^-$ system as an analogue of $(H_2OHOH_2)^+$.

The computed barriers to proton transfer in $(HOHOH)^-$ are exhibited in Fig. 5 as a function of the interoxygen distance R, and the bending angle α, as was done for $(H_2OHOH_2)^+$ in Fig. 4. Comparison of the data in these two figures illustrates that for linear configurations, i.e. $\alpha = 0°$, the barriers for the two systems are essentially identical over a range of H-bond lengths. However, the anionic system is much less sensitive to angular distortions. For example, for R = 2.75 Å, twisting the H-bond by 20° increases the barrier for $(H_2OHOH_2)^+$ by 30% but only half that for $(HOHOH)^-$. These observations have some interesting and potentially useful ramifications for the proton transport mechanism. If the hydrogen bonds contained in the chain are very close to linear then the conduction mechanism should show little pH dependence since the barriers of the different charge states are so similar. A great deal of pH sensitivity is expected, however, for a chain with bent H-bonds. At very high pH, where many hydroxyl-type groups may be present as anions, the barriers to proton transfer are substantially lower than for the neutral groups at low pH, leading to enhanced proton conduction at high pH. Of course, the actual pH required to deprotonate a hydroxyl group is rather high - in the range of 14 or so. It is likely though that many of the conclusions reached here for hydroxyl groups will be true also for groups of lower pK, closer to physiologically relevant pH. Work along these lines is currently in progress.

Validity of Model

There are two approximations made in the modeling of the biological system which are potentially very large sources of error. The first of these involves the representation of a long chain of 20 residues or more by a single H-bond between two molecules. It is possible in principle that the additional molecules along the chain may substantially alter the energetics of proton transfer. In order to test the magnitude of end effects in the dimer, calculations were carried out on a longer chain of four water molecules in which the transfer takes place between the two central units [23]. The transfer barriers in the tetramer $(H_2OH_2OHOH_2OH_2)^+$ were found to differ only slightly from those calculated for the dimer $(H_2OHOH_2)^+$ for a range of R(OO) distances between 2.55 Å and 2.95 Å. This near coincidence of results indicates that the dimer does in fact provide a satisfactory model with which to study proton transfer potentials in longer chains.

Table I. Barriers to proton transfer (in kcal/mol) in dimers of water and methanol. All H-bonds are linear.

R, Å	$(CH_3OH)_2H^+$	$(HOH)_2H^+$
2.55	1.9	1.4
2.75	8.9	7.6
2.95	19.6	16.9

A second question concerns the use of OH_2 in the calculations. While this small molecule will probably participate in the chain in the form of water of hydration, it may not offer a sufficiently realistic model of the larger and more complex hydroxyl-containing residues of the protein. Since each hydroxyl group in a protein is covalently bound to a C atom, it is possible to obtain a first approximation to the difference between water and protein residues with regard to proton transfers by studying $HOCH_3$. When methanol molecules are substituted for water in the dimer, the barriers to proton transfer listed in the first column of Table I are obtained [31]. When compared to the analogous barriers for the dimer of water presented in the next column, it is apparent that replacement of a hydrogen of water with a methyl group leads to an increase in the barrier to proton transfer. The magnitude of this increase is variable, being only 0.5 kcal/mol for R = 2.55 Å but 3.0 kcal/mol for R = 2.95 Å. In general, the barrier height increase seems to be on the order of 20-30%.

The results suggest that replacement of a hydroxyl-containing protein residue by a water of hydration may facilitate the proton transport process in a number of ways. First, the barriers to proton transfer are somewhat lowered for equivalent hydrogen-bond geometries. Second, the generally greater mobility of water allows it to H-bond to other links in the chain via stronger interactions. These stronger H-bonds are also associated with shorter and less bent configurations which, as we have seen here, correspond to lower barriers to proton transfer.

Involvement of Other Atoms

As mentioned above, oxygen is not the only atom that may be involved in the hydrogen-bonding chain. N atoms from peptide and amide moieties, amine groups of Lys and Arg, and imidazole groups from the His side chain are also expected to participate in these chains. It is therefore important to examine the question of nitrogen involvement in the proton transfer process. In analogy with the modeling of hydroxyl groups by HOH, the simplest repre-

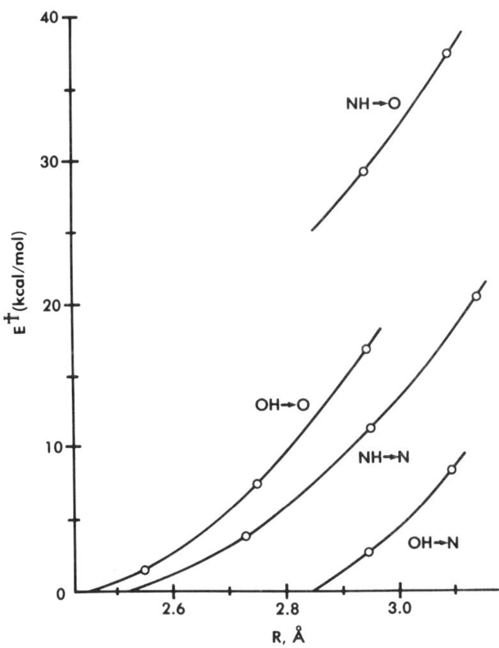

Fig. 6. Barriers to proton transfer, $E^{\dagger}$, in systems involving OH_2 and NH_3. Arrows indicate direction of motion of hydrogen; noncentral hydrogens are omitted from the nomenclature. For example, NH→O represents transfer of proton from N to O in $(H_3NHOH_2)^+$. In all cases, R is equal to the distance between first-row atoms; all H-bonds are linear.

sentative of the nitrogen atom in proteins is NH_3. Proton transfer potentials were computed [25,30] for transfers between two N atoms in the system $(H_3NHNH_3)^+$ as well as between N and O atoms in $(H_3NHOH_2)^+$. The barriers to transfer for linear configurations of the H-bond are presented in graphical form in Fig. 6. The curve labeled by OH→O represents the oxygen to oxygen proton trnasfer previously described for $(H_2OHOH_2)^+$ while the internitrogen transfer is labeled by NH→N. For the asymmetric transfer between N and O, the barrier for transfer in one direction is expected to be quite different than for transfer in the other direction. Thus, the barriers for transfer from N to O (NH→O) are quite a good deal higher than for OH→N transfer. This observation is not surprising since the nitrogen atom is much more basic than O and is expected to hold on to the proton more tightly.

Fig. 6 contains a good deal of information useful to the analysis of the proposed mechanism of proton transport in proteins. We note first that the interoxygen transfers require a bit higher activation energy than the NH→N transfers for the same hydrogen bondlength. This fact indicates that transfers of the latter type are somewhat more facile under the same geometry constraints. Conversely, the situation may be looked at in structural terms. For purposes of illustration, let us assume that the proper functioning of the transport mechanism requires a transfer barrier of, say, 5 kcal/mol at a particular interresidue H-bond. From the curves in Fig. 6, we see that an OH--O bond would need to be 2.68 Å in length whereas a longer distance of 2.78 Å is necessary for a bond of NH--N type. The transfer from O to N requires an OH--N distance of about 3.0 Å. It is particularly interesting that the lowest barrier for transfer from O to N is 25 kcal/mol. (Shorter bondlength than 2.85 Å leads to collapse of the potential into a single-well curve.) This fact allows an NH--O hydrogen bond to function as a one-way valve. The high OH→N transfer barriers minimize transfer in this direction but low barriers allow easy transfer in the opposite direction. This directional character is a desirable feature of the chain since transport in more than one direction is equivalent to "leakage" of protons through the membrane, dissipating the pH gradient and wasting the potential energy contained therein.

Analysis of the electronic structure provides some useful insights into the proton transfer process. Our calculations indicate that concurrent with the transfer of a proton from one molecule to another, there is a flow of electronic density in the opposite direction. This charge transfer appears to facilitate the motion of the proton as lower energy barriers to proton transfer are associated with larger amounts of charge shift from the proton acceptor molecule to the donor. The nitrogen atom, being less electronegative than oxygen, may release greater amounts of electronic charge and thus acts as a better proton-acceptor atom. Hence, the barriers to proton transfer to N are smaller than those to O in Fig. 6. Analogously, the greater electronegativity of oxygen makes it a better charge acceptor and therefore leads to lower barriers to proton transfer from O than from N (for equivalent proton acceptor molecules).

In addition to providing some explanation of the observed differences between these small molecules, the correspondence between transfer energetics and electronic properties offers an opportunity to draw inferences about similar quantities in larger and more complex systems. Since proton transfer is facilitated by a shift of charge in the reverse direction, it is anticipated that adding substituents to the proton-acceptor molecule that increase its ability to transfer electron density will make this molecule a better proton acceptor. Thus, replacement of one of the hydrogens of water with an electron-releasing substituent is expected to lead to lower barriers for proton transfer to this molecule than to water itself. Analogous arguments pertain to the proton donor.

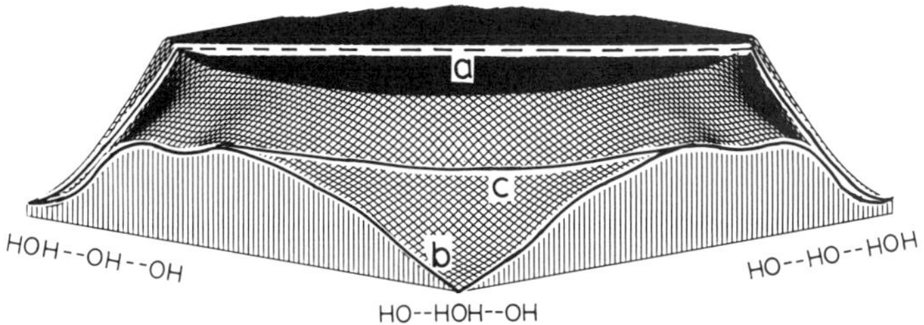

Fig. 7. Potential energy surface for motion of two protons between water molecules. Labels on the corners of the surface have been simplified by elimination of one H from each water. Thus HOH--OH--OH would more precisely be represented as H_2OH--OH_2--OH_2. Interoxygen distances are 2.95 Å. Energies above 25 kcal/mol were deleted from surface leaving the flat region at the top.

Synchronization of Transfers

The calculations described above have treated single transfers of a proton between a pair of molecules. However, the transport of one net proton across the membrane requires a large number of hops of protons from one residue to the next along the chain. In the proposed mechanism, each residue acts as proton acceptor in one transfer step and as donor in another. An important question concerns the concertedness of the two transfers in which a given residue is involved. That is, must this residue wait until it has fully accepted the first proton before it may begin to transfer the second proton or are the motions of the two protons synchronized to some extent?

To help provide an answer to this question, a chain of three* water molecules was considered [23]. These molecules are hydrogen-bonded to one another and an excess proton is added to the leftmost molecule: $(OH_3)^+(OH_2)(OH_2)$. This configuration corresponds to the left corner of Fig. 7 where it is denoted HOH--OH--OH. Transfer of a single proton from the first to the second water molecule leads to the $(OH_2)(OH_3)^+(OH_2)$ configuration, represented by HO--HOH--OH in the central lower portion of the figure. Transfer of two protons, one from molecule 1 to 2, and another from 2 to 3, yields HO--HO--HOH on the far right; i.e. $(OH_2)(OH_2)(OH_3)^+$. The surface in Fig. 7 represents the calculated potential energy for motion of the two protons. The height of

*The chain actually studied contained five water molecules. The additional water on each end was needed to minimize artifacts produced by end effects; proton transfers to the terminal waters were excluded.

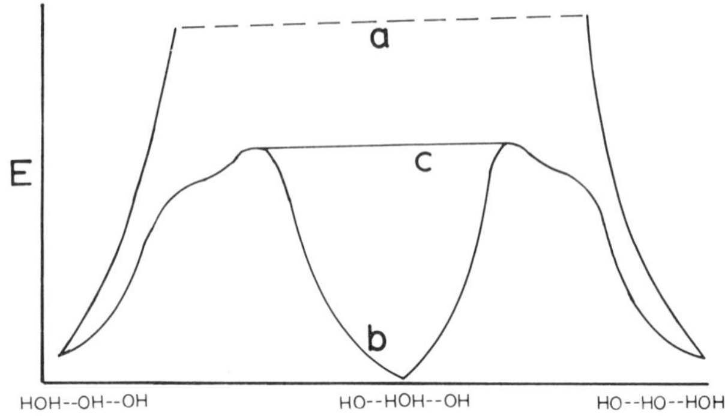

Fig. 8. Energy profiles along paths a, b, and c from Fig. 7. The broken section of path a indicates the energies greater than 25 kcal/mol along this path.

each point on the surface indicates its potential energy. The flat region at the top of the surface represents a ceiling of 25 kcal/mol placed on the surface; the energy of points in this region are actually considerably higher.

A fully concerted double transfer in which both protons move at the same rate between the respective water molecules corresponds to path a in Fig. 7. It is clear that the energy along this path rises rapidly and reaches the level of 25 kcal/mol quite early on. This point is illustrated in Fig. 8 which traces the energy along the reaction path from HOH--OH--OH to HO--HO--HOH. The energy along path a is actually greater than the 25 kcal/mol limit indicated by the dashed line. Such a fully synchronous mode of double proton transfer is obviously highly disfavored by energetic considerations.

Path b, on the other hand, corresponds to a stepwise trajectory in which proton 2 delays its transfer until proton 1 has completely transferred to molecule 2. The first half of path b passes first through a maximum and then to the intermediate HO--HOH--OH. Completion of the second half of the stepwise transfer to HO--HO--HOH is the mirror image of the first half, passing again through a maximum in the energy of around 16 kcal/mol.

A third possible mode of transfer, intermediate between the fully concerted and stepwise paths a and b, is represented by trajectory c in Figs. 7 and 8. This path initially follows b; that is, proton 2 remains fixed while the first proton begins its transfer. However, after the first proton has transferred approximately halfway, and the energy has reached its maximum of 16 kcal/mol, completion of the transfer of proton 1 is coupled to initial motion of proton 2. The energy of the system remains approximately constant while these two motions continue until, following completion of transfer of proton 1, the energy decreases as the second proton finishes its remaining transfer. By some synchronizing of the motions of these two protons, we may thus travel along path c which, as indicated in Fig. 8, involves no higher energy that that needed to reach the "transition state" involving half-transfer of only proton 1. In effect, we have transferred two protons for "the price of one"; once enough energy has been harnessed to achieve half transfer of the first proton, the second proton may be transferred as well at no further cost in energy, provided its motion does not progress fast enough to overtake proton 1 and take us out of the range between paths b and c. In fact, path c is not unique. Any path intermediate between b and c will fulfill the same requirements, thereby reducing the stringent requirements of a given single path.

Extension of these results to transfers of more than two protons leads to some interesting hypotheses concerning the proposed proton conduction mechanism. It is first possible to rule out a mechanism whereby all twenty or so protons transfer simultaneously and with a high degree of synchrony. Path a is energetically unfavorable for two protons and a similar mode is expected to be even more unlikely for a multiproton system. On the other hand, a certain amount of cooperativity between the various proton transfers may be capable of greatly speeding up the transport process as compared to a fully stepwise process. Specifically, once the first proton has reached the midpoint of its transfer, the second proton along the chain may begin its transfer without waiting for completion of the first transfer. Similar arguments may be applied to successive transfers as well. This cooperative mode has the additional advantage in that it may avoid dissipation of the "transfer energy" into other modes such as vibration of the protein structure.

ACKNOWLEDGMENTS

S. S. is the recipient of a Research Career Development Award from the National Institutes of Health (AM01059). This work was supported by research grants from the National Institute of General Medical Sciences (GM29391) and the Research Corporation. We are grateful to R. Gandour and S. Topiol for making available an IBM version of GAUSSIAN-80 and to Southern Illinois University for allocations of computer time.

REFERENCES

1. P. Mitchell, Nature 191:144 (1961).
2. P. Mitchell, Ann. Rev. Biochem. 46:996 (1977).
3. E. Racker, "A New Look at Mechanisms in Bioenergetics", Academic, New York (1976).

4. D. G. Nicholls, "Bioenergetics: An Introduction to the Chemiosmotic Theory", Academic, New York (1982).
5. V. P. Skulachev, P. C. Hinkle, Eds., "Chemiosmotic Proton Circuits in Biological Membranes", Addison-Wesley, Reading, MA (1981).
6. E. Racker, W. Stoeckenius, J. Biol. Chem. 249:662 (1974).
7. R. A. Bogomolni, R. A. Baker, R. H. Lozier, W. Stoeckenius, Biochem. 19:2152 (1980).
8. M. A. Marcus, A. Lewis, Science 195:1328 (1977).
9. D. M. Engelman, R. Henderson, A. D. McLachlan, B. A. Wallace, Proc. Nat. Acad. Sci., USA 77:2023 (1980).
10. J. Houstek, J. Kopecky, P. Swoboda, Z. Drahota, J. Bioenerg. Biomemb. 14:1 (1982).
11. R. H. Fillingame, Ann. Rev. Biochem. 49:1079 (1980).
12. Y. Kagawa, S. Ohta, M. Yoshida, N. Sone, Ann. NY Acad. Sci., 358:103 (1980).
13. R. S. Negrin, D. L. Foster, R. H. Fillingame, J. Biol. Chem. 255:5643 (1980).
14. Y. Kagawa, N. Sone, H. Hirata, M. Yoshida, J. Bioenerg. Biomemb. 11:39 (1979).
15. W. Stoeckenius in: "Membrane Transduction Mechanisms", R. A. Cone, J. E. Dowling Eds., Raven Press, New York (1979) pp. 39-47.
16. W. Stoeckenius, Sci. Amer., 234:38 (1976).
17. J. F. Nagle, H. J. Morowitz, Proc. Nat. Acad. Sci., USA 75:298 (1978).
18. J. F. Nagle, M. Mille, J. Chem. Phys. 74:1367 (1981).
19. J. F. Nagle, M. Mille, H. J. Morowitz, J. Chem. Phys. 72:3959 (1980).
20. H. F. Schaefer, Ed., "Applications of Electronic Structure Theory", Plenum, New York (1977).
21. P. Cársky, M. Urban, "Ab Initio Calculations: Methods and Applications in Chemistry", Springer-Verlag, Berlin (1980).
22. R. Ditchfield, W. J. Hehre, J. A. Pople, J. Chem. Phys., 54:724 (1971).
23. S. Scheiner, J. Am. Chem. Soc., 103:315 (1981).
24. S. Scheiner, L. B. Harding, J. Am. Chem. Soc., 103:2169 (1981).
25. S. Scheiner, J. Chem. Phys. 77:4039 (1982).
26. S. Scheiner, M. M. Szcześniak, L. D. Bigham, Int. J. Quantum Chem. (in press).
27. S. Scheiner, L. B. Harding, J. Phys. Chem. (in press).
28. W. J. Hehre, W. A. Lathan, R. Ditchfield, M. D. Newton, J. A. Pople, QCPE, GAUSSIAN-70, Prog. No. 236 (1974).
29. J. S. Binkley, R. A. Whiteside, R. Krishnan, R. Seeger, D. J. DeFrees, H. B. Schlegel, S. Topiol, L. R. Kahn, J. A. Pople, QCPE, GAUSSIAN-80, Prog. No. 406 (1981).
30. S. Scheiner, J. Phys. Chem., 86:376 (1982).
31. E. A. Hillenbrand, S. Scheiner (to be published).

H^+ GRADIENT CHANGES:

THEIR MEASUREMENT AND THEIR SIGNIFICANCE IN CELL STIMULATION

Elizabeth R. Simons, Nancy E. Norman and David B. Schwartz

Boston University School of Medicine
Boston, MA 02118 USA

As the sequence of events accompanying cell stimulation and oocyte fertilization becomes better known it has become clear that changes in the membrane potential as well as those in intracellular pH and pCa^{++} play an important role (1-21)Table I. Some of the possible ion gradient changes in the response of any cell possessing a receptor R for a given specific stimulus S can be depicted pictorially as in Figure 1. Although recognition and binding of the stimulus to its receptor are clearly the initiating steps, the subsequent sequence is not yet clear. Depolarization can be and in some cells has been (19-21) demonstrated to involve an influx of Na^+ ions. While there is some evidence that an accompanying H^+ outflow may be attributable to the stimulus induced opening of an Na^+ - H^+ antiport as depicted in Figure 1 (13,19), the temporal resolution has not been sufficiently great to allow one to conclude that the Na^+ and H^+ flux changes are simultaneous rather than sequential. A Na^+ influx undoubtedly stimulates a Na^+-K^+ ATPase in many types of cells (22), as well as a release of Ca^{++} from the membrane into the cytoplasm (23). Whether these changes in intracellular cation concentrations trigger membrane-bound enzymes (e.g. the lipases initiating the prostaglandin synthesis pathway in platelets (24)), enhance metabolic activity (25), act as signals for the fusion of organelles with the membrane or have other effects as "secondary" messengers remains the object of a number of studies.

Our own group's interest has been the initiating events themselves. Although we have also been involved in active studies of the nature and isolation of recognition (i.e. receptor) sites (26), we shall consider here only the next step(s), the induction of changes in the membrane permeability to Na^+ and to H^+. For most mammalian cells, stimulation is accompanied by a decrease in the

Table 1. Reported stimulus induced changes in pHi

SYSTEM	STIMULUS	REFERENCE
Oocytes	Fertilization	1, 2, 5-10
Muscle	Insulin	1, 11, 12
Platelets	Thrombin	13
Lymhocytes	Mitogens	14, 15
Amcebae	Pinocytosis	16
Bacterial Spores	Germination	17
E Coli	pH Gradient	18

Table 2. Stimulus response of secretory cells

1. BINDING OF STIMULUS TO CELL MEMBRANE RECEPTOR
2. MEMBRANE RESPONSE
   - A) ALTERED CATION PERMEABILITY (E. G., K^+, Na^+, H^+, CA^{++})
   - B) ACTIVATION OF MEMBRANE-BOUND ENZYMES (E.G., PHOSPHOLIPASES)
3. CYTOPLASMIC COUPLING FACTORS
   - A) ALTERED $[H^+]_i$
   - B) ALTERED $[Na^+]_i$, $[K^+]_i$
   - C) ALTERED $[Ca^{++}]_i$, $[Mg^{++}]_i$
   - D) ALTERED $[\text{CYCLIC NUCLEOTIDES}]_i$
4. CYTOPLASMIC RESPONSES
   - A) ACTIVATION OF METABOLIC CYCLES (E.G., GLYCOLITIC AND MITOCHONDRIAL PROCESSES)
   - B) ACTIVATION OF CONTRACTILE SYSTEM (E.G., MICROFILAMENTS, INTERMEDIATE FILAMENTS MICROTUBULES)
   - C) ACTIVATION OF ORGANELLAR SYSTEMS (E.G., SECRETION)

transmembrane potential (i.e. the inside of the cell becomes less negative) and in the intracellular concentration of H^+ (i.e. pH_i increases) although in bacteria a decrease in potential may be accompanied by a decrease in pH_i(18).

We shall here consider only the special case of secretory cells, whose interaction of S with R leads, by a series of steps hypothesized to follow the sequence depicted in Table II, to secretion of organellar contents. This hypothetical sequence has not yet been totally documented for any secretory cell, nor is it yet known whether all the steps are obligatorily sequential (and in what time frame) or whether several can exist in parallel rather than in series, or are simultaneous resultants of a single event. We (1,13, 19) as well as others (27-33) have demonstrated that secretory cells respond rapidly and that depolarization begins within less than 5 seconds. The fluorescence of certain cationic cyanine dyes is a good rapid indication of membrane potential, and can be monitored continuously (13,19,27-35), provided care and proper controls are provided. We have now attempted to devise a comparable, simple, continuously monitorable technique for measuring pH_i of small cells in suspension. Although the model cell system we have chosen is the human platelet, the methods described here should be applicable to any other cell in suspension provided the appropriate control experiments are performed.

As is true in membrane potential measurements (Table III), pH_i measurements on cells too small and/or mobile to permit microelectrode insertion have generally been made with distributive probes which are either isotopically labeled or optically detectable (Table IV). The conditions for such probes have been well defined (34,35,13): (a) The distribution between the interior of the cell and the external medium must be controlled by the property to be measured (i.e. by the potential or the pH) and not by the quantity of probe. Therefore, for the particular cell number to be used per measurement, one must determine experimentally the minimum probe concentration for which the distribution (i.e. the ratio inside: outside) ceases to be a function of that concentration; (b) The probe must not alter the resting state of the cells or their biological activity, or their ability to respond to a stimulus, when used at the optimal concentration determined in (a); (c) The distribution of the probe must be rapid relative to the response to be measured (i.e. a probe which requires 45 minutes for equilibration would not give a true measure of a transient response occurring in 15 seconds); (d) If a rapid response is to be observed, continuous measurements of the entire cell suspension (e.g. spectroscopic, fluorimetric) are easier than intermittent ones requiring separation of cells from their medium. Should the latter be necessary nevertheless, the use of a separating layer, e.g. silicone oil, between cell pellet and supernatant should be considered so that redistribution during centrifugation can be minimized; (e) In all cases it

Table III. Methods for Evaluation of pH Gradient Across Secretory Cell Plasma Membranes[a]

Method	Probe	Advantages	Disadvantages
Direct	Microelectrode	(1) Rapid (2) Continuous observation possible (3) Absolute $\Delta\Psi$ measurable	(1) Requires large immobile cells undamaged by micro-electrode penetration (2) Location within cell difficult to determine
Indirect	Lipophilic Cation Isotopically labeled	(1) Little instrumentation required (2) Absolute $\Delta\Psi$ measurable	(1) Requires separation of cells from super-natant (2) Difficult to use for rapid or transient changes (3) Measures average $\Delta\Psi$ as probe distributes extensively into organelles
	Lipophilic Cation Fluorescent	(1) Rapid response (2) Continuous Observation possible (3) Some (eg. $diSC_3(5)$) quenched inside cell	(1) Measures average $\Delta\Psi$ as probe distributes extensively into organelles (2) Absolute $\Delta\Psi$ cannot be calculated from this probe alone

(a) It is assumed that the necessary verifications of the probe (cf text) have been performed, including optimization of probe and of cell concentrations and verification of the viability of the probe-loaded cell.

Table IV. Methods for Evaluation of pH Gradient Across Secretory Cell Plasma Membranes[a]

Method	Probe	Advantages	Disadvantages
Direct	Microelectrode	(1) Rapid response (2) Continuous observation possible (3) Absolute pH measurable	(1) Requires large immobile cell undamaged by microelectrode penetration (2) Location within cell difficult to determine
Indirect	Isotopically labeled weak acid or base	(1) Rapid response (2) Little instrumentation required	(1) Requires separation of cells from supernatant (2) Difficult to use for rapid or transient changes (3) Measures average pH_i as probe also distributes extensively into organelles (4) pH range limited by pK of probe (5) Absolute pH_i cannot be calculated from this probe alone
	Fluorescent weak acid or base	(1) Rapid response (2) Continuous observation possible (3) Some (eg. 9aminoacridine) quenched inside cell	(1) Measures average pH_i as probe also distributes extensively into organelles (2) pH range limited by pK of probe (3) Absolute pH_i cannot be calculated from this probe alone
	Fluorescent in-situ probe	(1) Some (eg. 6 carboxyfluorescein) formed and trapped inside cell from nonfluorescent (diacetate) ester (2) Fluorescence highly sensitive to pH (3) Rapid response to pH_i changes (4) Continuous observation possible (5) Can obtain absolute pH_i	(1) Requires active non-specific cytoplasmic esterases (2) pH range limited by pH dependence of fluorescence (3) Need measurements at 2 wave lengths to obtain ratio independent of intracellular probe concentration

(a) It is assumed that the necessary verifications of the probe (cf text) have been performed, including optimization of probe and of cell concentrations and verification of the viability of the probe-loaded cell should be noted that a distributive probe enters into and partitions across the membranes of the internal organelles according to the latter's potential or pH_i respectively. Therefore one measures an overall average distribution and average change. In some cells such as the platelet (13) it is possible to detect the cytoplasmic probe content by using digitonin at a concentration low enough to permeabilize only the plasma membrane (13). Other cells, such as neutrophils, are stimulated by digitonin and no determination of actual cytoplasmic probe concentration can be made. (Simons, unpublished). The amount contained in each of the other intracellar organelles should eventually be measurable once separation of the various cell components on a routine basis becomes possible; but for the present no such information has been reported.

MATERIALS

Bovine serum albumin (BSA), adenosine diphosphate (ADP), and 9-aminoacridine (9AA) were purchased from Sigma Chemical Company (St. Louis, MO). 6-carboxyfluorescein diacetate (6 CFA), was purchased from Molecular Probes Inc. (Plano, TX). Silicone oil was purchased from Contour Chemical Company (Woburn, MA). Tetraphenylphosphonium bromide (Ph_4P^+Br) was purchased from Alpha Division (Ventron, Danvers, MA). $[H^3]PH_4P^+Br$ was the kind gift of Dr. W.R. VanPelt of Hoffman-LaRoche Inc. 3,3'-Dipropylthiocarbocyanine ($diSC_3(5)$) was the kind gift of Dr. A. Waggoner (Carnegie Mellon Univ., Pittsburgh, PA). Nigericin was the kind gift of Dr. Nils Bang, Eli Lilly Inc. Digitonin was purchased from Sigma Chemical Company and recrystallized three times from absolute ethanol. All other chemicals were of reagent grade, and were purchased from Fisher Scientific Co. Bovine α-thrombin was prepared by Dr. N.E. Larsen from Parke-Davis topical thrombin by the procedure of Lundblad _et al_. (44).

Buffers

0.01 M phosphate (KH_2PO_4/K_2HPO_4), pH 5.75 - 8.70; modified Tyrode's solution, pH 7.35: (0.14 M NaCl, 11.9 mM $NaHCO_3$, 0.1% D-glucose, 2.7 mM KCl, 1 mM $MgCl_2$ 6 H_2O, 0.5% serum albumin) or Hepes pH 7.0 and 7.35 (3.3 mM NaH_2PO_4, 2.7 mM KCl, 137.0 mM NaCl, 5.5 mM D-glucose, 3.8 mM Hepes, 0.98 mM $MgCl_2$ 6 H_2O). The Tyrode's solution was used and stored under a 5% CO_2 atmosphere and kept covered to maintain pH. For some calibration experiments the buffers were prepared from K^+ rather than Na^+ salts.

Platelets

Fresh platelets from normal human volunteers were prepared by gel filtration on Sepharose 2B in either modified Tyrode's or Hepes buffer containing 0.15 U apyrase/ml as previously described (13).

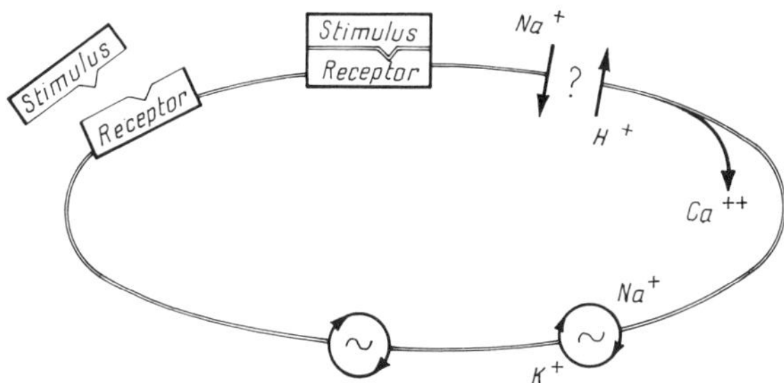

Figure 1: Hypothetical Schematic of Ion Movements in Stimulus Response.

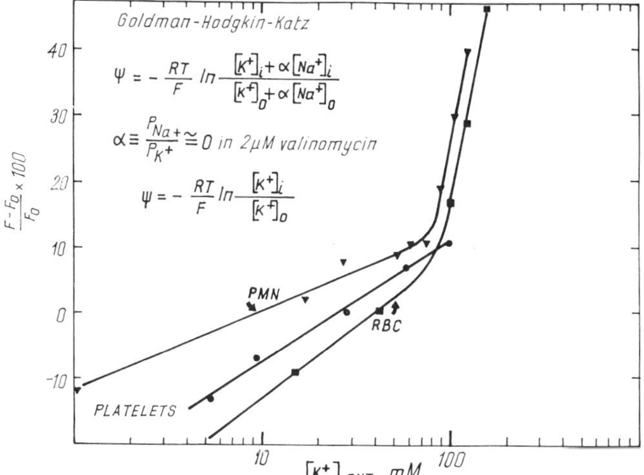

Figure 2: Effect of external potassium concentration on the relative change of di-S-C_3-(5) fluorescene upon addition of valinomycin

Fluorimetry

All fluorescence measurements were performed on a Perkin-Elmer MPF-2a or 650/10 spectrofluorimeter equipped with thermostating and stirring devices as previously described (13).

METHODS

Membrane Potential

We have chosen to use a fluorescent probe, 3,3'dipropylthiocarbocyanine ($diSC_3(5)$), prepared, described (37-39), and kindly supplied to us by Prof. Alan Waggoner of Carnegie Mellon University, Pittsburg, PA, as a probe of membrane potential. As Dr. Waggoner had indicated (37) and we have shown (13,19,30,40) this probe fulfills all the criteria described in the previous paragraph. Although some doubt of its long term (>10 to 20 min.) effect on mitochondria of resting cells has been raised (34,35,41-43), we have looked for and found no effect of 2 uM $diSC_3(5)$ on the potential of resting platelets or neutrophils in our buffer systems (all containing glucose) for periods of up to 30 minutes, an adequate leeway since all of our experiments are completed in less than 10 minutes. This probe's intracellular fluorescence is quenched by self association (37) and the observed fluorescence can safely be correlated with the concentration of extracellular probe. The Goldman-Katz equation is obeyed (Figure 2, reproduced from Ref. 30) for platelets and for neutrophils over a large range of $[K^+]$out, and the fluorescence can be correlated with actual membrane potential (40), using one of the isotopically labeled ions as indicator. Measurements of the resting membrane potential of platelets by several such probes have yielded the same result in our hands, - 50 mv (13), whether $^{36}Cl^-$, passively distributed, or ^3H tetraphenylphosphonium, $[^3H]Ph_4P^+$, a hydrophobic cation whose distribution is potential dependent (41), is used to measure that resting potential.

We therefore utilize the relative stimulus-induced change of 2 uM $diSC_3(5)$ fluorescence (λ_{exc} = 620, λ_{em} = 670 nm), $\Delta F = (F - Fo)/Fo$, as a measure of membrane potential change. In both of the systems we have studied depolarization begins virtually immediately - within the first 3 - 5 seconds - and is linear for 40-60 seconds at most. We therefore use either the initial slope $d(\Delta F)/dt$, or $\Delta F/Fo$ at 30 seconds as a measure of the stimulus-induced change in membrane potential. Furthermore we have demonstrated that this change is dose dependent and saturable, as receptor mediated events should be (13, 19,30-32,40,45). A "saturating" dose of stimulus does not fully depolarize the cell; for example the platelet membrane potential is reduced from -50 to -15 mV by α-thrombin doses equal to a greater than 0.02 U/ml of 5.5×10^7 platelets. The relative changes in potential for less than saturating doses, measured by either technique, fall on the same curve, (Figure 3, reproduced from reference 13).

pH Measurements

To date our pH_i measurements have been restricted to platelets (13,1). We have already shown, as discussed above, that platelet response to thrombin is extremely rapid. The isotopically labeled pH probes such as ^{14}C-5,5-dimethyl-2,4-oxazolidine-dione (DMO), ^{14}C methyl amine, or ^{14}C acetic acid, whose pH-dictated distribution must be evaluated by separate counting of cells (pellet) and external medium (supernatant) in centrifuged timed aliquots are difficult to use when time intervals of 5 or 10 seconds are desired. We therefore turned to fluorescent pH probes.

Several years ago Deamer (46) showed that 9 aminoacridine, a nonfluorescent organic base which becomes fluorescent in the protonated state (47), can be used to evaluate the pH of synthetic liposomes. Its applicability has been extended to intact cells (48, 49, 13) as well as to isolated cell components (50-52). Like any other lipophilic weak base, 9- aminoacridine traverses membranes freely in the uncharged base form, and becomes protonated in the external medium as well as in the intracellular sytoplasm to a degree dictated by their respective pH. Like any other distributive probe, this one will also penetrate into the intracellular organelles and will become trapped there in the protonated form. It will clearly be at highest concentrations in the most acidic compartment. Since the pK of this base is over 10 (46), very little of it will exist in the unprotonated form in normal mammalian cells whose pH=7. It has been shown (4,6,47) that the basic form of 9-aminoacridine is virtually nonfluorescent and that intracellular protonated 9-aminoacridine is essentially fully quenched and hence also nonfluorescent. The detectable fluorescence in a cell suspension is hence a measure of the external concentration of 9 aminoacridine. At concentrations below 10 uM, the relation between these two is linear, i.e. Beer's Law is obeyed. If the internal pH of that cell increases, less protonated probe will be trapped in the cell's interior and more fluorescence will be detected. One can use this change in fluorescence to calculate the apparent change in pH gradient between the external medium and the inside of the cell, an apparent value which can be converted to an absolute value by means of an appropriate calibration curve (1,13). If one expresses the above in equation form, the total concentration of probe, $[A]_T$, can be expressed as $[A]_T = [AH^+]_{out} + [AH^+]_{in} + [A]$ and therefore as $[A]_T = [AH^+]_{out} + [AH^+]_{in}$ where $[AH^+]_{out}$ and $[AH^+]_{in}$ are the probe concentrations in the external medium and in the cell's interior, respectively, and the unprotonated probe $[A] \approx 0$. The pH gradient across the membrane, ΔpH, can then be shown (46) to be a function of the relative volumes the external and internal probe permeates, V_{out} and V_{in}, i.e.

$$\Delta pH = \log \frac{[AH^+]_{in}}{[AH^+]_{out}} + \frac{V_{out}}{V_{in}} = \log \frac{[A]_T - [AH^+]_{out}}{[AH^+]_{out}} + \log \frac{V_{out}}{V_{in}}$$

For the measurement of stimulus-induced changes in pH_i it is not necessary to know the pH gradient across the membrane, but merely the stimulus-induced change in that gradient which is correlatable with the change in pH_i when the external medium is strongly buffered. If one assumes that a negligible change in V_{in} accompanies cell stimulation, the change in gradient, $\delta(\Delta pH)$, upon stimulation becomes independent of V_{out}/V_{in} (1,13,46):

$$\delta(\Delta pH) = \log \left[\frac{[A]_T - [AH^+]_{out}}{[AH^+]_{out}}\right]_{after} \div \left[\frac{[A]_T - [AH^+]_{out}}{[AH^+]_{out}}\right]_{before}$$

Since the fluorescence exhibited by 9-aminoacridine is proportional to the concentration of its protonated form, $F \cong [AH^+]_{out}$ at neutral pH. $[A]_T$ can be evaluated in a cell free system, since the total probe, $[A]_T$ will then be present in protonated form so that $F_T \cong [A]_T$. Therefore the apparent change in pH gradient, and thus the apparent change in internal pH, ΔpH_i, which results from stimulation of a cell can be calculated from the observed fluorescence of 9-aminoacridine (subscripts b and a refer to before and after stimulation respectively):

$$\delta(\Delta pH) = \log \left[\frac{F_T - F_a}{F_a}\right] \cdot \left[\frac{F_b}{F_T - F_b}\right]$$

RESULTS AND DISCUSSION

These relations have sufficed to evaluate relative changes in intraplatelet pH and to establish their dependence upon the dose of thrombin used to effect platelet stimulation (1,13) (Figure 3). Like the thrombin-induced initial platelet membrane potential changes and eventual serotonin secretion, the ΔpH_i becomes independent of dose at α-thrombin concentrations above 0,02 U/ml and the relative change, $\delta(\Delta pH_i)$, and serotonin release curves (defining 100% as the asymptotic value at doses >.025 U/ml in each case) are superimposed on those of the membrane potential change.

It should be re-emphasized that utilization of any distributive probe cannot yield an absolute value of the internal pH, nor of the gradient across the membrane, because of the amount of probe trapped in the organelles. The latter can be estimated by exposing probe-equilibrated platelets to a low concentration (80 uM) of digitonin, sufficient to permeabilize the plasma but not the organellar membranes (36). Approximately half (57%) of the platelet associated 9-aminoacridine was released (13), implying that the other half remained sequestered in the organelles. A saturating thrombin dose

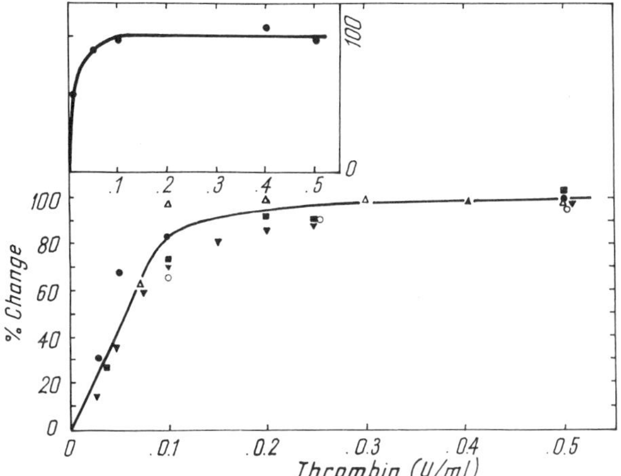

Figure 3: Normalized response to stimulation of washed human platelets by α-thrombin. Relative change in transmembrane potential as measured by fluorescence of di-S-C3-(5) ●; by distribution of ^{3}H-TPhP^{+}○; relative change in transmembrane pH gradient as measured by distribution of 9-aminoacridine (▲) or by fluorescence of 6-carboxyfluorescein (△); relative extent of release of [^{3}H]serotonin (■).

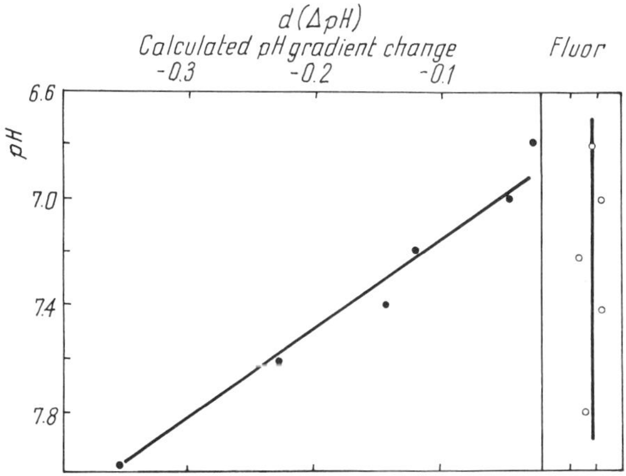

Figure 4: Fluorescence of 4 μM 9-aminoacridine (λ_{exc}=400nm, λ_{em}-456 nm). (A) In the absence of platelets in K^{+}-Hepes buffers of indicated pH; (B) δ(ΔpH) calculated (cf. text).

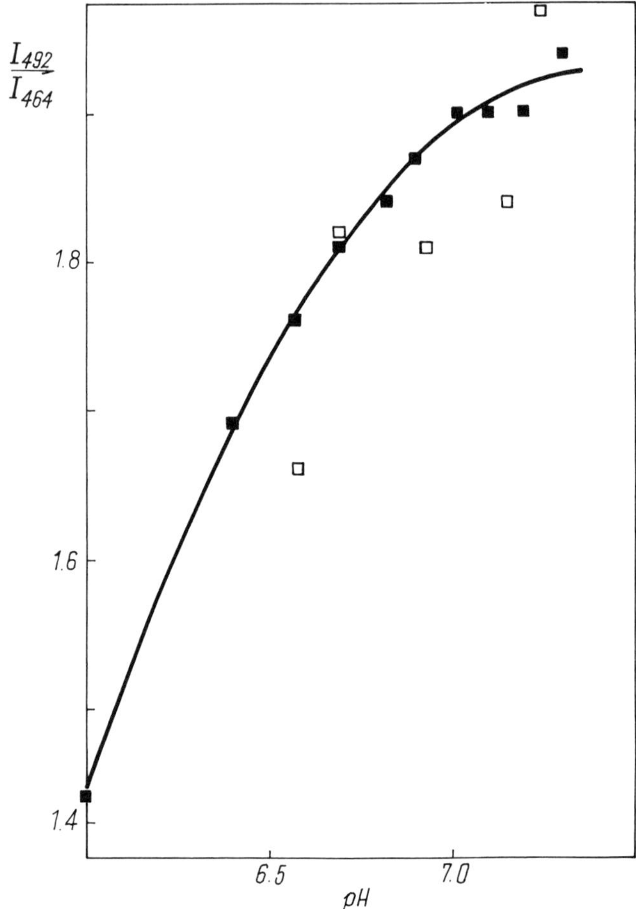

Figure 5: Fluorescence of 6-carboxyfluorescein; (I_{492}/I_{464} = $I(\lambda_{em}$=518, λ_{exc}=492nm) ÷ $I(\lambda_{em}$=518nm, λ_{exc}=464)). (■) buffer alone; (□) in the presence of 5.5 x 10^7 platelets/ml and 2 μM nigericin.

only caused 25% of the total platelet associated probe to be released to the external medium, another 32% to a total of 57% (but no more) being releasable by subsequent addition of digitonin to a final 80 uM concentration. Thus thrombin and digitonin cause outflow of 9-aminoacridine from the same compartment, the platelet cytoplasm. The 9-aminoacridine experiments therefore demonstrate that platelet pH_i rises upon thrombin stimulation. The maximal value of pH_i attained by that route with a saturating dose of thrombin remains, however, lower than that attained when the transplasma membrane pH gradient is fully collapsed with 80 uM digitonin, when $pH_{in} = pH_{out} = 7.40$.

The absolute value of the resting platelet's cytoplasmic pH cannot be determined by the method described so far, nor can one equate the calculated apparent $\delta(\Delta pH_i)^{app.}$ with the true $\delta(\Delta pH_i)$ because only a portion of the distributed 9-aminoacridine participates in the thrombin-induced change. We have shown that the actual $\delta(\Delta pH_i)$ is linearly proportional to the apparent $\delta(\Delta pH_i)^{app}$ between pH 7 and 7.9 (Figure 4, reproduced from Ref. 13). The lower pH limit is attributable to the low concentration of unprotonated 9-aminoacridine below pH 7 (pK > 10 (46)); the higher limit is attributable to our inability to keep platelets above pH 7.9 long enough to perform the calibration experiments. At pH below 7.9 the platelets' homeostatic system maintains pH_i (i.e. 9-aminoacridine fluorescence F^b constant) in a K^+-Hepes buffer long enough to permit measurement of F^b and of the fluorescence, F^a, after exposure to 2 uM nigericin to collapse the proton gradient. If the buffer's K^+ concentration, K^+_{out}, is equal to the known K^+_{in} of the human platelet, 120 mM, this addition of nigericin will bring $pH_{in} = pH_{out}$ since then $[K^+]_{out}/[K^+]_{in} = [H^+]_{out}/[H^+]_{in} = 1$. A calculation of $\delta(\Delta pH_i)^{app}$, by the equation given above, can then be made for each pH_{out} and the calibration curve (Figure 4) calculated. The actual increase in pH_i upon thrombin stimulation is therefore proportional but not equal to the $\delta(\Delta pH_i)^{app}$ calculated from the 9-aminoacridine fluorescence of a platelet suspension before and after thrombin addition.

In order to obtain a true absolute pH_i as well as an independent confirmation of our amine distribution experiments, we have adapted the method of Thomas et al. (1,54), the "trapped" probe technique. As they showed in Ascites cells (54) and we have now shown in human platelets (13), and in E coli (18) the nonfluorescent compound 6-carboxyfluorescein diacetate traverses plasma membranes with ease. Since Ascites cells contain a large concentration (and platelets a low one) of cytoplasmic nonspecific esterases, the nonfluorescent diacetate is converted into the highly fluorescent 6-carboxyfluorescein whose intensity of fluorescence is pH dependent up to pH 7.4 - 7.6. The probe is thus an in situ pH_i indicator. Furthermore, since it now is negatively charged its free passage through membranes is severely impeded, and the pH indicated is that of the cytoplasm. It should be noted that while we used the previously described digitonin treatment to demonstrate that 95% of the internalized hydrolyzed probe was in the platelet cytoplasm, one needs to confirm the location of the indicator for every cell utilized.

While Thomas et al. (53) report little leakage of probe from the Ascites cells, and negligible fluorescence outside the cell, we found major contribution (20-80%) by the external medium when platelets were prelabeled with 6-carboxyfluorescein diacetate. There were two sources of this fluorescence:spontaneous (i.e. non-enzymatic) external hydrolysis of intact diester at pH 7.4, and transmembrane leakage of cytoplasmic enzyme-lysed 6-carboxyfluorescein. The best solution to date has been continuous dialysis of the non-stimulated stock of 6-carboxyfluorescein-labeled platelets. The change in platelet probe content over the duration of a single determination, approximately 5-10 minutes, can thus be kept very small. In order to determine the contribution to fluorescence by the cytoplasmically trapped indicator, aliquots are centrifuged at the beginning and at the end of each determination. The fluorescence in the platelets, I_{492} platelet, can then be calculated from the difference, at the same excitation (492 nm) and emission (518 nm) wave lengths, between the fluorescence of the entire suspension and that of the supernatant after high speed centrifugation, I_{492} suspension - I_{492} supernatant.

For the observation of platelet response to any single dose of thrombin, it was therefore possible to observe I_{492} suspension continuously as thrombin was added remotely to a prelabeled platelet suspension. An increase in that fluorescence, denoting a rise in pH_i, was detectable within less than 5 sec. The relative thrombin dose dependence of this cytoplasmic alkalinization follows the same curve (Figure 3) as the 9-aminoacridine measured relative $\delta(\Delta pH)$, the $diSC_3(5)$-measured relative $\Delta\psi$, the $[^3H]Ph_4P^+$ measured relative $\Delta\psi$, and serotonin release.

An absolute pH_i before and after stimulation can be obtained from an appropriate calibration curve since 6-carboxyfluorescein is not a distributive probe. In order to obviate differences due to the extent of hydrolysis within the cell, we have used the ratio calculation described by Heiple and Taylor (54). Providing Beer's Law is obeyed, the observed intensity of fluorescence by the platelets (em. constant at 518 nm) when excitation is at 492 nm, I_{492}, is proportional to a constant molar coefficient E_{492}, the optical path length, l, and the concentration of the fluorescent species, c, i.e. $I_{492} = E_{492}\,lc$. For a different excitation wavelength, eg 464 nm, $I_{464} = E_{464}\,lc$. Since 464 nm is the absorbance isobestic wavelength for 6-carboxyfluorescein, the absorbance is independent of pH. The ratio I_{492} platelets/I_{464} platelets is then independent of the concentration of the fluorescent entity and is a function of pH as described above. Indeed a calibration curve of this ratio for pure 6-carboxyfluorescein in buffer (1,13, Figure 5 reproduced from reference 13) is superimposed on that of platelets treated with nigericin in K^+ Hepes as described above. The calculation of a resting platelet's pH_i from a large number of determinations of this ratio and the calibration curve allows us to deduce that resting human platelets have pH_i = 7.02 (1,13). Upon addition of a saturating dose of thrombin (>.02 U/ml) this pH rises to 7.3 when the external medium is at pH 7.4. If one uses pH_i = 7.02 and applies the calibration curve for 9-aminoacridine (Figure 4) and the thrombin induced calculated (ΔpH_i) values, one obtains a change of 0.3 pH units with a saturating thrombin dose. The agreement between the two methods, ΔpH_i = 0.28 and 0.30 respectively, is gratifying.

Thus we have described here in some detail two relatively simple techniques for evaluation of stimulus-induced intraplatelet pH changes, i.e. for stimulus induced H^+ flow. They should be adaptable with relative ease to other cellular systems.

REFERENCES

1. "Intracellular pH: Its Measurement, Regulation and Utilization in Cellular Functions." R. Nuccitelli and D.W. Deamer, eds., Alan R. Liss, Inc., New York (1982).
2. D. Epel, Mechanisms of Activation of Sperm and Egg during Fertilization of Sea Urchin Gametes, Curr. Top. Dev. Biol. 12:186-246 (1978).
3. A. Roos, and W.F. Boron, Intracellar pH, Physiol. Rev. 61:297-434 (1981).
4. R.J. Gillies, "Intracellular pH and Growth Control in Eukaryotic Cells" in "The Transformed Cell," I. Cameron, ed., Academic Press, New York 91981).
5. J.D. Johnson, D. Epel, and M. Paul, Intracellular pH and activation of sea urchin eggs after fertilization, Nature 262: 611-664 (1976).
6. S.S. Shen, and R.A. Steinhardt, Measurement of Intracellular pH during Metabolic Depression of the Sea Urchin Egg, Nature 272:253-254 (1978).
7. M.M. Winkler, and J.L. Grainger, Mechanism of Action of NH_4cl and Other Weak Bases in Activation of Sea Urchin Eggs, Nature 238-538 (1978).
8. T. Finkel, and D.P. Wolf, Membrane Potential, pH and the Activation of Surf Clam Oocytes, Gamete Research 3:299-304 (1980).
9. R. Nuccitelli, D.J. Webb, S.T. Lagier, and G.B. Watson, ^{31}P NMR Reveals an Increase in Intracellular pH after Fertilization in Xenopus Eggs, Proc. Natl. Acad. Sci. USA 78:4421-4425 (1981).
10. R.D. Moore, Elevation of Intracellular pH by Insulin in Frog Skeletal Muscle, Biochem. Biophys. Res. Commun. 91:900-904 (1979)

11. S. Lee, and R.A. Steinhardt, Observations on Intracellular pH during Cleavage of Eggs of Xenopus Laeirs, J. Cell Biol. 91:414-419 (1981).
12. R.D. Moore, Stimulation of NaiH Exchange by Insulin, Biophys. J. 33:203-210 (1981).
13. W.C. Horne, N.E. Norman, D.B. Schwartz, E.R. Simons, Changes in Cytoplasmic pH and in Membrane Potential in Thrombin-Stimulated Human Platelets, Eur. J. Biochem. 120:295-302 (1981).
14. D.F. Gerson, H . Kiefer, and W. Eufe, Intracellular pH of Mitogen-Stimulated Lymphocytes, Science 216:1009-1010 (1982).
15. D.F. Gerson, and H. Kiefer, High Intracellular pH Accompanies Mitotic Activity in Murine Lymphocytes, J. Cell Physiol. 112:1-4 (1982).
16. J.M. Heiple, and D.L. Taylor, pH Changes in Pinosomes and Phagosomes in the Ameba Choas Carolinensis, J. Cell Biol. 94:143-149 (1982).
17. B. Setlow, and P. Setlow, Measurements of the pH within Dormant and Germinated Bacterial Spores, Proc. Natl. Acad. Sci. USA 77:2774-2776 (1980).
18. E. Shechter, L. Letellier, and E.R. Simons, Fluorescence Dye as a Monitor of Internal pH in Escherichia Coli Cells, FEBS Letters, 139,121-124 (1982).
19. W.C. Horne, and E.R. Simons, Effects of Amiloride on the Response of Human Platelets to Bovine Thrombin, Thrombos. Res. 13:599-607 (1979).
20. R.I. Sha'afi, T.F.P. Molski, and P.H. Naccache, Chemotactic Factors Activate Differentiable Permeation. Pathways for Na^{+} and Ca^{++} in Rabbit Neutrophils, Biochem. Biophys. Res. Commun. 99:1271-1276 (1980).
21. T.F.P. Molski, P.H. Naccache, M. Volpi, L.M. Wolpert, and R.I. Sha'afi, Specific Modulation of the Intracellular pH of Rabbit Neutrophils by Chemotactic Factors, Biochem. Biophys. Res. Commun. 508:514 (1980).
22. E.W. Salzman, Some Basic Mechanisms in Platelet Physiology, Sem. Haemat. 8:3-49 (1976).
23. N.E. Owen, and G.C. LeBreton, Ca^{++} Mobilization in Blood Platelets as Visualized by Chlortetracycline Fluorescence, Am. J. Physiol. 241:H613-619 (1981).
24. M.J. Broekman, J.W. Ward, and A.J. Marcus, Phospholipid Metabolism in Stimulated Human Platelets, J. Clin. Invest. 66:275-283 (1980).
25. H. Holmsen, C.A. Setkowsky, and H.J. Day, Effects of antimycin and 2:deoxyglucose on Adenine Nucleotides in Human Platelets. Role of Metabolic ATP in Primary Aggregation, Secondary Aggregation and Shape Change of Platelets, Biochem. J. 144:385-396 (1974).
26. N.E. Larsen, and E.R. Simons, Preparation and Application of a Photoreactive Thrombin Analogue: Binding to Human Platelets, Biochem. 20:4141-4147 (1981).

27. B.E. Seligman, J.I. Gallin, Use of Lipophilic Probes of Membrane Potential to Assess Human Neutrophil Activation. J. Clin. Invest. 66:493-503 (1980).
28. H.M. Korchak, and G. Weissman, Changes in Membrane Potential of Human Granulocytes Antecede the Metabolic Responses to Surface Stimulation, Proc. Natl. Acad. Sci. USA 75:3818-3822 (1978).
29. K. Utsumi, K. Sugiyamam, M. Miyahara, M, Naito, M. Awai, and M. Inone, Effect of Concanavalin A on Membrane Potential of Polymorphonuclear Leukocytes Monitored by Fluorescent Dye, Cell Struct. Func. 2:203-209 (1977).
30. J.C. Whitin, C.E. Chapman, E.R. Simons, M.E. Chovaniec, and H.J. Cohen, Correlation Between Membrane Potential Changes and Superoxide Production in Human Granulocytes Stimulated by Phorbol Myristate Acetate, J. Biol. Chem. 255:1874-1878 (1980).
31. J.C. Whitin, R.A. Clark, E.R. Simons, and H.J. Cohen, Effects of the Myeloperoxidase System on Fluorescent Probes of Granulocyte Membrane Potential. J. Biol. Chem. 256:8904-8906 (1981).
32. H.J. Cohen, P.E. Newburger, M.E. Chovaniec, J.C. Whitin, and E.R. Simons, Opsonized Zymosan: Stimulated Granulocytes-Activation and Activity of the Superoxide-Generating System and Membrane Potential Changes, Blood 58:975-981 (1981).
33. G.S. Jones, K. Van Dyke, and V. Castrovana, Transmembrane Potentials Associated with Superoxide Release from Human Granulocytes. J. Cell Physiol. 106:75-83 (1981).
34. J.C. Freedman, and J.F. Hoffman, The Relation Between Dicarbocyanine Dye Fluorescence and the Membrane Potential of Human Red Blood Cells Set at Varying Donnan Equilibria, J. Gen. Physiol. 74:187-212 (1979).
35. J.C. Freedman, and J. Hoffman, Ionic and Osmotic Equilibria of Human Red Blood Cells Treated with Nystatin, J. Gen. Physiol. 74:157-185 (1979).
36. J.W.N. Akkerman, R.H.M. Ebberink, J.P.M. Lips, and G.C. Christiaens, Rapid Separation of Cytosol and Particle Fraction of Human Platelets, Br. J. Haematol. 44:291-297 (1980).
37. P.J. Sims, A.S. Waggoner, C.-H. Wang, and J. Hoffman, Studies on the Mechanism by which Cyanine Dyes Measure Membrane Potential in Red Blood Cells and Phosphatidylcholine Vesicles, Biochem. 13:3315-3330 (1974).
38. A. Waggoner, Optical Probes of Membrane Potential, J. Memb. Biol. 27:317-334 (1976).
39. A. Waggoner, Dye Indicators of Membrane Potential, Ann. Rev. Biophys. Bioeng. 8:47-68 (1979).
40. W.C. Horne, and E.R. Simons. Probes of Transmembrane Potentials in Platelets: Changes in Cyanine Dye Fluorescence in Response to Aggregation Stimulation, Blood 5:741-749 (1978).

41. T.C. Smith, J.T. Herlily, and S.C. Robinson, The Effect of the Fluorescent Probe 3,3'Diproplythiodicarbocyanine Iodide on the Energy Metabolism of Ehrlich Ascites Tumor Cells, J. Biol. Chem. 256:1108-1110 (1981).
42. R.M. Johnstone, P.C. Laris, and A.A. Eddy, The Use of Fluorescent Dyes to Measure Membrane Potentials: A Critique. J. Cell Physiol. 112:298-301 (1982).
43. J.C. Freedman, and P.C. Laris, Electrophysiology of Cells and Organelles: Studies with Optical Potentiometric Indicators, Int. Rev. Cytol. (Suppl.) 12:177-246 (1981).
44. R. Lundblad, L.C. Uhteg, C.N. Vogel, H.S. Kingdon, and K. Mann, Preparation and Partial Characterization of two forms of Bovine Thrombin, Biochem. Biophys. Res. Commun. 66:482-489, (1975).
45. N.E. Larsen, W.C. Horne, E.R. Simons, Platelet Interaction with Active and TLCK-inactivated Thrombin, Biochem. Biophys. Res. Commun. 87:403-409 (1979).
46. D.W. Deamer, R.C. Prince, and A.R. Crofts, The Response of Fluorescent Amines to pH Gradients across Liposome Membrane, Biochim. Biophys. Acta 274:323-335 (1972).
47. H. Rottenberg, T. Grunwald, and M. Avron, Determination of pH in Chloroplasts. Eur. J. Biochem. 25:54- (1982).
48. R. Casadio, A. Baccarini-Melandri, and B.A. Melandri. On the Determination of the Transmembrane pH Difference in Bacterial Chromatophores Using 9-aminoacridine, Eur. J. Biochem. 47: 121-128 (1974).
49. R.J. Gillies, and D.W. Deamer, Intracellular pH, Curr. Topics Bioenerg. 9:63-87 (1979).
50. W.S. Chow, and A.B. Hope. Light-indiced pH Gradients in Isolated Spinach Chloroplasts, Aust. J. Plant Physiol. 3: 141-153 (1976).
51. V. Pick, C.M. Avron, A Method for Measuring the Internal pH in Illuminated Chloroplasts Based on the Stimulation of Proton Uptake by Amines. Eur. J. Biochem. 70:569-576 (1976).
52. S. Shuldiner, H. Rottenberg, and C.M. Avron, Determination of pH in Chloroplasts. 2. Fluorescent Amines as a Probe for the Determination of pH in Chloroplasts, Eur. J. Biochem. 25:64-70 (1972).
53. J.A. Thomas, R.N. Buchsbaum, A. Zimniak, and E. Racker. Intracellular pH measurements in Ehrlich Ascites Tumor Cells Utilizing Spectroscopic Probes Generated in situ, Biochem. 18:2210 (1970).
54. J.M. Heiple, and D.L. Taylor, Intracellular pH in Single Motile Cells, J. Cell Biol. 86:885-890 (1982).

CALCIUM AND THE REGULATION OF CELLULAR PROCESSES WATER AND IONS IN MUSCLE

CALCIUM DEPENDENT REGULATION OF CONFORMATIONAL CHANGE IN THE SH_2 REGION OF SKELETAL MYOSIN

Tsuneo Kameyama

Department of Biochemistry, School of Medicine
Juntendo University
Hongo-2, Bunkyo-ku, Tokyo, 113, Japan

INTRODUCTION

Due to the finding of troponin,[1,2] actin-linked Ca^{2+}-dependent regulatory mechanism of muscle contraction was established.

On the other hand, since myosin-linked Ca^{2+}-dependent regulatory mechanism was widely accepted in molluscun muscle,[3,4] there have appeared reports[5-14] which demonstrate that physiological level of free Ca^{2+} concentration also affects the conformational changes in skeletal myosin, although these Ca^{2+}-sensitive conformational changes give no definite evidence for a significant contribution to physiological contraction.

The additional problems related to the Ca^{2+} regulation discussed here are based on the traditional studies in our laboratory. Myosin has a lot of thiol groups, most of which are located at the head of myosin (heavy meromyosin). The specific thiol groups, designated as SH_1 and SH_2, in the active site of myosin ATPase were selectively labeled with NEM. Their contribution to the enzymatic properties and the primary structures around them were also determined.[15-19] SH_1 modification with NEM caused activation of Ca^{2+}-ATPase activity, and successive modification of SH_2 caused inactivation of Ca^{2+}-ATPase to almost the original level (Fig. 1). Whereas SH_1 is extremely fast reactive, suggesting that SH_1 is exposed to the surface of myosin

---

Abbreviations:
SH_2 region, local area containing a specific thiol group, SH_2, of myosin; NEM, N-ethylmaleimide; myosin*, SH_1-NEM modified myosin; AMPPNP, adenylyl imidodiphosphate.

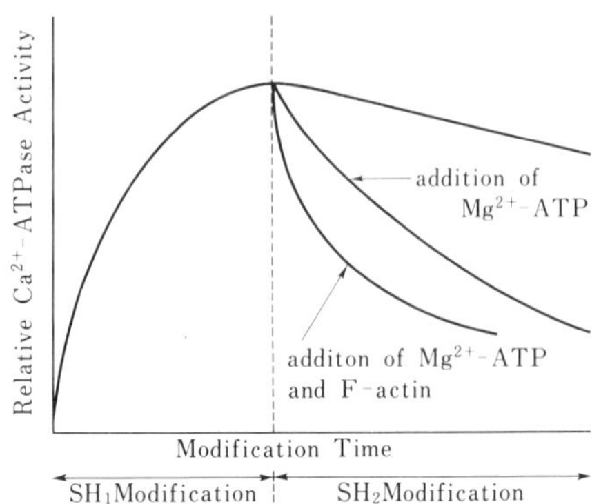

Fig. 1. Biphasic response of Ca^{2+}-ATPase activity to SH_1 and SH_2 modification with NEM.

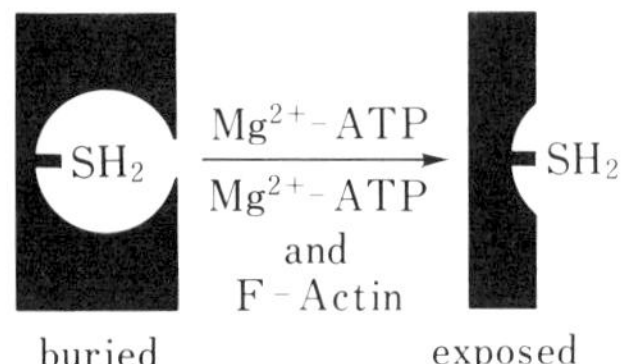

Fig. 2. Conformational change in the SH_2 region.

molecule, SH_2 is not reactive under the physiological ionic conditions, suggesting that SH_2 is in the folded state.[15] SH_2 become reactive in the presence of Mg^{2+}-ATP (Mg^{2+}-ATP-induced unfolding of the SH_2 region). This fact indicated that the physiological substrate, Mg^{2+}-ATP, or intermediates of its decomposition induced the conformational change around a specific thiol group, SH_2, (referred to as SH_2 region), leading to exposure of the SH_2 region to the surface of the molecule.[20] Furthermore, F-actin enhanced the Mg^{2+}-ATP-induced unfolding of the SH_2 region (Fig. 1,2).[13,21,22] Recently it was reported that SH_2 is located near the (subfragment-1)-(subfragment-2) junction in the head of myosin molecule.[23] In addition, taking the foregoing effects of Mg^{2+}-ATP and F-actin into consideration, SH_2 region is assumed to act as a hinge in the drive stroke action of the myosin head during the sliding process (Fig. 3).

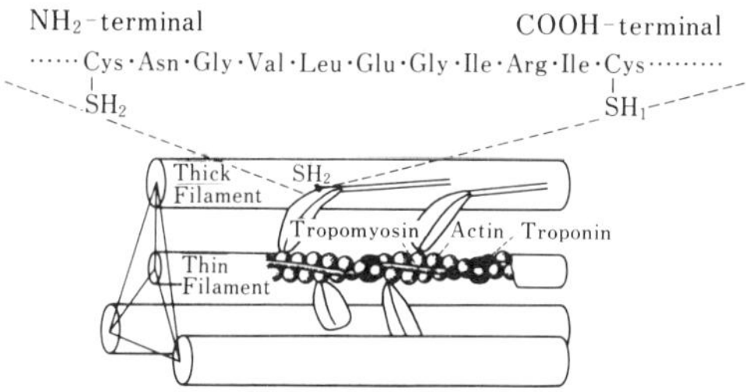

Fig. 3. A possible role of the SH_2 region during muscle contraction.

At present, an additional new finding which may relate to a physiological role of the SH_2 region was obtained. Namely, the present finding militates in favor of a possible regulatory mechanism by free Ca^{2+} through the SH_2 region in skeletal myosin, revealing that the physiological change in free Ca^{2+} concentration associated with the excitation-contraction coupling regulates the conformational change in the SH_2 region in the presence of Mg^{2+}-ATP.

EFFECT OF FREE Ca^{2+} CONCENTRATION ON THE SH_2 MODIFICATION IN THE PRESENCE OF Mg^{2+}-ATP

At present, it is impossible to modify SH_2 alone without modifying SH_1. Fortunately, SH_1-NEM modified myosin (myosin*) preserves the essential properties required for muscle contraction. Namely, myosin* has the ability to cause superprecipitation, actin-activated Mg^{2+}-ATPase activity,[24] and its actomyosin thread developes isometric tension similar to the normal actomyosin thread.[25] Therefore, all the experiments on SH_2 modification were carried out using myosin*. The extent of the conformational change (unfolding) in the SH_2 region has been evaluated from the rate constant of decrease in Ca^{2+}-ATPase activity which is caused by the reaction of SH_2 with NEM.[21]

As is already stated, physiological substances, e.g. Mg^{2+}-ATP, accelerated SH_2 modification, indicating that it induces the conformational change in the SH_2 region. As is shown in Fig. 4, the reactivity of SH_2 increased with increasing free Ca^{2+} concentration in the presence of Mg^{2+}-ATP, indicating that Mg^{2+}-ATP-induced conformational change increases with increased free Ca^{2+} concentration. The rate constants of the SH_2 modification, that is, the extents of the conformational changes in the SH_2 region were shown in Fig. 5. The curve relating this conformational change to pCa was a typical sigmoid shape, showing that the conformational change was most sensitive to the change in free Ca^{2+} concentration at pCa 5.

An essential property of this Ca^{2+} sensitivity was independent on the preparative methods of myosin (myosin prepared according to the method of Kielley and Bradely[26] or Perry[29]). Mg^{2+}-ADP had also a potent ability to induce the conformational change in the SH_2 region, however, there is no relation to the Ca^{2+} sensitivity. The conformational change was scarcely regulated in the presence of Mg^{2+}-ATP by free Sr^{2+} instead of free Ca^{2+}. Other nucleotides, 8-BrATP[30], GTP, ITP, or AMPPNP were all ineffective regardless of presence or absence of Ca^{2+} in the presence of Mg^{2+}. The conformational change did not occur in the absence of ATP and in the presence of Mg^{2+} or Mg^{2+} and Ca^{2+}.

Thus, it should be noted that the physiological level of free Ca^{2+} concentration regulates the conformational change in the SH_2 region of myosin molecule energized with Mg^{2+}-ATP and that both Ca^{2+} and Mg^{2+}-ATP have great specificity.[14,31]

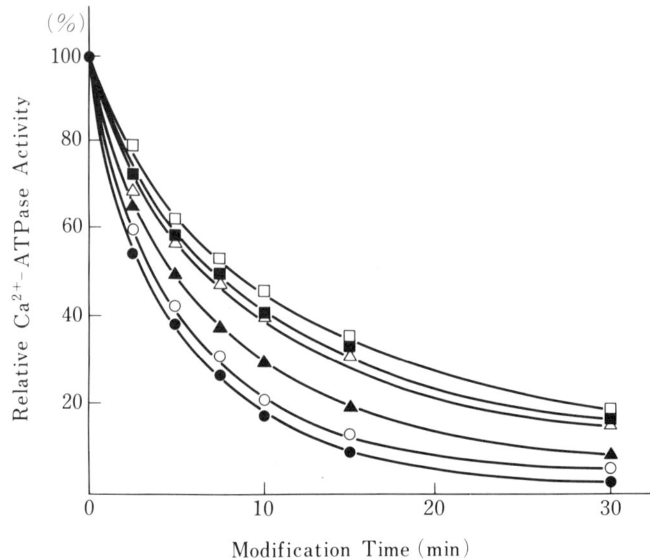

Fig. 4. Effect of free Ca^{2+} concentration on SH_2 modification of myosin* with NEM in the presence of Mg^{2+}-ATP. Myosin was prepared according to the method of Kielley and Bradely.[26] SH_2 modification with NEM was carried out at 0° C, in the reaction mixture containing 20 mM Tris-maleate buffer pH 6.8 (20 mM maleate was titrated with Tris solution), 0.5 M KCl, 2 mM $MgCl_2$, 1 mM ATP, Ca-EGTA buffer (0.1 mM EGTA),[27] 0.2 mM NEM and 1.2 mg of myosin*[21] in a total volume of 1.0 ml. The reaction was initiated by adding NEM. At the times indicated, in order to stop the reaction, aliquots of 0.1 ml were transferred to the solution of 1.9 ml of 0.5 M KCl containing 5 fold excess of β-mercaptoethanol with respect to NEM. The extent of SH_2 modification was estimated from the decrease in Ca^{2+}-ATPase activity. Assay of Ca^{2+}-ATPase activity was carried out at 37° C in the reaction mixture containing 0.1 ml of SH_2 modified myosin* solution (6 μg), 20 mM histidine buffer pH 7.6, 5 mM $CaCl_2$, 1 mM ATP and 0.55 M KCl in a total volume of 1.0 ml. Inorganic phosphate liberated from ATP was measured using the method by Fiske and SubbaRow.[28] 100% of Ca^{2+}-ATPase activity corresponded approximately to specific activity of 4.6 μmoles P_i/min/mg. pCa 3 (●), 4 (○), 5 (▲), 6 (△), 7 (■) and 8 (□).

IMPLICATIONS

In the head of myosin molecule (heavy meromyosin), there exists the mechanism of chemo-mechanical conversion which causes tension development.

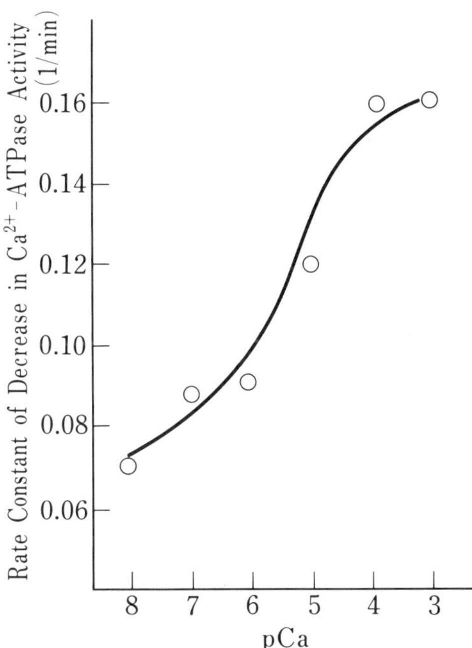

Fig. 5. The rate constants of decrease in Ca^{2+}-ATPase activities caused by SH_2 modification. The rate constants were obtained from linear parts of semilog plots of Fig. 4 (within 10 min from the start of the reactions).

On the other hand, it was known that there are a lot of thiol groups in myosin molecule which exist mainly in its head part. We have studied the conformational change in myosin in an effort to elucidate the mechanism of chemo-mechanical conversion, using the reactivity of these thiol groups as a structural probe. At present, our studies are focused on the teleological conformational change in the SH_2 region during muscle contraction.

Based on a series of our studies, my present view on the functional topology of myosin molecule will be discussed here. Physiological substrates, Mg^{2+}-ATP or Mg^{2+}-ADP, induced the conformational change (unfolding) in the SH_2 region.[15,17] The extent of the conformational change was greater in the presence of Mg^{2+}-ADP than in the presence of Mg^{2+}-ATP. So, it is reasonable to consider that this conformational change itself is not related to the energized state of myosin molecule due to ATP decomposition. However, the physiological level of free Ca^{2+} concentration regulated only Mg^{2+}-ATP-induced conformational change and failed to regulate Mg^{2+}-ADP-induced one (present report). F-actin also enhanced only Mg^{2+}-ATP-induced conformational change in the SH_2 region, not Mg^{2+}-ADP-induced one (without addition of Ca^{2+}).[21] In contrast to these conformational changes in the SH_2 region, Ca^{2+}-dependent reactivity of whole thiol

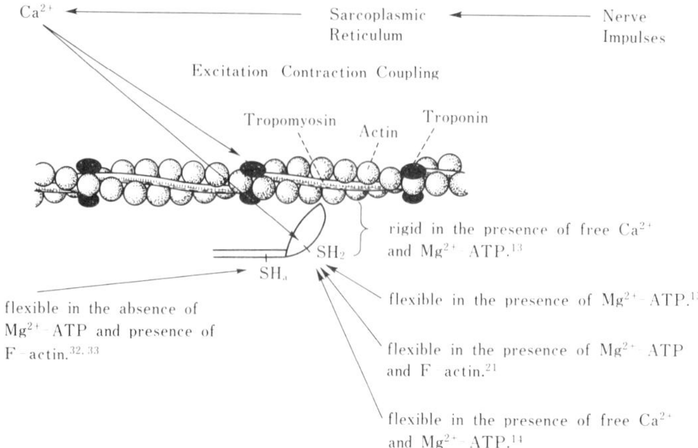

Fig. 6. Schematic diagram of functional topology in myosin molecule.

groups in myosin suggested that the physiological level of free Ca^{2+} concentration increased the rigidity of myosin head energized with Mg^{2+}-ATP with increasing the free Ca^{2+} concentration.[13] In addition, another local area containing a specific thiol group, SH_a, in light meromyosin was also reported to be unfolded in the presence of F-actin and the absence of Mg^{2+}-ATP.[32,33] These results were summarized diagrammatically in Fig. 6.

Thus it will be easily understood that the physiological level of free Ca^{2+} concentration associated with excitation-contraction coupling and F-actin may regulate the conformational change in the energized myosin molecule in favor of drive stroke action of myosin head to develope the tension during the sliding process.

REFERENCES

1. S. Ebashi, Third component participating in the superprecipitation of natural actomyosin, Nature 200:1010 (1963).
2. S. Ebashi, The Croonian Lecture, 1979. Regulation of muscle contraction, Proc. R. Soc. London. Ser. B 270:259 (1980).
3. A. G. Szent-Györgyi, E. M. Szentkiralyi, and J. Kendrick-Jones, The light chains of scallop myosin as regulatory subunits, J. Mol. Biol. 74:179 (1973).
4. J. Kendrick-Jones, Role of myosin light chains in calcium regulation, Nature 249:631 (1974).
5. E. A. Sugden and T. Nihei, The effects of calcium and magnesium ions on the adenosine triphosphatase and inosine triphosphat-

ase activities of myosin A, Biochem. J. 113:821 (1969).
6. K. Morimoto and W. F. Harrington, Evidence for structural changes in vertebrate thick filaments induced by calcium, J. Mol. Biol. 88:693 (1974).
7. M. M. Werber and A. Oplatka, Physico-chemical studies on the light chains of myosin. III. Evidence for a regulatory role of a rabbit myosin light chain, Biochem. Biophys. Res. Commun. 57:823 (1974).
8. J. C. Haselgrove, X-ray evidence for conformational changes in the myosin filaments of vertebrate striated muscle, J. Mol. Biol. 92:113 (1975).
9. S. S. Margossian, S. Lowey, and B. Barshop, Effect of DTNB light chain on the interaction of vertebrate skeletal myosin with actin, Nature 258:163 (1975).
10. W. Lehman, Thick-filament-linked calcium regulation in vertebrate striated muscle, Nature 274:80 (1978).
11. G. D. Rieser, R. A. Sabbadini, and P. J. Paolini, Calcium and pH-induced structural changes in skinned muscle fibers : Prevention by N-ethylmaleimide, Biochem. Biophys. Res. Commun. 90:179 (1979).
12. R. A. Sabbadini, G. D. Rieser, and P. J. Paolini, Calcium-induced structural changes in chemically skinned muscle fibers : Detection by optical diffractometry, Biochim. Biophys. Acta 578:526 (1979).
13. T. Kameyama, M. Komatsu, and T. Sekine, Actin-induced local conformational change in the myosin molecule. III. Reactivity of S_2 thiol and DTNB-reactive thiols of porcine cardiac myosin, J. Biochem. 87:587 (1980).
14. T. Kameyama, Calcium regulation of the conformational change around a specific thiol group, SH_2, of skeletal myosin, Proc. Jpn. Acad. Ser. B 58:191 (1982).
15. T. Sekine and M. Yamaguchi, Effect of ATP on the binding of N-ethylmaleimide to SH groups in the active site of myosin ATPase. J. Biochem. 54:196 (1963).
16. T. Sekine and W. W. Kielley, The enzymic properties of N-ethylmaleimide modified myosin, Biochim. Biophys. Acta 81:336 (1964).
17. M. Yamaguchi and T. Sekine, Sulfhydryl groups involved in the active site of myosin A adenosine triphosphatase. I. Specific blocking of the SH group responsible for the inhibitory phase in "biphasic response of the catalytic activity", J. Biochem. 59:24 (1966).
18. T. Yamashita, Y. Soma, S. Kobayashi, T. Sekine, K. Titani, and K. Narita, The amino acid sequence at the active site of myosin A adenosine triphosphatase activated by EDTA, J. Biochem. 55:576 (1964).
19. T. Yamashita, Y. Soma, S. Kobayashi, and T. Sekine, The amino acid sequence of SH-peptides involved in the active site of myosin A adenosinetriphosphatase, J. Biochem. 75:447 (1974).
20. T. Sekine, K. A. Kato, K. Takamori, M. Machida, and Y. Kanaoka,

Fluorescent thiol reagents. X. Fluorimetric estimation of the reactivity of thiols : Examples in small molecular compounds, Taka-amylase A and myosin A, Biochim. Biophys. Acta 354:139 (1974).

21. T. Kameyama, T. Katori, and T. Sekine, Actin-induced local conformational change in the myosin molecule. I. Effect of metal ions and nucleotides on the conformational change around a specific thiol group(S_2) of heavy meromyosin, J. Biochem. 81:709 (1977).
22. T. Kameyama, Actin-induced local conformational change in the myosin molecule. II. Conformational change around the S_2 thiol group related to the essential intermediate of ATP hydrolysis, J. Biochem. 87:581 (1980).
23. K. Sutoh, Location of SH_1 and SH_2 along a heavy chain of myosin subfragment 1, Biochemistry 20:3281 (1981).
24. T. Sekine and M. Yamaguchi, Superprecipitation of actomyosin reconstructed with F-actin and NEM-modified myosin, J. Biochem. 59:195 (1966).
25. S. Srivastava and J. Wikman-Coffelt, An investigation into the role of SH_1 and SH_2 groups of myosin in calcium binding and tension generation, Biochem. Biophys. Res. Commun. 92:1383 (1980).
26. W. W. Kielley and L. B. Bradely, The relationship between sulfhydryl groups and the activation of myosin adenosinetriphosphatase, J. Biol. Chem. 218:653 (1956).
27. Y. Ogawa, The apparent binding constant of glycoletherdiaminetetraacetic acid for calcium at neutral pH, J. Biochem. 64:255 (1968).
28. C. H. Fiske and Y. SubbaRow, The colorimetric determination of phosphorus, J. Biol. Chem. 66:375 (1925).
29. S. V. Perry, [94] Myosin adenosinetriphosphatase, in "Methods in Enzymology, Vol. 2" S. P. Colowick and N. O. Kaplan, ed., Academic Press, New York (1955).
30. M. Takenaka, M. Ikehara, and Y. Tonomura, Structure and function of myosin molecule. IV. Physiological functions of various reaction intermediates in myosin adenosinetriphosphatase, studied by the interaction between actomyosin and 8-Bromo-adenosinetriphosphate, J. Biochem. 80:1381 (1976).
31. T. Kameyama, Submitted for publication.
32. T. Yamashita, M. Kobayashi, and T. Horigome, The sulfhydryl groups involved in the active site of myosin B adenosinetriphosphatase, J. Biochem. 77:1037 (1975).
33. T. Horigome and T. Yamashita, The sulfhydryl groups involved in active site of myosin B adenosinetriphosphatase. IV. Structure around the S_a thiol group, J. Biochem. 83:49 (1978).

THE ROLE OF CALCIUM IONS IN THE CONFORMATIONAL CHANGES OF TROPONIN

Kozo Nagano

Faculty of Pharmaceutical Sciences
University of Tokyo
Hongo, Bunkyo-ku, Tokyo 113, Japan

INTRODUCTION

This paper aims to propose a new 3-dimensional model of troponin C, which can explain as many experimental observations on thin filaments in vertebrate striated muscles as possible. Thin filaments are composed of F-actin, tropomyosin (TM) and troponin. TM is believed to be a double-stranded coiled-coil.[1] Troponin consists of 3 components; Ca binding troponin C (TnC), actin-myosin interaction inhibiting troponin I (TnI), and TM binding troponin T (TnT).[2] The N-terminal fragment of TnT, viz. T_1, is mostly covered by troponins C and I, while the C-terminal fragment of TnT, viz. T_2, is more exposed and situated at the N-terminal side of TM.[3,4] Troponins C and I are both exposed, but it was not determined by immunoelectron microscopy on which side is actually TnC. The length of TnT is estimated to be longer than 90 Å.

NEUTRALIZATION OF TnI INHIBITION OF ACTOMYOSIN ATPASE ACTIVITY

There has been a controversy on whether TnC shows both high and low affinities for Ca in the system of all muscle proteins with ATP and Mg, or whether it reacts cooperatively with Ca ions. Kohama[5] determined the stability constants of troponin complex at various pHs, and found it to be $1.1 \times 10^7\ M^{-1}$ for the high affinity sites and $4 \times 10^5\ M^{-1}$ for the low affinity sites at pH 7.4 in the presence of 1.3 mM Mg. However, Ca-dependence of neutralization of TnI inhibition of actomyosin ATPase activity for TnC and its proteolytic fragments has suggested that the apparent stability constants of all 4 Ca binding sites of skeletal TnC in the thin filament are about $7 \times 10^6\ M^{-1}$ at pH 7.5 in the presence of 5 mM Mg.[6] The results have also shown that the N-terminal half of TnC binds strongly to TnT in the presence of Ca, but does not bind to it at all in the absence of Ca. On the other hand, it is known that the intact TnC can bind to TnT whether or not Ca ion is present. This suggests that the C-terminal half of TnC is important in binding to TnT in the absence of Ca, although the C-terminal fragments do not bind to it.

The equivalence of the stability constants of the 4 Ca binding sites of skeletal TnC in the system of actomyosin ATPase,[6] though the relative activity is different from one fragment to another, is quite contradictory to the assignment of high affinity Ca-Mg sites to the C-terminal half of TnC and the low affinity Ca-specific sites to its N-terminal half.[7] This indicates a cooperative phenomenon with respect to the conformational changes of TnC associated with binding of Ca ions. Why are two stability constants of the high affinity sites weakened in the system of actomyosin ATPase? Why are two of the low affinity sites strengthened to become equivalent to those of the high affinity sites? How to explain this problem in terms of a 3-dimensional model of TnC is the subject of this paper.

It was observed that calmodulin (CaM) can neutralize TnI inhibition of actomyosin ATPase activity in the presence of Ca, but not in the absence of Ca.[8] This suggests that the conformation of CaM would be quite similar to that of TnC in the presence of Ca, but different from that of TnC in the absence of Ca. The effect of parvalbumin on the experiment has never been published. I assume in this paper that parvalbumin would have no effect on the system of actomyosin ATPase.

Figure 1 shows a comparison of amino acid sequences of TnCs, CaM, parvalbumin and intestinal Ca binding protein. While parvalbumin and intestinal Ca binding protein have 2 Ca binding sequences composed of 12 to 14 residue loop flanked by two α-helices, TnCs and CaM have 4 such regions. However, the 1st region of cardiac TnC is not supposed to bind Ca. In a reasonable tertiary structure of TnC and CaM, the C-terminal T_2 region of TnT should interact with TnC, its proteolytic fragments, and CaM, but not with parvalbumin. The regions of rabbit skeletal muscle TnC around residues 47 to 54 and those 123 to 130 should face T_2, while the regions around residues 83 to 94 and both the N- and C-termini should be situated far from T_2. The N-terminal T_1 region including the long α-helix Tx is supposed to play an important role in determining the specific binding position and orientation of TnT against TM coiled-coil. The interaction between TM and Tx is mainly due to salt-bridges and a few hydrophobic contacts. Although the long α-helix Tx would be bent at a middle position unless it is bound to TM coiled-coil, the most reasonable binding mode of TM coiled-coil and Tx is a triple-stranded coiled-coil.[9,10] The C-terminal T_2 region is known to play an important role in binding TnI, regardless of Ca concentrations, as well as the N-terminal half of TnC, particularly when the Ca concentration is high.

	linker	helix	loop	helix	
			* * * * * *		
(a)	$_{0}$XMDDITKAAVEQLTEEQKNE	FKAAFDIF	VLGAEDG-CISTKE	LGKVMRML	48
(b)	$_{0}$XDTQQAEARSYLSEEMIAE	FKAAFDMF	-DADGGG-DISVKE	LGTVMRML	46
(c)	$_{0}$XADQLTEEQIAE	FKEAFSLF	-DKDGNG-TITTKE	LGTVMRSL	39
(a)	$_{49}$GQNPTPE-----E	LQEMIDEV	-DEDGSG-TVDFDE	FLVMMVRC	84
(b)	$_{47}$GQTPTKE-----E	LDAIIEEV	-DEDGSG-TIDFEE	FLVMMVRQ	82
(c)	$_{40}$GQNPTEA-----E	LQDMINEV	-DADGDG-TIDFPE	FLTMMARK	75
(d)	0	XAFAGVLN	DADIAAALEACKAA	DSFNHKAF	29
(a)	$_{85}$MKDDSKGKSEE-E	LSDLFRMF	-DKNADG-YIDLEE	LKIMLQAT	124
(b)	$_{83}$MKEDAKGKSEE-E	LAECFRIF	-DRNADG-YIDAEE	LAEIFRAS	122
(c)	$_{76}$MKDTDSEE----E	IREAFRVF	-DKDGNG-YISAAE	LRHVMTNL	112
(d)	$_{30}$FAKVGLTSKSADD	VKKAFAII	-DQDKSG-FIEEDE	LKLFLQNF	70
(e)	$_{1}$KSPEE	LKGIFEKY	AAKEGDPNQLSKEE	LKLLLQTE	35
(a)	$_{125}$GETITED-----D	IEELMKDG	-DKNNDG-RIDYDE	FLEFMKGV	E$_{161}$
(b)	$_{123}$GEHVTDE-----E	IESLMKDG	-DKNNDG-RIDFDE	FLKMMEGV	Q$_{159}$
(c)	$_{113}$GEkLTDE-----E	VDEMIREA	-DIDGDG-QVNYEE	FVQMMTAK	148
(d)	$_{71}$KADARALTDG--E	TKTFLKAG	-DSDGDG-KIGVDE	FTALVKA	108
(e)	$_{36}$FPSLLKGPS---T	LDELFEEL	-DKNGDG-EVSFEE	FQVLVKKI	SQ$_{75}$

Fig. 1. A comparison of amino acid sequences of troponin C and its related proteins: (a) bovine cardiac troponin C;[11] (b) rabbit skeletal troponin C; (c) bovine brain calmodulin;[12] (d) carp muscle parvalbumin; and (e) bovine intestine Ca binding protein. Reference should be made to Szebenyi et al.[13] for the sequences of (b), (d) and (e). Symbol * represents a residue coordinating to Ca in the EF hand structures.[14] The N-terminal sequence of parvalbumin in a box does not show any homology with other corresponding sequences. Number indicates a residue number at the end of the line. The ranges of helix and loop are those of the EF hand structure,[14] while the rest of the sequences is assigned to linker regions.

The 3-dimensional structures of two Ca binding proteins listed in Fig. 1 have been determined by X-ray crystallography. Those are carp muscle parvalbumin[15] and vitamin D-dependent Ca binding protein from bovine intestine.[13] Although the 1st region of the intestinal Ca binding protein was a little different from others, the conformations were very similar, and are called EF hand structure.[14] Two EF hand structures are hydrogen-bonded together at the 8th position of the 12 residue loops. Since no crystallographic results of TnC or CaM are available, we would like to predict the most reasonable tertiary structure of TnC. A famous model of tertiary structure of TnC with 4 bound Ca ions was predicted by Kretsinger and Barry.[16] The high affinity Ca ions are separated from the low affinity Ca ions by distances longer than 32 Å.

PREDICTED MODEL OF TnC WHICH EXPLAINS THE REGULATORY MECHANISM

If we try to fit this model to the predicted model of TM-TnT complex, the regions of the low affinity Ca-specific sites should come closer to T_2. It is contradictory to the observation that T_2 is also capable of binding to TnI regardless of Ca concentrations.[17] On the other hand, the region of rabbit skeletal TnC around residues 75 to 80 is the most conservatively hydrophobic and the most important in binding to TnI when the Ca concentration increases.[18] It suggests that the region would not be exposed in the absence of Ca. Although the predicted model of Kretsinger and Barry[16] for TnC exposes the region by making a close contact between two hydrophilic regions of residues 47 to 54 and those 123 to 130, it seems possible for the model to make this region less exposed by a slight modification of the spatial relationship between the high and the low affinity Ca binding domains. How could we explain the regulatory mechanism of muscle contraction due to TnI binding to the most conservatively hydrophobic region in the presence of Ca? Such a design of the contractile apparatus would not be efficient. The next question is why all α-helices of TnC do not run in parallel or antiparallel with the α-helices of the triple-stranded coiled-coil of TM-TnT complex. This is the reason why I did not adopt the separate domains of EF hand structures of the high and the low affinity Ca binding sites in the predicted model of the approximate quaternary structure of troponin complex bound to TM coiled-coil.[19] The final important question is how we could expect the same stability constants for all 4 Ca binding sites in the Kretsinger-Barry model? In particular, it is known that cardiac TnC can interact with TnI almost fully when 2 Ca ions are bound to it, because cardiac TnC has a very low affinity for the 3rd Ca.[20] In order to explain the cooperative phenomenon of Ca binding to TnC, the tertiary structure of skeletal TnC with 3 Ca ions bound to it should be almost the same as that binding 4 Ca ions.

The reasons for questioning the EF hand structures of the separate Ca binding domains of Kretsinger-Barry model led me to choose another possible tertiary structure, in which 4 Ca ions are situated in the centre of the molecule making a square or a lozenge, and separated from each other by 12 Å.[19] Although the type of coordination to Ca was similar to the EF hand type in the 4 Ca binding sites with some constraints in their main chain conformations, this type of model sits on the triple-stranded coiled-coil of TM-TnT complex, with all α-helices almost either parallel or antiparallel. Its length is about 45 Å, which is about a half of the length of TnT bound to TM coiled-coil. Its width is about 20 Å, which is about the diameter of TM-Tx triple-stranded coiled-coil. The T_2 region can recognize the turn regions of residues 47 to 54 and those 123 to 130, while both the N- and C-termini and the turn region of residues 83 to 94 are situated far from T_2. In the turn region CaM has 3 residues less, as shown in Fig. 1. Parvalbumin has 3 residues more in the turn region of residues 123 to 130.

The region binding to TnI independently of Ca concentrations is extended and in contact with both T_2 and actin. The region of residues 75 to 80 is the Ca-sensitive site of TnI binding, covering a part of the Ca-insensitive TnI binding site and protruding toward a myosin head. When the Ca concentration is very low, the upper two regions are supposed to be switched over so that the region of residues 75 to 80 would be buried deeply by the N-terminal α-helix of TnC. The movable end of TnI interacting with the region of residues 75 to 80 of TnC is also very close to the actin molecule next to the one in contact with the TnC molecule. This scheme is one of the simplest interpretations explaining why the reduction in Ca concentration triggers the dissociation of myosin head from the surface of the neighbouring actin molecule.

The type of coordination to Ca was modified so that the constraint in the main chain conformation could be lessened. In the EF hand structure at the high affinity Ca-Mg sites, the carboxylate oxygens of the 1st Asp, the 5th Asp, the 9th Asp and the 12th Glu coordinate to Ca from +X, +Z, -X, and -Z directions, respectively, the amide oxygen of the 3rd Asn coordinates to it from +Y direction, and the carbonyl oxygen of the 7th main chain peptide coordinates to it from -Y direction, as shown in Fig. 2(a). This type of coordination to Ca seems favourable for achieving the highest possible stability constants, because two acid pairs chelate a divalent cation.[21] This type of coordination, however, requires the angle of 100° to 110° between the axes of two α-helices flanking the Ca binding loop. This is quite unfavourable for binding to the surface of the triple-stranded coiled-coil of TM-TnT complex. In my predicted model of the improved tertiary structure of TnC, the carboxylate oxygen of the 1st Asp coordinates to Ca from +X direction, and that of the 12th Glu from -X direction. The other 4 coordinations to Ca are all made by main chain carbonyl oxygen atoms, as shown in Fig. 2(b). There are many examples in the Ca binding structures analysed so far by X-ray crystallography, in which more than two main chain carbonyl oxygens coordinate to Ca. However, this type of coordination has never been

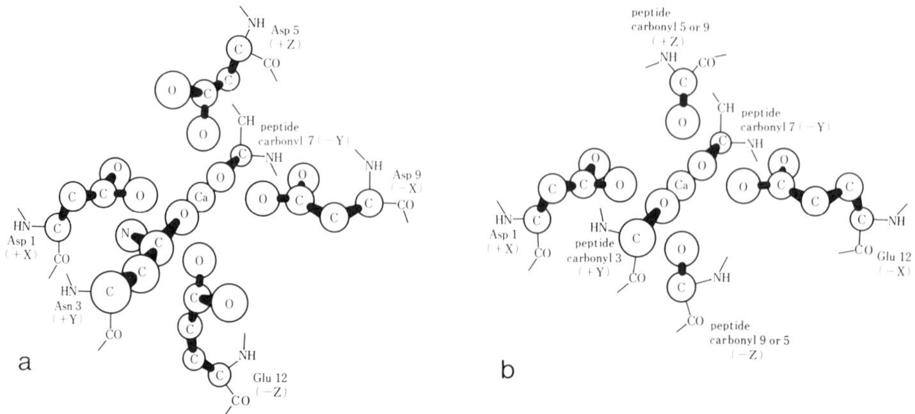

Fig. 2. (a) EF hand type coordination to Ca; (b) Nagano model type coordination to Ca.

observed. The stability constant of this type of Ca binding should be lower than that of the EF hand type, because only one acid pair chelates a Ca ion.[21] When a Ca ion is bound to one of the high affinity Ca-Mg sites, hydrogen-bondings to two neighbouring sites are very favourable. This is the reason for the cooperative phenomenon. Since 4 of the main chain carbonyls take part in chelating Ca in addition to the two carboxylate oxygens of both ends of the 12 residue loop, the stability constants for binding Ca can be expected to be equivalent in all sites of TnC and CaM.

CONFORMATIONAL CHANGE OF CaM AND TnC WHEN NO Ca ARE BOUND

One of the reasons why all sites of CaM and the N-terminal 2 sites of TnC have low affinity for Ca and no affinity for Mg might be that a specific Gly-X-Gly sequence could make a stable γ-turn,[22] which stabilizes a large antiparallel β-sheet in the case of CaM, as shown in Fig. 3(a). In the case of TnC, the CaM-like β-sheet, shown in Fig. 3(b) would not be so stable because of a salt linkage between Lys and Glu residues and also because of two sequences Gly-X-X at sites III and IV. Instead, two separate domains would be more favourable. One domain would easily reform into the high affinity Ca binding conformations in the absence of Mg, while the other keeps the low affinity Ca-specific sites. However, when Mg ions attack the high affinity Ca-Mg sites, the 9th Asp, the 5th Asp and the 3rd Asn could coordinate to Mg ions at the sites III and IV. Two acidic side chains of the 1st Asp and 12th Glu would be free because of the

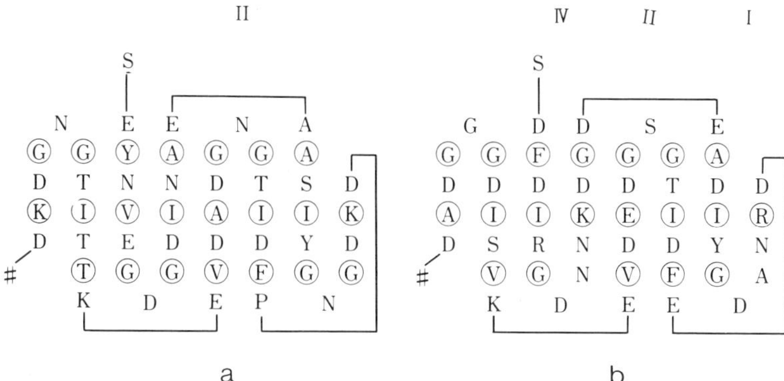

Fig. 3. (a) β-sheet of calmodulin without bound Ca; (b) β-sheet of skeletal troponin C. Side chains of residues in circles direct toward the viewer. # N-terminus. $ C-terminus.

negative charge repulsion. This is favourable for the extended and parallel arrangement of α-helices in my predicted model of TnC. The low affinity sites are extended so that the hydrophobic regions of flanking α-helices of the high affinity sites are covered by those of the low affinity sites. This is the reason for the enhancement of the stability constants of the low affinity sites.

This is how my model of TnC explains the cooperative phenomenon in the regulatory mechanism of muscle contraction.

REFERENCES

1. A. D. McLachlan and M. Stewart, The 14-fold periodicity in α-tropomyosin with actin, J. Mol. Biol. 103:271 (1974).
2. S. Ebashi, Regulatory mechanism of muscle contraction with special reference to the Ca-troponin-tropomyosin system, Essays in Biochemistry 10:1 (1974).
3. I. Ohtsuki, Molecular Arrangement of troponin-T in the thin filament, J. Biochem. 86:491 (1979).
4. I. Ohtsuki, Functional organization of the troponin-tropomyosin system, in:"Muscle Contraction: Its Regulatory Mechanisms," S. Ebashi et al. eds., Japan Sci. Soc. Press, Tokyo/Springer-Verlag, Berlin (1980).
5. K. Kohama, Role of the high affinity Ca binding sites of cardiac and fast skeletal troponins, J. Biochem. 88:591 (1980).
6. Z. Grabarek, W. Drabikowski, P. C. Leavis, S. S. Rosenfeld, and J. Gergely, Proteolytic fragments of troponin C - Interactions with the other troponin subunits and biological activty, J. Biol. Chem. 256:13121 (1981).
7. P. S. Leavis, S. S. Rosenfeld, J. Gergely, Z. Grabarek, and W. Drabikowski, Proteolytic fragments of troponin C - Localization of high and low affinity Ca^{2+} binding sites and inter-

actions with troponin I and troponin T, J. Biol. Chem. 253:5452 (1978).

8. G. W. Amphlett, T. C. Vanaman, and S. V. Perry, Effect of the troponin C-like protein from bovine brain (brain modulator protein) on the Mg^{2+}-stimulated ATPase of skeletal muscle actomyosin, FEBS Lett. 72:163 (1976).
9. K. Nagano, S. Miyamoto, M. Matsumura, and I. Ohtsuki, Possible formation of a triple-stranded coiled-coil region in tropomyosin-troponin T binding complex, J. Mol. Biol. 141:217 (1980).
10. K. Nagano, S. Miyamoto, M. Matsumura, and I. Ohtsuki, Prediction of a triple-stranded coiled-coil region in tropomyosin-troponin T complex, J. Theor. Biol. 94:743 (1982).
11. J.-P. van Eerd and K. Takahashi, Determination of the complete amino acid sequence of bovine cardiac troponin C, Biochemistry 15:1171 (1976).
12. H. Kasai, Y. Kato, T. Isobe, H. Kawasaki, and T. Okuyama, Determination of the complete amino acid sequence of calmodulin (phenylalanine-rich acidic protein II) from bovine brain, Biomed. Res. 1:248 (1980).
13. D. M. E. Szebenyi, S. K. Obendorf, and K. Moffat, Structure of vitamin D-dependent calcium-binding protein from bovine intestine, Nature 294:327 (1981).
14. R. M. Tufty and R. H. Kretsinger, Troponin and parvalbumin calcium binding regions predicted in myosin light chain and T4 lysozyme, Science 187:167 (1975).
15. R. H. Kretsinger and C. E. Nockolds, Carp muscle calcium-binding protein II. Structure determination and general description, J. Biol. Chem. 248:3313 (1973).
16. R. H. Kretsinger and C. D. Barry, The predicted structure of the calcium-binding component of troponin, Biochim. Biophys. Acta 405:40 (1975).
17. M. Tanokura, Y. Tawada, and I. Ohtsuki, Chymotryptic subfragments of troponin T from rabbit skeletal muscle. I. Determination of the primary structure, J. Biochem. 91:1257 (1982).
18. I. Ohtsuki and K. Nagano, Molecular arrangement of troponin-tropomyosin in the thin filament, Adv. Biophys. 15:93 (1982).
19. K. Nagano and I. Ohtsuki, Prediction of approximate quaternary structure of troponin complex, Proc. Japan Acad. 58B(3):73 (1982).
20. K. Kohama, Divalent cation binding properties of slow skeletal muscle troponin in comparison with those of cardiac and fast skeletal muscle troponins, J. Biochem. 86:811 (1979).
21. R. E. Reid and R. S. Hodges, Co-operativity and calcium/magnesium binding to troponin C and muscle calcium binding parvalbumin: An hypothesis, J. Theor. Biol. 84:401 (1980).
22. B. W. Matthews, The γ-turn. Evidence for a new folded conformation in protein, Macromolecules 5:818 (1972).

POTASSIUM LOSS AND WATER UPTAKE OF STIMULATED FROG MUSCLES

László Nagy and Péter Práger

Biophysical Institute
Medical University
Pécs, Hungary

INTRODUCTION

Ernst and Csucs (1929) were the first to demonstrate that when stimulating the muscle directly - i.e. when the stimulating current passes through the whole muscle - changes take place in the K^+-Na^+-Cl^- and P- content. Mond and Netter (1930) as well as Achelis (1932) reported similar results. Hodgkin and Horovitz (1959) demonstrated Na^+-K^+ exchange in single muscle fibers stimulated directly. Tigyi (1956) investigated isolated frog sartorii in a Ringer solution containing K^{42} radioactive isotope. Besides potassium loss, the directly stimulated muscles showed a significant increase in specific activity. Later different authors obtained conflicting results. We decided, therefore, to reinvestigate the problem. Changes of the K^+ and water content of frog muscle caused by the electric stimulating current were measured in order to obtain new data about the nature of these changes. We used blockers of ionic currents and of metabolism to find out their effect on the changes mentioned above.

METHOD

We performed our measurements on the m. gastrocnemius of the frog Rana esculenta. Square pulses were used as stimuli, their time duration was 0.1 ms, their amplitudes 0.15; 0.30; 0.45; and 0.75 mA. They were supplied by a current pulse generator of a high impedance. The positive pole of the generator was contacted with the muscle and the negative one was contacted with the test solution (10 ml). As a control we used a muscle kept in the same solution but it was not stimulated. The stimulus threshold was about 0.2 mA, and stimuli having amplitudes higher than 0.6 mA elicited an action potential of a maximum amplitude. Recording of action potentials of the muscle were carried out through a suction electrode, not shown in the figure (Fig. 1).

At each current intensity we performed measurements in 10 stimulated and 10 control muscles. The stimulation went on for 10 minutes with electric pulses of 10 cps. We calculated the water uptake from the increase in the mass of the muscle. The loss of potassium was determined with flame-photometer from the increase of K^+-concentration in the Ringer's solution. The sum of this increase and the potassium-content of the muscle ash gave the total potassium-content of the muscle.

The stimulus threshold of the motor axons among the muscle fibers is much lower than that of the muscle fibers. We could avoid the "indirect" stimulation of the muscle through these nerves by blocking the end-plates by a curare solution of a concentration of 10 mM/l. The various drugs (lidocaine, procaine, TEA, etc....) were applied in a concentration which was high enough to block action potentials and contraction.

The potassium and water content of the control muscles were not changed over a time range of four hours which was much longer than that of the test experiments.

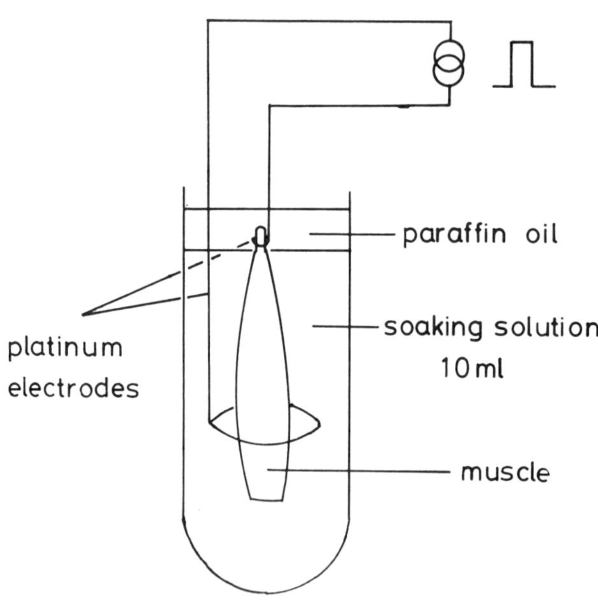

Fig. 1.

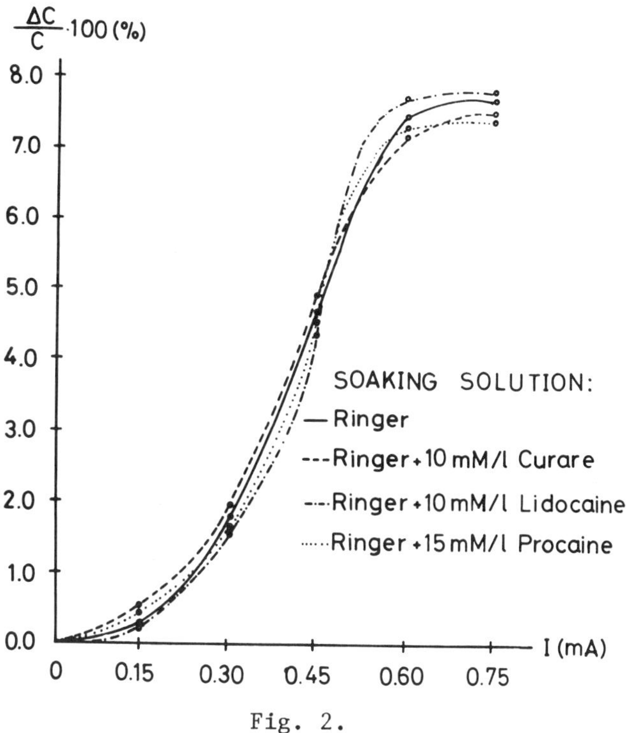

Fig. 2.

RESULTS

The K^+ loss from the muscles is plotted against the stimulating current intensity in Fig. 2 in case of different test solutions. The blockers used in our experiments did not cause any significant change ($p = 0.01$) in the slope of the curves.

As it was found in the case of the K^+ loss no significant difference can be observed between the relative water contents of muscles stimulated in normal Ringer's and those treated with local anesthetics under the influence of electric current pulses, though the action potential and the muscle contraction is blocked by the local anesthetics. One can see in the last two figures that the potassium loss and the water uptake increased when the amount of the charge driven by the stimulating current increased and they were saturated at a level of 7-8%.

The experiments showed that the K^+ loss was produced by the stimulating current in muscles inhibited by ouabain. We obtained the same results when examining the muscles in a Ringer's solution with a

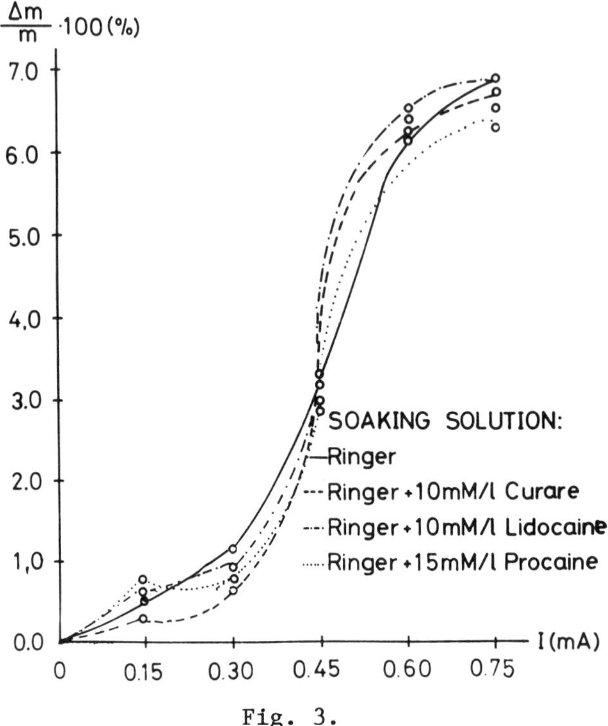

Fig. 3.

TEA content of a concentration of 10 mM/l, i.e. blocking selectively the K^+ channels of the membrane (Fig. 4).

The increase of the water content is not caused by an increased Na^+ influx because when stimulating the muscles in an Na^+-free choline-chloride Ringer's the increase of the water content does not differ from that demonstrated in Fig. 3.

The results of our experiments cannot support the assumption that the loss of K^+ and the water uptake in the muscle are caused by the effect of the electric current on the active transport.

The iontophoretic effect of the electric current does not explain the increased ion loss because when we changed the polarity of the stimulating current the effect remained the same (Fig. 5).

We made further calculations to learn how much K^+ could pass through the membrane under the effect of the applied amount of charge. The amount of K^+ calculated in this way proved to be considerably lower than that determined in the experiments.

ΔC/C ·100(%)
POTASSIUM LOSS FROM MUSCLE
10.0
9.0
8.0
7.0
6.0
5.0
4.0
3.0
2.0
1.0
0.0
0.15
0.45
0.75
I(mA)
SOAKING SOLUTION:
Ringer
Ringer+TFA
Ringer+Ouabain
Ringer+Choline Chloride

Fig. 4.

POLARITY OF THE MUSCLE ELECTRODE
+
—
+
—
7.7 %
7.8 %
6.7 %
6.8 %
POTASSIUM LOSS
WATER UPTAKE

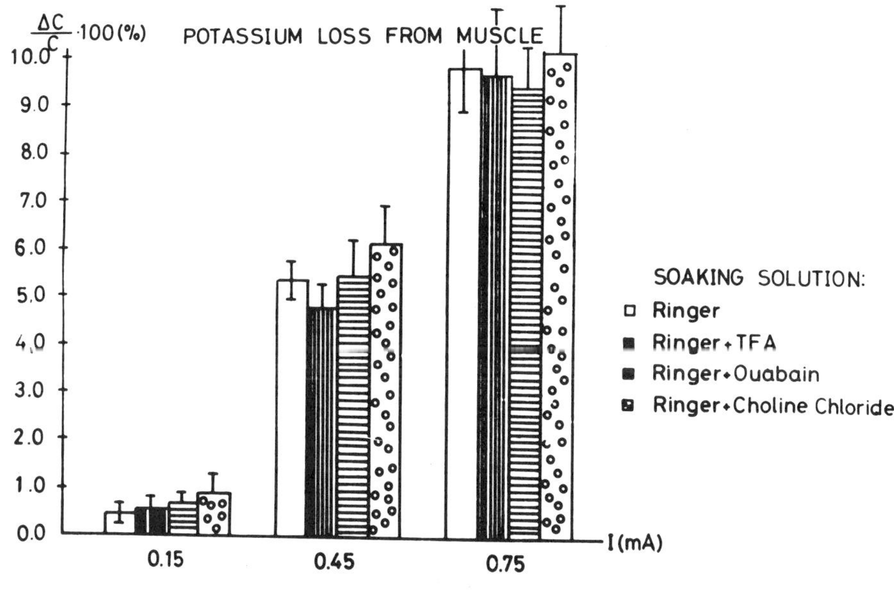

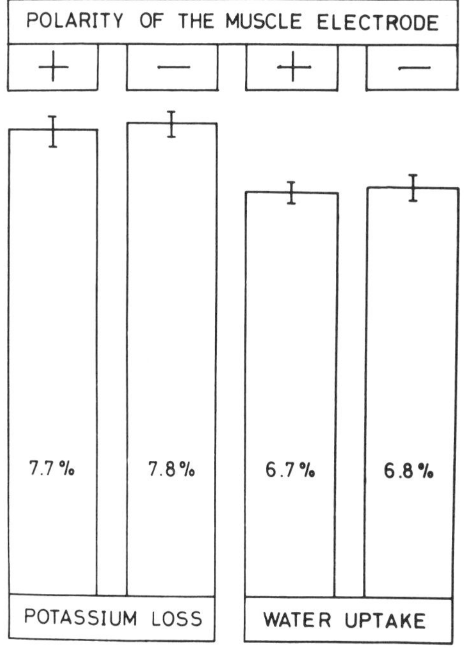

Fig. 5.

DISCUSSION

The K^+ loss found when stimulating current passes through the membrane cannot be explained by supposing that the excitatory processes bring it about, because the same loss was found when action potential was blocked by local anesthetics or the muscle was treated with TEA, known as a blocker of K^+-channels. Neither did the ouabain (an inhibitor of the potassium-ATPase) influence the K^+ loss which comes about during a "direct" stimulation.

The water uptake cannot be due to an increased Na^+ inward current, because the water uptake was not changed when Na^+ was replaced by non-permeating choline ions.

What remains is to suppose that the electric current opens non-selective ionic channels or it makes the membrane permeable in a way more or less repairable.

Apart from these speculations it seems certain that stimulating currents regarded as "physiological" can increase the ionic and water permeability of the membrane in a non-physiological way and this phenomenon must be taken into consideration when attempting to explain experimental results obtained by using "direct" stimulation by electric current.

REFERENCES

Achelis, J. D., 1932, Über die Polarisationskapacität ("Permeabilität") des Skeletmuskels bei indirekter Reizung, Pflügers Archiv für die gesamte Physiologie, 230:412.

Ernst, E., and Csucs, L., 1929, Untersuchungen über Muskelkontraktion, IX. Mitt. Permeabilität und Tätigkeit, Arch.Ges. Physiol., 223:663.

Hodgkin, A. L., and Horowitz, P., 1959, Movements of Na and K in single muscle fibers, J.Physiol., 145:505.

Mond, R., and Netter, H., 1930, Andert sich Ionpermeanilität des Muskels während seiner Tätigkeit? Pflügers Arch.Ges.Physiol., 224:702.

Tigyi, J., 1956, K-Na-P izotópcsere a müködö sartoriusban, Kisérletes Orvostudomány, 8:105 (in Hungarian).

DEPENDENCE OF THE ELECTRICAL AND CONTRACTILE ACTIVITIES OF THE GASTRIC SMOOTH MUSCLE ON CA IONS

M. Papasova and K. Boev

Institute of Physiology
Bulgarian Academy of Science
Bulgaria

In spite of the opinion of many authors that the excitation-contraction coupling in the smooth muscle is effected in a way similar to that characteristic of the striated muscle, recent data have shown that the smooth muscle has some peculiarities of the excitation-contraction coupling. Thus the smooth muscle contraction is not always connected with the appearance and the pattern of the action potentials. In the vascular smooth muscle Somlyo and Somlyo[1] have observed contractions which are not connected with spike potentials. Such contractions we have recorded from parts of the digestive tract. Of interest in this respect are the smooth muscles making up the different regions of the stomach. Whereas from the smooth muscles of the antrum and the corpus of the stomach of cat[2,3] and man[4] spontaneous rhythmic slow potentials, type plateau (we will call them plateau action potentials, PAP) are recorded, the smooth muscle of the stomach fundus is characterized by slow changes in the membrane potential connected with tonic contractions. The objective of the present investigation is the role of Ca^{2+} in the realization of the excitation-contraction coupling in the smooth muscles of the antrum and fundus of the cat stomach.

The electrical and contractile activities were recorded by the single or double sucrose gap technique[5,6].

The PAP recorded from the antrum smooth muscle consists of an initial fast component with amplitude of 20-30 mV/s and duration of 770±15 ms followed by a second slow component which determines the PAP plateau. The duration of PAP varies from 5 to 10 s. PAP are recorded from the antrum of cat as well as of dog and man. The PAP always triggers contractions even in the absence of spike potentials[7]. The amplitude and duration of the phasic contractions characteristic of the antrum of the stomach are determined by the amplitude and duration of the second component of the PAP. The appearance of the spike potentials on the second component of the PAP leads to an increase in the contraction amplitude. Na^+ and Ca^{2+} are necessary for the generation of PAPs. Furthermore, the first component of the PAP is mainly Na^+-dependent and the second component is Ca^{2+}-dependent[3]. The removal of Ca^{2+} from the nutrient solution decreases the amplitude and duration of the second component of the PAP and respectively the contraction amplitude. At the 5th minute after the Ca^{2+} removal both the second component of the PAP and the contractions are inhibited (Fig. 1).

A delay also occurs in the velocity of the increase of the first component of the PAP. This fact is of particular interest because earlier investigations have shown that excitation-contraction coupling in the antrum smooth muscles could be accomplished only when the velocity of the increase of the initial phase of the PAP is higher than 25 mV/s[8]. The PAP is very sensitive to Ca^{2+} antagonists. Thus nitroprusside sodium at a concentration of 10^{-6} M decreases the amplitude of the second component of the PAP and the phasic contraction and at a concentration of 10^{-5} completely abolished them. Mn^{2+} at a concentration of 2×10^{-3} M as early as the 5th minute inhibits the second component of the PAP and the related phasic contractions. Verapamil (10^{-6} M) or D600 (10^{-6} M) disturbs the coupling of the excitation and contractile processes, as at the 15-20th minute the contractions disappear while the PAP, is still present though with a decreased amplitude and duration. The fact that the PAP and the related contraction of the stomach antrum are

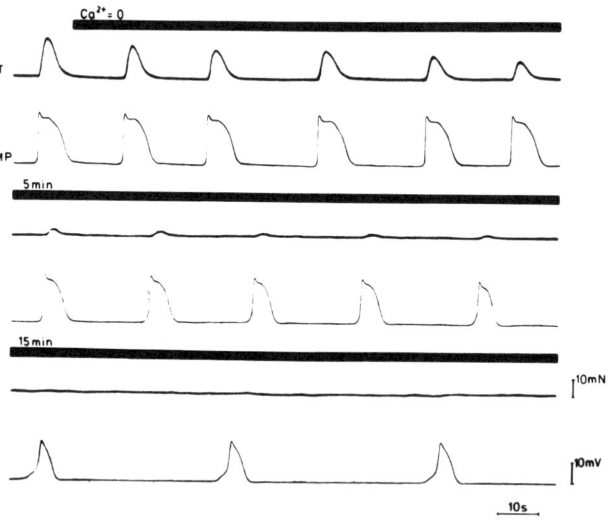

Fig. 1. Importance of Ca^{2+} for the generation of plateau action potentials and contractions from the antrum smooth muscle.

sensitive to Ca^{2+} and Ca antagonists suggests the existence of slow Ca channels in the smooth muscle cell membrane similar to the slow Ca channels in the cardial muscle cells. Probably as in the cardiac muscle[9] the slow Ca channels in the antrum smooth muscle are activated during the development of the PAP.

PAPs could be evoked by direct electrical stimulation. At threshold value of the current there appears a local response upon switching on of the depolarizing current as well as upon switching off the hypolarizing current. At suprathreshold value of the current (1,5 mA) the local response transforms into evoked PAP (Fig. 2). The amplitude of the evoked PAPs approaches that of the spontaneously generated PAPs and might reach 20-30mV. The evoked PAPs have the same shape and relation to the contractile process as the spontaneous PAPs. The voltage-current relationship is linear and shows abnormal rectification. The relation between the second component of PAP and the contractile process is illustrated on Fig. 3. Depolarizing current applied during the second component of PAP leads to membrane depolarization followed by an increase of the contraction (Fig. 3A). Of the same time membrane hyperpolarization causes a reduction of the second component of PAP and relaxation of the muscle (Fig. 3B). These findings could be explained by the activatioan of the slow Ca channels during membrane depolarization and their inactivation during membrane hyperpolarization. Thus the PAP controls the excitation-contraction coupling during contraction and relaxation in the antrum smooth muscle. In case of low extracellular Na the contraction disappears - only slow sinusoidal membrane potential changes are observed. On this background ACh (10^{-7} g/ml) leads to appearance of spike potentials. (Fig. 4).

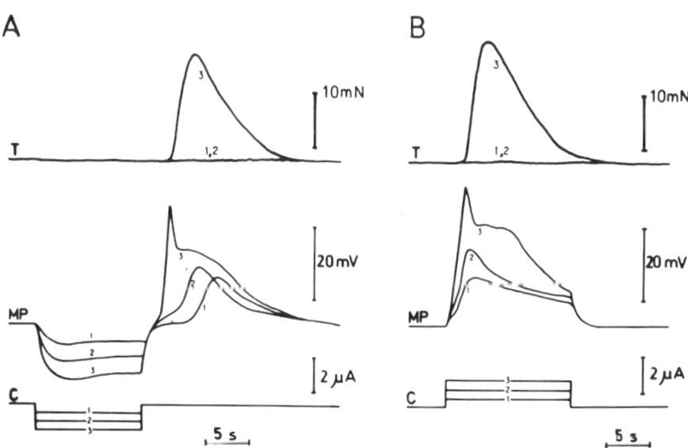

Fig. 2. Evoked action potentials elicited by inward (A) and outward (B) currents. 1 and 2 - local responses to subthreshold current; 3 - evoked plateau action potentials and contractions upon suprathreshold current; T - contractions; MP - membrane potential; C - current.

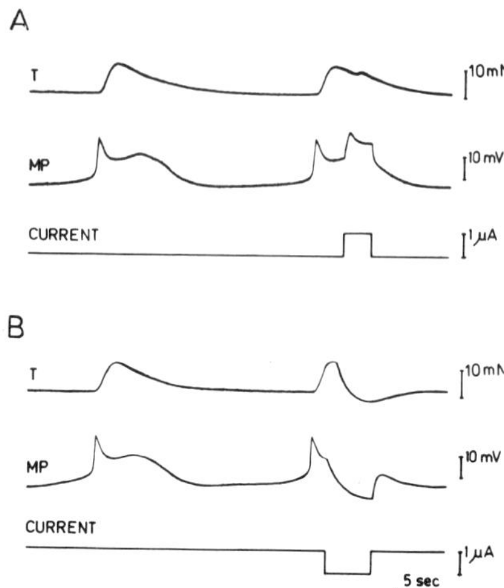

Fig. 3. Relationship between the second component of the potentials and contractions. A - inward current; B - outward current. Designations are as in Fig. 2.

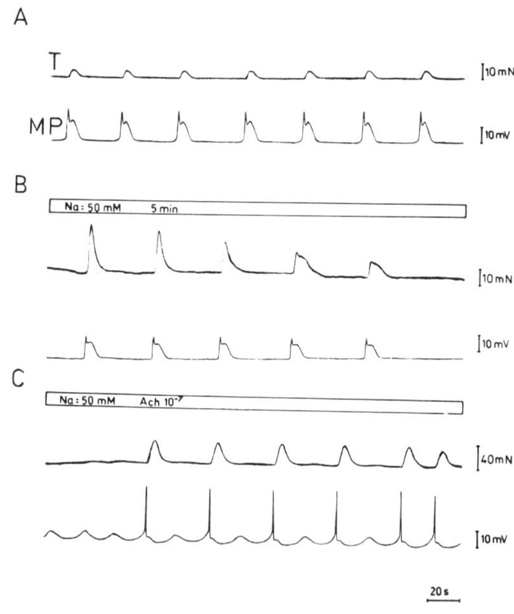

Fig. 4. Changes in the plateau action potentials by the decreased Na^+ concentration in the medium. A - background; B - 5 min. after 50 mM Na^+; C - 15 min. later there appeared sinusoidal slow waves. ACh produced spike potentials connected with contractions.

In Na-free medium the spontaneous electrical and contractile activities are completely inhibited, but the suprathreshold electrical stimulation induces spike potentials and verapamil - sensitive high - amplitude phasic contractions. Spike potentials could also appear on the PAPs after treatment of the antrum smooth muscle with TEA and 4-AP. It is known that these substances selectively block the K outward current and activate the Ca inward current[10,11]. Thus Fig. 5 shows that within two to five minutes after treatment with TEA (5×10^{-3} M) even with the background of Atropine (10^{-6} M) there appear spike potentials on the second component of PAP, leading to an increase of the contraction amplitude. Spike potentials with a relatively higher frequency are observed under the effect of 4-AP. All this suggests the existence not only of slow but also of fast Ca channels in the antrum smooth muscle cell membrane.

The smooth muscles making up the stomach fundus show essential differences in the character of the electrical and contractile activities as compared with the smooth muscles of the antrum. Some of the fundic preparations exhibit spontaneous electrical and contractile activities while others are spontaneously inactive. The latter are characterized by slow changes in the membrane potential and tonic contractions. The spontaneously active preparations are characterized by slow sinusoidal waves with a frequency of 1-2 cpm leading to contraction only when accompanied by spike potentials. Unlike the smooth muscles of the corpus and antrum of the stomach which respond according to the "all-or-none" low to electrical stimu-

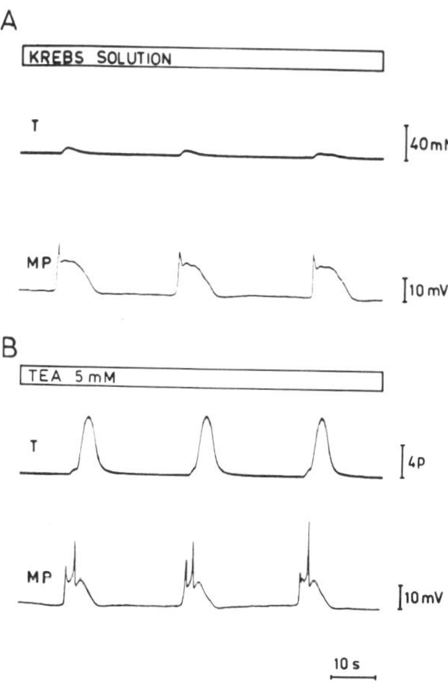

Fig. 5. Effect of TEA on the electrical and contractile activities of the antral smooth muscle. A - background; B - 5 min. after TEA.

lation, the fundic smooth muscles show a slight electrical excitability. Direct electrical stimulation of spontaneously inactive fundic preparations induces gradual depolarization as local responses are observed at higher current values. The local responses do not develop into action potentials even at high current values. The membrane depolarization is accompanied by tonic muscle contractions (Fig. 6A). These spike-free tonic contractions as well as the relaxation depend on the changes in the membrane potential. The spike-free contractions characteristic of the fundic smooth muscle disappear in the absence of Ca^{2+} as well as under the effect of the Ca antagonists D600 and nitroprusside sodium. In Ca^{2+} free medium the tonic contractions produced by direct electrical stimulation also disappear but the electrical response of the cell membrane to polarizing and depolarizing currents is not changed (Fig. 6B). This suggests that in the fundic smooth muscle extracellular Ca^{2+} is of essential importance for the realization of the excitation-contraction coupling. Most probably the tonic contraction characteristic of the fundic smooth muscle is effected by the Ca^{2+} influx into the cell membrane. On the other hand, as in the antrum smooth muscle TEA or 4-AP increases the excitability of the fundic smooth muscle. From Fig. 7 it is clear that with the background of TEA treatment direct electrical stimulation leads to the appearance of spike potentials and related contractions. This dependence is observed under the effect of 4-AP too. TEA or 4-AP in higher concentrations might induce spontaneous spike potentials and phasic contractions.

The fundic smooth muscle responds to ACh with spike free tonic contractions. These contractions are verapamil-resistant but are sensitive to nitroprusside sodium[12]. The different dependence of the fundic tonic contractions of different blockers as well as the possibility that fundic smooth muscles would trigger phasic contractions as well suggests that three types of Ca channels operate in the cell membrane of the fundic smooth muscle: (1) high threshold, fast, regenerative Ca channels, determining the fast inward Ca

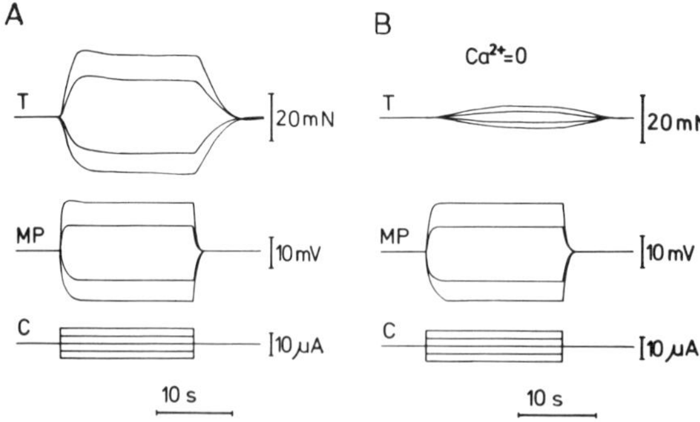

Fig. 6. Electrical excitability of the fundic smooth muscle. A - in Krebs solution; B - in Ca^{2+} free modium.

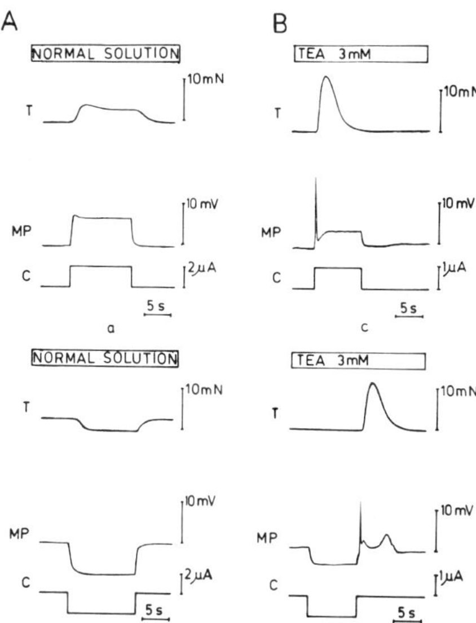

Fig. 7. Changes in the electrical excitability of the fundic smooth muscle under the effect of TEA. A - background; B - 10 min. after TEA.

current described by Bury and Boev[13]. It has been shown that similar to the other excitable membranes in the cell membrane of the fundic smooth muscle, too, these channels are activated by depolarization and are self-inactivating. The slight electrical excitability of the fundic smooth-muscle cells is determined by the early activation of the K outward current[13]. This is the reason for the absence of spontaneous electrical and contractile activity in the majority of fundic smooth-muscle preparations. Thus the inhibition of the K outward current under the effect of TEA, 4-AP leads to the appearance of spike potentials and phasic contractions of the spontaneously inactive fundic preparations; (2) membrane potential-dependent slow Ca channels which are not self-inactivating as are the fast regenerative Ca channels. They have a lower threshold of activation and a lower ion selectivity compared with the fast Ca channels. The slow Ca channels are verapamil-sensitive and determine the spontaneous myogenic tone of the fundic smooth muscle, the potential dependent tonic contractions and the verapamil-sensitive component of the ACh induced tonic contractions. Unlike the fast Ca channels the slow Ca channels are completely blocked by nitroprusside sodium; and (3) chemosensitive Ca channels or receptor-operating Ca channels according to Bolton[7].

Depending on the activation or inactivation of the different types of Ca channels characteristic of the cell membrane of the fundic smooth muscle there will appear either phasic or tonic contractions. Both types of contractions depend on the concentration of extracellular Ca^{2+}. The activation of the fast regenerative Ca channels and the release of intracellular Ca^{2+} as a results of membrane depolarization during the action potential lead to phasic contractions. The activation of the slow noninactivating Ca channels causes the occurrence of spike-free tonic contractions characteristic of the fundic smooth muscles. Furthermore, the verapamil-resistant component of the tonic contraction of the fundus is determined also by the membrane potential - independent Ca^{2+} influx through the receptor-operating Ca channels.

REFERENCES

1. A. P. Somlyo and A. V. Somlyo, J.Pharmacol.Exp.Ther., 159:129-145 (1968).
2. E. Daniel and K. Chapman, Am.J.Dig.Dis., 8:54-102 (1963).
3. M. Papasova, T. Nagai, and L. Prosser, Am.J.Physiol., 214:697-702 (1968).
4. M. Papasova, I. Altaparmoakov and K. Boev, C.R.Acad.Bulg.Sci., 25:545-548 (1972).
5. K. Boev, C.R.Acad.Bulg.Sci., 24:933-936 (1971).
6. K. Boev and K. Golenhofen, Pflügers Arch., 349:277-283 (1974).
7. M. Papasova and K. Boev, "Physiology of Smooth Muscle," E. Bülbring and M.F. Shuba, eds., Raven Press, New York, pp. 209-226 (1976).
8. V. Kochemasova, M. Shuba, and K. Boev, C.R.Acad.Bulg.Sci., 22: 1437-1440 (1969).
9. M. Kolhardt, B. Bauer, H. Krause, and A. Fleckenstein, Pflügers Arch., 335:309-322 (1972).
10. H. Inomata and T. Suzuki, 53:215-219 (1977).
11. M. Pelhate and Y. Pichon, J.Physiol.(London), 242:90P-91P (1974).
12. K. Boev, K. Golenhofen, and J. Lukanov, "Physiology of Smooth Muscle," E. Bülbring and M.F. Shuba, eds., Raven Press, New York, pp. 203-209 (1976).
13. V. Buryi and K. Boev, Experientia, 36:216-218 (1980).
14. T. B. Bolton, Physiol.Rev., 59:606-718 (1979).

CALCIUM-EXTRUSION PUMP

OF THE SMOOTH MUSCLE CELL MEMBRANE

L.M. Popescu and P. Ignat

Department of Cell Biology & Histology
Faculty of Medicine
Bucharest 35, Romania

STATEMENT OF THE PROBLEM

Ca^{++} Extrusion from the Smooth Muscle Cell is Obligatory because there is a (Dis)Continuous Ca^{++} Entry into the Smooth Muscle Cell and Ca^{++} Accumulation in Cellular Organelles is Limited

All Ca^{++}-flux studies have shown that the external membrane (sarcolemma) of the smooth muscle cell is permeable to Ca^{++} [1] and, therefore, due to the large electrochemical gradient ($\Delta\tilde{\mu}$ Ca^{++} of about 9 Kcal/mol Ca^{++}) an inward "basal" Ca^{++} leak appears obvious. This passive Ca^{++} leak is abruptly supplemented during activity, since the smooth muscle contraction is initiated by Ca^{++} entry into the cell through voltage-dependent and/or receptor-operated Ca^{++} channels. However, because there is a natural maximum limit of Ca^{++} sequestration in cellular organelles (Fig. 1), the maintenance of intracellular Ca^{++} homeostasis requires the existence of a sarcolemmal mechanism which ejects Ca^{++} out of the cell to balance Ca^{++} entry. The absolute necessity of Ca^{++} extrusion can be deduced as follows: (i) the contraction of one smooth muscle cell is associated with a Ca^{++} influx of about 0.05 femtomoles [2]; (ii) let us consider then, that there would be no Ca^{++} extrusion from the cell and that the relaxation would be promoted by excess-Ca^{++} accumulation into the sarcoplasmic reticulum; (iii) if this would be the case, then, after only about 100 successive contractions, when the Ca^{++}- storage capacity of the sarcoplasmic reticulum would be supersaturated, the smooth muscle would remain in a permanent state of contraction, which is simply biological nonsense.

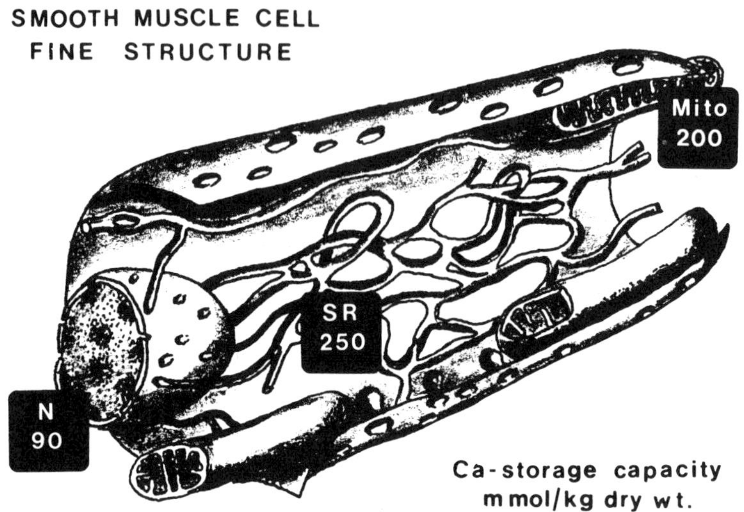

Fig. 1. The main cellular organelles which are able to sequester large amounts of Ca^{++}: SR, sarcoplasmic reticulum; Mito, mitochondria; N, nucleus. Their maximal capacity of Ca^{++} storage was determined after Ca^{++}-loading experiments, either in situ by electron probe microanalysis or on isolated subcellular fractions by atomic absorbtion spectrometry (Popescu et al.[3-5]). Somlyo et al.[6] also found by x-ray microanalysis a high Ca concentration of about 250 mmol/kg dry weight in some microdomains of the smooth muscle SR. Apparently, the Ca^{++}-storage capacity of the nucleus is (too) large. However, this could be ascribed to the newly-discovered nuclear Ca^{++}-binding proteins[7], besides the well-known Ca^{++}-binding to nucleic acids

To ensure that Ca^{++} does not accumulate intracellularly, two mechanisms might be considered in smooth muscle sarcolemma: (i) the Na^{+}/Ca^{++} exchange carrier, where the energy to move Ca^{++} out of the cell is derived from the electrochemically favoured Na^{+} entry and (ii) the ATP-driven Ca^{++}-extrusion pump, represented by a Ca^{++}-ATPase. However, it is becoming evident that Ca^{++} is ejected from the smooth muscle cell mainly by an ATP-dependent mechanism rather than via the Na^{+}/Ca^{++} exchanger[8,9]. We will review here briefly recent research in our laboratory focused on the isolation and characterization of the Ca^{++}-pump ATPase from smooth muscle sarcolemma.

Criteria to Define a Plasma Membrane Ca^{++}-Pumping ATPase

According to Penniston[10] there are three fundamental criteria: (1) the very origin from the plasma membrane: (ii) the high Ca^{++}-affinity of ATPase activity and (iii) the high Ca^{++}-affinity of Ca^{++} transport. In addition, some properties are now generally accepted as functional markers for the plasmalemmal Ca^{++}-pump ATPase[10,11]: inhibiton by low concentrations of vanadate, direct stimulation by calmodulin, activation by proteolysis, strictness of their requirement for ATP,etc.

EXPERIMENTAL APPROACH

Obtaining an Isolated Sarcolemmal Fraction

Our method for isolating myometrial sarcolemmae is evolved from the method described by Oliveira and Holzhacker[12] for the isolation of ileum smooth muscle cell membrane. In essence, the technique consists in the separation and disruption of smooth muscle cells to obtain "cell ghosts" (Fig. 2), followed by the evacuation of residual intracellular content to obtain empty sarcolemmal sheaths or sheets (Fig. 3). The details of our procedure are presented elsewhere[13]. However, it should be noted that the method produces open sarcolemmal sheaths (and sheets), which do not permit the study of Ca^{++} transport. The measurements of ATP-driven Ca^{++} transport would require the formation of closed plasmalemmal vesicles like those obtained in usual microsomal fractions[14,15], but it is obvious that in this case the SR vesicles, which have their own and quite different Ca^{++} pump, are contaminating such specimens*. To prepare calmodulin-depleted sarcolemmae, we used an adaptation of the procedure described by Caroni and Carafoli[17].

Solubilization of Sarcolemmal Ca^{++}-Pump ATPase

SDS was found the most efficient detergent when the solubilizing power of a detergent series was estimated as a function of the detergent-to-protein ratio[18]. We solubilized the isolated sarcolemmae with a high concentration of SDS for a short time period[13]. This treatment resulted in a partial solubilization of sarcolemmae, since the three-layered pattern of the membrane was still evident in the electron microscope.

---

* In smooth muscle cells the total surface area of SR over the total area of the cell membrane is about 0.5-1[16]. Therefore, it seems a personal option whether One looks at a crude microsomal fraction as being a plasma membrane fraction contaminated with SR or, vice versa, as being an SR fraction contaminated with plasma membrane !

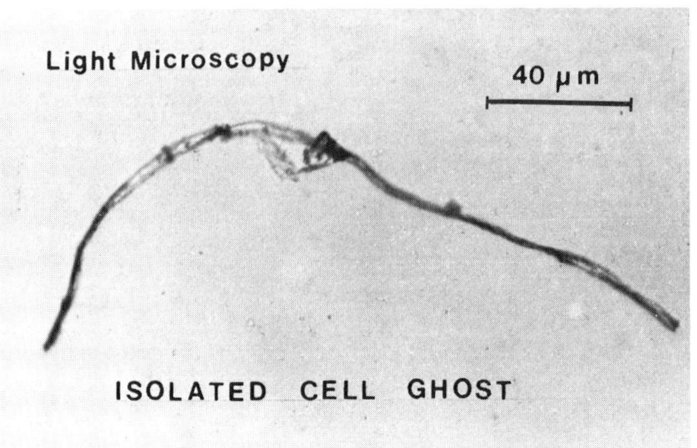

Fig. 2. Smooth muscle cell ghost isolated from human myometrium. Note the incomplete evacuation of the intracellular content. This is an intermediary step in obtaining sarcolemmal sheaths. The specimen was not fixed, sectioned or stained.

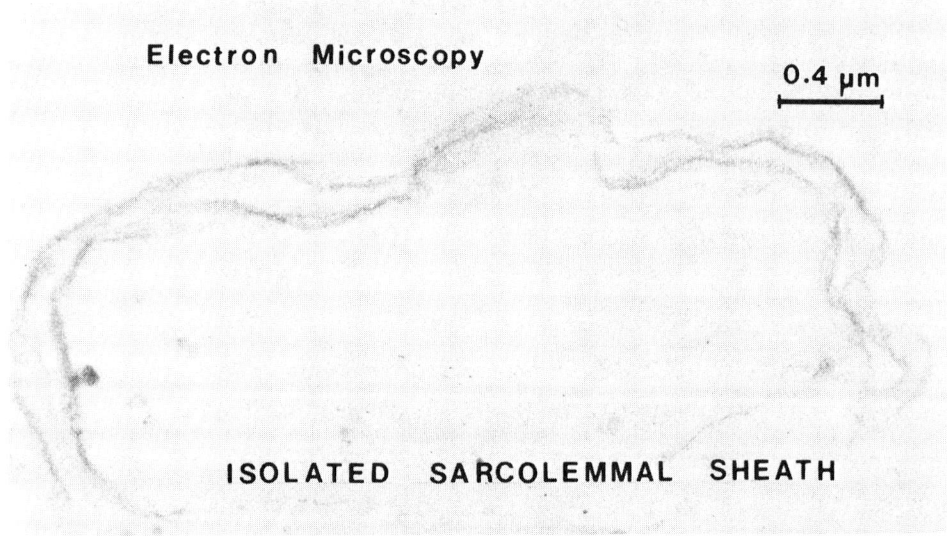

Fig. 3. Sarcolemmal sheath isolated from human myometrium. Compare with Fig. 2 and note the complete absence of intracellular content. Typically, such empty sarcolemmal tubes appear flattened and torsionated, and show membrane veils. The specimen was conventionally prepared for electron microscopy.

The Standard Assay Medium for Ca^{++}-ATPase:

a Compromise between the Ideal and Actual Conditions

From a physiological point of view, the main problem in making an assay medium for Ca^{++}-ATPase is obtaining as-good-as-possible a simulation of the in situ microenvironment of the enzyme. In practice, however, we achieved a reasonable compromise between the available information on the ionic composition of cytosol and the inevitable biochemical restrictions imposed by such an assay (e.g. the presence of EGTA to obtain definite pCa values). For calculating the total concentrations necessary to obtain specified free concentrations of Ca^{++}, Mg^{++}, K^{+} and Na^{+} in the presence of ATP and EGTA, we took advantage of the new computer programs[19] which transcended the considerable disagreement existing in the literature over the stability constants. Basically, two types of assay conditions were used: either a standard medium (pCa 6) or media with various free Ca^{++} concentrations, the other ionic species being kept constant. The free concentrations in the standard assay medium were (mM): Ca^{++}, 0.001; Mg^{++}, 0.5; K^{+}, 74; Na^{+}, 7.6; ATP, 0.4; pH 7.4. ATP hydrolysis was monitored spectrophotometrically according to Roos et al.[20].

IDENTIFICATION AND CHARACTERIZATION OF Ca^{++}-PUMP ATPase

To asses the purification of Ca^{++}-pump ATPase, the specific activity of the enzyme was successively measured in cell homogenates, cell ghosts, isolated sarcolemmae and sarcolemmal solubilisates, at pCa 6, using the standard assay medium. Indeed, a progressive increase of enzyme specific activity paralleled the isolation procedure. The overall purification factor for the detergent-solubilized enzyme was about 100 times.

Sarcolemmal Ca^{++}-pump ATPases isolated from myometria have a high Ca^{++}-affinity, expressed by apparent $K_m(Ca^{++})$ values in the submicromolar domain (Table 1). The maximum specific activity was found in the micromolar range of free-Ca^{++} concentrations. This fact seems to make perfect functional sense for an active Ca^{++}-extrusion pump because during the maximal contractions of smooth muscle myoplasmic Ca^{++} ions reach micromolar concentrations. The dependence of sarcolemmal Ca^{++}-ATPase activity on the ATP concentration is also functionally relevant. The apparent K_m(ATP) of 20-30 μM, found at pCa 6, suggests that the Ca^{++}-extrusion ATPase has the capacity to deal with increasing Ca^{++}-levels in cytosol even in the presence of very low ATP concentrations, which are clearly below the habitual intracellular concentration of ATP in smooth muscle. In contrast to the Ca^{++}-pump ATPase of SR, which can use efficiently CTP, GTP or UTP as substrates, the Ca^{++}-pump ATPase of smooth muscle sarcolemma is

Table 1. Kinetic Parameters of Sarcolemmal Ca^{++}-ATPases Isolated from Myometrium Smooth Muscle

Source of Ca^{++}-ATPase	Apparent K_m[a] (Ca^{++}) μM	Specific Activity[b] munits[c]
Human myometrium		
- nonpregnant	0.25 ± 0.04	20
- pregnant	0.19 ± 0.08	83
Monkey myometrium		
- nonpregnant	0.25 ± 0.17	17
- pregnant	0.31 ± 0.07	56

[a]In the presence of endogenous calmodulin.
[b]The average specific activity corresponds to pCa 6.
[c]Milliunits, nmoles P_i/mg protein/min.

virtually dependent on ATP. For instance, at pCa 6, CTP hydrolysis was about 1% of that displayed by ATP (L.M. Popescu and E. Toescu, unpublished observation).

To differentiate the sarcolemmal Ca^{++}-ATPase and the so-called basal Mg^{++}-ATPase (responsive to Mg^{++} alone) the ATPase activity was measured in the virtual absence of Ca^{++} (pCa 9), but in the presence of a total Mg concentration of 3.95 mM (free Mg^{++}: 0.5 mM). The remaining Mg^{++}-ATPase activity was less than 1% of the fully stimulated Ca^{++}-ATPase. The ouabain-sensitive component of the ATPase activity was defined as Na^{+}-K^{+}-ATPase. However, at pCa 6, ouabain 0.1 mM did not produce a significant inhibition.

Although no specific inhibitor of plasma membrane Ca^{++}-ATPase is available, orthovanadate is useful in distinguishing the sarcolemmal Ca^{++}-ATPase from the SR Ca^{++}-ATPase because the vanadate sensitivity of sarcolemmal Ca^{++}-ATPase is much higher[11,17]. Indeed, when the activity of sarcolemmal Ca^{++}-ATPases, isolated from human myometrium, was measured at pCa 6, the half maximal inhibition by vanadate was found at 0.8 μM. The concentration of 2 μM vanadate inhibited about 70% of Ca^{++}-ATPase activity and 10 μM vanadate produced the complete inhibition. These findings are to be compared with the following data: Ca^{++}-ATPase of skeletal muscle SR has a $K_{1/2}$ (vanadate) of 10 μM[21] and the $K_{1/2}$ (vanadate) of Ca^{++} uptake by SR vesicles isolated from smooth muscle is 12 μM[15].

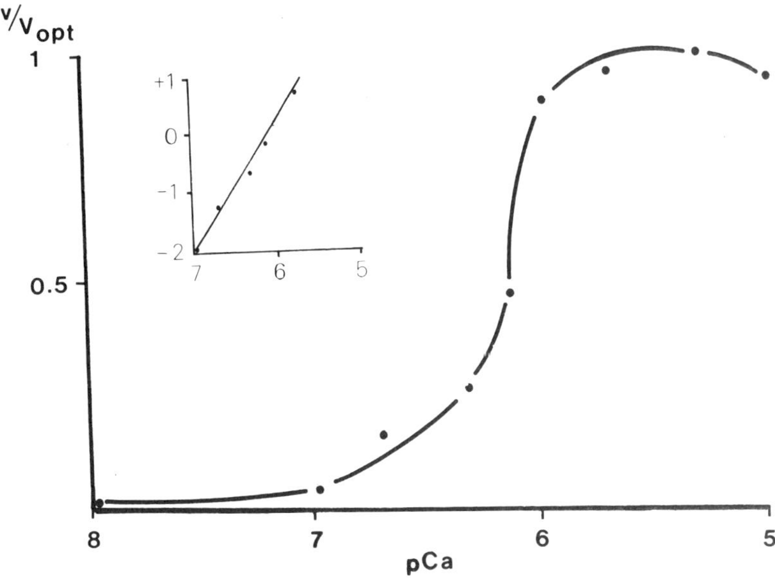

Fig. 4. The activity of sarcolemmal Ca^{++}-ATPase isolated from human pregnant myometrium expressed as a function of pCa: v, ATPase activity at given pCa; V_{opt}, maximal ATPase activity obtained at saturating free-Ca^{++} concentration. A Hill plot of the data is inserted: ordinate, log v/V_{opt} - v; abscissa, pCa. The slope shows a Hill coefficient of 1.5.

Calmodulin Regulation of Sarcolemmal Ca^{++}-Pump ATPase

Calmodulin-depletion experiments and the use of suitable anti-calmodulin concentrations of trifluoperazine provided evidence for the existence of calmodulin in isolated smooth muscle sarcolemmae[13]. Calmodulin dependence of Ca^{++}-ATPase was established according to the most recent formulation of Cheung's criteria[22].

Fig. 4 shows the effect, over 3 orders of magnitude, of increasing the free-Ca^{++} concentration on the solubilized Ca^{++}-ATPase activity. The sigmoidal resultant suggests the participation of an allosteric effector and the most probable candidate for such a Ca^{++}-dependent role is calmodulin. The modulator protein had a marked effect on the kinetic parameters of Ca^{++}-ATPase by shifting the enzyme from the low Ca^{++}-affinity (K_{Ca} in the micromolar range) to the high Ca^{++}-affinity (K_{Ca} in the submicromolar range) and increasing the V_{max} by a factor of about 10.

Recently, it was reported that the mild proteolysis activates Ca^{++}-ATPase from the erythrocyte membrane in a way which is similar to that caused by calmodulin[11,23]. Our results support the calmodulin -like activation of the isolated sarcolemmal Ca^{++}-ATPase by limited trypsin attack (L.M. Popescu and E. Toescu, in preparation).

CONCLUDING REMARKS

One can conclude that the high-affinity Ca^{++}-ATPase isolated from smooth muscle sarcolemma has all the important properties which are expected for an active Ca^{++}-extrusion pump. In addition, a comparison of the kinetic parameters of this Ca^{++}-extrusion pump with the kinetic

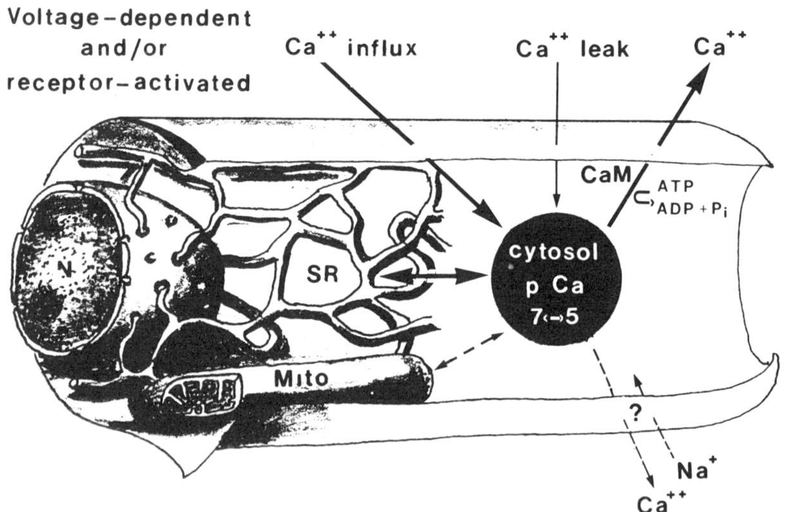

Fig. 5. Regulatory mechanisms of intracellular Ca^{++} homeostasis in smooth muscle. CaM, calmodulin.

parameters of Na^{+}/Ca^{++} exchanger[9,24] shows that the latter has only a limited physiological role in smooth muscle (Fig. 5).

Regulation of Ca^{++}-Extrusion Pump by Calmodulin

In our view, the direct involvement of calmodulin in the activation of Ca^{++}-pump ATPase is essential in the control of intracellular Ca^{++} homeostasis, because a calmodulin-operated "on-off" feed-back mechanism could be responsible for the adequate Ca^{++} extrusion. When the free cytoplasmic Ca^{++} is in the submicromolar range, the Ca^{++}-extrusion pump has a basal activity, and when the cytosolic Ca^{++} reaches micromolar levels calmodulin binds Ca^{++} thereby activating the Ca^{++}-extrusion ATPase. Then, Ca^{++} extrusion, together with Ca^{++} accumulation in SR (and mitochondria ?), return cytosolic Ca^{++} to the submicromolar steady-state level and the active calmodulin-Ca^{++}-pump complex will dissociate decreasing the pump activity again to its basal level. And the process goes on and on and on ... all life long.

REFERENCES

1. C. van Breemen, P. Aaronson, and R. Loutzenhiser, Sodium-calcium interactions in mammalian smooth muscle, Pharmacol. Rev. 30:167 (1979).
2. L. M. Popescu, Conceptual model of the excitation-contraction coupling in smooth muscle, Studia Biophys. 44:141 (1974).
3. L. M. Popescu and I. Diculescu, Calcium in smooth muscle sarcoplasmic reticulum in situ: conventional and x-ray analytical electron microscopy, J. Cell Biol. 67:911 (1975).
4. L. M. Popescu, Cytochemical study of the intracellular calcium distribution in smooth muscle, in: "Excitation-Contraction Coupling in Smooth Muscle", R. Casteels et al., eds., pp. 13-23, Elsevier/North-Holland, Amsterdam (1977).
5. L. M. Popescu, W. C. de Bruijn, U. Zelck, and N. Ionescu, Intracellular distribution of calcium in smooth muscle: facts and artifacts, Morphology & Embryology (Bucharest) 26:251 (1980).
6. A. P. Somlyo, A.V. Somlyo, and H. Shuman, Electron probe analysis of vascular smooth muscle, J. Cell Biol. 81:316 (1979).
7. A. Schibeci and A. Martonosi, Ca^{2+}-binding proteins in nuclei, Eur. J. Biochem. 113:5 (1980).
8. C. van Breemen, P. Aaronson, R. Loutzenhiser, and K. Meisheri, Ca^{2+} movements in smooth muscle, Chest 78:157 (1980).
9. N. Morel and T. Godfraind, Na-Ca exchange in heart and smooth muscle microsomes, Arch. int. Pharmacodyn. Ther. 258:319 (1982).

10. J. T. Penniston, Plasma membrane Ca^{2+} ATPases as active Ca^{2+} pumps, Physiol. Rev. (in press).
11. E. Carafoli and M. Zurini, The Ca^{2+}-pumping ATPase of plasma membranes, Biochim. Biophys. Acta 683:279 (1982).
12. M. M. Oliveira and S. Holzhacker, Isolation and characterization of smooth muscle cell membranes, Biochim. Biophys. Acta 332:221 (1974).
13. L. M. Popescu and P. Ignat, Calmodulin-dependent Ca^{2+}-pump ATP-ase of human smooth muscle sarcolemma, Cell Calcium (accepted).
14. F. Wuytack, G. De Schutter, and R. Casteels, The effect of calmodulin on the active calcium-ion transport and (Ca^{2+} + Mg^{2+})-dependent ATPase in microsomal fractions of smooth muscle compared with that in erythrocytes and cardiac muscle, Biochem. J. 190:827 (1980).
15. M. Wibo, N. Morel, and T. Godfraind, Differentiation of Ca^{2+} pumps linked to plasma membrane and endoplasmic reticulum in the microsomal fraction from intestinal smooth muscle, Biochim. Biophys. Acta 649:651 (1981).
16. G. Gabella, Structure of smooth muscle, in: "Smooth Muscle: an assessment of current knowledge", E. Bülbring et al., eds., pp. 1-47, Edward Arnold, London (1981).
17. P. Caroni and E. Carafoli, The Ca^{2+}-pumping ATPase of heart sarcolemma, J. Biol. Chem. 256:3263 (1981).
18. M. Klingenberg, The use of detergents for the isolation of intact carrier proteins, in: "Membrane and Transport", A. N. Martonosi ed., pp. 203-209, Plenum, New York (1982).
19. A. Fabiato and F. Fabiato, Calculator programs for computing the composition of the solutions containing multiple metals and ligands used for experiments in skinned muscle cells, J. Physiol.(Paris) 75:463 (1979).
20. I. Ross, M. Crompton, and E. Carafoli, The role of inorganic phosphate in the release of Ca^{2+} from rat-liver mitochondria Eur. J. Biochem. 110:319 (1980).
21. U. Pick, The interaction of vanadate ions with the Ca-ATPase from sarcoplasmic reticulum, J. Biol. Chem. 257:6111 (1982).
22. W. Y. Cheung, Calmodulin: an overview, Fed. Proc. 41:2253 (1982).
23. J. Stieger and H. J. Schatzmann, Metal requirement of the isolated red cell Ca-pump ATPase after elimination of calmodulin dependence by trypsin attack, Cell Calcium 2:601 (1981).
24. A. K. Grover, C. Y. Kwan, and E. E. Daniel, Na-Ca exchange in rat myometrium membrane vesicles enriched in plasma membranes, Am. J. Physiol. Cell Physiol. 9:C175 (1981).

REGULATION OF PINOCYTOSIS IN *AMOEBA PROTEUS* BY THE CALCIUM ION

Robert D. Prusch

Department of Life Sciences
Gonzaga University
Spokane, Washington 99258

Calcium and Pinocytosis

Pinocytosis in *Amoeba proteus* involves the uptake of surface bound solute and a portion of the bulk-phase medium by surface membrane infolding and vesiculation. The complex process of pinocytosis in the cell is initiated by the binding of a cationic inducer (Chapman-Andresen, 1962) to the cell surface and is terminated by the internalization of the membrane bound inducer into the cytoplasm. The calcium ion plays a major role in pinocytosis, as it does in a number of other physiological processes, participating in several distinct phases of the pinocytotic cycle. It was observed some time ago (Brandt and Freeman, 1967) that the induction of pinocytosis in the amoeba was correlated with a decrease in membrane resistance and an increase in membrane permeability. This change in membrane resistance associated with the initiation of pinocytosis in the amoeba could be reversed by increasing the level of Ca^{++} in the external medium. It has now been established that external Ca^{++} controls the overall solute permeability of the amoeba surface (Prusch and Dunham, 1972). Consequently, it has been suggested by a number of investigators (Cooper, 1968, Josefsson, 1976) that pinocytosis in the amoeba is initiated by the displacement of surface associated calcium upon the addition of inducer to the external medium. More recently, it has been demonstrated that pinocytotic intensity is dependent upon the level of Ca^{++} in the external medium (Josefsson, 1976; Prusch and Hannafin, 1979b).

In *Amoeba proteus* it has been determined that total cellular calcium is 4.59 mmole/kg cells. Of this total cellular calcium, 18% or approximately 0.84 mmole/kg cells is associated with the cell surface (Prusch and Hannafin, 1979a). This calcium is most likely associated with the cell surface by specific membrane binding sites and by a screening mechanism due to the negative surface charge of the external membrane. When a cationic inducer, e.g., Alcian blue or Na^+, is added to the external medium, a dose-dependent amount of surface calcium is displaced from the cell. The inducer presumably binds to some sites originally occupied by calcium, but given the wide variety of inducer substances (inorganic cations, amino acids, proteins, basic dyes, etc.) it is doubtful that they are all binding to identical sites on the amoeba surface or even binding in an identical fashion.

The initial interaction of pinocytotic inducers and calcium with the amoeba surface is most likely considerably more complex than a simple mass action competition for identical surface sites. For example, it has been established that ligand binding to receptors on the surface of some mammalian cells is dependent upon external Ca^{++} (Kaplan, 1981). In this case, the Ca^{++}- dependency is restricted only to those ligands which will be internalized by endocytosis, not those which serve only to transmit information from the cell surface to the cytoplasm. It may be that the relationship between pinocytotic intensity and external Ca^{++} may in part be a result of Ca^{++} interacting with the amoeba surface increasing the amount of inducer associated with the cell surface. In any event, it can be fairly certain that Ca^{++} is involved with the initial interaction of the inducer with the call surface.

Transmission of surface events or information to the cytoplasm during the initiation of pinocytosis in the amoeba most likely also involves calcium. That is, the information that a pinocytotic inducer is associated with the external cell surface must trigger off a response in the cytoplasm resulting in surface membrane invagination, vesiculation and inducer internalization. Under control conditions, calcium movements across the amoeba cell surface are very slow but there are indications that an increase in Ca^{++} influx is associated with the onset of pinocytosis. A difficulty here is separating specific ion influx from bulk-phase solute uptake associated with pinocytosis. This difficulty was overcome with the use of the Ca^{++} ionophore A23187 (Prusch, 1980a). This ionophore is an antibiotic which acts as a carrier for Ca^{++}. Structurally, A23187 is a relatively complex carboxylate compound with a molecular weight of about 523. The ionophore forms a complex with Ca^{++} only in its deprotonated anionic form (recalling that inducers of pinocytosis in the amoeba are all cations), with two A23187 anions taking up a pseudocyclic configuration around one Ca^{++} ion. The ionophore-Ca^{++} complex then migrates across the cell membrane and releases Ca^{++} into the cytoplasm. Calcium ionophores then represent an experimental method for increasing at least transiently the cytoplasmic Ca^{++} ion activity.

Application of A23187 to Amoeba proteus brings about an increase in the amount of calcium associated with the cell, i.e., an increase in unidirectional Ca^{++} influx. In addition to bringing about an increase in the amount of calcium associated with the cell, A23187 stimulates the uptake of labeled sucrose from the external medium, i.e., bulk-phase pinocytosis. Under normal circumstances the surface of Amoeba proteus is impermeable to sucrose and sucrose itself is not an inducer of pinocytosis (Chapman-Andresen and Holter, 1955). Sucrose uptake can therefore be used to measure the time course and intensity of pinocytosis in the amoeba. On the basis of these observations, i.e., an increase in the amount of calcium associated with the cell and a concomitant uptake of sucrose in the presence of A23187, it may very well be that it is the transmembrane movement of Ca^{++} itself which serves to transmit to the cytoplasm information of inducer association with the amoeba surface. It is suggested that the association of pinocytotic inducers with sites on the external amoeba surface brings about an increase in membrane permeability (among other things) which may then bring about an increase in the movement of Ca^{++} from outside the cell into the cytoplasm, bringing about an increase in the cytoplasmic Ca^{++} ion activity. The ionophore A23187 eliminates the need for the inducer and serves itself to move Ca^{++} into the cytoplasm bringing about pinocytosis.

Given that an increase in the cytoplasmic Ca^{++} ion activity is indeed associated with the onset of pinocytosis in the amoeba, it may be that the ionophore elevates the cytoplasmic Ca^{++} ion activity by entering the cell and releasing calcium from intracellular storage sites. It was extablished though that the ability of A23187 to elicit pinocytotic sucrose uptake in *Amoeba proteus* was dependent upon the presence of ecternal Ca^{++}-free medium, A23187 did not bring about sucrose uptake, but as the external Ca^{++} was slowly increased, pinocytotic sucrose uptake increased. The maximum effect of A23187 on observed pinocytotic sucrose uptake was when external Ca^{++} was approximately 10^{-4} M.

Under control conditions, the cytoplasmic Ca^{++} ion activity of the amoeba cytoplasm is maintained at very low levels, on the order of 10^{-7} M (Taylor, et al, 1973). The maintenance of this low cytoplasmic Ca^{++} ion activity is maintained in part by intracellular membrane systems which actively accumulate Ca^{++} from the cytoplasm (Reinold and Stockem, 1972) and by the active extrusion of Ca^{++} from the cytoplasm across the plasmalemma (Prusch, 1980b). Any increase in the cytoplasmic Ca^{++} ion activity associated then with the initiation or onset of pinocytosis would be transi-tory because of the functioning of these two calcium regulatory mechanisms. This transitory increase in the cytoplasmic Ca^{++} activity may serve in turn though to initiate the final phase of pinocytosis, surface membrane invagination and inducer uptake.

The area between the plasma membrane and the underlying ectoplasm is highly complex in the amoeba. In both *Chaos chaos* (Comly, 1973) and *Amoeba proteus* (Klein and Stockem, 1979) microfilaments have been found in association with the cell cortex. These microfilaments are particularly prevalent in the area of pinocytotic formation. In addition, it has been suggested (Wang, et al, 1982) that a distinct fraction of cellular actin in *Amoeba proteus* is associated with the cell membrane. It is quite conceibable that any increase in the cytoplasmic Ca^{++} ion activity in the localized vicinity of these actin filaments could bring about a contractile response in these filaments. The contraction of the filaments may then result in surface membrane invagination and pincytotic channel formation with subsequent inducer internalization.

The specificity of solute uptake during these initial phases of pinocytosis of pinocytosis does not reside at the level of the cell membrane (Prusch, 1981). A wide variety of external solutes are taken up by pinocytosis in Amoeba proteus, including solutes which may be beneficial to the cell as well as solutes which may be harmful. The only apparent uniform criterion for a solute to successfully initiate pinocytosis in the amoeba is that it must possess a net positive charge. Pinocytosis can be induced with substances such as proteins and amino acids, which presumably could be used by cell's metabolic processes, and basic dyes such as Alcian blue which are poisonous to the cell. When amoebae are exposed to 0.1 percent Alcian blue, the dye is quickly bound to the cell surface, apparently to localized areas, and the cells cease streaming. Within five to ten minutes after initial exposure to the dye, pinocytotic channels can be observed and inducer uptake beings, with dye accumulating in the cytoplasm over the course of the next thirty minutes. During the next thirty to sixty minutes, the cell begins to expel a portion of the internalized dye, along with ingested surface membrane, through a distinct channel leading from the cell interior to the cell surface. This excretory process is quite similar to exocytosis observed in a number of secretory cells in which it has also been established that calcium plays an important role (Allison and Davies, 1974). Induction of pinocytosis with inorganic cations (Na^+) or protein, which leads to the uptake of these solutes, does not elecit the solute extrusion mechanism noted with Alcian blue. This would suggest that the cell has an internal mechanism for sorting out solutes taken up by pinocytosis and then eliminating undesirable solutes. These observations reinforce the notion that pinocytosis is induced by a variety of solutes all of which possess a fundamental charge requirement allowing for some type of interaction involving calcium at the surface of the cell.

Arbitrarily dividing the process of pinocytosis into three distinct phases (Prusch, 1982), initial interaction of the inducer with the cell surface, transmission of surface information to the cytoplasm and inducer internalization, it is apparent that calcium plays an important role in several phases of the pinocytotic cycle. This central role of calcium in a number of diverse physiological processes ranging from protozoa to vertebrates emphasizes the underlying similarities inherent in a given cellular mechanism, regardless of cell type.

References

Allison, A. C. and Davies, P., 1974, Mechanisms of endocytosis and exocytosis, *Symp. Soc. Exp. Biol.*, 28:419.

Brandt, P. W. and Freeman, A. R., 1967, Plasma membrane: substructural changes correlated with electric resistance and pinocytosis, *Science*, 155:582-585.

Chapman-Andresen, D., 1961, Studies on pinocytosis in amoebae. C. R. Trav. Lab. Carlsberg, 33:73-264.

Chapman-Andresen, C. and Holter, H., 1955, Studies on the ingestion of ^{14}C-glucose by pinocytosis in *Amoeba proteus*. C. R. Trav. Lab. Carlsberg, 34:211-226.

Comly, L. T., 1973, Microfilaments in *Chaos carolinensis*: membrane association, distribution, and heavy meromyosin binding in the glycerinated cell. J. Cell Biol., 58:230-241.

Cooper, B. A., 1968, Quantitative studies of pinocytosis induced in *Amoeba proteus* by simple cations. C. R. Trav. Lab. Carlsberg, 36:385-403.

Josefsson, J. O., 1976, Studies on the mechanism of induction of pinocytosis in *Amoeba proteus*. Acta Physiol. Scand., Suppl. 432:1-65.

Kaplan, J., 1981, Polypeptide-binding membrane receptors: analysis and classification. Science, 212:14-20.

Klein, H. P. and Stockem, W., 1979, Pinocytosis and locomotion of amoeba. XII. Dynamics and motive force generation during induced pinocytosis in *Amoeba proteus*. Cell Tissue Res., 197:263-279.

Prusch, R. D., 1980a, Endocytotic sucrose uptake in *Amoeba proteus* induced with calcium ionophore A23187. Science, 209:691-692.

Prusch, R. D., 1980b, Active calcium extrusion by *Amoeba proteus*. J. Exp. Zool., 212:475-477.

Prusch, R. D., 1981, Bulk solute extrusion as a mechanism conferring solute uptake specificity by pinocytosis in *Amoeba proteus*. Science, 213:668-670.

Prusch, R. D., 1982, Calcium and pinocytosis. In, *The Role of Calcium in Biological Systems*, Vol I, eds., L. J. Anghileri and A. M. Tuffet-Anghileri, CRC Press, Boca Raton, Florida, Ch. 14, pp. 219-227.

Prusch, R. D., and Dunham, P. B., 1972, Ionic distribution in *Amoeba proteus*. J. Exp. Biol., 56:551-563.

Prusch, R. D. and Hannifin, J., 1979a, Calcium distribution in *Amoeba proteus*. J. Gen. Physiol., 74:511-521.

Prusch, R. D. and Hannafin, J., 1979b, Sucrose uptake by pinocytosis in *Amoeba proteus* and the influence of external calcium. J. Gen. Physiol., 74:523-535.

Reinold, M. and Stockem, W., 1972, Darstellung eines ATP-sensitiven membransystems mit Ca^{++}-transportierender funktion bei amoben. Cytobiologie 6:183-194.

Taylor, D. L., Condeelis, J. S., Moore, P. L. and Allen, R. D., 1973, The contractile basis of amoeboid movement. I. The chemical control of mobility in isolated cytoplasm. J. Cell Biol., 59:378-394.

Wang, U., Lanni, F., McNeil, P. L., Ware, B. R. and Taylor, D. L., 1982, Mobility of cytoplasmic and membrane-associated actin in living cells. Proc. Nat. Acad, Sci., 79:4660-4664.

HEAVY WATER EFFECTS IN LIVING SYSTEMS

STERIC *VERSUS* ELECTRONIC COMPONENTS OF SECONDARY DEUTERIUM ISOTOPE EFFECTS DEPEND ON THE SIZE OF ELECTROPHILIC PROBES

Alexandru T. Balaban

Polytechnic Institute, Organic Chemistry Department
Splaiul Independenţei 313 - TCH
76206 Bucharest 16, Roumania

INTRODUCTION

Isotope effects have become important for elucidating reaction mechanisms, for increasing the biological half-life of drugs, for studying metabolic pathways, etc. Several books[1] and reviews[2] have been devoted to isotope effects.

Deuterium Is Smaller Than Protium, and CD_3 Is Smaller Than CH_3

As a consequence of zero-point energy differences between deuterium and protium, the C-D bond length is shorter than the C-H bond length. From literature data,[3] the difference is about 0.4 % in mean bond lengths : C-H in CH_4 has 1.106 $\pm$ 0.001, C-D in CD_4 has 1.102 $\pm$ 0.001 ; C-H in C_2H_4 has 1.103 $\pm$ 0.001, and C-D in C_2D_4 has 1.099 $\pm$ 0.003 Å. The mean vibrational amplitude is 0.075 $\pm$ .002 for C-H in CH_4 and 0.066 $\pm$ .002 for C-D in CD_4.

Thus a CD_3 group has a slightly smaller Van der Waals envelope than a CH_3 group, therefore CD_3 groups are expected to exert a slightly lower steric hindrance than CH_3 groups.

Deuterium Is An Electron Donor Relative to Protium, and CD_3 Groups Are Donors Relative to CH_3 Groups

Also because of zero-point energy differences, the C-D bond is more polar than the C-H bond (inductive effect), and the C-D bond is less polarizable than the C-H bond (hyperconjugative effect). The overall result is that CD_3 groups are more electron-donating than CH_3 groups, as indicated by potentiometric titrations and by dipole moment measurements. Thus deuteration reduces the acidity of carboxylic acids (CD_3COOH vs. CH_3COOH gives $\Delta pK_a = 0.026 \pm 0.002$; CH_3CD_2COOH vs. CH_3CH_2COOH gives $\Delta pK_a = 0.034 \pm 0.002$), and increases the basicity of amines ($PhCD_2NH_2$ vs. $PhCH_2NH_2$ gives $\Delta pK_a = 0.048 \pm 0.005$; D_3C-NH-Ar vs. H_3C-NH-Ar gives $\Delta pK_a = 0.050 \pm 0.004$, with Ar = 2,4-dinitrophenyl or 2,4,6-trinitrophenyl). From these literature data,[4] the electronically-caused decrease of acidity or increase of basicity is 2-4 % per deuterium atom at an adjacent carbon.

It is therefore expected that by both mechanisms (inductive and hyperconjugative), replacement of CH_3 by CD_3 substituents will increase the basicity and nucleophilicity of methyl-substituted pyridines.

HOW IS IT POSSIBLE TO DISSECT SECONDARY ISOTOPE EFFECTS INTO STERIC AND ELECTRONIC COMPONENTS ?

We have selected secondary deuterium isotope effects (which can be kinetically either accelerating or retarding) instead of primary ones (which kinetically are always retarding) because the latter are strongly influenced not only by steric and electronic factors, but also by the molecular energetics and by the nature of the transition state. Deuterium was selected because it gives rise to larger isotope effects than any other stable isotope.

By working with one and the same type of substrate (in our case, regioselectively methyl-deuterated pyridines), as well as by involving in all cases an association between this substrate as nucleophile and verious electrophilic probes of varying size, it ought to be possible to dissect secondary isotope effects into their steric and electronic components, and to see the influence of the size of the probe. The basic underlying idea is that by deuterating α-standing methyl groups in pyridines, we facilitate the attack of the electrophile on the lone electron pair of the nitrogen heteroatom by both mechanisms : the electronic effect enhances the nucleophilicity of the nitrogen, and the smaller volume of α-CD_3 groups facilitates sterically the access of the electrophile by allowing it to come closer to the nitrogen ; when on the other hand, we deuterate the γ-methyl group, only the former mechanism is still operating but there is no longer any steric effect because the γ-methyl is too far from the nitrogen.

We have employed three electrophilic probes : (i) the smallest possible electrophile, the proton, in acid-base equilibria ; (ii) methyl iodide which alkylates the pyridine by the Menshutkin reaction, quaternizing the nitrogen heteroatom with a methyl group, which is appreciably larger than the proton involved in the preceding case ; (iii) europium chelates with a very large Van der Waals radius, which act as complexing lanthanide shift reagents.

In all three cases the nitrogen lone pair is involved : for forming a covalent bond in (i) and (ii), and a weak coordination in (iii). Whereas processes (i) and (iii) are equilibria, analyzed thermodynamically, process (ii) involves a kinetic analysis. Thus we measure pK_a values for (i), reaction rates for (ii), and molar induced chemical shifts in ^1H-NMR spectra for (iii).

<u>Preparation of Regioselectively-Deuterated Pyridines.</u>[5]

2,4,6-Trimethylpyridines deuterated at the α, at the γ, or at the α and γ methyl groups were obtained via the corresponding 2,4,6-trimethylpyrylium perchlorates :

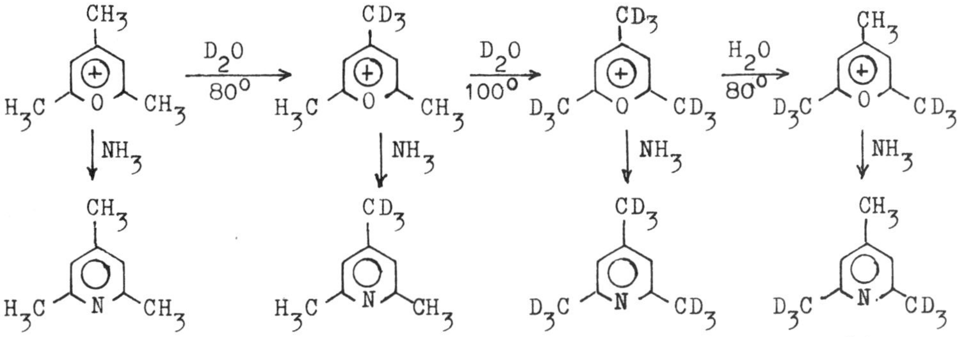

Similarly, 2,6-diethyl-4-methylpyrylium perchlorate can be deuterated in D_2O either rapidly in γ, or more slowly both in α and in γ. The latter product can be de-deuterated by H_2O to afford the α-deuterated pyrylium salt ; all these compounds are then converted into the corresponding pyridines by reaction with ammonia, without altering their deuteration pattern.

2,3,5,6-Tetramethylpyrylium is deuterated at the 2,6-methyl groups, while pentamethylpyrylium is deuterated in D_2O at the 2,4,6-methyl groups ; ammonia then converts these compounds into the corresponding pyridines, replacing heteroatom O^+ by N.

RESULTS AND DISCUSSION

(i) The Secondary Isotope Effect in the Protonation of 2,4,6-Trimethylpyridine Presents No Detectable Steric Component

By determining relative pK_a values of the regiospecifically deuterated 2,4,6-trimethylpyridines 1, 1A, 2A, and 2 (based on the literature pK_a value for the non-deuterated sym-collidine 1 i. e. 7.45 at 25° C [6]) it was found [7] that every replacement of

CH_3 by CD_3 increases the pK_a, irrespective whether this replacement occurs in γ or α : the pK_a values are 7.47 ± 0.01 for 1A, 7.49 ± 0.01 for 2A, and 7.53 ± 0.02 for 2. Thus the proton does not detect any steric component, the effect is purely electronic.

(ii) The Secondary Isotope Effect in the Quaternization of Pyridines with Methyl Iodide Contains Comparable Amounts of Steric and Electronic Components

The Menshutkin quaternization of pyridines is known [8] to be sensitive to steric hindrance. The Brønsted relationship :

$$\log(k^H/k^D) = \alpha(pK_a^H - pK_a^D)$$

allows to calculate on the basis of known (for 1, 1A, 2A, 2) or computed pK_a values (for other alkylpyridines) the **expected** electronic component of the secondary isotope effect. By measuring the total secondary isotope effect, one may determine by difference the steric component.

The data presented in the Table refer to two from the six selectively-deuterated pyridines which were investigated ; in all six cases, the secondary kinetic isotope effect was found to be accelerating, and to consist in comparable amounts of electronic and steric components when both α-positions were substituted with alkyl groups.[9]

Table 1. Kinetic secondary isotope effects and their constituent parts (in acetine at 25.0° unless otherwise stated)

Probe	Secondary isotope effect	2,3,5,6-Tetramethylpyridine (deuteration at the 2,6-methyls)		Pentamethylpyridine (deuteration at the 2,4,6-methyl groups)	
		pK_a	k	pK_a	k
CH_3I	H	7.91	2.92	8.75	3.5
	D	7.96	3.24	8.83	5.2 (at 30°)
	Electronic	4.9 %		-	
	Steric	6.1 %		-	
	Total	11.0 %		-	
CD_3I	H	7.91	3.31	8.75	4.0
	D	7.96	3.59	8.83	4.7
	Electronic	4.9 %		7.9 %	
	Steric	3.6 %		9.6 %	
	Total	8.5 %		17.5 %	

In this Table, rate constants k are expressed as $l\cdot mol^{-1}\cdot sec^{-1}$ times 10^{-6}.

A few remarks are needed to make the Table more easily understable. The quaternization rates were measured conductimetrically using a specially devised cell and carefully purified and dried reagents in an inert atmosphere during the purification and measurement.[9] The sensitivity of the method allowed rate measurements (about 20 experimental determinations for the linearization of the plot assuming bimolecular reaction kinetics) for final conversions of about 1-2 %. It can be noted from the Table that the rates of quaternization with CD_3I are about 12-15 % higher than those involving CH_3I ; it was, however, not possible to dissect these extra rate enhancements on replacing CH_3I by CD_3I, so that the Table separates into components only the effects due to deuteration of pyridinic methyl groups. For the other deuterated pyridines, not contained in the Table, as well as for a few examples described earlier in the literature,[10] the results are similar : the total kinetic secondary isotope effect for pyridines having 2,6-dimethyl substituents consists of approximately equal amounts of steric and electronic components.

The data for pentamethylpyridine were recorded at several temperatures and measured (with a lower accuracy) by graphical interpolation. This was due to the initial (up to about 0.5 % conversion) curvature of the kinetic plots, indicating the presence of an impurity with higher nucleophilicity than pentamethylpyridine ; separate experiments showed this impurity (amounting to less than 0.5 %) to be the Dewar-type valence isomer of pentamethylpyridine :

Me Me Me Me Me N $\underset{}{\overset{h\nu}{\rightleftharpoons}}$ Me Me Me Me Me N

(iii) The Secondary Isotope Effect in Lanthanide-Induced Shifts (^{1}H-NMR Spectra) for Alkylpyridines Is of Purely Steric Origin

Hinckley, who discovered that lanthanide chelates (lanthanide shift reagents, LSR) cause useful induced shifts in nuclear magnetic resonance (NMR) spectra, also discovered the secondary isotope effects caused by deuteration. He attributed the enhanced induced shift caused by α-deuteration of alcohols to electronic effects.[11]

We have investigated the ^{1}H-NMR spectra of many substituted pyridines, using as LSR tris(dipivaloylmethanato)europium, abbreviated as $Eu(dpm)_3$ or $Eu(thd)_3$, and the fluoro-substituted chelate $Eu(fod)_3$ or its deutero-counterpart which causes less line broadening and has a higher solubility. Though the results are expressed most accurately by the complexation equilibrium constants, a simpler, more convenient form (precise enough for most purposes) is through the molar induced shifts (MIS), i. e. the linearly extrapolated induced shift for one mole of LSR per mole of substrate (MIS values are expressed as parts per million, ppm).

A first observation is that the MIS values are extremely sensitive to steric factors of the α-alkyl groups. MIS values in CS_2, CCl_4 or $CDCl_3$ show that in 3 two α-ethyl groups reduce the complexation from 15.0 ppm in 1 to less than half (6.5 ppm) ; two α-isopropyl or t-butyl groups completely suppress any complexation (MIS = 0). One α-ethyl and one α-methyl give rise to a detectable steric hindrance (13.8 ppm), one α-isopropyl and one α-methyl cause about the same effect (6.6 ppm) as two α-ethyl groups (6.5 ppm), while one α-methyl and one α-t-butyl group almost totally reduce the complexation (MIS = 0.4 ppm). The above figures represent MIS values for protons attached to the α-carbon atoms in non-deuterated pyridines.[12]

The europium chelate, which is a very bulky electrophile, complexes more strongly with α-deuterated pyridines than with their non-deuterated counterparts : on mixing unequal amounts of corresponding deuterated and non-deuterated pyridines, and on adding increasing amounts of $Eu(dpm)_3$ as LSR, one obtains in certain cases induced shifts manifesting secondary isotope effects. It makes no difference whether the γ-methyl group is deuterated or non-deuterated, only the α-alkyl deuteration matters. Thus the MIS values for β-protons are 6.00 ppm for 1 and 1A, and 6.34-6.35 for 2A and 2 (Fig. 1).[13] For the 2,6-diethyl-4-methylpyridines 3-4 (Fig. 2) the MIS values are 2.52 and 2.80 (β-protons), 1.80 and 1.89 ppm (methyl protons of ethyl groups) when the α-methylene groups are non-deuterated and deuterated, respectively, irrespective whether the γ-methyl is or is not deuterated.[13] The only possible interpretation of these results is that the electronically-caused enhancement of nucleophilicity on deuteration plays no part in the complexation : the smaller CD_3 or CD_2CH_3 groups than CH_3 or CH_2CH_3 groups allow a closer approach of the LSR to the pyridinic lone pair of electrons.

CONCLUSION

We have shown that the relative importance of steric and electronic components in secondary isotope effects of deuterium depends on the steric requirements of the probe : with alkyl--substituted pyridines, small electrophiles (H^+) feel only the electronic component, bulky electrophiles on the other hand (lanthanide shift reagents) feel only the steric component, while both components are present with electrophilic probes of intermediate size such as methyl iodide in quaternizations.

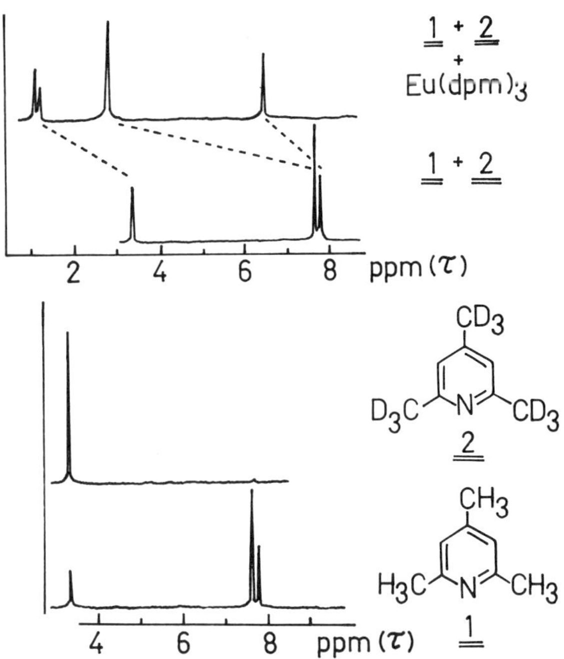

Fig. 1. On mixing a small amount of sym-collidine (1, bottom NMR spectrum) with a larger amount of totally methyl-deuterated counterpart (2, next spectrum) one obtains mixture 1 + 2 (second spectrum from top), which shows (top NMR spectrum) a splitting of the β-proton peak on addition of $Eu(dpm)_3$ evidencing thereby a secondary isotope effect. An exactly similar result is obtained with the mixture 1 + 2A, but no isotope effect (no splitting) is obtained with the mixture 1 + 1A. Since the taller peak moves faster on adding $Eu(dpm)_3$ the deuterated pyridine complexes more strongly.

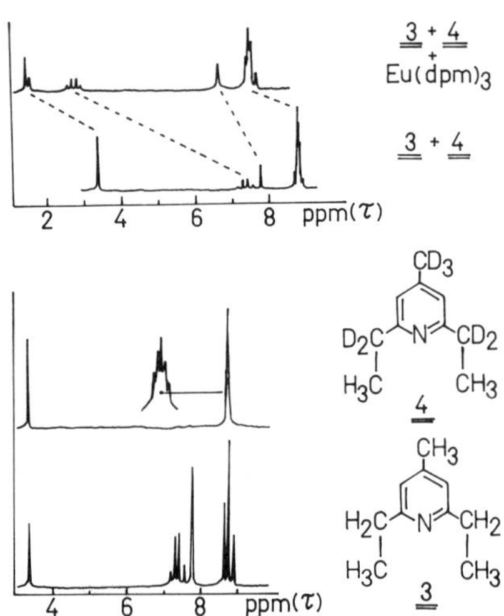

Fig. 2. On mixing a small amount of 2,6-diethyl-4-methylpyridine (3, bottom NMR spectrum) with a larger amount of its totally side-chain benzylic-position deuterated counterpart (4, next spectrum) one obtains the mixture 3 + 4 (its NMR spectrum is second from top) which shows (top spectrum) splittings of β-proton and CH_3 peaks on addition of LSR, evidencing thereby a secondary isotope effect. An identical result is obtained when only the ethyl CD_2CH_3 groups are present in the deuterated compound, but no isotope effect is observed when only the γ-methyl group is deuterated.

REFERENCES

1. L. Melander, "Isotope Effects on Reaction Rates", The Ronald Press Co., New York, 1960 ; C. J. Collins and N. S. Bowman (eds.) "Isotope Effects in Chemical Reactions", Van Nostrand - Reinhold, New York, 1970 ; A. F. Thomas, "Deuterium Labeling in Organic Chemistry", Appleton-Century-Crofts, New York, 1971, p.401.

2. C. J. Collins, Adv. Phys. Org. Chem. 2 : 63 (1964) ; E. R. Thornton, Ann. Rev. Phys. Chem. 17 : 349 (1968) ; E. A. Halevi, Progr. Phys. Org. Chem. 1 : 150 (1963).

3. L. S. Bartell, K. Kuchitsu and R. J. de Neui, J. Chem. Phys. 35 : 1211 (1961) ; L. S. Bartell, E. A. Roth, C. D. Hollowell, K. Kuchitsu and J. E. Young, Jr., ibid. 42 : 2693 (1965) ; L. S. Bartell, J. Am. Chem. Soc. 83 : 3567 (1971) ; K. Clausius and K. Weigand, Z. physik. Chem. B 46 : 1 (1940) ; P. J. Mitchell and L. Phillips, Chem. Commun. 908 (1975).

4. E. A. Halevi, N. Nussim and A. Ron, J. Chem. Soc. 866 (1963) ; B. D. Batts and E. Spinner, ibid. 789 (1968).

5. A. T. Balaban, J. Labelled Comp. Radiopharm. 18 : 1621 (1981) and further references therein.

6. A. Gero and J. J. Markham, J. Org. Chem. 16 : 1735 (1951) ; M. Tamres, S. Searles, E. M. Leighly and L. W. Mohrman, J. Am. Chem. Soc. 76 : 3983 (1954).

7. A. T. Balaban, I. I. Stănoiu and E. Gârd, Rev. Roumaine Chim. 22 : 1191 (1977).

8. H. C. Brown, J. Chem. Soc. 1248 (1956) ; L. W. Deady and J. Zoltewicz, J. Org. Chem. 37 : 603 (1972) ; J. A. Zoltewicz and L. W. Deady, Adv. Heterocyclic Chem. 22 : 71, 85 (1978) ; G. W. Duffin, ibid. 3 : 1 (1964).

9. A. T. Balaban, A. Bota, D. C. Oniciu, G. Klatte, C. Roussel and J. Metzger, J. Chem. Res. (S) 44 (1981) ; (M) 559 (1981).

10. H. C. Brown and G. J. Mc Donald, J. Am. Chem. Soc. 88 : 2514 (1966).

11. C. C. Hinckley, W. A. Boyd, G. V. Smith and F. Behbahany, in "Nuclear Magnetic Resonance Shift Reagents", R. E. Sievers, ed., Academic Press, New York, 1973, p. 1 ; G. V. Smith, W. A. Boyd and C. C. Hinckley, J. Am. Chem. Soc. 93 : 6319 (1971) .

12. A. T. Balaban and M. D. Gheorghiu, unpublished data.

13. A. T. Balaban, I. I. Stănoiu and F. Chiraleu, Chem. Commun. 984 (1976) ; Rev. Roumaine Chim. 23 : 187 (1978).

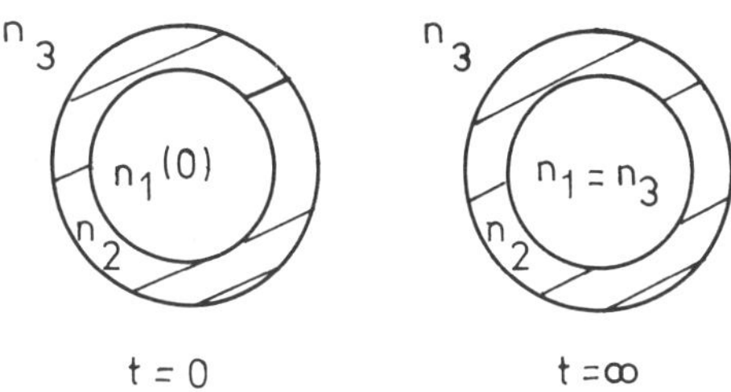

1. Schematic representation of the indices of refraction for a coated sphere (vesicle) at two times. n_2 is the index of refraction of the shell (lipid membrane). $n_1(0) \neq n_3$.

D

les prepared in H_2O are rapidly mixed with D_2O in a stopped-
apparatus. The transmitted light intensity I_t is observed as
ion of time after the short mixing period. The D_2O-molecules
ate through the membrane and enter into the intravesicular
rtment, while the H_2O-molecules move in the opposite direction
the isotopic compositions are equal on either side of the
ne. As consequence of the permeation process the initial
ence between the intra- ($n_1(t=0) = n(H_2O)$) and the extra-
lar index of refraction ($n_3 = 0.5\ (n(H_2O) + n(D_2O))$ is re-
to finally $n_1(t=\infty) = n_3$ (s. Fig. 1). The extravesicular
of refraction n_3 stays constant due to the excess of the
esicular volume. In the regarded case the intravesicular
of refraction n_1 decreases with time in a monoexponential
according to

$$ = n_3 + 0.5\ (n(H_2O) - n(D_2O))\ \exp(-k_{ex}\ t) \quad (1)$$

hange relaxation rate k_{ex} is related to the permeability
ient P_D by

$$ = k_{ex}\ V/S \quad (2)$$

and S are the vesicular volume and surface, respectively.
e-dependent change of the intravesicular index of refraction
an almost monoexponential change of the transmitted light
ty I_t or turbidity τ.

cribed method is restricted to large vesicles (no point
ers!). For that purpose the Mie scattering theory applied
d spheres[1] has to be used. The Mie coefficients a_n and b_n

D_2O/H_2O - PERMEABILITY OF MEMBRANES BY LIGHT SCAT

H.-P. Engelbert and R. Lawaczeck
Institute of Physical Chemistry
Marcusstr 9/11, 8700 Würzburg, FRG

ABSTRACT

The different indices of refraction of H_2O and
study the permeation of these molecules through
bilayers. Vesicles were prepared by sonication
mixed with D_2O in a stopped-flow apparatus. The
vesicular index of refraction decreases since
permeate into the interior of the vesicles. Co
transmitted light intensity increases until th
sition is equal on both sides of the membrane.
signal is a function of the vesicular size and
only for larger vesicles. Calculations based o
theory of coated spheres are in agreement with

INTRODUCTION

Vesicular lipid bilayers are a common model o
Studies on the permeation of molecules throug
of fundamental importance, since especially s
are basic properties of cellular membranes. I
difference of the indices of refraction of H
D_2O (n_D = 1.32844) the permeation of H_2O/D_2O
vesicular bilayers can be monitored in light
ments. The permeability coefficient P_D (cm/s
calculated from the measured exchange relaxa
if the vesicle-size is known. We present her
and first results obtained with this method

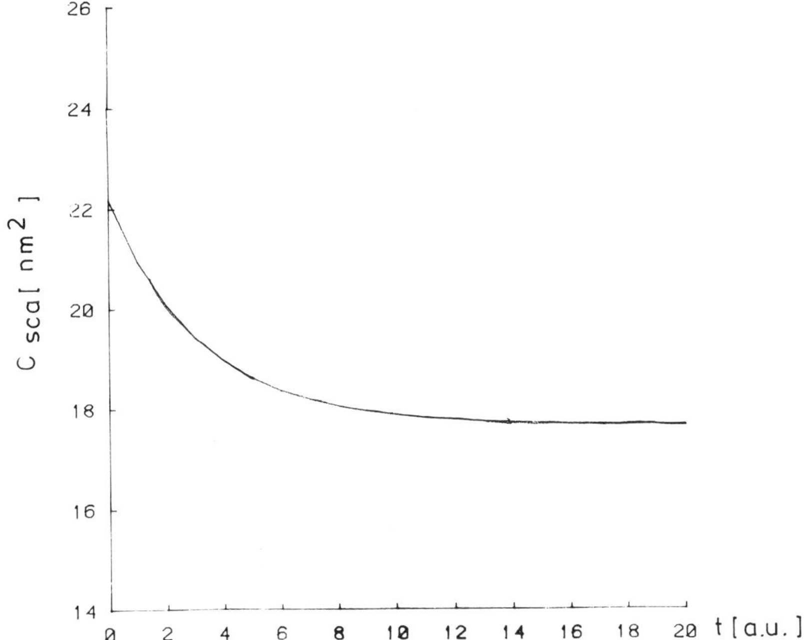

Fig. 2. Calculated scattering cross section C_{sca} for a coated sphere with outer radius 100 nm and shell thickness 5 nm as function of time. The interior index of refraction n_1 approaches n_3 according to eq. (1). k_{ex} = 0.3; λ = 405 nm; n_2 = 1.501; $n(H_2O)$ = 1.34274; $n(D_2O)$ = 1.33749.

are calculated as function of the wavelength λ, the outer and inner radii, the three indices of refraction n_1, n_2 (index of refraction of the vesicular bilayer) and n_3. The scattering cross section C_{sca} is obtained from the Mie coefficients a_n and b_n according to

$$C_{sca} = \lambda^2/2\pi \sum_n (2n+1)\,(|a_n|^2 + |b_n|^2) \tag{3}$$

The summation is performed until the sum converges ($n \approx 10$), depending mainly on the radius of the regarded sphere. In the limiting case of small spheres with n_1/n_3 and n_2/n_3 close to 1 eq. (3) reduces to the well-known Rayleigh-scattering.

We have calculated C_{sca} as function of time since the inner index of refraction n_1 was varied according to eq. (1). A typical result for one set of parameters is presented in Fig. 2. The difference in the scattering cross sections C_{sca}, i. e. $\Delta C_{sca} = C_{sca}(t=0) - C_{sca}(t=\infty)$, is illustrated in Fig. 3 as

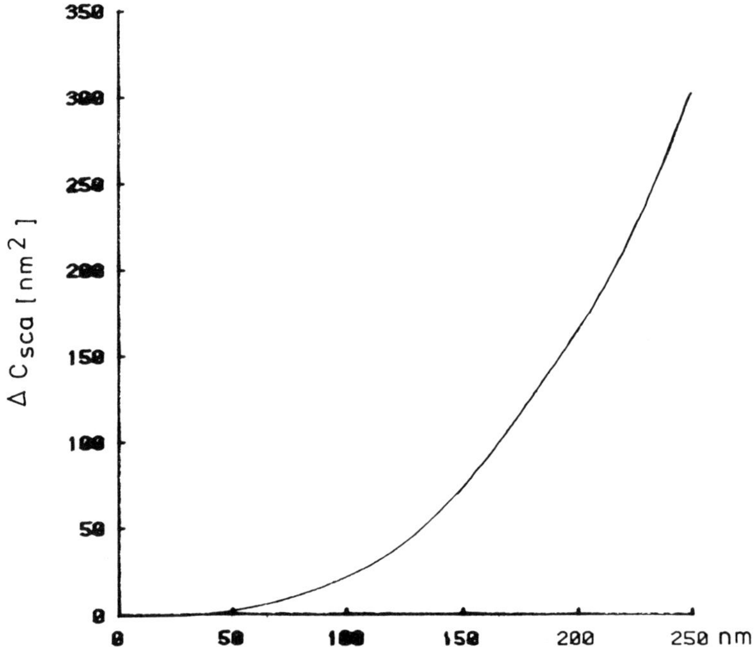

Fig. 3. Difference of the initial scattering cross section C_{sca} ($n_1 = n(H_2O)$) and the final C_{sca} ($n_1 = n_3$) vs. the outer radius of a coated sphere with shell thickness 5 nm.
$\Delta C_{sca} = |C_{sca}(t=0) - C_{sca}(t=\infty)|$; $\lambda = 405$ nm; $n_2=1.501$

function of the outer radius. ΔC_{sca} corresponds to the maximum experimental effect just after the short mixing period. From Fig. 3 it is evident that this effect becomes only measurable for larger vesicles.

In the absence of multiple scattering C_{sca} is related to the turbidity τ

$$\tau = C_{sca} N \tag{4}$$

with N the number of scatterers per unit volume[2]. From eq. (4) the intensity of the transmitted light I_t is obtained

$$I_t = I_o \exp(-\tau \cdot l) \tag{5}$$

where I_o is the incident light intensity and l the observation length.

For small values of $\tau \cdot l$ the exponential function in eq. (5) can be approximated

$$I_t = I_o (1 - C_{sca} Nl) \tag{6}$$

Under these conditions the time-dependence of I_t is equivalent to that of C_{sca}. Thus k_{ex} can easily be deduced from the measured intensity of the transmitted light I_t.

EXPERIMENTAL RESULTS

Large dipalmitoylphosphatidylcholine (DPPC) vesicles were prepared in H_2O (20 mM $CaCl_2$; 0.02% NaN_3). After interval sonication (10x30 sec) below T_c (41.5 °C for DPPC) the unannealed vesicles were incubated at 37 °C and allowed to fuse for 72 h. Prior to the measurement the vesicles were annealed at temperatures above T_c [3].

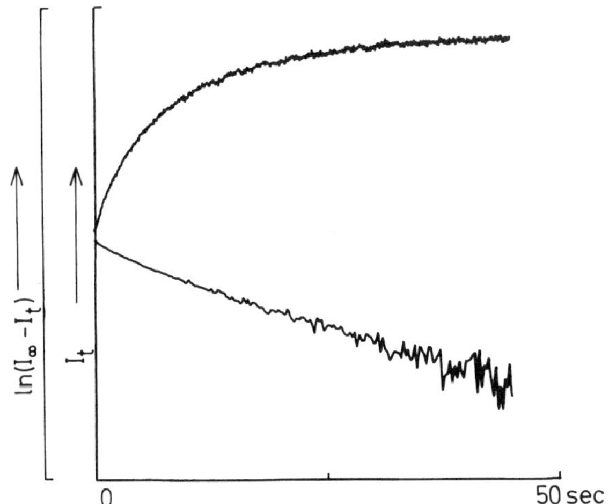

Fig. 4. Experimental curve of the transmitted light intensity I_t vs. time after mixing together with a logarithmic presentation (lower curve). DPPC-vesicles in H_2O were rapidly mixed with the deuterated solvent (t=0) in a stopped-flow apparatus. λ =405 nm; room temperature; large vesicles.

The large vesicles in the H_2O-buffer were rapidly mixed in a stopped-flow apparatus with the equivalent D_2O-buffer. The transmitted light intensity I_t was measured as function of time. In Fig. 4 a typical experimental result obtained with this method is shown together with a logarithmic presentation of the data points according to $\ln(I_\infty - I_t)$; $(I_\infty = I_t(t=\infty)$. Except for short times of observation the transmitted light intensity I_t approaches the new steady-state-signal in an almost mono-exponential manner thus allowing the direct deduction of k_{ex} from the experimental curve.

If vesicles were prepared in D_2O and mixed with H_2O the opposite tendency of the transmitted light intensity I_t is observed. In this case the intravesicular index of refraction increases with time and thus the transmitted light intensity I_t decreases. Control experiments where the isotopic composition is equal in both syringes of the stopped-flow apparatus show a constant value of I_t.

The described method can be extended to studies on biological membranes in a straightforward manner. First experiments with human erythrocytes were performed.

REFERENCES

1. A. L. Aden and M. Kerker, Scattering of electromagnetic waves from two concentric spheres, J.Appl.Phys. 22:1242 (1951)
2. M. Kerker, "The scattering of light", Academic Press, New York (1969)
3. R. Lawaczeck, M. Kainosho and S. I. Chan, The formation and annealing of structural defects in lipid bilayer vesicles, Biochim.Biophys.Acta 443:313 (1976)

Support by the Fonds der Chemischen Industrie and the Deutsche Forschungsgemeinschaft (DFG) is highly acknowledged (La 328/4-1).

THE ISOTOPE EFFECT OF D_2O ON THE FIRE-FLY BIOLUMINESCENCE SPECTRA

Constanţa Ganea and V. Vasilescu

Biophysical Department
Faculty of Medicine
Bucharest, Romania

INTRODUCTION

Fire-fly bioluminescent reaction represents the most efficient bioenergetical process, its quantum yield approaching unity. This is one reason for great interest in the study of the fire-fly luciferin-luciferase system. A large number of studies were carried out, mostly by McElroy and his coworkers,[1-4] at understanding the mechanism underlying the fire-fly bioluminescence process. This mechanism is still not completely elucidated.

In vitro, the process takes place in the presence of the following reagents: luciferin, luciferase, ATP, Mg^{2+} ions and molecular oxygen. The bioluminescence reaction implies two main stages, namely: luciferin activation by ATP resulting in the formation of adenyl-luciferin, and oxidation of this latter compound, accompanied by emission of light[1]:

$$E + LH_2 + ATP \underset{}{\overset{Mg^{2+}}{\rightleftharpoons}} E\ .\ LH_2 - AMP + PP$$

$$E\ .\ LH_2 - AMP + O_2 \overset{-CO_2}{\rightleftharpoons} [\rho\cdot - E.\ AMP] \longrightarrow h\nu + \text{products}$$

(E - luciferase; LH_2 - luciferin, PP - pyrophosphate, $\rho\cdot$ - reaction product which emits light when electronically excited).

Following a number of experiments performed in our Laboratory[2] concerning involvement of water and water protons in biological systems, we investigated the role of protons in this much simpler process which is a bioluminescent reaction. The method of quasi-total deuteration of the system was used and a comparison between the behavior of the deuterated and that of the non-deuterated bioluminescent systems was studied.

In a previous work[3] we reported some results concerning D_2O effects on the kinetic and thermodynamic parameters of the fire-fly bioluminescence in vitro.

Slower kinetics were found after deuteration, reflecting the increased value of half-time bioluminescence kinetics. A comparison between the thermodynamic parameters of the bioluminescence reaction in the deuterated and non-deuterated systems revealed the following: a) the values of the free energies of activation and denaturation were smaller in non-deuterated than in deuterated system, thus indicating an increased stability of luciferase on deuteration; b) the activation and denaturation entropies were lower in a deuterated system, which suggests a higher degree of structure of luciferase and its possible transition to a conformation more suitable for the occurrence of the bioluminescence reaction.

These investigations with measurements of D_2O effects on the fire-fly bioluminescence spectra are reported in this paper.

MATERIALS AND METHOD

Experiments were carried out on the fire-fly luciferin-luciferase system extracted from fire-fly dessicated tails obtained from SIGMA. The extraction was performed following McElroy method [4], improved in our laboratory, in order to obtain better sensitivity of the system[5].

The luciferin-luciferase system was extracted from fire-fly dessicated tails, kept at cool (0°-4°C) and introduced in a glycyl-glycine buffer solution (0.025M glycyl-glycine, 0.1M Mg^{2+}, pH = 7.5).

ATP solutions to be used were also prepared in the glycyl-glycine buffer solution. There were prepared in the same manner as the corresponding deuterated reagents, using as solvent D_2O. Equipment used consisted of a Shimadzu spectrofluorimeter, a SPECAN 101 analogic processor and a thermostat. A specially adapted spectrofluorimetric cuvette was used, which recorded the spectra very quickly following the rapid injection of reagents.

The samples contained 0.4 ml ATP (1.6×10^{-2}M) and 0.5 ml luciferin-luciferase solution (10 mg/ml) - both prepared in glycyl-glycine - a composition which was suitable for recording the spectra in the slower part of the kinetic curve.

We studied the bioluminescence spectra of deuterated and non-deuterated systems in two cases: a) at the constant temperature of 25°C (the optimum for fire-fly bioluminescence reaction) and a pH range of 7.15-7.64 in water and 6.8-7.24 in heavy water; b) at the constant pH value of 7.5 (the optimum pH value in this case) and a temperature range of 22-39°C in water and of 20-24°C in heavy water.

The pH correction for the deuterated system was applied according to Glasoe and Long[6].

RESULTS AND DISCUSSIONS

a) Spectra showed in all cases an asymmetry (Fig. 1) which suggested superimposing of two component spectra, one having a maximum at about 552 nm, and the other at about 610 nm.

These components have already been found by White et al.[7] according to whom the two spectra are due to the existence of two tautomeric forms of luciferin (Fig. 2): the enol form (λ_{max} = 556 nm) known as the greenish-yellow luminophore, and the keto-form (λ_{max} = 616 nm), known as the red luminophore.

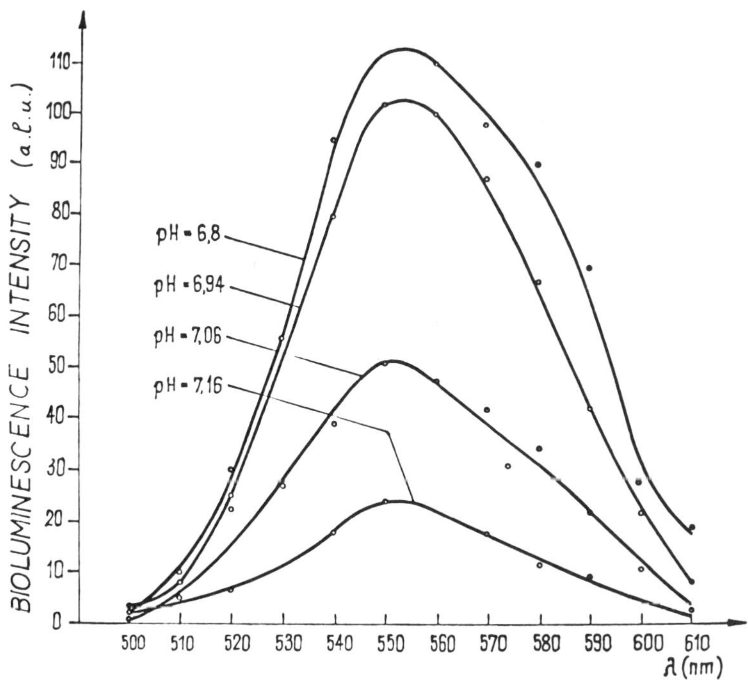

Fig. 1. Bioluminescence spectra in heavy water for various oH values.

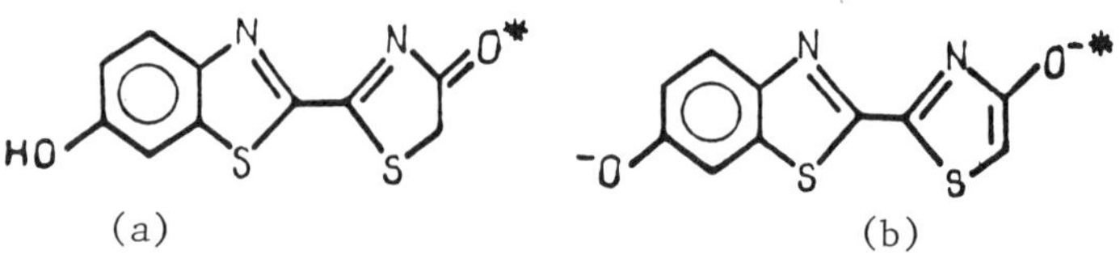

Fig. 2. a) Red luminophore; b) Greenish-yellow luminophore.

The λ_{max} of the resulting spectrum depends on the concentration of the two forms in the mixture.

b) The experiments, both in water and in heavy water, showed that λ_{max} of the bioluminescence spectra shifted to greater wavelengths as the pH of the system decreased (Fig. 3), that is, the equilibrium between the two forms shifted to the keto-form as the proton concentration increased[8].

The shift was greater in heavy water, the ratio of the two shifts (in heavy water and in water) being:

$$(\Delta\lambda_{max})_D/(\Delta\lambda_{max})_H \cong 3.2$$

a value which approaches that of the $\sqrt{k_H/k_D}$, where k_H, k_{Do} are the rate constants of dissociation of water and heavy water at 25°C.

c) Bioluminescence spectra at a constant pH value and at different temperatures showed in experiments performed both in non-deuterated and in deuterated systems, a shift of λ_{max}, as the temperature increased, which shift was stronger in D_2O. In order to give a

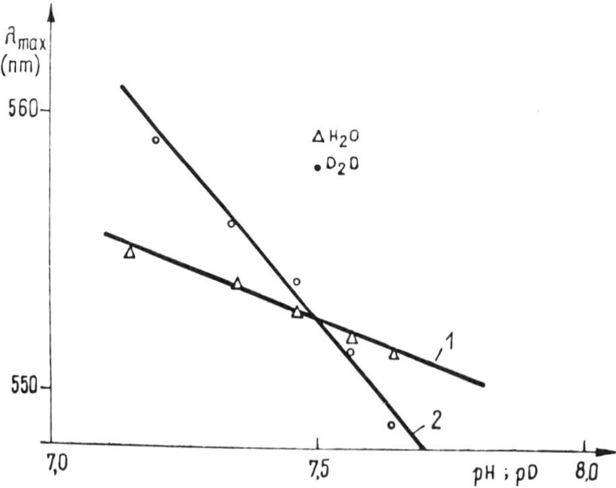

Fig. 3. Comparative graphic representation of λ_{max} decrease with pH increase, in water and in heavy water.

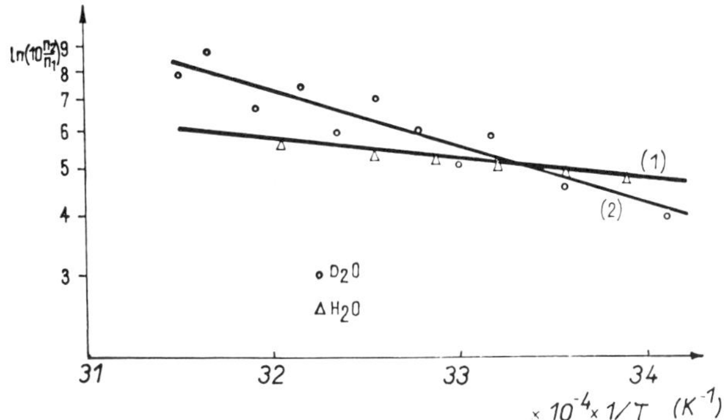

Fig. 4. In (n_2/n_1) variation with the inverse absolute temperature in water and in heavy water.

quantitative description of the respective broadening of the spectrum, we have calculated the ratio of the populations of two energy levels, namely those corresponding to the radiation with λ = 552 nm (n_1) and λ = 580 nm (n_2) respectively. Their ratio $R = n_2/n_1$ increased with temperature rise (Fig. 4), indicating a broadening of the spectrum towards greater wavelengths described by a Boltzmann distribution.

This broadening was enhanced in heavy water and we interpreted this as an isotope effect, expressed by an empirical relationship:

$$R_H/R_D = \exp[-5.9(1-T_m/T)] \quad (T_m = 298\ \text{K},$$

the optimum temperature for the bioluminescence reaction[8]).

Acknowledgement

The authors wish to thank Dr O. Hörer from the Institute of Virology "St. Nicolau" for valuable discussions and technical advice.

REFERENCES

1. M. De Luca and W. D. McElroy, Biochemistry, 13:921 (1974).
2. V. Vasilescu and E. Katona, "Frontiers of Bioorganic Chemistry and Molecular Biology," S.N. Ananchenko, ed., Pergamon Press, 4-15.

3. C. Ganea and V. Vasilescu, Studia Biophysica, 84:59-60 (1981).
4. W. D. McElroy, J.Cell Comp.Physiol., 21:95-116 (1956).
5. V. Vasilescu, E. Trutia, and C. Zaciu, Rev.Roum.Morphol.Embryol. Physiol., 12:87-90.
6. P. Glasoe and F. Long, J.Phys.Chem., 64:180 (1960).
7. E. H. White, E. Rapaport, H. H. Seliger, and T. A. Hopkins, Bioorgan.Chem., 1:92-98 (1971).
8. C. Ganea, Ph.D. Thesis (1980).

TWO DISTINCT ORDERS IN AGAROSE BIOSTRUCTURAL GELS : TRANSITION KINETICS AND THE ROLE OF TEMPERATURE AND ISOTOPIC SUBSTITUTION

M. Leone, S.L. Fornili, and M.B. Palma-Vittorelli

Istituto di Fisica dell'Universita' and
C.N.R. - Gruppo Nazionale Struttura della Materia
90123 - Palermo, Italy

INTRODUCTION

Aqueous agarose systems have been the object of much attention[1-4] as a consequence of their ability to form structural gels of high biological and practical interest. Thermoreversibility and hysterical behaviour of sol-gel transitions in these systems offer a further point of interest, since they evidence the existence of marked cooperative effects in the formation/ recognition/ aggregation/ stabilization of molecular and supramolecular patterns. Understanding the underlying physical processes may have very wide implications. In these systems gelation occurs already at a polysaccharide concentration of a fraction of a % w/v through the formation of a molecular order (double helices) and a supramolecular order (bundles)[1].

Certain similarities of these systems with polynucleotides add to their interest. From thorough studies of isotope exchange performed in the last few years, the existence and motion of "open states" in polynucleotides has been evidenced, perhaps due to highly non linear excitations such as solitary waves or even solitons[5]. These have an important biological role for accessibility of single strands. From the point of view of solvent-solute interactions, D-H substitution in the solvent causes similar effects in agarose and polynucleotides systems on the transition from the more ordered helix phase to the less ordered coil phase[2b,6]. This evidences in both cases a dynamical aspect of biomolecule-water interaction[2,6,7]. Further, computer simulation experiments show that solvation water seems to have a preference to form open and closed connectivity pathways in very similar ways around agarose and polynucleotides helices[8]. These considerations show the interest of a study of the nature of the ordering process of gelation, of the occurrence and thermal annealing of defective order, and of the possible role of the dynamical properties of molecules and solvent in the formation of the ordered state.

In this work, we report preliminary results of the first experimental study of the kinetics of gelation of a polysaccharide (Agarose) aqueous systems. The study was extended to different conditions of temperature, concentration and isotopic substitution in the solvent. Experimental findings show that a new markedly more ordered and stable gel state (henceforth LRO gel, for long-range-ordered gel) can be reached. This probably occurs when the formation of extended helical aggregates is favoured by the joint action of a slow formation of helically ordered polymers and their subsequent coalescence and annealing into bundles. These conditions are thought to be close to those occurring in biological conditions. Among other mechanisms of molecular dynamics[9], the diffusion of "open segments" similar to those operating in polynucleotides and similarly influenced by the dynamical properties of the solvent, are thought to be of importance in the studied process of ordering.

EXPERIMENTAL

In all extensive experimental studies available[1-4], the hysteretically reversible gel-sol cycle of agarose-aqueous systems has been studied by following progressive changes of some of its properties upon temperature scans whose rate was determined by practical circumstances. The occurrence of changes on longer time scales when samples were stored at constant temperature was observed in several occasions[2,4], yet never systematically studied. In the present work we have followed the changes in apparent optical density (due to turbidity) at different wavelengths, in a range where no electronic transitions are observable, in samples stored at constant temperature for times up to several weeks. A Jasco UVIDEC 320 Spectrophotometer was used, in connection with a fully computerized system for interactive data acquisition and analysis[10]. Agarose was Ultrapure Seakem HGT(P) from Marine Colloids, Inc., water was from Millipore Super Q with 0.22 μ filters, and D_2O (99.8%) from Prochem Ltd. Samples were dissolved by immersion in boiling water for 15 min, poured in stopped optical cuvettes (further sealed with parafilm), and stored at 70°C for about two hours. The cuvettes were then transferred into the cuvette holders, thermostatted at the wanted temperature. The time required for the start and completion of the gelation process was strongly dependent upon concentration and temperature. We disregarded kinetics in which a significant part of the process occured during thermalization of the samples, and those in which no appreciable growth of O.D. was observed after about 6 weeks. These time limitations set the limits of the temperature and concentration ranges of our experiments, which were 0.5%-5% w/v agarose concentration and 35-52°C.

Optical density (O.D.) growth curves were observed to differ

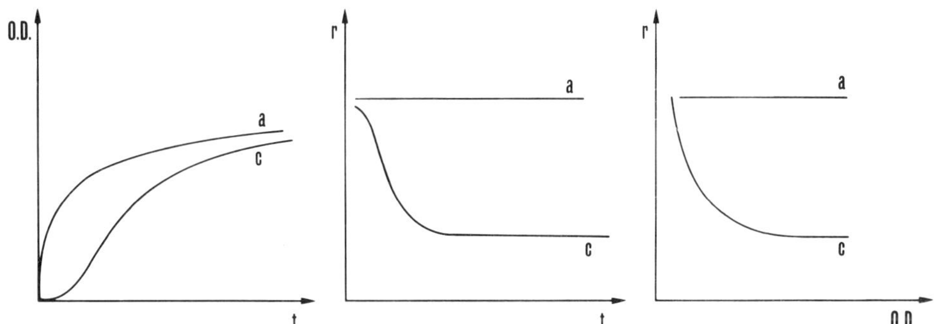

Fig. 1 Schematic comparison of two extreme gelling kinetics of Agarose aqueous systems. Left: apparent optical density (O.D.) vs. time; center: ratio r=(O.D. at 350 nm)/(O.D. at 450 nm) vs. time; right: r vs. O.D. upon elimination of the time variable. Curves of the a-type are observed at T<~40°C and curves of c-type at T>~45°C. Intermediate behaviour (not shown) is observed in the narrow intermediate temperature range.

markedly in different conditions. The same was true for the time evolution of the ratio (r) of O.D. values taken at different wavelengths. This ratio contains information on the correlation length of refractive index , i.e. the average distances along which the refractive index is fairly uniform. Detailed experimental data, including work still in progress will be published elsewhere[11]. In Fig.1 we show typical shapes only, of the dependence of O.D. and r upon time and, upon elimination of the time variable, of r vs. O.D. Curves of the type labeled _a_ (characterized by a monotonically decreasing time-derivative of O.D., a constant r, and a time scale from a few minutes to several hours) are observed in the lower temperature region (T<~40°C), while curves of the type labeled _c_ (characterized by a sigmoidal shape, a rapidly stepping down value of r and a time scale from several hours to several weeks and even much further) are observed in the higher temperature region (T>~45°C). Intermediate behaviours are observed in the narrow intermediate temperature range. Note that gels described in literature obtained with the usual rapid (or even non-equilibrium) process, can hardly be anything but type _a_, that is the type corresponding to a fast kinetic process. For this reason we refer to this type of gel (and process) as "normal". To the best of our knowledge, the slowly formed gel obtained in the higher temperature range had not been reported before: since the observed low value of r indicates a long correlation length, we will refer to it as "long range ordered" (or LRO) gel.

Comparison of the melting curves of normal and LRO gels (Fig.2), shows that the latter is characterized by a considerably higher melt-

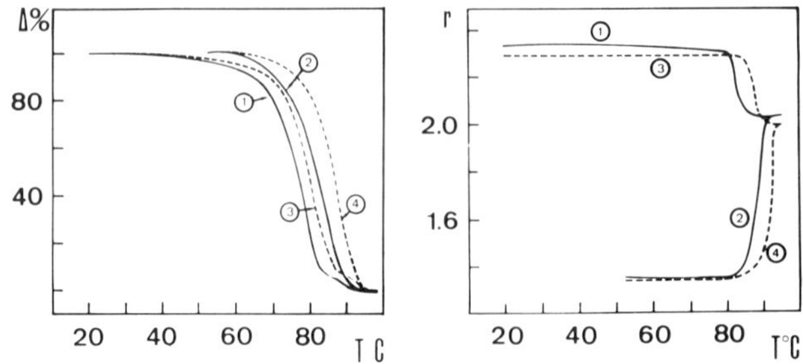

Fig. 2 Melting curves of 3% w/v agarose gels: 1)normal, in H_2O; 2)LRO, in H_2O; 3)normal, in D_2O; 4)LRO, in D_2O. Left: (O.D.)%; right: r=O.D.$_{350}$/O.D.$_{450}$

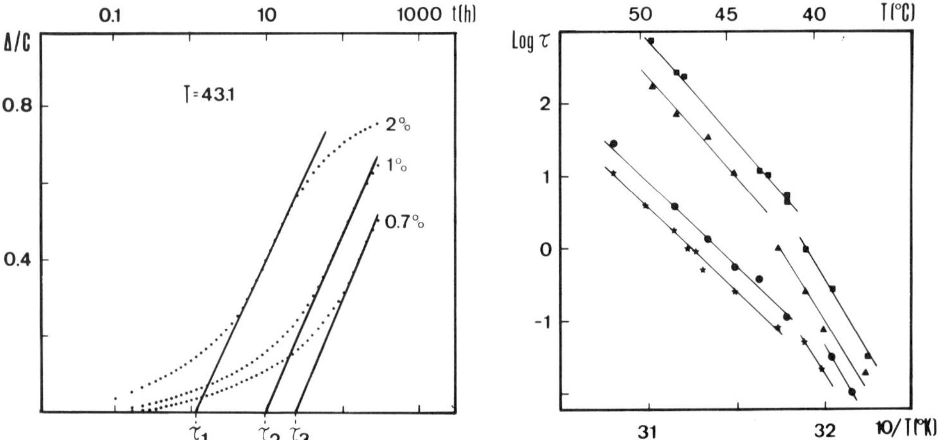

Fig. 3 Definition of a characteristic time for the kinetics, and its Arrhenius plots. Left: the tangent is drawn at the point of maximum slope and the intercept with the log(t) axis is taken. Note that on a logarithmic time scale all kinetic curves turn out to have a sigmoidal shape, so that their slope has a maximum in all cases. Right: dots: H_2O, 5% ; stars: D_2O, 5% ; squares: H_2O, 1% ; triangles: D_2O, 1% . Note that the effect of deuteration is a parallel shift towards higher temperature.

ing temperature (+6°C at 50% O.D. change) and a completely different behaviour of r.

A qualitative comparison of the growing rates due to the different shapes of kinetics can be made. We arbitrarily choose a characteristic time, defined as in Fig.3 (left). The temperature dependence of this characteristic time is shown in Fig.3 (right), for two typical concentrations. Results relative to D_2O are qualitatively similar to those for H_2O, with a systematic (yet not identical) upward shift of temperatures limiting the different ranges, and of the melting temperatures.

Finally, storage of normal gel at a temperature in the higher region causes a conversion into the LRO gel. A converse storage of LRO gel at lower temperatures has no effect on its properties. A study of the kinetics of these annealing processes, in comparison with the formation of the LRO gel from sol phase is in progress.

DISCUSSION

The average length of an helical strand of Agarose can be assumed to be some 300 nm[1c]. In the ordinary gel the observed r-value corresponds to a correlation length of refractive index of about 350 nm. This is consistent with the presence, in the ordinary gel, of bundles containing only a few strands[1].

The LRO type of gel is here reported and described for the first time: melting curves in Fig.2 show it to be appreciably stabler than the normal gel. The observed low value for the r parameter indicates a very large correlation length, corresponding to a much higher degree of order, hitherto unobserved in vitro. The long "incubation" times (up to several weeks, and more) observed in the sigmoidal growth curves of apparent O.D. suggest that the observed higher order is the result of a different and not only longer gelation process, possibly occurring through a different path in the phase space. The increase of correlation length, and the slowing-down of gelation times may deserve, however, more exhaustive attention and study, in the light of phase transitions theory.

Effects of D-H substitution in the solvent (and therefore also at all exchangeable hydrogen sites in the polymer) may appear trivial. Yet, as we are going to see, they provide hints as to differences between processes responsible for normal and LRO gelation. They can be summarized as a general, yet not equal, upward shift of all relevant temperatures (including the boundary region, above which LRO gelation is observed), and an acceleration of gelation kinetics. The latter can be counteracted by appropriate upward shifts of temperature.

Thermodynamic analysis, and comparison with similar effects in polynucleotides and model systems show that the sole consideration of a possibly stronger deuterium bonding is not sufficient to explain the experimental observations: a non electrostatic, isotopic modulation of the entropy term must play the central role[2b,6,7,12,13], and this derives from solute-solvent coupling. This is easily understood by considering that across a conformational transition, a solute biomolecule will alter the proportion of its hydrophilic and hydrophobic surface exposed to the solvent, and that hydrophobic interactions, entropic in nature, are stronger in D_2O than in H_2O[14,15]. From a microscopic point of view, we may more interestingly note that changing a dynamic parameter of solvent molecules (the hydrogen mass) causes alterations which cannot be explained in terms of an altered hydrogen bond strength, but must alter the overall entropy so as to entrain and shift in the scale of temperature the conformational transition(s) and process(es) in the biomolecular sub-system. This evidences a coupling between solvent molecular dynamics and molecular and supramolecular order of solute biomolecules[7,13].This view is in accord[7] with current views on liquid water as a network of uninterrupted hydrogen bond paths, rearranging itself in convex cages in the presence of hydrophobic surfaces[14], and with Monte Carlo calculations providing ample evidence for the existence of hydrogen bond paths pinned at hydrophilic sites only, in either DNA and Agarose double helices in water[8]. In other words, the structure of the solute biopolymer and that of the connectivity network in the solvent will depend on each other, while the latter happens to depend upon the dynamic parameters of solvent molecules[7,13,16].

All this offers viewing molecular and supramolecular stability in a much ampler volume in the phase space. Also, it offers a closer insight in processes responsible for LRO gelation and for the conversion of ordinary gels into LRO gels upon storage at sufficiently high temperature. These processes must be capable of annealing out localized defects of polymer ordering (such as hairpins), and of relaxing strains caused by the formation of bundles. So, they will probably include mechanisms such as unravelling, chain slippage and the like[8], and localized highly non-linear dynamical modes such as solitary waves[5], all implying the exposure of hydrophobic surfaces to the solvent and the role of highly non linear hydrophobic restoring "forces". In agreement with our observation, these mechanisms are expected to be more effective in the higher temperature range, where LRO gels are obtained. It is of interest to note in this connection that electron microscopic observations have shown that "pre-melting bubbles" in native DNA are affected by solvent deuteration just as expected on the basis of the foregoing considerations[17].

These considerations are being taken up as guidelines for work in progress on computer simulation of ordinary and LRO kinetics[18], following the conceptual framework of the Flory-Stockmayer theory[19,20] whereby different steps operate in cascade, each setting the necessary conditions for the occurrence of the next one. Results are so far encouraging towards a quantitative understanding of the kinetics[16].

ACKNOWLEDGEMENTS

Prof. M.U. Palma, Dr. M. Migliore and Dr. G. Guzzio took an active part in the early stages of this work and, although other committments have prevented their effective participation, we have benefited from a number of stimulating discussions with them. Mr. M. Lapis, p.i., had a major role in implementing the microelectronics and Mr. S. Pappalardo provided accurate and careful technical help. Financial support from Ministero Pubblica Istruzione (local funding) is acknowledged.

REFERENCES

1. a) I.C.M. Dea, A.A. Mc Kinnon and D.A. Rees, Tertiary and Quaternary Structure in Aqueous Polysaccharide Systems, J.Mol.Biol. 68:153 (1972)
   b) D.S. Reid, T.A. Bryce, A.H. Clark and D.A. Rees, Helix-Coil Transition in Gelling Polysaccharides, Faraday Discuss.Chem.Soc. 57:230 (1974)
   c) S. Arnott, A. Fulmer, W.E. Scott, I.C.M. Dea, R. Moorhouse and D.A. Rees, The Agarose Double Helix and Its Function in Agarose Gel Structure, J.Mol.Biol. 90:269 (1974)
2. a) P.L. Indovina, E. Tettamanti, M.S. Micciancio-Giammarinaro and M.U. Palma, Thermal Hysteresis and Reversibility of Sol-Gel Transition in Agarose-Water Systems, J.Chem.Phys. 70:2841 (1979)
   b) G. Vento, M.U. Palma and P.L. Indovina, Concentration and Isotope Effects in the Stability of Agarose Gel, ibid. 70:2848 (1979)
3. a) A. Hayashi, K. Kinoshita and M. Kuwano, Studies of the Agarose Gelling Systems by the Fluorescence Polarization Method. Part I, Polym.J. 9:219 (1977)
   b) A. Hayashi, K. Kinoshita, M. Kuwano and A. Nose, id. Part II, ibid. 10:485 (1978)
   c) A. Hayashi, K. Kinoshita and S. Yasueda, id. Part III, ibid. 12:447 (1980).
4. J.D. Aplin and L.D. Hall, Spin-Labelling Studies of the Agarose Gelling Systems, Carbohydr.Res. 75:17 (1979)
5. a) C. Mandal, N.R. Kallenbach and W. Englander, Base-Pair Opening and Closing Reactions in the Double Helix, J.Mol.Biol. 135:391 (1979)
   b) S.W. Englander, N.R. Kallenbach, A.J. Heeger, J.A. Krumhansl and S. Litwin, Nature of the Open State in Long Polynucleotide Double Helices: Possibility of Soliton Excitations, Proc.Natl.Acad.Sci.(USA) 77:7222 (1980)
6. A. Cupane, E. Vitrano, P.L. San Biagio, F. Madonia and M.U. Palma, Thermal Stability of Poly(A) and Poly(U) Complexes in H_2O and D_2O: Isotopic Effects on Critical Temperatures and Transition, Nucl.Acid.Res. 8:4283 (1980)

7. M.U. Palma, Isotope Effects and Collective Excitations, in: "Excitations in Biological Systems" ,F. Kremer, ed., Springer Verlag, Berlin (to appear)

8. a) E. Clementi and G. Corongiu, Solvation of DNA at 300 K: Computer Experiment, in: "Biomolecular Stereodynamics" vol. I, R.H. Sarma, ed., Adenine Press, New York (1981)

   b) S.L. Fornili, G. Corongiu and E. Clementi: to be published

9. M. Spodheim and E. Neumann, Kinetic Analysis of the Annealing Period in the Formation of the Poly(A).2Poly(U) Triple Helix, Biopolymers 16:289 (1977)

10. S.L. Fornili and M. Migliore, Microcomputer-Based System for Automation of Spectrophotometric Data Acquisition for Long Lasting Kinetics, J.Phys.E.:Sci.Instrum. 14:426 (1981)

11. M. Leone, S.L. Fornili and M.B. Palma-Vittorelli, to be published

12. S.L. Fornili, G. Sgroi and V. Izzo, Solvent Isotope Effect in the Monomer-Dimer Equilibrium of Methylene Blue, J.Chem.Soc. 77:3049 (1981)

13. M.U. Palma, Internal Dynamics of Biomolecules in Solution, in: "Structure and Dynamics: Nucleic Acids and Protein" E. Clementi and R.H. Sharma, eds., Adenine Press, New York (1983)

14. F. Franks, ed.,: "Water, a Comprehensive Treatise", Plenum Press, New York (seven volumes, 1972-1982)

15. F.H. Stillinger, Water Revisited, Science 209:451 (1980)

16. G. Aiello, M.S. Giammarinaro, M.B. Palma-Vittorelli and M.U. Palma, Behaviour of Interacting Protons: The Average Mass Approach to its Studies and Possible Biological Relevance, in: "Cooperative Phenomena", H. Haken and M. Wagner, eds., Springer-Verlag, Berlin (1973)

17. V. Izzo, S.L. Fornili and L. Cordone, Thermal Denaturation of B. Subtilis DNA in H_2O and D_2O Observed by Electron Microscopy, Nucl.Acid.Res. 2:1805 (1975)

18. M. Migliore, M. Leone and M.B. Palma-Vittorelli, to be published

19. P.J. Flory, "Principles of Polymer Chemistry", Cornell University, Ithaca, N.Y. (1953)

20. W.H. Stockamayer, Theory of Molecular Size Distribution and Gel Formation in Branched-Chain Polymers, J.Chem.Phys., 11:45 (1943)

PHYSICAL TECHNIQUES IN THE STUDY OF WATER IN BIOLOGICAL SYSTEMS CRYOBIOLOGY

ON THE STATE OF WATER IN BIOLOGICAL SYSTEMS

EVALUATION OF METHODS OF ITS INVESTIGATIONS

S.I. Aksyonov

Biology Department
Moscow State University
Moscow II7234, USSR

INTRODUCTION

Its obvious at present, that the fundamental problems of molecular biology and biology and biophysics - how high efficiency of biological reactions operating with the involvement of globular proteins is attained - cannot be solved without considering the role of water and its state. Questions regarding efficiency and energetics of biological reactions were not of much importance as long as much of research was aimed at elucidation of structural particularities of nucleic acids and mechanisms of their functioning. What is most important about nucleic acids is stability and precision of functioning, while contribution of processes depending on them is not great in the general energetics of the cell. At the same time, proteins are involved in practically all basic biological processes. Hence, the elucidation of questions concerning the mechanisms providing high efficiency of protein functioning can open the door to solving other problems of molecular biophysics. Solving this problem is also very important for practical application - synthesis of artificial catalysts with properties similar to biological ones. With due consideration of the role of water in biological systems we will probably be able to find the reason for the difference between calculated and observed rates of enzymatic reactions, which differ by 9-I2 orders of magnitude (Volkenstein, I975).

It is now believed that water has a direct influence on biological reactions associated with conformational changes of the macromolecules in the process and the resultant free energy of them is probably dependent on alterations in the structure of the water surrounding the macromolecules ("entropy-enthalpy compensation effects") (Lumry, 1974). If so, one should expect the presence of a considerable amount of water around biopolymer macromolecules which has structural and dynamical properties other than ordinary water. Moreover, the structure of the surrounding water must change while protein functioning because any biological reactions are accompanied by conformational changes of proteins. This view, however, is inconsistent with results of various experimental observations aimed at elucidating the state of biological water. Spin-echo NMR studies (Lubas and Wilczok, 1966, Abetsedarskaya et al, 1968), NMR measurements at low temperatures (Kuntz and Brassfield, 1971), calorimetry (Mrevlishvili and Privalov, 1969), dielectric spectroscopy (Schwan, 1965, Kashpur et al, 1976) and some other techniques (Petrochenko and Privalov, 1973, Kuntz and Kauzmann, 1974) indicate that in biological systems the amount of bound water with altered structure is not greater than a hydration monolayer surrounding the macromolecules. Besides, the amount of bound water changes little and in some cases decreases after denaturation of proteins (Petrochenko and Privalov, 1973). Physical modelling of systems shows that the amount of bound water observed in NMR measurements at low temperatures is correlated with the number of polar groups or centers on the boundary surface; they also show that such water lacks from areas adjacent to pure hydrophobic surfaces (Kurzaev et al, 1977).

If it is true that the structure of water is subject to alterations in the vicinity of polar groups only and that the amount of bound water does not practically vary even after denaturation of proteins, consideration of hydrophobic interactions and entropy-enthalpy compensation effects in the water-protein system does not make sense. The change in the free energy of the water component cannot occur without a change of its structure. Solving this contradiction and obtaining corrected data for water in biological systems in very important also because of a complex character of hydrophobic groups distribution on the surface and within the interior of a protein globule (Klotz, 1970, Janin, 1979, Keshavarz and Nakai, 1979). This distribution is an individual characteristics of each protein depending on the regulatory action of water on the structure of protein, when water determines the balance of forces within globule, (Aksyonov, 1981) and its influence on the energetics of processes occuring with the involvement of proteins. Besides, some theoretical evaluations of hydrophobic effects that have been made were based on erroneous assumptions. In particular, hydrophobic interactions in proteins are interpreted as being due an additional coordination bond formed at the surface of globules in which the water molecules are H-bonded to 4 nearest neighbour water molecules (Scheraga, 1965). However, such bonded structure may be realized near the surface in rare cases because the bonding in a structure with quadrupole bound water is continuous in space.

For all these reasons a more critical evaluation is needed of some interpretations previously put forth basing on results obtained through use of various methods.

RESULTS AND DISCUSSION

Let us consider results of NMR measurements at low temperatures. Bound water content is determined here as non-freezing fraction of water at 30°C and it accounts for a narrow NMR line (Kuntz and Brassfield, 1971). However, as we have mentioned already (Aksyonov, 1977), this line can be discernibly observed when molecules of water are in separate liquid or quasiliquid phase for which the lifetime is greater than the inverse of resonance line width (about 10^{-3}s). In the surrounding of charged groups, this is true for the water molecules located in a monomolecular layer, for as it follows from the relation:

$$\nu \sim kT/h \cdot e^{-E/RT}$$

(ν is the frequency of molecule transition); at an energy of binding E higher than 12 kcal/mole the lifetime of the water molecules in this layer is as long as 10^{-3}s. (Entropy contribution is not discussed here). Water molecules with steric restrictions too should have a long lifetime (Aksyonov, 1977). All other water molecules with altered structure should exchange with bulk water within this time which will find its reflection on the wide NMR line belonging to freezed bulk water. It is known, that width of this line gradually approaches its limit width on lowering the temperature to 210°K (Kurzaev et al, 1977).

Fast transitions of water molecules from one state to another find their reflection in IR spectra too. As the lifetime of H-bonds in liquid water is about 1,5 10^{-12}s (Yukhnevich, 1973) and the rate of proton transfer is greater than the rate of diffusion in water (Kaldin, 1964), one may expect a broadening of some separated lines and disappearance of some other from IR spectra in the range to several dozens of cm^{-1}.

Consider now some results of calorimetric measurements. The lack of a heat absorption peak in the biopolymer-water system at about 0°C is attributed to bound water. This peak appears immediately after the water content in the system becomes greater than the amount of water surrounding the polar groups and some other regions of the macromolecules (Mrevlishvili and Privalov, 1969). From this it has been concluded that all water which is recorded after the appearance of this peak is free water. However, from the appearance of the peak of heat capacity of about 0°C it follows only that some of the absorbed water forms four H-bonds. Moreover, these investigators did not take into consideration the sequence of water absorption in various sites of macromolecules as water content increases (Aksyonov, 1977). The location of water in the layer adjacent to the hydrophobic regions is energetically an unfavourable situation. Water molecules localize in this layer only after the bulk water has filled regions adjacent to the absorbed water in polar groups and then the excess water in the macromolecules has entered into contact with the hydrophobic regions of the macromolecule surface. In this connection it is worth mentioning that with increasing water content, in DNA and protein samples, heat absorption increases at low temperatures (down to -30°C) (Mrevlishvili and Privalov, 1969), a fact which cannot be explained in terms of bulk water only. The above mentioned reasoning must be taken into consideration in interpreting data of gravimetry, low temperature NMR studies of protein crystals etc.

Another method, dielectric spectroscopy, which is frequently used in water investigations (Schwan, 1965, Kashpur et al, 1976) is concerned with absorption of high frequency electromagnetic radiation. It is thought that bound water does not absorb in the mm wave range which belongs to free water absorption. However, the reorientation time of water molecule dipoles differs from the lifetime of molecules of bound water in various states. NMR data indicate that there is no isotopic substitution effects for water in biological systems (Abetsedarskaya et al, 1968, Aksyonov and Kharchuk, 1977, Hallenga and Koenig, 1976, Lubas and Wilczok, 1966). This textifies a fast rotation of bound water molecules about a axis beyond the reorientation within the charge field. Water molecules are limited in freedom of reorientation definitely only when a charge field or steric restrictions are imposed on them. There are a number of other situations with restricted freedom of movement of water which must be studied additionally.

First NMR spin-echo results were interpreted on the assumption that there is a fast exchange between the fractions of bound and free water and that bound water is homogeneous (Abetsedarskaya et al, 1968, Lubas and Wilczok, 1966). From two measured parameters, relaxation times T_I and T_2, one can, in principle, determine correlation time τ_c for bound water and its content. However, the view that bound water is homogeneous is in contradiction with some results obtained by this method. Theory predicts that T_I should decrease with temperature increasing, when T_I is longer than T_2. NMR measurements show an opposite dependency (Abetsedarskaya et al, 1968, Aksyonov and Kharchuk, 1977, Hallenga and Koenig, 1976, Lubas and Wilczok, 1966). The NMR theory also predicts that the ratio of T_I/T_2 should be proportional to the square of resonance frequency, while NMR observations give not support to these theoretical expectations. For the temperature dependence of T_I and T_2 for water protons, the scattering of activation energy values appeared to be within 0 to 6 kcal/mole, this value also being dependent on biopolymer concentration, pH etc. We feel, therefore, that the above approach to the interpretation of NMR spin-echo data is incorrect.

In later works, models involving two states of bound water were suggested, belonging to weakly and strongly bound water, with each state having its own correlation time (Kuntz and Kauzmann, 1974, Grosch and Noack, 1976). It has been reported that there is a good correlation between calculated and experimental values of T_I for protons of water in SA solutions of different concentrations in the frequency range from I6 kHz to 70 MHz (Grosch and Noack, I976). Note, however, it does not seem appropriate to speak of "a good correlation" bearing in mind that the amount of strongly bound water per one protein macromolecule changed from 4,5 to 242 for different SA concentrations and the amount of weakly bound water by about 5 times (Grosch and Noack, I976).

The approach taken in another model based on the fact there is a correlation between values of T_I for water protons and deutrons at the different resonance frequencies and mobility of biopolymer macromolecules (Hallenga and Koenig, I976). The T_I frequency dependencies that have been found for water solutions of proteins of different molecular weight indicate that there is a broad spectrum of motions of bound water molecules. Analysis of the data has led to the conclusion that for different proteins there is some correlation between the position of the center point on the I/T_I vs resonance frequency curve, where is a sharp decline of the curve and protein molecular weight (Hallenga and Koenig, I976). Using the values of correlation times for bound water molecules, determined from this relation, one can calculate that the amount of strongly bound water constitutes about 0,I% of protein molecular weight M. Moreover, one can readily see that there is a similar or even better correlation between positions of the above mentioned centers and corresponding values of $M^{2/3}$ for them or macromolecule surface size. Clearly, correlations of the type discussed here do not reflect the real dynamics of bound water molecules in biopolymer solutions.

To obtain a true picture of the state of water in biological systems, one should take into consideration the following factors. First, the direct dependency of water dynamics on biopolymer dynamics exist only in cases, where the lifetime of bound water molecules is much greater than the timescale of biopolymer motion. This condition is fulfilled as a rule with strongly bound water which has a lifetime of IO^{-3} to IO^{-7}s. Furthermore, as water is bound on the boundary surface of biopolymers first of all, the motion of the lateral groups which have higher frequencies, compared with the motion of the macromolecule, must determine the dynamic characteristics of the strongly bound water and relaxation times T_I and T_2 from protons of water in biopolymer solutions. In view of this one can expect a much greater amount of strongly bound than the above mentioned value. The extent of influence of other fractions of strongly bound water, which are not in contact with the boundary layer of the solution (water molecules within the subsurface layers of macromolecules) on T_I and T_2 shall be limited by the time of their exchange with the bulk water. The presence of two mentioned types of states of strongly bound water also follows from the scale of T_2 values for water protons in SA solutions of various concentrations at the different temperatures, pH and ionic strength and from comparison of these values with T_2

values for protons of lateral groups of protein in such solutions (Aksyonov and Kharchuk, I977). The observed deviations from linearity of the concentration dependence of I/T_2 for water protons correspond quantitatively to the slowing-down of motion of the lateral groups of globules which is considerable because of a change in interactions between the protein macromolecules with increasing concentration or with a variation of pH, ionic strength etc. (Aksyonov and Kharchuk, I977). If the situation is as suggested here, one can easily explain the broad frequency dispersion of T_I for water protons in protein solutions (Hallenga and Koenig, I976) whose frequencies of motion are higher and lower than those of a protein globule.

Strongly bound water is only part of the water which is bound by the biopolymer macromolecule. The temperature dependencies of T_I obtained by us and other investigators for water protons in solutions of SA, TMV and other biopolymers which have strongly different dynamic characteristics (Aksyonov and Kharchuk, I977, Hallenga and Koenig, I976, Lubas and Wilczok, I966 etc) show the presence of two types of temperature dependencies belonging to solutions of biopolymers with great and little mobility. The activation energy was found to be within from 3 to 5 kcal/mole for the former and close to O for the latter. Analysis of these data indicates the presence of at least 2 states of weakly bound water (Aksyonov and Kahrchuk, I977). The temperature dependencies that have been obtained suggest that the frequencies of motions of its both states exceed the resonance frequency in solutions of mobile biopolymers, while in solutions of biopolymers with little mobility one of the two states of weakly bound water has a lower frequency than the resonance frequency. The two states of bound water have opposite temperature dependencies of T_I for water protons and in sum both determine that this value has a weak temperature dependence (Aksyonov and Kharchuk, I977). This conclusion is in good agreement with experimentally obtained frequency dependencies of T_I for water protons in various protein solutions (Hallenga and Koenig, I976). Experiments have revealed a high frequency component for bound water motion which is not in correlation with the motion of macromolecules. Second high frequency component of motion of bound water falls within the tail region of the frequency dispersion observed in a solution of heavy macromolecules with a molecular weight of about I million daltons (Hallenga and Koenig, I976). The motion of this fraction of weakly bound water depends but only to some extent on macromolecule motion. These two components have opposite signs of the T_I temperature dependence and are thought to belong to above mentioned two states of weakly bound water.

Energy contribution of weakly bound water approaches to one of strongly bound water, that follows for example, from the data of direct calorimetric study of DNA dehydration effects by adding polyethylenglicole (Platonov et al, I976).

Thus, the real picture of the state of water in biological system is more complex than usually postulated on the basis of results that are now available with the aid of different physical methods. It does not contradict to the view that water has great influence on the structure and functioning of biopolymers. A quantitative evaluation of weakly bound water in biopolymer solutions has not been made as yet and little is known about its alteration in biological reactions. Solving this question is an urgent task for molecular biophysics. Extension of studies to frequencies corresponding to frequencies of motion of bound water molecules using NMR frequencies above IOO MHz is one of possible ways to carry out such investigations.

More complicated problems arise in investigations of water in biological tissues. Markedly shorter values of T_I and T_2 for water protons were observed in such investigations, compared with protein solutions, when in both cases the concentrations of the dry substance were equal. Moreover, with tissues changes in water proton relaxation times were seen in pathological cases, although in solution even protein denaturation shows a weak influence on water proton relaxation time (Damadian, I97I, Mathur-de-Vre, I979, Murza et al, I978). Data that have been obtained in model solutions of biopolymers suggest that consideration of interactions with structures having little mobility is important for proper evaluation of water proton relaxation times for tissues. Also, special experiments with biological tissues show that the intensity of slow components of the spinecho decay curve from protons on partially dried fir needles exceeds markedly the waited one from water protons only. Besides, we observe a lack of change of decay curve shape after partial H_2O-D_2O substitution. Similar results were obtained too on the model solutions of starch. All this shows a marked contribution to T_2 for tissues water protons from the exchange between the protons of water and NH- and OH-groups of biopolymers, including little mobile macromolecules of polysaccharides (Aksyonov and Kharchuk, I978). This conclusion was confirmed by an observation of a considerable difference in the T_I/T_2 ratio for protons and deutrons of water compared with the one seen in ^{I7}O of water molecules in tissues (Chivan et al, I978). At the same time the contribution from the interactions of water with solute biopolymers constitutes only a small part of the relaxation rate for water protons in tissues. A change in hydration of solute biopolymers, associated with the processes that occur in pathological tissues gives even a smaller contribution to T_I and T_2. All this shows that magnitudes of T_I and T_2 for intracellular water are not sufficient indicators of the state of water in a biological tissues.

The conclusion that interactions with structures having little mobility give the greatest contribution to the relaxation of water in tissues is important, for consideration of them allows us to interpret correctly NMR data for tumours which show greater values of T_I for water protons compared with normal tissues (Damadian, I97I, Mathur-de-Vre, I979, Murza et al, I978). Increasing T_I for water in tumours may be associated with a decrease in the content of bound water in them compared with normal tissues (Damadian, I97I). However, calirimetry of cancer tissues shows an increased content of unfrozed water in them (Andronikashvili and Mrevlishvili, I968). This contradiction may be explained in terms of destruction of structures in cancer having little mobility. Due to destruction the number of biopolymer groups which interact with water increases and can be registered by the calorimetry method. At the same time, the destruction leads to greater mobility of water which is bonded through such structures, especially by lateral groups. Mobility of lateral groups of biopolymers must be very sensitive, for example, to the change of intermolecular interactions. Increasing mobility of water is observed as increased T_I for water protons in cancer. The view that in tumours there occurs a destruction of structures having little mobility correlates with the appearance of narrow proton resonance lines in such tissues after D_2O substitution, while such effect was not observed in normal tissues. (Zenin et al, I976). Destruction of structures with a small degree of mobility leads also to an increase in water content in tumours compared with normal tissues (Mathur-de-Vre, I979, Murza et al, I978) because of increase of the osmotic pressure in such cells.

Thus, values of T_I and T_2 for protons and deutrons in intracellular water is an indicator of the state of cellular structures rather than an indicator of the state of water in biological tissues.

REFERENCES

Abetsedarskaya, L. A., Miftakhutdinova, F. G., and Fedotov V. D., I968, On the state of water in living tissues, Biofizika (USSR), I3:630

Aksyonov, S. I., I977, On the evaluations of the state of water in biological systems with aid of various physical methods, Biofizika (USSR), 22:923

Aksyonov, S. I., I98I, Water as a regulator in biological systems, Studia biophysica, 84:37

Aksyonov, S. I., and Kharchuk, O. A., I977, On the state of water in solutions of proteins and viruses, In: "Bound water in heterogeneous systems", vol. 4, V. F. Kisselev, V. I. Kvlividze, and R. I. Zlochevskaya, eds, Moscow Univ. Ed., Moscow (in Russian).

Aksyonov, S. I., and Kharchuk, O. A., I978, On the NMR characteristics for water protons and the state of water in plants, In: "Water relations of plants in various ecological conditions", N. S. Petinov, ed., Kazan Univ. Ed., Kazan, (in Russian).

Andronikashvili, E. L., and Mrevlishvili, G. M., I968, Study of state of water in tumours by calorimetry method, Doklady AN SSSR, I83:463

Chivan, M. M., Achlama, A. N., and Shporer, V., 1978, The relationship between the transverse and longitudinal NMR relaxation rate of muscle water, Biophys. J., 21:127
Damadian, R., 1971, Tumour detection by NMR, Science, 171:1151
Damadian, R., Zaner, K., Hot, D., DiMaio, T., Minkoff, L., and Goldsmith, M., 1973, Ann, N. Y., Acad. Sci., 222:1048
Grosch, L., and Noack, F., 1976, NMR relaxation investigation of water mobility in aqueous bovine serum albumin solutions, Biochim, Biophys, Acta, 453:218
Hallenga, K., and Koenig, S. H., 1976, Protein rotation relaxation as studied by solvent ^{1}H and ^{2}H magnetic relaxation, Biochemistry, 15:4255
Janin, J., 1979, Surface and inside volumes in globular proteins, Nature, 277:491
Kaldin, E. F., 1964, "Fast reactions in solutions", Blackwell Sci. Publ., Oxford
Kashpur, V. A., Maleev, V. Ya., and Schegoleva, T. Yu., 1976, Study of globular protein hydration by aid of differential dielectric spectroscopy, Molec. biologiya (USSR), 10:568
Keshavarz, E., and Nakai, S., 1979, The relationship between hydrophobicity and interfacial tension of proteins, Biochim. Biophys. Acta, 576:269
Klotz, I. M., 1970, Comparison of molecular structure of proteins: helix content, distribution of apolar residues, Arch. Biochem. Biophys., 138:704
Kuntz, I. D., and Brassfield, T. S., 1971, Hydration of macromolecules. II. Effects of urea on protein hydration, Arch. Biochem. Biophys., 142:660
Kuntz, I. D., and Kauzmann, W., 1974, Hydration of proteins and polypeptides, Adv. Protein Chemistry, 28:239
Kurzaev, A. B., Kvlividze, V. I., and Kisselev, V. F., 1977, in: Specifics of the phase transition of water on the surface of biological and model dispersed bodies at low temperatures, in: "Bound water in heterogeneous systems," vol. 4, V. F. Kisselev, V. I. Kvlividze, and R. I. Zlochevskaya, eds, Moscow Univ. Ed., Moscow, (in Russian).
Lubas, B., and Wilczok, T., 1966, Spin-echo technique study of the non-rotational hydration of DNA, Biochim. Biophys. Acta, 120-427
Lumry, R., 1974, Participation of water in protein reactions, Ann. N. Y. Acad. Sci., 227:471
Mathur-De-Vre, R., 1979, The NMR studies of water in biological systems, Progr. Biophys. Molec. Biol., 35-103
Mrevlishvili, G. M., and Privalov, P. L., 1969, Calorimetric investigation of macromolecular hydration, in: "Water in biological systems", L. P. Kayushin, ed., Consultants Bureau, N. Y.
Murza, L. I., Bunto, T. V., Naidich, V. I., and Emanuel, N. M., 1978, Proton magnetic relaxation in investigation of tumour influence on organism, Doklady AN SSSR, 238:474

Petrochenko, S. I., and Privalov, P. L., I973, Change of globular protein hydration after thermal denaturation, Biofizika (USSR), I8:555

Platonov, A. L., Protassevich, I. I., Evdokimov, Yu. M., Akimenko, N. M., Chebotareva, N. A., Varshavsky, Ya. M., I976, Thermal effect preceeded to compactization of two-chain DNA on polythylenglicole-water-salt solutions, Molec. biologiya, (USSR), IO:32I

Scheraga, H. A., I965, The effects of solutes on the structure of water and its implications for protein structure, Ann. N. Y. Acad. Sci., I25:253

Schwan, H. P., I965, Electrical properties of bound water. Ann. N. Y. Acad. Sci., I25:344

Volkenstein, M. V., I975, "Molecular Biophysics", Nauka, Moscow, (in Russian)

Yukhnevich, G. V., I973, "IR-spectroscopy of water". Nauka, Moscow, (in Russian)

Zenin, S. V., Chupina, G. I., and Vinnik, L. A., I976, NMR study of normal and pathological tissues of human, Studia biophysica, 52:53

AN ESR STUDY OF THE OXIDATION AND REDUCTION OF BISULFITE (HYDRATED SULFUR DIOXIDE) IN BIOLOGICAL SYSTEMS

Colin F. Chignell, Carolyn Mottley, Kandiah Sivarajah, Thomas E. Eling, and Ronald P. Mason
Laboratory of Environmental Biophysics
National Institute of Environmental Health Sciences
Research Triangle Park, N.C. 27709 USA

INTRODUCTION

Sulfur dioxide is recognized as a major air pollutant, particularly near large cities (Rall, 1974), while the ionized forms, bisulfite and sulfite, are found as preservatives in food and wine. In the lung, sulfur dioxide is hydrated rapidly according to the following equation,

$$H_2O + SO_2 \rightleftharpoons HSO_3^- + H^+$$

with the equilibrium constant 1.07 x 10^2 mole/liter. The bisulfite anion is a weak acid which dissociates according to the reaction,

$$HSO_3^- + H_2O \rightleftharpoons H_3O^+ + SO_3^{-2}$$

with an equilibrium constant of 1.07 x 10^7 mole/liter. At pH values above 7 the equilibrium lies to the right and sulfite predominates, although there is always an equilibrium between sulfite and bisulfite. In this paper the terminology (bi)sulfite will be used when it cannot be determined which species is involved in a given reaction.

OXIDATION OF (BI)SULFITE

Horseradish Peroxidase

The addition of (bi)sulfite to horseradish peroxidase at pH 8.6 results in the generation of a single line ESR spectrum (Fig. 1). The species giving rise to this line has been identified as the $SO_3^{\cdot-}$ radical based on a comparison of the experimental g-value (2.0031) with the reported g-value for the sulfur trioxide anion free radical (Behar and Fessenden, 1972). It may be seen from Fig. 1 that all of the components of the system - (bi)sulfite, H_2O_2, and horseradish peroxidase - must be present in order for the $SO_3^{\cdot-}$ spectrum to be observed. The probable mechanism for the generation of the $SO_3^{\cdot-}$ radical in this system is as follows:

$$\text{HRP} \xrightarrow{H_2O_2} \text{HRP-Compound I}$$

$$\text{HRP-Compound I} + \text{(bi)sulfite} \longrightarrow \text{HRP-Compound II} + SO_3^{\cdot-}$$

$$\text{HRP-Compound II} + \text{(bi)sulfite} \longrightarrow \text{HRP} + SO_3^{\cdot-}$$

Prostaglandin synthase

Prostaglandin synthase consists of two enzymatic components, a fatty acid cyclo-oxygenase which converts arachidonic acid to the hydroperoxide prostaglandin G_2 and a hydroperoxidase which reduces the hydroperoxide group of prostaglandin G_2 to an alcohol to yield prostaglandin H_2. The hydroperoxidase often catalyzes reactions identical to those catalyzed by horseradish peroxidase.

No $SO_3^{\cdot-}$ is observed when ram seminal vesicle microsomal prostaglandin synthase are incubated with (bi)sulfite and arachidonic acid. However, when the spin trap 5,5-dimethyl-1-pyrroline-N-oxide (DMPO) is present a strong ESR spectrum is observed (Fig. 2).

This signal is derived from the DMPO spin adduct of $SO_3^{\cdot-}$ (1) formed according to the following equation:

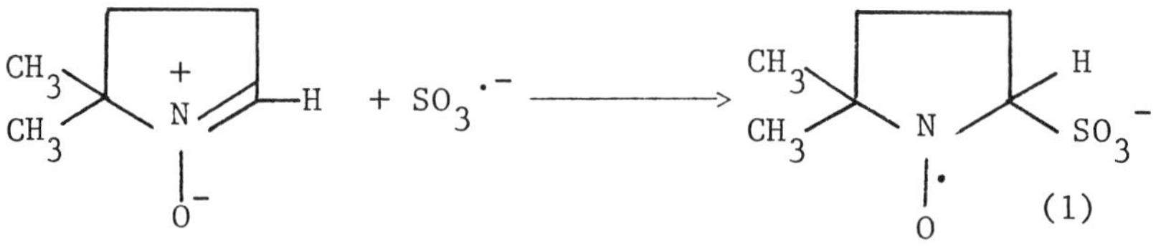

$$\text{DMPO} + SO_3^{\cdot-} \longrightarrow \text{DMPO-}SO_3^{-} \text{ adduct} \quad (1)$$

The same adduct is formed in the horseradish peroxidase, hydrogen peroxide system. Indomethacin, a prostaglandin cyclooxygenase inhibitor, which prevents prostaglandin G_2 formation, significantly decreases the ESR signal intensity (Fig. 2B). There is some peroxidase activity in the microsomes which is independent of arachidonic acid (Fig. 2C). This activity is largely driven by H_2O_2 as shown by inhibition with catalase (Fig. 2D). The

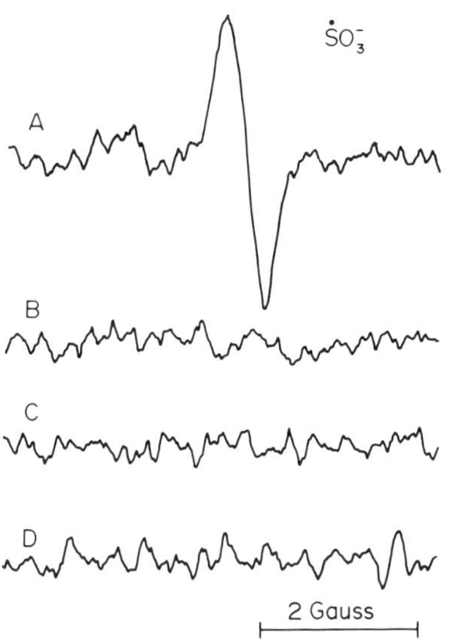

Fig. 1. ESR spectra of the HRP/H_2O_2/sulfite system A, 10 mM Na_2SO_3, 10 μM H_2O_2 and Type VI HRP (0.25 mg/ml, 83 units/ml) in pH 8.6 boric acid/sodium borate buffer; B, 10 mM Na_2SO_3 and 10 μM H_2O_2 in buffer; C, 10 mM Na_2SO_3 and HRP (0.25 mg/ml) in buffer; D, 10 μM H_2O_2 and HRP (0.25 mg/ml) in buffer. Reproduced from Mottley et al. (1982b) with permission.

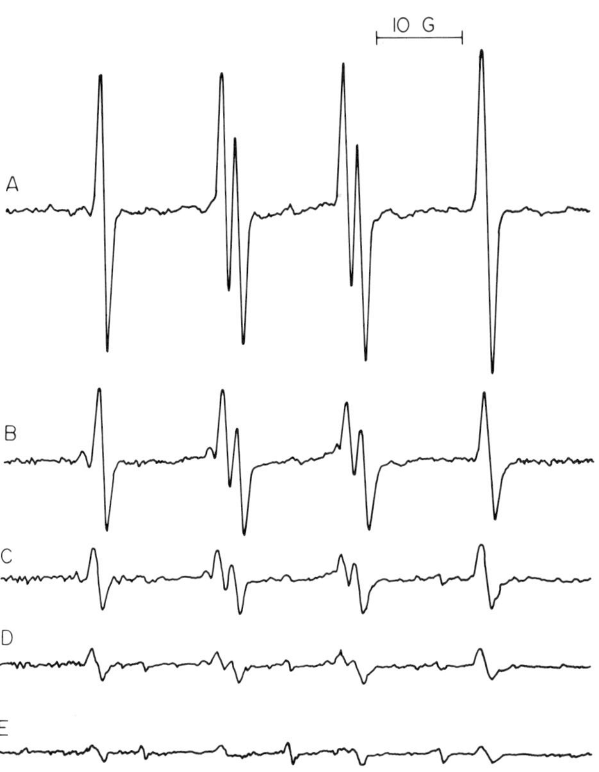

Fig. 2. The ESR spectrum of DMPO-trapped SO_3^- produced by ram seminal vesicle microsomes. A, incubation containing 1 mM Na_2SO_3, 1 mg of ram seminal vesicle microsomal protein/ml of incubation, 410 μM arachidonic acid, and 270 mM DMPO. B, same as A but microsomes were preincubated with 500 M indomethacin for 10 min at 4°C. C, same as A but with no arachidonic acid. D, same as C with 8000 units of catalase/ml of incubation. E, same as A but microsomes were heat-denatured by heating for 2 min in boiling water. Reproduced from Mottley et al. 1982a with permission.

H_2O_2 is probably derived from the autoxidation of the (bi)sulfite stock solution. The entire signal is abolished by heat denaturation of the microsomes. The peroxidatic activity of the ram seminal vesicles can also be initiated by either hydrogen peroxide or 15-hydroperoxy-arachidonic acid. Guinea pig lung microsomes may also be used as a source of prostaglandin synthase, even though the activity of this preparation is 200 times less than that of ram seminal vesicles.

REDUCTION OF (BI)SULFITE

Although the biological oxidation of (bi)sulfite is certainly more common than reduction, under anerobic conditions (bi)sulfite is reduced to dithionite by either reduced flavodoxins or a mixture of paraquat, H_2 and hydrogenase (Mayhew, 1978). In the latter case the paraquat cation free radical was found to reduce (bi)sulfite to form the sulfur dioxide anion free radical which is in equilibrium with dithionite,

$$2SO_2^{\cdot -} \rightleftharpoons S_2O_4^{2-}$$

Cytochrome-P450

The incubation of (bi)sulfite with rat liver microsomes and an NADPH-generating system results in a weak single line ESR spectrum which increases linearly for several hours (Fig. 3). The signal is completely dependent on the presence of all components in the system and is not observed under aerobic conditions. Heat denaturation of the microsomes results in no detectable signal. The ESR spectrum is charactized by a g-value of 2.0056 ± 0.0001 and a peak-to-peak line width of 0.95 G.

These findings suggest that cytochrome P-450 reduces (bi)sulfite to the sulfur dioxide anion radical. If dithionite is added to a microsomal incubation the same ESR spectrum is observed (Fig. 3D). Thus, the well known reduction of cytochrome P-450 by dithionite is reversible,

$$\text{P-450}(Fe^{2+}) + SO_2 \rightleftharpoons \text{P-450}(Fe^{3+}) + SO_2^{\cdot -} \rightleftharpoons \text{P-450}(Fe^{3+}) + \tfrac{1}{2}S_2O_4^{2-}$$

and a redox equilibrium is established between dithionite/(bi)sulfite and NADPH/$NADP^+$ with cytochrome P-450 and NADPH-cytochrome P-450 reductase as the catalysts.

CONCLUSION

From these studies it is obvious that horseradish peroxidase, prostaglandin hydroperoxidase and probably other peroxidases can

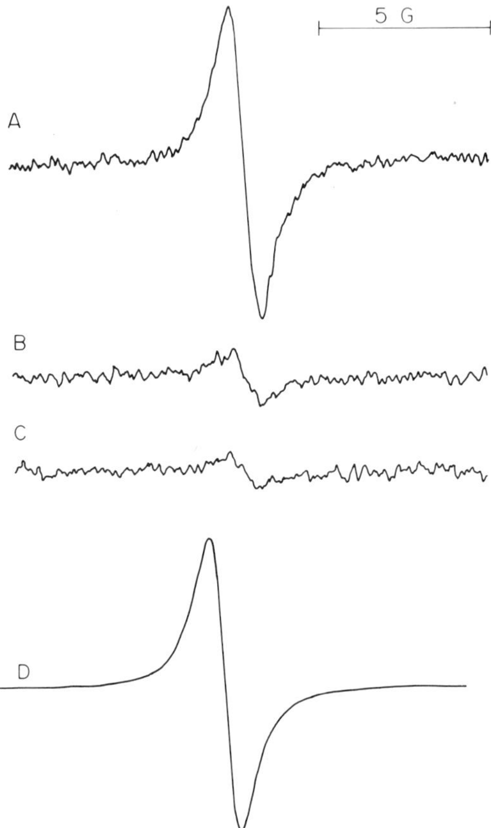

Fig. 3. Electron spin resonance spectra of liver microsomal incubations (1 mg/ml) containing (bi)sulfite (10 mM) and NADH-generating system. A, The microsomal incubation was under a 100% nitrogen atmosphere. B, Identical to A, but under a 100% carbon monoxide atmosphere. C, Identical to A, but containing 5 mM metyrapone. D, Identical to A, but sodium dithionite was added instead of the NADH generating system.

oxidize (bi)sulfite to $SO_3^{\bullet -}$. Other experiments have shown that $SO_3^{\bullet -}$ can react with oxygen to form superoxide, hydrogen peroxide and other sulfur oxides. The reaction of $SO_3^{\bullet -}$ with DMPO is direct proof that this radical can add across double bonds. This propensity, together with the strong oxidizing properties of $SO_3^{\bullet -}$, may account for the toxicity of this radical. Known reactions include the oxidation of diphosphopyridine nucleotide (Klebanoff, 1961) and methionine (Yang, 1970), the destruction of tryptophan (Yang, 1973) and β-carotene (Peiser and Yang, 1979), reaction with nucleic acids (Hayatsu, 1976), the peroxidation of fatty acids (Lizada and Yang, 1981) and the cleavage of DNA (Hayatsu and Miller, 1972).

Anaerobic incubations of (bi)sulfite with rat hepatic microsomal protein and NADPH or NADH generate the sulfur dioxide anion radical. Since it is not known whether this radical is formed in the presence of air, the toxicological implications of this reaction are unknown.

REFERENCES

Behar, D and Fessenden, R.W., 1972, Electron spin resonance studies of inorganic radicals in irradiated aqueous solutions. I. Direct observations, J. Phys. Chem. 76:1706-1710.

Hayatsu, H., 1976, Bisulfite modification of nucleic acids and their constituents, in: "Progress in Nucleic Acid Research and Molecular Biology", Vol. 16, W.E. Cohn, ed., Academic Press, New York.

Hayatsu, H., and Miller, R.C., 1972, The cleavage of DNA by the oxygen-dependent reaction of bisulfite. Biochem. Biophys. Res. Commun., 46:120-124.

Inouye, B., Ikeda, M., Ishida, T., Ogata, M., Akiyama, J. and Utsumi, K., 1978, Participation of superoxide free radical and Mn^{2+} in sulfite oxidation, Toxicol. Appl. Pharmacol., 46: 29-38.

Klebanoff, S.J., 1961, The sulfite-activated oxidation of reduced pyridine nucleotides by peroxidase, Biochim. Biophys. Acta, 48:93-103.

Lizada, M.C.C. and Yang, S.F., 1981, Sulfite-induced lipid peroxidation, Lipids, 16:189-194.

Mayhew, S.G., 1978, The redox potential of dithionite and SO_2^- from equilibrium reactions with flavodoxins, methyl viologen and hydrogen plus hydrogenase, Eur. J. Biochem., 85:535-547.

Mottley, C., Mason, R.P., Chignell, C.F., Sivarajah, K. and Eling, T.E., 1982a, The formation of sulfur trioxide radical anion during the prostaglandin hydroperoxidase-catalyzed oxidation of bisulfite (hydrated sulfur dioxide), J. Biol. Chem., 257:5050-5055.

Mottley, C., Trice, T.B. and Mason, R.P., 1982b, Direct detection of the sulfur trioxide radical anion during the horseradish peroxidase-hydrogen peroxide oxidation of sulfite (aqueous sulfur dioxide, _Mol. Pharmacol._, 22:732-737.

Peiser, G.D. and Yang, S.F., 1979, Sulfite-mediated destruction of β-carotene, _J. Agric. Food Chem._, 27:446-449.

Rall, D.P., 1974, Review of the health effects of sulfur oxides, _Environ. Health Perspect._, 8:97-121.

Yang, S.F., 1970, Sulfoxide formation from methionine or its sulfide analogs during aerobic oxidation of sulfite, _Biochemistry_, 9:5008-5014.

Yang, S.F., 1973, Destruction of tryptophan during the aerobic oxidation of sulfite ions, _Environ. Res._, 6:395-402.

NMR STUDIES OF THE ROLE OF INTRACELLULAR SODIUM IONS IN THE MECHANISM OF INSULIN ACTION ON AN AMPHIBIAN OOCYTE

Raj K. Gupta, Adele B. Kostellow and Gene A. Morrill

Department of Physiology & Biophysics
Albert Einstein College of Medicine
New York, New York 10461 U.S.A.

INTRODUCTION

^{23}NMR spectroscopy provides a noninvasive technique for the study of ^{23}Na ions in living cells and tissues, permitting a direct observation of the cations via their own resonance absorption [1-10]. Rana oocytes constitute a particularly favorable cellular system for NMR study. A Rana female contains up to 2-3 thousand large (1.8 mm diameter) oocytes arrested in first meiotic prophase. Interestingly, insulin can release the block at prophase arrest in the amphibian oocyte and reinitiate the meiotic divisions in vitro, although its physiological role in vivo remains unclear. Insulin action on the plasma membrane causes a rapid change in ion permeability and electrical properties and we have carried out measurements to determine its effect on intracellular Na^+ concentration [6,7].

Until recently, a major problem in the application of the NMR technique to the study of intracellular ^{23}Na ions had been that the resonances of intra- and extracellular Na^+ ions occur at the same NMR frequency. The difficulty in separating these resonances prevented accurate quantitation of NMR-visible intracellular Na^+ concentrations. To overcome this problem, Gupta and Gupta introduced the highly anionic complex of dysprosium (III) with tripolyphosphate ($Dy(PPP_i)_2^{7-}$) [3] as an extracellular paramagnetic shift reagent. The detection of frequency-resolved resonances from intra- and extracellular ^{23}Na ions in cell suspensions depends on the fact that this anionic paramagnetic reagent and its components do not penetrate the cell membrane over the time scale of NMR measurements and therefore the reagent remains only on the outside of the cells [3-10]. Thus only extracellular ^{23}Na ions experience the resonance shift. Since the anionic shift reagent causes effective resonance separation at sufficiently low concentrations (about 5 ppm with 2 mM reagent) at a physiological level of extracellular Na^+, Ca^{2+} and pH, and since all of the Dy^{3+} is complexed to PPP_i, a biometabolite, it causes no significant perturbation of the cellular system [5-10].

We have used dysprosium bis(tripolyphosphate) to separate and study the resonances of intra- and extracellular Na^+ ions in *Rana* oocytes. This technique makes possible a non-invasive analysis of intracellular Na^+ levels in-situ and allows detection of small changes in cell Na^+. The present work applies it to examine the effect of insulin on intracellular ^{23}Na ion levels in the amphibian oocyte.

MATERIALS AND METHODS

Materials Sexually mature *Rana pipien* females were obtained from the New England region of the United States and maintained in artificial hibernation at 4^oC. Ovaries were removed from a pithed animal, rinsed with Ringer's solution, and the follicles were dissected from fresh ovaries [11]. The outer theca cell-containing epithelium was removed with fine-tipped forceps, and remaining follicle cells were detached by a 15-20 min exposure to Ca,Mg-free Ringer's solution containing 1.0 mM EDTA. These oocytes are termed "denuded" in the following text. They were rinsed several times and allowed to re-equilibrate in Ringer's solution for 1 h before use. Sodium insulin was obtained from Eli Lilly Company and dissolved in Ringer's solution immediately before use. Dysprosium bis(tripolyphosphate) $Dy(PPP_i)_2^{7-}$ was prepared as described below.

Water Content and Total Na^+ Analysis The water content of 10-20 follicles or denuded oocytes from each preparation was measured as described elsewhere [11]. Ten-20 follicles or denuded oocytes were rinsed with choline Ringer's solution or 0.24 M sucrose containing 1.1 mM $CaCl_2$ and digested at room temperature overnight in 0.3 ml fuming nitric acid, and the digestion was completed the next day by heating in a boiling water bath for 30 min. The digests were diluted to 10 ml with ion-free water and analyzed for total sodium using a Perkin Elmer Model 360 atomic absorption spectrometer.

Preparation of Shift Reagent The anionic paramagnetic shift reagent dysprosium bis(tripolyphosphate) was prepared by titrating $DyCl_3$ (Ventron, Alfa Division, Danvers, MA) with $Na_5(PPP_i)$ (Sigma, St. Louis, Mo) to obtain the complex $Dy(PPP_i)_2$, the formation of which is indicated by the visual disappearance of a white precipitate in the solution. At PPP_i levels less than two mole equivalents for each mole of Dy^{3+}, the resulting complexes are partly insoluble and give rise to white precipitate [3].

Electrophysiological Measurements Membrane potential measurements were made using a W-P Instruments M-707 microprobe system (W-P Instruments, New Haven, CT) and standard 2.5 M KCl-filled glass micropipettes. Oocytes were voltage-clamped by applying the membrane potential signal into a negative feedback circuit, the output of which was fed back into the oocyte via a second micropipette [7].

Measurement of Intracellular pH by NMR The ^{31}P NMR technique was used to measure intracellular pH in isolated follicles. In contrast to pH microelectrodes, this technique provides a noninvasive method to study intracellular pH by following the chemical shift of the intracellular P_i resonance [6,7]. Since it is possible to measure a chemical shift difference more accurately than an absolute chemical shift, we carried out direct measurements of the intracellular P_i resonance relative to that of the intracellular phosphocreatine [6,7]. ^{31}P NMR spectra were recorded at 81 MHz with a Varian XL-200 NMR instrument using the techniques described previously [7].

Measurement of Intracellular $^{23}Na^+$ Ion Concentration by NMR Unless otherwise specified, all NMR measurements were made at 53 MHz and 21°C using a Varian XL-200 FT NMR spectrometer with a 5/10 mm diameter concentric combination of sample tubes. About 50 oocytes or follicles were placed in the inner tube and D_2O (for field-frequency locking) in the annular space between the inner and outer tubes. NMR free induction decay signals were time-averaged for about 3 min to obtain the spectra presented in this paper. Comparable spectral signal to noise could be obtained in only 10 sec with a 10 mm diameter sample tube containing about 500 oocytes. However, since follicles and oocytes have low oxygen uptake and develop normally in packed condition for one or more hours, the smaller sample tube which required fewer oocytes for obtaining adequate signal to noise ratio was used. The paramagnetic shift reagent dysprosium bis(tripolyphosphate) was added to a final concentration of about 4 mM to the Ringer's solution to separate the intra- and extracellular ^{23}Na signals. Once separation of the resonances of intra- and extracellular Na^+ was achieved using the shift reagent, a comparison of the intensity of the resonance of extracellular ions (A_{out}) with that of a cell-free control (A_o) containing the same concentration of ^{23}Na ions as present in the extracellular medium (Na_{out}) in an identical sample geometry directly yielded the fractional space in the NMR window that was extracellular (S_{out}). The intensities of the ^{23}Na resonances of intracellular (A_{in}) and extracellular ions and a knowledge of the fractional space in the NMR window that is extracellular then directly yielded the concentration of intracellular Na^+ that contributes to the observed resonance signal. The following straightforward equations provide the relationship between the observed resonance intensities and the NMR-visible intracellular ^{23}Na ion concentration [3]:

$$S_{out} = \frac{A_{out}}{A_o} \qquad [1]$$

$$[Na_{in}] = \left[\frac{A_{in} \quad S_{out}}{A_{out} \quad (1-S_{out})} \right] \frac{[Na_{out}]}{w} \qquad [2]$$

w is the fractional water content of the oocytes and k is a constant which is 2.5 or 1 depending upon whether first order nuclear quadrupolar interactions cause a splitting of the ^{23}Na resonance into a broad and a narrow component, leading to a 60% loss in the observable intensity of the intracellular ^{23}Na resonance. $[Na_{in}]$ obtained in this way is the concentration expressed on the basis of cell water content.

RESULTS

Figure 1 illustrates the ^{23}Na NMR spectra of a gently packed suspension of isolated Rana follicles (trace a) and denuded oocytes (trace b) in Ringer's solution containing 4 mM of the shift reagent $Dy(PPP_i)_2$ at pH 7.4. Two well-resolved Na^+ resonances are directly observable in each spectrum showing the spectral separation of intra- and extracellular ^{23}Na ions in the follicle and denuded oocyte suspensions by the shift reagent. As described elsewhere [3], the resonance at right (upfield) corresponds to extracellular Na^+ which interacts with the shift reagent while the smaller resonance at left (downfield) arises from intracellular ^{23}Na ions that are not accessible to the shift reagent. The chemical shift of the intracellular Na^+ resonance is essentially unaffected by the presence of the paramagnetic shift reagent. In the absence of shift reagent, oocytes exhibit only a single ^{23}Na resonance due to overlap of signals from intra- and extracellular Na^+ ions. The intensity of the intracellular ^{23}Na resonance is also not altered by variations in the concentration of the paramagnetic shift reagent.

We have found that 4 mM dysprosium bis(tripolyphosphate) has no effect on oocyte membrane potential or conductance, or on the level of phosphocreatine as observed by ^{31}P NMR for periods of at least 2 h. Similarly, the reagent did not inhibit insulin-induced nuclear breakdown in Rana oocytes. Thus, this shift reagent appears to be non-toxic to amphibian oocytes.

Previous work on ion uptake and exchange kinetics with denuded oocytes [12], yielded values of 1.3 and 0.15 mM/h for Na^+ and K^+, respectively. This indicates that no significant changes would occur in intracellular and extracellular Na^+ and K^+ concentrations in the packed cells over the time scale of NMR measurements (200 sec).

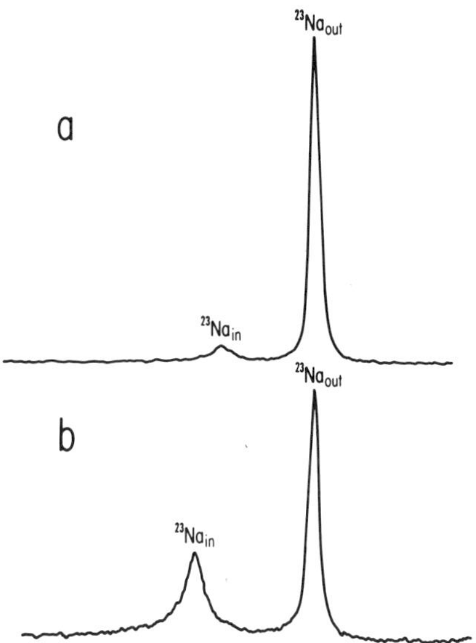

Figure 1. ^{23}Na NMR spectra (53 MHz, 21°C) of gently packed *Rana* follicles (trace a) or denuded oocytes (trace b) in a physiological Ringer's solution (pH 7.4) with 4 mM dysprosium tripolyphosphate. To obtain each spectrum 1,000 transients of free induction decay signal following 90° pulses with a pulse recycle time of 0.2 sec were time-averaged over a period of 200 sec, and an artificial line-broadening of 5 Hz was used to improve the spectral signal-to-noise ratio. The resonances of intra- and extracellular Na^+ ions are labelled as $^{23}Na_{in}$ and $^{23}Na_{out}$, respectively.

A comparison of the resonance intensity of extracellular ^{23}Na ions (A_{out}) with that of a cell-free control (A_o) containing the same concentration of Na^+ ions as present in the extracellular medium ($^{23}Na_{out}$), in an identical sample geometry, directly yields the extracellular space (S_{out}) of the cell suspension. The extracellular space calculated using equation 1 was 38 $\pm$ 4% and 30 $\pm$ 3% (N=6) for gently packed follicles and denuded oocytes, respectively. It should be mentioned that these estimates are for the extracellular space as seen by Na^+ ions themselves. Knowledge of fractional extracellular space and the relative intensities of intra- and extracellular resonances are used to yield the NMR-visible intracellular Na^+ ion concentrations in accordance with equation 2. As reported previously [11], both follicle-enclosed oocytes as well as denuded oocytes are relatively anhydrous and contain only about 50% water. Therefore, a w value of 0.5 was used in equation 2 for the calculation of intracellular Na^+ concentrations on cell water basis.

In follicles from six females, the NMR-visible Na^+ accounted for only 17 ± 3% of the total follicle Na^+. Total follicle Na^+ was measured by atomic absorption. Follicles were rinsed for <1 min with Na^+-free Ringer's solution to prevent loss of a rapidly exchanging Na^+ fraction that appears to be associated with the epithelial cell layers [13]. This rapidly exchanging Na^+ fraction represents about 10 mmol/liter follicle water [12,13]. The calculation of fractional NMR-visibility of oocyte Na^+ implicitly assumes that the shift reagent penetrates the inulin space of the isolated *Rana* follicle. If not, the true NMR-visible Na^+ for follicles would be even lower. In contrast to the follicles, about 40% of the total Na^+ of the denuded oocytes was calculated to be visible in accordance with equation 2. The higher Na^+ content of denuded oocytes compared to follicles reflects a gain in $[Na^+]_i$ during removal of the follicle cells in Ca-free Ringer's solution and most of this added Na^+ appears to be NMR-observable.

To study the role of Na^+ in the mechanism of insulin action in reinitiation of the meiotic divisions, NMR-observable and total intracellular Na^+ levels in untreated and insulin-treated follicles and denuded oocytes were compared. NMR-observable Na^+ (Table 1) and total Na^+ were essentially unchanged (< ± 1 mM) following a 1 h incubation of denuded prophase oocytes in Ringer's solution without insulin but both increased somewhat over the same time period in insulin-treated oocytes. Total Na^+ measured by the atomic absorption technique exhibited a net increase in oocyte Na^+ from 101 ± 1 to 110 ± 1 (N=3) mmol/liter oocyte water over the first 1 h in the presence of 10 μM insulin. When follicles were allowed to accumulate Na^+ in a Ca,Mg-free medium containing 1 mM EDTA for 1.5 h, the NMR-

Table 1. Changes in NMR-Observable Na^+ in *Rana* Follicles and Denuded Prophase Oocytes in Response to Insulin

Preparation	Pretreatment	Treatment	Time[a] h	NMR-Na^+ mmols/l	Total Na^+ cell water
Denuded Oocytes	None	None	0	42	101
"	None	None	1	41	100
"	None	10 μM Insulin	1	44	110
Follicles	None	None	0	12	91
"	Ca-free, 1.5 h	None	0	74	...
"	"	None	3	33	...
"	"	10 μM Insulin	3	46	...

[a]Incubation in calcium containing Ringer's solution at 21°C for the time indicated in column 4.

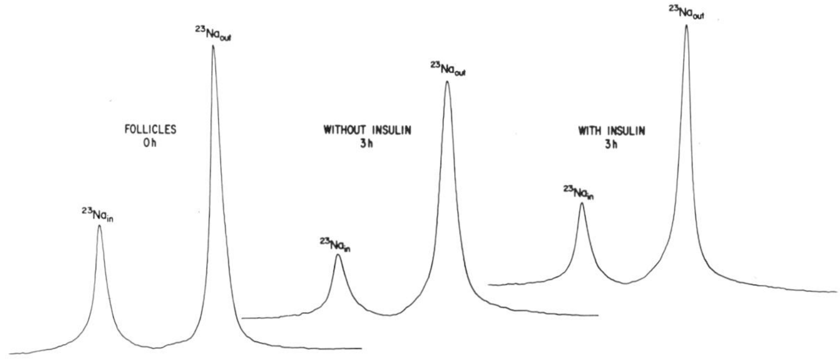

Figure 2. ^{23}Na NMR spectra, obtained at 53 MHz and 21°C, showing the effect of insulin (10 μM) on *Rana* follicles. Follicles were loaded with ^{23}Na by 1.5 h pre-incubation in a Ca,Mg-free Ringer's solution without insulin and subsequently transferred to the standard Ca,Mg-containing Ringer's solution for 3 h with or without insulin. At the end of the incubations, the intracellular ^{23}Na resonance was significantly larger in oocytes from insulin-containing Ringer's solution.

visible intracellular Na^+ level increased from 12 to 74 mmol/liter cell water. The follicles lost preloaded Na^+ when returned to a Ca, Mg-containing Ringer's solution but the decrease in Na^+ occurred at a slower rate in the presence of 10 μM insulin (figure 2). Thus, 3 h after being transferred to a Ca,Mg-containing medium, insulin-treated follicles retained 46 mmol of NMR-visible intracellular Na^+/liter cell water whereas untreated follicles exhibited a significantly lower level of NMR-visible Na^+ (33 mmol/liter cell water, Table 1).

The effect of insulin on the *Rana* oocyte plasma membrane potential has also been investigated. Insulin (10 μM) produced a sizable negative-going hyperpolarization of the plasma membrane in denuded prophase oocytes. Hyperpolarization generally began in 10 min, the potential reached a new steady state in 60-90 min and the oocyte remained hyperpolarized for several hours [7]. Membrane potential of untreated oocytes was 47 ± 3 mV (N=12). In oocytes from several females, the magnitude of insulin-induced mean maximal hyperpolarization was 17.5 ± 1.1 mV [7]. These values were typical for denuded oocytes from fall through winter. By late spring, the resting potential of denuded prophase oocyte had increased to 75 ± 5 mV (inside negative) and addition of insulin no longer produced hyperpolarization.

The hyperpolarizing effect of insulin was compared in both low (2.5 mM) Na^+ and in Li^+-Ringer's solutions. In low (2.5 mM) Na^+-Ringer's solution the oocyte plasma membrane hyperpolarized by about 30 mV and subsequent addition of 10 µM insulin did not significantly affect the membrane potential. Amiloride, an inhibitor of Na^+ channels and Na^+/H^+ exchange at high concentrations, produced a 25 mV hyperpolarization that was insensitive to insulin. In contrast, substitution of Li^+ for Na^+ resulted in a 10 mV depolarization and subsequent addition of 10 µM insulin produced an 18 mV hyperpolarization [7]. The addition of the serine proteinase inhibitor phenylmethylsulfonyl fluoride (PMSF) 30-60 min prior to the addition of insulin completely blocked both hyperpolarization and meiosis (nuclear breakdown), whereas PMSF added following insulin addition had no inhibitory effect on the denuded oocyte.

The effect of insulin on intracellular pH has also been followed using ^{31}P NMR. In experiments with 6 females, the mean pH_i of isolated ovarian follicles was 7.38 ± 0.03. pH_i rose to 7.66 ± 0.15 within 2 h after exposure to 10 µM insulin and remained elevated for 1-2 h. PMSF blocked the insulin-induced elevation in pH_i if added 0.5 h prior to insulin but had no significant effect on pH_i if added 0.5 h after insulin. PMSF alone produced a 0.1 unit decrease in pH_i but had no effect on intracellular phosphocreatine or ATP levels, as measured by ^{31}P NMR, indicating that it did not poison the energy metabolism in the oocyte [7].

DISCUSSION

The results presented here and elsewhere indicate that insulin reinitiates the meiotic divisions in *Rana* oocytes, increases intracellular Na^+ in Ca-free medium and produces a sizable hyperpolarization of the plasma membrane as well as an increase in intracellular pH during the first 1-2 h after exposure [7]. The membrane hyperpolarization seen in oocytes appears to be due, in large part, to a decrease in membrane Na^+ conductance although stimulation of the electrogenic Na^+ pump may contribute a small part [7]. Our results would indicate that Na^+ is a major conducting ion and that insulin acts to reduce Na^+ conductance of the oocyte plasma membrane. This is apparent from hyperpolarization arising from decreased conductance seen for oocytes incubated in low Na^+ medium or normal Na^+ medium containing amiloride. Our finding that insulin will hyperpolarize oocytes in Li^+-substituted media where the electrogenic Na^+/K^+-ATPase in the oocyte is inhibited suggests that hyperpolarization is not primarily due to stimulation of an electrogenic pump [7].

Insulin-treated oocytes exhibited a markedly (about 40%) higher NMR-visible intracellular Na^+ in Ca,Mg-free medium in comparison to untreated controls. In a Ca,Mg-containing medium, however, only a slight (7%) increase in NMR-visible Na^+ relative to controls was apparent. These observations suggest that insulin stimulates a non-conducting pathway for Na^+ influx in the oocyte plasma membrane. In a Ca,Mg-containing medium the cytoplasmic level of intracellular Na^+ is low and the Na^+ pump may not be saturated with this cation. Any transient increase in internal Na^+ due to increased influx would shift the operating point resulting in enhanced activation of the Na^+ pump which would tend to return the intracellular Na^+ towards the level found in the absence of insulin. In a Ca,Mg-free medium, however, the level of intracellular Na^+ is much higher and the Na^+ pump is likely to be oparating close to its maximal efficiency. Therefore, an increased influx would result in a markedly higher intracellular Na^+ level.

Part of the increased uptake of Na^+ is apparently due to an increase in fluid phase turnover via endocytosis. About 2% of the total oocyte fluid volume (1.4 µl/oocyte) exchanges per hour via endocytosis [6]. The fluid uptake has been estimated to be 25 ± 2 nl/oocyte/h [6]. Since the medium contains about 110 mM Na^+, fluid uptake via endocytosis could account for about 2 mmols Na^+ uptake per kg per h in the prophase-arrested oocyte. Thus about one-third of the Na^+ uptake by the prophase oocyte may be via endocytotic vesicles. Insulin increases fluid uptake and turnover by stimulating endocytotic pathway 2 to 3-fold over the first hour after treatment. This increased endocytosis could be contributing significantly to the insulin-induced increase in NMR-visible Na^+ (Table 1). About half of the increase in the total Na^+ level in response to insulin (4-5 mmol/kg/h) may be due to enhanced Na^+ uptake during increased endocytosis. Insulin, however, also causes an increase in intracellular pH [6,14-16]. In Rana oocytes insulin-induced elevation of pH_i was Na^+ dependent indicating that intracellular pH may be regulated by Na^+/H^+ exchange and that, as proposed for muscle [15], insulin acts at least in part by stimulating the Na^+/H^+ exchange system in the oocyte plasma membrane. Thus, the increase in NMR-visible Na^+ may be due in part to increased Na^+ uptake in exchange for H^+ as well as increased fluid phase turnover. Insulin stimulation of Na^+/H^+ exchange in the oocyte plasma membrane may occur via a decrease in the activation energy of this system which would allow Na^+ moving down its free energy gradient into the cell to provide energy to move protons outward, against their free energy gradient, with a resulting increase in intracellular pH. Since the buffering capacity of the oocyte is relatively large, the increase in pH_i by 0.25 units [16] may be accompanied by significant Na^+ uptake. Like Na^+ pump, Na^+/H^+ exchange in Rana oocytes may also not have a measurable conductance.

Earlier studies have indicated that insulin acts to generate a peptide(s) that stimulates a multitude of enzymes affecting a variety of cellular processes [17,18]. It has been postulated that, in fat cells, the serine proteinase trypsin mimics insulin action via mediator peptide formation and also that serine proteinase inhibitors block insulin action [19]. In our studies the serine proteinase inhibitor PMSF blocked the insulin-induced hyperpolarization and increase in intracellular pH as well as insulin-initiation of the meiotic divisions. This inhibition occurred if PMSF was added 0.5 h before treatment with insulin but not if added o.5 h thereafter. It is thus possible that activation of a serine proteinase may be one of the steps in the sequence of ionic events associated with release of the prophase block.

ACKNOWLEDGEMENTS

This work was supported in part by NIH research Grants AM-32030 and HD-10463, and by NCI Core Grant CA-13330.

REFERENCES

1. F. W. Cope, J. Gen. Physiol. 50:1353 (1967).
2. M. M. Civan and M. Shporer, Biol. Magn. Reson. 1:1 (1978).
3. R. K. Gupta and P. Gupta, J. Mag. Res. 47:344 (1982).
4. R. K. Gupta, P. Gupta, and W. Negendank, in: "Ions, Cell Proliferation and Cancer," A. L. Boynton, W. L. McKeehan, and J. E. Whitfield, eds., Academic Press, N.Y., p. 1 (1982).
5. R. K. Gupta, P. Gupta, and R. D. Moore, Annu. Rev. Biophys. Bioeng. 13:221 (1984).
6. G. A. Morrill, A. B. Kostellow, S. P. Weinstein, and R. K. Gupta, Physiol. Chem. Phys. Med. NMR 15:357 (1983).
7. G. A. Morrill, A. B. Kostellow, S. P. Weinstein, and R. K. Gupta, Biochim. Biophys. Acta, in press (1985).
8. R. K. Gupta, A. B. Kostellow, and G. A. Morrill, J. Biol. Chem., in press (1985).
9. B. A. Wittenberg and R. K. Gupta, J. Biol. Chem. 260:2031 (1985).
10. B. M. Rayson and R. K. Gupta, J. Biol. Chem., in press (1985).
11. D. H. Ziegler and G. A. Morrill, Develop. Biol. 60:318 (1977).
12. G. A. Morrill and D. H. Ziegler, Develop. Biol. 74:216 (1980).
13. G. A. Morrill, D. H. Ziegler, and V. S. Zabrenetsky, J. Cell Sci. 26:311 (1977).
14. R. K. Gupta and R. D. Moore, J. Biol. Chem. 255:3987 (1980).
15. R. D. Moore and R. K. Gupta, Int. J. Quant. Chem. Quant. Biol. Symp. 7:83 (1980).
16. G. A. Morrill, A. B. Kostellow, S. Mahajan, and R. K. Gupta, Biochim. Biophys. Acta 804:107 (1984).
17. L. Jarett and J. T. Seals, Science 206:1407 (1979).
18. J. Larner, G. Galasko, K. Cheng, A. DePaoli-Roach, L. Huang, P. Daggy, and J. Kellogg, Science 206:1408 (1979).
19. J. Larner, K. Cheng, C. Schartz, K. Kikuchi, S. Creacy, R. Dubler, G. Galasko, C. Pullin, and M. Katz, Fed. Proc. 41:2724 (1982).

THE LOSS OF INTRACELLULAR WATER DURING FREEZING IN PRESENCE OF HYDROXYETHYL STARCH

Christoph Körber, Klaus Wollhöver, and Max-Werner Scheiwe

Helmholtz-Institut für Biomedizinische Technik an der RWTH Aachen
Goethestr. 27-29
D-5100 Aachen, West-Germany

INTRODUCTION

According to Mazur's two-factor-hypothesis (Mazur 1965), the kinetics of the loss of intracellular water during freezing play a governing role concerning the survival of biological cells in suspensions exposed to low temperatures. If cooling is too slow or too fast, the cell may either be damaged by osmotical effects or intracellular ice formation, respectively, while an intermediate range of cooling rates generally yields a relative maximum in the survival curve. The absolute value of the corresponding "optimal" cooling rate is closely related to the hydraulic membrane permeability (being the limiting factor in volume shrinkage), and hence varies - in some cases orders of magnitude - from one particular type of cell to another (Rapatz et al., 1968). This qualitative interpretation of freezing injury has become widely accepted, and Mazur was also the first to give a quantitative description of the shrinkage of the cell volume due to the osmotically induced water loss resulting from freezing (Mazur 1963). In the meantime, the model has been picked up by many authors who modified and refined it in order to take into account particular aspects (e.g. Ling and Tien, 1969; Mansoori, 1975; Silvares et al., 1975; Levin et al., 1978; Hua et al, 1982). More recently, experimental results obtained by means of cryomicroscopy have been correlated with the model (e.g., Knox and Diller, 1978; Diller, 1982; Scheiwe and Körber, 1982 and 1983). The agreement is not always satisfactory, among other reasons possibly because of the rather rough approximation of the extracellular medium obeying phase diagram conditions during the freezing process while considerable deviations from equilibrium have been shown experimentally (Körber et al.,1982 and 1983).

It has also been tried to extend the model to ternary systems, i.e. water-salt-glycerol (Fahy 1981, Lynch and Diller 1981), on the basis of the respective phase diagram (Shepard et al., 1976), but difficulties arise from the ability of the cryoprotective additive to penetrate the cell wall, requiring the treatment of the (cross-coupled) fluxes of two species. In the case of a polymeric and hence non-penetrating agent, on the other hand, the direct influence must be limited to the extracellular medium (or to the outer surface of the membrane itself, a case which will not be considered in this contribution). But the changes induced extracellularly may alter the driving force of the water transport across the cell membrane, and hence also the intracellular composition during freezing which may result in a shift of optimal cooling rates to different values. In the following, such effects will be discussed by the example of hydroxyethyl starch (HES).

EXTRACELLULAR EFFECTS OF HES

The cryoadditive HES has the advantage of being non-toxic, thus allowing direct application after thawing without dilution procedures (Scheiwe et al., 1979). Although its cryoprotective properties have not yet been fully understood, it is one of the most interesting extracellular agents for the low temperature preservation of erythrocytes and granulocytes (Allen et al.,1976, Lionetti et al.,1980). From the physico-chemical behavior in aqueous solution it has been derived that the stability of the completely amorphous state is of negligible importance in the case of HES (Körber et al., 1982). Calorimetric studies, however, revealed that water absorption seems to play a major role (Franks et al., 1977, Körber et al., 1982). The water bound to HES is not subject to freezing and may hence be considered as thermally inert. On that basis it is also possible to understand phase diagram data obtained for the HES-NaCl-H_2O ternary system (Körber and Scheiwe, 1980): HES does not cause a significant depression of the salt-water binary eutectic temperature but a shift of the isothermal eutectic trough which is related to its water absorptive capacity. It has been proposed that this behavior also offers an explanation of the cryoprotective properties: by selecting the proper value of the initial HES concentration it may be avoided that the salt concentration exceeds a lethal level when the freezing path of the solution in the ternary diagram reaches the eutectic trough.

This effect may be understood more clearly from Fig.1. The diagram shows isoplethal sections through the liquidus surface of the ternary phase diagram, i.e. sections along lines of constant ratio of starch and salt (remaining unchanged because of the withdrawal of pure water during solidification). A linear approximation was used in this case for the NaCl-H_2O binary liquidus (labelled 0% HES). Under these circumstances the curves resulting from the addition of HES may simply be calculated from the following expression:

$$T(c_{NaCl}) = T_0 - \frac{T_0 - T_E}{\left(\frac{1}{c_{NaCl}} - (1+\varphi)\cdot\frac{c^0_{HES}}{c^0_{NaCl}}\right) c^E_{NaCl}} \qquad (1)$$

In this equation, T_0 =273.15 K represents the freezing temperature of pure water, T_E =251.95 K the eutectic temperature of the H_2O-NaCl system, φ =0.412 the fraction of water bound per mass unit of HES, c denotes ternary mass

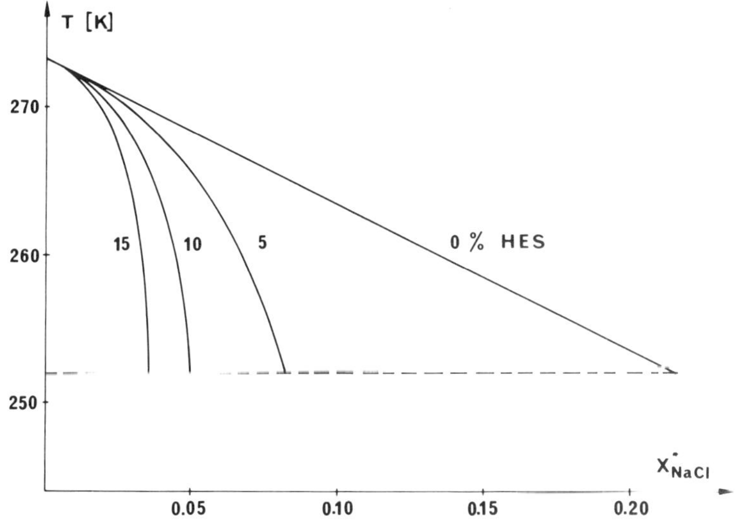

Fig.1: Effect of HES on the liquidus curve (linear approximation) shown by isoplethal sections through ternary phase diagram. c_{NaCl}: ternary mass fraction of sodium chloride. Dashed line represents eutectic temperature.

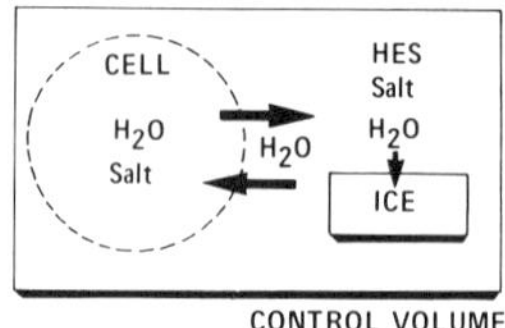

Fig.2: Schematic of water transport model

fractions of starch and water with superscripts o and E referring to initial (isotonic) and eutectic compositions, respectively, with $c^o_{NaCl}=0.009$, $c^E_{NaCl}=0.233$ and $c^o_{HES}=0.05$, 0.10 and 0.15 as indicated in the diagram. It can be seen that the presence of HES drastically increases the "steepness" of the liquidus and reduces the final concentration reached at the eutectic trough (dashed line). These alterations representing the state of the extracellular medium can then be inserted into the model describing the loss of intracellular water during freezing.

WATER TRANSPORT MODEL

The model which is, except for the effects caused by HES, basically similar to those mentioned above, envisages a situation as sketched in Fig.2. The cell be composed of only NaCl and water (a certain fraction of which is osmotically inactive) and surrounded by an ideally semipermeable membrane. The extracellular medium be a solution of NaCl and HES in water which is partly bound to HES and partly converted into ice according to the conditions described above. These and several other assumptions are summarized in Tab.1.

The driving force for the water efflux is the difference of the chemical potential of water at both sides of the cell wall, expressed by the logarithm of the mole fractions of water extra- and intracellularly, x^{ex}_w and x^{in}_w, according to assumption #5:

$$\frac{dn^{in}_w}{dT} = L_p(x^{ex}_w,T)\cdot\frac{ART}{\bar{v}^2_w B}\left(\ln x^{in}_w - \ln x^{ex}_w\right) \quad (2)$$

n^{in}_w denotes the number of moles water in the cell, A

Tab.1 Assumptions of water transport model

1. two phase system
2. no extracellular supercooling
3. no intracellular ice formation
4. isotonic initial conditions
5. ideal solutions
6. no density changes
7. conservation of matter
8. membrane permeable to water only
9. constant cell surface area
10. no hydrostatical pressure difference
11. intracellular medium partially inactive osmotically
12. no interaction between cells
13. homogeneous temperature
14. no concentration gradients extra- and intracellularly

the cell surface area, R is the universal gas constant, $\bar{v}_w$ the average partial molar volume of water. Introducing the (constant) cooling rate B, the differential in time was replaced by that in temperature. The water permeability L_p was considered to be both temperature and concentration dependent as expressed by the coefficients α_1, and α_2 according to (Scheiwe et al., 1980; Scheiwe and Körber, 1983):

$$L_p(x_w^{ex}, T) = L_p^o \cdot \exp\left[-\alpha_1\left(\frac{1}{T} - \frac{1}{T_o}\right) + \alpha_2\left(\frac{1}{1-x_w^{ex}} - \frac{1}{1-x_w^{oex}}\right)\right] \quad (3)$$

The index o refers to the reference values before freezing. In order to take into account the change of the extracllular composition with temperature represented by Eq.(1), the mass fractions of salt have to be converted into the corresponding mole fractions of water.

$$x_w^{ex}(T) = \left[1 + \frac{2M_W}{M_{NaCl}} \cdot \frac{c_{NaCl}(T)}{1-c_{NaCl}(T)}\right]^{-1} \quad (4)$$

The set of equations (1) through (4) can be solved with standard numerical methods.

RESULTS AND DISCUSSION

Fig.3 shows shrinkage curves (i.e. normalized cell colume V/V_o vs. temperature T) calculated from the model described above for the case of the human red blood cell. The influence of HES (concentrations of 0, 5, 10, and 15 wt%) is demonstrated for two different cooling rates, 50 and 1000 K/min. Two general trends may be seen: first, the final water content reached asymptotically during the cooling process becomes larger if HES is added. This effect is due to the reduction of the maximum salt content reached extracellularly (cf. concentrations at dashed line in Fig.1). Second, the intracellular water content at a given temperature is always higher than without HES, i.e., deviations from osmotical equilibrium become larger. That retardation of the water exchange is related to the increasing slope of the liquidus curves in Fig.1. It can be seen that these effects produce a shift of the above mentioned optimal cooling rate (yielding a maximum recovery) to lower values. Such a reduction is highly desired in order to get into a range of cooling conditions which can be achieved technically in large samples as needed for cryopreservation purposes (Geiser and Scheiwe, 1981). With 5wt% HES added, for instance, the optimal value of about 4800 K/min without cryoprotective agent (Scheiwe et al., 1979 and 1982) is reduced to about 1000 K/min. This tendency is qualitatively in good agreement with experimental results obtained from a systematic investigation of the interdependence between the survival of red blood cells and the parameters cooling rate, HES concentration, and hematocrit (Scheiwe et al., 1982). It is hoped that the correlation can also be quantified by further extension and refinement of the model. It is also planned to obtain more direct experimental evidence from cryomicroscopical studies which are currently in progress.

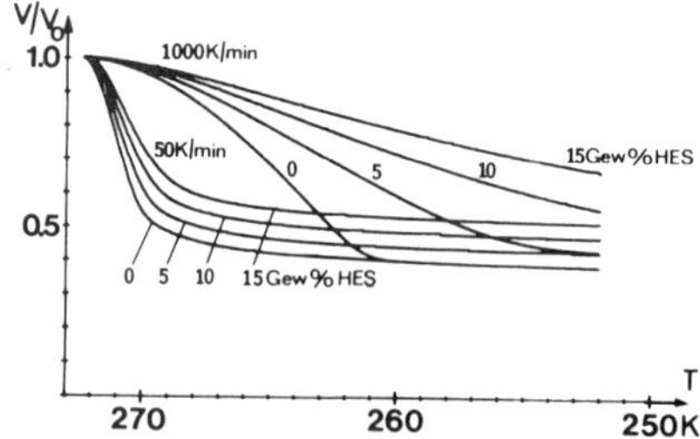

Fig.3: Calculated erythrocyte shrinkage curves (normalized volume V/V_0 vs. temperature T) for various cooling rates and HES contents (0, 5, 10, and 15 wt%).

REFERENCES

Allen, E. D., Weatherbee, L., Spencer, H. H., Lindenauer, S. M., and Permoad, P. A., 1976, Large unit red cell cryopreservation with hydroxyethyl starch, Cryobiol., 13:500.

Diller, K. R., 1982, Quantitative low temperature optical microscopy of biological systems, J. Micros., 126:9.

Fahy, G. M., 1981, Simplified calculation of cell water content during freezing and thawing in non-ideal solutions of cryoprotective agents and its possible application to the study of "solution effects" injury, Cryobiol., 18:473.

Franks, F., Asquith, M. H., Hammond, C. C., Le B.Skaer, H., and Echlin, P., 1977, Polymeric cryoprotectants in the preservation of biological ultrastructure, I Low temperature states of aqueous solutions of hydrophylic polymers, J. of Microscopy, 110:223.

Geiser, T., and Scheiwe, M. W., 1981, Design of freezing containers for submerging into LN_2: the temperature field and its influence on the recovery of hydroxyethyl starch preserved red blood cells, Cryo-Lett., 2:291.

Hua, T. C., Cravalho, E. G., and Jiang, L., 1982, The temperature difference across the cell membrane during freezing and its effect on water transport, Cryo-Lett., 3:255.

Knox, J. M., and Diller, K. R., 1978, Volumetric changes induced in living cells during freezing and thawing in:"Advances in Biomedical Engineering", R. C. Eberhard, A. H. Burnstein, eds., ASME, New York.

Körber, C., and Scheiwe, M. W., 1980, The cryoprotective properties of hydroxyethyl starch investigated by means of DTA, Cryobiol., 17:54.

Körber, C., Wollhöver, K., and Scheiwe, M. W., 1981, The redistribution of solute in front of the advancing ice-liquid interface, in:"Refrigeration Science and Technology Cryosurgery and Medical Applications of Refrigeration - Current Situation and Perspectives", International Institute of Refrigeration, ed., I.I.R., Paris.

Körber,.C., Scheiwe, M. W., and Boutron, P., 1982, Can the stability of the completely amorphous state explain the cryoprotection of HES? Cryo-Lett., 3:83

Körber, C., Scheiwe, M. W., Boutron, P., and Rau, G., 1982, The influence of hydroxyethyl starch on ice formation in aqueous solutions, Cryobiol., 19:478

Körber, C., Wollhöver, K., and Scheiwe, M.W., 1983, Solute polarization during planar freezing of aqueous salt solutions, Int. J. Heat Mass Transfer, (in press).

Körber, C., Wollhöver, K., and Scheiwe, M. W., 1983, Observations on the non-planar freezing of aqueous salt solutions, J. Crystal Growth, (in press).
Levin, R. L., Cravalho, E. G., and Huggins, C. E., 1978, The concentration polarization effect in a multicomponent electrolyte solution - the human erythrocyte, J. theor. Biol., 71:225.
Ling, G. R., and Tien, C. L., 1969, Analysis of cell freezing and dehydration, ASME paper No 69 WA/HT 31.
Lionetti, F. J., Luscincas, F. W., Hunt, S. M., Valeri, C. R.,and Callaghan, A. B., 1980, Factors affecting the stability of cryogenically preserved granulocytes, Cryobiol., 17:297
Lynch, M. E., and Diller, K. R., 1981, Analysis of the kinetics of cell freezing with cryophylactic additives, in:"1981 Advances in Bioengineering", ASME, New York, p. 229.
Mansoori, G. A., 1975, Kinetics of water loss from cells at subzero centigrade temperatures, Cryobiol., 12:34
Mazur, P., 1963, Kinetics of water loss from cells at subzero temperatures and the likelihood of intracellular freezing, J. Gen. Physiol., 47:347.
Mazur, P., 1965, Causes of injury in frozen and thawed cells, Fed. Proc., 24:175.
Rapatz, G., Sullivan, J. J., and Luyet, B., Preservation of erythrocytes in blood containing various cryoprotective agents, frozen at various rates and brought to a given final temperature, Cryobiol., 5:18.
Scheiwe, M. W., Körber, C., and Nick, H. E., 1979, Physical and chemical aspects of cryopreservation of human RBC's with HES 450/0.7, Vox Sanguinis, 37:354.
Scheiwe, M. W., Körber, C., and Wollhöver, K., 1980, Ermittlung der optimalen Kühlrate roter Blutkörperchen (RBK) aufgrund thermodynamischer Modellbetrachtungen, Biomed. Tech., 25(suppl.):432.
Scheiwe, M. W., and Körber, C., 1982, Thermally defined cryomicroscopy and some applications on human leukocytes, J. Microscopy, 126:29.
Scheiwe, M. W., Nick, H. E., and Körber, C., 1982, An experimental study on the freezing of red blood cells with and without HES, Cryobiol., 19:461
Scheiwe, M. W., and Körber, C., 1983, Basic investigations on the freezing of human lymphocytes, Cryobiol., 20(in press).
Shepard, M. L., Goldstone, C. S., and Cocks, F. H., 1976, The H_2O-NaCl-glycerol phase diagram and its application in cryobiology, Cryobiol., 13:9.
Silvares, O. M., Cravalho, E. G., Toscano, W. M., and Huggins, C. E., 1975, The thermodynamics of water transport from biological cells during freezing, J. Heat Transfer, 97:582.

ERYTHROCYTES IN ALTERNATING ELECTRIC FIELDS

Vasile V. Morariu [x], Alexandra Chifu [x],
Titus Simplaceanu [x], and Petre T. Frangopol [xx]

[x] Institute of Isotopic and Molecular Technology
P.O.Box 243, R-3400 Cluj-Napoca, ROMANIA
[xx] Institute of Physics and Nuclear Engineering
P.O. Box Mg-6, R-76900 Magurele-Bucuresti, Romania

INTRODUCTION

Cells exposed to relatively mild alternating field intensities of the order of hundreds of volts per centimeter show the phenomenon known as dielectrophoresis. This is the displacement of the cells towards regions of higher field intensities[1]. When a few short pulses of higher field intensities were applied, fusion of the cells or lysis was demonstrated to occur[2,3].

We here present results obtained with human erythrocytes at intermediate alternating field intensities, when an important deformation of the cells is evident. We suggest that the investigation of erythrocytes in alternating fields is a convenient method to study the flexibility of a cell. The deformability of erythrocytes is usually estimated by microfiltration methods. They refer either to the bulk of the sample[4,5] or to individual cells[6].

We have also noticed that there is a striking similarity between the morphological changes of the erythrocytes following exposure to alternating electric fields and those observed when subjected to a shear stress, although different mechanisms are operative.

MATERIALS AND METHODS

Human blood was collected by venipuncture on heparine and used within one hour. The blood was washed several times in

150 mM mannitol solution until the conductivity of the washings reached a value of $1X10^{-5}$ ohm^{-1} cm^{-1} or less. Finally a diluted suspension of erythrocytes was prepared for examination at the optical microscope.

The microslide for dielectrophoresis was prepared by vacuum deposition of a gold film. A copper wire of 80 μm diameter was fixed on the microslide prior to deposition. As a result a gold-free gap was obtained, having a relatively nonuniform width. The slide was subsequently subjected to a thermal treatment at 350°C in vacuum for three hours in order to increase the adhesivity of the metallic film. The edges of the gold-free gap served as electrodes. If a voltage is applied to the electrodes, a non-uniform field is obtained in the gap. The field is more intense nearby the edges of the gap. Further nonuniformity of the field is brought about by the fact that the width of the gap is not constant along the slide.

The cells were examinated with an optical microscope equipped with a set-up for phase contrast at X300 magnification. All experiments were carried out at room temperature while the microslide was covered by a cover slip to avoid evaporation. Microphotographs were taken successively before and after applying the alternating voltage.

The experiments were carried out at 1.4 MHz frequency and an output voltage of 10 V. The estimated value of the field intensity for our microslide was 1.2 kV/cm, although significant variations around this value must have occured along and across the gold-free gap for reasons already explained.

RESULTS

Red blood cells subjected to moderate field intensities show the usual dielectrophoretic motion and pearl-chain formation without deformation[1,2]. However, when local field intensities are higher, elongation of the cells becomes evident. This process occurs within seconds after coupling the field and the cells return to the normal shape at an apparent similar rate. This kind of deformations was termed as being elastic. We found that red blood cells take an oval shape while at higher field intensities they transform to shuttles. Only discocytes and stomatocytes showed this kind of deformation while spherocytes appeared to be rigid. Echinocytes showed little or no deformation while nearby discocytes or stomatocytes were significantly elongated.

The deformation of the cell can be conveniently described by the ratio of the maximum to the minimum diameter D/d. Discocytes and stomatocytes showed a range of elastic deformation extending up to D/d = 3.

Non-elastic deformations were noticed, too. We found that strongly elongated shuttle cells only recovered to oval cells and this shape remained unchanged even after fifteen minutes. Some echinocytes located at the gap edges were so strongly elongated that the spikes disappeared and the cell became a shuttle. The spikes recovered within seconds after decoupling the field but the cell remained elongated.

It was not possible to precisely establish the border between elastic and inelastic deformations because of the technical limitations of our device. However this border seems to be located at $3 < D/d < 4$.

Discocytes and stomatocytes attached to the gold-free gap edges did not show symmetric deformation, unlike the cells situated within the gap. Their typical shape was that of a bullet with the tip pointing to the gap.

Eventually higher local field intensities caused fusion in a matter of seconds. Pearl-chains of various lengths were typically converted into what appeared to be tubular cells.

DISCUSSION

These results suggest that alternating electric fields can be used to investigate the flexibility of the red blood cells. We have found that discocytes and stomatocytes show comparable deformabilities in the elastic range although the surface-volume ratio of the stomatocytes appears to be smaller than for discocytes. This ratio and the high bending capacity of the membrane are considered to be mainly responsible for the flexibility of the erythrocyte[6].

Spherocytes were never seen to elongate in our field conditions.

Echinocytes behaved as rigid bodies too, but only in the elastic range of discocytes. At higher field intensities, when inelastic deformation occured, the spikes appeared to have elastic properties. This clearly suggests that spikes in the absence of the field represent a local increased flexibility of the membrane and this flexibility is not altered after exposure to the field even if the cell as a whole is irreversibly deformed.

It is interesting to note that the deformation of the cells in the radiofrequency field appears to be similar to the deformation caused by a shear stress. The cells in our field conditions have $1.5 < D/d < 3$ while similar values were found for a shear stress ranging between 25 and 250 s^{-1} [7]. The value of D/d tends to level off beyond 600 s^{-1} where $D/d \approx 3.7$. (We have estimated the values of D/d at various values of the shear stress from the microphotographs given in Ref. 7). This is in the range suggested by us as the limit between elastic and inelastic deformation. Despite of such an apparent similarity, the inelastic deformation caused by the penetration into capillaries having diameters of 2.0 - 2.5 μm (where the equivalent deformation is $3.0 \lesssim D/d \lesssim 3.5$) is qualitatively different. The normal cells are ejected as echinocytes, while an equivalent shuttle cell becomes an oval cell after decoupling the field. Therefore it seems likely that the inelastic deformation of the cells induced by the field is qualitatively different from a mechanical stress induced deformation.

The elastic deformation of a pearl-chain extending across the gap can offer an image of the field non-homogeneity. This is illustrated in Fig. 1. Obviously the cells need to be of similar morphology, preferably discocytes. It can be seen that the cells are strongly elongated at the edges of the poles while in the mid of the gap they are less deformed.

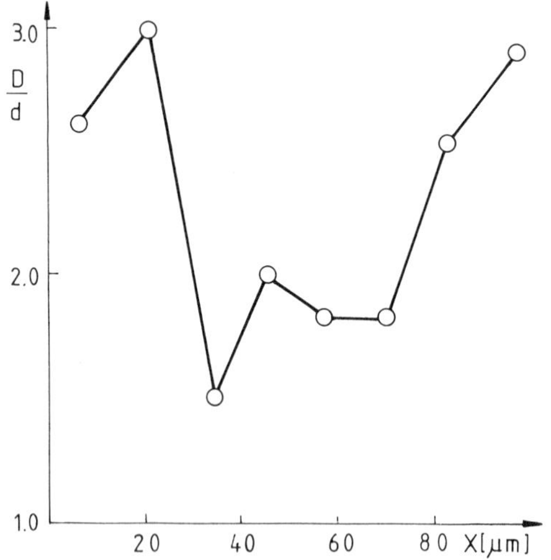

Fig. 1. Deformation of human erythrocytes exposed to 1.4 MHz frequency, 10 V, as a function of their position in the gold-free gap of the microslide for dielectrophoresis.

We suggest that the elastic deformation of the erythrocytes could be used for the characterization of the field distribution in the microslide for dielectrophoresis.

Fusion or hemolysis in our field conditions could be regarded as the final stage of the cells undergoing inelastic deformation. Fusion of red blood cells has also been achieved by exposing cells to a few short fiels pulses of higher voltage [2,3]. The fused cells were spherical, therefore that kind of fusion gave quite different results compared to ours where the field is continuously applied for much longer times (seconds compared to microseconds).

The deformation of cells in alternating electric fields clearly suggests that a strong polarization force acts upon the cell. Polarization of the cell without deformation may occur as well and the dielectrophoretic movement of the cells and the formation of pearl-chains are the typical results commonly observed by other authors[1,2]. Polarization of a cell may be caused by interfacial polarization (charge separation), dielectric polarization (dipole orientation within the membrane), or by the movement of counter charges close to the charged membrane surface[8,9]. The last process is considered to be important only up to about 1 kHz which is far below our working frequency. The first two contributions are considered to be effective at 1.4 MHz[8,9] while, based on the cell rotational resonance, charge separation seems to be the most probable operating mechanism[10].

The fact that the cells are not deformed by short microsecond pulses but only by much longer exposure to the field is in apparent agreement with the charge separation hypothesis. Indeed, short intense field pulses may cause electrical breakdown but not a significant charge separation which could become effective at only longer field exposure. The electrical breakdown caused by the dipole orientation within the membrane is a process much faster than the charge separation. It is possible that the mobile charge concentrations inside the cell should be higher than outside as a result of repeated washing. Therefore we suggest that the polarization of the cell is mainly due to these internal charges.

Finally, the inelastic deformation of the cell may be the result of two contributions: a) elongation of the cell beyond the mechanical limits accepted by the membrane, and b) exposure of local regions of the cell to a high concentration of charges representing an additional damaging factor. This second contribution could possibly explain why a simple mechanical stress gives a different result.

In conclusion, the main finding of this work is the elastic and inelastic deformation of erythrocytes induced by alternating fields and the suggestion that moderate field intensities when continuously applied can cause lysis by a different mechanism compared to the action of short intense field pulses. Therefore these different experimental conditions can be used to approach various properties of the membrane such as those related to the dielectric polarization of the membrane or to the interfacial polarization, leading to the inelastic deformation of the cells.

REFERENCES

1. H. A. Pohl, "Dielectrophoresis", Cambridge University Press, London and New York (1979).
2. U. Zimmermann, P. Scheurich, G. Pilwat and R. Benz, Cells with manipulated functions: new perspectives for cell biology, medicine, and technology, Angew. Chem. Int. Ed. Engl., 20:325 (1981).
3. U. Zimmermann and J. Vienken, Electric field-induced cell-to-cell fusion, J. Membrane Biol., 67:165 (1982).
4. J. Unger and G. Geyer, CDM-Untersuchungen an Erythrocyten. II. Methodologische Grundlagen, Wiss. Z. FSU Jena, Math.-naturwiss. R., 26:521 (1977).
5. G. Geyer, K. J. Halbhuber, D. Stibenz, C. Scheven, J. Unger, A. Benser, R. Fröber, J. Makovitzky, D. Geiling and H. G. Geiling, Alteration by procaine of spectrin cross-links, deformability, and fluidity related properties of the erythrocyte membrane, Folia Haematol. Leipzig, 107:472 (1980).
6. A. W. L. Jay, Viscoelastic properties of the human red blood cell membrane. I. Deformation, volume loss, and rupture of red cells in micropipettes, Biophys. J., 13:1166 (1973).
7. H. Schmid-Schönbein, Blood rheology and physiology of microcirculation, in "Research in Clinic and Laboratory", vol. XI, Supplement No 1, p. 13-33.
8. C. Holzapfel, J. Vienken and U. Zimmermann, Rotation of cells in an alternating electric field: theory and experimental proof, J. Membrane Biol., 67:13 (1982).
9. H. P. Schwan, Dielectric properties of biological tissues and biophysical mechanisms of electromagnetic field interaction, in ACS Symposium Series No 157, "Biological Effects of Nonionizing Radiation", Karl H. Illinger, Editor, (1981).
10. H. H. Hub, H. Ringsdorf and U. Zimmermann, Rotation of polymerized vesicles in an alternating electric field, Angew. Chem. Int. Ed. Engl., 21:134 (1982).

WATER IN POLYMERS AND ARTEMIA CYSTS: RELATION OF NEUTRON SCATTERING AND NMR RESULTS

H. E. Rorschach

Physics Department
W. M. Rice University
Houston, Texas 77251

INTRODUCTION

Water is one of the major constituents in viable biological systems. An understanding of its physical properties is important to a quantitative treatment of chemical reaction rates, protein conformations, ionic distributions and other essential cellular processes. A large number of different techniques have been used to measure the properties of water in association with macromolecular systems. Some of these measurements suggest that the properties of water in these systems differ little from those of pure water, while others suggest that there are substantial changes (Drost-Hansen and Clegg, 1979). A proper interpretation of the results obtained from these techniques requires an appreciation of the heterogenous nature of the cell and the scale of distance over which variations in physical properties occur.

Two important quantities that characterize the scale of the heterogeneity of the cell are presented in Table I. The dimensions a that characterize the cell size (as determined by the membrane structures) and the scale of heterogeneity within the cell (determined by the organelle size and separation) as well as the times t_D required for the diffusion of water molecules through these characteristic distances, are tabulated.

Table I. Dimension & Diffusion Time for Cell Structures

Structure	Dimension a	Diffusion time $t_D=a^2/D$
Cell membranes	1-10 microns	1 ms
Organelle/Macro-molecular spacing	0.01 microns	0.1 μs

NMR METHOD

Nuclear magnetic resonance (NMR) methods have been used to study the properties of water in a variety of biological systems (Hazlewood, 1979). Measurements of the relaxation times T_1 and T_2 and the diffusion coefficient D under a wide range of measuring conditions have been reported and indicate that the values of these parameters differ markedly from those of pure water. These measurements are strongly influenced by the heterogeneous nature of the cell. Most methods employ pulse techniques in which the measuring time is $\gtrsim$ 10 ms. This time is larger than t_D, the characteristic diffusion time for the structures of the cell. The observed parameters (T_1, T_2 & D) will thus be averages over the cellular environment and will be influenced by the interaction of the water molecules with the macromolecules and adsorbed water fractions. The influence of these structures on the relaxation times T_1 and T_2 will be especially strong, and models in which rapid exchange occurs amongst the various cellular environments have been developed to describe this influence (Zimmerman and Brittin, 1957).

The diffusion coefficient D is less sensitive to the properties of the adsorbed fractions, but it is influenced by the macromolecular obstructions. These obstructions will decrease the NMR-measured diffusion coefficient when the measuring time is comparable to or larger than t_D. Modern methods, which make use of pulsed field gradients (Stejskal, 1972), allow a variation of the measuring time over a range that permits an evaluation of the influence of the membrane structures on the diffusion coefficient (Tanner, 1975 & 1980). It is, however, not possible to reduce the NMR measuring time sufficiently to study directly the influence of the organelles and macromolecules on the diffusion coefficient.

QNS METHOD

A powerful method of growing importance in studying the dynamics of water in association with clays, polymers and biological structures is that of quasi-elastic neutron scattering (QNS) (Touret-Poinsignon and Timmins, 1981; Springer, 1972). Measurements of the neutron scattering law $S(\vec{Q},\omega)$ give direct information on the Fourier transform of the space-time correlation function $G(\vec{r},t)$ (Van Hove, 1954; Marshall and Lovesey, 1971). The quasi-elastic scattering for neutron energy transfers $\hbar\omega$ near zero is determined by the dynamical properties of the diffusive motion of the scattering molecules. The most important feature of the QNS measurements is that they permit the study of this motion on a scale of a few angstrom, thus extending the diffusion measurements to a region, inaccessible to NMR, in which the influence of barriers and obstructions becomes negligible.

The scattering law $S(\vec{Q},\omega)$ is an experimentally determined quantity that is proportional to the scattering cross-section for each value of neutron momentum transfer $\hbar\vec{Q}$ and energy transfer $\hbar\omega$. Most theories of diffusive motion predict that the scattering law is composed of one or more Lorentzian lines that can be characterized by a width $\Gamma(Q)$. This width can be related to the microscopic parameters that characterize the diffusive motion (Springer, 1972). The diffusive properties of water are often described by a model in which jump-diffusion of the molecule is combined with rotational Brownian motion. The parameters that characterize this model are the translational diffusion coefficient D, the residence time between translational jumps τ, and the rotational diffusion coefficient D_r. This method of analyzing the dynamics of water in biological and polymer systems has been described in more detail by Trantham, et al. (1982).

The application of the QNS method to the study of water in biological systems is relatively new. In the next section we present some results of this method for pure water, and water in an agarose gel, a polyox gel and *Artemia* cysts. NMR results on these systems are also summarized, if available.

SOME EXPERIMENTAL RESULTS

NMR Method

The relaxation mechanism in pure water is well understood and can be characterized in terms of the fluctuating fields associated with the translational and rotational motion (Abragam, 1961). The relaxation rate for pure water is given by

$$\left(\frac{1}{T_1}\right)^{o} = \left(\frac{1}{T_1}\right)^{o}_{trans.} + \left(\frac{1}{T_1}\right)^{o}_{rot.} \qquad (1)$$

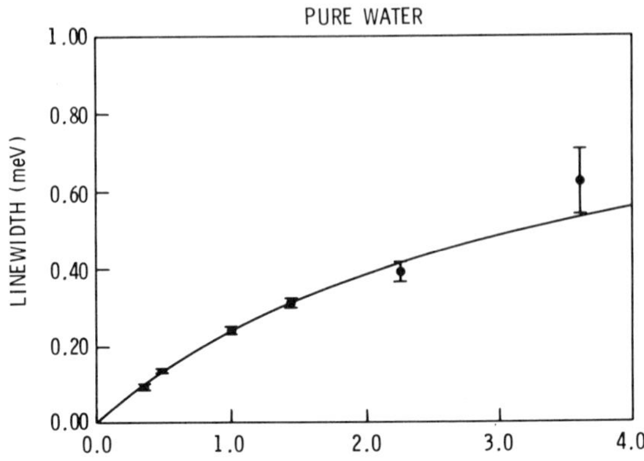

Fig. 1. Line width Γ(meV) vs Q^2(A^{-2}) for pure water.

where $(\frac{1}{T_1})^{\circ} \simeq 0.33\ s^{-1}$ and

$$(\frac{1}{T_1})^{\circ}_{rot.} \simeq 2(\frac{1}{T_1})^{\circ}_{trans.} \qquad (2)$$

(The zero superscript indicates the value for pure water.)

Measurements on polymer and biological systems show that the relaxation rate for water is substantially increased in these systems. For an agarose gel of hydration 4g H_2O/g dry solids, a relaxation rate $(1/T_1)_{agarose} \simeq 2.5\ s^{-1}$ has been measured (Derbyshire & Duff, 1974). Biological systems generally show an even larger relaxation rate. Measurements on Artemia cysts of hydration 1.2 g/g dry solids give $(1/T_1)_{artemia} \simeq 4.0\,s^{-1}$ (Seitz,et al.,1980). These increases in relaxation rate have been usually explained by models in which the major component of the cytoplasmic water is in rapid exchange with a minor component that is associated with the macromolecules and is characterized by a long correlation time τ_c (Edzes & Samulski, 1978).

QNS Method

This method has been used to study the diffusive properties of water in agarose gels and Artemia cysts (Trantham et al., 1982).

The technique will be described in more detail elsewhere. Some typical results for $\Gamma(Q)$ are given in Figure 1 which shows results for pure water, and in Figure 2 which shows results for a 4 g/g agarose gel and 1.2 g/g *Artemia* cysts. Figure 3 shows some preliminary results for a 1.8 g/g gel of polyethylene oxide. These results for $\Gamma(Q)$ can be used to derive the parameters D, τ & D_r that characterize the diffusive motion in these systems. Such an analysis requires a careful consideration of all contributions to the line shape, including the influence of the tightly bound protons and the resolution function of the spectrometer, which will be reported elsewhere. In Table II, we give the values of the diffusive parameters obtained by this analysis.

COMPARISON OF NMR AND QNS RESULTS

In this section, we want to show that the values of T_1 obtained from NMR measurements on *Artemia* cysts are consistent with the values of D and D_r obtained from QNS studies, and that the T_1 relaxation mechanism is thus primarily a *bulk* mechanism.

In a previous section, the relaxation rate for pure water was characterized by contributions from the translational and rotational motion. We assume that the relaxation mechanisms in *Artemia* can be described in the same way:

$$\left(\frac{1}{T_1}\right)_{\text{artemia}} = \left(\frac{1}{T_1}\right)_{\text{trans.}} + \left(\frac{1}{T_1}\right)_{\text{rot.}} \qquad (3)$$

In the short correlation time limit, each relaxation rate is inversely proportional to the corresponding diffusion coefficient $(1/T_1)_i \propto 1/D_i$ (Abragam, 1961). In *Artemia* cysts, both D and D_r are reduced, and Table II shows that the relaxation rate due to translational motion is greater by a factor of 3.4 than that in pure water while the rate due to rotational motion is greater by a factor of $\sim$ 13. If we assume

Table II. Diffusion Parameters from QNS Studies

System	(D/D_o)	(τ/τ_o)	$D_r\,(10^9\ s^{-1})$
Pure H_2O	1^a	1^a	b
Agarose (4 g/g)	0.80	1.4	--
Artemia Cysts (1.2 g/g)	0.29	3.9	6.0

$^aD_o = 2.4 \times 10^{-5}\ cm^2/s$; $\tau_o = 1.2 \times 10^{-12}$ s

$^bD_r = 8.1 \times 10^{10}\ s^{-1}$ given by dielectric relaxation studies (Hasted, 1972).

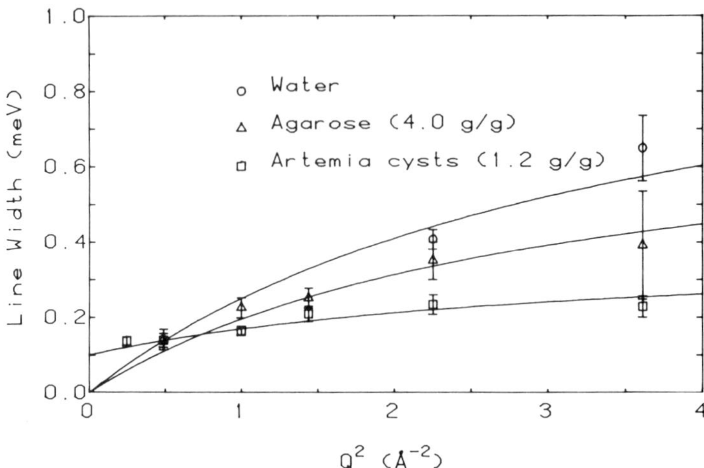

Fig. 2. Line width Γ(meV) vs $Q^2(A^{-2})$ for pure water, the 4 g/g agarose-H_2O gel and the 1.2 g/g *Artemia* cysts. The curves drawn through the data are obtained from diffusion models with the parameters listed in Table II.

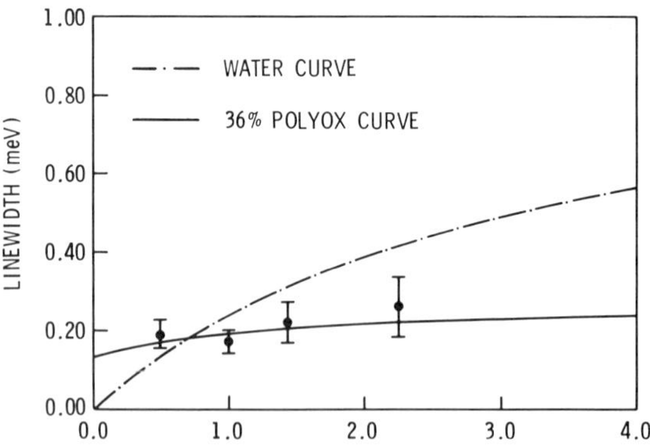

Fig. 3. Line width Γ(meV) vs $Q^2(A^{-2})$ for the 1.8 g/g poly(ethylene) oxide gel.

that the other parameters (e.g. atomic distances) that determine the relaxation rate are unchanged, then we can estimate $(1/T_1)_{artemia}$ from equations (1) & (3):

$$\left(\frac{1}{T_1}\right)_{artemia} = 3.4 \left(\frac{1}{T_1}\right)^{o}_{trans.} + 13 \left(\frac{1}{T_1}\right)^{o}_{rot.}$$

and Eq. (2) then gives

$$(1/T_1)_{artemia} \simeq 10(1/T_1)^{o} = 3.3 \ s^{-1}.$$

This rate agrees rather well with the value obtained by Seitz, et al. (1980). (The difference could be due to the neglect of the rapid exchange mechanism.)

A similar analysis cannot be carried out for the agarose gel, since D_r is not known for this system. A value of D_r that is less than that of pure water by a factor of ~ 8 is required to produce the observed value of T_1 with the same assumptions regarding relaxation mechanisms as those used for Artemia.

Analysis of the polyox data has not yet been completed. The QNS properties are quite similar to those of the Artemia cysts, which is not surprising since the hydrations are low in both cases, and the "structuring" of the water would be expected to be similar.

ACKNOWLEDGEMENTS

The author wishes to acknowledge support of this work from the Office of Naval Research through ONR Contract N00014-79-C-0492 and from the Department of Energy through ORAU Participation agreement S-2016. The author thanks Drs. J. C. Clegg, C. F. Hazlewood R. M. Nicklow, E. C. Trantham, and D. B. Heidorn for important suggestions and for use of the unpublished QNS data.

REFERENCES

1. Abragam, A., 1961, "The Principles of Nuclear Magnetism" Clarendon Press, Oxford, pp. 289-305.
2. Derbyshire, W. and Duff, I. D., 1974, NMR of Agarose Gels. Faraday Discussions of the Chemical Society, 57:243.
3. Drost-Hansen, W. and Clegg, J.S., editors, 1979, "Cell-Associated Water," Academic Press, New York.
4. Edzes, H. T. and Samulski, E. T., 1978, The Measurement of Cross-Relaxation Effects in the Proton NMR Spin-Lattice Relaxation of Water in Biological Systems: Hydrated Collagen and Muscle. J. Mag. Res. 31:207.
5. Hasted, J. B., 1972, Liquid Water: Dielectric Properties, in: "Water Vol. 1," F. Franks, ed., Plenum Press, NY, Chapt. 7.
6. Hazlewood, C. F., 1979, A View of the Significance and Understanding of the Physical Properties of Cell-Associated Water, in:

"Cell-Associated Water," W. Drost-Hansen and J. Clegg, ed., Academic Press, NY, p. 165.
7. Marshall, W. and Lovesey, S. W., 1971, "Theory of Thermal Neutron Scattering," Clarendon Press, Oxford.
8. Seitz, P., Hazlewood, C.F., and Clegg, J., 1980, Proton Magnetic Resonance Studies on the Physical State of Water in *Artemia* Cysts, *in*: "The Brine Shrimp Artemia, Vol 2. Physiology, Biochemistry and Molecular Biology," G. Persoone, P. Sorgeloos, O. Rolls, and E. Jaspers, ed., Universa Press, Wetteren, Belgium, p. 545.
9. Springer, T., 1972, "Quasi-Elastic Neutron Scattering for the Investigation of Diffusive Motions in Solids and Liquids. Springer Tracts in Modern Physics," Vol. 64, G. Höhler, ed. Springer-Verlag, NY.
10. Stejskel, E.O., 1972, Spin-echo Measurement of Self-Diffusion in Colloidal Systems. *Adv. Mol. Relax. Proc*. 3:27.
11. Tanner, J. E., 1975, "Self-diffusion in Cells and Tissues," Office of Naval Research Report NWSC/CR/RDTR-6, Division of Medical and Dental Science, Arlington, VA.
12. Tanner, J. E., 1980, NMR Measurements of Self-Diffusion in Cells, Abstract #813, 24th Annual Meeting of the Biophysical Society.
13. Touret-Poinsignon, C., and Timmins, P., eds., 1981, "Workshop on Water at Interfaces," Institute Laue-Langevin Report 81T055S, Oct. 12-13, Grenoble Cedex-France.
14. Trantham, E. C., Rorschach, H. E., Clegg, J. C., Hazlewood, C. F., and Nicklow, R. M., 1982, QNS Measurements on Water in Biological and Model Systems, *in*: "Neutron Scattering--1981," John Faber, Jr., ed., American Institute of Physics, NY.
15. Van Hove, L., 1954, Correlations in Space and Time and Born Approximation Scattering in Systems of Interacting Particles. *Phys. Rev*. 95:249.
16. Zimmerman, J. P., and Brittin, W. F., 1957, Nuclear Magnetic Resonance Studies in Multiple Phase Systems: Lifetime of a Water Molecule in an Absorbing Phase on Silica Gel. *J. Phys. Chem*. 61:1328.

WATER, IONS AND DRUG ACTION, INCLUDING ANAESTHETICS

THE EFFECT OF INHALATION ANESTHETICS ON EPITHELIAL WATER AND ION TRANSPORT IN VIVO: STUDIES IN RATS

Heribert Konder, Rüdiger Dennhardt*, and Herbert Lennartz

Abteilung für Anaesthesie und interdisziplinäre Intensivtherapie der Philipps-Universität
D-3550 Marburg
*Klinik fur Anästhesie und operative Intensivmedizin im Klinikum der Freien Universität
D-1000 Berlin

Until now most studies of general anesthetics have been made on their effects on the central nervous system. Claude Bernard pointed out that narcosis causes alteration in tissues as well as in cells. There is little information available about the effect of modern inhalation anesthetics like Fluothane and Enflurane on epithelial transport systems[1-7].

From our point of view epithelial tissues possessing unexcitable membranes are most suitable for investigations on basic mechanisms of anesthetic action on membrane transport. For our experiments we examined rat jejunum because this epithelial tissue shows good correlation in its main transport functions to other epithelial tissues [8].

In former in vitro transport-kinetic studies using the micro-everted-sac technique of Semenza and Mühlhaupt[9] carried out on rat jejunum with sorbose, glucose and 3-0-methyl-glucose as substrates we could establish that fluothane and enflurane diminish passive membrane diffusion and slow down active transport. The loss in membrane transport capacity can be compensated in part by an increase of the cellular metabolism of glucose[3].

At the same time we studied the effect of 3 Vol% of fluothane and enflurane on jejural water content in vitro (Fig. 1). These experiments have been carried out in Krebs-Henseleit bicarbonate buffer at 37°C with an incubation period of 4 minutes in the absence of sugars. The water content after the incubation has been determined as difference between wet and dry weight after vacuum drying for 24 hours at 100°C[10]. The data have been compared to experiments without inhalation anesthetics. The results show that 3 Vol% of fluothane and enflurane significantly augment water content of jejumum by about 4.2% or 3.7% respectively compared to the control experiments.

These results prompt the examination of the effect of enflurane on the water and ion transport in rat jejunum in vivo. For our studies we used the in vivo perfusion technique of rat intestine developed by Haberich et al.[11] (Fig. 2).

The animals were operated on at least three days before the beginning of the experiments. An inflatable cuff was placed at each end of a chosen intestinal segment, so that a jejunal segment of about 10 cm in length could be temporarily isolated and subsequently perfused by means of a tube placed at the proximal end and an outlet tube anastomosed to the distal end of the gut segment. All tubes came out on the animal's back and were protected by a semicircular piece of metal. When no experiments were being carried out the cuffs are evacuated, the tubes closed and attached to the neck, and the rats had free access to food and water.

In order to perform an experiment we placed the animal into a plastic tube, blew up the cuffs to isolate 10 cm of jejunum and washed out the ingesta through the implanted catheters. During the

Jejunum	Control	3Vol.% Fluothane	3Vol.% Enflurane
	(% wet weight)		
$\bar{x}$	74.21	77.3	76.98
s	2.09	2.6	1.68
n	15	15	15

Control vs. Fluothane: $p<0.002$
Control vs. Enflurane: $p<0.001$

Incubation: 37°C; 4min. in Krebs-Henseleit-bicarbonate-buffer; Carbogen

Drying: 100°C; 24 hours; in vacuo

Fig. 1. Effect of 3 Vol% of Fluothane and Enflurane on jejunal water content in vitro.

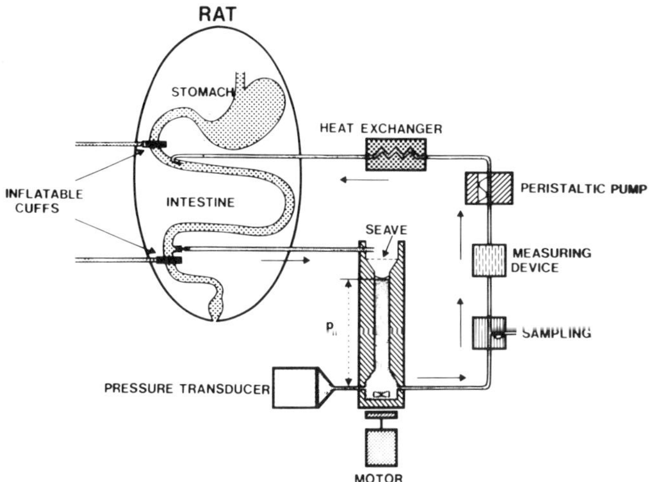

Fig. 2. Illustration of experimental procedures for perfusion of rat's intestine in an extracorporal closed circular course. Changes in net water fluxes are recorded immediately by pressure transducer at the ground of the buret.

experiment jejunum was perfused with Krebs-Henseleit bicarbonate buffer with an initial osmolality of either 290 mosmol/kg isotonic, 150 mosmol/kg hypotonic or 400 mosmol/kg hypertonic. Different osmolalities were prepared by changes in the starting sodium concentration. The in vivo experiments are carried out in an extracorporal closed circular course where the perfusate is pumped through a heat exchanger into the jejunum and subsequently dropped into a buret. Changes in net water transport due to absorption or secretion are continuously registered as hydrostatic pressure differences at the ground of the buret. The unidirectional water fluxes are measured by liquid scintillation counting of tritium water added to the perfusate. At the same time changes in the amount of net sodium, potassium, chloride and osmotic transport are analyzed over the duration of the experiment.

Firstly every rat had to undergo a control perfusion over 60 min without enflurane inhalation followed by a second perfusion period over 60 min under the influence of enflurane. We started the perfusion experiment when a steady state of anesthesia with enflurane had been reached. The minimal anesthetic concentration (MAC) was proved by gripping the tail of the rat. The individually proved MAC for rats is about 2 Vol% of enflurane.

Fig. 3 shows the mean values of 10 experiments of net sodium transport in µmol per 10 cm of jejunum and 60 min of experimental time. Chloride transport is equimolecular to sodium transport. Absorption(-) means net transport from the intestinal lumen to the blood. Net sodium absorption is significantly diminished by enflurane during isotonic perfusion and significantly increased under hypotonic conditions whereas there is no difference during hypertonic perfusion according to control experiments.

In the same animal we simultaneously analyzed the effect of enflurane on epithelial net and unidirectional water fluxes (Fig. 4). Influx means the unidirectional flux from the blood into the intestinal lumen and efflux the opposite unidirectional flux. Subtraction of both unidirectional fluxes results in the net water flux which we determined immediately at the ground of the buret. The net water fluxes are comparable to the net sodium fluxes and follow passively the osmotic gradient existing across the intestinal epithelial barrier. When looking for the unidirectional water fluxes it is seen

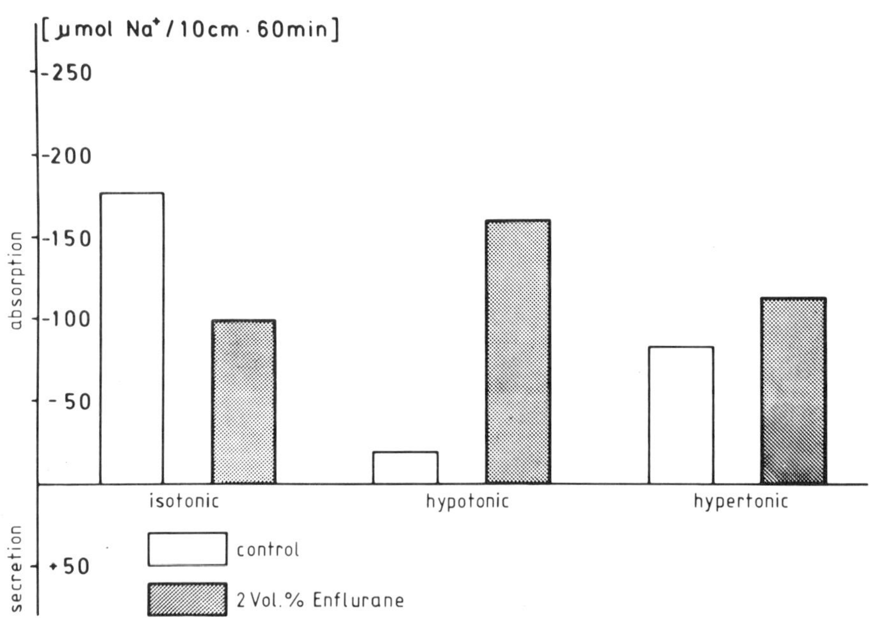

Fig. 3. Net sodium absorption in rat's jejunum with and without (control) inhalation of enflurane (mean values of 10 experiments resp.)

that there is a relative increase in the influx - blood to lumen - during inhalation of enflurane.

Fig. 5 demonstrates the results of the net potassium transport. It is wellknown that there is no active potassium transport in the

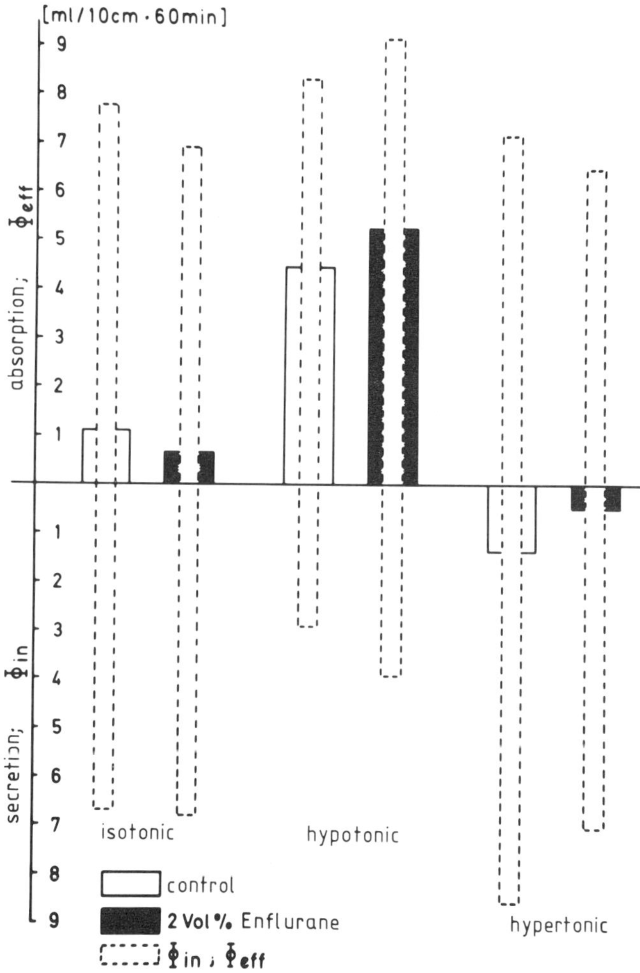

Fig. 4. Net water fluxes and unidirectional water fluxes with and without inhalation of enflurane (n = 10).

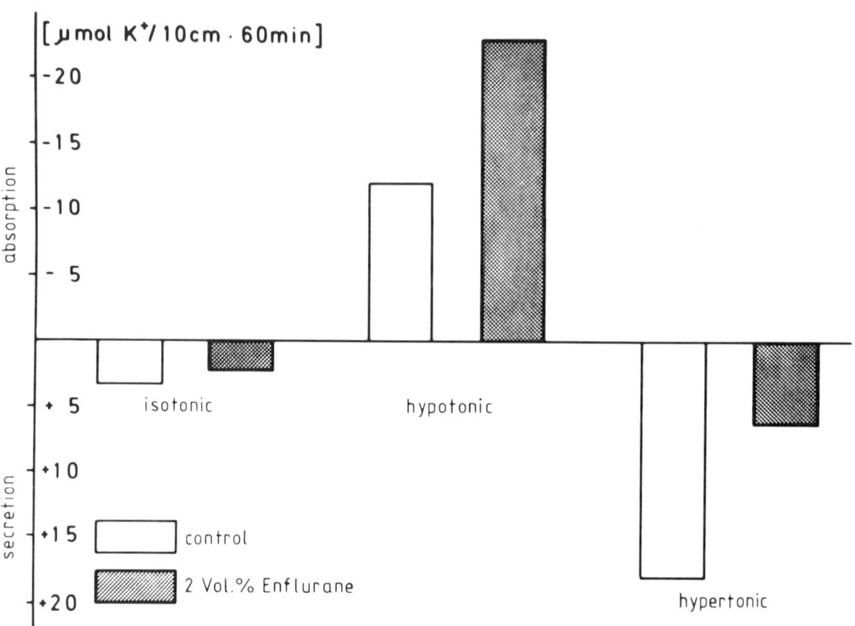

Fig. 5. Net potassium fluxes with and without inhalation of enflurane (n = 10).

jejunum; potassium follows water transport rather like a "bulk flow". These results emphasize that enflurane affects epithelial membrane permeability.

In Fig. 6 the sodium-water-equivalent can be seen as linear significant regression. The straight lines are the result of the simultaneously obtained data of net sodium and net water transfer in each experiment according to the different osmotic respectively sodium concentrations in the intestinal lumen.

During isotonic perfusion (full lines) isotonic net water transport occurs, i.e. absorption of 140 µmol of sodium is coupled with a net water absorption in the amount of 1 ml per 60 min and 10 cm of jejunum. Enflurane does not alter isotonic transport. If we apply hypertonic solution to the jejunal lumen we find increased sodium absorption due to a higher concentration of sodium in the perfusate. At the same time a raised net water secretion into the jejunal lumen occurs in order to get osmotic balance compared with blood. The regression line under hypertonic perfusion runs parallel with that under isotonic perfusion demonstrating that the sodium-water equivalent is unchanged. The regression line during enflurane inhalation runs parallel with that during hypertonic control experiments, which means that the net sodium absorption and net water secretion are

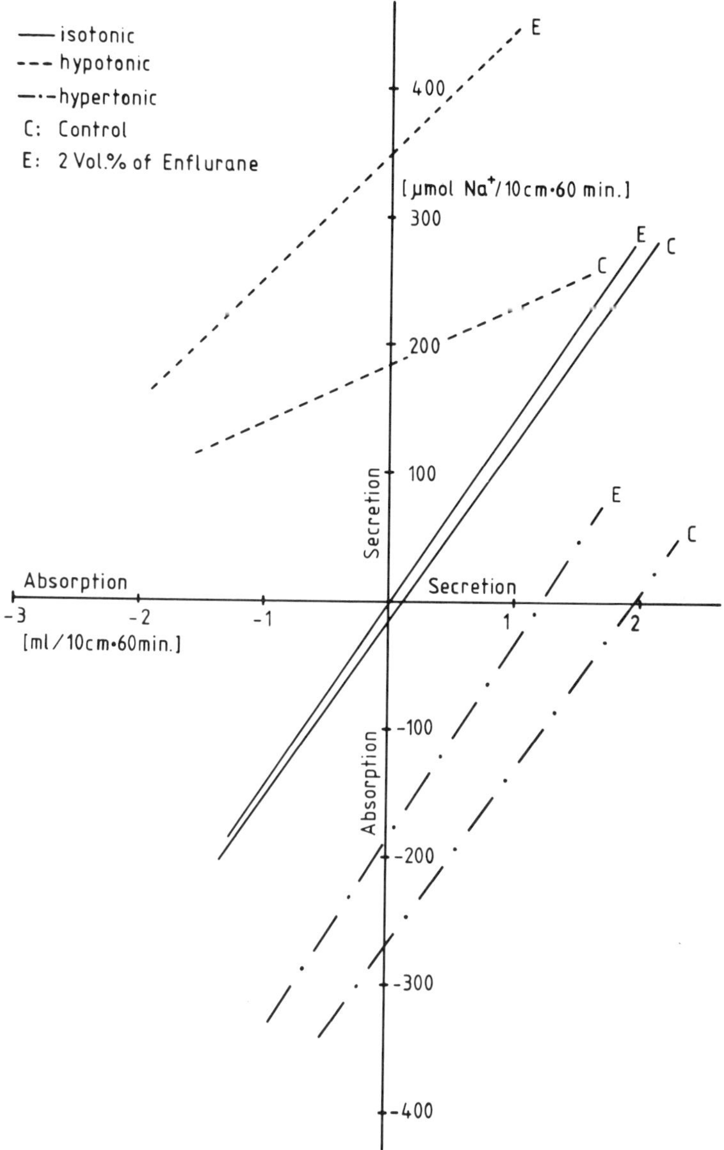

Fig. 6. Sodium(y)-water(x)-equivalents of the data demonstrated in Fig. 3 and 4 during perfusion of the jejunum with different osmotic gradients.
Isotonic: Control y = -16.9 + 139.9 x; r = 0.989;
Enflurane y = -5.1 + 145.4 x; r = 0.988;
Hypotonic: Control y = 183.5 + 45.5 x; r = 0.732;
Enflurane y = 349.9 + 96.4 x; r = 0.965;
Hypertonic: Control y = 269.4 + 137.2 x; r = 0.817;
Enflurane y = 187.4 + 153.6 x; r = 0.907.

diminished in the same ratio. The sodium-water equivalent is untouched.

For the hypotonic perfusion experiments we should expect net sodium secretion and net water absorption. This is supported by the regression line under control perfusion but the steepness of the regression line is much smaller than under isotonic control experiments. The sodium-water equivalent under hypotonic perfusion has been shifted in favor of an enhanced sodium-absorption which counteracts the secretion induced by the low concentration of sodium in the lumen. This phenomenon is the result of the active sodium absorption in the jejunum. Enflurane alters the sodium-water equivalent during hypotonic perfusion and moves the quotient to isotonic conditions. At the same time secretion is raised by enflurane similar to the results obtained under hypertonic perfusion.

CONCLUSIONS

Our experiments indicate that enflurane in MAC causes inhibition of active sodium absorption in epithelial membranes. This finding is in agreement with the results we established on active epithelial hexose absorption. Besides this enflurane diminishes the permeability of membranes located at the serosal side of the epithelium, so that in part an inversion of net transport occurs from absorption to secretion. This is well illustrated by the experiments during perfusion with different osmotic gradients.

Our data are in agreement with theoretical considerations about the mechanism of anesthetic action. The critical volume theory[12] points out that anesthetics act by taking a partial molal volume in a membrane so producing either a hydrostatic effect or an inhibition of conformational changes of proteins[13], thus influencing active membrane transport as well as membrane permeability. Trudell[14] established a theory of anesthesia based on lateral phase separations in membranes illustrating that changes in membrane ion transport are due to alterations in membrane channels and carriers.

Trudell's considerations are in agreement with the critical volume theory. Our findings support these theoretical considerations.

REFERENCES

1. L. Amaranath and N. B. Andersen, The effect of anesthetics on permeability to water of the inactivated toad bladder, Anesthesiology, 40:168 (1974).
2. N. B. Andersen and L. Amaranath, Anesthetic effects on transport across cell membranes, Anesthesiology, 39:126 (1973).

3. R. Dennhardt, H. Konder, and H. Lennartz, Beeinflussung des Hexose-Transports an epithelialen Strukturen durch Inhalationsanaesthetika, in: "Anaesthesiology and Intensive Care Medicine," 141, Vol. 3, pp. 32-36, B. Haid and G. Mitterschiffthaler, eds., Springer-Verlag, Berlin, Heidelberg, New York (1981).
4. H. Konder, R. Dennhardt, and H. Lennartz, Beeinflussung des Elektrolyt- und Wassertransports am Dünndarm durch Enfluran, in: "Anaesthesiology and Intensive Care Medicine," 141, Vol. 3, pp. 26-31, B. Haid and G. Mitterschiffthaler, eds., Springer-Verlag Berlin, Heidelberg, New York (1981).
5. A. G. Macdonald and K. T. Wann, "Physiological Aspects of Anaesthetics and Inert Gases," Academic Press, London, New York, San Francisco (1978).
6. E. M. Nemoto, S. W. Stezoshi, and B. S. D. MacMurdo, Glucose transport across the rat blood-brain barrier during anesthesia, Anesthesiology, 49:170 (1971).
7. P. Seeman and S. Roth, The general anesthetics expand cell membranes at surgical concentrations, Biochim.Biophys.Acta, 255:171 (1972).
8. K. J. Ullrich, E. Frömter, and H. Murer, Prinzipien des epithelialen Transportes in Niere und Darm. Klin.Wochenschr., 57:977 (1979).
9. O. Semenza and E. Mühlhaupt, Studies on intestinal sucrase and sugar transport. VII. A method for measuring intestinal uptake, The absorption of the anomeric forms of some monosaccharides, Biochim.Biophys.Acta, 173:104 (1969).
10. R. K. Crane and P. Mandelstam, The active transport of sugars by various preparations of hamster intestine, Biochim.Biophys. Acta, 45:460 (1960).
11. F. J. Haberich, R. Herzer, O. Aziz, and R. Dennhardt, Resorptions- und Sekretionsstudien am Darm. I. Technik der extrakorporalen Perfusion beliebiger, vorübergehend funktionell isolierter Darmabschnitte an der wachen Ratte, Z.Ges.Exp. Med., 148:223 (1968).
12. L. J. Mullins, Anaesthetics, in: "Handbook of Neurochemistry," Vol. 6, Plenum Press, New York (1971).
13. B. P. Schoenborn and R. M. Featherstone, Molecular forces in anaesthesia, Adv.Pharmacol., 5:1 (1967).
14. J. R. Trudell, A unitary theory of anesthesia based on lateral phase separations in nerve membranes, Anesthesiology, 46:5 (1977).
15. R. Dennhardt, B. Lingelbach, and F. J. Haberich, Intestinal absorption under the influence of vasopressin: studies in unanesthetized rats, GUT, 20:107 (1979).

WATER AND DNA-DRUG INTERACTION

Günter Löber and Renate Klarner

Central Institute of Microbiology and Experimental Therapy, Acad. of Sci. of the GDR, Department of Drugs and Isotopes, DDR-69 Jena

INTRODUCTION

Several biologically active compounds are capable of forming intermolecular non-covalent complexes with deoxyribonucleic acid (DNA) under free energy changes of less than 40 kJ/mol (approximately 10 kcal/mol)[1-4]. These complexes are in the first place investigated on isolated DNA. There is, however, indication that they are also present in biological systems, where the nucleic acids exist in their natural state[5,6].

The number of papers using absorption and fluorescence spectroscopic techniques in analyzing the interaction of drugs with DNA is great and still increases[7,8]. There are overwhelming indications that the drug binding occurs by two principal modes of binding, termed as type I and type II[8]. Type I corresponds to the monomer binding, consisting of various possible subtypes including intercalation, while type II corresponds to different modes of binding which are based on the mutual interaction of bound drug molecules [8-10].

This paper is concerned with the question as to how the binding is affected when the aqueous medium, in which the drug and the DNA are solved, is partially substituted by organic solvents. Thus far such studies are capable of yielding information on the role of water in drug-DNA interaction, the nature of the interaction forces and the spectroscopic properties of type I complexes.

INFLUENCE OF ORGANIC SOLVENTS ON THE INTERACTION OF DRUGS WITH ISOLATED DNA

Influence of Organic Solvents on Type I Complexes

The effect of various organic solvents on the type I complex of the mutagenic drug proflavine with DNA was studied by means of absorption and fluorescence spectra and by the thermal denaturation technique[11].

The binding of proflavine to DNA is accompanied by a decrease of the optical density as well as by a red shift of the characteristic long-wave absorption band of the drug. It follows from the behavior of the absorption spectrum of the drug in the presence of ethanol that the tendency for the interaction is significantly lowered with increasing quantities of the organic solvent present in the drug-DNA system (Figure 1). A more detailed study using different kinds of organic solvents showed that the effectiveness of solvents in destabilizing the type I complex of DNA with proflavine increases in the order: water<glycerol<eythylene glycol<methanol<formamide<ethanol <isopropanol<n-propanol<p-dioxane<dimethylsulfoxide.

In several previous contributions it was shown that the fluorescence intensity of a drug is very strongly altered upon binding to the DNA (review,[8]). Addition of DNA to solutions of proflavine leads to quenching of the dye fluorescence due to the binding. Increasing quantities of the organic solvents decreases the dye binding to the DNA and as a consequence the changes of the fluorescence intensity become less expressive. As shown in Figure 2 addition of ethanol evoked a significant decrease in DNA-induced quenching and at 50-60% (v/v) of ethanol the lack of fluorescence quenching indicates that the degree of the binding is below the threshold of the fluorimetric detection.

Similar effects were also found for other solvents like methanol, n-propanol, isopropanol, formamide, dimethylsulfoxide, p-dioxane, glycerol and ethylene glycol. To demonstrate the decrease of dye binding more quantitatively the relative fluorescence intensity, f/f , of proflavine at 9×10^{-5} M DNA phosphorus was measured as a function of the volume percentages of various organic solvents. The relative effect of the solvents was characterized by the volume percentage which caused reduction of f/f to one half. Those data listed in Table 1 give some insight into the nature of the binding. Thus it is obvious that the observed solvent effect cannot be related to the dielectric constant of the added solvent. Even though the tested alcohols have a considerably smaller and formamide a higher dielectric constant compared with water, all reduce the interaction between proflavine and DNA. On the other hand, the effectiveness of the organic solvents increases with increasing hydrocarbon content (propylalcohols>ethanol>methanol).

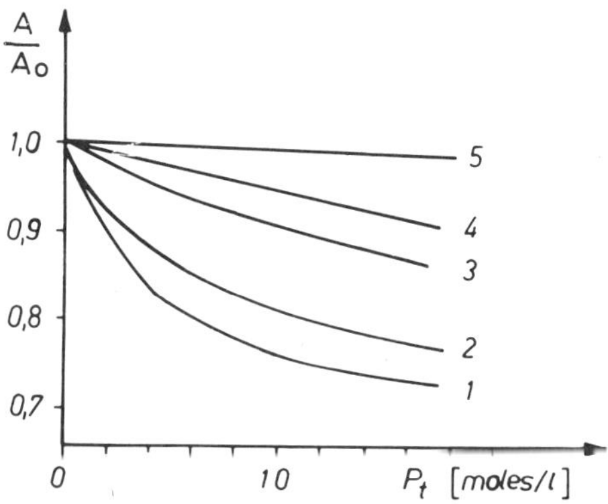

Fig. 1. DNA-induced changes of the optical density of the long-wave absorption band of proflavine in the presence of increasing amounts of ethanol; drug concentration is 2×10^{-5} M; A_o, optical density without DNA; A, optical density in the presence of 0% (1), 10% (2), 20% (3), 30% (4) and 50% (5) of ethanol (% given as v/v); P_t is the total concentration of DNA phosphorus.

These results demonstrate the importance of hydrophobic forces for the formation of the complexes of the acridine derivative with DNA in aqueous solution. In particular, Table 1 demonstrates that the binding ability is strongly reduced at organic solvent concentrations which are not sufficient to cause DNA denaturation[12].

Independent information on the binding of drug to DNA was obtained by thermal denaturation experiments following the method described in[13]. The stabilization effect of the bound proflavine, expressed as the difference between the melting temperature of the complex and pure DNA, ΔT_m, is reduced in the presence of the organic solvents. The reduction of the stabilization effect in the presence of organic solvents can be attributed to the decrease of the total amount of the bound dye and the relative decrease of the proportion of proflavine bound by the stronger binding type 1 (melting experiments were done by Dr. V. Kleinwächter, Brno, Czechoslovakia)[11].

Similar to proflavine, the stability of the complexes of DNA with a number of drugs, e.g., phenosafranine, acriflavine, acridine orange, ethidium bromide, quinacrine, and netropsin could only be

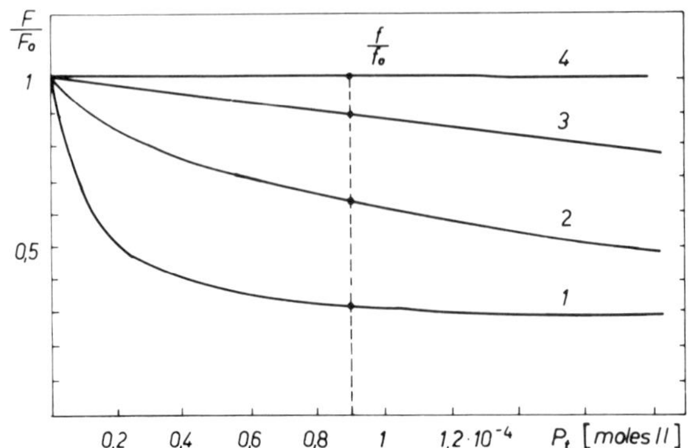

Fig. 2. DNA-induced fluorescence changes of proflavine in the presence of different amounts ethanol; drug concentration is 1×10^{-6} M; F_o, fluorescence intensity without DNA: F, fluorescence intensity in the presence of 0% (1), 20% (2), 30% (3) and 60% (4) ethanol; f/f_o, value of F/F_o at a constant concentration of DNA chosen arbitrarily.

Table 1. Solvent Effect on Denaturation and Proflavine Binding of DNA.

Solvent	Dielectric constant	Vol % Solvent: Denaturation midpoints taken from Herskovits[12]	Vol % Solvent: Reduction of the binding to one half
Methanol	32.0	80	30
Ethanol	25.8	80	24
n-Propanol	22.0	80	17
i-Propanol	19.0	80	20
Formamide	110.5	75	25
Dimethyl-sulfoxide	45.0	62	14
p-Dioxane	2.24	-	15
Glycerol	47	-	67
Ethylene glycol	38	93	39

ensured in the presence of water of generally more than 50% (unpublished results of authors).

Glycosidic antitumor antibiotics of the anthracycline type contain an aglycone part and a sugar residue. In the case of the anthracycline derivative violamycin BI (VBI), the glycosidic part consists of two positively charged rhodosamine residues one attached at position 7 and the other at position 10 of the anthracycline skeleton. This drug was shown to display an extremely high binding affinity to DNA[14]. The effect of ethanol or formamide on the absorption and fluorescence spectroscopic properties of the VBI bound to DNA by type I was investigated. Up to 60% (v/v) of organic solvent no significant destabilization of the complex was observed. Much stronger decrease was found in the binding of the aglycone to DNA, e.g., 40% (v/v) ethanol decreases the binding below the threshold of detection in fluorescence quenching experiments (Figure 3). It is apparent from these results that the aglycone behaves basically in the same way as proflavine.

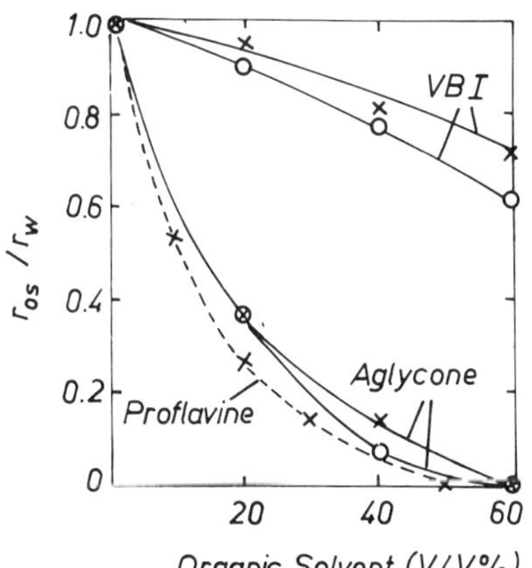

Fig. 3. Influence of ethanol (ooo) and formamide (xxx) on the binding of the anthracycline derivative violamycin BI (VBI), its aglycone and proflavine (for comparison) to calf thymus DNA; r_{OS} number of bound drug molecules per DNA phosphorus in the presence of organic solvents; r_W, number of bound drug molecules per DNA phosphorus in 0.01 M NaCl. Total concentrations of VBI and DNA phosphorus were $2x20^{-5}$ M and $1x10^{-4}$ M, respectively. The values of r were determined fluorimetrically.

The two charged sugar residues, however, evidently cause the high binding ability of VBI to DNA, which is reflected in the relatively high stability of the VBI-DNA complex. The smaller influence of organic solvents can be interpreted by a higher contribution of electrostatic forces in the binding (the role of different interaction forces in the binding of anthracyclines was discussed recently)[15].

Spectral Properties of Type I Complexes

The most characteristic effect of the interaction of nucleic acids with drugs is the displacement of the electronic absorption spectrum to longer wavelengths, when the drug molecules are bound by binding type I. This spectral shift was confirmed for a great number of substances[1,4,6,8]. Thus, the absorption technique became one of the standard procedures for qualitative and quantitative determination of the drug-DNA binding. Various hypotheses have been offered in order to explain this spectral behavior[8]. One of them is strongly connected with the key role of water in the formation and for the stability of drug-DNA complexes. It can be seen from Figure 4 that by increasing amounts of the organic solvent content in an ethanol-water mixture or by adding DNA to the solution of proflavine the absorption spectrum of the drug becomes red-shifted, while the fluorescence spectrum becomes blue-shifted. The maxima of absorption and fluorescence spectra of proflavine, acridine orange and ethidium bromide measured in water, seven organic solvents, and in the DNA-bound state are schematically represented in Figure 5. Moreover, absorption and fluorescence maxima of eosin found in similar solvent systems and complexed with human serum albumin (HSA) are shown[16]. In any case the absorption maxima of the drugs dissolved in water appear at the shortest wavelength. The lack of an isosbestic point in the sets of absorption spectra in the organic solvent-water mixtures excludes a specific interaction between the solute and the solvent molecules. Thus, one inclines to consider that a perturbation of the hydration shell around the solved drug molecules by organic solvent molecules opens a possibility for the latter to interact with the drug. This gives rise to the observed spectral changes which are found for all used solvents. Importantly, the direction of the red-shift in the absorption band of the drug by DNA binding, or in the case of eosin by HSA binding, essentially agrees with the solvent effects. Similarly, the maxima of the fluorescence spectra of proflavine, acridine orange and eosin upon binding with DNA and HSA, respectively, move towards the fluorescence maxima found in a number of organic solvents. This holds, however, not true for ethidium bromide, where, unexpectedly, the fluorescence maximum of the ethidium-DNA complex is blue-shifted, while the fluorescence maxima of the drug in all the checked organic solvents are red-shifted. Nevertheless, the major part of the data (and also the absorption spectrum of ethidium bromide) supports the notion that the

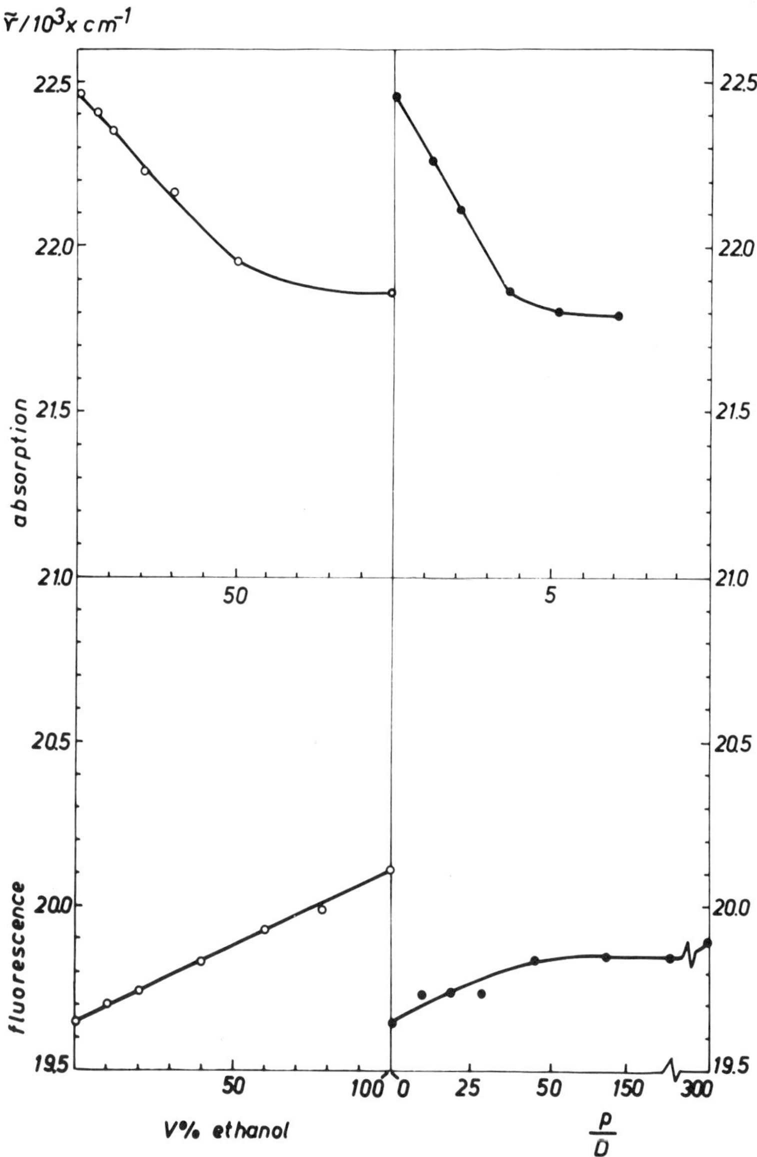

Fig. 4. Changes in the wavenumbers, $\tilde{\nu}$, of absorption and fluorescence maxima of proflavine on addition of ethanol (left) or DNA (right). Dye concentration is $2x10^{-5}$ M; P/D, ratio of DNA phosphorus to dye.

changes of environment, when a dye molecule goes from its hydrated state to the bound state where it is surrounded by the organic moieties of the biopolymer, are with respect to the position of the electronic transition similar to those accompanying the change from

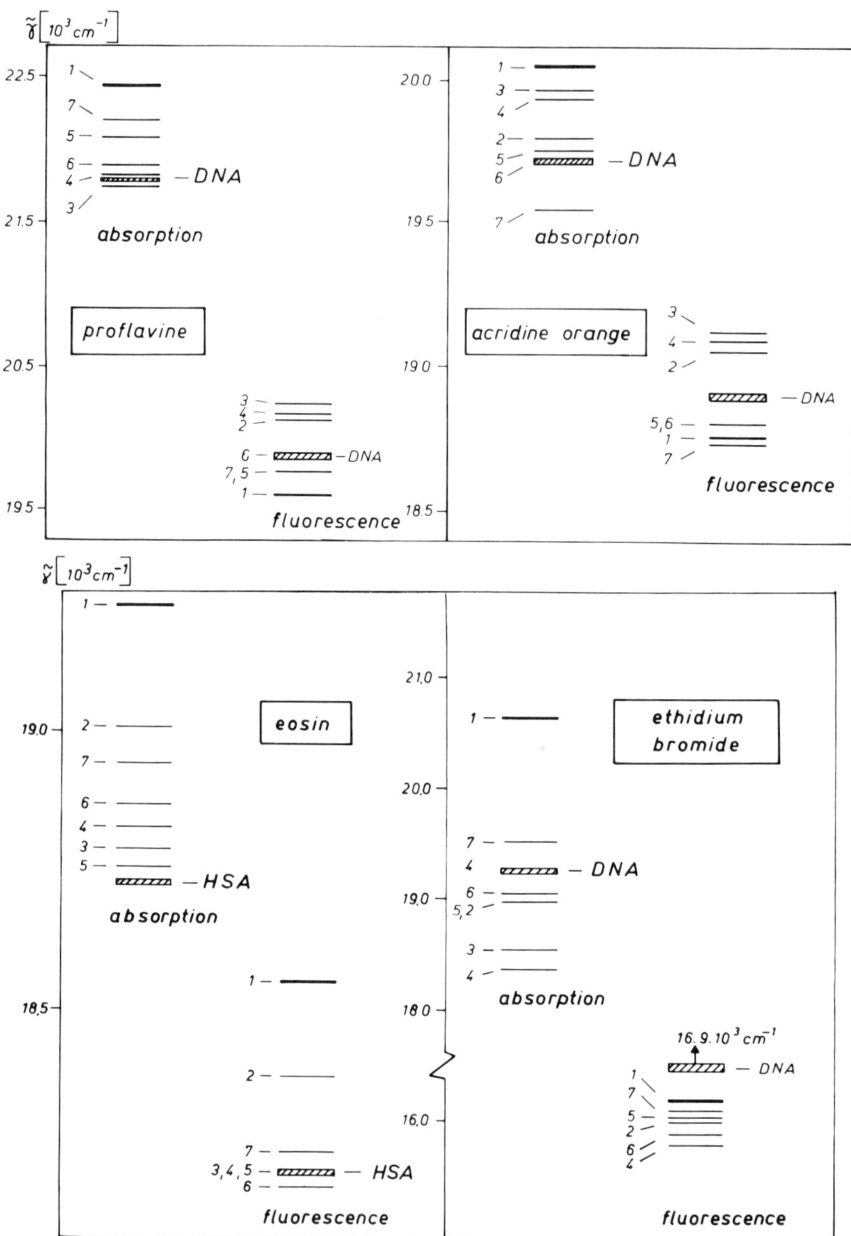

Fig. 5. Schematic representation of the wavenumbers of absorption and fluorescence maxima of proflavine, acridine orange, ethidium bromide and eosin in various solvents; water (1), methanol (2), n-propanol (3), isopropanol (4), formamide (5), ethylene glycol (6), glycerol (7); wavenumbers upon binding of the drugs with deoxyribonucleic acid and human serum albumin are indicated by "DNA" and "HSA", respectively.

water to an organic solvent. Very good approximations are the absorption spectra measured in ethanol solution (Table 2).

Interpretative energy diagrams and potential curves appropriate to the binding of dye molecules by type I should be based on the following findings (energy differences between the Franck-Condon state and the equilibrium state are neglected. This seems reasonable, when the dipole moment does not change with excitation, so that reorientation effects are small. In the case of proflavine and acridine orange this condition is fulfilled[17]).

(a) The change of the binding enthalpy is mostly negative. This results in a lowering of the energy of the ground state in the dye-DNA complex, when compared with the free dye in water;

(b) the red-shift of the 0.0 transition in going from the free to the bound state of dye is small. Thus, the level of the excited singlet state is only slightly more lowered than the level of the singlet ground state;

(c) the long-wave absorption band of a bound dye is stronger red-shifted than the 0.0 transition, which is attributed to the retention

Table. 2. DNA and Solvent-Induced Wavenumber Shifts in 10^3 cm^{-1} for Various DNA Binding Active Dyes[a].

Dye	Wavenumber in Water	Wavenumber in Ethanol
Proflavine	22.5	21.8
Acriflavine	22.2	21.5
Phenosafranine	19.3	18.8
Ethidium bromide	20.6	18.7
Pinacyanol	18.3; 16.8	17.9; 16.5

DNA-bound state	Change in the wavenumber in going from: Water to ethanol	Water to DNA-bound state
21.8	0.7	0.7
21.5	0.7	0.7
18.7	0.5	0.6
19.3	1.9	1.3
18.0;16.6	0.4;0.3	0.3;0.2

[a] Order of drugs in the lower part as in the upper part of the table.

of light absorption at the short-wave side (narrowing of the absorption band); Figure 6 represents a tentative scheme of the cross sections potential curves.

This scheme explains:

(i) the bathochromic shift of the long-wave absorption band and its narrowing in the complex with DNA;

(ii) the hypsochromic shift of the fluorescence band without narrowing in the complex with DNA;

(iii) the bathochromic shift of the long-wave absorption band, the hypsochromic shift of the fluorescence band, and narrowing of both in substituting the aqueous with an organic environment (ethanol);

(iv) the increased probability of the 0.0 transition, when the dye is bound to DNA or dissolved in ethanol, as compared with the aqueous environment (for details see [8]).

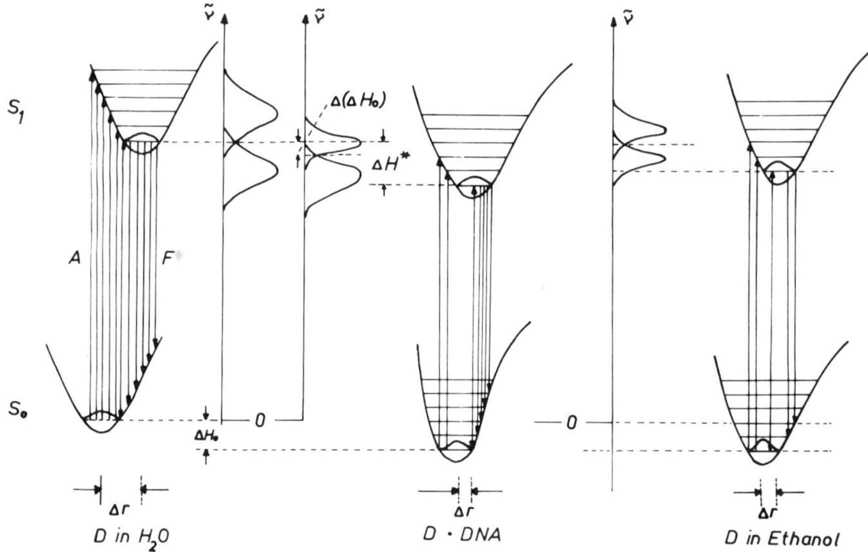

Fig. 6. Scheme of the cross sections of potential hyperfaces along an unspecified nuclear coordinate r responsible for the main progression in the spectra of proflavine (D) in H_2O, in ethanol and bound to DNA by type I; S_0 is the singlet ground state, S_1 is the first excited singlet state; A is absorption; F is fluorescence; ΔH_0 and ΔH_0* are the apparent binding enthalpies in S_0 and S_1, respectively; $\Delta(\Delta H_0)=\Delta H_0^*-\Delta H_0$; $\tilde{\nu}$ is wave number. The potential curves of D in H_2O and ethanol are of approximately same shape in S_0 and S_1. The potential curve of D in complex with DNA is steeper in S_0. Δr being smaller in ethanol and in complex.

Influence of Organic Solvents on Type II Complexes

In this chapter we attempt to offer data on the role of water in drug binding by type II, exploiting the effect of non-aqueous solvents on the surface binding, which involves the stacking properties of the drug along a polymer chain. Spectroscopic and equilibrium dialysis measurements were performed with acridine orange, i.e., an acridine dye which exhibits a high tendency for type II binding. By using polyphosphate instead of DNA the occurrence of type I binding is a priori widely excluded. The influence of different organic solvents on the binding of acridine orange to polyphosphate (Graham salt) was investigated at a phosphorus-to-drug ratio of about 500, where in the absence of organic solvents the drug is completely bound and stacked. By adding organic solvents to the solution of the complexes, an unstacking takes place with increasing organic solvent content which leads to free or perhaps only weakly, non-cooperatively bound monomeric dye molecules[18,19].

In Figure 7 the percentage of the stacked dye to the total dye concentration is plotted versus the concentration of organic solvents. As may be seen from Figure 7 and Table 3, the effectiveness of the solvents decreased in the order: formamide>tertiary butanol> isopropanol>n-propanol>ethanol>methanol>ethylene glycol>glycerol. The order of the solvents decreasing the stacking and binding ability of the drug bound by type II is nearly the same as that reported above for binding type I. Apparently, there is no correlation between the effectiveness of the solvents in lowering the binding and their dielectric constants. The cooperative binding of the drug molecules at the surface of the polyphosphate, which is controlled by dye-dye interactions, is favored in aqueous solution, demonstrating the importance of hydrophobic forces in determining the structure of the complexes.

From the spectrophotometric measurements in bulk solution it is not possible to decide, whether the solvent effect consists in an unstacking process, leading to randomly oriented but still bound dye molecules at the polyphosphate chain, or if dye dissociation takes place. Table 4 presents data obtained from spectroscopic and equilibrium dialysis measurements showing that complete dissociation of acridine orange from the polyphosphate requires a higher concentration of ethanol than is necessary for the complete change of the absorption spectrum due to the unstacking process.

Interestingly, stacking at the polyphosphate chain does also occur for anthracyclines. Those compounds become probably primary bound by the sugar part and in a second step the interaction between the drug chromophores takes place[14]. A study of the effect of ethanol on the binding of violamycin BI (VBI) by type II along the polyphosphate chain yielded a destabilization effect, however, less prominent than for acridine orange. The two charged sugar residues

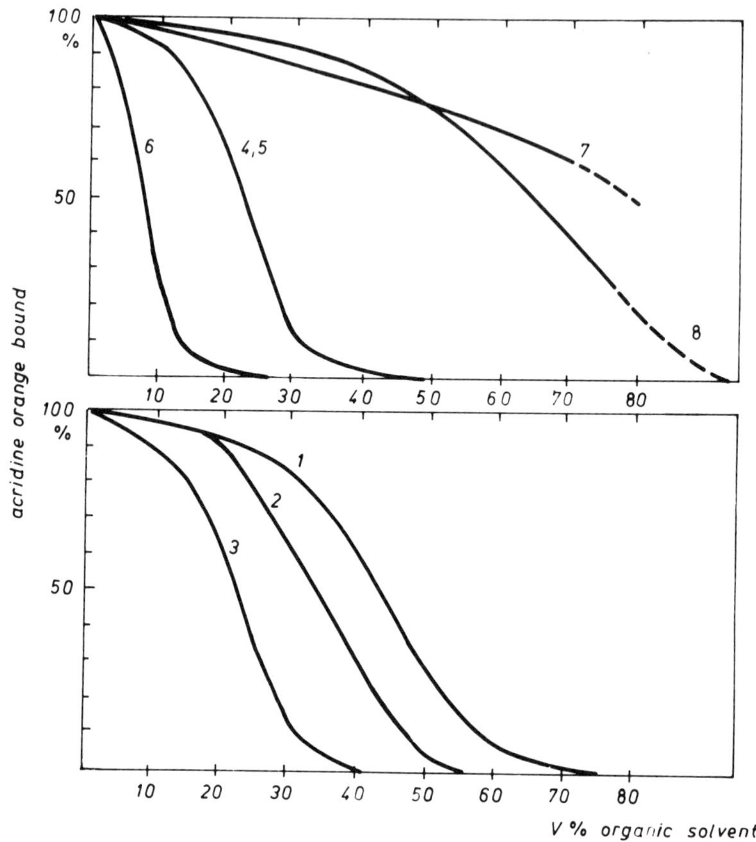

Fig. 7. Percentage of acridine orange bound to polyphosphate (Graham salt) in dependence on the volume % of various organic solvents; for the numbering of the curves see Table 3.

Table 3. Solvent Effect on the Cooperative Binding of Acridine Orange to Polyphosphate (Graham Salt).

No.	Solvent	Dielectric constant	Solvent concentration corresponding to 50% unstacking of the dye
1	Methanol	32.0	43
2	Ethanol	25.8	34
3	Tert. butanol	12.2	22
4	n-Propanol	22.0	22
5	Isopropanol	19.0	22
6	Formamide	110.5	8
7	Glycerol	47.0	80
8	Ethylene glycol	38	65

Table 4. Comparison of the Dissociation and Unstacking of Acridine Orange Bound to Polyphosphate (Graham Salt) in the Presence of Ethanol.

Ethanol concentration volume %	Dissociated dye[a] %	Unstacked dye[b] %
0	0	0
10	10	5
20	18	8
30	24	40
40	35	69
50	45	97
60	78	100
80	98	100

[a] data obtained from equilibrium dialysis; [b] data obtained from spectrophotometric measurements in bulk solution.

evidently cause the increased binding affinity to the polyanion, which is reflected in the relatively high stability of the complex in the presence of the organic solvent.

The results presented here give evidence that the presence of water is also essential for the cooperative binding process type II.

WATER AND FLUORESCENCE STAINING OF CHROMOSOMES

A very intense application of the fluorescence of dye-nucleic acid complexes is found in chromosome research. Binding of cationic dyes by chromosomal nucleic acids is probably the most important step in the process of chromosomal staining. Each metaphase chromosome stained with the fluorescent dye quinacrine shows its own specific fluorescence banding pattern. Positive and negative Q bands (where Q stands for quinacrine) mean chromosomal regions displaying high and low fluorescence intensities, respectively (for a review see [8]). In order to understand the processes resulting in chromosomal banding after staining with appropriate fluorescent dyes it is necessary to know the factors that produce a highly localized brilliant fluorescence in certain regions as well as less intense fluorescence in other regions of a chromosome. Probably of importance are the spectroscopic processes of fluorescence quenching and fluorescence enhancement of dyes bound to G.C- and A. T-rich DNA sequences, respectively[20-22]. It has been widely accepted that quenching of the fluorescence of quinacrine by guanine in G.C-containing regions is responsible for the negative Q bands (Figure 8). On the other side binding of quinacrine to A.T-rich DNA sequences enhances the dye fluorescence giving rise to the occurrence of the positive Q band.

The striking similarities in spectroscopic properties of dyes bound to DNA and the same dyes dissolved in organic solvents have been shown above. With respect to the fluorescence it was pointed out that the fluorescence intensity of acridine orange[23], quin-

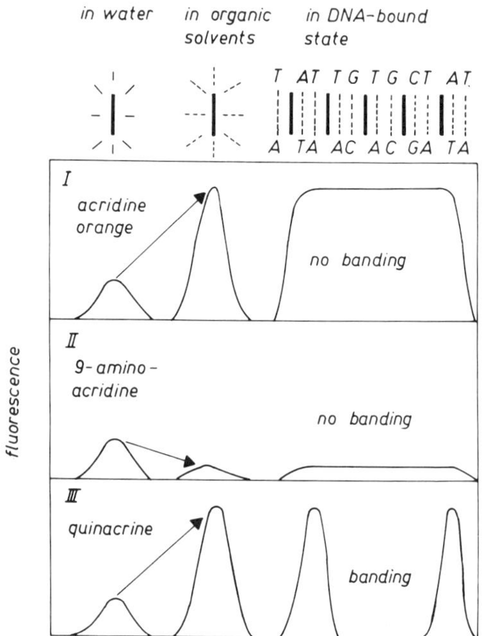

Fig. 8. Model for the dye-induced fluorescence banding pattern in eukaryotic metaphase chromosomes; schematic representation of the fluorescence changes (indicated by the arrow) of different dyes in passing from an aqueous to an organic environment (dye dissolved in organic solvent or attached to DNA).

acrine[23] and proflavine[16] dramatically increases with the amount of organic solvents in an organic solvent-water mixture. Depending on the nature of the solvent the increase of the fluorescence intensity of the quinacrine varies by a factor which ranges between 2 and 5 (Figure 9). Whatever the spectroscopic background for this behavior may be, water comes into play in as far as it reduces the probability of a fluorescence emission of dyes more than any of the organic solvents. A plausible hypothesis to explain this phenomenon has been developed[24]. It is based on a differential lowering of the corresponding singlet and triplet states of dyes in that a decrease of the singlet triplet-energy split in water facilitates intersystem crossing so that part of the excitation energy gets lost in a radiationless process.

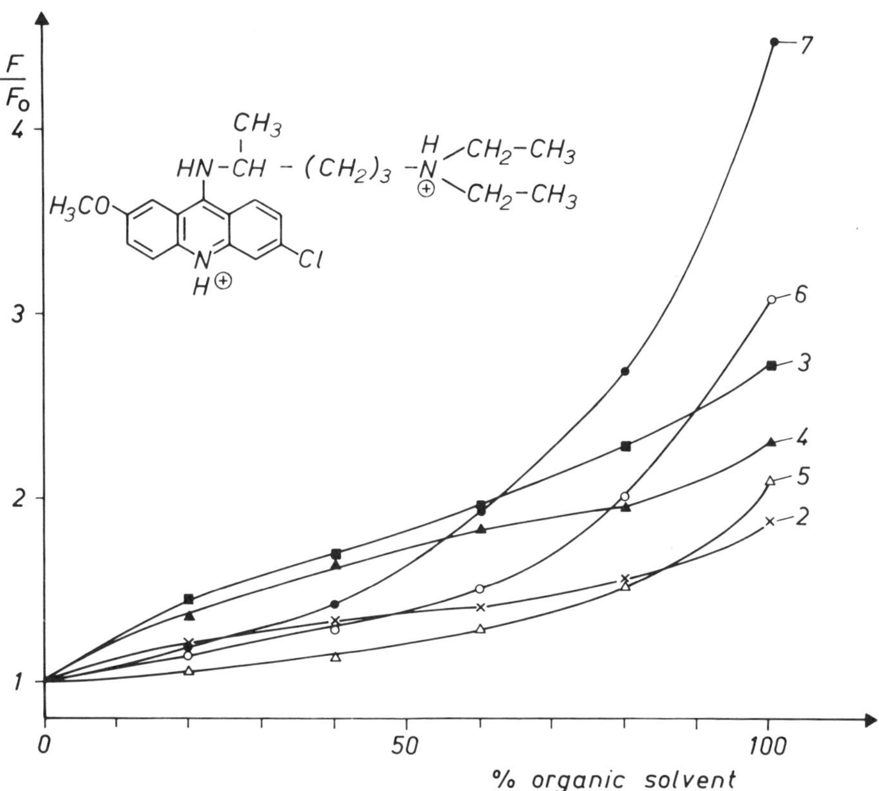

Fig. 9. Changes of the relative fluorescence intensity, F/F_0, of quinacrine ($1x10^{-6}M$) excited with light of 440 nm wavelength in various organic solvent-water mixtures; fluorescence maximum in water at 496 nm, the small solvent-induced wavelength shift was neglected; methanol (2), n-propanol (3), isopropanol (4), formamide (5), ethylene glycol (6), glycerol (7).

With this background in mind three cases can be distinguished in Figure 8:

Case I, (e.g. realized by acridine orange), is characterized by a strong increase of the fluorescence when going from an aqueous to an organic environment. This drug, however, is not very well suited for banding, since no base-specific quenching occurs (the metachromatic effect was left out of consideration).

Case II, (e.g. valid for 9-aminoacridine) involves neither an increase of the fluorescence in organic medium nor any base-specific fluorescence quenching. Drugs possessing such properties are extremely unfavorable for chromosomal banding.

Banding, however, appears for Case III, which is characterized by an increase in the fluorescence intensity of the dye, e g., of quinacrine, in passing from water to an organic medium or to the DNA-bound state and were G. C-specific quenching takes place.

Investigations on the direct action of organic components on the fluorescence banding pattern of chromosomes have not come to the authors' knowledge. However, no significant deviation from the behavior of the Giemsa-banding pattern is expected, for which a destaining of the chromosomes in the presence of dimethylsulfoxide was found[25].

SUMMARY

Organic solvents, e. g. methanol, ethanol, n-propanol, isopropanol, formamide, dimethylsulfoxide, p-dioxane, ethylene glycol and glycerol decrease the binding ability of drugs with DNA. This is valid for type I binding (monomeric binding) studied for proflavine, acriflavine, phenosafranine, ethidium bromide and the anthracycline derivative violamycin BI as well as for type II binding (aggregate binding) studied for acridine orange and violamycin BI. The effectiveness of the solvents increases with their hydrocarbon content, but can hardly be related to their macroscopic dielectric constant. The complex formation is effectively suppressed below the threshold of the spectroscopic detection by solvent concentrations of 50% (v/v), in which DNA still preserves its double-helical conformation. These results demonstrate the importance of water and of hydrophobic interaction forces in the formation of the complexes. The high contribution of electrostatic interaction forces for violamycin BI, due to the two positively charged sugar moieties, explains the stability of the violamycin BI-DNA complex at solvent concentrations greater than 50% (v/v).

Spectroscopic changes occurring, when a drug molecule goes from its hydrated state to the type-I bound state are similar to those accompanying the change from water to an organic solvent. Appropriate spectroscopic term schemes are presented.

One important feature in fluorescence labelling of metaphase chromosomes by quinacrine is the enhanced fluorescence emission in A.T-rich DNA sequences, which was discussed on grounds of an enhanced probability of the emission process in an less polar environment, when compared with water.

Taken together, we consider that water is essential for the formation of DNA-drug complexes. The markedly different properties of drugs in aqueous solution on one side and in organic solvents or in the DNA-bound state on the other side help to interpret a number of spectroscopic findings.

REFERENCES

1. A. Blake and A. R. Peacocke, The interactions of aminoacridines with nucleic acids, Biopolymers 6:1225 (1968).
2. G. Löber, On the complex formation of acridine dyes with DNA-IV. The equilibrium constants of substituted proflavine and acridine orange derivatives, Photochem.Photobiol., 8:23 (1968).
3. G. Löber and G. Achtert, On the complex formation of acridine dyes with DNA-VII. Dependence of the binding on the dye structure, Biopolymers 8:595 (1969).
4. G. Löber, Zur Komplexbindung von Farbstoffen mit Desoxyribonucleinsäuren, Z.Chem., 9:252 (1969).
5. G. Löber, W. Fleck, H.-E. Jacob, and K. Rost, Beziehungen zwischen der Komplexbindung mit DNS und einigen biologischen Wirkungen von Acridinfarbstoffen, in Wirkungsmechanismen von Fungiziden, Antibiotika und Cytostatika, H. Lyr and W. Rawald, eds., Akademie-Verlag, Berlin (1970) p.39.
6. G. Löber, Acridine - ihre physikochemische und biochemische Bedeutung. Eine Betrachtung anläBlich der Entdeckung des Acridins vor 100 Jahren. Teil II. Z.Chem. 11:135 (1971).
7. G. Löber, and L. Kittler, Selected topics in photochemistry of nucleic acids, Recent results and perspectives, Photochem.-Photobiol. 25:215 (1977).
8. G. Löber, The fluorescence of dye-nucleic acid complexes. J.Luminescence 22:221 (1981).
9. Z. Balcarovä, V. Kleinwächter, J. Koudelka, G. Löber, K. E. Reinert, L. P. G. Wakelin, and M. J. Waring, Interaction of phenosafranine with nucleic acids and model polyphosphates. II. Characterization of phenosafranine binding to DNA. Biophys.Chem. 8:27 (1978).
10. G. Löber, L. Kittler, R. Klarner, Z. Hradecna, V. Kleinwächter, Z. Balcarová, M. Skalka, J. Koudelka, E. Smékal, L. Popa and V. Beensen, DNA-drug interactions (A minireview). Studia biophysica 88:1 (1982).

11. G. Löber, H. Schütz, and V. Kleinwächter, Effect of organic solvents on the properties of the complexes of DNA with proflavine and similar compounds. Biopolymers 11:2439 (1972).
12. T. T. Herskovits, Nonaqueous solution of DNA: factors determining the stability of the helical configuration in solution. Arch.biochem.Biophys., 97:474 (1962).
13. V. Kleinwächter and J. Koudelka, Thermal denaturation of deoxyribonucleic acid acridine orange complex. Biochim. Biophys.Acta 91:539 (1964).
14. G. Löber, R. Klarner, E. Smékal, T. Räim, Z. Balcarová, J. Koudelka, and V. Kleinwächter, Spectroscopic investigations on the interaction of the anthracycline antibiotic violamycin BI with deoxyribonucleic acid, Int.J.Biochem., 15:663-673 (1983).
15. U. Katenkamp, E. Stutter, I. Petri, F. A. Gollmick, and H. Berg, Interaction of authracyline antibiotics with biopolymers. VIII. Binding parameters of aclacinomycin A to DNA. J.Antibiotics, 36:1222-1227 (1983).
16. G. Löber, V. Kleinwächter, J. Koudelka, and E. Smékal, On spectral properties of type I complexes of dyes with deoxyribonucleic acid and human serum albumin. Studia biophysica 45:91 (1974).
17. H. Lang and G. Löber, Die Lösungsmittelabhängigkeit der Elektronenspektren von kationischen Acridinfarbstoffen, Ber.Bunsenges.physik.Chem., 73:710 (1969).
18. G. Löber, and V. Kleinwächter, Effect of organic solvents on the properties of the complex polyphosphate-acridine orange (preliminary note), Studia biophysica 33:73 (1972).
19. G. Löber, V. Kleinwächter, and H. Berg, Effect of organic solvents on the properties of the complexes of a polyphosphate with acridines, Studia biophysica 35:29 (1973).
20. G. Löber, V. Kleinwächter, and J. Koudelka, Staining of chromosomes with basic dyes, Studia biophysica 55:49 (1976).
21. G. Löber, V. Kleinwächter, J. Koudelka, Z. Balcarová, J. Filkuka, P. Krejci, P. Döbel, V. Beensen, and R. Rieger, Molecular and spectroscopic aspects of chromosome banding, Biol.Zbl. 95:169 (1976).
22. G. Löber, V. Beensen, Ch. Zimmer, and H. Hanschmann, Changes of quinacrine staining of human chromosomes by the competitive binding of A.T-and G.C-specific substances, Studia biophysica 69:237 (1978).
23. G. Löber, On the spectroscopic basis of acridine-induced fluorescence banding patterns in chromosomes. Studia bio physica 48:109 (1975).
24. C. J. Seliskar and L. Brand, Electronic spectra of 1-aminonaphthalene-6-sulfonate and related molecules, J.Am. Chem.Soc., 93:5414 (1971).
25. R. D. G. McKay, The mechanism of G- and C-banding in mammalian metaphase chromosomes. Chromosoma (Berlin) 44:1 (1973).

MAPPING BY THE MTD METHOD OF A CATECHOLAMINE RECEPTOR IMPLIED IN THE IONIC PERMEABILITY OF CELL MEMBRANES

Z.Simon, N.Dragomir, G.I.Mihalaş,
M.G.Plauchithiu and Tamara Bânzaru

Institute of Medicine, Department of Biophysics
and Department of Pharmacology
1900 Timişoara, Romania

INTRODUCTION

The acţion of adrenaline and adrenaline agonists, upon the alpha adrenergic receptors, produces an increase of the permeability for K^{+}, which is related to an increase of Ca^{2+}- concentration on the inner side of the membrane. The alpha adrenergic receptor can exist in two, relatively stable conformations - the antagonistic, inactive conformation and the agonistic conformation, induced by adrenaline and its agonists[1]. The ionophoric function of the agonistic receptor conformation is, probably, responsible for the Ca^{2+} increase[2].

X-ray crystallography data are largely used in the analysis of protein structures, but the active receptor conformation is, presumably, not the lowest energy conformation of the receptor protein. Here, we try to infer about the structures of the active and inactive conformations of the alpha adrenergic receptors from quantitative atructure activity relations (QSAR's), established for the inhibition of tritiated dihydroergocryptin (^{3}H-DHE)- binding to this receptor by alphy adrenergic antagonists and, separately, by alpha adrenergic antagonists.

METHOD

Relatively large series of data for the inhibition of ^{3}H-DHE binding to uterine membrane by alphy adrenergic agonists and by antagonists are listed by Williams and Lefkovits[3] and by Ariens[4]. As ^{3}H-DHE must dislocate agonists and respectively antagonists from the receptor, the inhibition data must be pertinent to the agonistic and respectively antagonistic receptor conformation. The biological activity used here is defined as A $\equiv$log KD, with K_D- the agonist (or antagonist) -receptor dissociation constant. There is a good parallelism between receptor binding and physicologic activity for agonistic drugs[3].

Classical QSAR methods and the minimal steric difference (MTD)-method[5,6] will by used to establish the structure-activity relations. The MTD-method allows to account for the molecular stereochemistry by means of the hypermolecule. The hypermolecule is a network obtained by an approximative atom per atom superposition of the considered M_i-molecules (i=1,2,...,N), with neglect of the small, hydrogen-atoms. If the j-vertex of the hypermolecule (j=1,2,...,M) is occupied by the M_i-molecule, this situation is described by the x_{ij}=1 value; if j is not occupied by M_i, we have x_{ij}=O. The minimal steric difference, MTD_i, for M_i is defined by the formula:

$$MTD_i = s + \sum_{j=1}^{M} \varepsilon_j x_{ij} \qquad (1)$$

whence $\varepsilon_j = -1$ for the vertices supposed to belong to the receptor cavity, $\varepsilon_j = +1$, for the vertices within the receptor walls and $\varepsilon_j = O$, for the vertices situated in the exterior of the receptor; s is the number of cavity-vertices. The receptor - M_i - affinity is supposed to decrease linearly with MTD_i, corresponding to a regressional equation of the type:

$$\hat{A}_i = \alpha - \beta MTD_i \qquad (2)$$

The assignment of $\varepsilon_j = -1$, O or +1-values to the j-vertices is performed according to an optimisation procedure which is slightly modified[7] as compared to the procedure used in previous papers[5,6].

As "classical" structural variables, π_i - hydrophobicities of molecular fragments are used (calculated according to Rekker's fragmental constants[8]) and indicator variables, which take the value δ_i=1 if a certain structural characteristic is present in the M_i-molecule and δ_i=O, if it is absent.

STRUCTURE - ACTIVITY RELATIONS

Clonidine and seventeen agonistic β- phenylethylamine derivatives are listed in Table 1 (nr.1-18). The rigid skeleton of ergonovine, an active, partial adrenergic agonist, was used for the relative orientation of molecules, in the construction of the hypermolecule, depicted in Figure 1. The Phe-C-C-N^+- moiety of the agonistic molecules occupies always the same, unnumerotated, vertices of the hypermolecule. The imidazole ring of clonidine occupies also vertices j : 10,11 and 12, its two chlorine atoms occupy vertices j : 4 and 6; the β-OH-group of phenylethylamines, in L-configuration, occuples j :8, if in D-configuration,9 etc. The initial assignments of vertices to cavity, wall, or exterior space are done according to some qualitative structure-activity relations for agonists, as given by Ariens[4]. The final vertex assignments, as resulted by the optimisation procedure of the MTD-method, the so called optimised receptor map, S^*, and the corresponding regressional equation are:

$$S^* \begin{cases} j(\varepsilon=-1): 1,2,4,6,8,10,12,14 \\ j(\varepsilon=0): 3,5,7,9 \\ j(\varepsilon=+1): 11,13,15 \end{cases}$$

$$\hat{A}_{AG} = 10.320-0.916\ \text{MTD};\ r = 0.860,\ s = 0.418,\ N=18 \quad (3)$$

All the vertices of the common molecular core were considered as cavity ($\varepsilon_j = -1$) vertices.

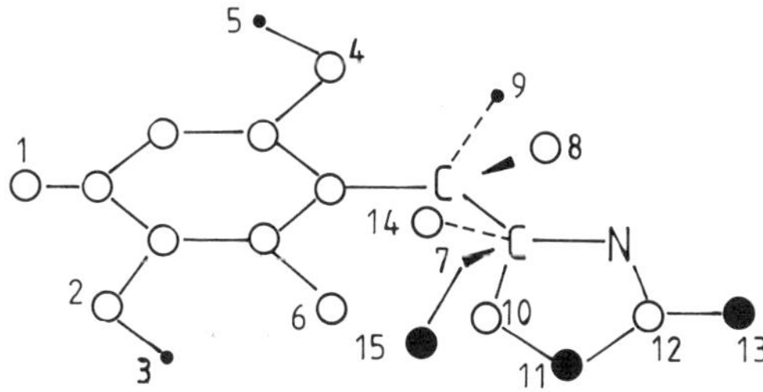

Figure 1. Hypermolecule and optimised receptor map for agonists. Open circles, $\varepsilon_j = -1$ - vertices, black circles, $\varepsilon_j = +1$ - vertices, dots - $\varepsilon_j = 0$ - vertices

Table 1. Structure and activity data for β- phenyl-ethyl amines with agonistic and antagonistic alpha adrenergic activities, of the general formula (at pH≃7)

$$R_1,R_2\text{-}C_6H_3\text{-}\overset{R_3}{\underset{|}{C}}H\text{-}\overset{R_4}{\underset{|}{C}}H\text{-}\overset{+}{N}H_2\text{-}R_5$$

Nr.	R_1	R_2	R_3	R_4	R_5	A_{AG}	A_{ANT}
	a) Agonists:						
1	Clonidine					6,63	-
2	OH	OH	L-OH	H	CH_3	6.59	-
3	OH	OH	L-OH	H	H	6.19	-
4	OH	OH	L-OH	CH_3	H	5.94	-
5	H	OH	L-OH	CH_3	H	5.60	-
6	H	OH	L-OH	H	CH_3	5.46	-
7	2,5 di-	OCH_3	L-OH	CH_3	H	5.46	-
8	OH	OH	D-OH	H	CH_3	5.30	-
9	H	H	H	α-di-CH_3	CH_3	4.90	-
10	H	OH	L-OH	H	C_2H_5	4.90	-
11	OH	OH	D-OH	H	H	4.70	-
12	OH	H	L-OH	H	H	4.70	-
13	OH	OH	H	H	H	4.68	-
14	H	H	L-OH	CH_3	CH_3	4.61	-
15	OH	OH	L-OH	H	iso-C_3H_7	4.37	-
16	OH	H	H	H	H	4.30	-
17	OH	OH	L-OH	C_2H_5	H	4.13	-
18	OH	H	L-OH	H	iso-C_3H_7	3.90	-
	b) Antagonists:						
19	OH	H	OH	H	$-CH(CH_3)CH_2OC_6H_5$	-	6.3
20	OH	H	OH	CH_3	$-CH(CH_3)CH_2CH_2C_6H_5$	-	6.0
21	OH	OH	OH	H	$-C(CH_3)_2CH_2C_6H_4$-OH (para)	-	5.4
22	OH	OH	OH	H	$-CH(CH_3)CH_2C_6H_4OH$ (para)	-	5.0
23	Cl	Cl	OH	H	iso-C_3H_7	-	4.6
24	OH	OH	OH	H	$-C(CH_3)_2CH_2C_6H_5$	-	5.0
25	OH	OH	OH	H	$-CH(CH_3)CH_2CH_2C_6H_5$	-	5.1
26	OH	OH	OH	H	$-CH(CH_3)CH_2C_6H_5$	-	4.4
27	OH	OH	OH	H	$-C(CH_3)_2CH_2C(CH_3)_3$	-	3.5
28	OH	OH	OH	H	nC_4H_9	-	2.0

A_{AG} and A_{ANT} are equal to - log K_D, with K_D-dissociation constant for the receptor-agonist and respectively receptor-antagonist complex, according to data listed by Williams and Lefkovits[3] (Tab.6-2A) and by Aries[4](Tab.III).

Data for antagonistic β-phenyl ethyl amines are also listed in Table 1 (nr.19-28). Several other drugs have also strong antagonistic activities (A_{ANT} - according to[3], in paranthesis): phenoxybenzamine (7.75), phentolamine (7.82), yohimbine (6.66) and especially ergot alkaloid-derivatives; dihydroergotamine (7.82), ergotamine (7.68), dihydroergocristine (7.38), ergocristine (6.86), dihydroergocriptine (8.oo), ergocriptine (6.92), dihydroergocornine (7.22), ergocornine (7.oo) and ergonovine (6.35). The wide structural variety of the antagonists suggests, that they may have different orientations in the complex with the receptor, or bind to different receptor conformations; the only condition is that they prevent the formation of the active, agonic, receptor conformation.

For the antagonic β-phenyl-ethylamines following regressional equation could be established:

$$\hat{A}_{ANT} = -0.088 + 1.156\pi_{14} + 0.675\pi_5 + 2.112\delta$$
$$r = 0.925,\ s = 0.477,\ N = 10 \qquad (4)$$

whence π_{14} is the calculated hydrophobicity for the β-phenylethyl moiety (e.g. π_{14}=1.74 for $C_6H_5CH_2CH_2$, π_{14} = o,59 for the trihydroxylated analogue), π_5 for the R_5-substituent (e.g. π_5=o,7o for CH_3, 2.29 for nC_4H_9, 3.88 for $-CH(CH_3)CH_2CH_2C_6H_5$) and $\delta = 1$ for the presence of a phenyl ring in R_5.

For the antagonic ergot alcaloid derivatives:

$$\hat{A}_{ANT} = 6.840 - 0.520\,\delta_1 - 0.490\,\delta_2 + 1.155\,\delta_3 \qquad (5)$$
$$r = 0.898,\ s = 0.230,\ N = 9$$

was obtained; δ_1=1 for the presence of a CH_3-group, δ_2= 1 for the presence of a double bond in the ergonovine cycle and δ_3=1 for the presence of the tripeptidic cycle.

THE STRUCTURE OF THE ALPHA ADRENERGIC RECEPTOR

The QSAR's obtained for alpha adrenergic agonists (eq.3 and the optimised map, S^{*}) and antagonists (eqs.4 and 5), suggest different structures for the agonicactive and the antagonic - inactive receptor conformations.

The multitude of antagonic structures suggests a rather planar receptor surface for the antagonic receptor conformation, with one anionic and two hydrophobic sites for binding the antagonist molecules. Aromatic groups potentiate this binding, possibly by stacking to aromatic aminoacidic side chains, of the hydrophobic sites. A third site binds, possibly, the tripeptidic cycle of ergot alcaloids.

The agonist molecule binds to the anionic site and to one of the hydrophobic sites, but "covers" itself with a second receptor moiety, which forms hydrogen bonds with hydroxyl groups at C_{β} (if in L-configuration) and at the benzene ring. The aminic N^{+}-atom is now surrounded by the receptor cavity and large R_5 substituents, here, inhibit the formation of the closed, agonic receptor conformation.

Figure 2 depicts the proposed relation between the two receptor conformations. Antagonists should be any molecules binding strongly to the receptor and inhibiting the formation of the agonic conformation.

The existance of two different receptor conformations[1] is difficult to reconcile with the alpha adrenergic receptor model of Belleau[9,10] in which adrenaline (the agonist) plays the role of a "coferment" enhancing the binding of Ca^{2+} and the cleavage of the ATP-molecule. Nevertheless, these last two components could be included into the "second" receptor moiety

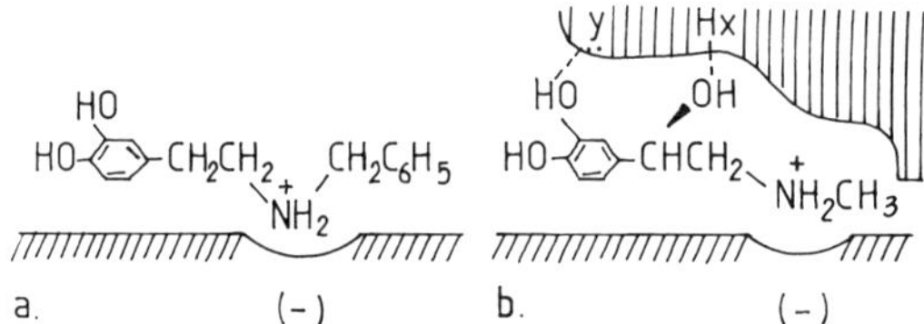

Figure 2. Alpha adrenergic receptor model (a) - in the antagonic conformation with an antagonist molecule; (b) in the agonic conformation with an agonist molecule. Hachures, ////, mark hydrophobic receptor zones.

REFERENCES

1. D.C.U.Pritchard, D.A.Greenberg and S.H.Snyder: Mol.Pharmacol. 13, 45 (1977)

2. D.H.Jenkinson, D.G.Haylett and K.Koller, in "Cell Membrane Receptors for Drugs and Hormones. A Multidisciplinary Approach", R.W.Straub and Liana Bolis, editors, Raven Press, New York 1978, pp 89-1o5
3. L.T.Williams and R.J.Lefkovitz: "Receptor Binding Studies in Adrenergic Pharmacology", Raven Press, New York 1978, Chap.6
4. E.J.Ariens: Adv.Drug Res. 3,235(1966)
5. A.T.Balaban, A.Chiriac, I.Moţoc and Z.Simon: "Steric Fit in Quantitative Structure Activity Relations", Lecture Notes in Chem Ser., vol.15.,Springer, Heidelberg, 198o, Chap.5
6. Z.Simon, N.Dragomir, M.G.Plauchithiu, H.Glatt and F.Kerek: Eur.J.Medic. Chem 15,521 (198o)
7. S.Holban, D.Ciubotariu and G.I.Mihalaş: in preparation
8. R.F.Rekker: "The Hydrophobic Fragmental Constant",Elsevier, Amsterdam, New York, 1977, Appendix
9. B.Belleau: Ann.N.Y.Acad.Sci.139,586 (1967)
10. M.G.Plauchithiu, M.Rocsin and Z.Simon: Rev.Roumaine Biochim.15,67 (1978)

Abstract Structure-activity relations are established for the binding of alpha adrenergic agonists to the receptor, by the MTD-method and for antagonists, by classical QSAR-methods. A model for the alpha adrenergic receptor is put forward, in which the lowest energy, inactive, antagonic receptor conformation should present a rather planar combing site, while the active, agonic conformation should enclose the agonist-molecule into a "closed" combining site.

SUPEROXIDE ANION IN ACUTE CCl_4 INTOXICATION: THE PROTECTIVE ACTION OF $ZnCl_2$

E. Truţia, Rodica Bartock and Veronica Dinu

Department of Biochemistry
Medical Faculty
Bucharest, Romania

Recent years have witnessed a large development of the research on biochemical mechanisms involved with CCl_4 toxicity, as a result of the more frequent presence of this compound in the atmosphere of certain industrial centers. Thus CCl_4, known for its high toxic potential, appears to be one of the substances polluting the atmosphere[1].

According to a large number of studies, CCl_4 toxicity depends on the previous metabolic activation by the hepatic mixed function oxidase system (MFOS)[2,3]. This process is thought to be started by uptake of one electron (e) in the NADPH - cytochrome P_{450} system, after which the resulting anion dissociates and the trichlormethyl radical is formed[4].

$$CCl_4 + e \longrightarrow (CCl_4^-) \longrightarrow CCl_3 + Cl^-$$

The trichlormethyl radical is a species sufficiently reactive for interaction with various components of the endoplasmic reticulum[5]. CCl_4 hepatotoxicity would be - as suggested by many authors - the consequence of an uncontrolled peroxidation induced by the trichlormethyl radical and other species formed during CCl_4 activation[6,7].

Our attention was centered on the superoxide anion, a compound of radical nature, produced during CCL_4 activation and which, under certain conditions, may detach from the complex $CCl_3O_2^-$ [8] exposing the cell to an increased flux of O_2^-. In these circumstances response of the superoxide dismutase, an enzyme involved in O_2^- splitting ($2\ O_2^- + 2\ H^+ \longrightarrow H_2O_2 + O_2$) is expected[9]. The enhanced production of O_2 favors formation of OH radicals and singlet oxygen ($^1O_2$)[10], species of high oxidation potential which can initiate the peroxidation processes. The present work deals with the response of antioxidant systems, namely of superoxide dismutase, gluthathione peroxidase and gluthathione reductase in severe intoxication by CCl_4[11].

We have studied also the effect of zinc chloride on these enzymatic systems, taking into account the ability of this compound to reduce CCl_4 induced hepato-toxicity[12].

MATERIAL AND METHOD

Experiments were performed on a number of 60 Wistar male rats weighing 100g each, distributed into three groups. The animals in the first group were intraperitoneally injected with 600 µl CCl_4/kg bodyweight, while those in the second group, with saline solution; the rats in the third group received 20 mg Cl_2Zn/kg bodyweight administered 24 h before injecting CCl_4. Animals were killed 24 h after CCl_4 administration and their liver was rapidly excised. Determinations were carried out on the total homogenate after centrifugation at 600g.

SOD activity was measured by its ability to inhibit autooxidation of epinephrine at alkaline pH[13].

We appreciated the gluthathione reductase activity from the decrease in the extinction of NADPH at 340 nm as a result of the reduction of oxidized gluthathione (GSSG)[14].

The activity of gluthathione peroxidase was determined by the conjugation of the gluthathione peroxidase reaction with the gluthathione reductase reaction and by watching the decrease of the extinction of NADPH.

$$2\ GSH + ROOH \xrightarrow{GP_x} GSSG \xrightarrow[NADPH\ \ NADP]{GR} 2\ GSH$$

The quantity of protein was appreciated by the Lowry technique[15].

RESULTS AND DISCUSSION

The analysis of the results showed in the group treated with CCl_4 a severe decrease of the SOD activity from 17.4 enzymatic units to 9.2 eu (Table 1).

The activity of GP_x increases significantly in the intoxicated rats compared to the control ones (Table 2).

Table 1. Activity of Superoxide in Liver of Rats Treated by CCl_4 (600 μl/kg bodyweight) and $ZnCl_2$ + CCl_4 as Compared to a Control Group.

Group	No. animals	Average values (Eu)*	SD	t	P
Control	20	17.4	0.45	-	-
Injected with CCl_4	20	9.2	0.68	14.6	<0.01
Injected with $ZnCl_2$ + CCl_4	20	13.9	0.78	3.2	<0.01

* Eu = quantity of enzyme producing an inhibition of 50% in adrenaline oxidation to adrenochrome/min/mg protein.

Table 2. Activity of Gluthathione Peroxidase in Liver of Injected Rats with CCl_4 (600 μl/kg bodyweight) and $ZnCl_2$ + CCl_4 as Compared to a Control Group.

Group	No. animals	Average values (Eu)*	SD	t	P
Control	20	1.76	0.24	-	-
Injected with CCl_4	20	2.50	0.42	3.1	<0.01
Injected with $ZnCl_2$ + CCl_4	20	2.12	0.20	2.96	<0.01

* Eu = 1 μmol GSSG/min/mg protein at 30°C and pH = 7.4.

The activity of gluthathione reductase shows a spectacular increase in the intoxicated animals presenting average values of 9.2 eu/mg protein as against the average value of 3.3 eu/mg protein in the control group (Table 3).

The $ZnCl_2$ effect proved to be quite important. Animals treated by CCl_4 under $ZnCl_2$ protection showed a substantially diminished activity of GP_x and GR and increased activity of SOD, which reflects their tendency to normalization (Tables 1,2 and 3).

As mentioned above, it is generally agreed that the metabolic activation of CCl_4 by the hepatic mixed function oxidase system is essential for the toxic action of this substance[2,3].

During the metabolic activation of CCl_4, it is thought that the superoxide anion (O_2^-) arises as a reactive species involved in the very oxidizing process of CCl_4[4]. The products derived from O_2^- as

Table 3. Activity of Gluthathione Reductase in Liver of Injected Rats with CCl_4 (600 µl/kg bodyweight) and $ZnCl_2$ + CCl_4 as Compared to a Control Group.

Group	No. animals	Average values (Eu)*	SD	t	P
Control	20	3.30	0.43	-	-
Injected with CCl_4	20	9.20	0.64	3.06	<0.01
Injected with $ZnCl_2$ + CCl_4	20	3.15	0.33	5.4	<0.01

* Eu = 1 nmol GSSG reduced/min/mg protein at 30°C and pH = 7.4.

for example hydroxyl radical ($\cdot OH$) resulted in the Haber Weiss reaction ($H_2O + O_2^- \longrightarrow HO^\cdot + HO^- + {}^1O_2$) and singlet oxygen (${}^1O_2$) produced by the spontaneous dismutation of O_2^- ($2\ O_2^- + 2\ H^+ \longrightarrow H_2O_2 + {}^1O_2$), represent high oxidizing species, able to initiate peroxidation processes[10].

Under common conditions - SOD - an enzyme representing the first antioxidizing defence line limits the evolution of O_2^- to $\cdot OH$ and ${}^1O_2$.

The increase of both the activity of gluthathione peroxidase and of gluthathione reductase (results which agree with those of other authors[16]) represents the adaptive response of the cell to the selfpropagation of the radical oxidation chains resulting in the limitation of the $\cdot OH$ and ${}^1O_2$ effects.

The increase of the intracellular redox potential, noticed by Prager and coworkers[17], as well as the accumulation of lipid peroxides clearly demonstrate that the response capacity of gluthathione peroxidase is surpassed.

The exceeding of the antioxidizing defence capacity by SOD, GP_X and GR produces a severe oxidizing stress and leads to the alteration of the structure and function of the essential chemical components of the cell, and finally to necrosis and carcinogenesis[2,3].

Since $\cdot OH$ radical has a higher mobility and a reduced discriminatory action towards biological macromolecules, it initiates radical reaction outside the endoplasmic reticulum too, as demonstrated by Sowkin and Jakobson[18] in mitochondrial and cytosol fraction. We consider that the decrease of superoxide dismutase activity, producing subsequently some species of high oxidizing potential exceeding the response capacity of cellular antioxidizing enzymatic system could be significant for the accumulation of lipid peroxides responsible for the CCl_4 hepatotoxicity.

In our experiment (Tables 1,2,3) zinc chloride was found to significantly increase SOD activity, the first line of antioxidizing protection, and to decrease at the same time GP_x and GR activity.

Membrane protection by $ZnCl_2$ is considered to be due to MFOS inhibition, entailing a diminished metabolic activation of CCl_4 to radical species with oxidation potential[12,19].

It is quite possible that $ZnCl_2$ also stimulates SOD activity resulting in a diminution of $^{\cdot}OH$ and $^1O_2$ production and implicity of peroxidation. The lowered activity of GP_x and GR supports this hypothesis.

$ZnCl_2$ treatment, by re-establishing the redox equilibrium disturbed in CCl_4 intoxication, could be efficient in all necrotic-degenerative lesions of the hepatic cell.

REFERENCES

1. P. V. Cappurro, Clin.Toxic., 6:109 (1973).
2. T. F. Slater, Nature, 209:36 (1966).
3. P. Goli Giuseppe, E. Albano, and M. Dianzani, Exp.Molec.Pathol., 30:116 (1979).
4. T. F. Slater, M. Ahmed, C. Benedetto, K. Cheeseman, J. E. Packer, and R. L. Willson, Int.J.Quant.Chem., Quant.Biol. Symposium 7, 347 (1980).
5. P. B. McCay, M. Margaret King, J. Lee Poyer, and K. Lai, Annals New York, Acad.Sci., 37:23 (1982).
6. R. O. Recknagel and A. K. Ghoshal, Lab.Invest., 15:132 (1966).
7. T. F. Slater, "Free Radical Mechanisms in Tissue Injury," Pion, London (1972).
8. A. E. Ahmed, V. L. Kubic, J. L. Stevens, and M. W. Anders, Fed.Proc., 39:3150 (1980).
9. J. McCord and I. Fridovich, J.Biol.Chem., 244:6049 (1969).
10. S. Kong and A. L. Davison, Arch.Biochem.Biophys., 204:18 (1980).
11. V. Dinu, E. Trutia, and D. D. Belloiu, Rev.Roum.Bioch., 10:93 (1983).
12. M. Chvapil, Proc.Soc.Exp.Biol.Med., 141:150 (1972).
13. H. P. Misra and I. Fridovich, J.Biol.Chem., 287:3170 (1972).
14. R. E. Pinto and N. Bartley, Biochem.J., 112:109 (1969).
15. O. H. Lowry, N. J. Rosebrough, A. L. Farr, and R. J. Randall, J.Biol.Chem., 193:265 (1951).
16. B. Matkovics and R. Novak, Gen.Pharmacol., 93:92 (1978).
17. P. Prager, P. Andras, M. Tibor, and T. Andras, Kiserl.Orvostud., 33.337 (1981).
18. A. Sowkin and G. S. Jakobson, Vop.Med.Himii., 25:433 (1979).
19. M. Chvapil, Exp.Molec.Pathol., 19:186 (1973).

IONIC CHANNELS

THE ORDERED WATER ION CHANNEL MODEL

Donald T. Edmonds

The Clarendon Laboratory
Parks Road
Oxford OX1 3PU U.K.

INTRODUCTION

A lipid bilayer presents a formidable energy barrier to the passage of ions which is largely electrostatic. The electrostatic self-energy in joules of an ion of charge Q coulombs and radius R metres immersed in a continuous fluid of relative dielectric constant ε is given by

$$U = (1/4\pi\varepsilon\varepsilon_o)(Q^2/2R)$$

where $\varepsilon_o = 8.85 \times 10^{-12}$Fm. From this formula one would calculate that a sodium ion needs to surmount an energy barrier of 6×10^{-19}J (144 k_BT) to leave water with $\varepsilon = 80$ and enter a lipid layer with $\varepsilon = 2$. To treat water and lipid as continuous fluids with their bulk dielectric constant is clearly a poor approximation in these circumstances but more realistic calculation[1] and experiment still predict that the barrier is so large as to completely preclude thermally activated transit at normal temperatures.

To reproduce the ion transfer rates measured across real biological membranes special structures for facilitated transport are needed which are called channels. To be plausible model channels must mimic most of the measured properties of real channels listed below.

(i) Transfer Rate. Many real channels achieve transfer rates in excess of 10^6 ions/sec.

(ii) Selectivity. The sodium channel in the apical membrane of toad bladder has a permeability that discriminates[2] 1000:1 in favour of Na^+ over K^+ and channels in nerve and muscle often display[3] permeability ratios in excess of 100:1 between similar ions.

(iii) Gating. Many real channels switch from high to low permeability in response to changes in membrane voltage difference or the binding of messenger molecules.

The transfer rate quoted above allows us to rule out shuttling ionophores such as valinomycin or nonactin as models for most biological channels as the maximum plausible transfer rate for such carriers is about 10^3 ions/sec. The known high transfer rates also mean that the ions must surmount an energy barrier very much less than the full electrostatic barrier discussed above so that some mechanism of reducing the electrostatic self energy of ions during transit is needed. An apparent conflict exists between this need and selectivity. Thus a fully hydrated ion has the low energy required but it is difficult to conceive of a selectivity mechanism that can sharply discriminate between like ions when they are surrounded by layers of water.

Most current models of selective channels depend upon a mechanical size filter[4] for a partially hydrated ion or passage between charged sites[5] within the pore. Room temperature thermally activated motion of individual atoms in compact proteins have typical[6,7] root mean square amplitudes of $\pm$ 0.05 nm for the main chain and about $\pm$ 0.1 nm for some side chain atoms even within a solid crystal. For a mechanical filter to discriminate between Na^+ (R = 0.095 nm) and K^+ (R = 0.133 nm) demands quite exceptional rigidity, particularly if the pore size is to be determined by binding between membrane penetrating α-helical strands of a folded protein or even binding between olgomeric sub units. Exposed charged sites in the low dielectric environment of a channel have prohibitively high electrostatic energies, as we have discussed, unless surrounded by highly polarizable groups which screen the charged site. Like charged sites may not be sited close together because of like charge repulsion and thus high electrostatic energy. It is possible to conceive of selective charge sites guarded by mobile screening that can be displaced by the ion, but again a conflict exists between tight binding for high selectivity and the high transfer rate required.

Current models of gated channels postulate a separate gate, often mechanical, in series with the selective filter. As we shall see below all the requirements for transfer rate, selectivity and gating can be simulatneously satisfied by postulating an electrically ordered array of water molecules lining a cylindrical protein defined pore and using only the known properties of these molecules.

A BRIEF DESCRIPTION OF THE MODEL

Detailed treatments of individual features are to be found in the papers quoted and a more complete description is currently in press.[8]

The channel is supposed to consist of a ring or net structure of water molecules which lines a pore about 1 nm in diameter defined by parallel helical protein rods which span the membrane perpendicularly. Helical protein rods support the water array by providing hydrogen bonding sites that match some of the unsaturated hydrogen bonds of the array so that the water array and the protein rods together form the channel protein or group of proteins. The water array is thought to be composed largely of planar pentagon and puckered hexagon rings which are two small ring structures that preserve the tetrahedral bonding angles required by a water molecule in its ground state. The ions find low energy sites at the centres of the water rings which are selective[9] for unhydrated radius, valence state and the algebraic sign of the charge. The ions move over the water structure by passing from water ring to contiguous water ring.

Due to the peculiar low dielectric constant and mechanically shielded environment found within the membrane and also to the elongated shape of the water channel, the water molecule electric dipoles are ordered.[10] Any water molecule in a static tetrahedrally bonded array may point in one of six directions while maintaining the directions of its four hydrogen bonds to its neighbours. In the two equally low energy ground states of the channel water array the water dipoles have a predominantly positive or negative projection on the outward normal to the membrane.

An ordered array leads to the presence of very large electric fields within the channel so that ion transfer may be controlled by these electric fields alone[11] and there is no need of mechanical gating particles. A switch between the two oppositely polarized states with the lowest energy involves no translation but only rotation of the water molecules. It can be brought about by the electric field across the membrane (electrical gating) or by the proximity of a suitably charged group (chemical gating). Thus each ordered water channel has two distinct ground states each with a different current v. voltage relationship depending on the direction of the polarization within the channel. An individual channel in a given polarized configuration would not conduct continuously although an assembly of such channels would display the current v. voltage characteristics of that polarization. The conduction of an individual channel would depend for example on freedom from the electrically charged hydrogen bonding defects which would block the channel and so conduct only in randomly timed bursts. A distinction needs to be made between activation, which corresponds to the dipole polarization switch and subsequent conduction of ions as sketched in figure 1.

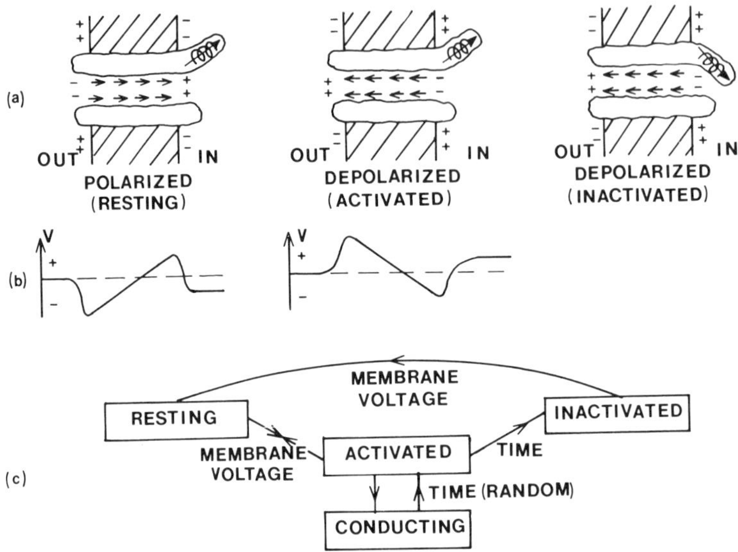

Figure 1. (a) A sketch of possible protein support structures with the axial component of the polarization of the water array indicated by short arrows. (b) The resultant voltage as a function of distance experienced within the channel. (c) A possible kinetic scheme incorporating inactivation due to the time delayed but spontaneous motion of an α-helical protein tail due to its attractive interaction with the electric field of a polarized water array when depolarized. Repolarization is delayed by this interaction but when achieved results in the repulsion of the tail to the non-inactivated position.

The degree to which a channel is selective to particular ions will depend on the extent to which the protein rods provide hydrogen bonding sites that discriminate in favour of one particular type of water molecule ring structure that provides particularly low energy sites for one particular ion. No configurational change is required of the protein support rods when a channel switches as the hydrogen bonds between the water array and the protein do not change. The protein rods do however have a vital role in providing specific binding sites for chemical messengers at both ends of the channel. For example the specific Tetrotoxin and Amiloride binding sites on the outer ends of sodium channels in nerve and tight epithelia respectively would be provided by particular protein rod structures.

The action of internal Pronase[12] suggests that inactivation may be the result of a conformational change of the protein structure protruding from the inner surface of the membrane. The fact that activation but not necessarily conduction must precede inactivation of the sodium channel in nerve[13] suggests that this conformational change cannot take place until the inner end of the water channel has become negatively charged by the rotation of the water dipole column resultant from activation of the configuration attained by membrane depolarization. One such possibility based upon an α-helical tail of one of the protein rods with its known large electrical dipole moment[14] is sketched in figure 1.

PREDICTIONS OF THE MODEL

(a) A given channel can exist in two configurations each with a distinct current versus voltage characteristic. In contrast to a mechnically gated model channel which is either open or shut both configurations will conduct[15,11] at suitable membrane voltages. One configuration is made more likely by depolarization and the other by hyperpolarization of the membrane from a particular membrane voltage known as the "switching voltage". A channel switches about a membrane voltage such that the average external electric field acting upon the water molecules of the channel is zero. This membrane voltage may deviate from zero because of local charges and the electric dipole moments of the protein helices that form the pore[11]. The model thus predicts that pairs of experimentally determined channels now thought of as distinct may in fact be the same channel in its two configurations. Channels are often pronounced as distinct when they are blocked by different internally or externally applied agents. Such categorization may not be valid for the reasons given in (b) below.

(b) When a channel switches the electric field at the channel ends due to the polarized water molecules reversed sign as sketched in figure 1(a). The channel ends and the adjacent protein structures become "modulated receptors" and the attractiveness of a given end for a charged molecule seeking to bind changes when the channel switches.[8] Evidence for this is already at hand. Externally applied Zn^{++} is known[16] to delay the onset of both Na^+ current and the gating charge transfer on activation but not to effect the decay of the Na^+ current or back transfer of the gating charge on repolarization of sodium channels of myelinated nerve fibres. This can be explained by Zn^{++} binding near to the negative outer end of a resting channel but not at the same site when the outer end becomes positive (see figure 1(a)) in the activated configuration. Similar effects are observed in the binding of internally applied local anaesthetics which probably act in their cationic form.[17] For the potassium channel similar effects are observed in the squid axon[18] and the node of Ranvier in frog nerve[19] for internally applied tetraethylammonium cations and derivatives.

(c) By abstracting the calculated voltage gradient within an ordered water array to a linear stepped ramp it is possible[11] to calculate easily the approximate expected current versus voltage characteristics of the channels. Good agreement is obtained with the results measured for the sodium and potassium channels of the squid axon both for physiological and also non-physiological and artificially applied external and internal free ion concentrations[11,15]. The behaviour of the sodium channel in the apical membrane of tight epithelia where the unidirectional inward flux of sodium is controlled inversely by the internal free concentration of Na^+ is reproduced[20] by the model.

(d) The model channel should conduct protons[21] which would be "gated" by channel switching in the same sense as cations. The first evidence for such gated proton currents in heavy ion channels has recently been obtained[22] in molluscan neurons. A water molecule array is uniquely suited to conduct both alkali metal ions and protons.

(e) When an ordered water array reverses its polarization the calculated[10,21] energy change is given by qV_M where V_M is the voltage across the membrane and q is a positive charge between 1.2 and 1.6 times the proton charge. This is in good agreement with the measured[23] gating charge transfer in the sodium channel of the squid axon of approximately 1.4 times the proton charge.

(f) The stability of the water molecule array should be affected by the inclusion of small uncharged chemically inert molecules which can occupy and stabilize cages of water molecules as in the clathrate hydrates.[24] Dodecahedral water cages have an internal free diameter of about 0.5 nm and so could include N_2O, CF_4, CO_2, H_2S, Xe, N_2 or O_2.

ACKNOWLEDGEMENT

I am most grateful to the Royal Society for the award of a Senior Research Fellowship which has allowed me time to pursue these interests.

REFERENCES

1. A.D. Buckingham, Discuss. Faraday Soc., 24 : 151 (1957).
2. L.G. Palmer, J. Membrane Biol., 67 : 91 (1982).
3. C. Edwards, Neuroscience, 7 : 1335 (1982).
4. B. Hille, in : "Membranes", Vol 3, G. Eisenman, ed., (1975).
5. G. Eisenman, Biophys. J., 2 : 259 (1962).
6. H. Frauenfelder, G.A. Petsko and D. Tsernoglu, Nature, 280 : 558 (1979).
7. P.J. Artymink, C.C.F. Blake, D.E.D. Grace, S.J. Oatley, D.C. Phillips and M.J.E. Sternberg, Nature, 280 : 563 (1979).

8. D.T. Edmonds, in : "Biological Membranes", Vol. 5, D. Chapman, ed., Academic Press, London (1983).
9. D.T. Edmonds, Proc. R. Soc., B211 : 51 (1980).
10. D.T. Edmonds, Chem. Phys. Lett., 65 : 429 (1979).
11. D.T. Edmonds, Proc. R. Soc., B214 : 125 (1981).
12. C.M. Armstrong, F. Bezanilla and E. Rojas, J. Gen. Physiol. 62 : 375 (1973).
13. R. Horn, J. Patlak and C.F. Stevens, Nature, 291 : 426 (1981).
14. W.G.J. Hol, P.T. van Duijnen and H.J.C. Berendsen, Nature 273 : 443 (1978).
15. D.T. Edmonds, Trends in Biochem. Sci., 6 : 92 (1981).
16. C.M. Armstrong and W.F. Gilly, J. Gen. Physiol. 74 : 691 (1979).
17. K.R. Courtney, J. Phar. Exp. Ther., 195 : 225 (1975).
18. C.M. Armstrong, J. Gen. Physiol., 58 : 413 (1971).
19. C.M. Armstrong and B. Hille, J. Gen. Physiol., 59 : 388 (1972).
20. D.T. Edmonds, Proc. R. Soc., B217 : 111 (1982).
21. D.T. Edmonds, Biochem. Soc. Symp., 46 : 91 (1980).
22. R.C. Thomas and R.W. Meech, Nature, 299 : 826 (1982).
23. R.D. Keynes, N.G. Greeff and D.T. van Helden, Proc. R. Soc., B215 : 391 (1982).
24. D.W. Davidson, in : "Water a Comprehensive Treatise", Vol. 2, Chap. 4., F. Franks, ed., Plenum Press, London (1973).

GATING KINETICS IN IONIC CHANNELS

L. Goldman and J.L. Kenyon

Department of Physiology
School of Medicine, University of Maryland
Baltimore, Maryland 21201, U.S.A.

INTRODUCTION

It is now well established that the rising phase of the action potential seen in many nerve, muscle and cardiac cells generates from a selective increase in the membrane permeability to Na ions, P_{Na}. When studied under voltage clamp, the changes in P_{Na} which underlie the upstroke of the action potential are measured as changes in the electrical conductance, g_{Na}. For a positive step change in the clamped membrane potential g_{Na} is seen to, with a delay, rise to a peak value (activate) and then, even though the potential is held constant, to decline to a small value (inactivate). A central issue in the problem of the molecular organization of this Na channel gating machinery is the nature of the relation between the activation and inactivation processes.

In their classic voltage clamp analysis Hodgkin and Huxley (1952) proposed a kinetic model in which activation and inactivation are mediated by two separate gating systems, each independently detecting and responding to changes in the membrane potential. The activation system in their scheme responds rapidly and the inactivation system more slowly to the membrane potential producing a transient increase in g_{Na}. However, since that time there has accumulated a considerable body of experimental evidence both from the kinetics of g_{Na} itself and from the gating current (an asymmetrical portion of the capacitative current which is at least largely associated with the actual operation of the gate) demonstrating that the activation and inactivation processes are coupled together in some way (see Goldman, 1976 for a review).

For the gating current the relevant observation is "charge

immobilization" (Armstrong and Bezanilla, 1974) in which clamp protocols which produce Na inactivation also reduce the gating current. The point is that a charge displacement with the kinetics of the activation process (Armstrong and Gilly, 1979) is affected by channel inactivation. Less clearly established is the nature of the coupling. Specifically, there is very little evidence available indicating that the coupling is of the sequential sort in which channels must enter the conducting state before they can inactivate.

There are available kinetic data on g_{Na} from *Myxicola* and other preparations which are most simply accounted for if inactivation follows sequentially on channel opening. For example, the steady state inactivation curve (in which the effect of long lasting conditioning polarizations of various values on the peak g_{Na} during a fixed test pulse is determined) shifts to the right along the voltage axis as test pulses with a larger maximum g_{Na} are selected (Goldman and Schauf, 1972). This is the sort of result expected if it is primarily those channels that have activated that inactivate. The inactivation curve shifts to the right as the channels inactivated during the conditioning potential become a smaller fraction of those maximally activatable during the test pulse. Again, in *Myxicola* (Goldman and Schauf, 1973) the Na current, I_{Na}, during moderate steps in potential declines to negligible values. However, when probed with strongly depolarizing test steps during and subsequent to the decline, inactivation is found not to be complete. Peak I_{Na} during the test step is again high. Such results are expected if it is only those channels that have activated that can inactivate. Similar effects are seen in lobster (Oxford and Pooler, 1975) and crab (Connor, 1976) axons and in cultured heart cells (Ebihara and Johnson, 1980).

These data are consistent with sequential activation and inactivation, but the evidence is not direct. And, the gating current experiments do not seem to bear on the issue of sequential coupling. Moreover, in currents recorded from single Na channels in cultured rat muscle Horn, Patlak and Stevens (1981) found the rate of inactivation to be the same for channels that opened late as compared to those that opened early during a potential step, suggesting that for this preparation inactivation proceeds in parallel with channel opening. Hence the parallel mode of channel organization does exist in nature. The question addressed here is whether the opposite mode - inactivation sequential to channel opening - also exists in biological channels.

To approach this question, a detailed examination of the time course of inactivation development was made in *Myxicola* giant axons. The point of the experiments was to look for a delay preceding inactivation development. Delays are in general expected if inactivation develops subsequent to a precursor process.

METHODS

All methods are given in Goldman and Kenyon (1982). The protocol is to present a series of conditioning steps in potential of fixed amplitude, but varying duration, each followed by a fixed test step in potential. When the peak I_{Na} values (obtained by tetrodotoxin substraction) during the test steps are examined as a function of conditioning step duration, the time course of the inactivation that develops during the conditioning potential is traced out. A gap (i.e. a step back to the holding potential) that is long relative to the time constant of activation is always included between the conditioning and test pulses. In this way the measured time course of inactivation is uncontaminated by the time course of activation that develops during brief conditioning pulses.

In a voltage clamp experiment the potential of the clamp is not the true membrane potential but differs from it by the quantity $I_m R_s$ where I_m is the membrane current and R_s is the resistance between the potential sensing electrodes and the membrane (i.e. in series with the membrane capacitance). To correct this problem R_s is electronically reduced by the use of compensated feedback, and I_m is reduced by conducting the experiments in bathing media with the Na concentration reduced by substitution with an impermeant cation.

All potentials are reported as absolute membrane potentials (inside minus outside) and have been corrected for liquid junction potentials.

RESULTS AND DISCUSSION

Figure 1 shows a typical determination. Peak value of I_{Na} during the test pulse is shown as a function of the duration of the conditioning potential. Inactivation always developed with an initial delay followed by a simple exponential decline of time constant τ_c (solid curve). Identical results were obtained in 57 determinations from 27 axons. The inset shows the early part of the curve on an expanded time scale to illustrate how the inactivation delays were operationally determined. They are taken as the time at which the unconditioned I_{Na} value intersects the exponential.

Note that to demonstrate a delay an initial plateau (i.e. initial slope of zero) in the inactivation curve is not required. The initial slope can be either positive, zero or even negative depending on the relative values of the time constants for any precursor (delay producing) and follower processes and on the relative values of the coefficients on these two exponential terms. To demonstrate a delay one only needs to show that the unconditioned I_{Na} value falls below the exponential.

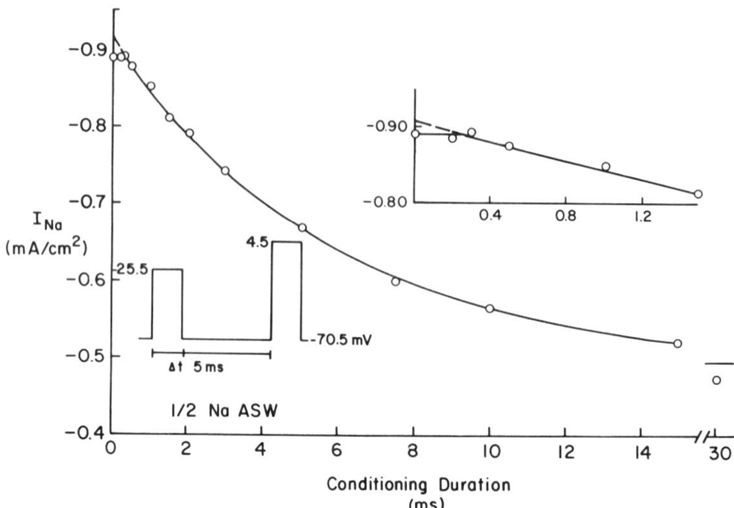

Fig. 1. Inactivation time course. Peak I_{Na} during a 4.5 mV test pulse is shown as a function of the duration of a -25.5 mV conditioning pulse. The curve is a simple exponential with a time constant (τ_c) of 5.71 ms. Inset shows the initial part of the curve on an expanded time scale. Delay is 285 μs. Pulse schedule is shown in lower left, where Δt is the conditioning duration. From Goldman and Kenyon (1982).

The inactivation time course, then, is as expected if inactivation develops subsequent to a precursor process. However, it is important to insure that these experiments are not affected by technical artifact. We first consider the resistance in series with the membrane, R_s.

Note that the results of figure 1 are qualitatively different from those expected from Hodgkin-Huxley kinetics in the presence of an uncompensated R_s (Gillespie and Meves, 1980) in that we see delays (i.e. deviations from a simple exponential) only for brief conditioning potentials where g_{Na} is little activated. Hodgkin-Huxley kinetics in the presence of some R_s predict deviations from a simple exponential only for conditioning pulse durations long enough to substantially activate g_{Na}.

Nevertheless we have put the issue to direct experimental test. Two determinations were made in each of four axons with the same holding potential, gap width and conditioning and test pulse amplitudes. One of each pair of determinations was made in 1/2 Na ASW and the other in 1/4 Na ASW. In each case the two τ_c exponentials were the same. In fact the 1/2 Na curves could always be fitted by those in 1/4 Na scaled up by the ratio of the two unconditioned I_{Na} values. Correspondingly the delays in each case were just the same also, and reducing the current density by up to five-fold had no effect on either τ_c or the delay. Inactivation delays are therefore not produced by R_s errors.

The width of the gap between conditioning and test pulses was 6 ms. This is long relative to the slowest activation time constant seen in Na tail currents in Myxicola at these same potentials and temperature (about 1.0 ms, Goldman and Hahin, 1978). Correspondingly, for the experiments reported here, in each case the tail currents flowing during the gap were seen to decay to zero well before the start of the conditioning pulse.

However, we have put this issue to direct experimental test also. Two determinatons were made in each of three axons, with holding potential, conditioning and test pulse amplitudes and solution composition all held constant. For one determination the gap width was the usual 6 ms while for the other it was reduced to 2 ms. For each of the three axons the results were the same. τ_c was just the same for both determinations, and the delays were correspondingly found to be just the same also. Hence even a 2 ms gap is sufficient to avoid any significant distortion of the inactivation time course by the time course of activation development. Note that if the delay seen for the 6 ms gap had been produced by such distortions then the delay would necessarily have increased with the 2 ms gap.

The results presented so far establish unambiguously that there is a process preceding the development of inactivation (i.e. inactivation is not a simple two state process). Similar results have been reported in well controlled experiments in crayfish (Bean, 1981) and squid (Gillespie and Meves, 1980) axons. However, these experiments provide little information as to the natue of the precursor process. Those described below address this issue.

Figure 2A shows the results of two determinations on the same axon. In this experiment the holding potential, gap width, test pulse amplitude and the solution composition were all held constant. As indicated, the conditioning potential was -26 mV for the determination shown in the upper and -11 mV for that shown in the lower part of the figure. τ_c decreased with more positive conditioning potential in the usual way (5.51 ms for the upper and 3.37 ms for the lower curve). However as shown in fig. 2B, which is the early portion of both parts of fig. 2A on an expanded time scale, the delay also decreased with more positive conditioning potential (525 μs for -26 and 309 μs for -11 mV). This is just as expected if activation and inactivation are sequentially coupled owing to the reduced time for activation to develop at -11 mV. When collected values of delay vs conditioning potential were examined, delay was found to decrease steeply with more positive potential in the negative potential range and tended to saturate at positive potentials.

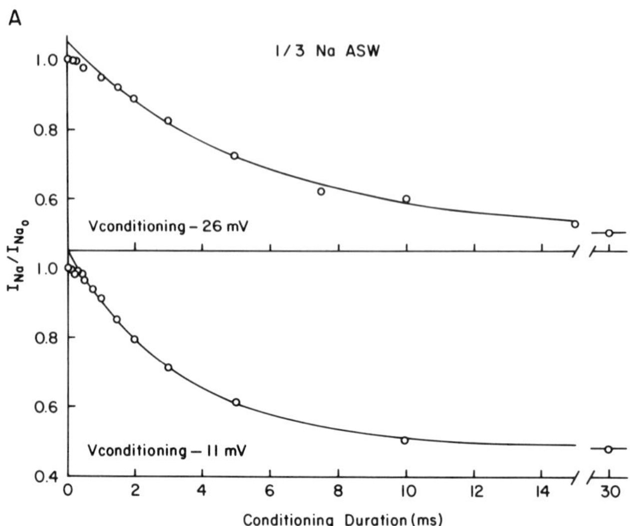

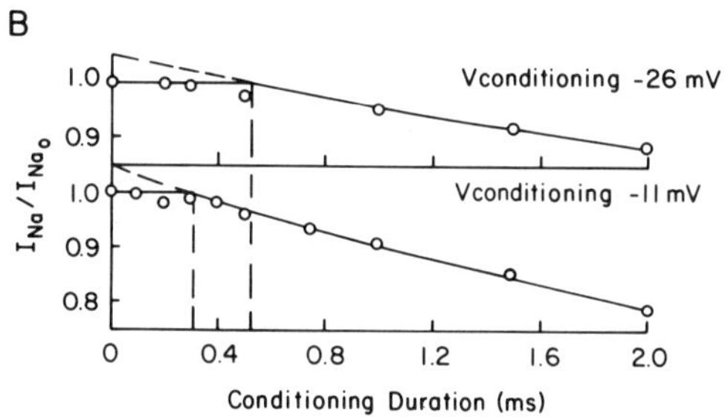

Fig. 2. A. Two inactivation determinations on the same axon. Test potential was 4 mV for both determinations. Upper curve determined with a -26 mV conditioning potential (τ_c = 5.51 ms), and lower with -11 mV (τ_c = 3.37 ms). B. Initial portions of the two curves in A on an expanded time scale. Delay was 525 μs for the upper and 309 μs for the lower curve (indicated by the vertical lines). From Goldman and Kenyon (1982).

Somewhat more information about the delay process is gained by plotting inactivation delay as a function of the time to peak g_{Na} during the conditioning pulse. The collected results are shown in fig. 3. Delay increases with time to peak g_{Na}, and the two variables may be described as roughly proportional. If activation and inactivation are sequentially coupled then the inactivation delay as well as time to peak g_{Na} will both vary with the activation time constant, and proportionality between the two is expected for the simplest sort of sequential coupled process. Considering the operational way in which the delay is determined, the agreement between the data of

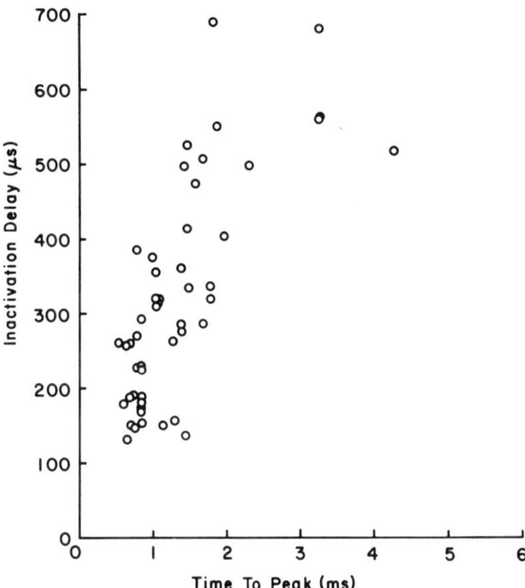

Fig. 3. Inactivation delay as a function of time to peak g_{Na} during the conditioning pulse. Data pooled from 24 axons. From Goldman and Kenyon (1982).

fig. 3 and the expectations for a sequential coupled process is quite reasonable.

In 7 of the determinations from 6 different axons, the difference between the experimental Na currents for brief conditioning pulses and the extrapolated τ_c exponential were large enough to measure reliably. These difference values could also be described as an exponential, and in these few cases the time constant of the delay process, τ_{delay}, could be determined. τ_{delay}, values are plotted as a function of conditioning potential as the filled triangles in fig. 4. Also included (circles) are the Myxicola τ_m (V) values (Goldman and Schauf, 1973). τ_{delay} (V) is in reasonable agreement with τ_m (V) again as expected for sequential activation and inactivation.

τ_m is the activation time constant for an independent kinetic model. However, no difficulties are raised by the agreement between τ_m (V) and τ_{delay} (V). Fig. 4 also includes (open triangles) activation time constants determined by fitting a sequential coupled scheme to Myxicola voltage clamp data (Goldman and Hahin, 1978 fig.5). The scheme used was:

$$A \rightleftharpoons B \rightleftharpoons C \rightleftharpoons D \rightleftharpoons E$$

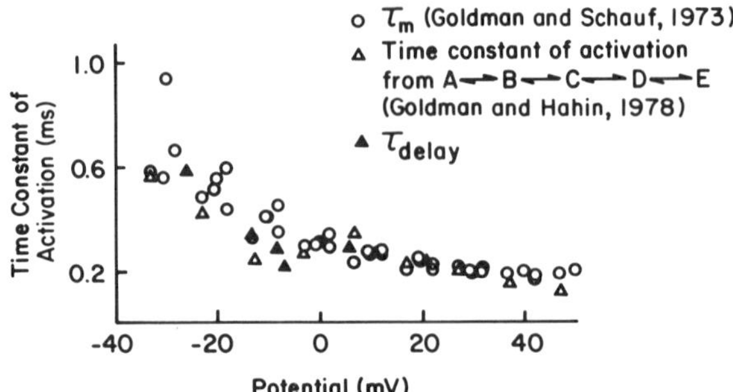

Fig. 4. Activation time constant as a function of membrane potential. Circles are τ_m values for Myxicola (Goldman and Schauf, 1973). Open triangles are activation time constants fitted to clamp data from a coupled model (Goldman and Hahin, 1978), and the filled triangles are τ_{delay} values. From Goldman and Kenyon (1982).

where A is the resting, B and C activated but not conducting, D the conducting and E the inactivated state. This scheme displays three activation time constants consistent with the three relaxations seen in Na tail currents in Myxicola, and the slowest of these (τ_3 from Goldman and Hahin, 1978) is plotted in fig. 4. Goldman and Hahin presented evidence from initial conditions experiments that it is the τ_3 relaxation that is most closely associated with the filling of the conducting state. τ_3 (V) also agrees with τ_m (V), and the agreement between τ_{delay} (V) and τ_m (V) is fully consistent with a sequential coupled process.

Na channels in Myxicola, then, display a delay in the development of inactivation whose time course parallels activation kinetics. Either inactivation, at least in part, sequentially follows the activation process, or inactivation sequentially follows a process whose kinetics are identical to the activation process. We suggest that the precursor process is likely to be the activation process as: (i) previous kinetic results in Myxicola (Goldman and Schauf, 1972; 1973) were most simply accounted for if inactivation sequentially followed the activation process; (ii) a specific prediction from this view - that there should be a delay in inactivation development - was borne out experimentally; (iii) in an extensive series of experiments, the potential dependency of the time constant of the delay process and that of the activation process were found to be the same (fig. 3); (iv) in a smaller number of determinations, the actual value of the time constant of the delay process was found to be that of the activation process (fig. 4). Na channels in Myxicola, then behave in a number of respects just as expected if they, at least in part, must open before they can inactivate.

ACKNOWLEDGEMENTS

We are grateful to Mr. Mark Pohl for excellent technical assistance. This work was supported by National Institutes of Health research grants NS 07734 and NS 14800. Figures 1-4 have been reproduced from the Journal of General Physiology 80:83-102 (1982) by copyright permission of the Rockefeller University Press.

REFERENCES

Armstrong, C.M., and Bezanilla, F., 1974, Charge movement associated with the opening and closing of the activation gates of the Na channels, J. Gen. Physiol., 63:533.
Armstrong, C.M., and Gilly, W.F., 1979, Fast and slow steps in the activation of sodium channels. J. Gen. Physiol, 74:691.
Bean, B.P., 1981, Sodium channel inactivation in the crayfish giant axon. Must channels open before inactivating? Biophys. J., 35:595.
Connor, J.A., 1976, Sodium current inactivation in crustacean axons, Biophys. J., 16:24a (Abstr.).
Ebihara, L., and Johnson, E.A., 1980, Fast sodium current in cardiac muscle. A quantitative description, Biophys. J., 32:779.
Gillespie, J.I., and Meves, H., 1980, The time course of sodium inactivation in squid giant axons, J. Physiol. (Lond.), 299:289.
Goldman, L., 1976, Kinetics of channel gating in excitable membranes, Q. Rev. Biophys., 9:491.
Goldman, L., and Hahin, R., 1978, Initial conditions and the kinetics of the sodium conductance in Myxicola giant axons. II. Relaxation experiments, J. Gen. Physiol., 72:879.
Goldman, L., and Kenyon, J.L., 1982, Delays in inactivation development and activation kinetics in Myxicola giant axons, J. Gen. Physiol., 80:83.
Goldman, L., and Schauf, C.L., 1972, Inactivation of the sodium current in Myxicola giant axons. Evidence for coupling to the activation process, J. Gen. Physiol., 59:659.
Goldman, L., and Schauf, C.L., 1973, Quantitative description of sodium and potassium currents and computed action potentials in Myxicola giant axons, J. Gen. Physiol., 61:361.
Hodgkin, A.L., and Huxley, A.F., 1952, A quantitative description of membrane current and its applications to conduction and excitation in nerve, J. Physiol. (Lond.), 117:500.
Horn, R., Patlak, J., and Stevens, C.F., 1981, Sodium channels need not open before they inactivate, Nature (Lond.), 291:426.
Oxford, G.S., and Pooler, J.P., 1975, Selective modification of sodium channel gating in lobster axons by 2,4,6-trinitropohenol. Evidence for two inactivation mechanisms, J. Gen. Physiol., 66:765.

THRESHOLD FUNCTION OF THE NERVE EXCITABLE MEMBRANE

AND THE NOISE ASSOCIATED WITH IONIC CURRENTS

Cornelia Zaciu

Department of Biophysics
Faculty of Medicine
Bucharest 76241, Romania

INTRODUCTION

Certain of our previous studies performed by means of spectral analysis[1,3] showed that the nerve membrane of all or none type, processes stimuli so that they might have optimal characteristics from the viewpoint of the excitation process.

This analysis was carried out for stimuli more frequently used in physiological research, namely; rectangular pulses, sinusoidal and ramp stimuli.

In the case of pulse stimuli, nerve membrane limits the duration of these stimuli to $\tau = \tau_{excit}$ (the time constant of the excitation process), a way through which the optimal covering of the frequency range of sodium current noise is realized; for the short pulse stimulus the energetic condition is more severe as compared with that of the stimulus of long duration.

The sinusoidal stimulus is processed so as to provide both the energetic condition at low frequencies and the covering of a spectral range which even in a liminal case must be between 0 and 200-300 Hz. Spectral analysis of these stimuli enables the pararesonance phenomenon to be explained.

Ramp stimuli enables us to study accommodation phenomena which, from a physiological viewpoint, characterize the ability of a cell to memorize temporal characteristics of the stimulus. Starting from the experimental evidence obtained for liminal ramp stimuli, the spectral covering was calculated to be 0-200 Hz for giant axon and 0-250 Hz for the Ranvier node membrane in a motor nerve fibre of Rana

Esculenta sciatic nerve. Therefore the spectrum of a liminal ramp seems to exceed the spectrum of I_K noise and partly cover that of I_{Na} noise.

MATHEMATICAL MODEL OF THE THRESHOLD FUNCTION OF EXCITABLE NERVE MEMBRANE

The main conclusion of our studies is that the threshold function of excitable nerve membrane may be defined by two conditions: [2,3] - the energetic condition at low frequencies which can be written as:

$$A^2 \tau^2_{excit} = \frac{A^2}{(2\pi f_{o1})^2} = \frac{A^2 \alpha^2_m}{4} = \frac{A^2 T^2_{o2}}{(4)^2} \qquad (1)$$

where A is the rehobasic threshold; τ_{excit} the time constant of excitation process; f_{01}, the maximum frequency of a sinusoidal stimulus with $\tau = T_0/4$; α_m, the minimal duration of a ramp stimulus; f_{02}, the maximum frequency of a sinusoidal stimulus with $\tau = T_0/2$.

- the condition of spectral covering, whose mathematical expression is:

$$\frac{p}{100} \frac{\int_{-\infty}^{\infty} |S_{zg}(f)| df}{|S_{zg\,K}|_M} < \frac{p}{100} \frac{\int_{-\infty}^{\infty} |S_s(f)|^2 df}{|S_s|^2_M} \leqslant \frac{p}{100} \frac{\int_{-\infty}^{\infty} |S_{zg\,Na}(f)| df}{|S_{zg\,Na}|_M} \qquad (2)$$

where $p = 10$; $\int_{-\infty}^{\infty} |S_s(f)|^2 df$ is the signal energy, and $|S_s|^2_M$ its spectral energy density in the origin; $|S_{zg\,Na}(f)|$ the spectral power density of the fluctuations associated with the sodium current and $|S_{zg\,Na}|_M$ the maximum value of this density; $|S_{zg\,K}(f)|$ the spectral power density of the fluctuations associated with the potassium current and $|S_{zg\,K}|_M$ the maximum value of this density.

On the basis of this model suggested by us one can make a number of predictions verifiable through experimental data obtained by us or by other authors, and of which the most important would be: a) the slope of a ramp stimulus which, applied in voltage clamp measurements, gives the characteristics I_{Na}-V; b) the optimal frequency in the pararesonance phenomenon; c) the behavior of the axon membrane, only on hyperpolarization as a filter adapted to signals commonly used as physiological stimuli[2,4].

The last of the above predictions would possibly cast further light on the correlations between macroscopic and microscopic phenomena, by opening a way to two theoretical objectives formulated as the following questions:

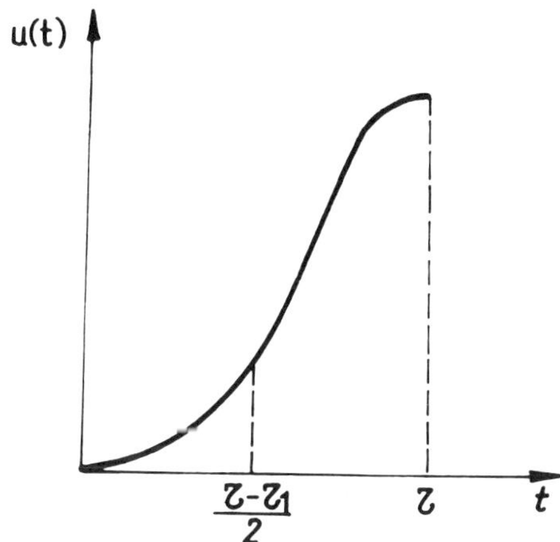

Fig. 1. Signal shape representing the rising phase of the action potential.

- which is the functional model of a filter adapted to a signal having the shape of an action potential?

- which is the signal adapted to a filter with a frequency characteristic similar to that obtained when applying white noise to the membrane near the excitation threshold?

Our recent studies enabled us to solve the first of these two problems, and the answer was given in the form of a functional model representing the axon membrane as a filter adapted to a signal like the rising phase of an action potential.

FUNCTIONAL MODEL REPRESENTING THE AXON MEMBRANE AS FILTER ADAPTED TO A SIGNAL SIMILAR IN SHAPE TO THE ACTION POTENTIAL

If the signal is considered as an action potential, the synthesis of a filter adapted to this signal may be performed. By definition, the transfer characteristic $H(\omega)$ of the adapted filter can be determined by calculating the conjugated spectral function $S(\omega)$ of the signal.

$$H(\omega) = k\, S^*(\omega) \tag{3}$$

In the excitation process it is the rising phase of the action potential which is important. The signal equivalent to this rising phase (Figure 1) can be assimilated to two fragments of parabola:

$$u_1(t) = UAt^2 \qquad \text{for } 0 < t < \frac{\tau - \tau_1}{2}$$
$$u_2(t) = UB + UC(t - \frac{\tau}{2})^2 \qquad \text{for } \frac{\tau - \tau_1}{2} < t < \tau/2 \tag{4}$$

The spectral function of the signal given by the relations (4) will be:

$$S(\omega) = \frac{8U}{\tau(\tau-\tau_1)} \frac{1}{(j\omega)^3} - \frac{8U}{\tau(\tau-\tau_1)} \frac{1}{(j\omega)^3} e^{-j\omega(\frac{\tau-\tau_1}{2})}$$
$$+ \frac{8U}{\tau\tau_1} \frac{1}{(j\omega)^3} e^{-j\omega\tau/2} (1 - e^{j\omega\tau_1/2}) - U\frac{1}{(j\omega)} e^{-j\omega\tau/2} \tag{5}$$

According to the definition, the transfer characteristic of the adapted filter is:

$$H(\omega) = k \left[- \frac{8U}{\tau(\tau-\tau_1)} \frac{1}{(j\omega)^3} + \frac{8U}{\tau(\tau-\tau_1)} \frac{1}{(j\omega)^3} e^{j\omega(\frac{\tau-\tau_1}{2})} \right.$$
$$\left. - \frac{8U}{\tau\tau_1} \frac{1}{(j\omega)^3} e^{j\omega\tau/2}(1 - e^{-j\omega\tau_1/2}) + U\frac{1}{(j\omega)} e^{j\omega\tau/2} \right] \tag{6}$$

The functional model of a filter having the frequency characteristics given by the equation (6) appears (with the notation $\alpha = \frac{8kU}{\tau(\tau-\tau_1)}$ and $\beta = \frac{8kU}{\tau\tau_1}$) in Figure 2. In this mode one can observe:

a) a way with three integrators which, in the mathematical model H-H, may be assimilated to sodium activation $m(I_{Na} = m^3h\bar{I}_{Na}$ where $h \cong 1$ at the initial moment and $\bar{I}_{Na}$ is the maximum permeability to the sodium current).

b) a way with only one integrator, which may be assimilated to the capacitive component of the membrane impedance.

DISCUSSION

A useful discussion on the model in Figure 2 may be generated when

a) $\tau \to \tau_1$ which represents an action potential with small latency;
b) $\tau_1 \to 0$ which represents an action potential with large latency.

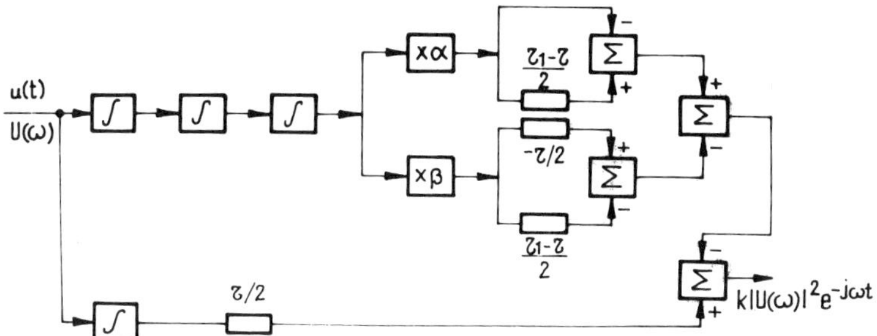

Fig. 2. Functional model of a filter adapted to a signal representing the rising phase of the action potential.

In the case a) the transfer function of the adapted filter becomes:

$$H(\omega) = k\left[4\frac{U}{\tau}\frac{1}{(j\omega)^2} - \frac{8U}{\tau^2}\frac{1}{(j\omega)^3}e^{j\omega\tau/2}(1-e^{-j\omega\tau/2}) + U\frac{1}{(j\omega)}e^{j\omega\tau/2}\right] \quad (7)$$

The functional model of an adapted filter with the frequency characteristic given by formula (7) is illustrated (with the notations $\gamma = 4kU/\tau$ and $\delta = 8kU/\tau^2$) in Figure 3.

In the case b) the transfer function of an adapted filter will be:

$$H(\omega) = k\left[\frac{-8U}{\tau^2}\frac{1}{(j\omega)^3} + \frac{8U}{\tau^2}\frac{1}{(j\omega)^3}e^{j\omega\tau/2} - \frac{4U}{\tau}\frac{1}{(j\omega)^2} + \frac{U}{(j\omega)}e^{j\omega\tau/2}\right] \quad (8)$$

With the same notions as in case a), the functional model of an adapted filter having the frequency characteristic given by equation (8) may be seen in Figure 4. The conclusion we reached, on the basis of the models conceived for the two limiting cases (Figure 3 and Figure 4), is that the two ways (with two and three integrators) of ionic conduction are identical but, in the final summation, signals are applied with inverse phase: the way with three integrators is added in phase for the case a) and in antiphase for the case b) and, inversely the way with two integrators is added in phase for the case b) and in antiphase for the case a).

Therefore, the regenerative process which occurs at the excitation threshold may be described as a resonance phenomenon.

Our model enables calculation of the resonance frequency which is conditioned by a zero signal output. But this is a problem which, with that mentioned at point 2, will form the object of our subsequent studies.

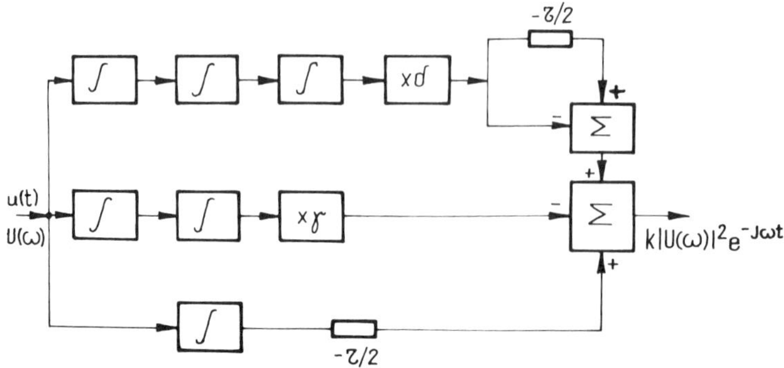

Fig. 3. Functional model of a filter adapted to a signal representing the action potential, in the case of great latency of the excitation process ($\tau_1 \rightarrow 0$).

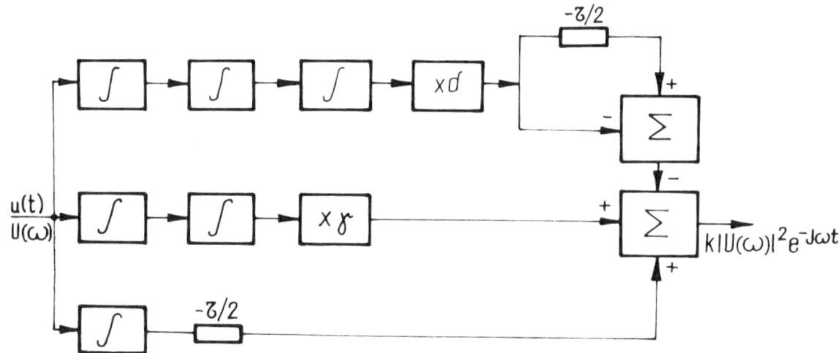

Fig. 4. Functional model of a filter adapted to a signal representing the action potential in the case of a small latency of the excitation process ($\tau_1 \rightarrow \tau$).

Acknowledgements

Many thanks are due to Prof. L. Goldman (Dept. of Physiol. Univ. of Maryland, Baltimore), whose generous discussions during the Conference prompted the author to continue this study and thoroughly develop it in the future.

REFERENCES

1. C. Zaciu and D. G. Margineanu, Membrane noise and microscopical significance of critical slopes of ramp voltages, J.Theor. Biol., 80:589-593 (1979).
2. C. Zaciu, Microscopical significance of the critical characteristics of stimuli in excitation processes as revealed by spectral analysis, J.Theor.Biol., 91:241-254 (1981).

3. C. Zaciu, Information transmission through nerve fibre: noise and memory of the myelinated nerve fibre: Information processes and systems 12:243-249 (Publishing House of the Romanian Academy, Bucharest, 1982) (in Romanian).
4. C. Zaciu, Information transmission through nerve fibre: the behavior of axonal membranes as adaptive systems (in press) (Publishing House of the Romanian Academy, Bucharest, 1983) (in Romanian).

INDEX